The Biology of Scorpions - Polis - 595.46
B615 (SR)

Arachnology   595.4406
                               161a (SR)

striped Blister Beetle pg 292
              cantharidin

Entomology

Antlions . . Pg 274
Water Scorpions .. Pg 211
Ticks (Life cycle)
      The Animal MIND - Gould .. Pg 9·10
           591. 94 (SR)
Ticks - Coevolution of parasites ARTHROPods
         and Mammals ... 599.0249 (SR)

The Encyclopedia of Land Invertebrate Behavior
      595.20451 (SR)

The Manual of Forensic Entomology - smith
         614.1 (SR)

Bugs in the System . . . 595.7

Entomology

Cedric Gillott

University of Saskatchewan
Saskatoon, Saskatchewan, Canada

Plenum Press · New York and London

Library of Congress Cataloging in Publication Data

Gillott, Cedric.
  Entomology.

  Bibliography: p.
  Includes index.
  1. Entomology. I. Title.
QL463.G54   595.7   79-21675
ISBN 0-306-40514-8 (pbk)

©1980 Plenum Press, New York
A Division of Plenum Publishing Corporation
227 West 17th Street, New York, N.Y. 10011

Printed in the United States of America

Preface

The idea of writing this book was conceived when, in the late 1960s, I began teaching a senior undergraduate class in general entomology. I soon realized that there was no suitable text for the class I intended to give. The so-called "general" or "introductory" texts reflected the traditional taxonomic approach to entomology and contained relatively little information on the physiology and ecology of insects. This does not mean that there were no books containing such information. There were several, but these were so specialized and detailed that their use in an introductory class was limited. I hold a strong belief that an undergraduate general entomology course should provide a balanced treatment of the subject. Thus, although some time should be devoted to taxonomy, including identification (best done in the laboratory, using primarily material which students themselves have collected, supplemented with specimens from the general collection), appropriate time should be given also to discussion of the evolution, development, physiology, and ecology of insects. In the latter category I include the interactions between insects and Man because it is important to stress that these interactions follow normal ecological principles. Naturally, the format of this book reflects this belief.

The book has been arranged in four sections, each of which necessarily overlaps with the others. Section I (Evolution and Diversity) deals with the evolution of the Insecta both in relation to other arthropods (Chapter 1) and in terms of the individual orders within the class (Chapter 2). Chapter 3 serves two purposes: it provides a description of external structure, which remains the basis on which insects can be classified and identified, while stressing diversity by reference to mouthpart and limb modifications. In Chapter 4 the principles of classification and identification are discussed, and a key to the orders of insects provided. Diversity of form and habits is again emphasized in Chapters 5 to 10, which deal with the orders of insects.

Section II (Anatomy and Physiology) deals with the homeostatic systems of insects; that is, those systems which serve to keep an insect "in tune" with its environment. The goal of these physiological systems is to enable insects to grow and reproduce optimally. The section begins with a discussion of the integument (Chapter 11), since this has had such a profound effect on the success of insects. Chapter 12 examines sensory systems, whose form and function are greatly influenced by the cuticular nature of the integument. In Chapter

13 neural and chemical integration are discussed, and this is followed, in Chapter 14, by a discussion of muscle structure and function, including locomotion. Chapter 15 reveals the remarkable efficiency of the tracheal system in gaseous exchange. Chapter 16 deals with the acquisition and utilization of food, Chapter 17 with the structure and functions of the circulatory system, and Chapter 18 with nitrogenous waste removal and salt/water balance.

In Section III are discussed reproduction (Chapter 19), embryonic development (Chapter 20), and postembryonic development (Chapter 21). The final section of the book (Ecology) examines those factors which affect the distribution and abundance of insects. In Chapter 22 abiotic (physical) factors in an insect's environment are discussed. Chapter 23 deals with biotic factors that influence insect populations and serves as a basis for the final chapter in which the specific interactions of insects and Man are discussed.

As can be inferred from the opening paragraph of this Preface, the book is intended as a text for senior undergraduates taking their first class in entomology. Such students probably will have an elementary knowledge of insects, gleaned from an earlier class in general zoology, as well as a basic understanding of animal physiology and ecological principles. With such a background, students should have no difficulty in understanding the present text.

On occasions, several of my colleagues have questioned the wisdom (ability?) of a single author attempting to cover the entire field of entomology; the more outspoken ones have plainly told me to "Get help!" I have resisted this advice, partly, I suppose, through selfishness, but mainly on the grounds that multiauthored treatises tend to lack continuity. However, to guard against errors of fact and interpretation resulting from my own ignorance, I have had each chapter reviewed by a specialist.

During preparation of the book, there were moments when I wondered whether the finished product would ever see the light of day or, indeed, whether I should have undertaken the task in the first place! Now that the book is finished, I can truly say that its preparation has been a rewarding and pleasant experience. My hope is that its readers will now receive the same pleasure.

Cedric Gillott

April 1979

Acknowledgments

Though this book has single authorship, its preparation would not have been possible but for the contributions of numerous individuals and organizations, to whom I am most grateful.

These include Miss Shirley Shepstone, who prepared all the original drawings and who redrew a large number of figures from their original sources; Mr. David Wong Mr. Dennis Dyck, and Mr. Alex Campbell, who photocopied numerous figures; Mrs. Evelyn Peters, who typed the bulk of the manuscript, as well as Mrs. Denise Nowoselski, Mrs. Peggy Baird, and Mrs. Joan Ryan for their smaller, but nevertheless important, typographical contributions.

I also thank the following individuals who reviewed specific chapters: Dr. G. G. E. Scudder (Chapters 1–3, 8), Dr. D. M. Lehmkuhl (Chapters 4, 6, 9, 22–24), Dr. D. K. McE. Kevan (Chapters 5, 7), Dr. G. E. Ball (Chapter 10), Dr. M. Locke (Chapter 11), Dr. R. Y. Zacharuk (Chapters 12 13), Dr. P. J. Mill (Chapters 14, 15), Dr. J. E. Steele (Chapter 16), Dr. J. C. Jones (Chapter 17), Dr. J. E. Phillips (Chapter 18), Dr. K. G. Davey (Chapter 19), and Dr. B. S. Heming (Chapters 20, 21). An especial debt of gratitude is owed to Dr. Scudder, who most willingly served also as general reviewer for the entire manuscript.

Thanks are also extended to the large number of publishers, editors, and private individuals who allowed me to use material for which they hold copyright. In particular, I thank Mr. Charles S. Papp, who generously supplied me with negatives for a large number of his drawings. The source of each figure is acknowledged individually in the text.

I am also grateful to Professor E. J. W. Barrington, former Head, Zoology Department, The University of Nottingham, for provision of facilities in his department, where preparation of this book began, during tenure of a sabbatical leave; and to the Plenum Publishing Corporation, especially Mr. Kirk Jensen (Editor), Mr. John Matzka (Managing Editor), and Mr. Geoffrey Braine (Production Editor), for their patience and assistance in seeing this project through to completion.

And finally, the unceasing encouragement, assistance, and patience of my wife, Anne, must be acknowledged. To her fell such jobs as translating my hieroglyphics into a first typewritten draft, proofreading at all stages of the book's progress, and a variety of clerical work. It is to her that this book is dedicated.

Contents

I. Evolution and Diversity

II. Anatomy and Physiology

17

**The
Circulatory
System**

18

**Nitrogenous
Excretion and
Salt and Water
Balance**

III. Reproduction and Development

19

Reproduction

20

**Embryonic
Development**

21

Postembryonic Development

IV. Ecology

22

The Abiotic Environment

23

The Biotic Environment

24

Insects and Man

I

Evolution and Diversity

1

Arthropod Evolution

1. Introduction

Despite their remarkable diversity of form and habit, insects possess several common features by which the group as a whole can be distinguished. They are generally small arthropods whose bodies are divisible into cephalic, thoracic, and abdominal regions. The head carries one pair of antennae, one pair of mandibles, and two pairs of maxillae (the hind pair fused to form the labium). Each of the three thoracic segments bears a pair of legs and, in the adult, the meso- and/or metathoracic segments usually have a pair of wings. Abdominal appendages, when present, generally do not have a locomotory function. The genital aperture is located posteriorly on the abdomen. With few exceptions eggs are laid, and the young form may be quite different from the adult; most insects undergo some degree of metamorphosis.

Although these may seem initially to be an inauspicious set of characters, when they are examined in relation to the environment it can be seen quite readily why the Insecta have become the most successful group of living organisms. This aspect will be discussed in Chapter 2.

In the present chapter we shall examine the possible origins of the Insecta, that is, the evolutionary relationships of this group with other arthropods. In order to do this meaningfully it is useful first to review the features of the other groups of Arthropoda. As will become apparent below, the question of arthropod phylogeny is controversial, and various theories have been proposed. For details of these, consult Tiegs and Manton (1958), Manton (1973, 1977), Sharov (1966), and Anderson (1973).

2. Arthropod Diversity

Despite their widely different structures and habits all Arthropoda possess certain features with which the group can be defined. These are as follows: segmented body covered with a chitinous exoskeleton, paired jointed appendages on a variable number of segments, dorsal heart with paired ostia, pericardium present, hemocoelic body cavity, nervous system comprised of a dorsal brain and ventral ganglionated nerve cord, muscles almost always striated, and epithelial tissue almost always nonciliated.

3

2.1. Onychophora

Modern Onychophora (Figure 1.1) are terrestrial animals living in tropical regions of the world. They are generally confined to humid habitats and are found beneath stones, in rotting logs and leaf mold, etc. They possess a combination of annelidan and arthropodan characters and, as a result, are always prominent in discussions of arthropod evolution. Although possessing a chitinous cuticle, the body wall is annelidan, as are the method of locomotion, unjointed legs, the excretory system, and the nervous system. Their arthropodan features include a hemocoelic body cavity, the development and structure of the jaws, the possession of salivary glands, an open circulatory system, a tracheal respiratory system, and claws at the tips of the legs. Of the various groups of arthropods Onychophora resemble the myriapods most closely: their body form is similar, tagmosis (the grouping of segments into functional units) is restricted to the three-segmented head, exsertile vesicles are present in some species as well as in Diplopoda and Symphyla, a "liver" is absent, the midgut is similar, the genital tracts resemble those of myriapods, the gonopore is subterminal, and certain features of embryonic development are common to both groups (Tiegs and Manton, 1958).

2.2. Trilobita

The trilobites (Figure 1.2) are marine fossils which reached a peak (in terms of their diversity) in the Cambrian and Ordovician periods (500–600 million years ago). Despite their antiquity they were, however, not primitive but highly specialized arthropods. In contrast with modern arthropods the trilobites as a whole show a remarkable uniformity of body structure. The body, usually oval and dorsoventrally flattened, is divided transversely into three regions (head, thorax, and pygidium) and longitudinally into three lobes (two lateral pleura and a median axis). The head, which bears a pair of antennae, compound eyes, and four pairs of biramous appendages, is covered by a carapace. A pair of identical biramous appendages is found on each thoracic segment. The basal segment of each limb bears a small inwardly-projecting endite which is used to direct food toward the mouth.

Much about the habits of trilobites can be surmised from examination of their remains and the deposits in which these are found. In general they appear to have been bottom dwellers, feeding either as scavengers or, like earthworms, by taking in mud and digesting the organic matter from it. On the basis of X-ray studies of pyritized trilobite specimens, which show that trilobites possess a combination of chelicerate and crustacean characteristics, Cisne (1974) concludes that the Trilobita, Chelicerata, and Crustacea form a natural group with a common ancestry. Their ancestor would have a body form similar to that of trilobites.

Although the decline of trilobites (and their replacement by the crusta-

FIGURE 1.1. *Peripatopsis* sp. (Onychophora). [From A. Sedgwick, 1909, *A Student's Textbook of Zoology*, Vol. III, Swan, Sonnenhein and Co., Ltd.]

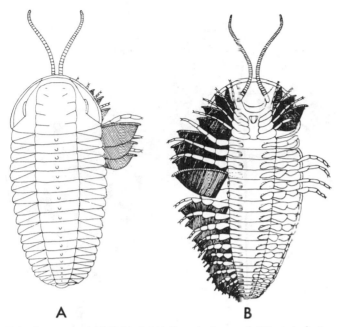

A   B

FIGURE 1.2. *Triarthrus eatoni* (Trilobita). (A) Dorsal view and (B) ventral view. [From R. D. Barnes, 1968, *Invertebrate Zoology*, 2nd ed. By permission of the W. B. Saunders Co., Philadelphia, Pa.]

ceans as the dominant aquatic arthropods) is a matter solely for speculation, Tiegs and Manton (1958) suggest that their basic, rather cumbersome body plan may have prohibited the evolution of fast movement at a time when highly motile predators such as fish and cephalopods were becoming common. In addition, the many identical limbs presumably moved in a metachronal manner, which is a rather inefficient method in large organisms.

2.3. The Chelicerate Arthropods

The next four groups are often placed together under the general heading of Chelicerata because their members possess a body that is divisible into cephalothorax and abdomen, the former usually bearing a pair of chelicerae (but lacking antennae), a pair of pedipalps, and four pairs of walking legs. Although there is little doubt of the close relationship between the Merostomata, Eurypterida, and Arachnida, the position of the Pycnogonida is uncertain. Hedgepeth (1954) and Tiegs and Manton (1958) recommend that they be considered as an entirely separate group, with the rank of a subphylum.

Merostomata (Xiphosurida). *Limulus polyphemus*, the king or horseshoe crab (Figure 1.3), is one of the few surviving species in this previously extensive class of arthropods. King crabs occur in shallow water along the eastern coasts of North and Central America. Other species occur along the coasts of China, Japan, and the East Indies. Like their trilobite relatives they are bottom feeders, stirring up the substrate and extracting the organic material from it. In *Limulus* the cephalothorax is covered with a horseshoe-shaped

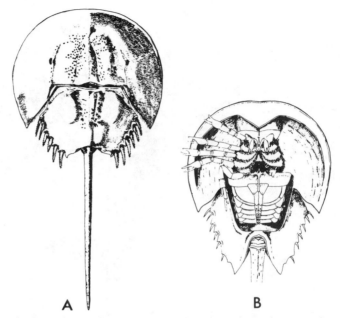

FIGURE 1.3. The horseshoe crab, *Limulus polyphemus*. (A) Dorsal view and (B) ventral view. [From R. D. Barnes, 1968, *Invertebrate Zoology*, 2nd ed. By permission of the W. B. Saunders Co., Philadelphia, Pa.]

carapace. The abdomen articulates freely with the cephalothorax and at its posterior end carries a long telson. On the ventral side of the cephalothorax are six pairs of limbs. The most anterior pair are the chelicerae, and these are followed by five pairs of legs. Each leg has a large gnathobase, which serves to break up food and pass it forward to the mouth. Six pairs of appendages are found on the abdomen. The first pair fuse medially to form the operculum. This protects the remaining pairs, which bear gills on their posterior surface.

The larva of *Limulus* is without the long telson to the abdomen and in general form looks very similar to a trilobite.

Eurypterida. The Eurypterida (giant water scorpions) are an extinct group of predatory arthropods which existed from the Cambrian to the Permian periods. Because of their sometimes large size (up to 2.5 meters) they are also known as Gigantostraca. They are believed to have been important predators of early fish, providing selection pressure for the evolution of dermal bone in the Agnatha. In body plan they were rather similar to the xiphosurids. Six pairs of limbs occur on the cephalothorax, but, in contrast to those of king crabs, the second pair is often greatly enlarged and chelate forming pedipalps, which presumably served in defense and to capture and tear up prey. The trunk of eurypterids can be divided into an anterior "preabdomen" on which appendages (concealed gills) are retained and a narrow taillike "postabdomen" from which appendages have been lost.

Arachnida. Scorpions, spiders, ticks, and mites belong to the class Arachnida whose members are more easily "recognized" than "defined." Members of the group are terrestrial (although a few mites are secondarily aquatic) and have respiratory organs in the form of lung books or tracheae. In contrast to the two aquatic chelicerate groups described earlier most arachnids take only liquid food, extracted from their prey by means of a pharyngeal sucking pump, often after extraoral digestion.

Pycnogonida. The Pycnogonida (sea spiders) [Figure 1.4) are a group of arthropods whose affinities are unclear. Sea spiders are found at varying depths in all the oceans of the world, but they are particularly common in the shallower waters near the North and South Poles. They live on the sea floor and feed on coelenterates, bryozoans, and sponges. On the cephalothorax is a large proboscis, a raised tubercle bearing four simple eyes a pair of chelicerae and an associated pair of palps, and five pairs of legs. The legs of the first pair differ from the rest in that they are small and positioned ventrally. These ovigerous legs are used in the male for carrying the eggs. The abdomen is very small and lacks appendages.

The Pycnogonida were traditionally thought of as being related to the Chelicerata (and sometimes even the Crustacea). More recently they have been placed separately and given the rank of class or subphylum (Hedgepeth, 1954; Tiegs and Manton, 1958). Although the presence of chelicerae, the structure of the brain, and the nature of the sense organs are chelicerate characters (Barnes, 1968), the structure and innervation of the proboscis, the similarity between the intestinal diverticula and those of annelids, the multiple paired gonopores, and the suggestion that the pycnogonids have a true coelom show that they must have left the main line of arthropod evolution at a very early date (Sharov, 1966). Other nonchelicerate features which they possess are (1) the partial segmentation of the leg-bearing part of the body, (2) the reduction of the opisthosoma to a small abdominal component, and (3) the presence, in the male, of ovigerous legs.

2.4. The Mandibulate Arthropods

All the remaining groups of arthropods are frequently grouped together as the Mandibulata (Snodgrass, 1938), so named because its members possess a pair of mandibles as the primary masticatory organs. However, some authors,

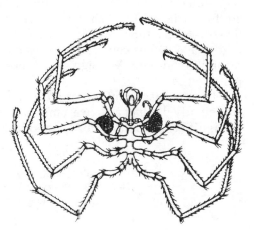

FIGURE 1.4. *Nymphon rubrum* (Pycnogonida). [From R. D. Barnes, 1968, *Invertebrate Zoology*, 2nd ed. By permission of the W. B. Saunders Co., Philadelphia, Pa.]

particularly Manton (1964), believe that the mandible of the crustaceans is not identical with that of the myriapods and insects. The term Mandibulata should not therefore be used to imply a phylogenetic relationship but only a common level of advancement reached by several groups independently (Tiegs and Manton, 1958; Manton, 1964).

Crustacea. To the Crustacea belong the crabs, lobsters, shrimps, prawns, barnacles, and woodlice. The Crustacea are a successful group of arthropods, some 26,000 species having been described (Barnes, 1968). They are primarily aquatic, and few have managed to successfully conquer terrestrial habitats. They exhibit a remarkable diversity of form; indeed, many of the parasitic forms are unrecognizable in the adult stage. Typical Crustacea, however, usually possess the following features: body divided into cephalothorax and abdomen; cephalothorax with two pairs of antennae, three pairs of mouthparts (mandibles and first and second maxillae), and at least five pairs of legs; biramous appendages.

The reason for the success of Crustacea (and perhaps the reason why they replaced trilobites as the dominant aquatic arthropods) is their adaptability. Like their terrestrial counterparts, the insects, crustaceans have exploited to the full the advantages conferred by possession of a segmented body and jointed limbs. Primitive crustaceans, for example, the fairy shrimp (Figure 1.5), have a body which shows little sign of tagmosis and limb specialization. In contrast, in a highly organized crustacean such as the crayfish (Figure 1.6) the appendages have become specialized so that each performs only one or two functions, and the body is clearly divided into tagmata. In the larger (bottom-dwelling) Crustacea specialized defensive weapons have evolved (for example, chelae, the ability to change color in relation to the environment, and the ability to move at high speed over short distances by snapping the flexible abdomen under the thorax). In the smaller, planktonic Crustacea, which are rather defenseless against predators, the reproductive capacity is generally high and the life span short.

The members of four classes of mandibulate arthropods (Chilopoda, Diplopoda, Pauropoda, and Symphyla) have a body composed of a five- or six-segmented head and elongate trunk which bears many pairs of legs, and they are found in similar habitats (in leaf mold, loose soil, rotting logs, etc.). For these reasons they were traditionally placed in a single large group, the Myriapoda. Other common features are the single pair of antennae, a tracheal respiratory system, Malpighian tubules for excretion, and the absence of compound eyes.

Despite these similarities most modern zoologists consider the four groups sufficiently distinct as to warrant class status and do not attach any phylogenetic significance to the term Myriapoda. Manton (1964, 1973) believes, however,

FIGURE 1.5. *Branchinecta* sp., a fairy shrimp. [From R. D. Barnes, 1968, *Invertebrate Zoology*, 2nd ed. By permission of the W. B. Saunders Co., Philadelphia, Pa.]

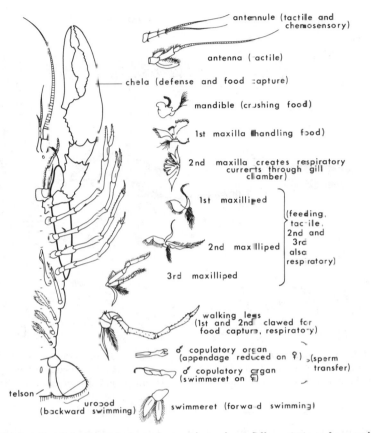

antennule (tactile and chemosensory)

antenna (tactile)

chela (defense and food capture)

mandible (crushing food)

1st maxilla (handling food)

2nd maxilla creates respiratory currents through gill chamber)

1st maxilliped

} (feeding, tactile, 2nd and 3rd also respiratory)

2nd maxilliped

3rd maxilliped

walking legs (1st and 2nd clawed for food capture, respiratory)

♂ copulatory organ (appendage reduced on ♀)
♂ copulatory organ (swimmeret on ♀)
} (sperm transfer)

telson

uropod (backward swimming)

swimmeret (forward swimming)

FIGURE 1.6. Crayfish. Ventral view of one side to show differentiation of appendages.

that the classes show enough affinity for the term to be reinstated and given the rank of subphylum.

Chilopoda. The chilopods (centipedes) (Figure 1.7A) are probably the commonest myriapods. They are typically active, nocturnal predators whose bodies are flattened dorsoventrally. The first pair of trunk appendages (maxillipeds) are modified into poison claws that are used to catch prey. In most centipedes the legs increase in length from the anterior to the posterior of the animal to facilitate rapid movement.

Diplopoda. In contrast with the centipedes the diplopods (millipedes) (Figure 1.7B) are slow-moving, primarily herbivorous animals. The distinguishing feature of the class is the presence of diplosegments, each bearing two pairs of legs, formed by fusion of two originally separate somites. It is believed that the diplosegmental condition enables the animal to exert a strong pushing force with its legs while retaining rigidity of the trunk region. Since they cannot escape from would-be predators by speed, many millipedes have evolved such protective mechanisms as the ability to roll into a ball and the secretion of defensive chemicals.

Pauropoda. These are minute arthropods (0.5–2 mm in length) that live in soil and leaf mold. Superficially they resemble centipedes, but detailed

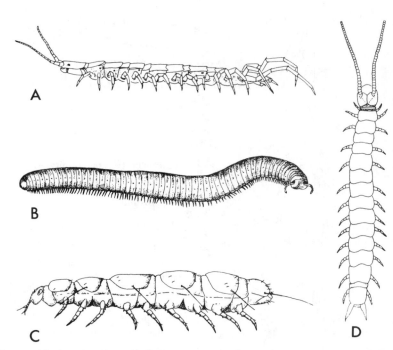

FIGURE 1.7. Myriapoda. (A) *Lithobius* sp. (centipede), (B) *Julus terrestris* (millipede), (C) *Pauropus silvaticus* (pauropod), and (D) *Scutigerella immaculata* (symphylan). [From R. D. Barnes, 1968, *Invertebrate Zoology,* 2nd ed. By permission of the W. B. Saunders Co., Philadelphia, Pa.]

examination reveals that they are more closely related to the millipedes. This affinity is confirmed by such common features as the position of the gonopore, the number of head segments, and the absence of appendages on the first trunk segment (Sharov, 1966). A characteristic feature are the large tergal plates on the trunk, which overlap adjacent segments (Figure 1.7C). It is believed that these large structures prevent lateral undulations during locomotion.

Symphyla. Symphylans (Figure 1.7D) are small arthropods that differ from other myriapods in the possession of a labium (the fused second maxillae) and the position of the gonopore (on the eleventh body segment). Although forming only a very small class of arthropods, the Symphyla are of particular interest to entomologists because they appear to be the myriapods most closely related to insects. The symphylan and insectan heads have an identical number of segments and, according to some zoologists, the mouthparts of symphylans are insectan in character, although Manton (1964) disputes this. At the base of the legs of symphylans are eversible vesicles and coxal styli. Similar structures are found in some apterygote insects.

Hexapoda. The Protura, Collembola, Diplura, Thysanura, and pterygote insects were included originally, mainly because of their six-legged condition, in the class Insecta (=Hexapoda). With increasing knowledge it has become apparent that the first three groups are less closely related to insects than was believed and perhaps deserve a more elevated status (see Section 3.3). In contrast with the thysanurans and winged insects that live on the surface, the

Protura, Collembola, and Diplura are specialized to varying degrees for existence in soil. Because of this they exhibit a good deal of similarity with some of the myriapodan groups, especially the Symphyla. Since their taxonomic status is controversial, the Protura, Collembola, and Diplura have been included with the Thysanura in Chapter 5 (Apterygote Hexapods) where details of their biology are presented.

3. Evolutionary Relationships of Arthropods

3.1. The Problem

In determining the evolutionary relationships of a group of animals the zoologist uses evidence from a variety of sources. The comparative morphology, embryology, and physiology of living members of the group provide clues about the evolutionary trends that have occurred within the group. It is, however, only the fossil record that can provide the *direct* evidence for such processes. Unfortunately, in the case of arthropods the fossil record is poor. By the time the earth's crust became suitable for preservation of dead organisms, in the Cambrian period (about 600 million years ago), the arthropods had already undergone a wide adaptive radiation. Trilobites, crustaceans, and eurypterids were abundant at this time. Even after this time the fossil record is incomplete mainly because conditions were unsuitable for preserving rather delicate organisms such as myriapods and insects. The remains of such organisms are only preserved satisfactorily in media which have a fine texture, for example, mud, volcanic ash, fine humus, and resins (Ross, 1965). Therefore, arthropod phylogeneticists have had to rely almost entirely on comparative studies. Their problem then becomes one of determining the relative importance of similarities and differences that exist between organisms. Evolution is a process of divergence, and yet, paradoxically, organisms may evolve toward a similar way of life (and hence develop similar structures). A distinction must therefore be made between *parallel* and *convergent* evolution. As we shall see below, the difficulty in making this distinction has led to the development of very different theories for the origin of and relationship between various arthropod groups.

3.2. Theories of Arthropod Evolution

As Manton (1973, p. 111) noted, "It has been a zoological pastime for a century or more to speculate about the origin, evolution, and relationships of Arthropoda, both living and fossil." Many famous zoologists have, at various times, expounded their views on this subject. Unfortunately, for the reasons noted above, these views have been widely divergent. Within the last 30 years, however, much evidence has been accumulated, especially in the areas of functional morphology and comparative embryology, so that a clearer, more plausible theory can now be proposed (but see Patterson, 1978). This theory suggests that the major arthropod groups [Chelicerata, Crustacea, and Uniramia (Onychophora–Myriapoda–Hexapoda)] originated independently; that is, the

Arthropoda is a polyphyletic assembly of organisms. Notwithstanding, it is not without historical interest first to briefly trace the development of earlier monophyletic and diphyletic theories (see also Scudder, 1973).

3.2.1. Mono- and Diphyletic Theories

The first scheme for arthropod evolution was devised by Haeckel (1866),* who saw the arthropods as a monophyletic group, divisible into the Carides (Crustacea, which included Xiphosura, Eurypterida, and Trilobita) and the Tracheata (Myriapoda, Insecta, and Arachnida).

In the 1870s new information became available on *Peripatus* (Onychophora), which Haeckel had included in the Annelida. Moseley (1894)* showed that *Peripatus* had a number of arthropodan features (including a tracheal system), and he envisaged it as being the ancestor of the Tracheata, with the Crustacea having evolved independently. Here, then, was the first diphyletic theory for the origin of arthropods. During the latter part of the nineteenth century more of the biology of *Peripatus* became known, especially the internal anatomy and embryology, all of which served to emphasize the close link between Onychophora and Tracheata. As a result, Haeckel (1896)* published a revised, diphyletic scheme similar to the one hinted at earlier by Moseley. It should be emphasized that in this scheme the Arachnida were placed in the Tracheata and were believed to have had a terrestrial origin.

Lankester's (1881)* careful reexamination of *Limulus* showed that this creature, which had previously been included in the Crustacea, was in fact an aquatic arachnid. This discovery provided support for Claus' (1876, 1880)* proposal that the aquatic Gigantostraca were the ancestors of all the terrestrial arachnids. As a result the eurypterid–xiphosuran–arachnid group emerged as an evolutionary line entirely separate from the myriapod–insect line and having perhaps only very slight affinities with the crustaceans. Thus emerged the first example of convergence in the Arthropoda, namely, a twofold origin of the tracheal system.

In 1904 Lankester,* convinced that the arthropods were a unique group with their own distinctive features, produced a fundamentally different, monophyletic scheme based on the number of segments anterior to the mouth, which, he suggested, had moved posteriorly during evolution. In this arrangement the "Hyparthropoda," which had no preoral segment, gave rise to the "Proarthropoda" (for example, *Peripatus*), which had one preoral segment. From such organisms arose the "Euarthropoda" whose earliest members still possessed one preoral segment (Diplopoda, Pauropoda, and Symphyla). Later euarthropods had either two preoral segments (Trilobita and arachnids) or three (Crustacea and, evolving from them, the Hexapoda and the Chilopoda). Lankester's proposal was never taken seriously as it contained several major errors: it was based on a single character of questionable value; in the scheme the Myriapoda, an apparently natural group, disappeared; the Crustacea, already present in the Precambrian period, were preceded by the Diplopoda whose earliest fossils appear in the Upper Silurian; the idea necessitated at

*Cited from Tiegs and Manton (1958).

least a fivefold origin of the tracheal system; because they evolved from Crustacea, the earliest insects were aquatic and, Lankester suggested, the wings evolved from the lateral tracheal gills.

Handlirsch (1908, 1925, 1937)* saw the Trilobita as the group from which all other arthropod classes arose separately. *Peripatus*, clearly not ancestral to the trilobites, was returned to the Annelida, its several arthropod features presumably being due to convergence. The greatest difficulty with Handlirsch's scheme is the idea that the pleura of trilobites became the wings of insects. In other words, in both Handlirsch's and Lankester's schemes it is inferred that the apterygote insects must have evolved from winged forms. This is contrary to all available evidence.

It was at about this time that the fossil *Aysheaia pedunculata* was discovered. This is a *Peripatus*-like creature but having a number of primitive features (six claws at the tip of each leg, a terminal mouth, first appendages postoral, second and third appendages are legs). The associated fauna suggested that this creature was from a marine or amphibious habitat. This and other discoveries led Snodgrass (1938) to suggest another monophyletic scheme of arthropod evolution. In this scheme the hypothetical ancestral group were "lobopod" annelids that had lobelike outgrowths of the body wall serving as legs. Following chitinization of the cuticle and loss of all except one pair of tentacles (which formed the antennae), the lobopods gave rise to the Protonychophora. From the protonychophorans developed, on the one hand, the Onychophora and, on the other, the Protarthropoda in which the cuticle became sclerotized and thickened. Such organisms lived in shallow water near the shore or in the littoral zone. The Protarthropoda gave rise to the Protrilobita (from which the trilobite–chelicerate line developed) and the Protomandibulata (Crustacea and Protomyriapoda). From the protomyriapods arose the myriapods and hexapods. In other words, two essential features of Snodgrass' scheme are that the Onychophora play no part in arthropod evolution and that the mandibulate arthropods (Crustacea, Myriapoda, and Hexapoda) form a natural group, the Mandibulata. The theory has certain advantages over earlier ideas: it supposes a monophyletic origin for the group, it retains the unity of the myriapod–insect line, and it avoids much of the convergence inferred in earlier schemes. However, there are several major difficulties in accepting it. First, there is a lack of "centipedelike" (protomandibulate) fossils in the Cambrian period. The first such fossils do not appear until the Upper Silurian and Devonian periods, and these are clearly myriapodan. Second, the theory implies a unity of all mandibulate arthropods based on a single character. There is, however, reason to believe that the mandible has evolved convergently in Crustacea and in the Myriapoda–Insecta line (Tiegs and Manton, 1958; Manton, 1964; also see Section 3.2.2). A third difficulty is the supposed homology of the seven- to nine-segmented, biramous appendage of Crustacea with the five-segmented, uniramous appendage of Insecta. Those who favor the Mandibulata concept, for example, Matsuda (1970), derive the insect leg from the ancestral crustacean type by proposing that the "leftover" segments were incorporated into the thorax as subcoxal components. This difficulty does not arise in polyphyletic

---

*Cited from Tiegs and Manton (1958).

schemes where the Crustacea are considered unrelated to the Insecta (Manton, 1973; Cisne, 1974).

Another proponent of a monophyletic theory was the Russian paleontologist Sharov (1966) who, like Snodgrass (1938), believed that the Arthropoda and the Onychophora arose from primitive crawling annelids during the Precambrian period. Such organisms were soft-bodied and segmented, moved by means of lobopodia, and captured food by means of an eversible proboscis. Sharov considered that the common features of the Arthropoda and Onychophora are due to convergence and stated (p. 24) that "Tiegs and Manton plainly exaggerate the importance of the features which they claim indicate the close affinity of the Onychophora and atelocerates" (myriapods and insects) (see following discussion). Thus, the Onychophora had no part in Sharov's scheme of arthropod evolution (Figure 1.8).

Sharov suggested that in the earliest arthropods (the Proboscifera) the proboscis was retained and formation of the exoskeleton and a labrumlike structure occurred. He considers the fossil *Opabinia regalis* (Figure 1.9) to be an example of this group. From such organisms arose creeping bottom feeders which strained organisms from the mud or ingested mud directly and digested the organic matter within it, much as do earthworms today. These bottom feeders (the Tetracephalosomita) gave rise to two main lines, one leading to the trilobites and later to the Chelicerata, the other to the Crustacea. Sharov suggested that the ancestral crustaceans retained a bottom-feeding habit which led to the transformation of the first postantennal appendages into tactile organs (the second antennae) and the limbs of the second postantennal segment into masticatory organs (the mandibles). Presumably these changes would increase the efficiency of food gathering. According to Sharov the early crustaceans belonged to the class Gnathostraca, whose modern representatives include water fleas, fairy shrimps, and brine shrimps, and it was from this class that the other classes of Crustacea arose. In Sharov's theory the Atelocerata (meaning "without the second antenna") evolved from primitive Malacostraca (the class which today includes crabs, lobsters, shrimps, and prawns) that lived in shallow water along the shore. The earliest atelocerates were supposedly primitive myriapods, which gave rise to the Monomalata ("one-jawed"— Chilopoda, Diplopoda, and Pauropoda) and Dimalata ("two-jawed"—Symphyla,

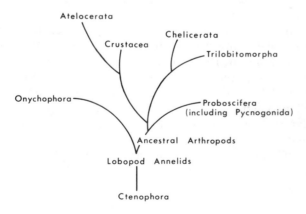

FIGURE 1.8. Sharov's view of the evolutionary relationships between the Annelida, Onychophora, and Arthropoda. [After D. T. Anderson, 1966, Arthropod arboriculture, *Ann. Mag. Nat. Hist. Ser. 13* **9:**445–456. By permission of Taylor and Francis Ltd. and the author.]

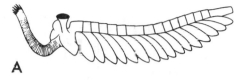

A

FIGURE 1.9. *Opabinia regalis.* (A) Lateral view and (B) dorsal view. [From A. G. Sharov, 1966, *Basic Arthropodan Stock.* By permission of Pergamon Press Ltd., Oxford.]

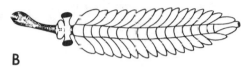

B

Protura, Collembola, Diplura, and Insecta, including Thysanura). As can be seen from his scheme (Figure 1.10) Sharov did not consider the hexapodous groups to have evolved as an offshoot from the line leading to the modern Symphyla, a feature which is implicit in the "Symphylan theory" of Imms (1936) (see Figure 1.11). Like Manton (1964) Sharov interpreted the common features of Symphyla and Hexapoda as being due to parallel evolution between the two groups, which are only distantly related.

Although to the nonexpert Sharov's proposal may seem quite reasonable, it has been criticized severely by several arthropod phylogeneticists. Anderson (1966), for example, notes (p. 445) that "Crude comparison of isolated segments of different animals, no longer acceptable as a basis for a phylogenetic argument, is the cardinal method employed here in establishing conclusions." Anderson also states that Sharov's argument contains many factual errors, selective misquotations, and already disproved hypotheses. The fossil *Opabinia*, which Sharov suggested was close to the ancestral arthropod, is considered by most authorities to be an aberrant, pelagic trilobite. The Tetracephalosomita, which Sharov considers to have arisen from *Opabinia*-like organisms, are in Anderson's view a pure invention for which there is no supporting evidence. In considering the possible origin of myriapods and insects Sharov ignores the argument of Tiegs and Manton (1958) for an affinity with the Onychophora on the grounds that the evidence for such an affinity greatly outweighs that which is against it. Another major error in Sharov's argument, according to Anderson,

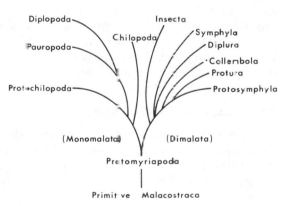

FIGURE 1.10. Sharov's scheme for the evolutionary relationships of Atelocerata.

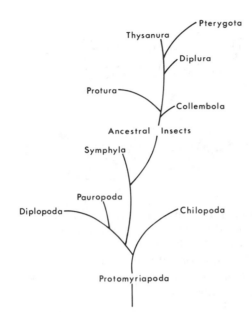

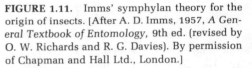

FIGURE 1.11. Imms' symphylan theory for the origin of insects. [After A. D. Imms, 1957, *A General Textbook of Entomology,* 9th ed. (revised by O. W. Richards and R. G. Davies). By permission of Chapman and Hall Ltd., London.]

is the assumption that the Mandibulata are a monophyletic group as a basis for deriving the atelocerates from crustaceans. This is a direct contradiction of the work of Manton (1964) demonstrating the nonhomologous nature of crustacean and myriapodan–insectan mandibles. Furthermore, Sharov completely ignores the fundamental differences in embryonic and larval development between the crustaceans and atelocerates. Manton (1973, p. 128) in her review of arthropod phylogeny devotes two sentences to the dismissal of Sharov's work "with its abundant phantasies and errors of fact."

We must turn, then, to the polyphyletic theory as being a more likely theory for the evolution of arthropods, despite the high degree of convergence that must be assumed to have occurred. Some comfort is gained from the knowledge that convergence is a fairly common phenomenon in evolution. Indeed, on theoretical grounds alone, it could be expected that two unrelated groups of animals would evolve toward the same highly desirable situation and, as a result, develop almost identical structures serving the same purpose. As Manton (1964, p. 106) has noted, "Of great importance is the recognition of the limited range in form which is serviceable for certain organ systems. For example, the constancy of nine peripheral and two central fibrils in cilia and flagella throughout the living world would hardly have been maintained had divergencies from this plan been equally advantageous."

3.2.2. The Polyphyletic Theory

Tiegs and Manton (1958), Anderson (1973), and Manton (1973, 1977) have presented strong arguments in favor of a polyphyletic origin of the arthropods. According to these authors the evidence, derived largely from comparative embryology and functional morphology, weighs heavily in support of division of the arthropods into at least three natural groups, each with the rank of phylum. Thus in this scheme the term "Arthropoda" no longer has taxonomic status but merely indicates a particular grade of advancement. The phyla are the Chelicerata (excluding the trilobites whose relationships remain in doubt), the Crustacea, and the Uniramia (Onychophora–Myriapoda–Hexapoda).* Any polyphyletic theory implies that extensive convergent evolution has occurred. Tiegs and Manton argue that this could be expected in view of the similar functions that structures perform. There is little difficulty in accepting that the tracheal system and Malpighian tubules have evolved at least twice (in the Arachnida and Myriapoda–Insecta group), but convergence in the compound eyes, mandibles, and even arthropodization itself is more difficult to establish. Tiegs and Manton argue that the identical structure of the compound eye in arthropods can be explained simply in terms of the similar function that it performs in all groups. Manton (1973, p. 119) notes that "the detailed knowledge . . . concerning structure and function of compound eyes does not support the original idea that these organs could have been evolved once only." Because of their apparent likeness in structure, musculature, and innervation, it has been argued by Snodgrass (1950) that the mandibles of Crustacea, Myriapoda, and Insecta could have evolved but once. However, Tiegs and Manton again argue that convergence has occurred and their claim receives support from the work of Manton (1964). This author has shown that the mandibles of Crustacea are not homologous with those of Myriapoda and Insecta. In crustaceans the jaw is nonsegmented and formed from the limb base (coxal endite). In myriapods and insects the mandible is formed from the whole limb and is therefore segmented, and these arthropods bite with the tip of the limb. Tiegs and Manton advance the argument that, since arthropodization confers such advantages on organisms, it could be expected to have evolved more than once. They point out that neither chitin nor sclerotin is restricted solely to arthropods, but it is only in arthropods that the two molecules combine to produce cuticle. This process, they claim, could have occurred several times.

3.3. Relationships within the Uniramia

Tiegs and Manton (1958) and Manton (1973) review the evidence in favor of uniting the Onychophora, Myriapoda, and Hexapoda as subphyla of the Uniramia (Figure 1.12). These authors agree with the nineteenth century zoologists Haeckel and Moseley that the many similarities between onychophorans and myriapods indicate true affinity and are not due to convergence. Further, Manton (1949) had demonstrated convincingly that the soft body wall of onychophorans (i.e., lack of arthropodization) is correlated with their habit of passing through narrow spaces so as to escape predators. Tiegs and Manton visualize the evolution of myriapods and insects from onychophoranlike ancestors as a process of progressive cephalization. To the original three-segmented head (seen in modern Onychophora) were added progressively, mandibular, first maxillary, and second maxillary (labial) seg-

---

*As noted earlier, Cisne (1974) concludes on the basis of his X-ray studies of trilobites that there are only *two* natural arthropod groups, the Chelicerata–Trilobita–Crustacea assemblage and the Uniramia.

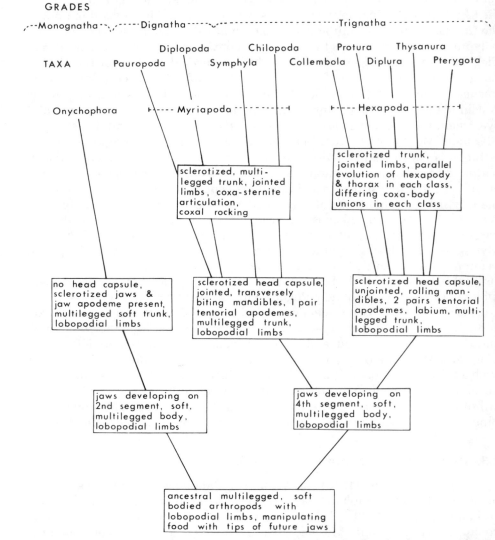

FIGURE 1.12. Manton's view of the probable courses of evolution within the Uniramia. [After S. M. Manton, 1973, Arthropod phylogeny—A modern synthesis, *J. Zool. (London)* **171**:111–130. By permission of the Zoological Society of London.]

ments, giving rise to the "monognathous," "dignathous," and "trignathous" conditions, respectively. Of the monognathous condition there has been found no trace. The dignathous condition is found in the Pauropoda and Diplopoda, and the trignathous condition is seen in the Chilopoda (in which the second maxillae remain leglike) and the Symphyla and Hexapoda (in which the second maxillae fuse to form the labium). Thus, on this criterion alone, the Symphyla, among the myriapods, stand closest to the insects. Indeed, Tiegs and Manton (1958) argued strongly in support of the "Symphylan theory" for the origin of insects proposed by Calman (1936) and Imms (1936). The theory suggested that the insects are derived from symphylanlike creatures and was based on the many resemblances between symphylans and the dipluran *Campodea*. These include the insectan head, beaded antennae, abdominal styli, exsertile vesicles, terminal cerci, segmented mandibles, the 14-segmented trunk, and the dorsal organ. That insects were derived from many-legged ancestors is confirmed by studies of the embryonic development of the most primitive insects. The weakness of the theory is that the comparison is between Symphyla and *Campodea*, a dipluran rather than a typical insect. It is important to emphasize that modern Symphyla are not thought of as the ancestors of insects. As Tiegs and Manton (1958, p. 277) concluded, "The implication is that it [the Symphylan theory] presents the Symphyla as modified survivors of a group of myriapods that had achieved a grade of organization foreshadowing the insects, and that it was from among these 'Protosymphyla' (Imms, 1936), by assumption of longer legs and a hexapodous gait, that the insect stem took shape." Manton (1964) later changed her mind and, in agreement with Snodgrass (1951), recommended abandonment of the Symphylan theory. She claims that many of the similarities, for example, possession of a labium, are superficial and due to parallel evolution resulting from distant common inheritance.

Although there are five groups of hexapods, the wingless Collembola, Protura, Diplura, and Thysanura, and the winged Pterygota, considerable difference of opinion exists as to their relationships. Traditionally, the wingless forms were placed in the subclass Apterygota (Ametabola) of the class Insecta. It is now clear, however, that this is a rather heterogeneous assembly of organisms, with only the Thysanura having a close affinity with the Pterygota. Indeed, Manton (1973) suggests that these two groups may have had a common ancestry. The Collembola and Protura are probably the least related to winged insects. Noninsectan features of Collembola include an abdomen with only six segments, a myriapodan gonad, and total cleavage in the egg. Protura lack antennae, have 11 abdominal segments (3 of which are added by anamorphosis*) and have vestigial appendages on the first three abdominal segments. The position of the Diplura is even more questionable; some authors consider them closely related to the Thysanura (even to the point of grouping the two within the same order of the class Insecta), whereas others believe they are closer to myriapods than insects and accordingly merit the rank of class. Their differences from typical insects include the distinctive entognathous† mouthparts, antennae whose flagellar segments are each equipped with muscles, and certain features of the respiratory system. Manton (1973) considers that each hexapod group should be given the rank of class. Her detailed study of leg articulation (Manton, 1972) has shown that members of the five hexapod groups have entirely distinctive and mutually exclusive leg mechanisms. In other words, hexapody has evolved on five separate occasions.

---

*ANAMORPHOSIS is an increase in the number of segments after emergence from the egg. Segments are added by intercalary growth between the telson and the last segment.

†In the ENTOGNATHOUS condition the mandibles and maxillae lie horizontally beneath the head, enclosed within a pouch formed by the fusion of the labium with the lateral margins of the head.

4. Summary

The Arthropoda are a very diverse group of organisms whose evolution and interrelationships are obscured by a lack of fossil evidence. However, recent functional morphology and comparative morphology strongly suggest that the group had a polyphyletic origin and includes three distinct phyla: Chelicerata, Crustacea, and Uniramia. Thus the term "Arthropoda" does not indicate a taxonomic status but merely a grade of organization achieved (independently) by the three groups. The major problem associated with any theory suggesting a polyphyletic origin is acceptance of the high degree of convergence that must be assumed to have occurred.

Within the Uniramia are three subphyla—Onychophora, Myriapoda, and Hexapoda—distinguished by their jaw mechanisms, associated head structures, and limb articulation. The Hexapoda evolved from many-legged soft-bodied, Onychophora-like organisms through a process of progressive cephalization. Hexapody has evolved in five groups: Collembola, Protura, Diplura, Thysanura, and Pterygota. The first three groups are sufficiently distinct from each other and from typical insects as to each merit the rank of class. However, the Thysanura and Pterygota show enough affinity to suggest that they had a common ancestor and should be united in the class Insecta.

5. Literature

Anderson, D. T., 1966, Arthropod arboriculture, *Ann. Mag. Nat. Hist. Ser. 13* **9**:445–456.

Anderson, D. T., 1973, *Embryology and Phylogeny of Annelids and Arthropods*, Pergamon Press, Oxford.

Barnes, R. D., 1968, *Invertebrate Zoology*, 2nd ed., Saunders, Philadelphia, Pa.

Calman, W. T., 1936, The origin of insects, *Proc. Linn. Soc. London* **1935–1936**:193–204.

Cisne, J. L., 1974, Trilobites and the origin of arthropods, *Science* **186**:13–18.

Hedgepeth, J. W., 1954, On the phylogeny of the Pycnogonida, *Acta Zool. (Stockholm)* **35**:193–213.

Imms, A. D., 1936, The ancestry of insects, *Trans. Soc. Br. Entomol.* **3**:1–32.

Manton, S. M., 1949, Studies on Onychophora. VII. The early embryonic stages of *Peripatopsis*, and some general considerations concerning the morphology and phylogeny of the arthropods, *Philos. Trans. R. Soc. London Ser. B.* **233**:483–580.

Manton, S. M., 1964, Mandibular mechanisms and the evolution of arthropods, *Philos. Trans. R. Soc. London Ser. B* **247**:1–183.

Manton, S. M., 1972, The evolution of arthropod locomotory mechanisms. Part 10. Locomotory habits, morphology and evolution of the hexapod classes, *Zool. J. Linn. Soc.* **51**:203–400.

Manton, S. M., 1973, Arthropod phylogeny—A modern synthesis, *J. Zool. (London)* **171**:111–130.

Manton, S. M., 1977, *The Arthropoda: Habits, Functional Morphology and Evolution*, Oxford University Press, London.

Matsuda, R., 1970, Morphology and evolution of the insect thorax, *Mem. Entomol. Soc. Can.* **76**:431 pp.

Patterson, C., 1978, Arthropods and ancestors, *Antenna* **2**:99–103.

Ross, H. H., 1965, *A Textbook of Entomology*, 3rd ed., Wiley, New York.

Scudder, G. G. E., 1973, Recent advances in the higher systematics and phylogenetic concepts in entomology, *Can. Entomol.* **105**:1251–1263.

Sharov, A. G., 1966, *Basic Arthropodan Stock*, Pergamon Press, Oxford.

Snodgrass, R. E., 1938, Evolution of the Annelida, Onychophora, and Arthropoda, *Smithson. Misc. Collect.* **97**:159 pp.

Snodgrass, R. E., 1950, Comparative studies on the jaws of mandibulate arthropods, Smithson. Misc. Collect. **116**:85 pp.

Snodgrass, R. E., 1951, *Comparative Studies on the Head of Mandibulate Arthropods,* Comstock, New York.

Tiegs, O. W., and Manton, S. M., 1958, The evolution of the Arthropoda, *Biol. Rev.* **33**:255–337.

2

Insect Diversity

1. Introduction

In this chapter we shall examine the evolutionary development of the tremendous variety of insects that we see today. From the limited fossil record it would appear that the earliest insects were wingless, thysanuranlike forms which abounded in the Silurian and Devonian periods. The major advance made by their descendants was the evolution of wings, facilitating dispersal and, therefore, colonization of new habitats. During the Carboniferous and Permian periods there was a massive adaptive radiation of winged forms, and at this time most of the modern orders had their beginnings. Although members of many of these orders retained a life history similar to that of their wingless ancestors, in which the change from juvenile to adult form was gradual (the hemimetabolous or exopterygote orders), in other orders a life history evolved in which the juvenile and adult phases are separated by a pupal stage (the holometabolous or endopterygote orders). The great advantage of having a pupal stage (although this was not its original significance) is that the juvenile and adult stages can become very different from each other in their habits, thereby avoiding competition for the same resources. The evolution of wings and development of a pupal stage have had such a profound effect on the success of insects that they will be discussed as separate topics in some detail below.

2. Monura, Microcoryphia, and Zygentoma

The earliest wingless insects to appear in the fossil record are thysanuran-like creatures (Figure 2.1) of the order Monura from the Upper Carboniferous and Permian periods. It is, however, generally acknowledged that these are but

FIGURE 2.1. Reconstruction of *Dasyleptus brongniarti* [Monura). [After A. G. Sharov, 1966, *Basic Arthropodan Stock.* By permission of Pergamon Press Ltd., Oxford.]

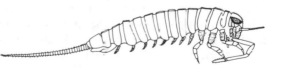

a few remnants of an originally extensive apterygote fauna that existed in the Silurian and Devonian periods. Primitive features of the Monura include the lack of differentiation of the thoracic segments, the clearly visible terga of the mandibular, maxillary, and labial segments of the head, the rather uniform segmentation of the legs, which have a one-segmented tarsus, and the well-developed styli on the abdominal segments. Despite such features Sharov (1966) suggested that the Monura were not ancestral to thysanurans, but rather the two groups had a common ancestry. The absence of anal cerci was regarded as secondary by Sharov. Carpenter (1977) does not share this opinion; he regards the absence of cerci as a primitive feature and proposes that the Monura represent a stage in the evolution of the thysanurans.

Fossil thysanurans resembling modern Machilidae (see Chapter 5) are known from the Triassic period. It is likely, however, that by this time a major split had taken place in the evolution of the group, giving rise to the Microcoryphia (bristletails) and Zygentoma (silverfish, firebrats). Indeed, many authors consider the two groups sufficiently distinct as to warrant ordinal status. Of particular interest in the present discussion is the species *Tricholepidon gertschi,* a living thysanuran discovered in California in 1961. This species is sufficiently different from other Thysanura that it is placed in a separate family, the Lepidotrichidae. *Tricholepidon* possesses a number of primitive features common to both Microcoryphia and Monura (see Chapter 5) which leaves no doubt that the family to which it belongs is closer than any other to the thysanuranlike ancestor of the Pterygota (Sharov, 1966).

3. Evolution of Winged Insects

3.1. Origin and Evolution of Wings

The origin of insect wings has been one of the most debated subjects in entomology for more than a century and a half, and even today the question remains far from being answered. Most authors agree, in view of the basic similarity of structure of the wings of insects, both fossil and extant, that wings are of monophyletic origin; that is, wings arose in a single group of ancestral apterygotes. Where disagreement occurs is with regard to (1) the position(s) on the body at which wing precursors (pro-wings) developed (and, related to this, how many pairs of pro-wings originally existed); (2) the nature of the original functions of pro-wings; (3) the nature of the selection pressures that led to the formation of wings from pro-wings; and (4) the nature of the ancestral insects, that is, whether they were terrestrial or aquatic and whether the pro-wings arose in larval or adult forms.

Oken (1811, cited in Wigglesworth, 1973) apparently made the first suggestion as to the origin of wings, namely, that they were derived from gills. Woodworth (1906, cited in Wigglesworth, 1973), having noted that gills are soft, flexible structures perhaps not easily converted (in an evolutionary sense) into rigid wings, modified the GILL THEORY by suggesting that wings were more likely formed from accessory gill structures, the movable gill plates which protect the gills and cause water to circulate around them. The gill plates, by their very functions, would already possess the necessary rigidity and strength. This proposal receives support from embryology, which has shown abdominal segmental gills of larval Ephemeroptera to be homologous with legs not wings. Wigglesworth (1973, 1976) has recently resurrected, and attempted to extend, the gill theory by proposing that in terrestrial apterygotes the homologues of the gill plates are the coxal styli, and it was from the thoracic coxal styli that wings evolved. Wigglesworth, a firm believer in a monophyletic origin for the arthropods (see Chapter 1), attempts to strengthen his argument by pointing out the enormous diversity in structure and function that has evolved from the primitive exite (supposedly homologous with the coxal stylus) and endite of the ancestral biramous limb. Kukalova-Peck (1978) states that the homology of the wings and styli as proposed by Wigglesworth is not acceptable and points out that wings are always located above the thoracic spiracles, whereas legs (even if ancestrally biramous) always articulate with the thorax below the spiracles. However, in support of Wigglesworth's (1973, 1976) proposal, it should be noted that primitively wings are moved by muscles attached to the coxae (see Chapter 14, Section 3.3.3) and are tracheated by branches of the leg tracheae.

The SPIRACULAR FLAP THEORY of Bocharova-Messner (1971, cited in Kukalova-Peck, 1978) suggests that wings evolved from integumental folds which initially protected the spiracles. The folds enlarged, became hinged, and began to function as a ventilating mechanism for the tracheal system. Improved efficiency of ventilation would be achieved by a flattening of the flaps and an increase in their surface area. At the same time the flapping motion would also assist with locomotion over the substrate and, it is proposed, leg muscles might be used to coordinate leg and flap movements. Further selection pressure would lead to the development of wings from the flaps.

Bradley's FIN THEORY (1942, cited in Kukalova-Peck, 1978) proposed that wings arose in a group of terrestrial insects that became amphibious, that is, spent much of their time in water, returning to land for mating and dispersal. These insects originally possessed nonarticulated pro-wings that had, perhaps, a protective function. As the insects spent increasing amounts of time in water the pro-wings took on new functions as fins in propulsion and attitudinal control. In other words, they would become articulated, strengthened by the development of veins, and would develop suitable musculature. Selection pressure would also lead to enlargement of the fins to the point at which they would be also capable of moving an insect through air.

The most popular theory for the origin of wings is undoubtedly the PARANOTAL THEORY, first proposed by Woodward (1876, cited in Hamilton, 1971), and supported by Crampton (1916), Sharov (1966), Hamilton (1971), Wootton (1976), and others. Essentially the theory proposes that wings arose from rigid, lateral outgrowths (paranota) of the thoracic terga which became enlarged and, eventually, articulated with the thorax. Although large paranotal lobes are found on the prothorax of many fossil insects (Figure 2.5), fully developed and articulated wings occur only on the meso- and metathoracic segments. Presumably, whereas three pairs of paranotal lobes were ideal for attitudinal control (see below), only two pairs of flapping wings were necessary to provide the most mechanically efficient system for flight. Indeed, as insects have evolved there has been a trend toward the reduction of the number of functional wings to one pair (see Chapter 3, Section 4.3.2). This freed the prothorax for other functions such as protection of the membranous neck and serving as a base for attachment of the muscles that control head movement.

Various suggestions have been made to account for the development of paranotal lobes. For example, Alexander and Brown (1963) proposed that the lobes functioned originally as organs of epigamic display or as covers for pheromone-producing glands. Most authors believe, however, that the paranota arose to protect the insect, especially, perhaps, its legs or spiracles. Enlargement and articulation of the paranotal lobes were associated with movement of the insect through the air. Packard (1898, cited in Wigglesworth, 1973) suggested that wings arose in surface-dwelling, jumping insects and served as gliding planes that would increase the length of the jump. However, the almost synchronous evolution of insect wings and tall plants (Hocking, 1957) supports the idea that wings evolved in insects living on plant foliage. Wigglesworth (1963a,b) proposed that wings arose in small aerial insects where light cuticular expansions would facilitate takeoff and dispersal. The appearance later of muscles for moving these structures would help the insect to land the right way up. Hinton (1963a), on the other hand, argued that they evolved in somewhat larger insects and the original function of the paranotal lobes was to provide attitudinal control in falling insects. There is an obvious selective advantage for insects that can land "on their feet," over those which cannot, in the escape from predators. As the paranotal lobes increased in size, they would become secondarily important in enabling the insect to glide for a greater distance. Flower (1964) has examined the hypotheses of both Wigglesworth and Hinton. Flower's calculations show that small projections (rudimentary paranotal lobes) would have no significant advantage for very small insects in terms of aerial dispersal. However, such structures would confer great advantages in attitudinal control and, later, glide performance for insects 1–2 cm in length. The final step would be the transition from gliding to flapping flight, that is, the development of a hinge so that the wings became articulated with the body. Some authorities believe that the hinge evolved initially in order that the projections could be folded along the side of the body, thereby enabling the insect to crawl into narrow spaces and thus avoid capture. Movement of the wings during flight was therefore a secondary adaptation. Perhaps a more likely explanation is that the earliest movements of wings during flight were simply to improve attitudinal control. Only later did the movements become sufficiently strong as to make the insect more or less independent of air currents for its distribution. In this hypothesis the earliest flying insects would rest with their wings spread at right angles to the body, as do modern dragonflies and mayflies. The final major step in wing evolution was the development of wing flexing, that is, the ability to draw the wings when at rest over the back. This ability would be strongly selected for, as it would confer considerable advantage on insects that possessed it, enabling them to hide in vegetation, in crevices, under stones, etc., thereby avoiding predators and desiccation. An implicit part of the paranotal theory is that this ability evolved in the adult stage.

The paranotal theory is based on three pieces of evidence: the occurrence of rigid tergal outgrowths (wing pads) of modern larval exopterygotes (ontogeny recapitulating phylogeny), the occurrence in fossil Paleodictyoptera of rigid pronotal expansions whose venation is homologous with that of wings, and the assumed homology of wing pads and lateral abdominal expansions, both of which have rigid connections with the terga and, internally, are in direct communication with the hemolymph.

This evidence has been severely criticized by Kukalova-Peck (1978). In a thought-provoking paper Kukalova-Peck argues that the fossil record supports none of this evidence. Rather, it indicates just the opposite sequence of events, namely, that the primitive arrangement was one of freely movable pro-wings on all thoracic and abdominal segments of juvenile insects, and it was from this arrangement that the fixed wing-pad condition of modern exopterygote larvae evolved. According to Kukalova-Peck, numerous fossilized juvenile insects have been found which possessed articulated thoracic pro-wings. However, with few exceptions even in the earliest fossil insects, both juvenile and adult, the abdominal pro-wings are already fused with the terga and frequently reduced in size. Some juvenile Protorthoptera with articulated abdominal pro-wings have been described, and in Ephemeroptera the abdominal pro-wings are retained as movable gill plates.

In proposing her ideas for the origin and evolution of wings, Kukalova-Peck emphasizes that these events probably occurred in "semiaquatic" insects living in swampy areas and feeding on primitive terrestrial plants, algae, rotting vegetation, or, in some instances, other small animals. It was in such insects that pro-wings developed. The pro-wings developed on all thoracic and abdominal segments, were present in all instars, and at the outset were hinged to the sides of the body (not the terga). With regard to the selection pressures that led to the origin of pro-wings, Kukalova-Peck uses ideas expressed by earlier authors. She suggests that pro-wings may have functioned initially as spiracular flaps to prevent entry of water into the tracheal system when the insects became submerged or to prevent loss of water via the tracheal system as the insects climbed vegetation in search of food. Alternatively, they may have been plates that protected the gills and/or created respiratory currents over them, or tactile organs comparable to (but not homologous with) the coxal styli of thysanurans. Initially, the pro-wings were saclike and internally confluent with the hemocoel. Improved mechanical strength and efficiency would be gained, however, by flattening and by restricting hemolymph flow to specific channels (vein formation). Kukalova-Peck speculates that eventually the pro-wings of the thorax and abdomen became structurally and functionally distinct, with the former growing large enough to assist in forward motion, probably in water. This new function of underwater rowing would create selection pressure leading to increased size and strength of pro-wings, improved muscular coordination, and better articulation of the pro-wings, making rotation possible. These improvements would also improve attitudinal control, gliding ability, and therefore survival and dispersal for the insects if they jumped or fell off vegetation when on land. The final phase would be the development of pro-wings of sufficient size and mobility that flight became possible.

Probably more interesting than her views on wing origin are Kukalova-Peck's suggestions as to the evolution of fused wing pads and wing flexing. As noted above, the earliest flying insects had wings that stuck out at right angles to the body. Thus, as they developed (in an ontogenetic sense) the insects would be subjected to two selection pressures. One, exerted in the adult stage, would be toward improvement of flying ability; the other, which acted on juvenile instars, would promote changes that enabled them to escape or hide more easily under vegetation, etc. In other words, it would lead to a streamlining of body shape in juveniles. In most Paleoptera streamlining was achieved through the evolution of wings, which in early instars were curved so that the tips were directed backward. At each molt, the curvature of the wings became less until the "straight-out" position of the fully developed wings was achieved. Two other groups of paleopteran insects became more streamlined as juveniles through the evolution of a wing-flexing mechanism, a feature which was also advantageous to, and was therefore retained in, the adult stage. The first of these groups, the order Diaphanoptera, remained primitive in other respects and is included therefore in the infraclass Paleoptera (Table 2.1 and Figure 2.6). The second group, whose wing-flexing mechanism was different from that of Diaphanoptera, contained the ancestors of the Neoptera. The greatest selection pressure would be exerted on the older juvenile instars, which could neither fly nor hide easily. In Kukalova-Peck's scheme, the older juvenile instars were eventually replaced by a single metamorphic instar in which the increasing change of form between juvenile and adult could be accomplished. To further aid streamlining and, in the final juvenile instar, to protect the increasingly more delicate wings developing within, the wings of juveniles became firmly fused with the terga and more sclerotized, that is, wing pads. This state is comparable to that in modern exopterygote (hemimetabolous) insects. Further reduction of adult structures to the point at which they exist until metamorphosis as undifferentiated embryonic tissues (imaginal discs) beneath the juvenile integument led to the endopterygote (holometabolous) condition, that is, the evolution of the pupal stage.

In concluding this discussion on the origin of wings, the differences and similarities between the paranotal theory and Kukalova-Peck's proposal may be summarized. In the paranotal theory wings arose in truly terrestrial insects and environmental water played no part either in their origin or evolution. Kukalova-Peck's theory is that wings arose in semiaquatic insects, and both their origin and evolution were water-related. Both theories suggest that the original function of pro-wings was not movement-related but one of protection. The paranotal theory suggests that the legs may have been the structures protected, hence the development of pro-wings only on thoracic segments. Kukalova-Peck proposes that spiracles or gills were protected by the pro-wings which thus arose on all thoracic and abdominal segments. A major difference between the two theories concerns the site of origin and the onset of articulation of the pro-wings. The paranotal theory proposes that pro-wings arose as rigid outgrowths of the thoracic terga, and articulation came later; Kukalova-Peck argues that pro-wings developed between the spiracles and terga and were hinged from the outset. According to the paranotal theory, enlargement of the pro-wings was associated with improved gliding ability in insects living on terrestrial plants; the evolution of articulated pro-wings led to improved attitudinal control. In Kukalova-Peck's theory, the increase in size of the already

TABLE 2.1. The Major Groups of Pterygota.

		Divisions within Neoptera	
Infraclass	Orders	Martynov's scheme	Hamilton's scheme
Paleoptera	Paleodictyoptera[a] Megasecoptera[a] Diaphanoptera[a] Protodonata[a] Protephemeroptera[a] Odonata (dragonflies, damselflies) Ephemeroptera (mayflies)		
Neoptera	Protoblattoidea[a] Protorthoptera[a] Dictyoptera (cockroaches, mantids) Isoptera (termites) Orthoptera (grasshoppers, locusts, crickets) Miomoptera[a] Protelytroptera[a] Dermaptera (earwigs) Grylloblattodea (grylloblattids) Phasmida (stick and leaf insects) Embioptera (web spinners)	Polyneoptera	Pliconeoptera
	Hadentomoida[a] Caloneurodea[a] Protoperlaria[a] Plecoptera (stoneflies) Zoraptera (zorapterans)		
	Glosselytrodea[a] Psocoptera (booklice) Phthiraptera (biting and sucking lice) Hemiptera (bugs) Thysanoptera (thrips)	Paranecptera	Planoneoptera
	Megaloptera (dobsonflies, alderflies) Raphidioptera (snakeflies) Neuroptera (lacewings, mantispids) Mecoptera (scorpion flies) Lepidoptera (butterflies, moths) Trichoptera (caddis flies) Diptera (true flies) Siphonaptera (fleas) Hymenoptera (bees, wasps, ants, ichneumons) Coleoptera (beetles) Strepsiptera (stylopoids)	Oligoneoptera	

[a] Fossil orders.

articulated pro-wings was related to their use as oars for rowing an insect through water. Both theories are in agreement that the further increase in size and improved articulation of the wings was associated with their use in actively propelling insects through air. Although both theories agree that the selection pressure leading to the evolution of wing flexing was the need to escape from predators or to avoid desiccation, they differ as to the life stage on which this pressure was exerted. The paranotal theory implies that wing flexing evolved in adult insects. Kukalova-Peck argues that the need for concealment arose in juvenile stages, and it was in these that the wing-flexing mechanism developed, though its advantage would also apply to adults. Thus, in her view, the well-sclerotized, nonarticulated wing pads of modern exopterygotes are secondarily derived.

Regardless of their origin, the wings of the earliest flying insects were presumably well-sclerotized, heavy structures with ill-defined veins. Slight traces of FLUTING (the formation of alternating concave and convex veins for added strength) may have been apparent (Figure 2.2A) (Hamilton, 1971). The wings (and flight efficiency) were improved by a reduction in sclerotization, giving rise to the condition seen in Paleoptera (Figure 2.2B). Only the basal part of a wing and the integument adjacent to tracheae within the wing remained sclerotized, forming definite basal plates and veins, respectively. Fluting of the wing was accentuated in the Paleoptera, and the distal portion of the wing was additionally strengthened by the formation of nontracheated intercalary veins and numerous crossveins (Hamilton, 1971, 1972c). This is known as the ARCHEDICTYON arrangement (Figure 2.4).

Primitively, and still seen in Ephemeroptera, the base of each wing articulated at three points with the tergum, the three axillary sclerites running in a straight line along the body (Figure 2.3A). In the evolution of Neoptera the axillary sclerites altered their alignment; specifically, the second axillary sclerite moved away from the tergum and became articulated with the pleural wing

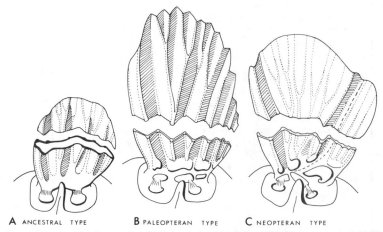

A ANCESTRAL TYPE   B PALEOPTERAN TYPE   C NEOPTERAN TYPE

FIGURE 2.2. Basic types of wings. (A) Supposed ancestral type, heavily sclerotized and with slight fluting; (B) paleopteran type, less sclerotized but strongly fluted; and (C) neopteran type, slightly sclerotized and limited fluting. [After K. G. A. Hamilton, 1971, The insect wing, Part I, *J. Kans. Entomol. Soc.* **44**(4):421–433. By permission of the author.]

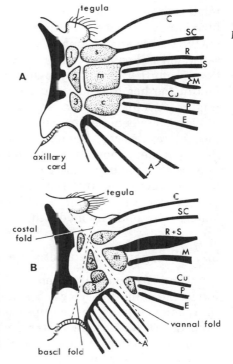

FIGURE 2.3. Basal sclerites of wings. [A) Theoretical ancestral type (and still found in Ephemeroptera) with three points of articulation with tergum, and (B) generalized neopteran type with first and third axillary sclerites articulated with tergum. [After K. G. A. Hamilton 1971, The insect wing. Part I, *J. Kans. Entomol. Soc.* **44**(4):421–433. By permission of the author.]

process (see Figure 3.18B) so that each wing articulated with the tergum at only two points (Figure 2.3B). This alteration of alignment made WING FLEXING possible. It will be seen from Figure 2.3B that the two axes along which folding occurs (the basal and vannal folds) cross in the vicinity of the first axillary sclerite. Contraction of the wing flexor muscle which runs from the pleuron to the third axillary sclerite (see Figure 14.8B) pulls the wing dorsally and posteriorly over the body.

A second important consequence of the altered articulation of the wing was a further improvement in flight efficiency. In Ephemeroptera and, presumably, most or all fossil Paleoptera the wing beat is essentially a simple up-and-down motion; in Neoptera each wing twists as it flaps and its tip traces a figure-eight path. In other words, the wing "rows" through the air, pushing against the air with its undersurface during the downstroke yet cutting through the air with its leading edge on the upstroke. To carry out this rowing motion effectively necessitated the loss of most of the wing fluting. Only the costal area (see Figure 3.27) needs to be rigid as this leads the wing in its stroke, and fluting is retained here (Hamilton, 1971).

Another evolutionary trend, again leading to improved flight, has been toward a reduction in wing weight, permitting both easier wing twisting and an increased rate of wing beating (see also Chapter 14, Section 3.3.4). Concomitant with this reduction in weight has been a fusion or loss of some major veins and the loss of crossveins. The extent and nature of fusion or loss of veins follow certain well-definable patterns which, together with other structural features, are important characters on which conclusions about the evolutionary relationships of neopteran insects can be based. Hamilton (1972c) recognizes three primary venational types, in addition to the archedictyon. The POLYNEUROUS arrangement (Figure 2.4), which has the crossveins more widely separated and parallel compared with those of the archedictyon, is characteristic of Protorthoptera, many fossil and extant orthopteroid orders, and the ancestors of the hemipteroids and endopterygotes. In the COSTANEUROUS pattern (Figure 2.4) crossveins remain only between the costal and subcostal veins (and in Plecoptera between the medial and plical veins). In addition to the wings of Plecoptera, those of Isoptera, Megaloptera, Neuroptera, and the fossil order Protelytroptera show the costaneurous condition. All hemipteroids, most endopterygotes, and members of the orthopteroid orders Dermaptera and Embioptera have wings that almost entirely lack crossveins, a condition described as OLIGONEUROUS. In some insects the basic pattern for the order to which they belong is secondarily modified, especially by the addition of crossveins.

3.2. Phylogenetic Relationships of the Pterygota

There are about 26 orders of living insects and 12 containing only fossil forms, the number varying according to the authority consulted. Elucidation of the relationships of these groups has been hampered by a lack of fossil evidence from the Devonian and Lower Carboniferous periods during which a great adaptive radiation of insects occurred. By the Permian period, from which many more fossils are available, almost all the modern orders had been established. Misidentification of fossils by early paleontologists led to their drawing incorrect conclusions about the phylogeny of certain groups and the construction of rather confusing nomenclature. For example, *Eugereon*, a Lower Permian fossil with sucking mouthparts, was placed in the order Protohemiptera. It is now known that this insect is, in fact, a member of the order Paleodictyoptera and is not related to the modern order Hemiptera as was originally concluded. Likewise the Protohymenoptera, whose wing venation superficially resembles that of Hymenoptera, were thought originally to be ancestral to the Hymenoptera. It is now appreciated that these fossils are paleopteran insects, most of which belong to the order Megasecoptera (Hamilton, 1972c).

To aid subsequent discussion of the evolutionary relationships within the Pterygota the various orders referred to in the text are listed in Table 2.1.

It is generally assumed that the Paleoptera and Neoptera had a common ancestor in the Devonian period [see, for example, Ross (1965) and Sharov

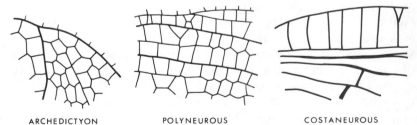

ARCHEDICTYON   POLYNEUROUS   COSTANEUROUS

FIGURE 2.4. Primary types of crossvenation. [After K. G. A. Hamilton, 1972c, The insect wing, Part IV, *J. Kans. Entomol. Soc.* **45**(3):295–308. By permission of the author.]

(1966)], although there is no fossil record of this ancestor. By the Upper Carboniferous period when conditions became suitable for fossilization, a number of paleopteran and neopteran orders had evolved.

Among the Paleoptera at least two major evolutionary lines evidently appeared. The more primitive of these, the ephemeropteroid series, includes the Protephemeroptera (Upper Carboniferous) and Ephemeroptera (Permian–Recent). The fossil *Triplosoba* looks very much like the modern mayflies except that it has two identical pairs of wings. It is therefore placed in a separate order, Protephemeroptera and considered to be slightly off the line of evolution leading to the Ephemeroptera. The second group, the odonatoid series, contains three, possibly four, fossil orders and the extant Odonata. The Paleodictyoptera (Upper Carboniferous–Permian) and Megasecoptera (Upper Carboniferous–Permian) were closely related orders, the former possibly ancestral to the latter (Hamilton, 1972c), whose members were large and heavily built. The Paleodictyoptera (Figure 2.5) were phytophagous or saprophagous and usually equipped with chewing mouthparts. However, one group, which includes *Eugereon*, may have had a sucking proboscis and fed on plant juices. The Megasecoptera were rather like the modern dragonflies in their habits; that is, they were carnivorous, catching their prey on the wing. Venational and other morphological features indicate that the Paleodictyoptera and Megasecoptera were allied to the Protodonata (Upper Carboniferous–Permian) and Odonata (Permian–Recent). The last two orders are closely related, and it is probable that the Odonata evolved from primitive Protodonata. Members of both orders were predaceous, and it is in the Protodonata that the largest known insects (the Meganeuridae) occur, some of which had a wingspan of up to 75 cm. Though not known as fossils, it is presumed that juvenile Protodonata

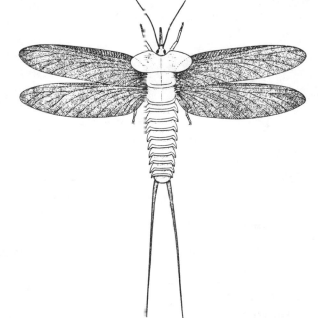

FIGURE 2.5. *Stenodictya* sp. (Paleodictyoptera). [From A. G. Sharov, 1966, *Basic Arthropodan Stock.* By permission of Pergamon Press Ltd., Oxford.]

were aquatic (Carpenter, 1977). The remaining fossil order of Paleoptera, the Diaphanoptera (Upper Carboniferous–Permian), is of interest because its members were able to fold their wings over the body at rest, though the mechanism used was not apparently the same as in Neoptera. Features of their venation and mouthparts indicate that Diaphanoptera were related to the Paleodictyoptera–Megasecoptera, assemblage (Carpenter, 1977).

In contrast with the Paleoptera, which were inhabitants of open spaces, the Neoptera evolved toward a life among overgrown vegetation where the ability to fold the wings over the back when not in use would be greatly advantageous. The early fossil record for Neoptera is poor, but from the great diversity of fossil forms discovered in Permian strata it appears that the major evolutionary lines had become established by the Upper Carboniferous period.

Two major schools of thought exist with regard to the origin and relationships of these evolutionary lines. The more widely accepted view, proposed by Martynov (1938), is that, shortly after the separation of ancestral Neoptera from Paleoptera, three lines of Neoptera became distinct from each other (Table 2.1 and Figure 2.6A). Based on his studies of fossil wing venation Martynov arranged the Neoptera in three groups, Polyneoptera (orthopteroid orders), Paraneoptera (hemipteroid orders), and Oligoneoptera (endopterygotes). In a modification of this view Sharov (1966) proposed that the Neoptera and Paleoptera had a common ancestor (that is, the former did not arise from the latter) and, more importantly, that the Neoptera may be a polyphyletic group. In his scheme (Figure 2.6B) each of the three groups arose independently, a consequence of which must be the assumption that wing folding arose on three separate occasions.

Ross (1955), from studies of body structure, and Hamilton (1972c), who examined the wing venation of a wide range of extant species as well as that of fossil forms, concluded that there are two primary evolutionary lines within the Neoptera, the Pliconeoptera and the Planoneoptera (Figures 2.6C and 2.7). Ac-

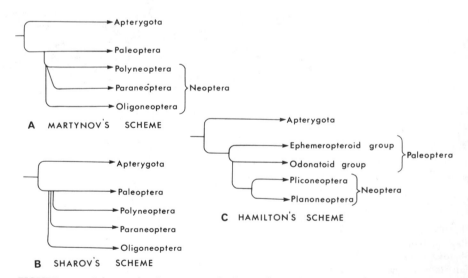

FIGURE 2.6. Schemes for the origin and relationships of the major groups of Neoptera.

cording to Hamilton, the ancestral neopteran group was the Protoblattoidea from which arose the Protorthoptera. The latter is the largest and most diverse of the fossil orders and may even be polyphyletic (Carpenter, 1977). From this group evolved the plicopteran line, the Caloneurocea and the Hadentomoida. The latter order contained the ancestors of the Planoneoptera (Figure 2.7 and Table 2.1). As the figure and table indicate, the Pliconeoptera of Hamilton

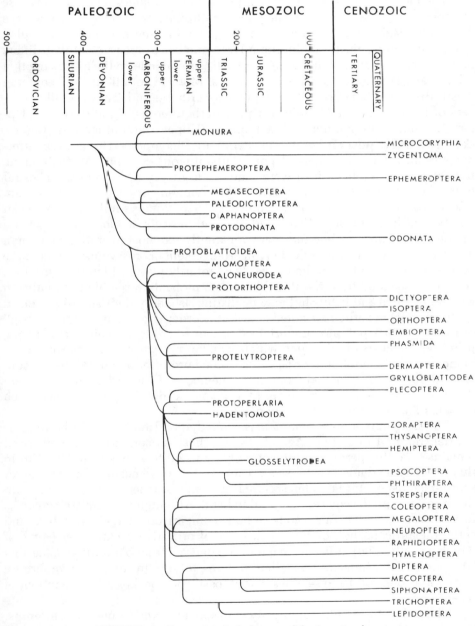

FIGURE 2.7. A suggested phylogeny of the insect orders.

corresponds approximately to the Polyneoptera of Martynov, that is, includes the orthopteroid orders except for the Plecoptera, Zoraptera, and a few fossil orders considered planoneopteran by Hamilton, and the Planoneoptera includes two main groups that are approximately equivalent to the Paraneoptera and Oligoneoptera in Martynov's scheme. In other words, both schools agree that there are three major groups within the Neoptera but differ with regard to the relationships among these groups.

The most primitive group of Neoptera is the orthopteroid complex, whose members are characterized by chewing mouthparts, presence of cerci, a large number of Malpighian tubules, separate ganglia in the nerve cord, complex wing venation (typically including a large number of crossveins) that differs between fore wings and hind wings, and a large anal lobe on the hind wing that folds like a fan along numerous anal veins. Most, if not all, of the neopteran fossils discovered in Carboniferous strata are orthopteroid. They include members of the orders Protoblattoidea (Upper Carboniferous), Protorthoptera (Upper Carboniferous–Permian), Miomoptera (Upper Carboniferous–Permian), Caloneurodea (Upper Carboniferous–Permian), Dictyoptera (Upper Carboniferous–Recent), and Orthoptera (Upper Carboniferous–Recent). As their name indicates, the Protoblattoidea were thought originally to have been ancestral to the Dictyoptera or at least to have had a common ancestry with this group. According to Hamilton (1972c), however, venational similarities between the two orders are superficial and do not indicate a close phylogenetic relationship. The Protoblattoidea are apparently the most primitive neopteran group as indicated by the archedictyon arrangement of crossveins and the more or less complete fluting of the wings. The Protorthoptera, which in Hamilton's scheme descended from the Protoblattoidea, probably resembled the latter in general features and habits but may be distinguished by their slightly reduced (polyneurous) venation. As noted above, this order is probably a "mixed bag" of insects with a polyphyletic origin (Carpenter, 1977). Probably it was from primitive members of this order, which soon became extinct, that the ancestors of most recent orthopteroid orders arose. Some protorthopteran fossils are considered sufficiently distinct from other members of the order Protorthoptera that Hamilton (1972c) places them in a separate order, Hadentomoida, which contains the ancestors of the Planoneoptera.

The Dictyoptera underwent a massive radiation (the Upper Carboniferous is often referred to as the "Age of Cockroaches" so common are their fossil remains from this period) and the order remains quite extensive today. Within this order two trends can be seen. The cockroaches became omnivorous or saprophagous, nocturnal, often secondarily wingless insects, whereas the mantids remained predaceous and diurnal in habit. Although Isoptera are known as fossils only from the Cretaceous period, comparison of their structure and certain features of their biology with those of primitive cockroaches (some of which are subsocial) indicates that they are derived from blattoidlike ancestors (Weesner, 1960). Indeed, the method of wing folding in the primitive termite family Mastotermitidae resembles that of fossil cockroaches rather than that of recent ones.

The order Orthoptera was already widespread by the Upper Carboniferous, its members being readily recognizable by their modified hind legs. Early in the evolution of the Orthoptera a split occurred, one l ne leading to the Ensifera (long-horned grasshoppers and crickets), the other to the Caelifera (short-horned grasshoppers and locusts).

The affinities of the two remaining orders of fossils found in the Upper Carboniferous, the Miomoptera and Caloneurodea, are problematical. The Miomoptera were small to very small insects with some orthopteroid features (for example, chewing mouthparts and cerci) but with a wing venation somewhat like that of the hemipteroid order Psocoptera. Members of the Caloneurodea had chewing mouthparts, short cerci, and orthopteroid wing venation on which basis Carpenter (1977) assigns the order to the orthopteroid group.

The affinities of several orders of orthopteroids are debatable. The Protelytroptera (Permian) were small insects whose fore wings were hard protective sheaths rather similar to the elytra of beetles, under which the hind wings folded at rest. It is possible that they were ancestral to the Dermaptera, which are first known as fossils from the Jurassic period. The Dermaptera have been placed by some authors close to the Orthoptera, Plecoptera, Embioptera, and even the nonorthopteroid Coleoptera. Other authors have suggested that they split off early from either Protoblattoidea or Protorhoptera and are only very distantly related to the other modern orthopteroid orders (Giles, 1963). As a result of his comparative morphological study, Giles concludes that the nearest living relatives of earwigs are the Grylloblattodea, and the two groups probably had a common ancestor. The Grylloblattodea show an interesting mixture of orthopteran, dictyopteran, phasmid, and dermapteran features, which has led several authorities to consider them as living remnants of a primitive stock from which both Orthoptera and Dictyoptera evolved. However, the presence of several specialized features combined with the fact that they are not known as fossils, probably precludes their being directly ancestral to any of the other orthopteroid orders (Imms, 1957). Kamp's (1973) numerical analysis has shown that considerable similarity exists between the Dermaptera and Grylloblattodea, which supports the conclusion reached by Giles (1963) of common ancestry for the two groups. The fossil record of the Phasmida is poor, specimens being known only from the Triassic onward. Kamp's (1973) study indicates that the order Phasmida has a close phenetic affinity with the Dermaptera and Grylloblattodea, and this author suggests that the three orders may form a natural group.

Although apparently a very ancient group with a fossil record extending back to the Permian period, the relationships of the Embioptera are obscure. They do not appear to be close to any other orthopteroid group and are generally assumed to be an early offshoot of protorthopteran stock (Illies, 1965).

Plecoptera (Permian–Recent) are well represented from the late Permian and probably had, as their ancestors, members of the Protoperlaria (Permian), which themselves are believed by most authorities to have evolved from Protorthoptera. Hamilton (1972b) notes, however, that unlike that of typical orthopteroids, the anal lobe of Plecoptera has only four or five veins and, as the hind wing folds, the lobe rolls under the rest of the wing rather than folding fanwise. Thus, Hamilton places the plecopteroid orders in the Planoneoptera as an early offshoot of the Hadentomoida.

The phylogenetic position of the order Zoraptera is uncertain. Members of the order are not known as fossils, but in some respects they resemble cockroaches, which has led to placement of the order in the orthopteroid group. On the other hand, other morphological features, including wing venation, led Ross (1955) and Hamilton (1972c) to suggest that the affinities of the order lie with the hemipteroids, perhaps as an offshoot of the Hadentomoida.

The hemipteroid orders, which with one possible exception have living representatives, share a number of features. They possess suctorial mouthparts, a fusion of ganglia in the ventral nerve cord, and few Malpighian tubules. They lack cerci, and the anal lobe of the hind wing is reduced, never having more than five veins. When the hind wing is drawn over the abdomen, it folds once along the anal or jugal fold, not between the anal veins as in the orthopteroids. The wing venation of hemipteroids is much reduced due to fusion of primary veins and almost complete loss of crossveins and, when both fore wings and hind wings are present, is basically similar in each. The Glosselytrodea (Lower Permian–Triassic) are generally considered to have been endopterygotes, probably having affinities with the Neuroptera (Carpenter, 1977). Hamilton (1972c) concludes, however, that members of the order had no venational features that would link them with neuropteroids but do have features in common with Hemiptera. As a result, Hamilton considers them as ancestral to the remaining hemipteroids. Both Hemiptera and Psocoptera are known from the Lower Permian. These early Hemiptera belonged to the suborder Homoptera, which radiated considerably in the Permian, Triassic, and Jurassic periods. Undoubted fossils of the suborder Heteroptera are known from the Jurassic period, but opinion differs as to the point at which they separated from the Homoptera. Although fossil Thysanoptera are first seen in the Upper Permian period, the record is generally poor. Ross (1955) suggests that the order stands closest to the Hemiptera. Fossil Phthiraptera have not been found, but they probably evolved from psocopteran ancestors as both Psocoptera and the more primitive, chewing lice possess a unique, rather complex hypopharynx (Imms, 1957). Members of the suborder Anoplura (sucking lice) apparently evolved in the Cretaceous from ancestral chewing forms, and are found only on placental mammals.

The main feature which unites members of the endopterygote orders is the presence of a pupal stage between the larval and adult stages in the life history. The wing venation of endopterygotes, though planoneopteran, is quite diverse and shows no obvious evolutionary trends among the orders (Hamilton, 1972c). Despite the occurrence of the pupal stage there has been considerable difficulty in deciding whether the group had a monophyletic or polyphyletic origin. The difficulty arises because, whereas the orders Megaloptera, Raphidioptera, Neuroptera, Mecoptera, Lepidoptera, Trichoptera, Diptera, and Siphonaptera show affinities with each other and form the "panorpoid complex" of Tillyard (1918–1920, 1935), the remaining orders (Coleoptera, Hymenoptera, and Strepsiptera) are quite distinct, each bearing little similarity to any other endopterygote group. The consensus of opinion is that the endopterygotes form a monophyletic group, though the major subgroups were formed at a very early date (Hinton, 1958; Mackerras, 1970; Hamilton, 1972c).

Many endopterygote orders were well established in the Lower Permian period. At that time, the Mecoptera was a diverse group that included many more families than presently exist. The neuropteroid orders (Megaloptera, Neuroptera, and Raphidioptera) were clearly distinguishable. Trichoptera were abundant, and primitive Coleoptera (suborder Archostemata) existed (Carpenter, 1977). Both the Mecoptera and neuropteroid group have on different occasions been considered as ancestral to the remaining endopterygotes on the grounds that some living representatives of both groups possess a number of very primitive features. Martynova (1961) suggested that it was from the neuropteroid group that the remaining endopterygotes arose, and this suggestion is supported by Hamilton's (1971, 1972a–c) studies of wing venation.

The neuropteroid orders were originally included by Tillyard in the panorpoid complex, in view of the several primitive features that they share with the Mecoptera. The modern view is that the neuropteroid group is a heterogeneous assembly of insects belonging to three distinct orders and, further, should be excluded from the panorpoid complex (see Hinton, 1958). Probably the Mecoptera and neuropteroids evolved from a common ancestor in the Upper Carboniferous period.

The extant Mecoptera are the survivors of a formerly quite extensive group from which two distinct lines arose in the Permian period, one leading to the Diptera and the other to the Trichoptera and Lepidoptera. The close link between Diptera and Mecoptera has been confirmed by the discovery of Permian fossils having a mixture of dipteran and mecopteran features. These fossils, although having a wing venation rather similar to the modern Tipulidae (crane flies), still possess four wings. The metanotum is, however, reduced, which suggests that the trend toward a single pair of wings seen in the modern Diptera had already begun (Jeannel, 1960).

The phylogenetic position of the Siphonaptera is unclear. Fossil fleas are known only from the Lower Cretaceous onward, and comparative studies of living fleas are of little use, since these insects are so highly modified for their ectoparasitic mode of life. Fleas resemble primitive Diptera in certain larval features and in the nature of their metamorphosis. For this reason some authors have suggested a dipteran ancestry. Hinton (1958) and Holland (1964), however, argue that the Siphonaptera probably evolved from ancestors which were similar to the modern Boreidae, a rather distinct family of Mecoptera. The argument is based on a number of similarities in larval structure and the fact that, within the panorpoid group, only the adults of Siphonaptera and Boreidae have panoistic ovarioles.

The Permian fossil *Belmontia* has wings with a blend of trichopteran and mecopteran features, indicating the affinity of these two orders (Ross, 1967). Although the Trichoptera is a long-established order and most modern families can be traced back to the Triassic and Jurassic periods, the use of "cases" in which to shelter seems to have been a relatively recent development. These structures are known as fossils only from the end of the Cretaceous period (Jeannel, 1960). Fossil Lepidoptera are known with certainty from the Cretaceous, although it is generally believed that the group split off the trichopteran line during the early Triassic period. There is no doubt that the Lepidoptera and Trichoptera are closely related, and this is attested by their many common structural features and habits (Ross, 1967). Comparison of primitive Trichoptera with primitive Lepidoptera has led Ross to suggest that the common terrestrial ancestor was in the adult stage trichopteran and in the larval stage lepidopteran in character. In the subsequent evolution of the Trichoptera the larva became specialized for an aquatic existence, but the adult remained primitive. Along the line leading to Lepidoptera the larva retained its primitive features, but the adult became specialized, especially in the development of the suctorial proboscis. The primitive lepidopteran family Micropterygidae is a relic group close to the point of separation of the Trichoptera and Lepidoptera. Members of this group have, as adults, functional biting mouthparts. In addition some species have a wing venation very similar to that of Trichoptera (Imms, 1957). The paucity of fossils makes it difficult to establish the point at which the trichopteran and lepidopteran lines separated. However, the variety of structure seen among the Eocene fossils and the distribution of certain modern groups indicate that the order is a very ancient one which in its infancy had a rather restricted distribution. The great adaptive radiation of the group probably came at the end of the Cretaceous period and was correlated with the evolution of the flowering plants (Jeannel, 1960).

Remains of genuine Coleoptera are known from the Upper Permian period. Somewhat earlier elytralike remains which were assigned to the order Protocoleoptera are now known to be the fore wings of Protelytroptera, an orthopteroid group (see above). Although in the earliest beetles the elytra were not yet thickened and strengthened for their modern function as protective sheaths, they were still sufficiently specialized as to be of little use in phylogenetic studies. The origin of Coleoptera is, therefore, uncertain. Handlirsch and Zeuner (1938 and 1933, respectively, cited in Crowson, 1960) concluded that they evolved from protoblattoid ancestors, which implies that the endopterygotes have at least a diphyletic origin. Crowson (1960) and others claim on the basis of larval form and mode of development that the Coleoptera are derived from neuropteroid stock. According to Crowson this claim is substantiated by the Lower Permian fossil *Tshekardocoleus*, which is intermediate in form between primitive Coleoptera and Megaloptera.

Fossil Hymenoptera are known from the Triassic period, but these were already quite specialized individuals, clearly recognizable as belonging to the suborder Symphyta. The more advanced Hymenoptera of the suborder Apocrita, which contains the parasitic and stinging forms, are not known as fossils until the Jurassic and Cretaceous periods. The great adaptive radiation of the Hymenoptera was, like that of the Lepidoptera, clearly associated with the evolution of the flowering plants. Although early paleontologists suggested that the Hymenoptera evolved from Protorthopteran stock, a more likely origin is from neuropteroid ancestors, judging by the similarity between the larvae of Symphyta and those of panorpoid insects, and the ease with which the wing venation of Hymenoptera can be derived from the pattern found in Megaloptera (Imms, 1957).

The position of the Strepsiptera, highly modified endoparasitic insects, is uncertain. The earliest fossils are already well-developed forms from the Oligocene period and therefore speculation on their origin is based on comparative morphology. Jeannel (1960) argues that the Strepsiptera arose from an ancient stock derived from very primitive Hymenoptera. However, most authorities, including Crowson (1960), place the Strepsiptera near (or even within) the Coleoptera by virtue of their several common features, of which the most obvious is the use of the hind wings in flight. In both groups extensive sclerotization of the sternum (rather than the tergum) occurs. The first instar larva of Strepsiptera resembles closely the triungulin larva of the beetle families Meloidae and Rhipiphoridae. In addition the endoparasitic forms of the Rhipiphoridae have habits that are generally similar to those of Strepsipera.

The foregoing discussion on the evolutionary relationships within the Insecta is summarized in Figure 2.7.

3.3. Origin and Function of the Pupa

As noted in the previous section the Oligoneoptera (endopterygote orders) are characterized by the presence of a pupal stage between the juvenile and adult phases in the life history. The development of this stage, which serves a variety of functions, is a major reason for the success (that is, diversity) of endopterygotes. Several theories have been proposed for the origin of the pupal stage (Figure 2.8).

Berlese's Theory. During its development the insect embryo passes through three distinct stages. In the first (protopod) stage no appendages are visible; this is followed by the polypod stage (in which appendages are present on most segments; and finally the oligopod stage (when the appendages on the abdomen have been resorbed). The Berlese theory suggests that the eggs of exopterygotes, by virtue of their greater yolk reserves, hatch in a postoligopod stage of development, whereas the eggs of endopterygotes, which have less yolk, hatch in the polypod or oligopod stages. Thus in this theory the abdomi-

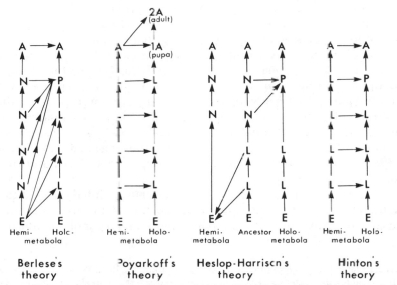

FIGURE 2.8. Theories for the origin of the pupal stage. Abbreviations: A, adult; E, egg; L, larva; N, nymph; P, pupa. [Partly after H. E. Hinton, 1963b, The origin and function of the pupal stage, *Proc. R. Entomol. Soc. London Ser. A* **38**:77–85. By permission of the Royal Entomological Society.]

nal prolegs must be considered homodynamous with the thoracic legs. The larva of endopterygotes corresponds to a free-living embryonic stage and the pupa represents the compression of the (postoligopod) nymphal stages into a single instar.

Berlese's theory has been criticized by Hinton (1948, 1955, 1963b) on the grounds that there is no evidence that the eggs of exopterygotes have a better supply of yolk than those of endopterygotes. Hinton also claims that the abdominal prolegs are not homodynamous with the thoracic legs but are secondary larval structures. Heslop-Harrison (1958), in proposing a modified version of Berlese's ideas, argues that Hinton's claim is not justified, especially in view of the polypodous nature of the ancestors of insects.

Novak's Theory. Novak's (1966) theory is based on Berlese's proposal that the eggs of endopterygotes hatch at an earlier stage of embryonic development than do those of exopterygotes. According to Novak in exopterygotes, which supposedly hatch in the postoligopod stage, differentiation of adult tissues begins before an effective concentration of juvenile hormone is reached in larval life so that, specifically, the wing rudiments can develop on the body surface. (Juvenile hormone inhibits adult tissue differentiation. See Chapter 21, Section 6.1.) In endopterygotes, however, which hatch earlier, production of juvenile hormone begins before adult tissues have started to differentiate. Thus, the adult tissues cannot begin to differentiate until the hormone titer falls to a sufficiently low level in the final instar larva. Thus, two molts are necessary; one to evaginate the undifferentiated wing discs so that the wings can develop in a suitably large space, and another to permit expansion of the fully developed wings.

Poyarkoff's Theory. This theory has been supported by several authorities, including Hinton (1948), Snodgrass (1954), and DuPorte (1958). The advantages of this theory over Berlese's is that it provides a causal explanation for the origin and function of the pupal stage. According to this theory, the eggs of both endopterygotes and exopterygotes hatch at a similar stage of development; the adult stage in the exopterygote ancestors of the endopterygotes became divided into two instars, the pupa and the imago; the subimago of Ephemeroptera and the "pupal" stage of some exopterygotes are equivalent to the endopterygote pupal stage; the pupal stage evolved in response to the need for a mold in which the adult systems, especially flight musculature, could be constructed. The second (pupal–imaginal) molt was then necessary in order that the new muscles could become attached to the exoskeleton. In Poyarkoff's theory there is no difference between the endopterygote larva and the exopterygote nymph.*

In supporting Poyarkoff's theory, Hinton (1948) pointed out that the pupa (especially that of primitive endopterygotes such as Neuroptera) resembles the adult form rather than that of the larva, supernumerary molts could be artifi-

---

*Because they were considered originally to be quite distinct, the juvenile stages of exopterygotes and endopterygotes were referred to as "nymph" and "larva," respectively. The modern view (see Hinton's theory) is that nymphal and larval stages are homologous and that pterygote juvenile stages should be called larvae. For clarity of discussion, however, in this chapter only, the traditional distinction has been retained.

cially induced under certain conditions in exopterygotes, and in Ephemeroptera a "natural" adult molt (from subimago to imago) occurs. Unfortunately no evidence was available to support the idea that the function of the pupal stage was to serve as a mold for development of the adult wing musculature.

This theory and Hinton's (1948) support of it have been criticized by Heslop-Harrison (1958) and Hinton (1963b). Heslop-Harrison claimed that the explanation given for the origin of the pupal stage is teleological; in other words, Poyarkoff's explanation is that the pupal stage arose in fulfillment of a "need." It is further suggested by Heslop-Harrison that Hinton's (1948) supporting evidence was carefully selected and does not represent the true situation. For example, Hinton's claim that the pupal stage resembles generally the adult was refuted by Heslop-Harrison, who pointed out several similarities between the pupal and larval instars. Hinton (1963b) stated that there is direct evidence against Poyarkoff's idea concerning the function of the pupal stage. First, it has been shown that tonofibrillae (microtubules within the epidermal cells which attach muscles to the exoskeleton) can be formed long after the pupal–adult molt has occurred. Second, even in highly advanced endopterygotes the fiber rudiments of the wing muscles are present at the time of hatching. These develop in the larva in precisely the same way as the flight muscles of many primitive exopterygotes. In other words, no molt is required.

Heslop-Harrison's Theory. Implicit in the theories of Berlese, Poyarkoff, and Hinton (1963b) is the evolution of endopterygotes from exopterygote ancestors. Heslop-Harrison believed, however, that the earliest forms of both groups were present at the same time and evolved from a common ancestor. This ancestor had a life history similar to that of modern Isoptera and Thysanoptera, namely, EGG → LARVAL INSTARS (showing no sign of wings) → NYMPHAL INSTARS (having external wing buds) → ADULT. Heslop-Harrison proposed that in the evolution of exopterygotes the larval instars were suppressed, and the modern free-living juvenile stages correspond to the nymphal instars of the ancestors. In the evolution of endopterygotes the nymphal stages were compressed into the quiescent prepupal and pupal stages of modern forms. (The prepupal stage is a period of quiescence in the last larval instar prior to the molt to the pupa. In other words the prepupal stage is not a distinct instar.) Thus, Berlese's original concept that the pupa comprised the ontogenetic counterparts of nymphal instars was supported by Heslop-Harrison. The basis of Heslop-Harrison's theory was his comparative study of the life history of various Homoptera in which the last nymphal instar is divided into two phases. In the most primitive condition the first of these phases is an active one where the insect feeds and/or prepares its "pupal" chamber. In the most advanced condition both phases are inactive, and there are, for all intents and purposes, distinct prepupal and pupal stages, as in true endopterygotes.

The main general criticism of Heslop-Harrison's theory is that it lacks supporting evidence. More specific criticisms are that (1) the fossil record indicates that the earliest insects were only exopterygote, and most authorities believe that the endopterygotes evolved from exopterygote stock (perhaps the Protorthoptera); (2) the Isoptera and Thysanoptera on which Heslop-Harrison's "primitive life history" was based are two highly specialized exopterygote orders; and (3) the implied homology of the endopterygote pupa and the last juvenile instars of the exopterygote Homoptera studied by Heslop-Harrison is not justified [see discussion in Hinton (1963b)].

Hinton's Theory.   Perhaps the attraction of Hinton's (1963b) theory is its simplicity. It avoids the "suppression of larval," "compression of nymphal" and "expansion of imaginal" stages, found in the earlier theories and provides a simple functional explanation for the evolution of a pupal stage.

In Hinton's theory the pupa is homologous with the *final* nymphal instar of exopterygotes, and the terms "larva" and "nymph" are synonymous. Hinton proposes that, during the evolution of endopterygotes, the last juvenile stage (with external wings) was retained to complete the link between the earlier juvenile stages (larvae with internal wings) and the adult, hence the general resemblance between the pupa and adult in modern endopterygotes. Initially, the pupa would also resemble the earlier instars (just as the final instar nymph of modern exopterygotes resembles both the adult and the earlier nymphal stages). Once this intermediate stage had been established it is easy to visualize how the earlier juvenile stages could have become more and more specialized (for feeding and accumulating reserves) and quite different morphologically from both the pupa and the adult (the reproductive and dispersal stage). At the same time the pupa itself became more specialized. It ceased feeding actively, became less mobile, and was concerned solely with the metamorphosis from the juvenile to the adult form.

Concerning the functional significance of the pupal stage, Hinton suggested that, as the endopterygote condition evolved, there was insufficient space in the thorax to accommodate both the "normal" contents (muscles and other organ systems) and the wing anlage. Thus, the function of the larval–pupal molt was to evaginate the wings. This would permit not only considerable wing growth (as greatly folded structures within the pupal external wing cases) but also the enormous growth of the imaginal wing muscles within the thorax. The latter is facilitated, of course, by histolysis of the larval muscles (a process which is often not completed for many hours after the pupal–adult molt). The function of the pupal–adult molt is simply to effect release of the wings from the pupal case. The original function of the pupal stage was, then, to create space for wing and wing muscle development. But, once a stage had been developed in the life history in which structural rearrangement could take place, the way was open for increasing divergence of juvenile and adult habits and, subsequently, a decrease in the competition for food, space, etc. between the two stages.

4. The Success of Insects

The degree of success achieved by a group of organisms can be measured either as the total number of organisms within the group or, more commonly, as the number of different species of organisms which comprise the group. On either account the insects must be considered highly successful. Success is dependent on two interacting factors: (1) the potential of the group for adapting to new environmental conditions, and (2) the degree to which the environmental conditions change. Since success measured as the number of different species is a direct result of evolution, the environmental changes that must be considered are the long-term climatic changes that have occurred in different parts of the world over a period of several hundred million years.

4.1. The Adaptability of Insects

The basic feature of insects to which their success can be attributed must surely be that they are arthropods. As such they are endowed with a body plan that is superior to that of any other invertebrate group. Of the various arthropodan features the integument is the most important, since it serves a variety of functions. Its lightness and strength make it an excellent skeleton for attachment of muscles as well as a "shell" within which the tissues are protected. Its physical structure (usually including an outermost wax layer) makes it especially important in the water relations of arthropods. Because they are generally small organisms, arthropods in almost any environment face the problem of maintaining a suitable salt and water balance within their bodies. The magnitude of the problem (and, therefore, the energy expended in solving it) is greatly reduced by the impermeable cuticle (see Chapters 11 and 18). Arthropods are segmented animals and therefore have been able to exploit the advantages of tagmosis to the full extent. Directly related to this is the adaptability of the basic jointed limb, a feature exploited fully by different groups of arthropods (see Figure 1.6 and descriptions of segmental appendages in Chapter 3).

Since all arthropods possess these advantageous features, the obvious question to ask is "Why have only insects been so successful?" or, put differently, "What features do insects possess that other arthropods do not?" Answering this question will provide only a partial answer, since, as was stressed above, the environmental changes that take place are also very important in determining degree of success. Take, for example, the Crustacea. Compared to other invertebrate groups they must be regarded as successful (at least 26,000 species have been described), yet in comparison to the Insecta they come a distant second. Although this is due partly to their different features, it must be due also to the different habitats in which they evolved. The Crustacea are a predominantly marine group. In other words, they evolved under relatively stable environmental conditions. Furthermore, it is likely that when they were evolving the number of niches that were available to them would be quite limited because most were occupied by already established groups. The insects, on the other hand, evolved in a terrestrial environment subject to great changes in physical conditions. They were one of the earliest groups to "venture on land" and, therefore, had a vast number of niches available to them in this new adaptive zone.

Most insects, modern and fossil, are small animals. A few early forms achieved a large size but became extinct presumably because of climatic changes and their inability to compete successfully with other groups. Small size confers several advantages on an organism. It facilitates dispersal, it enables the organism to hide from potential predators, and it allows the organism to make use of food materials that are available in only very small amounts. The great disadvantage of small size in terrestrial organisms is the potentially high rate of water loss from the body. In insects this has been successfully overcome through the development of an impermeable exoskeleton.

The ability to fly was perhaps the single most important evolutionary development in insects. With this asset the possibilities for escape from predators and for dispersal were greatly enhanced. It would lead to colonization of new habitats, geographic isolation of populations, and, as a result, formation of new species. Wide dispersal is particularly important for those species whose food and breeding sites are scattered and in limited supply.

The reproductive capacity and life history are two related factors that have contributed to the success of insects. Production of large numbers of eggs, combined with a short life history, means a greater amount of genetic variation can occur and be tested out rapidly within a population. The net result is (1) rapid adaptation to changes in environmental conditions, and (2) rapid attainment of genetic incompatibility between isolated populations and formation of new species. The evolution of a pupal stage between the larval and adult stages has led to a more specialized (and, in a sense, a more "efficient") life history. The main function of the larva is accumulation of metabolic reserves, whereas the adult is primarily concerned with reproduction and dispersal. This means that insects can utilize food sources that are available for only short periods of time. Although the pupa has as its main function the transformation of the larval to the adult form, it has in many species become a stage in which insects can resist unfavorable conditions. This development, and the restriction of feeding activity to one phase of the life history, have facilitated the expansion of the insect fauna into some of the world's most unfriendly habitats.

Several features of insects have contributed therefore to their success (diversification). It is important to realize that these features have acted in combination to effect success, and, furthermore, little of this success would have been possible except for the unstable environmental conditions in which the insects evolved.

4.2. The Importance of Environmental Changes

The importance of environmental changes in the process of evolution, acting through natural selection, is well known. These changes can be seen acting at the population or species level on a short-term basis, and many examples are known in insects, perhaps the best two being the development of resistance to pesticides and the formation of melanic forms of certain moths in areas of industrial air pollution. Of greater interest in the present context are the long-term climatic changes that have taken place over millions of years, for it is these that have controlled the evolution of insects both directly and indirectly through their influence on the evolution of other organisms, especially plants.

Although life began at least 2.5 billion years ago, it was not until the Silurian period (about 425–500 million years ago) that the first terrestrial organisms appeared, an event probably correlated with the formation of an ozone layer in the atmosphere, which reduced the amount of ultraviolet radiation reaching the earth's surface (Berkner and Marshall, 1965). The earliest terrestrial organisms were simple low-growing land plants that reproduced by means of spores. They were soon followed by mandibulate arthropods (scorpions, myriapods, and wingless insects) which presumably fed on the plants, their decaying remains, their spores, or on other small animals. In these early plants, spores were produced on short side branches of upright stems. An important evolutionary development was the concentration of the sporangia (spore-producing structures) into a terminal spike. Whether these spikes were particularly attractive as food for insects and whether, therefore, they may have been important in the evolution of flight is a matter for speculation (Smart and Hughes, 1973).

During the Devonian and Lower Carboniferous periods, a wide radiation of plants occurred. Especially significant was the development of swamp forests that contained, for example, tree lycopods, calamites, and primitive gymnosperms. The evolution of treelike form, though in part due to the struggle for light, may also have enabled the plants to protect (temporarily) their reproductive structures against spore-feeding insects and other animals. In contrast to the humid or even wet conditions on the forest floor, the air several meters above the ground was probably relatively dry. Thus, the evolution of trees with terminal sporangia several meters above the ground may have been an important stimulant to the evolution of a waterproof cuticle, spiracular closing mechanisms, and eventually, flight (Kevan et al., 1975).

The trees, together with the ground flora, would provide a wide range of food material. As noted earlier, winged insects appeared in the Lower Carboniferous. Many of these were mandibulate (for details see previous section) and fed on soft parts of plants or litter on the forest floor. Others, for example, some Paleodictyoptera, had mouthparts in the form of a proboscis which has led to the suggestion that these insects were adapted to feeding on liquids either free-standing or as sap in plant tissues. Smart and Hughes (1973) believe, however, that the proboscis might have been used as a probe for extracting pollen and spores from the plants' reproductive structures. They argue that not until the Upper Carboniferous did plants evolve which had phloem close enough to the stem surface that it was accessible to Hemiptera. Yet other insects such as the Protodonata, Megasecoptera and, later, Odonata, were predators. Some of these were very large, and it has been suggested that the evolution of large size was a result of competition between these insects and the earliest terrestrial vertebrates, the Amphibia. Certainly large size would be favored by the year-round, uniform growing conditions (Ross, 1965). In addition to the forest ecosystem, there were presumably other ecosystems, for example, the edges of swamps and higher ground, that had their complement of insects. However, such ecosystems did not apparently favor fossilization, and their insect fauna is practically unknown.

Toward the end of the Carboniferous period other climatic changes of importance in insect evolution took place. At this time extensive mountain ranges were formed and over many parts of the earth the climate became cooler and drier. These changes created not only many new habitats but also barriers that prevented gene flow between populations. By the end of the Permian period most of the older orders were extinct and had been replaced by representatives of the modern orders.

In the Triassic and early Jurassic periods the gradual radiation of insects continued but was largely overshadowed by evolution of the reptiles. The latter occurred in such large numbers and in such a variety of habitats that the Triassic period is generally known as the "Age of Reptiles." Many of them were insectivorous, and this acted as a selection pressure favoring small size, which is a general feature of fossil insects from this period.

Early in the Triassic period the first bisexual flowers appeared. The occurrence together of male and female structures immediately leads to the possibility of a role as pollinators for insects that feed on the reproductive parts of plants. Because of the risk of having their reproductive structures eaten, these early plants probably produced a large number of small ovules and much pollen (Smart and Hughes, 1973). The insect fauna of the Triassic period still included the "orthodox" plant feeders such as Orthoptera and some Coleoptera, the plant-sucking Hemiptera, and predaceous species (Odonata and Neuroptera). However, there were also a large number of primitive endopterygotes, mostly belonging to the panorpoid complex, Hymenoptera, and Coleoptera, which as adults were mandibulate and therefore were potential "mess-and-soil" pollinators (Smart and Hughes, 1973).

By the middle of the Jurassic period the decline of the reptiles had begun and the earth's climate had become generally warmer. As a result the insect fauna increased both in mean body size and in variety. A good deal of mountain formation occurred at the end of the Jurassic, creating new climatic conditions in various parts of the world including, for the first time, winters.

The Cretaceous period was an extremely important phase in the evolution of insects, for it was during this time that vast adaptive radiations of several endopterygote orders took place. In some instances it must be assumed that these radiations directly paralleled the evolution of angiosperms, although, it must be emphasized, the fossil record of angiosperm flowers is sparse. As a result of this coevolution, some extremely close interrelationships have evolved between plants and their insect pollinators (see Chapter 23, Section 2.3). For other orders, the radiation was only indirectly due to plants; for example, a large variety of parasitic Hymenoptera appeared, correlated with the large increase in numbers of insect hosts.

The Cretaceous period was also rather active, geologically speaking, for a good deal of mountain making, lowland flooding (by the sea), and formation and breakage of land bridges took place. All these processes would assist in the isolation and diversification of the insect fauna.

The decline of the reptiles became accelerated during the late Cretaceous period, and they were succeeded by the mammals and birds. These groups became very widespread and diverse during the Tertiary period. Paralleling this diversification was the evolution of their insect parasites. Throughout the Tertiary period the climate seems to have alternated between warm and cold. In the Paleocene epoch the climate was cooler than that of the Cretaceous. Thus cold-adapted groups became widely distributed. A warming trend followed in the Eocene so that cold-adapted organisms became restricted to high altitudes, while the warm-adapted types spread. It appears that by the end of the Eocene period (approximately 36 million years ago) most modern tribes or genera of insects had evolved (Ross, 1965). In the Oligocene the climate became cooler and remained so during the Miocene and Pliocene epochs. However, in these two epochs new mountain ranges were formed, and some already existing ones were pushed even higher. The Tertiary period ended about one million years ago and was followed by the Pleistocene epoch of the Quaternary period. Temperatures in the Pleistocene (Ice Age) were generally much lower than in the Tertiary, and four distinct periods of glaciation occurred, at which time most of the North American and European continents had a thick covering of ice. Between these periods warming trends caused the ice to recede northward. Accompanying these ice movements were parallel movements of the fauna and flora. However, the overall significance of the Ice Age in terms of insect evolution is uncertain at the present time (Ross, 1965).

5. Summary

From their beginning as primitively wingless thysanuranlike creatures, the insects have undergone a vast adaptive radiation to become the world's most successful group of living organisms. Undoubtedly, a large measure of this success can be attributed to the evolution of wings. The latter evolved from paranotal lobes whose original function appears to have been attitudinal control, enabling falling insects to land on their feet, thus facilitating escape from predators. Wing articulation, a later development, was probably concerned initially with improving attitudinal control.

The Pterygota probably had a monophyletic origin in the Devonian or early Carboniferous periods, and soon split into two major evolutionary lines, the Paleoptera and the Neoptera. By the late Carboniferous three distinct neopteran groups were established. These were the orthopteroids, hemipteroids, and endopterygotes. Of the three the endopterygotes have been by far the most successful. This is due, in large part, to the evolution of a pupal stage within the group. Various theories have been advanced for the origin of the pupal stage. The most likely theory proposes that it is equivalent to the final nymphal instar of the original exopterygote ancestor. Its initial function was to provide space for wing and wing muscle development.

Insect success (that is, diversity) is due not only to the group's adaptability but also to the environment in which they have evolved. Being arthropods, insects possess a body plan which is superior to that of other invertebrates. They are generally small and able to fly. They usually have a high reproductive capacity, often coupled with a life history which is short and contains a pupal stage. Insect evolution was coincident with the evolution of land plants. The insects were among the first invertebrates to establish themselves on land. By virtue of their adaptability they were able to colonize rapidly the new habitats formed as a result of climatic changes over the earth's surface.

6. Literature

For further information students should consult Jeannel (1960), Ross (1965), Mackerras (1970), Riek (1970), Hamilton (1971, 1972a–c), and Carpenter (1977) for details of the geological history and phylogenetic relationships of insects; Wigglesworth (1963a,b, 1973, 1976), Hinton (1963a), and Kukalova-Peck (1978) [discussion on origin of wings]; Heslop-Harrison (1958) and Hinton (1963b) [theories on the origin and function of the pupal stage]; and Becker (1965), Smart and Hughes (1973), and Kevan *et al.* (1975) [coevolution of insects and plants].

Alexander, R. D., and Brown, W. L., Jr., 1963, Mating behavior and the origin of insect wings, *Occas. Pap. Mus. Zool. Univ. Mich.* **628**:1–19.

Becker, H. F., 1965, Flowers, insects, and evolution, *Nat. Hist.* **74**:38–45.

Berkner, L. V., and Marshall, L. C., 1965, On the origin and rise of oxygen concentration in the earth's atmosphere, *J. Atmos. Sci.* **22**:225–261.

Carpenter, F. M., 1977, Geological history and evolution of the insects, *Proc. XV Int. Congr. Entomol.*, pp. 63–70.

Crampton, G. C., 1916, Phylogenetic origin and the nature of the wings of insects according to the paranotal theory, *J. N.Y. Entomol. Soc.* **24**:1–39.

Crowson, R. A., 1960, The phylogeny of Coleoptera, *Annu. Rev. Entomol.* **5**:111–134.

DuPorte, E. M., 1958, The origin and evolution of the pupa, *Can. Entomol.* **90**:436–439.

Flower, J. W., 1964, On the origin of flight in insects, *J. Insect Physiol.* **10**:81–88.

Giles, E. T., 1963, The comparative external morphology and affinities of the Dermaptera, *Trans. R. Entomol. Soc. London* **115**:95–164.

Hamilton, K. G. A., 1971, 1972a–c, The insect wing, Parts I–IV, *J. Kans. Entomol. Soc.* **44**:421–433; **45**:54–58, 145–162, 295–308.

Heslop-Harrison, G., 1958, On the origin and function of the pupal stadia in holometabolous Insecta, *Proc. Univ. Durham Philos. Soc. Ser. A* **13**:59–79.

Hinton, H. E., 1948, On the origin and function of the pupal stage, *Trans. R. Entomol. Soc. London* **99**:395–409.

Hinton, H. E., 1955, On the structure, function and distribution of the prolegs of the Panorpoidea, with a criticism of the Berlese–Imms theory, *Trans. R. Entomol. Soc. London* **106**:455–545.

Hinton, H. E., 1958, The phylogeny of the panorpoid orders, *Annu. Rev. Entomol* **3**:181–206.

Hinton, H. E., 1963a, Discussion: The origin of flight in insects, *Proc. R. Entomol. Soc. London Ser. C* **28**:23–32.

Hinton, H. E., 1963b, The origin and function of the pupal stage, *Proc. R. Entomol. Soc. London Ser. A* **38**:77–85.

Hocking, B., 1957, Aspects of insect flight, *Sci. Month.* **85**:237–244.

Holland, G. P., 1964, Evolution, classification, and host relationships of Siphonaptera, *Annu. Rev. Entomol.* **9**:123–146.

Illies, J., 1965, Phylogeny and zoogeography of the Plecoptera, *Annu. Rev. Entomol.* **10**:117–140.

Imms, A. D., 1957, *A General Textbook of Entomology*, 9th ed., (revised by O. W. Richards and R. G. Davies), Methuen, London.

Jeannel, R., 1960, *Introduction to Entomology*, Hutchinson, London.

Kamp, J. W., 1973, Numerical classification of the orthopteroids, with special reference to the Grylloblattodea, *Can. Entomol.* **105**:1235–1249.

Kevan, P. G., Chaloner, W. G., and Savile, D. B. O., 1975, Interrelationships of early terrestrial arthropods and plants, *Palaeontology* **18**:391–417.

Kukalova-Peck, J., 1978, Origin and evolution of insect wings and their relation to metamorphosis, as documented by the fossil record, *J. Morphol.* **156**:53–126.

Mackerras, I. M., 1970, Evolution and classification of the insects, in: *The Insects of Australia* (I. M. Mackerras, ed.), Melbourne University Press, Carlton, Victoria.

Martynov, A. V., 1938, Etudes sur l'histoire géologique et de phylogénie des ordres des insectes (Pterygota), 1–3e. partie, Palaeoptera et Neoptera-Polyneoptera, *Trav. Inst. Paléont., Acad. Sci. URSS*, pp. 1–150.

Martynova, O. M., 1961, Palaeoentomology, *Annu. Rev. Entomol.* **6**:285–294.

Novak, V. J. A., 1966, *Insect Hormones*, Methuen, London.

Riek, E. F., 1970, Fossil history, in: *The Insects of Australia* (I. M. Mackerras, ed.), Melbourne University Press, Carlton, Victoria.

Ross, H. H., 1955, Evolution of the insect orders, *Entomol. News* **66**:197–208.

Ross, H. H., 1965, *A Textbook of Entomology*, 3rd ed., Wiley, New York.

Ross, H. H., 1967, The evolution and past dispersal of the Trichoptera, *Annu. Rev. Entomol.* **12:**169–206.

Sharov, A. G., 1966, *Basic Arthropodan Stock*, Pergamon Press, Oxford.

Smart, J., and Hughes, N. F., 1973, The insect and the plant: Progressive palaeoecological integration, *Symp. R. Entomol. Soc. London* **6:**143–155.

Snodgrass, R. E., 1954, Insect metamorphosis, *Smithson. Misc. Collect.* **122:**124 pp.

Tillyard, R. J., 1918–20, The panorpoid complex. A study of the phylogeny of the holometabolous insects with special reference to the subclasses Panorpoidea and Neuropteroidea, *Proc. Linn. Soc. N.S.W.* **43:**235–284, 395–408, 626–657; **44:**533–718; **45:**214–217.

Tillyard, R. J., 1935, The evolution of the scorpion flies and their derivatives, *Ann. Entomol. Soc. Am.* **28:**1–45.

Weesner, F. M., 1960, Evolution and biology of termites, *Annu. Rev. Entomol.* **5:**153–170.

Wigglesworth, V. B., 1963a, Origin of wings in insects, *Nature (London)* **197:**97–98.

Wigglesworth, V. B., 1963b, Discussion: The origin of flight in insects, *Proc. R. Entomol. Soc. London Ser. C* **28:**23–32.

Wigglesworth, V. B., 1973, Evolution of insect wings and flight, *Nature (London)* **246:**127–129.

Wigglesworth, V. B., 1976, The evolution of insect flight, *Symp. R. Entomol. Soc.* **7:**255–269.

Wootton, R. J., 1976, The fossil record and insect flight, *Symp. R. Entomol. Soc.* **7:**235–254.

3

External Structure

1. Introduction

The extreme variety of external form seen in the Insecta is the most obvious manifestation of this group's adaptability. To the taxonomist who thrives on morphological differences, this variety is manna from Heaven; to the morphologist who likes to refer everything back to a "basic type," it can be a nightmare! Paralleling this variety is, unfortunately, a massive terminology, even the basics of which an elementary student may find difficult to absorb. He may receive some consolation from the fact that "form reflects function." In other words, seemingly minor differences in structure may reflect important differences in the functional capabilities of two groups. It is impossible to deal in a text of this kind with all the variation in form that exists, and only the basic structure of an insect and its most important modifications will be described.

2. General Body Plan

Like other arthropods insects are segmented animals whose bodies are covered with cuticle. Over most regions of the body the outer layer of the cuticle becomes hardened (tanned) and forms the exocuticle (see Chapter 11). These regions are separated by areas (joints) in which the exocuticular layer is missing, and the cuticle therefore remains membranous, flexible, and often folded. The presence of these cuticular membranes facilitates movement between adjacent hard parts (SCLERITES). The degree of movement at a joint depends on the extent of the cuticular membrane. In the case of intersegmental membranes there is complete separation of adjacent sclerites, and therefore movement is unrestricted. Usually, however, especially at appendage joints, movement is restricted by the development of one or two contiguous points between adjacent sclerites; that is, specific articulations are produced. A MONOCONDYLIC joint has only one articulatory surface, and at this joint movement may be partially rotary (for example, the articulation of the antennae with the head). In DICONDYLIC joints (for example, most leg joints) there are two articulations, and the joint operates like a hinge. The articulations may be either INTRINSIC, where the contiguous points lie within the membrane [Figure

3.1A), or EXTRINSIC, in which case the articulating surfaces lie outside the skeletal parts (Figure 3.1B).

In many larval insects (as in annelids) the entire cuticle is thin and flexible, and segments are separated by invaginations of the integument (the INTERSEG-MENTAL FOLDS) to which longitudinal muscles are attached (Figure 3.2A). Animals possessing this arrangement (known as PRIMARY SEGMENTATION) have almost unlimited freedom of body movement. In the majority of insects, however, there is heavy sclerotization of the cuticle to form a series of dorsal and ventral plates, the TERGA and STERNA, respectively. As shown in Figure 3.2B, these regions of sclerotization do not correspond precisely with the primary segmental pattern. The tergal and sternal plates do not cover entirely the posterior part of the primary segment, yet they extend anteriorly slightly beyond the original intersegmental groove. Thus the body is differentiated into a series of secondary segments (SCLEROMATA) separated by membranous areas (CON-JUNCTIVAE) that allow the body to remain flexible. This is termed SECONDARY SEGMENTATION. Each secondary segment contains four exoskeletal components, a tergum and a sternum separated by lateral, primarily membranous, PLEURAL AREAS. Each of the primary components may differentiate into several sclerites to which the general terms TERGITES and STERNITES are applied; small sclerites, generally termed PLEURITES, may also occur in the pleural areas. The primitive intersegmental fold becomes an internal ridge of cuticle, the AN-TECOSTA, seen externally as a groove, the ANTECOSTAL SULCUS. The narrow strip of cuticle anterior to the sulcus is the ACROTERGITE (when dorsal) or ACROSTER-NITE (when ventral). The posterior part of both the tergum and sternum is primitively a simple cuticular plate, but this undergoes considerable modification in the thoracic region of the body. The pleurites are usually secondary sclerotizations but in fact may represent the basal segment of the appendages. The pleurites may become greatly enlarged and fused with the tergum and sternum in the thoracic segments. In the abdomen the pleurites may fuse with the sternal plates.

The basic segmental structure is frequently obscured as a result of tag-

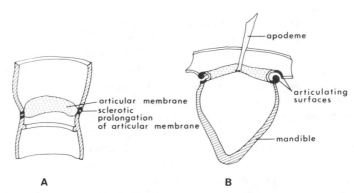

A   **B**

FIGURE 3.1. Articulations. (A) Intrinsic (leg joint) and (B) extrinsic (articulation of mandible with cranium). [From R. E. Snodgrass, *Principles of Insect Morphology.* Copyright 1935 by McGraw-Hill, Inc. Used with permission of McGraw-Hill Book Company.]

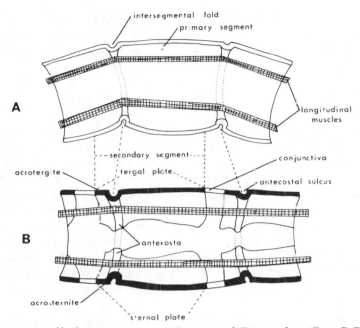

FIGURE 3.2. Types of body segmentation. (A) Primary and (B) secondary. [From R. E. Snodgrass, *Principles of Insect Morphology*. Copyright 1935 by McGraw-Hill, Inc. Used with permission of McGraw-Hill Book Company.]

mosis. In insects three tagmata are found: the head, the thorax, and the abdomen. In the head almost all signs of the original boundaries of the segments have disappeared, though, for most segments, the appendages remain. In the thorax the three segments can generally be distinguished, although they undergo profound modification associated with locomotion. The anterior abdominal segments are usually little different from the typical secondary segment described above. At the posterior end of the abdomen a variable number of segments may be modified, reduced, or lost, associated with the development of the external genitalia.

Examination of the exoskeleton reveals the presence of a number of lines or grooves whose origin is variable. If the line marks the union of two originally separate sclerites it is known as a SUTURE. If it indicates an invagination of the exoskeleton to form an internal ridge of cuticle (APODEME), the line is properly termed a SULCUS (Snodgrass, 1960). Pits may also be seen on the exoskeleton. These pits mark the sites of internal, tubercular invaginations of the integument (APOPHYSES). Secondary discontinuations of the exocuticular component of the cuticle may occur, for example, the ecdysial line along which the old cuticle splits during molting, and these are generally known as sutures.

Primitively each segment bore a pair of appendages. Traces of these can still be seen on almost all segments for a short time during embryonic development, but on many segments they soon disappear, and typical insects lack abdominal appendages on all except the posterior segments. The primitive segmental appendage of insects was five-segmented, and this arrangement may still be seen in the thoracic legs of many extant species, which comprise a basal

COXA, TROCHANTER, FEMUR, TIBIA, and TARSUS. The remaining appendages of the head and abdomen have become highly modified, and homologizing their segmental components with those of the leg may be extremely difficult.

3. The Head

The head, being the anterior tagma, bears the major sense organs and the mouthparts. Considerable controversy still surrounds the problem of segmentation of the insect head, especially concerning the number and nature of segments anterior to the mouth.* At various times it has been argued that there are from three to seven segments in the insect head, although it is generally agreed that there are three postoral (gnathal) segments, the MANDIBULAR, MAXILLARY, and LABIAL. It is not feasible to discuss here the many theories concerning the segmental composition of the insect head, and the reader is referred to the work of Rempel (1975) for a recent review of the subject. The major points of contention appear to have been (1) whether a preantennal segment occurs between the acron and the antennal segment and what appendages are associated with such a segment, and (2) whether the antennae are segmental appendages or merely outgrowths of the acron. Recent embryological studies (for example, those of Rempel and Church, 1971) have demonstrated convincingly that the head comprises the ACRON (which is nonsegmental and homologous with the annelid prostomium), three preoral segments, and the three gnathal segments. The first preoral segment is the LABRAL segment whose appendages fuse to form the labrum. The second preoral (ANTENNAL) segment bears the antennae, which are therefore true segmental appendages. The appendages of the third preoral (INTERCALARY) segment appear briefly during embryogenesis, then are lost.

3.1. General Structure

Primitively the head is oriented so that the mouthparts lie ventrally (the HYPOGNATHOUS condition) (Figure 3.3B). In some insects, especially those that pursue their prey or use their mouthparts in burrowing, a PROGNATHOUS condition exists in which the mouthparts are directed anteriorly (Figure 3.4A). In many Hemiptera the suctorial mouthparts are posteroventral in position (Figure 3.4B), a condition described as OPISTHOGNATHOUS (OPISTHORHYNCOUS).

The head takes the form of a heavily sclerotized capsule, and only the presence of the antennae and mouthparts provides any external indication of its segmental construction. A pair of COMPOUND EYES is usually situated dorsolaterally on the cranium, and usually three OCELLI occur between them on the

*Perhaps the most interesting conclusion of all is drawn by Snodgrass (1960, p. 51) who states "... it would be too bad if the question of head segmentation ever should be finally settled; it has been for so long such fertile ground for theorizing that arthropodists would miss it as a field for mental exercise"!

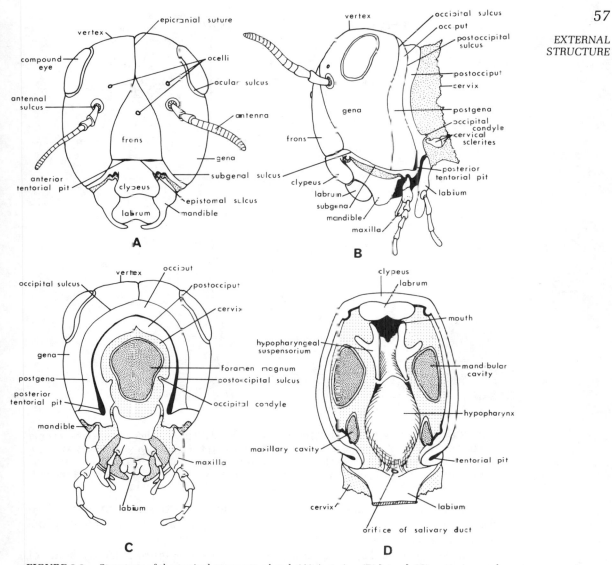

FIGURE 3.3. Structure of the typical pterygotan head. (A) Anterior, (B) lateral, (C) posterior, and (D) ventral (appendages removed). [From R. E. Snodgrass, *Principles of Insect Morphology.* Copyright 1935 by McGraw-Hill, Inc. Used with permission of McGraw-Hill Book Company.]

anterior face (Figure 3.3A). The two posterior ocelli are somewhat lateral in position; the third ocellus is anterior and median. The ANTENNAE vary in location from a point close to the mandibles to a more median position between the compound eyes. On the posterior surface of the head capsule is an aperture, the OCCIPITAL FORAMEN, which leads into the neck. Of the mouthparts, the LABRUM hangs down from the anterior edge of the cranium, the LABIUM lies below the occipital foramen, and the paired MANDIBLES and MAXILLAE occupy ventrolateral positions (Figure 3.3B). The mouth is situated behind the base of the

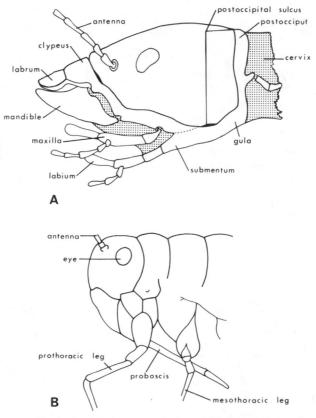

FIGURE 3.4. Diagrams illustrating (A) the prognathous and (B) the opisthognathous types of head structure. [A, from R. E. Snodgrass, *Principles of Insect Morphology.* Copyright 1935 by McGraw-Hill, Inc. Used with permission of McGraw-Hill Book Company. B, after R. F. Chapman, 1971, *The Insects: Structure and Function.* By permission of Elsevier North-Holland, Inc., and the author.]

labrum. The true ventral surface of the head capsule is the HYPOPHARYNX (Figure 3.3D), a membranous lobe that lies in the preoral cavity formed by the ventrally projecting mouthparts.

There are several grooves and pits on the head (Figure 3.3A–C), some of which, by virtue of their constancy of position within a particular insect group, constitute important taxonomic features. The grooves are almost all sulci. The POSTOCCIPITAL SULCUS separates the maxillary and labial segments and internally forms a strong ridge to which are attached the muscles used in moving the head and from which the posterior arms of the TENTORIUM arise (see following paragraph). The points of formation of these arms are seen externally as deep pits in the postoccipital groove, the POSTERIOR TENTORIAL PITS. The EPI-CRANIAL SUTURE is a line of weakness occupying a median dorsal position on the head. It is also known as the ECDYSIAL LINE, for it is along this groove that the cuticle splits during ecdysis. In many insects the epicranial suture is in the shape of an inverted Y whose arms diverge above the median ocellus and pass ventrally over the anterior part of the head. The OCCIPITAL SULCUS, which is commonly found in orthopteroid insects, runs transversely across the posterior part of the cranium. Internally it forms a ridge which strengthens this region of the head. The SUBGENAL SULCUS is a lateral groove in the cranial wall running slightly above the mouthpart articulations. That part of the subgenal sulcus lying directly above the mandible is known as the PLEUROSTOMAL SULCUS; that part lying behind is the HYPOSTOMAL SULCUS, which is usually continuous with the postoccipital suture. In many insects the pleurostomal sulcus is continued across the front of the cranium (above the labrum), where it is known as the EPISTOMAL (FRONTOCLYPEAL) SULCUS. Within this sulcus lie the ANTERIOR TENTORIAL PITS, which indicate the internal origin of the anterior tentorial arms. The ANTENNAL and OCULAR SULCI indicate internal cuticular ridges bracing the antennae and compound eyes, respectively. A SUBOCULAR SULCUS running dorsolaterally beneath the compound eye is often present.

The TENTORIUM (Figure 3.5) is an internal, cranial-supporting structure whose morphology varies considerably among different insect groups. It is appropriately considered at this point, since it is produced by invagination of the exoskeleton. Generally it is composed of the anterior and posterior tentorial arms that may meet and fuse within the head. Frequently, additional supports in the form of dorsal arms are found. The latter are secondary outgrowths of the anterior arms and not apodemes. The junction of the anterior and posterior arms is often enlarged and known as the TENTORIAL BRIDGE or CORPCROTENTORIUM. In addition to bracing the cranium, the tentorium is also a site for the insertion of muscles controlling movement of the mandibles, maxillae, labium, and hypopharynx.

The grooves described above delimit particular areas of the cranium which, although of no functional significance, are useful in descriptive or taxonomic work. The major areas are as follows. The FRONTOCLYPEAL AREA is the facial area of the head, between the antennae and the labrum. When the

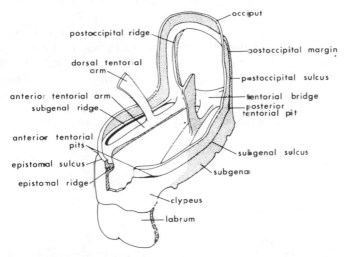

FIGURE 3.5. Diagram showing the relationship of the tentorium to grooves and pits on the head. Most of the head capsule has been cut away. [From R. E. Snodgrass, *Principles of Insect Morphology.* Copyright 1935 by McGraw-Hill, Inc. Used with permission of McGraw-Hill Book Company.]

epistomal sulcus is present the area becomes divided into the dorsal FRONS and the ventral CLYPEUS. The latter is often divided into a POSTCLYPEUS and an ANTECLYPEUS. The VERTEX is the dorsal surface of the head. It is usually delimited anteriorly by the arms of the epicranial suture and posteriorly by the occipital sulcus. The vertex extends laterally to merge with the GENA, whose anterior, posterior, and ventral limits are the subocular, occipital, and subgenal sulci, respectively. The horseshoe-shaped area lying between the occipital sulcus and postoccipital sulcus is generally divided into the dorsal OCCIPUT, which merges laterally with the POSTGENAE. The POSTOCCIPUT is the narrow posterior rim of the cranium surrounding the occipital foramen. It bears a pair of OCCIPITAL CONDYLES to which the ANTERIOR CERVICAL SCLERITES are articulated. Below the gena is a narrow area, the SUBGENA, on which the mandible and maxilla are articulated. The labium is usually articulated directly with the neck membrane (Figure 3.3C), but in some insects a sclerotized region separates the two. This sclerotized area develops in one of three ways: as extensions of the subgenae which fuse in the midline to form a SUBGENAL BRIDGE, as extensions of the hypostomal areas to form a HYPOSTOMAL BRIDGE, or (in most prognathous heads) through the extension ventrally and anteriorly of a ventral cervical sclerite to form the GULA. At the same time the basal segment of the labium may also become elongated (Figure 3.4A).

3.2. Head Appendages

3.2.1. Antennae

A pair of antennae is found on the head of the pterygote insects and the apterygote groups with the exception of the Protura. However, in the larvae of many higher Hymenoptera and Diptera they are reduced to a slight swelling or disc.

In a typical antenna (Figure 3.6) there are three principal components: the basal segment (SCAPE) by which the antenna is attached to the head, the second segment (PEDICEL) containing Johnston's organ (Chapter 12), and the FLAGELLUM, which is usually long and made up of many subsegments. The scape is set in a membranous socket and surrounded by the antennal sclerite on which a single articulation may occur. In the majority of insects movement of the whole antenna is effected by muscles inserted on the scape and attached to the

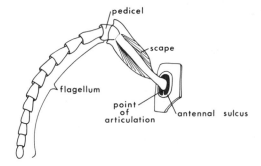

FIGURE 3.6. Structure of an antenna. [From R. E. Snodgrass, *Principles of Insect Morphology.* Copyright 1935 by McGraw-Hill, Inc. Used with permission of McGraw-Hill Book Company.]

cranium or tentcrium. In Collembola and Diplura all the antennal segments have intrinsic muscles.

Although retaining the basic structure outlined above, the antennae take on a wide variety of shapes (Figure 3.7). These are important diagnostic features in insect taxonomy.

3.2.2. Mouthparts

The mouthparts consist of the labrum, a pair of mandibles, a pair of maxillae, the labium, and the hypopharynx. The latter structure is not, of course, an appendage in the strict sense, but simply part of the ventral surface of the head. In Collembola, Protura, and Diplura the mouthparts are enclosed within a cavity formed by the ventrolateral extension of the genae which fuse in the midline (the ENTOGNATHOUS condition). In Thysanura and Pterygota the mouthparts project freely from the head capsule, a condition described as ECTOGNATHOUS. The form of the mouthparts is extremely varied (see below), and it is appropriate to describe first their structure in the more primitive chewing condition.

Typical Chewing Mouthparts. In a typical chewing insect the LABRUM (Figure 3.3A) is a broadly flattened plate hinged to the clypeus. Its ventral (inner) surface is usually membranous and forms the lobelike epipharynx, which bears chemosensilla.

The MANDIBLE (Figure 3.8A) is a heavily sclerotized, rather compact struc-

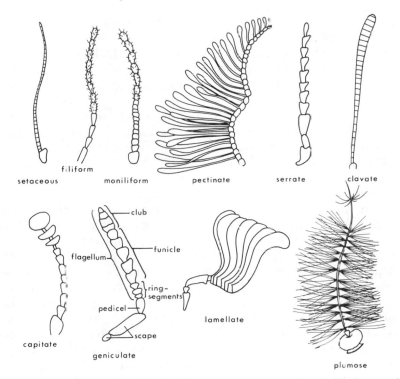

FIGURE 3.7. Types of antennae. [After A. D. Imms, 1957, *A General Textbook of Entomology*, 9th ed. (revised by O. W. Richards and R. G. Davies). By permission of Chapman and Hall Ltd.]

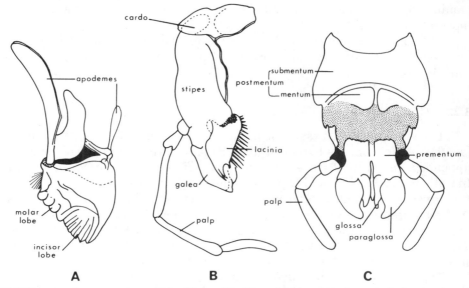

FIGURE 3.8. Structure of mandible (A), maxilla (B), and labium (C) of a typical chewing insect. [From R. E. Snodgrass, *Principles of Insect Morphology.* Copyright 1935 by McGraw-Hill, Inc. Used with permission of McGraw-Hill Book Company.]

ture having usually a dicondylic articulation with the subgena. Its functional area varies according to the diet of the insect. In herbivorous forms there are both cutting edges and grinding surfaces on the mandible. In carnivorous species the mandible possesses sharply pointed "teeth" for cutting and tearing. In apterygotes (except the Lepismatidae) and larval Ephemeroptera the mandible has a single articulation with the cranium and, as a result, much greater freedom of movement.

Of all the mouthparts the MAXILLA (Figure 3.8B) retains most closely the structure of the primitive insectan limb. The basal segment is divided by a transverse line of flexure into two subsegments, a proximal CARDO and a distal STIPES. The cardo carries the single condyle with which the maxilla articulates with the head. Both the cardo and stipes are, however, attached on their entire inner surface to the membranous head pleuron. The stipes bears an inner LACINIA and outer GALEA, and a MAXILLARY PALP. This basic structure is found in both apterygotes and the majority of chewing pterygotes, although in some forms reduction or loss of the lacinia, galea, or palp occurs.

The LABIUM (Figure 3.8C) is formed by the medial fusion of the primitive appendages of the postmaxillary segment, together with, in its basal region, a small part of the sternum of that segment. The labium is divided into two primary regions, a proximal POSTMENTUM corresponding to the maxillary cardines plus the sternal component, and a distal PREMENTUM homologous with the maxillary stipites. The postmentum is usually subdivided into SUBMENTUM and MENTUM regions. The prementum bears a pair of inner GLOSSAE and a pair of outer PARAGLOSSAE, homologous with the maxillary laciniae and galeae, respectively, and a pair of LABIAL PALPS. When the glossae and paraglossae are fused they form a single structure termed the LIGULA.

Arising as a median, mainly membranous, lobe from the floor of the head

capsule and projecting ventrally into the preoral cavity is the HYPOPHARYNX
(Figures 3.3D and 3.9). It is frequently fused to the labium. In some insects
(Thysanura, Ephemeroptera, and Dermaptera) a pair of lobes, the SUPERLIN-
GUAE, which arise embryonically in the mandibular segment, become associated with the hypopharynx. The hypopharynx divides the preoral cavity
into anterior and posterior spaces, the upper parts of which are the CIBARIUM
(leading to the mouth) and SALIVARIUM (into which the salivary duct opens),
respectively.

Mouthpart Modifications. The typical chewing mouthparts described
above can be found with minor modifications in Odonata, the orthopteroid
orders, Neuroptera and Coleoptera (with the exceptions mentioned below),
Mecoptera, primitive Hymenoptera, and larval Ephemeroptera, Trichoptera,
and Lepidoptera. However, the basic arrangement may undergo great modification associated with specialized feeding habits (especially the uptake of liquid
food) or other, nontrophic functions. Suctorial mouthparts are found in members of the hemipteroid orders, and adult Siphonaptera, Diptera, higher
Hymenoptera, and Lepidoptera. The mouthparts are reduced or absent in nonfeeding or endoparasitic forms.

Examination of the structure of the mouthparts provides information on an
insect's diet and feeding habits, and is also of assistance in taxonomic studies.
Some of the more important modifications for the uptake of liquid food are
described below It will be noted that all sucking insects have two features in
common. Some components of their mouthparts are modified into tubular
structures, and a sucking pump is developed for drawing the food into the
mouth.

Coleoptera and Neuroptera In certain species of Coleoptera and
Neuroptera the mouthparts of the larvae are modified for grasping, injecting,
and sucking. In the beetle *Dytiscus*, for example, the laterally placed mandibles

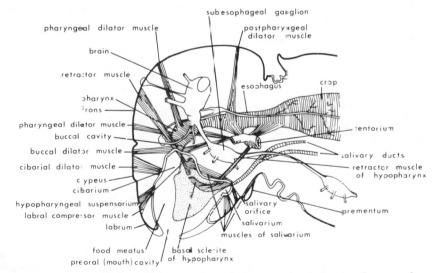

FIGURE 3.9. Simplified sectional diagram through the insect head showing the general arrangement of the parts. [From R. E. Snodgrass, *Principles of Insect Morphology*. Copyright 1935 by
McGraw-Hill, Inc. Used with permission of McGraw-Hill Book Company.]

are long, curved structures with a groove having confluent edges on their inner surface (Figure 3.10). The labrum and labium are closely apposed so that the cibarium is cut off from the exterior. When prey is grasped digestive juices from the midgut are forced along the mandibular grooves and into the body. After external digestion liquefied material is sucked back into the cibarium. In *Dytiscus* the suctorial pump is constructed from the cibarium, the pharynx, and their dilator muscles (see Figure 3.9).

Hymenoptera. In adult Hymenoptera a range of specialization of mouthparts can be seen. In primitive forms, such as sawflies, the mandible is a typical biting structure, and the maxillae and labium, though united, still exhibit their component parts. In the advanced forms, such as bees, the mandibles become flattened and are used for grasping and molding materials rather than biting and cutting. The maxillolabial complex is elongate, and the glossae form a long flexible "tongue," a sucking tube capable of retraction and protraction (Figure 3.11). The laciniae are lost and the maxillary palps reduced, but the galeae are much enlarged, flattened structures, which in short-tongued bees are used to cut holes in the flower corolla to gain access to the nectary. When the food is easily accessible, the glossae, labial palps, and the galeae form a composite tube up which the liquid is drawn. When the food is confined in a narrow cavity such as a nectary, only the glossae are used to obtain it.

The sucking mechanism of the Hymenoptera includes the pharynx, buccal cavity, and cibarium, and their dilator muscles.

Lepidoptera. In the adults of only one family of Lepidoptera, the Micropterygidae, are functional biting mouthparts retained. In all other groups the mouthparts (Figure 3.12) are considerably modified in conjunction with the insect's diet of nectar. The mandibles are usually completely lost, the labrum is reduced to a narrow transverse sclerite, and the labium is a small flap (though its palps remain quite large). The long, suctorial proboscis is formed from the interlocking galeae, whose outer walls comprise a succession of narrow sclerotized arcs alternating with thin membranous areas: presumably this arrangement facilitates coiling. Extension of the proboscis is brought about by a local increase in blood pressure.

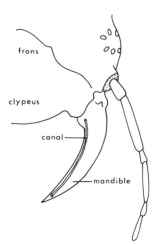

FIGURE 3.10. Left mandible of *Dytiscus* larva, seen dorsally, showing the canal on its inner side. [From R. E. Snodgrass, *Principles of Insect Morphology*. Copyright 1935 by McGraw-Hill, Inc. Used with permission of McGraw-Hill Book Company.]

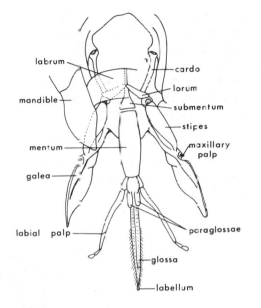

FIGURE 3.11. Mouthparts of the honeybee. [After R. E. Snodgrass, 1925, *Anatomy and Physiology of the Honeybee*, McGraw-Hill Book Company.]

The sucking pump of Lepidoptera comprises the same elements as that of Hymenoptera. In Lepidoptera that do not feed as adults all mouthparts are greatly reduced and the pump is entirely absent.

Diptera. In both larval and adult Diptera the form and function of the mouthparts have diverged considerably from the typical chewing condition. Indeed, in extreme cases [exemplified in some of the larvae (maggots) of Cyclorrhapha] it appears that not only a new feeding mechanism but an entirely new functional head and mouth have evolved, the true mouthparts of the adult fly being suppressed during the larval period. This remarkable modification of the head and its appendages is, of course, the result of the insect living entirely within its food.

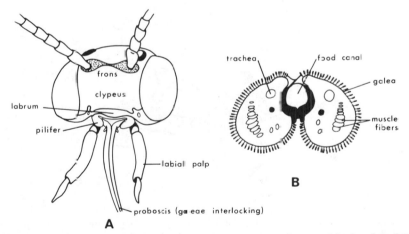

FIGURE 3.12. Head and mouthparts of Lepidoptera. (A) General view of the head, and (B) cross section of the proboscis. [From R. E. Snodgrass, *Principles of Insect Morphology*. Copyright 1935 by McGraw-Hill, Inc. Used with permission of McGraw-Hill Book Company.]

Larva. In many of the more primitive Orthorrhapha the head is retracted into the thorax and enclosed within a sheath formed from the neck membrane. The mandibles and maxillae possess the typical biting structure (though the palps are small or absent). The labrum is large and overhanging. The labium is rudimentary and often confused with the HYPOSTOMA, a toothed, triangular sclerite on the neck membrane (Figure 3.13A–C).

In maggots the true head is completely invaginated into the thorax and the conical "head" is, in fact, a sclerotized fold of the neck. The functional "mouth" is the inner end of the preoral cavity, the ATRIUM, from which a pair of sclerotized hooks protrude. The cibarium is transformed into a massive sucking pump, and the true mouth is the posterior exit from the pump lumen (Figure 3.13D).

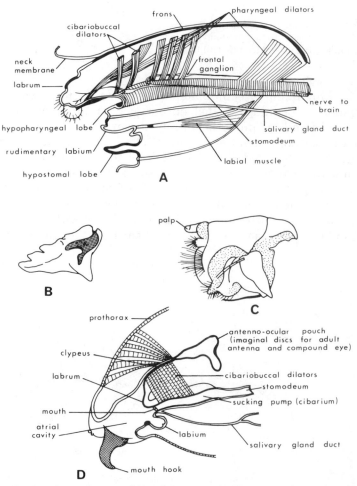

FIGURE 3.13. Head and mouthparts of larval Diptera. (A) Diagrammatic section through the retracted head of *Tipula*, (B) right mandible of *Tipula*, (C) left maxilla of *Tipula*, and (D) diagrammatic section through the anterior end of a maggot. [From R. E. Snodgrass, *Principles of Insect Morphology.* Copyright 1935 by McGraw-Hill, Inc. Used with permission of McGraw-Hill Book Company.]

Adult. No adult Diptera have typical biting mouthparts, although, of course, many are said to "bite" when they pierce the skin. The mouthparts can be divided functionally into those that only suck and those that first pierce and then suck. In the latter the piercing structure may be the mandibles, the labium, or the hypopharynx.

In the Diptera which merely suck or "sponge" up their food (for example, the housefly and blowfly) the mandibles have disappeared and the elongate feeding tube, the PROBOSCIS, is a composite structure that includes the labrum, hypopharynx, and labium (Figure 3.14). The proboscis is divisible into a basal ROSTRUM bearing the maxillary palps, a median flexible HAUSTELLUM, and two apical LABELLA. The latter are broad sponging pads, equipped with PSEUDO-TRACHEAE along which food passes to the oral aperture. The latter is not the true mouth, which lies at the upper end of the food canal. As in other Diptera, the sucking apparatus is formed from the cibarium and its dilator muscles which are inserted on the clypeus.

In the same family (Muscidae) as the housefly and blowfly are several flies (the tsetse fly, stable fly, and horn fly) that feed on blood and therefore have piercing and sucking mouthparts. Like their "nonpiercing" relatives these flies have a composite proboscis. However, the haustellum is elongate and rigid, and the distal labellar lobes are small but bear rows of PRESTOMAL TEETH on their inner walls. The labrum and labium interlock to form the food canal within which lies the hypopharynx enclosing the salivary duct (Figure 3.15).

Other blood-feeding flies (for example, horse- and deerflies, blackflies, and mosquitoes) use the mandibles for piercing the host's skin. The mouthparts of the horsefly *Tabanus* may be taken as an example (Figure 3.16). The labrum is dagger-shaped but flexible and blunt at the tip. On its inner side is a groove closed posteriorly by the mandibles to form the food canal. The mandibles are long and sharply pointed. The maxillae retain most of the components of the typical biting form (except the laciniae) but the galeae are long bladelike structures. The hypopharynx is a styletlike structure and contains the salivary duct.

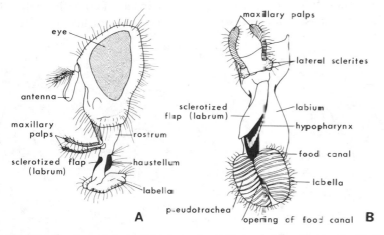

FIGURE 3.14. Head and mouthparts of the housefly. (A) Lateral view of the head with the proboscis extended and (B) anterodistal view of the proboscis. [From R. E. Snodgrass, *Principles of Insect Morphology.* Copyright 1935 by McGraw-Hill, Inc. Used with permission of McGraw-Hill Book Company.]

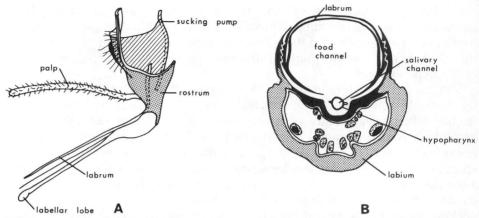

FIGURE 3.15. Mouthparts of the tsetse fly. (A) Lateral view and (B) cross section of the proboscis. [From R. E. Snodgrass, *Principles of Insect Morphology.* Copyright 1935 by McGraw-Hill, Inc. Used with permission of McGraw-Hill Book Company.]

The labium is a large, thick appendage with a deep anterior groove into which the other mouthparts normally fit. Distally it bears two large labellar lobes. Blood flows along the pseudotracheae to the tip of the food canal.

    Hemiptera. The major contributor to the hemipteran proboscis (Figure 3.17) is the labium, a flexible segmented structure with a deep groove on its anterior surface. Within this groove are found the piercing organs, the MAN-DIBULAR and MAXILLARY BRISTLES. The two maxillary bristles are interlocked within the labial groove and form the food and salivary canals. Because of the great enlargement of the clypeal region of the head associated with the opisthognathous condition, the cibarial sucking pump is entirely within the head.

4. The Neck and Thorax

    The thorax is the locomotory center of the insect. Typically each of its three segments (pro-, meso-, and metathorax) bears a pair of legs, and in the adult

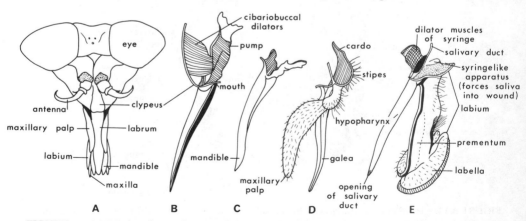

FIGURE 3.16. Head and mouthparts of the horsefly. (A) Anterior view of the head and (B–E) lateral views of the separated mouthparts. [From R. E. Snodgrass, *Principles of Insect Morphology.* Copyright 1935 by McGraw-Hill, Inc. Used with permission of McGraw-Hill Book Company.]

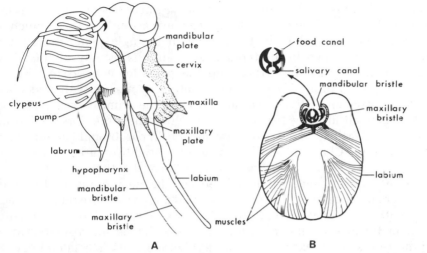

FIGURE 3.17. Head and mouthparts of Hemiptera. (A) Head with the mouthparts separated and (B) cross section of the proboscis. [From R. E. Snodgrass, *Principles of Insect Morphology.* Copyright 1935 by McGraw-Hill, Inc. Used with permission of McGraw-Hill Book Company.]

stage of the Pterygota the meso- and metathoracic segments each have a pair of wings. Between the head and thorax is the membranous neck (CERVIX).

4.1. The Neck

Study of its embryonic development shows that the neck contains both labial and prothoracic components and therefore the primary intersegmental line must be within the neck membrane. The muscles that control head movement arise on the postoccipital ridge and are attached to the antecosta of the prothoracic segment. Thus they must include the fibers of two segments. This modification of the basic segmental structure was apparently necessary to provide sufficient freedom of head movement.

Usually supporting the head and articulating it with the prothorax are the cervical sclerites, a pair of which occurs on each side of the neck (Figure 3.3B). Either one or both sclerites may be absent. When only one occurs it is often fused with the prothorax. Occasionally additional dorsal and ventral sclerites are found.

4.2. Structure of the Thorax

In the evolution of the typical insectan body plan there have been two phases associated with the development of the thorax as the locomotory center; in the first the walking legs became restricted to the three thoracic segments, and in the second articulated wings were formed on the meso- and metathoracic terga. Accompanying each of these developments were major changes in the basic structure of the secondary segments of the thoracic region. These changes were primarily to strengthen the region for increased muscular power.

In the apterygotes and many juvenile pterygotes the thoracic terga are little different from those of the typical secondary segment described in Section 2. In the adult the terga of the wing-bearing segments are enlarged and much modified (Figure 3.18). Although it may remain a single plate, the tergum (or NOTUM, as it is called in the thoracic segments) is usually divided into the anterior wing-bearing ALINOTUM, and the posterior POSTNOTUM. These are firmly supported on the pleural sclerotization by means of the PREALAR and POSTALAR ARMS, respectively. The antecostae of the primitive segments become greatly enlarged forming PHRAGMATA, to which the large dorsal longitudinal muscles are attached. Since wing movement is in part brought about by flexure of the terga (see Chapter 14), which is itself caused by contraction and relaxation of the dorsal longitudinal muscles, it is clear that the connection between the mesonotum and metanotum and between the metanotum and first abdominal tergum must be rigid. The intersegmental membranes are therefore reduced or absent. Additional supporting ridges are developed on the meso- and metanota, the most common of which are the V-shaped SCUTOSCUTELLAR RIDGE and the TRANSVERSE (PRESCUTAL) RIDGE (Figure 3.18A). The lateral margins of the alinotum are constructed for articulation of the wing. They possess both ANTERIOR and POSTERIOR NOTAL PROCESSES, to which the FIRST and THIRD AXILLARY SCLERITES, respectively, are attached. Further details of the wing articulation are given in Section 4.3.

The originally membranous pleura have been strengthened to varying degrees by sclerotization and the formation of internal cuticular ridges. In some

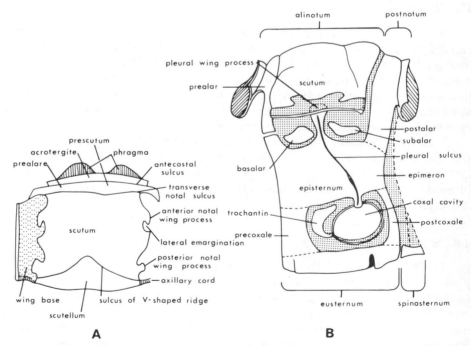

FIGURE 3.18. (A) Dorsal view of a generalized alinotum and (B) lateral view of a typical wing-bearing segment. [From R. E. Snodgrass, *Principles of Insect Morphology.* Copyright 1935 by McGraw-Hill, Inc. Used with permission of McGraw-Hill Book Company.]

apterygotes, for example, two small, crescent-shaped pleural sclerites may be seen above the coxa, though the rest of the pleuron is membranous. In the prothorax of Plecoptera there are likewise two sclerites, but these are much larger than those of apterygotes and occupy more than half the pleural area. In the thoracic segments of all other pterygotes the pleura are fully sclerotized and are additionally strengthened by the formation of an internal PLEURAL RIDGE which extends dorsally into the PLEURAL WING PROCESS (Figure 3.18B). Articulating with this process is the SECOND AXILLARY SCLERITE. Each pleural ridge is extended inwardly as a PLEURAL ARM (Figure 3.19) which meets and may fuse with similar apophyses from the sternum. The pleural ridge is seen externally as the PLEURAL SULCUS (Figure 3.18B) above the coxa. This groove divides the pleuron into an anterior EPISTERNUM and posterior EPIMERON. Often these sclerites are divided secondarily into dorsal and ventral areas, the SUPRAEPISTERNUM and INFRAEPISTERNUM, and SUPRAEPIMERON and INFRAEPIMERON. Derived from the episternum and epimeron and appearing above them usually as distinct, articulated sclerites are the BASALAR and SUBALAR, to which important wing muscles are attached.

In the thorax the acrosternite of the typical secondary segment forms an independent intersegmental plate or INTERSTERNITE. The intersternites which are found between the pro- and mesothorax and between the meso- and metathorax are known as SPINASTERNA because each bears an internal spine to which a few ventral muscles are attached. Frequently the spinasterna fuse with the segmental plate, EUSTERNUM, of the preceding segment. The eusternum is a composite structure, comprising the primary sternal plate and the sternopleurite. The eusternum may be divided secondarily into an anterior BASISTERNUM and posterior STERNELLUM by the STERNACOSTAL SULCUS (Figure 3.20). The latter is the result of an invagination to form the STERNACOSTA, a ridge of cuticle

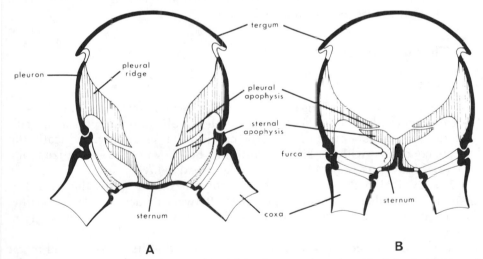

A   **B**

FIGURE 3.19. Diagrammatic cross sections of the thorax to show the endoskeleton. (A) Normal condition and (B) condition when furca present. [From R. E. Snodgrass, *Principles of Insect Morphology*. Copyright 1935 by McGraw-Hill, Inc. Used with permission of McGraw-Hill Book Company.]

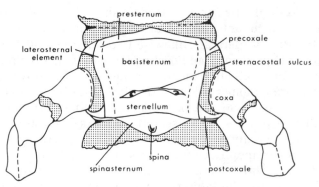

FIGURE 3.20. Ventral view of a generalized thoracic sternum. [From R. E. Snodgrass, *Principles of Insect Morphology*. Copyright 1935 by McGraw-Hill, Inc. Used with permission of McGraw-Hill Book Company.]

that unites the STERNAL APOPHYSES (Figure 3.19). In the higher pterygotes these apophyses are borne on a median internal ridge and form a Y-shaped FURCA (Figure 3.19). As noted earlier these apophyses combine with the pleural arms to form a rigid internal support. The latter provides attachment for the major longitudinal ventral muscles and certain muscles of the leg.

It must be emphasized that many variations occur from the rather general description of a thoracic segment provided above. In all insects the prothorax is modified by the development of the neck region. The pronotum especially is different, lacking the antecostal region and phragma through neck membranization. In some groups (for example, Orthoptera, Hemiptera, and Coleoptera) the pronotum is greatly enlarged; in others it is reduced to a narrow band between the head and mesothorax. In those orders whose members have a single pair of functional wings, the tergal plates of the segment from which the wings are absent are usually reduced in size.

4.3. Thoracic Appendages

4.3.1. Legs

In the vast majority of insects each thoracic segment bears a pair of legs. In the cases where legs are absent, for example, in all dipteran, and many coleopteran and hymenopteran larvae, the condition is secondary. Typically the legs are concerned with walking and running, but they may be specialized for a variety of functions, some of which are described below.

The Typical Walking Leg. The leg consists of five parts, the coxa, trochanter, femur, tibia, and tarsus (Figure 3.21). Between adjacent parts is a narrow, annulated membrane, the CORIUM, and usually a mono- or dicondylic articulation.

The coxa is a short, thick segment strengthened at its proximal end by an internal ridge, the BASICOSTA (Figure 3.22). The coxa usually has a dicondylic articulation with the pleuron. In some orders the BASICOSTAL SULCUS is U- or V-shaped over the posterior half of the coxa (Figure 3.22). The sclerite thus demarcated becomes thickened and is known as the MERON. The trochanter is a

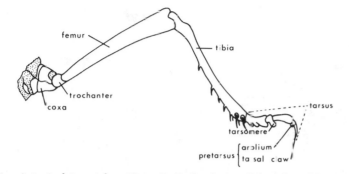

FIGURE 3.21. A typical insect leg. [From R. E. Snodgrass, *Principles of Insect Morphology*. Copyright 1935 by McGraw-Hill, Inc. Used with permission of McGraw-Hill Book Company.]

small segment. It always has a dicondylic articulation with the coxa but is usually firmly fixed to the femur, which is generally the largest leg segment. Following the slender tibia is the tarsus, a segment that is usually subdivided into between two and five TARSOMERES and a PRETARSUS. The pretarsus, in most insects, takes the form of a pair of TARSAL CLAWS and a median lobe, the AROLIUM (Figure 3.23).

Leg Modifications. The functions for which the legs have become modified include jumping, swimming, grasping, digging, sound production, and cleaning.

In Orthoptera and a few Coleoptera (for example, flea beetles) the femur on the hindleg is greatly enlarged to accommodate the extensor muscles of the tibia used in jumping. In swimming insects, the tibia and tarsus of the hindlegs (occasionally also the middle legs) are flattened and bear rigid hairs around the periphery (Figure 3.24A). Legs modified for grasping are found in predaceous insects such as the mantis and giant water bug, in the ectoparasitic lice, and in the males of various species where they are used for hanging onto the female during mating. In the mantis, the tibia and femur of the foreleg are equipped

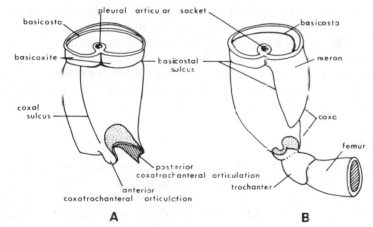

FIGURE 3.22. Structure of the coxa. (A) Lateral view and (B) coxa with a well-developed meron. [From R. E. Snodgrass, *Principles of Insect Morphology*. Copyright 1935 by McGraw-Hill, Inc. Used with permission of McGraw-Hill Book Company.]

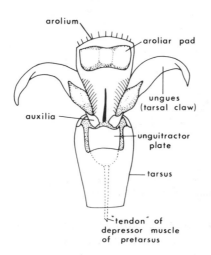

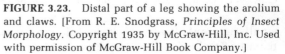

FIGURE 3.23. Distal part of a leg showing the arolium and claws. [From R. E. Snodgrass, *Principles of Insect Morphology.* Copyright 1935 by McGraw-Hill, Inc. Used with permission of McGraw-Hill Book Company.]

with spines and operate together as pincers (Figure 3.24B). The foreleg of a louse is short and thick and has at its tip a single, large tarsal claw which folds back against the tibial process (Figure 3.24C). Suctorial pads have been developed on the forelimbs in the males of many beetle species. In *Dytiscus*, for example, the first three tarsomeres are flattened and possess large numbers of cuticular cups, two of which are extremely enlarged (Figure 3.24D). The forelegs of soil-dwelling insects such as the mole cricket (Figure 3.24E), the cicadas, and various beetles, are modified for digging. The legs are large, heavily sclerotized, and possess stout claws. The tarsomeres are reduced in number or may disappear entirely in some forms. In many Orthoptera sounds are produced when the hind femora, which are equipped with a row of cuticular pegs on their inner surface, are rubbed against ridged veins on the fore wing. Modifications to the forelegs for cleaning purposes are found in many insects. In certain Coleoptera and Hymenoptera, for example, the honeybee (Figure 3.25A), a notch lined with hairs occurs on the metatarsus of the foreleg through which the antenna can be drawn and cleaned. The hindlegs of the bee are modified for pollen collection (Figure 3.25B). Rows of hairs, the COMB, on the inner side of the first tarsomere scrape pollen off the abdomen. The RAKE, a fringe of hairs at the distal end of the tibia, then collects the pollen from the comb on the opposite leg and transfers it to the pollen PRESS. When the press is closed the pollen is pushed up into the pollen BASKET, where it is stored until the bee returns to its nest.

4.3.2. Wings

The majority of adult Pterygota have one or two pairs of functional wings. The complete absence of wings is a secondary condition, associated with the habits of the group concerned, for example, soil-dwelling or endoparasitism. The wings may be modified for a variety of purposes other than flight.

Development and General Structure. The wing is a flattened evagination of the dorsal body wall. As such it comprises the usual integumental elements (cuticle, epidermis, and basement membrane) and its lumen, being an exten-

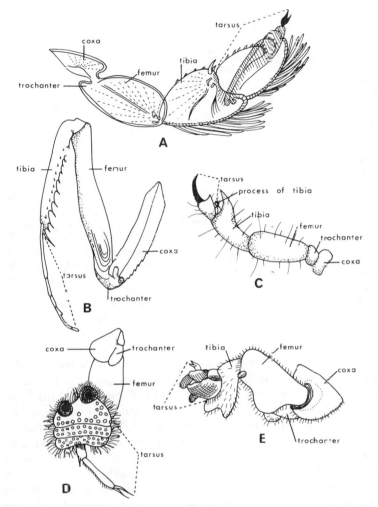

FIGURE 3.24. Leg modifications. (A) Hindleg of *Gyrinus* (swimming), (B) foreleg of a mantis (grasping prey), (C) foreleg of a louse (attachment to host), (D) foreleg of *Dytiscus* (holding onto female), and (E) foreleg of a mole cricket (digging). [A, after L. C. Miall, 1922, *The Natural History of Aquatic Insects*, published by Macmillan Ltd. B, D, E, after J. W. Folsom, 1906, *Entomology with Special Reference to Its Ecological Aspects*, published by Blakiston's Son and Co., Philadelphia, Pa. By permission of McGraw-Hill Book Company. C, after A. D. Imms, 1957, *A General Textbook of Entomology*, 9th ed. (revised by O. W. Richards and R. G. Davies). By permission of Chapman and Hall Ltd.]

sion of the hemocoel, contains tracheae, nerves, and hemolymph. As the wing develops the dorsal and ventral integumental layers become closely apposed over most of their area forming the WING MEMBRANE. The remaining areas form channels, the future VEINS, in which the nerves and tracheae may occur. The cuticle surrounding the veins becomes thickened to provide strength and rigidity to the wing.

It is generally considered that the extremely varied arrangement of veins found among the different insect orders is derived from a primitive common pattern, the ARCHETYPE VENATION. Details of the latter are not known with

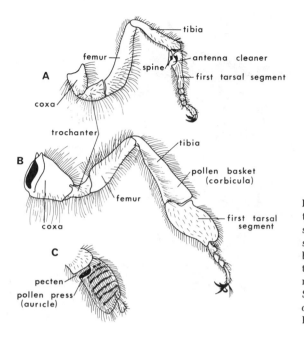

FIGURE 3.25. Leg modifications in the worker honeybee. (A) Foreleg showing the cleaning notch, (B) outer surface of hindleg showing the pollen basket, and (C) inner surface of hind tarsus and tip of hind tibia showing rake and pollen press. [After R. E. Snodgrass, 1925, *Anatomy and Physiology of the Honeybee*, McGraw-Hill Book Company.]

certainty because even the earliest fossil insects have a highly complex venation. However, by comparing the wing venation of members of many orders and noting the constant features and by considering features of the venation of paleopteran orders, it is possible to propose a hypothetical venational type that approximates the archetype venation (Hamilton, 1972a). Determination of the homologies of veins is of great importance in phylogenetic and taxonomic studies. The usual method of determining homology is direct comparison of the position and form of veins. In addition, there are other, more subtle features of wing venation with which to determine homologies (Ragge, 1955; Hamilton, 1972a). Thus, certain veins are always associated with particular axillary sclerites. Some veins have associated with them a row of hairlike mechanosensilla. In many wings this row persists even when the original vein has disappeared. The Comstock–Needham system of wing venation was based on the assumptions that tracheae are always present in particular veins and that the pattern of tracheae develops in a characteristic manner and thereby determines the venational pattern. Though these assumptions are generally valid, there are exceptions (see Hamilton, 1972a, for details). Other authors have based their studies of vein homology on wing fluting, the alternation of concave and convex veins. This approach is of limited use, however, and is not applicable to branched vein systems where fluting is practically nonexistent, or to wings that are secondarily fluted (Hamilton, 1972a). Alone, each method is deficient in some respect, and it is necessary, therefore, to use them in combination as was done by Hamilton (1972a,b). In the system of Hamilton (1972a) the following arrangement of veins is proposed (Figure 3.26). The first three veins (COSTA, SUBCOSTA, and RADIUS) are unbranched. The costa is a stout vein that runs along the anterior margin of the wing and proximally is associated with the SUBCOSTAL PLATE (Figure 2.3). The base of the subcosta is fused with the sub-

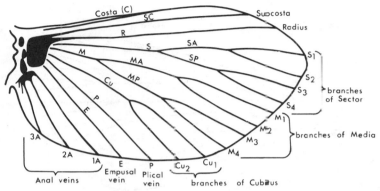
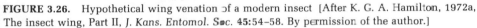

FIGURE 3.26. Hypothetical wing venation of a modern insect [After K. G. A. Hamilton, 1972a, The insect wing, Part II, *J. Kans. Entomol. Soc.* 45:54–58. By permission of the author.]

costal plate, which articulates with the FIRST AXILLARY SCLERITE. The radius is generally the strongest wing vein. It fuses with the anterior edge of the median plate, which articulates with the SECOND AXILLARY SCLERITE. In Hamilton's system the next three veins (SECTOR, MEDIA, and CUBITUS) are branched. The sector and media each have four branches, the cubitus two. (In the archetype venations proposed by other authors the sector is called the "radial sector" and is shown as a proximal branch of the radius.) Proximally, the sector and media are associated with the MEDIAN PLATE, while the cubitus fuses with the CUBITAL PLATE. The latter articulates with the THIRD AXILLARY SCLERITE. The five remaining veins are unbranched and in some earlier schemes were all designated "anal" veins. The first two are, however, quite distinct from the last three; both arise from the cubital plate and are anterior to the VANNAL (ANAL) FOLD. Hamilton designates these veins as the PLICAL and EMPUSAL veins. The last three veins, which have a common base, usually a common tracheal supply, and articulate with the third axillary sclerite, are true ANAL VEINS in Hamilton's scheme. In concluding this description it must be emphasized that this venation is hypothetical; it has not been found in any insect, fossil or extant. The venation observed in extant species deviates greatly from this proposed ancestral type as a result of reduction (fusion or degeneration) of veins or addition (through further subdivision of existing veins or secondary development of new veins). The position and articulation of the basal plates may also alter and further complicate the question of homology.

Various terms are used to describe the different parts of a wing (Figure 3.27). The wing is approximately triangular in shape and, as such, has three margins: the anterior COSTAL MARGIN, lateral APICAL MARGIN, and posterior ANAL MARGIN. These margins form three angles: the HUMERAL ANGLE at the wing base, the APEX between the costal and apical margins, and the ANAL ANGLE (TORNUS) between the apical and anal margins. The two major areas of the wing are the REMIGIUM, the anterior, more rigid part, and the posterior, flexible VANNAL (ANAL) LOBE. In some insects a small JUGAL LOBE occurs proximally to the vannal lobe. The lines of flexure between these three areas are the (VANNAL) ANAL and JUGAL FOLDS. In certain orders (Psocoptera, Hymenoptera, and Odonata) an opaque area, the PTEROSTIGMA, is found near the costal margin of the wing. The areas between adjacent veins are referred to as CELLS, which may

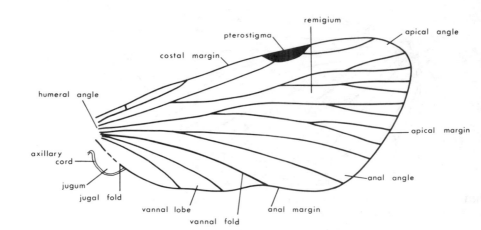

FIGURE 3.27. Diagram showing the major areas, folds, and angles of a generalized wing. [After R. F. Chapman, 1971, *The Insects: Structure and Function.* By permission of Elsevier North-Holland, Inc., and the author.]

be open (when extending to the wing margin) or closed (when entirely surrounded by veins). The cells are named after the longitudinal vein which forms their anterior edge.

Wing Modifications. There are many modifications of the typical condition in which the insect has two pairs of triangle-shaped flapping wings. The modifications fall, however, into two broad categories: (1) those that lead, directly or indirectly, to improved flight; and (2) those in which a wing takes on a function entirely unrelated to movement of the insect through the air. Both types of modification are, of course, possible in the same insect.

It seems that the two-winged condition is aerodynamically more efficient than the four-winged condition, and in a number of insect orders wing-coupling mechanisms have evolved that link together the fore and hind wings on the same side of the body (see Tillyard, 1918). The precise nature of the coupling mechanism varies, but usually it consists of groups of hairs, the FRENULUM, on the anterior basal margin of the hind wing that interlock with hairs or curved spines, the RETINACULUM, attached to various veins of the fore wing (Figure 3.28). The other way in which the two-winged condition has been achieved is through the loss, functionally speaking, of either the fore wings (in Coleoptera and male Strepsiptera) or the hind wings (Diptera, and some Ephemeroptera and Hemiptera). In these insects the wings, no longer used directly in flight (that is, to create lift through flapping), are still present but modified for other functions. The modified hind wings, HALTERES, of Diptera and male Coccoidea (Hemiptera) are highly developed sensory structures used in attitudinal control (see Figure 14.15). The heavily sclerotized ELYTRA (fore wings) of Coleoptera are mainly protective in function, though they may be secondarily important in control of attitude.

Even in insects that retain two functional pairs of wings there are frequently modifications of these structures for other functions. In many Lepidoptera, for example, the wing margins are irregular and the wings appropriately colored so that when the insect is at rest it is camouflaged. The wings of male Orthoptera are commonly modified for sound production. In the cric-

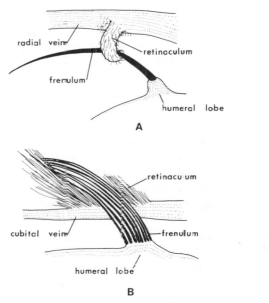

FIGURE 3.28. Wing-coupling mechanism in *Hippotion scrota* (Lepidoptera). (A) Male and (B) female. [After R. J. Tillyard, 1918, The panorpoid complex. Part I. The wing-coupling apparatus with special reference to the Lepidoptera, *Proc. Linn. Soc. N.S.W.* **43**:286–319. By permission of the Linnean Society, N.S.W.]

kets and long-horned grasshoppers the hardened fore wings possess a toothed "file" (the modified cubital vein) and a "scraper" (a sclerotized ridge at the wing margin). Rapid opening and closing of the wings causes the file on one wing to be dragged over the scraper of the other wing and sound to be produced. In the short-horned grasshoppers the file is on the hind femur; the scraper takes the form of ridged veins on the fore wings.

5. The Abdomen

The abdomen differs from the head and thorax in that its segments generally have a rather simple structure, they are usually quite distinct from each other, and most of them lack appendages. The number of abdominal segments is variable. The primitive number appears to be 12, though this number is found today in only the Protura. Most insects have 10 or 11 segments, but several of these are reduced. The reduction occurs primarily at the posterior end of the abdomen, but in some endopterygotes the first segment is reduced and intimately fused with the metathorax. For the purpose of discussion the abdominal segments may be considered to form three groups: the PREGENITAL SEGMENTS, the GENITAL SEGMENTS, and the POSTGENITAL SEGMENTS.

5.1. General Structure

The more anterior (pregenital) segments are little different from the typical secondary segment described in Section 2. In the first segment, however, the antecosta of the tergum bears internally a pair of phragmata to which the metathoracic dorsal longitudinal muscles are attached. Furthermore, the acrotergite of this segment forms the postnotal plate of the metathorax. Frequently the antecostal region and acrotergite are clearly separated from the rest of the tergum and form part of the metanotum. In the higher Hymenoptera the entire first segment, the PROPODEUM, is fused with the metathorax and the conspicuous "waist" of these insects occurs between the first and second abdominal segments.

The genital opening (GONOPORE) is located, in the majority of Pterygota, on or behind the eighth or ninth sternum in the female, and behind the ninth sternum in the male. In Ephemeroptera and most male Dermaptera paired gonopores occur. The genital segments are modified in various ways for oviposition or sperm transfer. In female Diptera, Lepidoptera, and Coleoptera the posterior segments lack appendages and form smooth cuticular cylinders often telescoped into the anterior part of the abdomen. When extended (Figure 3.29) they form a long narrow tube which facilitates egg laying in inaccessible places. Sometimes the tip of the abdomen is sclerotized for piercing tissues. In other orders the tergum and sternum of the genital segments remain as distinct cuticular plates. In the female the sternum of the eighth segment is sometimes enlarged to form the FEMALE SUBGENITAL PLATE, in which case the sternum of the ninth segment is reduced to a membranous sheet. In the male the tergum and sternum of the ninth segment are distinct but may be greatly modified. The genital segments retain their appendages, which are modified to serve in the reproductive process. In the female they form the OVIPOSITOR, and in the male, CLASPING and INTROMITTENT organs.

The postgenital segments include the tenth and, when present, the eleventh abdominal segments. In the lower orders where both postgenital segments are present, the tenth segment is usually united with the ninth or eleventh segments and it never bears appendages. The eleventh segment comprises a somewhat triangular tergal plate, the EPIPROCT, and a pair of ventrolateral plates, the PARAPROCTS. It bears appendages, the cerci, inserted in the membranous area between the tenth and eleventh segments on each side of the body (Figure 3.30).

In the majority of endopterygotes there is only one postgenital segment,

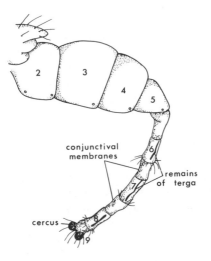

FIGURE 3.29. Abdomen of the housefly extended. The segments are numbered. [Reprinted from L. S. West, *The Housefly*. Copyright © 1951 by Comstock Publishing Co., Inc. Used by permission of the publisher, Cornell University Press.]

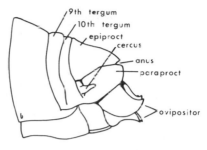

9th tergum
10th tergum
epiproct
cercus
anus
paraproct
ovipositor

FIGURE 3.30. Postgenital segments of a female grasshopper. [From R. E. Snodgrass, *Principles of Insect Morphology.* Copyright 1935 by McGraw-Hill, Inc. Used with permission of McGraw-Hill Book Company.]

the tenth abdominal segment. It is usually much modified and has no true appendages. When appendiculate structures are present they are secondary developments.

5.2. Abdominal Appendages

Generally the only appendages seen on the abdomen of an adult pterygote insect are those on the genital segments (the EXTERNAL GENITALIA) and the cerci. In the apterygote groups there are, in addition, appendages on a variable number of pregenital segments. All these may be considered as "primary" appendages; that is, derived from the primitive insectan limb.

Many endopterygote larvae possess segmentally arranged prolegs or gills on the abdomen. According to some authorities (for example, Hinton, 1955), these are secondary structures and are not derived from the primitive limb. Finally, a number of insects possess nonsegmental appendages on the abdomen, which clearly are of secondary origin.

5.2.1. External Genitalia

The morphology of the external genitalia is extremely varied, especially in the male, and it is sometimes extremely difficult to homologize the different components. In the female the appendages combine to form an ovipositor, used for placing eggs in specific sites. The appendages of the male are used as clasping organs during copulation. Because of their specific form, the external genitalia are widely used by taxonomists for identification purposes.

Female. According to Scudder (1961) the primitive structure of the pterygote insect ovipositor can be seen in the thysanuran *Lepisma* (Figure 3.31A). The ovipositor typically has a basal part and a shaft. The basal part consists of two pairs of GONOCOXAE (VALVIFERS) on the eighth and ninth segments, homologous with the coxa of the leg. Projecting ventroposteriorly from each gonocoxa is an elongate process, the GONAPOPHYSIS (VALVULA). Together the four gonapophyses make up the shaft. The first gonapophyses are ventral, the second dorsal. In most pterygote orders the second gonocoxa has a second posterior process, the third VALVULA or GONOPLAC, attached to it (Figure 3.31B). The gonoplac may form part of the ovipositor or serve as a protective sheath. The GONANGULUM is a small sclerite, articulated with the second gonocoxa (from which it is derived) and the ninth tergum.

Among Pterygota an ovipositor is found in Grylloblattodea, Dictyoptera,

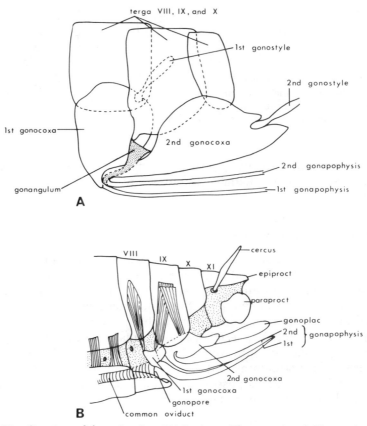

FIGURE 3.31. Structure of the ovipositor. (A) *Lepisma* (Thysanura) and (B) a typical pterygote insect. [A, after G. G. E. Scudder, 1961, The comparative morphology of the insect ovipositor, *Trans. R. Entomol. Soc. London* **113**:25–40. By permission of the Royal Entomological Society. B, from R. E. Snodgrass, *Principles of Insect Morphology.* Copyright 1935 by McGraw-Hill, Inc. Used with permission of McGraw-Hill Book Company.]

Orthoptera, and Hymenoptera, some Odonata and most Hemiptera, Thysanoptera, and Psocoptera. In each of these groups it is more or less modified from the general condition described above. The ovipositor of Odonata and Grylloblattodea is almost identical with that of *Lepisma*, except that a gonoplac is present. In Dictyoptera and Orthoptera the gonangulum fuses with the first gonocoxa. In viviparous cockroaches the ovipositor is considerably modified. The gonoplac, in Orthoptera, is a large sclerotized plate that forms part of the ovipositor shaft. The second gonapophyses, when present, are usually median and concealed. These structures are much reduced in Acrididae.

In Hemiptera, Thysanoptera, and Psocoptera, an ovipositor may or may not be present. When present the gonangulum is fused with the ninth tergum, and the second gonapophyses are fused, making the shaft a three-part structure. The gonoplacs normally ensheath the shaft.

The ovipositor of Hymenoptera may be considerably modified for boring, piercing, sawing, and stinging (Figure 3.32). However, only when modified for stinging does it no longer participate in egg laying, though it still retains the

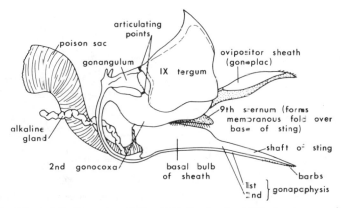

FIGURE 3.32. Sting of the honeybee. [After R. E. Snodgrass, 1925, *Anatomy and Physiology of the Honeybee*, McGraw-Hill Book Company.]

basic components, with the exception of the first gonocoxae, which have disappeared. The first gonapophyses are attached to the gonangulum, which is articulated with the ninth tergum and second gonocoxa. As in Hemiptera, the second gonapophyses are fused. The shaft is typically very elongate, and the eggs are considerably compressed as they are squeezed along it. In the stinging Hymenoptera the eggs are released at the base of the ovipositor. The fused second valvulae form an inverted groove that is enlarged proximally into the BASAL BULB. When an insect stings the POISON GLAND releases its fluid into the basal bulb. The poison is caused to flow along the groove by back and forth movements of the LANCETS (modified first gonapophyses).

   Male. "The great structural diversity in the male genitalia of insects is the delight of taxonomists, the despair of morphologists," wrote Snodgrass (1957, p. 11). Paralleling this diversity of structure is a mass of taxonomic terms (see Tuxen, 1970). The genitalia are composed of two sets of structures. First, there are basic structures which, though variable in form, are common to all insects, and second there are structures which are peculiar to a group or even a species. Space does not permit description of the latter, or even the variation that occurs within the former, set of structures. The following paragraphs will, therefore, be limited to a discussion of the basic form of the genitalia.

   Comparative study indicates that in all insects the basic genitalia are derived from a pair of PRIMARY PHALLIC LOBES, ectodermal outgrowths belonging to the tenth segment (Figure 3.33A). However, only in Ephemeroptera do they remain as separate lobes, through the posterior tips of which open the gonopores (Figure 3.33B). In Thysanura the lobes meet in the midline to form a short, tubular structure, the "PENIS." The latter is, in fact, a misnomer, since this structure is not an intromittent organ. In Odonata secondary copulatory structures are developed on the ventral surface of the anterior abdominal segments (see Chapter 6), and the genitalia on the tenth abdominal segment are much reduced. In all the remaining orders the primary lobes are each divided into two secondary lobes, the PHALLOMERES. Between the median pair, MESOMERES, the ectoderm invaginates to form the EJACULATORY DUCT. The outer pair, PARAMERES, elongate and develop into clasping organs. The mesomeres unite medially to form a tubular intromittent organ, the AEDEAGUS, whose inner

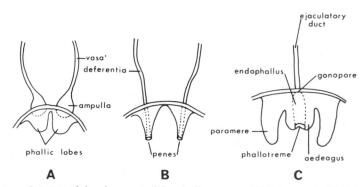

FIGURE 3.33. Origin and development of the phallic organs. (A) Primary phallic lobes, (B) paired penes of Ephemeroptera, and (C) formation of the aedeagus. [Reproduced by permission of the Smithsonian Institution Press from *Smithsonian Miscellaneous Collections*, Volume 135, "A revised interpretation of the external reproductive organs of male insects," Number 6, Dec. 3, 1957, 60 pages, by R. E. Snodgrass: Figures 1A–C, page 3. Washington, D.C., 1958, Smithsonian Institution.]

passage is termed the ENDOPHALLUS (Figure 3.33C). The distal opening of the endophallus is the PHALLOTREME. Occasionally the parameres and aedeagus are united basally, this region being known as the PHALLOBASE. Usually, however, the parameres are placed laterally on the ninth segment.

Two assumptions that are generally made concerning the male genitalia are (1) that they are modified limb appendages, and (2) that they belong to the ninth abdominal segment. Snodgrass (1957) has questioned the validity of both these assumptions. Snodgrass presents a convincing argument against assumption (1) and suggests, instead, that the genitalia are derived from primitive paired penes. According to Snodgrass, the evidence from embryonic and postembryonic development does not support assumption (2) but indicates that the genitalia arose primitively on the tenth abdominal segment.

5.2.2. Other Appendages

Cerci. Paired CERCI occur in Diplura, thysanurans, Ephemeroptera, Zygoptera (a suborder of the Odonata), the orthopteroid orders, and Mecoptera. It is generally agreed that the cerci (Figure 3.31) are true appendages of the eleventh segment, although frequently all traces of the latter have disappeared. Typically, they are elongate multisegmented structures which function as sense organs. They may, however, be considerably modified. In nearly all Dermaptera the cerci form unjointed forceps. The cerci of nymphs of Zygoptera are modified to form the LATERAL CAUDAL LAMELLAE. The latter, along with the MEDIAN CAUDAL LAMELLA (developed from the epiproct), are accessory respiratory structures. In adult male Zygoptera the cerci form CLASPERS for grasping the female during copulation. In most male Embioptera the basal segment of the left cercus forms a hook with which the insect can clasp his mate.

Styli and Eversible Vesicles. STYLI occur on most abdominal segments of thysanurans and Diplura, and on the ninth sternum of some male orthopteroid insects. In some thysanurans the styli are articulated with a distinct coxal plate (Figure 3.34), but generally the original coxal segment is fused with the ster-

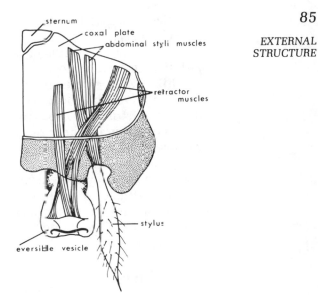

FIGURE 3.34. Stylus and eversible vesicle of a thysanuran. Part of the wall of the plate has been removed to show the musculature. [From R. E. Snodgrass, *Principles of Insect Morphology.* Copyright 1935 by McGraw-Hill, Inc. Used with permission of McGraw-Hill Book Company.]

num. In thysanurans, at least, the styli serve to raise the abdomen off the ground during locomotion.

The EVERSIBLE VESICLES (Figure 3.34) are short cylindrical structures found on some pregenital segments of apterygotes. They are closely associated with the styli when present, but their homology is unclear. They are believed to have the ability to take up water from the environment.

Prolegs. Segmentally arranged, leglike structures are present on the abdomen of many larvae (Figure 3.35A). They are known as PROLEGS, PSEUDOPODS, or LARVAPODS. Their structure is variable, though in a "typical" form (for example, that of a caterpillar) three regions can be distinguished: a basal membranous articulation, followed by a longer section having a

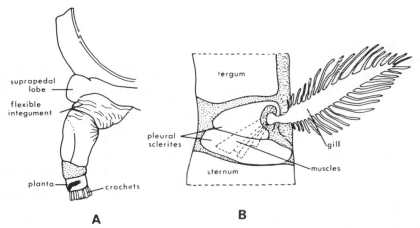

A   B

FIGURE 3.35. Secondary segmental appendages. (A) Proleg of a caterpillar and (B) gill of a mayfly larva. [From R. E. Snodgrass, *Principles of Insect Morphology.* Copyright 1935 by McGraw-Hill, Inc. Used with permission of McGraw-Hill Book Company.]

sclerotized plate on the outer wall, and an apical protractile lobe, the PLANTA, which bears claws, CROCHETS, peripherally. The planta is protracted by means of blood pressure. Immediately above the leg there is a swollen area in the body wall, the SUPRAPEDAL LOBE.

Gills. A large number of aquatic larvae possess segmentally arranged GILLS on a variable number of abdominal segments. These are flattened, filamentous structures, frequently articulated at the base (Figure 3.35B).

Nonsegmental Appendages. In many insects nonsegmental structures are present. These are typically a mediodorsal projection on the last abdominal segment. Examples of such structures are the median lamella of zygopteran larvae and the CAUDAL FILAMENT of Thysanura and Ephemeroptera. Occasionally these structures are paired (for example, the UROGOMPHI of some larval Coleoptera) and are easily mistaken for cerci. The ANAL PAPILLAE of certain dipteran larvae also fit in this category.

6. Literature

There are many books and review articles that deal with more or less specific aspects of external structure, for example, Snodgrass (1935) [general morphology]; DuPorte (1957), Snodgrass (1960), Matsuda (1965) [head]; Matsuda (1963) [thorax]; Schneider (1964) [antennae]; Matsuda (1976) [abdomen]; Snodgrass (1957), Scudder (1971) [male external genitalia]; Scudder (1961, 1971) [ovipositor]. Care must be taken when reading papers that deal with morphology because authors may use differing terminology and have differing views on the homology of structures.

DuPorte, E. M., 1957, The comparative morphology of the insect head, *Annu. Rev. Entomol.* **2:**55–70.

Hamilton, K. G. A., 1972a,b. The insect wing, Parts II and III, *J. Kans. Entomol. Soc.* **45:**54–58, 145–162.

Hinton, H. E., 1955, On the structure, function, and distribution of the prolegs of Panorpoidea, with a criticism of the Berlese–Imms theory, *Trans. R. Entomol. Soc. London* **106:**455–545.

Matsuda, R., 1963, Some evolutionary aspects of the insect thorax, *Annu. Rev. Entomol.* **8:**59–76.

Matsuda, R., 1965, Morphology and evolution of the insect head, *Mem. Am. Entomol. Inst. (Ann Arbor)* **4:**334 pp.

Matsuda, R., 1976, *Morphology and Evolution of the Insect Abdomen*, Pergamon Press, New York.

Ragge, D. R., 1955, *The Wing-Venation of the Orthoptera Saltatoria*, British Museum, London.

Rempel, J. G., 1975, The evolution of the insect head: The endless dispute, *Quaest. Entomol.* **11:**7–25.

Rempel, J. G., and Church, N. S., 1971, The embryology of *Lytta viridana* Le Conte (Coleoptera: Meloidae). VII. Eighty-eight to 132 h: The appendages, the cephalic apodemes, and head segmentation, *Can. J. Zool.* **49:**1571–1581.

Schneider, D., 1964, Insect antennae, *Annu. Rev. Entomol.* **9:**103–122.

Scudder, G. G. E., 1961, The comparative morphology of the insect ovipositor, *Trans. R. Entomol. Soc. London* **113:**25–40.

Scudder, G. G. E., 1971, Comparative morphology of insect genitalia, *Annu. Rev. Entomol.* **16:**379–406.

Snodgrass, R. E., 1935, *Principles of Insect Morphology*, McGraw-Hill, New York.

Snodgrass, R. E., 1957, A revised interpretation of the external reproductive organs of male insects, *Smithson. Misc. Collect.* **135:**60 pp.

Snodgrass, R. E., 1960, Facts and theories concerning the insect head, Smithson. Misc. Collect. 142:61 pp.

Tillyard, R. J., 1918, The panorpoid complex Part 1—The wing coupling apparatus, with special reference to the Lepidoptera, Proc. Linn. Soc. N.S.W. 43:285–319.

Tuxen, S. L., 1970, Taxonomist's Glossary of Genitalia in Insects, 2nd ed., Munksgaard, Copenhagen.

4

Classification and Identification

1. Introduction

Classification and identification are part of the more inclusive term systematics. Systematics is "the scientific study of the kinds and diversity of organisms and of any and all relationships among them" (Simpson, 1961, p. 7). Within this broad definition are thus included classification ["the ordering of organisms into groups (or sets) on the basis of their relationships" (Sneath and Sokal, 1973, p. 3)], identification ["the allocation or assignment of additional unidentified objects to the correct class once a classification has been established" (Sneath and Sokal, 1973, p. 3)], and taxonomy ["the theoretical study of classification, including its bases, principles, procedures and rules" (Simpson, 1961, p. 11)]. It must be emphasized that these definitions are not adopted by all authors. Taxonomy and classification are used by some to denote systematics (as defined above). Others consider taxonomy to be the more inclusive term, with systematics a part of it. Classification is also rather loosely used as a synonym of identification by some writers.

The classification and identification of insects, like those of most other groups of living organisms, are based primarily on external structure, though increasing use is being made (sometimes of necessity) of physiological, developmental, behavioral, and cytogenetic data. The purpose of this chapter is to provide a short introduction to classification and identification, as a basis for Chapters 5 to 10 inclusive, which deal with individual insect orders.

2. Classification

Biological systems of classification are hierarchical; that is, the largest groups (TAXA, singular TAXON) of organisms are subdivided into successively smaller groups. Thus, each taxon has a particular level (RANK) within the system. Groups of the same rank are said to belong to the same taxonomic category, to which a particular name is given. Some of these categories are obligate (capitalized in the example below), while others are optional. To show the hierarchical arrangement and to introduce the names of the various categories, let us take as an example the classification of the honeybee, *Apis mellifera*:

KINGDOM	Animalia
PHYLUM	Uniramia
Subphylum	Hexapoda
CLASS	Insecta
Subclass	Pterygota
Infraclass	Neoptera
Division	Oligoneoptera
ORDER	Hymenoptera
Suborder	Apocrita
Superfamily	Apoidea
FAMILY	Apidae
Subfamily	Apinae
Tribe	Apini
Subtribe	—
GENUS	*Apis*
Subgenus	—
SPECIES	*Apis mellifera*
Subspecies	—

[In zoology, the subspecies is the lowest category considered valid; in botany, variety, form, and subform are recognized (and given latinized names).]

Classification, then, is a means of more efficiently storing (and retrieving) information about organisms. In other words, it is not necessary to describe all the characteristics of a species each time that species is referred to. For example, as standard practice, a large proportion of entomological research articles include in their titles, after the name of the species being studied, the family, (superfamily), and order to which the species belongs. In this way, a reader can immediately gain some insight into the nature of the insect being studied, even though he may not be familiar with the species. Related to this last point, classification is also important in that it enables predictions to be made about incompletely studied organisms. For example, organisms are almost always classified first on the basis of their external structure. However, once an organism has been assigned to a particular taxon using structural criteria, it may then be possible to predict, in general terms, its habits (including life history), internal features, and physiology, on the basis of what is known concerning other, better studied, members of the taxon.

A classification may be either artificial or natural. It is possible, for example, to arrange organisms in groups according to their habitat or their economic importance. Such classifications may even be hierarchical in their arrangement. Artificial classifications are usually designed so that organisms belonging to different taxa within the system can be separated on the basis of single characters. As a result, such schemes have extremely restricted value and, usually, can be used only for the purpose for which they were initially designed. More importantly, artificial classifications provide no indication of the "true" or "natural" relationships of the constituent species.

Almost all modern classifications are natural; that is, they indicate the affinity (degree of similarity) between the organisms within the classification. Organisms placed in the same taxon (showing the greatest affinity) are said to form a natural group. There is, however, considerable controversy among systematists over the meaning of "degree of similarity," "natural group," and "natural classification." Essentially systematists fall into three major groups, according to their interpretation of the above terms. These are the phyleticists, cladists, and pheneticists. The two latter groups represent the extreme views regarding classification, whereas the phyleticists, which are the most common, may be considered as forming the "middle-of-the-road" group. To the cladistic group, led by Hennig (see Hennig, 1965, 1966), belong those systematists who base classification entirely on the recency of common ancestry, ignoring similarities and differences between existing organisms. It follows that such a classification can be erected only when the evolutionary history of a group is thoroughly known. In contrast, pheneticists devise classifications entirely on the basis of characters of extant organisms, without any direct reference to their evolutionary history. In principle, as many characters as possible are compared, chosen from all parts of an organism, so that a measure of overall similarity (or difference) can be obtained. Because of the large number of characters that may be considered, pheneticists commonly resort to mathematical analyses of their data and are thus known also as numerical taxonomists. The major proponents of phenetic classification are Sokal and Sneath (see Sokal and Sneath, 1963; Sneath and Sokal, 1973). Phyleticists employ both cladistic and phenetic information in formulating their classifications. The proportions of cladistic and phenetic information used may vary widely. Because, for many groups, the fossil record is poor, a phyletic classification will be largely phenetic in origin; that is, based primarily on comparative studies of extant species.

Cogent arguments both for and against the three systems have been made. However, it is neither possible nor appropriate here to do more than summarize some of the major points of contention. For a full discussion of the three systems the reader is referred to the works of Sokal and Sneath, and of Hennig, mentioned above, and to Wagner (1969), Michener (1970), Mayr (1969), and Ross (1974). Pheneticists argue that the schemes devised by cladists and phyleticists are too arbitrary or subjective; that is, the cladists and phyleticists "intuitively" decide which features of the organisms being studied are primitive (fundamental) and specialized, and then assign appropriate weight to these features as a basis for determining the relationships of the organisms. In fact, it has been argued that phyletic schemes are equally subjective. Obviously, for no organisms are *all* characters known. Similarity (difference) can be estimated only on the information available, and this information may be inadequate for a "true" measure of affinity. Further, pheneticists "select" their data, that is, use characters which they consider important and/or weight some characters more heavily than others. For example, should an eye be weighted equally with a leg, a part of a leg, or a hair on a part of a leg? They also "select" the method to be used in assessment of the degree of similarity, introducing even more subjectivity into their analysis. The pheneticists counter this by pointing out that their methods, being quantitative, are perfectly repeatable (say, on other groups of organisms), whereas the "value judgments" of phyleticists and cladists are not. Another criticism of phenetic schemes is their apparent inability to account for parallel and convergent evolution, though according to Sneath and Sokal (1973) this criticism is not justified.

2.1. The History of Insect Classification

Wilson and Doner (1937) have fully documented the many schemes that have been devised for the classification of insects, and it is from their account that the following short history is mainly compiled. [Papers marked with an asterisk (*) are cited from Wilson and Doner's review.] Only the major developments (that is, those which have had a direct bearing on modern schemes) have been included, though it should be realized that a good many more systems have been proposed.

Insect systematics may be considered to have begun with the work of Aristotle, who, according to Kirby and Spence (1815–1826),* included the Entoma as a subdivision of the Anaima (invertebrates). Within the Entoma Aristotle placed the Arthropoda (excluding Crustacea), Echinodermata, and Annelida. Authors who have examined Aristotle's writings differ in their conclusions regarding this author's classification of the insects, but it does appear clear that Aristotle realized that there were both winged and wingless insects and that they had two basic types of mouthparts, namely, chewing and sucking. Amazingly, it was not for almost another 2000 years that further serious attempts to classify insects were made. Aldrovanus (1602)* divided the so-called "insects" into terrestrial and aquatic forms and subdivided these according to the number of legs they possessed and on the presence or absence and the nature of the wings. In Aldrovanus' classification the term "insect" encompassed other arthropods, annelids, and some molluscs. The work of Swammerdam (1669)* is of particular interest because it represents the first attempt to classify insects according to the degree of change that they undergo during development. Although Swammerdam's concept of development was inaccurate, he distinguished clearly between ametabolous, hemimetabolous, and holometabolous insects. A more elaborate scheme of classification, still based primarily on the degree of metamorphosis but also incorporating such features as number of legs, presence or absence of wings, and habitat, was that of Ray and Willughby (1705).* Ray was the first naturalist to form a concept of a "species," a term which was to take on more significance following the introduction, by Linnaeus, of the binomial system some 30 years later. Between 1735 and 1758 Linnaeus* gradually improved on his system for the classification of insects, based entirely on features of the wings. Linnaeus recognized seven orders of "insects," namely, the Aptera, Neuroptera, Coleoptera, Hemiptera, Lepidoptera, Diptera, and Hymenoptera. Of the seven, the first four orders each contained a heterogeneous group of insects (and other arthropods) that today are separated into many different orders. The Diptera, Lepidoptera, and Hymenoptera have remained, however, more or less as Linnaeus envisaged them more than 200 years ago. Like earlier authors, Linnaeus included in the Aptera (wingless forms) spiders, woodlice, myriapods, and some nonarthropodan animals. He failed also to distinguish between primitively and secondarily wingless insect groups.

Surprisingly, perhaps, up to this time no one had made a serious attempt to classify insects according to the mouthparts. However, the Danish entomologist Fabricius, who was a student of Linnaeus, produced several "cibarian" or "maxillary" systems for classification during the period 1775–1798.* The primary subdivision was into forms with biting mouthparts and forms with sucking mouthparts. Like Linnaeus, however, Fabricius included a variety of noninsectan arthropods in his system and, furthermore, based his systems on a single anatomical feature.

De Geer (1778),* who also studied under Linnaeus, appears to have been one of the earliest systematists to realize the importance of using a combination of features as a basis for classification. Such an approach was used by the French entomologist Latreille, who, during the period 1796–1831,* gradually produced what he considered to be a natural arrangement of the Insecta. In 1810 Latreille separated the Crustacea and Arachnida from the "Insecta," in which he included still the Myriapoda. The latter group was not given class status until 1825. In the final version of his system Latreille distinguished 12 insect orders. The Linnaean order Aptera was split into the orders Thysanura, Parasita (=Anoplura), and Siphonaptera, although Latreille did not appreciate that the first group was primitively wingless, while the other two were secondarily so. The order Coleoptera of Linnaeus was subdivided into Coleoptera (*sensu stricto*), Dermaptera, and Orthoptera. The Phiphiptera (=Strepsiptera), believed to be related to the Diptera in which order they had been included, were separated as a distinct group by Latreille. The Frenchman was also among the earliest systematists to appreciate the heterogeneity of the Linnaean order Neuroptera, splitting the group into three tribes, the Subulicarnes (=modern Odonata and Ephemeroptera), Planipennes (=modern Plecoptera, Isoptera, Mecoptera, and neuropteroid insects†) and Plicipennes (=modern Trichoptera).

During the first half of the nineteenth century a large number of systematists produced their version of how insects should be classified. A majority argued, like Latreille, that the wings (presence or absence, number and nature) were the primary feature on which a classification should be established. Yet others, such as Leach (1815)* and von Siebold (1848),* considered that the nature of metamorphosis was the first-order character, with wings, mouthparts, etc. of secondary importance. If nothing else, the use of metamorphosis as a separating character drew further attention to the heterogeneity of the neuropteroid group, which contained both hemi- and holometabolous forms. Indeed, in his classification von Siebold adopted Erichson's (1839)* arrangement in which the termites, psocids, embiids, mayflies, dragonflies, and damselflies were removed from the Neuroptera and placed together as the suborder Pseudoneuroptera in the order Orthoptera.

The foundations of modern systems of classification were laid by Brauer (1885),* who appears to have been greatly influenced by the principles of comparative anatomy and paleontology established by the French zoologist Cuvier, and by the work of Charles Darwin. Brauer divided the Insecta into two subclasses, the APTERYGOGENEA, containing the primitively wingless Thysanura and Collembola, the latter having been given ordinal status by Lubbock (1873),* and the PTERYGOGENEA, containing 16 orders, in which he placed the winged

†Insects which are included in the modern orders Neuroptera. Megaloptera, and Raphidioptera.

and secondarily wingless forms. Three major divisions were established in the Pterygogenea: (1) MENOGNATHA AMETABOLA and HEMIMETABOLA (insects with biting mouthparts in both juvenile and adult stages, or mouthparts atrophied in the adult and with no or partial metamorphosis) containing the orders Dermaptera, Ephemerida, Odonata, Plecoptera, Orthoptera (including Embioptera), Corrodentia (which included the termites, psocids, and lice), and Thysanoptera; (2) MENORHYNCHA (insects with sucking mouthparts in both the juvenile and adult stages), containing the order Rhynchota (=Hemiptera); and (3) MENOGNATHA METABOLA and METAGNATHA METABOLA (insects having a complete metamorphosis, and with biting mouthparts in the juvenile stage and biting, sucking, or atrophied mouthparts in the adult), containing the neuropteroid insects, and the orders Panorpatae (=Mecoptera), Trichoptera, Lepidoptera, Diptera, Siphonaptera, Coleoptera, and Hymenoptera. Thus, Brauer appreciated the heterogeneity of the "Neuroptera" and correctly separated the Plecoptera, Odonata, and Ephemerida from the neuropteroids, Mecoptera and Trichoptera. He failed, however, to recognize the heterogeneity of the orders Orthoptera and Corrodentia.

Between 1885 and 1900 a number of modifications to Brauer's system were suggested. Most of these were concerned solely with the subdivision or aggregation of orders according to the author's views on the affinity of the groups. There were, however, two proposals that have a more direct bearing on modern systems. In 1888 Lang* proposed that the terms Apterygota and Pterygota be substituted for Apterygogenea and Pterygogenea, respectively. Sharp (1899) refocused attention on the importance of metamorphosis, but, claiming that the terms Ametabola, Hemimetabola, and Holometabola were not sufficiently definite for taxonomic purposes, proposed new terms describing whether the wings developed internally or externally. His arrangement was as follows: Apterygota (primitively wingless forms); Anapterygota (secondarily wingless forms); Exopterygota (forms in which the wings develop externally); Endopterygota (forms in which the wings develop internally). Sharp was criticized for grouping together the secondarily wingless orders (Mallophaga, Anoplura, Siphonaptera), since these contained both hemi- and holometabolous forms, and the term Anapterygota was discarded. The terms Exopterygota and Endopterygota were widely accepted, however, and became synonymous with Hemimetabola and Holometabola, respectively. It was not until the work of Crampton and Martynov in the 1920s (see below) that it was realized that these terms had no phylogenetic significance but were merely descriptive, indicating "grades of organization." Sharp recognized 21 orders of insects. His system improved on Brauer's mainly in the splitting of the Corrodentia and Orthoptera, thereby giving ordinal status to the Isoptera, Embioptera, Psocoptera. Mallophaga, and Siphunculata.

Toward the end of the nineteenth century the full force of Darwin's ideas on evolution and the importance and usefulness of fossils began to make themselves felt in insect classification. Gone was the old idea that evolution was a single progressive series of events, and in its place came the appreciation that evolution was a process of branching. Thus, insect classification entered, at the beginning of the twentieth century, the phylogenetic phase of its development, although Haeckel (1866)* had been the first to use a "phylogenetic tree" to indicate the relationships of the Insecta. Unfortunately his ideas on genealogy were incorrect. Most recent systems have been influenced to some degree by the work of an Austrian paleoentomologist, Handlirsch, who criticized earlier workers for their one-sided systems, in which a single character was used for separation of the major subdivisions. Another failure of the nineteenth-century authors was, he claimed, their inability to distinguish between parallel and convergent evolution of similar features. Finally, he pointed out that almost no one had taken into account fossil evidence. Handlirsch's first scheme, produced in 1903, was, at the time, regarded as revolutionary. He raised the Collembola, Campodeoidea (=Diplura), and Thysanura, each to the level of class. (Prior to this the Diplura had been considered usually as a suborder of the Thysanura.) He also raised the Pterygogenea of Brauer to the level of class and arranged the 28 orders of winged insects in 11 subclasses. His second scheme, published in 1908, was identical with the first except for some slight changes in the names of orders. In 1925 Handlirsch published his modified views on insect classification. In this scheme he reintroduced Brauer's two subclasses, Apterygogenea and Pterygogenea. In the former group he placed the orders Thysanura, Collembola, Diplura, and the recently discovered Protura. In the Pterygogenea he listed 29 orders (including the Zoraptera, first described in 1913) arranged in 11 superorders (his former subclasses). The most significant point in Handlirsch's work was his recognition of the heterogeneous nature of the Orthoptera, the contents of which he split into orders and regrouped with other orders in two superorders, Orthoptera (containing the orders Saltatoria, Phasmida, Dermaptera, Diploglossata, and Thysanoptera) and Blattaeformia (containing the Blattariae, Mantodea, Isoptera, Zoraptera, Corrodentia, Mallophaga, and Siphunculata). He did not appreciate however, the orthopteroid nature of the Plecoptera and placed the group in a superorder of its own. Handlirsch was also in error in regarding the Corrodentia, Mallophaga, and Siphunculata as orthopteroid groups. They are undoubtedly more closely related to the Hemiptera. Handlirsch's arrangement was strongly criticized by Börner (1904), who said that it did not express the true phylogenetic relationships of the Insecta. Börner considered that fossil wings did not have much value in insect systematics, and, in any case, there were far too few fossils for paleontology to have much bearing on classification. Comparative anatomical studies of recent forms, Börner argued, would give a more accurate picture. Börner, whose system was widely accepted, arranged the 19 orders of winged insects that he recognized in five sections. Three of these correspond with the "paleopteran orders," "orthopteroid orders," and ' hemipteroid orders" recognized today. In other words, Börner correctly assigned the Corrodentia, Mallophaga, and Siphunculata with the Hemiptera. The two remaining sections contained the endopterygote orders, though Börner's ideas on their affinities were to be shown by Tillyard (see below) to be incorrect.

Comstock (1918, and earlier), an American entomologist, supported Brauer's arrangement as a result of his comparative studies of the wing venation of living insects. Comstock was the first person to make extensive use of wing venation in determining affinities. He emphasized, however, that classifications should be based on many characters and not wings alone.

During a period of more than 20 years, beginning in 1917, Tillyard expounded his views on insect phylogeny, stemming from his extensive research into the fossil insects of Australia and North America. Although he made important contributions concerning the origin and relationships of many insect orders, Tillyard's (1918–20) work on the endopterygotes is particularly well known. In this work he showed that the Hymenoptera and Coleoptera (with the Strepsiptera) form two rather distinct orders, only distantly related to the other endopterygote groups which collectively formed the panorpoid complex. Within the complex, the Mecoptera, Trichoptera, Lepidoptera, Diptera, and Siphonaptera form a well-defined group, with the neuropteroid orders clearly distinct from them. In fact, as noted in Chapter 2, Hinton (1958) has made a strong case for excluding these orders entirely from the panorpoid complex and placing them closer to the Coleoptera.

While Tillyard was concentrating on the phylogeny of the endopterygotes, his American contemporary, Crampton, was directing his efforts toward solution of the problems of exopterygote relationships, especially the position of the Zoraptera, Embioptera, Grylloblattidae, and Dermaptera. Following his anatomical study on the newly discovered winged zorapteran *Zorotypus hubbardi*, Crampton (1920) concluded that the Zoraptera were related to the orthopteroid orders, and he placed them in a group (superorder Panisoptera) that also contained the Isoptera, Blattida, and Mantida. However, the following year Crampton revised his views and transferred the Zoraptera to the psocoid (hemipteroid) superorder, after consideration of their wing venation. In 1922 Crampton placed the Zoraptera in the order Psocoptera and suggested that it was from psocoidlike ancestors that the modern hemipteroid orders evolved. Originally, Crampton (1915) had placed the Grylloblattidae in a separate order, Notoptera, in the orthopteroid group. Five years later he concluded that the grylloblattids were closer to the Orthoptera (*s. str.*) than the blattoid groups and made the Grylloblattodea a suborder of the Orthoptera. The modern view is that the grylloblattids are probably survivors of the protorthopteran stock from which both the orthopteran and blattoid lines developed. Crampton considered that the closest relatives of the Embioptera were the Plecoptera, placing the two groups in the superorder Panplecoptera. In his early schemes Crampton also placed the Dermaptera in the Panplecoptera. He later changed this view and included them in the orthopteroid superorder, at the same time pointing out that the Diploglossata (Hemimerida) are parasitic Dermaptera. It is currently held that the Grylloblattodea are the nearest relatives of the Dermaptera.

Almost simultaneously in 1924 Crampton and the Russian paleoentomologist Martynov proposed an apparently natural division of the winged insects on the basis of the ability to flex the wings horizontally over the body when at rest. In the Paleoptera (=Paleopterygota=Archipterygota) are the orders Ephemeroptera and Odonata whose members do not possess a wing-flexing mechanism. It must be emphasized, however, that the two orders are only very distantly related through their paleodictyopteran ancestry. The remaining orders, whose members are able to flex their wings over the body, are placed in the Neoptera (=Neopterygota). The latter contains three natural subdivisions, the Polyneoptera (orthopteroid orders), Paraneoptera (hemipteroid orders), and Oligoneoptera (endopterygote orders).

Although modern writers seem to be generally agreed on the major subdivisions of the class Insecta, there is still controversy with regard to the position of certain orders such as the Zoraptera, which show a mixture of orthopteroid and hemipteroid characters, and the rank that certain groups should be given, for example, ordinal versus subordinal status for the Grylloblattodea. At the present time these are subjective matters open to an individual's whim, and only when fresh data are produced will they be resolved. The system adopted in the present volume is given below:

Phylum Uniramia
1. CLASS and ORDER. Collembola
2. CLASS and ORDER. Protura
3. CLASS and ORDER. Diplura
4. CLASS. Insecta
   I. SUBCLASS. Apterygota
      ORDERS. Microcoryphia and Zygentoma
  II. SUBCLASS. Pterygota
    A. INFRACLASS. Paleoptera
      ORDERS. Ephemeroptera and Odonata
    B. INFRACLASS. Neoptera
      a. DIVISION. Polyneoptera (orthopteroid orders)
      ORDERS. Orthoptera, Grylloblattodea, Dermaptera, Plecoptera,
          Embioptera, Dictyoptera, Isoptera, and Zoraptera
      b. DIVISION. Paraneoptera (hemipteroid orders)
      ORDERS. Psocoptera, Phthiraptera, Hemiptera, and Thysanoptera
      c. DIVISION. Oligoneoptera (endopterygote orders)
      ORDERS. Mecoptera, Lepidoptera, Trichoptera, Diptera, Siphonaptera, Neuroptera, Megaloptera, Raphidioptera, Coleoptera, Strepsiptera, and Hymenoptera

3. Identification

In principle the identification of insects is the same as that of any other animal. In practice it is more difficult, for two major reasons. First, the enormous number of species that occur means that often very minor differences in structure must be used to distinguish between forms, and second, the small size of most insects frequently means that the identifying characters are not easily seen. There are various methods for identifying organisms: (1) the specimen may be sent to an expert, (2) it may be compared with the specimens in a labeled collection, (3) it may be compared with pictures or descriptions, or (4) it may be identified by use of a key. Frequently only the last of these alternatives is available.

There are different ways of arranging a key, though all involve the same general principle, namely, the stepwise elimination of characters until a name is reached. Keys may be devised so as to reflect the evolutionary relationships between the taxa identified. However, because character differences between closely related taxa may be slight, the use of a phylogenetic key with "weak" or "difficult" couplets may make identification difficult. Thus, most keys are quite arbitrary, since they have as their only objective, ease of identification. In this arbitrary system the same taxon may key out at several points in the key, whereas in a phylogenetic key, the taxon would appear only once.

Typically, a key is in the form of a series of couplets (occasionally triplets may be included) of contrasting features. For maximum usefulness, the couplets should present clear-cut alternatives for the characters under consideration. (Ideally, in dichotomous keys, the two branches of the couplet should, where possible, use opposites for the characters.) The simplest form of sequence within a key is one in which each couplet includes only a single character. The danger of such monothetic keys is that they do not work for organisms in which a character does not follow the norm. The alternative is a polythetic key in which at least some couplets include several statements, each about a different character. Sneath and Sokal (1973) suggest three reasons for using polythetic keys: (1) one or more characters may not be observable (for example, if the specimen is incomplete, damaged, or at the "wrong" life stage), (2) some species may be exceptional for a particular character, and (3) the user of a key may err in deciding about a character. By having several characters in each couplet with which to work, a user can operate on a "majority vote" basis, that is, select the branch of the couplet which overall most closely describes the characters of his specimen. A disadvantage of such an arrangement is that a decision on which branch to select may not be clear-cut (especially if the specimen is exceptional in one of the characters listed). Further the "rules" to be followed in a polythetic key must be carefully stated [i.e., do all characters in a couplet have equal value, or does one (the first) or more carry greater weight—and, if so, how much?].

A serious drawback to many keys is that in order not to become unwieldy they are constructed either specifically for identification of specimens in a particular geographic area or for identification of specimens to a higher taxonomic level only, for example, family. This is especially true of insect keys because of the great diversity of the insect fauna. In short, their use may be rather limited. The arrangement in this text is the provision of a polythetic key for identification of insects to the level of order, rarely the suborder. A list of keys for identification beyond the ordinal level is then provided under the description of each order (see Chapters 5 to 10). This list is by no means exhaustive and it is anticipated that instructors will direct students to useful keys for the geographic area or insect group of interest.

3.1. Key to the Orders of Insects

The following key, modified from Brues, Melander, and Carpenter (1954), is in accordance with the classification used in this book. A few comments are necessary regarding its use. The key is suitable for use with the adult and most larval forms of insects. However, the larval forms of the endopterygote orders are often difficult to identify and, if at all possible, they should be allowed to metamorphose to the adult stage. In some cases it is important to know the original habitat of the specimen, and care should be taken to note this when the collection is made. The apterygote hexapods, Protura, Collembola, and Diplura, are included for completeness even though their insectan status is in doubt.

Orders marked with an asterisk (*) are unlikely to be encountered in a general collection.

Key to the Orders of Insecta

1. Wings developed 2
   Wingless, or with vestigial wings, or with rudimentary wings not
   suitable for flight (wingless adults and immature stages) 31
2. Fore wings horny, leathery, or parchmentlike, at least at base;
   hind wings membranous (occasionally absent). Prothorax
   large and not fused with mesothorax (except in Strepsiptera) ...3
   Fore wings membranous 11
3. Fore wings containing veins, or at least hind wings not folded
   crossways when hidden under fore wings 4
   Fore wings veinless, of uniform horny consistency; hind wings,
   when present, folded crossways as well as lengthwise when
   at rest and hidden beneath fore wings; mouth mandibulate 10
4. Mouthparts forming a jointed beak, fitted for piercing and
   sucking. Bugs ....HEMIPTERA   (Page 191)
   Mouthparts with mandibles fitted for chewing and moving
   laterally 5
5. Hind wings not folded, similar to fore wings; thickened basal
   part of wings very short, separated from rest of wing by a
   preformed transverse suture; social species, living in colonies.
   Termites ....ISOPTERA   (Page 152)
   Hind wings folding, fanlike, broader than fore wings 6
6. Usually rather large or moderately large species; antennae usually lengthened and threadlike; prothorax large and free from
   mesothorax; cerci present; fore wings rarely minute, usually
   long 7
   Very small active species; antennae short with few joints, at
   least one joint bearing a long lateral process; no cerci; fore
   wings minute; prothorax small. Rare short-lived insects,
   parasites of other insects, usually wasps and bees ....
   ....Males of STREPSIPTERA*   (Page 296)
7. Hind femora not larger than fore femora; body more or less
   flattened with wings superposed when at rest; tergites and
   sternites subequal 8
   Hind femora almost always much larger than fore femora,
   jumping species, if not (Gryllotalpidae) front legs broadened
   for burrowing; species usually capable of chirping or making
   a creaking noise; body more or less cylindrical. wings held
   sloping against sides of the body when at rest, tergites usually
   larger than sternites. Grasshoppers, katydids. crickets ....
   ....ORTHOPTERA   (Page 171)
8. Body elongate; head free, not concealed from above by the
   prothorax; deliberate movers 9
   Body oval, much flattened; head nearly concealed beneath the

oval pronotum; legs identical, coxae large and tibiae noticeably spiny or bristly. Cockroaches ....
............DICTYOPTERA, Suborder BLATTODEA   (Page 145)
9. Prothorax much longer than mesothorax; front legs almost
always heavily spined, formed for seizing prey; cerci usually
with several joints, Mantids ....
...........DICTYOPTERA, Suborder MANTODEA   (Page 145)
Prothorax short; legs similar, formed for walking; cerci un
jointed. Stick and leaf insects ....PHASMIDA   (Page 168)
10. Abdomen terminated by movable, almost always heavily chitinized forceps; antennae long and slender; fore wings short,
leaving most of abdomen uncovered, hind wings nearly circular, delicate, radially folded from near the center; elongate
insects. Earwigs ....DERMAPTERA   (Page 164)
Abdomen not terminated by forceps; antennae of various forms
but usually 11-jointed; fore wings usually completely sheath-
ing the abdomen; generally hard-bodied species. Beetles ....
.....................................COLEOPTERA   (Page 276)
11. With four wings 12
With but two wings (the mesothoracic) usually outspread when
at rest 29
12. Wings long, very narrow, the margins fringed with long hairs,
almost veinless; tarsi 1- or 2-jointed, with swollen tips;
mouthparts asymmetrical without biting mandibles, fitted for
lacerating and sucking plant tissues; no cerci; minute
species. Thrips .................THYSANOPTERA   (Page 213)
Wings broader and most often supplied with veins; if wings rarely
somewhat linear, tarsi have more than two joints and last
tarsal joint is not swollen 13
13. Wings, legs, and body clothed, at least in part, with elongate
flattened scales (often intermixed with hairs) that nearly
always form a color pattern on the wings; mouthparts (rarely
vestigial) forming a coiled tongue composed of the maxillae;
biting mandibles present only in Micropterygidae. Moths and
butterflies ....LEPIDOPTERA   (Page 228)
Wings, legs, and body not clothed with scales, although sometimes hairy and having a few scales intermixed; sometimes
clothed with bristles, especially on legs, or rarely with waxen
flakes or dust; color pattern when present extending to wing
membrane 14
14. Hind wings with anal area separated, folded fanlike when at
rest, nearly always wider and noticeably larger than fore
wings; antennae prominent; wing veins usually numerous 15
Hind wings without a separated anal area, not folded and not
larger than fore wings 17
15. Tarsi 5-jointed; cerci not pronounced 16
Tarsi 3-jointed; cerci well developed, usually long and many-
jointed; prothorax large, free; species of moderate to large
size. Stoneflies ....PLECOPTERA   (Page 137)

16. Wings with a number of subcostal crossveins; prothorax rather large; species of moderate to large size. Alderflies ....
   ....MEGALOPTERA (Page 269)
   Wings without subcostal crossveins, with surface hairy; prothorax small; species of small to moderate size. Caddis flies .....
   ....TRICHOPTERA (Page 221)

17. Antennae short and inconspicuous; wings netveined with numerous crossveins; mouthparts mandibulate 18
   Antennae larger if antennae of rather small size, wings have few crossveins or mouthparts form a jointed sucking beak........19

18. Hind wings much smaller than fore wings; abdomen ending in long threadlike processes; tarsi normally 4- or 5-jointed; sluggish fliers. Mayflies .............EPHEMEROPTERA (Page 121)
   Hind wings nearly like fore wings; no caudal setae; tarsi 3-jointed; vigorous, active fliers, often of large size. Dragonflies, damselflies ....ODONATA (Page 128)

19. Head elongated ventrally forming a rostrum, at tip of which are mandibulate mouthparts; hind wings not folded; wings usually with color pattern, crossveins numerous; male genitalia usually greatly swollen, forming a reflexed bulb. Scorpion flies .... MECOPTERA (Page 217)
   Head not drawn out as a mandibulate rostrum; male abdomen not forcipate 20

20. Mouthparts modified for sucking (occasionally reduced or absent); mandibles absent or in form of long bristles; no cerci; crossveins few 21
   Mouthparts for biting [occasionally for sucking (higher Hymenoptera)]; mandible always present and having typical biting form 22

21. Wings not covered with scales, not outspread when at rest; prothorax large; antennae with few joints; mouthparts forming a jointed piercing beak. Bugs .... .HEMIPTERA (Page 191)
   Wings and body covered with colored scales that form a definite pattern on wings; antennae many-jointed; mouthparts when present forming a coiled tongue. Moths and butterflies ....LEPIDOPTERA (Page 228)

22. Tarsi 5-jointed; if rarely 3- or 4-jointed, hind wings are smaller than front ones and wings lie flat over body; no cerci 23
   Tarsi 2-, 3-, or 4-jointed; veins and crossveins not numerous 26

23. Prothorax small or only moderately long. (In Mantispidae prothorax is very long, but front legs are strongly raptorial.) 24
   Prothorax very long and cylindrical, much longer than head; front legs normal; antennae with more than eleven joints; crossveins numerous. Snakeflies... RAPHIDIOPTERA (Page 270)

24. Wings similar, with many veins and crossveins; prothorax more or less free.. 25
   Wings with relatively few angular cells, costal cell without crossveins; hind wings smaller than fore pair; prothorax fused with mesothorax; abdomen frequently constricted at

base and ending in a sting or specialized ovipositor. Ants,
wasps, bees, etc. ....HYMENOPTERA (Page 299)

25. Costal cell, at least in fore wing, almost always with many
crossveins. Lacewings, antlions .... NEUROPTERA (Page 272)
Costal cell without crossveins. Scorpion flies ....
......................................MECOPTERA (Page 217)

26. Wings equal in size, or rarely hind wings larger, held superposed on top of abdomen when at rest; media fused with
radial sector for a short distance near middle of wing; tarsi
3-, 4-, or 5-jointed 27
Hind wings smaller than fore wings; wings held at rest folded
back against abdomen; radius and media not fusing; tarsi
2- or 3-jointed 28

27. Tarsi apparently 4-jointed; cerci usually minute; wings with a
transverse preformed suture near the base; social species,
living in colonies. Termites ....ISOPTERA (Page 152)
Tarsi 3-jointed, front metatarsi swollen; cerci conspicuous;
usually solitary species. Webspinners ....
.................................EMBIOPTERA* (Page 143)

28. Cerci absent; tarsi 2- or 3-jointed; wings remaining attached
throughout life; radial sector and media branched, except
when fore wings are much thickened. Book lice ....
......................................PSOCOPTERA (Page 183)
Cerci present; tarsi 2-jointed; wings deciduous at maturity,
venation greatly reduced; radial sector and media simple,
unbranched ....ZORAPTERA* (Page 181)

29. Mouthparts not functional; abdomen with a pair of caudal
filaments 30
Mouthparts forming a proboscis, only exceptionally vestigial;
abdomen without caudal filaments; hind wings replaced by
knobbed halteres. True flies ....DIPTERA (Page 246)

30. No halteres; antennae inconspicuous; crossveins abundant.
A few rare mayflies .............EPHEMEROPTERA (Page 121)
Hind wings represented by minute hooklike halteres; antennae
evident; venation reduced to a forked vein; crossveins lack-
ing; minute delicate insects. Males of scale insects ....
..............HEMIPTERA, Suborder HOMOPTERA (Page 200)

31. Body with more or less distinct head, thorax and abdomen, and
jointed legs; capable of locomotion 32
Without distinct body parts or without jointed legs, or incapable
of locomotion..75

32. Terrestrial, breathing through spiracles; rarely without
special respiratory organs 33
Aquatic, usually gill-breathing, larval forms 62
Parasites on warm-blooded animals 70

33. Mouthparts retracted into head and scarcely or not at all visible;
underside of abdomen with styles or other appendages; less
than three joints on maxillary palpi if antennae present; delicate, small or minute insects 34

Mouthparts conspicuously visible externally; if mouthparts
mandibulate, maxillary palpi more than 2-jointed; antennae
always present; underside of abdomen rarely with styles 36

34. Antennae absent. no long cerci, pincers, springing apparatus,
or anterior ventral sucker on abdomen; head pear-shaped ....
....PROTURA   (Page 113)
Antennae conspicuous; pincers, long cerci or basal ventral
sucker present on abdomen 35

35. Abdomen consisting of six segments or less, with a forked
sucker at base of abdomen below; no terminal pincers or long
cerci; usually with conspicuous springing apparatus near
end of abdomen. Springtails ....COLLEMBOLA   (Page 110)
Abdomen consisting of more than eight visible segments, with
long multiarticulate cerci or strong pincers at the end; eyes
and ocelli absent ....DIPLURA   (Page 115)

36. Mouthparts mandibulate, formed for chewing 37
Mouthparts haustellate, formed for sucking 59

37. Body usually covered with scales; abdomen with three prominent caudal filaments and bearing at least two pairs of ventral
styles 38
Body never covered with scales; never with three caudal filaments; ventral styles absent on abdomen 39

38. Head with large compound eyes and ocelli; legs with three
tarsal segments; paired styli present on each abdominal
segment. Bristletails .............MICROCORYPHIA   (Page 117)
Compound eyes small or absent; legs with two to four tarsal
segments; paired styli on abdominal segments 7 to 9 (rarely
2 to 9). Silverfish, firebrats ....ZYGENTOMA   (Page 119)

39. Underside of abdomen entirely without legs 40
Abdomen bearing false legs beneath, which differ from those of
thorax; body caterpillarlike, cylindrical; thorax and abdomen
not distinctly separated; larval forms 57

40. Antennae long and distinct 41
Antennae short, not pronounced; larval forms 54

41. Abdomen terminated by strong movable forceps; prothorax free.
Earwigs ....DERMAPTERA   (Page 164)
Abdomen not ending in forceps 42

42. Abdomen strongly constricted at base; prothorax fused with
mesothorax. Ants, etc. ....HYMENOPTERA   (Page 299)
Abdomen not strongly constricted at base; broadly joined to
thorax 43

43. Head not elongated ventrally 44
Head elongated ventrally forming a rostrum, at tip of which are
mandibulate mouthparts. Scorpion flies ....
.................................MECOPTERA   (Page 217)

44. Very small species; body soft and weakly sclerotized; tarsi
2- or 3-jointed 45
Usually much larger species; tarsi usually with more than three

joints, or, if not, body is hard and heavily sclerotized and
cerci are absent 46
45. Cerci absent. Book lice . . . . .PSOCOPTERA (Page 183)
Cerci unjointed, prominent . . . . .ZORAPTERA* (Page 181)
46. Hind femora enlarged; wing pads of larva when present in
inverse position, that is, metathoracic overlapping meso-
thoracic . . . . .ORTHOPTERA (Page 171)
Hind legs not enlarged for jumping; wing pads, if present, in
normal position 47
47. Prothorax much longer than mesothorax; front legs fitted for
grasping prey. Mantids . . . .
. . . . .DICTYOPTERA, Suborder, MANTODEA (Page 145)
Prothorax not greatly lengthened 48
48. Cerci present; antennae usually with more than fifteen joints,
often many-jointed 49
No cerci; body often hard-shelled; antennae usually with eleven
joints. Beetles . . . . .COLEOPTERA (Page 276)
49. Cerci with more than three joints 50
Cerci short, with one to three joints 52
50. Body flattened and oval; head inflexed; prothorax oval. Cockroaches . . . . .DICTYOPTERA, Suborder BLATTODEA (Page 145)
Body elongate; head nearly horizontal 51
51. Cerci long; ovipositor chitinized, exserted; tarsi 5-jointed . . . .
. . . . .GRYLLOBLATTODEA* (Page 162)
Cerci short; no ovipositor; tarsi 4-jointed; social forms, living
in colonies. Termites . . . . .ISOPTERA (Page 152)
52. Tarsi 5-jointed; body usually very slender and long. Stick
insects . . . . .PHASMIDA (Page 168)
Tarsi 2- or 3-jointed; body not linear 53
53. Front tarsi with 1st joint swollen, containing a silk-spinning
gland, producing a web in which the insects live; body long
and slender. Webspinners . . . . .EMBIOPTERA* (Page 143)
Front tarsi not swollen, without silk-spinning gland; body
much stouter; social species. Termites . . . .ISOPTERA (Page 152)
54. Body cylindrical, caterpillarlike 55
Body more or less depressed, not caterpillarlike 56
55. Head with six ocelli on each side; labium with spinnerets;
antennae inserted in membranous area at base of mandibles . . . . .
. . . . .Larvae of some LEPIDOPTERA (Page 229)
Head with more than six ocelli on each side; metathoracic legs
distinctly larger than prothoracic legs . . . .
. . . . .Larvae of Boreidae (MECOPTERA) (Page 218)
56. Mandibles united with corresponding maxillae to form sucking
organs .Larvae of NEUROPTERA (Page272)
Mandibles almost always separate from maxillae . . . .
. . . . .Larvae of COLEOPTERA (Page 277);
RAPHIDIOPTERA (Page 271); STREPSIPTERA* (Page 297);
DIPTERA (Page 247)
57. False legs (prolegs) numbering five pairs or less, located on

various abdominal segments, but not on 1st, 2nd, or 7th; false legs tipped with many minute hooks (hookless prolegs rarely on 2nd and 7th segments) . . . .
. . . . .Larvae of most LEPIDOPTERA   (Page 229)
False legs numbering from six to ten pairs, one pair of which occurs on 2nd abdominal segment; prolegs not tipped with minute hooks 58

58. Head with a single ocellus on each side. . . . .
. . . . .Larvae of some HYMENOPTERA   (Page 300)
Head with several ocelli on each side. .. . .
. . . . .Larvae of MECOPTERA   (Page 218)

59. Body bare or with few scattered hairs, or with waxy coating 60
Body densely clothed with hair or scales; proboscis if present coiled under head. Moths . . . . .LEPIDOPTERA   (Page 228)

60. Last tarsal joint swollen; mouth consisting of a triangular unjointed beak; minute species. Thrips . . . .
. . . . .THYSANOPTERA   (Page 213)
Tarsi not bladderlike at tip, and with distinct claws 61

61. Prothorax distinct. Bugs . . . . .HEMIPTERA   (Page 191)
Prothorax small, hidden when viewed from above. True flies . . . . .DIPTERA   (Page 246)

62. Mouthparts mandibulate 63
Mouthparts suctorial, forming a strong pointed inflexed beak . . . . . .
. . . . .Larvae of HEMIPTERA   (Page 191)

63. Mandibles exserted straight forward and united with the corresponding maxillae to form piercing jaws . . . .
. . . . .Larvae of NEUROPTERA   (Page 273)
Mandibles normal, moving laterally to function as biting jaws . . .64

64. Body not encased in a shell made of sand, pebbles, leaves, etc. . .65
Case-bearing forms; tracheal gills usually present. Caddis-worms . . . . .Larvae of TRICHOPTERA   (Page 221)

65. Abdomen furnished with external lateral gills of respiratory processes (a few Coleoptera and Trichoptera larvae here also) 66
Abdomen without external gills 67

66. Abdomen terminated by two or three long caudal filaments . . . .
. . . . .Larvae of EPHEMEROPTERA   (Page 122)
Abdomen with short end processes . . . .
. . . . .Larvae of MEGALOPTERA   (Page 270)

67. Labium strong, extensile, and furnished with a pair of opposable hooks . . . . .Larvae of ODONATA   (Page 129)
Labium not capable of being thrust forward and not hooked 68

68. Abdomen without false legs 69
Abdomen bearing paired false legs on several segments . . . .
. . . . .A few larvae of LEPIDOPTERA   (Page 229)

69. The three divisions of thorax loosely united; antennae and caudal filaments long and slender . . . .
. . . . .Larvae of PLECOPTERA   (Page 138)
Thoracic divisions not constricted; antennae and caudal fila-

ments short (also some aquatic larvae of Diptera and a few
Trichoptera here) ....Larvae of COLEOPTERA (Page 277)

70. Body flattened (or larval maggots) 71
Body strongly compressed; mouth formed as a sharp inflexed
beak; jumping species. Fleas .......SIPHONAPTERA (Page 264)

71. Mandibulate mouthparts 72
Mouthparts formed for piercing and sucking 73

72. Mouth inferior; cerci long; ectoparasites of bats or rodents ....
..............................Rare DERMAPTERA* (Page 164)
Mouth anterior; no cerci; generally elongate-oval insects with
somewhat triangular head; ectoparasites of birds (occa-
sionally mammals.) Chewing lice ....
.........................PHTHIRAPTERA, in part (Page 186)

73. Antennae exserted, visible, though rather short 74
Antennae inserted in pits, not visible from above (also larval
maggots, without antennae) ....
............................Pupiparous DIPTERA (Page 262)

74. Beak unjointed; tarsi formed as a hook for grasping hairs of the
host (Figure 3.24C); permanent parasites. Sucking lice ....
.......................PHTHIRAPTERA, in part (Page 186)
Beak jointed; tarsi not hooked; temporary parasites ....
...............................Some HEMIPTERA (Page 191)

75. Legless grubs, maggots or borers; locomotion effected by a
squirming motion ....Larvae of STREPSIPTERA* (Page 297);
SIPHONAPTERA (Page 265); and of some COLEOPTERA
(Page 277) (see also couplet 56); DIPTERA (Page 247);
LEPIDOPTERA (Page 229); and HYMENOPTERA
(Page 300). (If living in body of wasps and bees, with
flattened head exposed, compare females of STREPSIPTERA*
(Page 297); if aquatic wrigglers, see larvae and pupae of
mosquitoes, etc.)
Sedentary forms, incapable of locomotion 76

76. Small degraded forms bearing little superficial resemblance to
insects, with long slender beak, and usually covered with a
waxy scale, powder, or cottony tufts; living on various plants.
Scale insects ....HEMIPTERA (Page 200)
Body quiescent, but able to bend from side to side; not capable
of feeding, enclosed in a skin which is tightly drawn over all
members, or which leaves limbs free but folded against body;
sometimes free; sometimes enclosed in cocoon or in shell
formed from dried larval skins 77

77. Skin encasing legs, wings, etc., holding members tightly against
body; prothorax small; proboscis showing 78
Legs, wings, etc., more or less free from body; biting mouthparts
showing 79

78. Proboscis usually long, rarely absent; four wing cases; sometimes in cocoon ....Pupae of LEPIDOPTERA (Page 229)

Proboscis short; two wing cases, pupa often enclosed in oval
shell (puparium) formed of hardened larval skin ....
................................Pupae of DIPTERA   (Page 247)
79. Prothorax small, fused into one piece with mesothorax; some-
times enclosed in loose cocoon ....
........................Pupae of HYMENOPTERA   (Page 300)
Prothorax larger and not closely fused with mesothorax ... .80
80. Wing cases with few or no veins ....
...........................Pupae of COLEOPTERA   (Page 277)
Wing cases with several branched veins . ...
.........................Pupae of NEUROPTERA   (Page 273)

4. Literature

Börner, C., 1904, Zur Systematik der Hexapoda, *Zool. Anz.* **27**:511–533.

Brues, C. T., Melander, A. L., and Carpenter, F. M., 1954, *Classification of Insects*, 2nd ed:, Museum of Comparative Zoology, Cambridge, Mass.

Comstock, J. H., 1918, *The Wings of Insects*, Comstock, New York.

Crampton, G. C., 1915, The thoracic sclerites and the systematic position of *Grylloblatta campodeiformis* Walker, a remarkable annectant orthopteroid insect, *Entomol. News* **26**:337–350.

Crampton, G. C., 1920, Some anatomical details of the remarkable winged zorapteran, *Zorotypus hubbardi* Caudell, with notes on its relationships, *Proc. Entomol. Soc. Wash.* **22**:98–106.

Crampton, G. C., 1922, Evidences of relationship indicated by the venation of the fore wings of certain insects with especial reference to the Hemiptera–Homoptera, *Psyche* **29**:23–41.

Crampton, G. C., 1924, The phylogeny and classification of insects, *Pomona J. Entomol. Zool.* **16**:33–47.

Handlirsch, A., 1903, Zur Phylogenie der Hexapoden, *Sitz. Nat. Klasse Acad. Wiss.* **112**:716–738.

Handlirsch, A., 1906–08, *Die fossilen Insecten und die Phylogenie der rezenten Formen*, Engelman, Leipzig.

Handlirsch, A., 1925, Geschichte, Litteratur, Technik, Paläontologie, Phylogenie und Systematik der Insecten, in: *Handbuch der Entomologie* (C. Schröder, ed.), Vol. 3, pp. 1–1201, Fischer, Jena.

Hennig, W., 1965, Phylogenetic systematics, *Annu. Rev. Entomol.* **10**:97–116.

Hennig, W., 1966, *Phylogenetic Systematics*, University of Illinois Press, Urbana, Ill.

Hinton, H. E., 1958, The phylogeny of the panorpoid orders, *Annu. Rev. Entomol.* **3**:181–206.

Martynov, A. V., 1924, The interpretation of the wing venation and tracheation of the Odonata and Agnatha, *Russk. Ent. Obozr.* **18**:145–174. [Original in Russian. Translated, with an introductory note, by F. M. Carpenter in *Psyche* **37**(1930):245–280.]

Mayr, E., 1969, *Principles of Systematic Zoology*, McGraw-Hill, New York.

Michener, C. D., 1970, Diverse approaches to systematics, *Evol. Biol.* **4** 1–38.

Ross, H. H., 1974, *Biological Systematics*, Addison-Wesley, Reading, Mass.

Sharp, D., 1899, Some points in the classification of the Insecta Hexapoda, *Proc. 4th Int. Congr. Zool.*, pp. 246–249.

Simpson, G. G., 1961, *Principles of Animal Taxonomy*, Columbia University Press, New York.

Sneath, P. H. A., and Sokal, R. R., 1973, *Numerical Taxonomy—The Principles and Practice of Numerical Classification*, W. H. Freeman and Co., San Francisco, Calif.

Sokal, R. R., and Sneath, P. H. A., 1963, *Principles of Numerical Taxonomy*, W. H. Freeman and Co., San Francisco, Calif.

Tillyard, R. J., 1918–20, The panorpoid complex. A study of the holometabolous insects with special reference to the sub-classes Panorpoidea and Neuropteroidea, *Proc. Linn. Soc. N.S.W.* **43**:265–284, 395–408, 626–657; **44**:533–718; **45**:214–217.

Wagner, W. H., Jr., 1969, The construction of a classification, in: *Systematic Biology*, National Academy of Science, Washington, D.C., Publ. #1692.

Wilson, H. F., and Doner, M. H., 1937, *The Historical Development of Insect Classification*, Madison, Wis.

5

Apterygote Hexapods

1. Introduction

Traditionally the groups included in the term "apterygote hexapods," namely, the Collembola, Protura, Diplura, and Thysanura (including Microcoryphia and Zygentoma), were considered as orders of primitively wingless insects and placed in the subclass Apterygota (Ametabola). They exhibit the following common features: lack of wings, lack of a pleural sulcus on the thoracic segments, presence of pregenital abdominal appendages, slight or absent metamorphosis, and indirect sperm transfer. As more information on their structure and habits has become available, it has become apparent that (1) their status as insects (except for Thysanura) is doubtful, and (2) the relationship of the groups with each other is more distant than originally believed. Several authors have therefore recommended that the insectan subclass Apterygota be reserved solely for the Thysanura and that the Collembola, Protura, and Diplura each be given the rank of class. These three groups differ fundamentally from insects in several features, for example, they are entognathous, have intrinsic musculature in the antennae, and lack compound eyes which are characteristic of most insects, at least in the adult stage. The Collembola are probably furthest removed from the winged insects. They possess only six abdominal segments, a postantennal sensory organ similar to the organ of Tömösvary found in myriapods, gonads with lateral (rather than apical) germaria, and eggs in which there is total cleavage. Noninsectan features of the Protura are the absence of antennae (perhaps a secondary condition associated with their soil-dwelling habit), the occurrence of anamorphosis, and a genital aperture that opens behind the eleventh segment. Diplura are superficially similar to Thysanura, with which they were formerly grouped by some authors. However, in addition to the features mentioned above, they differ from typical insects in having unusual respiratory and reproductive systems. Even though all Thysanura are considered insectan, it is now apparent that the group contains two distinct subgroups, the Microcoryphia and the Zygentoma (=Thysanura sensu stricto), to which some authors accord ordinal status. The primary basis for this distinction concerns the mouthparts. In the Microcoryphia (such as *Machilis* and its allies, the bristletails) the mandibles have a single articulation with the head and bite with a rolling motion. In the Zygentoma (which includes silverfish and firebrats), on the other hand, there is a dicondylic articulation of the mandible, which thus bites transversely as in other insects. As noted in Chapter 1, differences in the structure and operation of the mouthparts are of fundamental phylogenetic importance.

2. Collembola

SYNONYMS: Oligentoma, Oligoentomata   COMMON NAME: springtails

Small to minute wingless hexapods; head pro- or hypognathous, antennae 4- to 6-segmented, compound eyes absent, mouthparts entognathous; abdomen 6-segmented, typically with three medially situated pregenital appendages (collophore on segment 1, retinaculum on segment 2, furcula on segment 4), gonopore on 5th segment.

The Collembola have a worldwide distribution. About 2000 species have been described, including about 250 from the United Kingdom and more than 300 from North America. However, the collembolan fauna of the latter region has not been extensively studied, and the species total for North America is probably much greater. Individual species may be quite cosmopolitan.

Structure

Collembola vary in length from about 0.2 to 10 mm. They are generally dark, but many species are whitish, green, or yellowish, and some are striped or mottled. The body is either elongate (suborder Arthropleona) (Figure 5.1B) or more or less globular (suborder Symphypleona) (Figure 5.1A). The head is pro- or hypognathous, and the mouthparts are enclosed within a pouch formed by the ventrolateral extension of the head capsule. The mouthparts are typically of the chewing type, though in many species they are rasping or suctorial. The 4- to 6-segmented antennae vary greatly in length. Immediately behind the antennae is a structure of variable form, the postantennal organ, which appears to have a sensory (olfactory?) function. Compound eyes are never present, but a variable number of ocelli (up to eight) are found on each side of the head. Generally the three thoracic segments are distinct and identical, but in the Symphypleona they are intimately fused with each other and with the abdomen, and the intersegmental margins are largely obscured. The legs have no true tarsus but terminate in one or two claws which arise from the tibia. No more than six abdominal segments can be distinguished at any time (even during embryonic development), and frequently fusion of segments occurs. The first abdominal segment bears the collophore (ventral tube), which arises by fusion and differentiation of the embryonic appendages. The tube contains a pair of vesicles that can be protruded by means of blood pressure. The function of the tube is unknown, though roles in respiration, water resorption, or adhesion to the substrate have been suggested. Most Collembola have a springing organ (furcula) on the fourth abdominal segment. The furcula is held under the body by a hooklike structure, the retinaculum (tenaculum), formed from the appendages of the third abdominal segment. When released from the re-

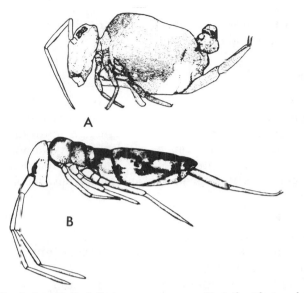

FIGURE 5.1. Collembola. (A) *Sminthurus purpurescens* (Sminthuridae) and (B) *Entomobrya nivalis* (Entomobryidae). [Reprinted from Elliott A. Maynard, *A Monograph of the Collembola or Springtail Insects of New York State.* Copyright © 1951 by Cornell University. Used by permission of the publisher, Cornell University Press.]

tinaculum, the furcula is forced downward and backward. As it strikes the ground, the animal is thrown forward through the air. Abdominal appendages may be greatly reduced in small subterranean forms. Cerci are absent in Collembola.

Noteworthy features of the internal structure of Collembola are the absence of Malpighian tubules and tracheal system, and the lateral germaria in the gonads of both sexes. The nervous system is specialized, the ganglia of the abdominal segments having moved anteriorly to fuse with the ganglion of the metathorax.

Life History and Habits

Springtails occupy a variety of humid or moist habitats. They are extremely abundant in moist soils and are found in leaf mold, under the bark of rotting logs, in the fleshy parts of some fungi, in the nests of termites or ants, and on the seashore. Several species occur on the surface of standing water (including, rarely, tidal pools). Some species occasionally form large aggregations on the surface of snow, though the significance of this is not clear. Collembola are saprophagous, fungivorous (including some spore feeders), or phytophagous (including some pollen feeders). Only rarely do they feed on animal matter. A few species are occasionally of economic importance because of the damage they cause to plants (for examples, see under families below).

Eggs, generally pale, spherical structures, are laid singly or in small clusters. Several females may oviposit in one place to form a large egg cluster. For an optimum rate of development the eggs require high humidity and low temperatures. There is little change in external form as the young animal develops, and sexual maturity is reached usually after four or five molts. Many additional molts occur during the adult phase.

Classification

The striking difference in body form appears to provide a natural subdivision of the Collembola into the suborders Arthropleona and Symphypleona. Classification beyond this level is variable, though most authors follow the arrangement of Gisin (1960) given below. However, Salmon (1964–65) has proposed that the group should be divided into four suborders, mainly on the position (hypognathous or prognathous) and form (chewing or suctorial) of the mouthparts:

Suborder Arthropleona
    Superfamily Hypogastruroidea
        Families ONYCHIURIDAE and HYPOGASTRURIDAE
    Superfamily Entomobryoidea
        Families ENTOMOBRYIDAE and ISOTOMIDAE
Suborder Symphypleona
    Superfamily Sminthuroidea
        Family SMINTHURIDAE

Of the five families, the Hypogastruridae, Entomobryidae, and Sminthuridae are the most common.

Sminthuridae (Figure 5.1A)

Members of the large family Sminthuridae are usually between 1 and 3 mm in length. They are characterized by a roundish body, broad fusion of the thorax and abdomen, and an abdomen which shows little sign of segmentation, except at the posterior end. Many are either pale yellow or white with pink markings. Generally they have a pair of conspicuous black ocelli. They are found, often in large numbers, on vegetation, and a number of species are economically important for the damage they do; for example, *Sminthurus viridis*, the lucerne flea, a European species introduced into Australia, has become an important pest on alfalfa (lucerne) and other leguminous crops. Other species may do considerable damage in greenhouses and to many garden vegetables when the latter are in the seedling stage.

The Hypogastruridae and Entomobryidae may be distinguished from the sminthurids by their elongate body, distinct thorax and abdomen, and obvious abdominal segmentation.

Hypogastruridae

The Hypogastruridae are generally 1 to 3 mm in length, whitish or darkly colored Collembola. They differ from Entomobryidae in having an obvious prothorax, a granular cuticle and short, 4-segmented antennae; postantennal organs are absent. They are found in a wide range of habitats. Probably most species live among decaying vegetation, in soil, in cracks in the bark of trees, and like habitats. Some, however, including the very common collembolan *Podura aquatica*, live near the shoreline of ponds and lakes. *Anurida maritima*, another common species, lives in the intertidal zone on the seashore. Many Hypogastruridae have the common name of snow fleas through their being found, sometimes in immense numbers, jumping about on snow, usually shortly after a period of mild weather. One widely distributed species, *Hypogastrura armata*, sometimes becomes a pest in mushroom cellars.

Entomobryidae (Figure 5.1B)

In Entomobryidae are included many of the larger Collembola, which reach 5 mm or more in length. Other features include the reduced prothorax, which is hidden beneath the mesothorax, a smooth cuticle, and 4- to 6-segmented antennae of variable length. Postantennal organs may or may not be present. Species are found in much the same type of habitat as those of the previous family, namely, in leaf litter, under bark, among moss, etc. They are very variable in color.

Literature

Accounts of the biology of Collembola are provided by Maynard (1951), Christiansen (1964), Hale (1967), and Butcher *et al.* (1971). Keys for the identification of Collembola are to be found in Maynard (1951) and Scott (1961) [North American forms], Salmon (1951) [genera of the World], Gisin (1960) [European species] (in German), and Richards (1968) [World Sminthuridae].

Butcher, J. W., Snider, R., and Snider, R. J., 1971, Bioecology of edaphic Collembola and Acarina, *Annu. Rev. Entomol.* **16**:249–288.

Christiansen, K., 1964, Bionomics of Collembola, *Annu. Rev. Entomol.* **9**:147–178.

Gisin, H., 1960, *Collembolenfauna Europas*, Museum d'histoire naturelle, Genève.

Hale, W. G., 1967, Collembola, in: *Soil Biology* (A. Burges and F. Raw, eds.), Academic Press, New York.

Maynard, E. A., 1951, *A Monograph of the Collembola or Springtail Insects of New York State*, Comstock, New York.

Richards, W. R., 1968, Generic classification, evolution, and biogeography of the Sminthuridae of the World (Collembola), *Mem. Entomol. Soc. Can.* **53**:54 pp

Salmon, J. T., 1951, Keys and bibliography to the Collembola, *Zool. Publ. Victoria Univ. Coll. N.Z.* **8**:1–82.

Salmon, J. T., 1964–65, *An index to the Collembola*, *Bull. R. Soc. New Zealand*, No. 7: Vols. 1 and 2 (1964), Vol. 3 (1965).

Scott, H. G., 1961, Collembola: Pictorial keys to the nearctic genera, *Ann. Entomol. Soc. Am.* **54**:104–113.

3. Protura

SYNONYM: Myrientomata   COMMON NAME: proturans

Minute wingless hexapods; head cone-shaped, compound eyes, ocelli, and antennae absent, mouthparts entognathous and suctorial; abdomen 11-segmented with appendages on first 3 segments, gonopore posterior to 11th segment.

About 200 species of proturans have been described which are distributed throughout the World. Of these, approximately 20 are North American and 20 are British.

Structure

Proturans (Figure 5.2) are elongate, generally pale arthropods 2 mm or less in length. The head is cone-shaped and bears anteriorly the styliform entognathous mouthparts. Photoreceptor organs are absent from the head, as are typical antennae. However, a pair of "pseudoculi" occur dorsolaterally that may be homologous with the antennae and serve as humidity receptors. The thoracic segments are distinct, though the first is greatly reduced. The six identical legs have an unsegmented tarsus. The forelegs are generally not used in locomotion but are held aloft and probably act as sense organs. In adult proturans the abdomen is 11-segmented; in the newly hatched animal there are only eight abdominal segments, three being added anamorphically during postembryonic development. Short, unsegmented or 2-segmented appendages with eversible vesicles are found on the first three abdominal segments. Cerci are absent.

Internally there are no distinct Malpighian tubules, but six papillae occur at the junction of the midgut and hindgut, and these may serve an excretory function. A tracheal system may or may not be present. The nervous system is generalized, with ganglia found in the first six abdominal segments.

Life History and Habits

Like springtails, proturans are found in a variety of moist habitats. Although frequently overlooked because of their small size, they are quite numerous in certain situations particularly in soil and litter. Few details of their biology are available. They are thought to be fungivorous. It is reported that five juvenile stages are passed through before sexual maturity is reached, but it is not known whether the animal molts as an adult. There are probably several

FIGURE 5.2. *Acerentulus barberi*, a proturan. [From H. E. Ewing, 1940, The Protura of North America, *Ann. Entomol. Soc. Am.* **33**:495–551. By permission of the Entomological Society of America.]

generations per year with, in cooler climates, the adults spending the winter in a dormant condition.

Classification

Tuxen (1964) recognizes two superfamilies and three families of Protura. These are superfamily Eosentomoidea (family EOSENTOMIDAE) and superfamily Acerentomoidea (families ACERENTOMIDAE and PROTENTOMIDAE). The Eosentomidae, considered to be the most primitive family, possess a tracheal system opening to the exterior via two pairs of thoracic spiracles. Tracheae are absent in the Acerentomoidea. In the Acerentomidae at least two pairs of the abdominal appendages bear eversible vesicles; in the Protentomidae only the first pair of abdominal appendages possess vesicles.

Literature

Most literature on Protura is taxonomic in nature. The best general account of their biology is that of Ewing (1940), although Tuxen (1964) provides a good bibliography of papers that contain ecological and biological data. Tuxen's monograph contains details of the external morphology of Protura as a basis for his key to their identification. Womersley (1927, 1928) and Ewing (1940) deal with the British and North American species, respectively.

Ewing, H. E., 1940, The Protura of North America, *Ann. Entomol. Soc. Am.* **33**:495–551.
Tuxen, S. L., 1964, *The Protura. A Revision of the Species of the World with Keys for Determination,* Hermann, Paris.
Womersley, H., 1927, Notes on the British species of Protura, with descriptions of new genera and species, *Entomol. Mon. Mag.* **63**:140–148.
Womersley, H., 1928, Further notes on the British species of Protura, *Entomol. Mon. Mag* **64**:113–115.

4. Diplura

SYNONYMS: Dicellura, Entotrophi, Entognatha   COMMON NAME: diplurans

Elongated apterygotes; head with long many-segmented antennae, compound eyes and ocelli absent, mouthparts entognathous; thoracic segments distinct, legs with unsegmented tarsus; abdomen 10-segmented, most pregenital segments with styli, eversible vesicles on some abdominal segments, cerci present as either long multi-segmented or forcepslike structures, gonopore between segments 8 and 9.

More than 600 species of this widely distributed, though mainly tropical or subtropical, order have been described. Most holarctic forms belong to the family Campodeidae, including the 11 British species. At least 25 species of Diplura have been described from North America.

Structure

In general form Diplura (Figure 5.3) resemble Thysanura but differ in being entognathous and lacking a median process on the last abdominal segment.

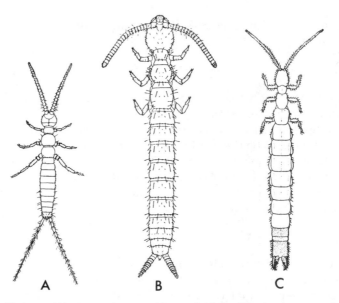

FIGURE 5.3. Diplura. (A) *Campodea* sp. (Campodeidae), (B) *Anajapyx vesiculosus* (Pro-japygidae), and (C) *Heterojapyx* sp. (Japygidae). [From A. D. Imms, 1957, *A General Textbook of Entomology*, 9th ed. (revised by O. W. Richards and R. G. Davies). By permission of Chapman and Hall Ltd.]

Most species are a few millimeters long, but a few may reach 50 mm in length. The roundish or oval head carries the multisegmented antennae, whose segments (except the most distal) are provided with muscles. The reduced biting mouthparts occupy a pouch on the ventral surface. The six identical legs have an unsegmented tarsus. Two to four lateral spiracles occur on the thorax. Ten abdominal segments are distinguishable. The sterna of segments 2 to 7 bear styli, and eversible vesicles occur on a variable number of segments. The conspicuous cerci vary in structure and provide the main basis for division of the order into families (see Classification). In most species seven pairs of abdominal spiracles can be found.

As in Protura, Malpighian tubules are represented usually by a variable number of papillae at the junction of the midgut and hindgut. The tracheal system is developed to a variable extent. Tracheae leading from one spiracle never anastomose with those from other spiracles, and they lack cuticular supporting rings characteristic of insectan tracheae. The nervous system is not specialized, the ventral nerve cord containing seven or eight abdominal ganglia. The reproductive system is greatly variable within the Diplura, although in all species the germaria are apical. In *Japyx* (Japygidae) there are seven pairs of segmentally arranged ovarioles, in *Anajapyx* (Projapygidae) two, and in *Campodea* (Campodeidae) one. One or two pairs of testes occur in the male.

Life History and Habits

Because of their rather secretive habits little is known of the life history of diplurans. They are found in damp habitats, for example, leaf litter, under stones and logs, and in soil. The campodeids are vegetarian, but the japygids are carnivorous, catching prey with their forceps (modified cerci). The eggs are laid in groups, and in some species the young are guarded by the female. Development is slow and molting continues through life.

Classification

Three families of Diplura can be distinguished, though only members of the Campodeidae are likely to be encountered. In the CAMPODEIDAE (Figure 5.3A) the cerci are multisegmented and usually as long as the abdomen. At maturity campodeids are usually more than 4 mm in length. In contrast, the PROJAPYGIDAE (=ANAJAPYGIDAE) (Figure 5.3B) are minute arthropods whose cerci are short (less than half the length of the abdomen) and have less than 10 segments. The cerci of the JAPYGIDAE (Figure 5.3C) take the form of strongly sclerotized, unsegmented forceps.

Literature

General information on Diplura is given by Imms (1957) and Wallwork (1970). The British species can be identified from Delaney (1954). Smith (1960) provides a key to North American Diplura.

Delaney, M. J., 1954, Thysanura and Diplura, R. Entomol. Soc. Handb. Ident. Br. Insects 1 (2):1–7.
Imms, A. D., 1957, *A General Textbook of Entomology*, 9th ed. (revised by O. W. Richards and R. G. Davies), Methuen, London.
Smith, L. M., 1960, The family Projapygidae and Anajapygidae (Diplura) in North America, *Ann. Entomol. Soc. Am.* **53**:575–583.
Wallwork, J. A., 1970 *Ecology of Soil Animals*, McGraw-Hill, London.

5. Microcoryphia

SYNONYMS: Archaeognatha, Ectotrophi   COMMON NAME: bristletails
          (in part), Ectognatha (in part)

Small or moderately sized apterygote insects; head with long multisegmented antennae, large compound eyes and ocelli, ectognathous chewing mouthparts, mandibles with single articulation; thorax with large paranotal lobes, legs with three tarsal segments; abdomen 11-segmented, though 10th segment reduced and tergum of 11th forming median caudal filament, paired styli present on each abdominal segment, long multisegmented cerci present.

As noted in the Introduction to this chapter, Microcoryphia and Zygentoma (see Section 6) were originally united in the order Thysanura. However, fundamental differences in their structure (compare the definitions of the orders) has led to their separation as distinct orders.

Structure

Microcoryphia (Figure 5.4A) are elongate insects up to 20 mm in length. Their body is strongly convex dorsally and generally tapered posteriorly. The head is hypognathous, in some species prognathous, and carries prominent

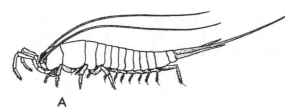

FIGURE 5.4. Microcoryphia and Zygentoma. (A) *Machilis* sp. (Machilidae) and (B) *Lepismodes inquilinus* (Lepismatidae). [A, reprinted from R. E. Snodgrass, *A Textbook of Arthropod Anatomy.* Copyright © 1952 by Cornell University. Used by permission of the publisher, Cornell University Press. B, from A. D. Imms, 1957, *A General Textbook of Entomology,* 9th ed. (revised by O. W. Richards and R. G. Davies). By permission of Chapman and Hall Ltd.]

chewing mouthparts, the maxillary palps being particularly conspicuous. Each mandible has a single articulation with the head. The antennae are filiform and comprise 30 or more segments which lack intrinsic musculature. Compound eyes are well developed and may be contiguous (meet in a middorsal position). Median and paired lateral ocelli are also present. The legs have three tarsal segments and, in some species, those of the mesothorax and metathorax bear coxal styli. Abdominal styli occur on segments 2 to 9, and eversible vesicles are almost always found on abdominal segments 1 to 7. In females an ovipositor is formed from the appendages of abdominal segments 8 and 9. In males the appendages of the 9th abdominal segment fuse to form a median penis.

Internally Microcoryphia exhibit features that might be expected of primitive insects. They have 12–20 Malpighian tubules, a nervous system that includes paired longitudinal connectives and eight abdominal ganglia, nine pairs of spiracles and tracheae which do not anastomose with those of adjacent segments, and a primitive reproductive system. In females there are seven panoistic ovarioles on each side of the body, and in some species these are arranged in a more or less segmental manner. Females lack a spermatheca. In males each testis comprises several lobes, though these are not segmentally arranged.

Life History and Habits

Microcoryphia lead a concealed life in rotting wood, among leaves, under stones, etc., and are restricted to such habitats by their inability to resist desiccation. They appear to be omnivorous or phytophagous.

Details of the life history of Microcoryphia are not well known. They apparently lay eggs singly or in groups of up to 30 in crevices or holes dug with the ovipositor. The absence of a spermatheca suggests that they have to mate frequently. Postembryonic development is slow (generally taking several months from hatching to adulthood) and includes at least five juvenile instars in *Machilis.* Molting continues in adults.

Classification

The order includes a single superfamily Machiloidea, which contains three families, MACHILIDAE, PRAEMACHILIDAE, and MEINERTELLIDAE. Only Machilidae (bristletails) are commonly encountered. The family is widely distributed, though primarily in the Northern Hemisphere. Bristletails are quick-running insects and are capable of jumping for short distances when disturbed.

Literature

See under Section 6 (Zygentoma).

6. Zygentoma

SYNONYMS: Thysanura (*sensu stricto*), COMMON NAMES: silverfish, firebrats
Ectotrophi (in part),
Ectognatha (in part)

Small or moderately sized apterygote insects; head with long multisegmented antennae, compound eyes small or absent, ocelli absent, ectognathous chewing mouthparts, mandible with dicondylic articulation; legs with 2 to 4 tarsal segments; abdomen 11-segmented but with 10th segment reduced and tergum of 11th segment forming median caudal filament, paired styli on abdominal segments 7 to 9 (rarely 2 to 9), long multisegmented cerci present.

Structure

Zygentoma (Figure 5.4B) are broadly similar to Microcoryphia, and only the more important differences in structure will be noted here. The body of Zygentoma is dorsoventrally flattened. The head bears much-reduced compound eyes but lacks ocelli. The mandibles have a dicondylic articulation with the head, as do those of pterygote insects. The legs include two to four tarsal segments Abdominal styli normally occur only on segments 7 to 9, though in Nicoletiidae they are found on segments 2 to 9. Eversible vesicles are absent in many Lepismatidae but occur on abdominal segments 2 to 7 in Nicoletiidae.

Four to eight Malpighian tubules are present. Ten pairs of spiracles occur and the tracheal system is well developed compared to that of Microcoryphia. On each side of the body a longitudinal trunk links the tracheae originating from the spiracles, and transverse segmental tracheae link one side with the other. The female reproductive system includes five panoistic ovarioles, a spermatheca, and accessory glands. The number of lobes comprising each testis is greater than in Microcoryphia (eight lobes in Lepismatidae, many in Nicoletiidae).

Life History and Habits

In their general habits the Zygentoma resemble Microcoryphia. They lead a secluded existence, often in places with a high relative humidity, though some are extremely resistant to desiccation. Like Microcoryphia, they are omnivorous or phytophagous.

Eggs are laid singly or in groups, in crevices, etc. Individuals may live for several years, and as many as 60 molts have been recorded. About a dozen molts occur before sexual maturity is reached. Because the lining of the spermatheca is shed at each molt, females must mate between molts in order to produce fertilized eggs. An apparent scarcity of males suggests that parthenogenesis may occur in some species.

Classification

Members of the order are arranged in a single superfamily, Lepismatoidea, containing three families, LEPISMATIDAE, NICOLETIIDAE, and LEPIDOTRICHIDAE. Only members of the Lepismatidae are likely to be encountered. Some species of this family are very common. They include a number of domiciliary species (found in buildings), the two most common being the common silverfish, *Lepismodes inquilinus* (=*Lepisma saccharina*) and the firebrat, *Thermobia domestica*. The former prefers warm and humid environments and is often found in places such as bookcases, cupboards, and bathrooms. Firebrats, in contrast, live in hot, dry environments, for example,in the vicinity of fireplaces, furnaces, boilers, and in bakeries. They are highly resistant to desiccation. Both species may cause considerable damage to books, clothing, and foods that contain starch or cellulose, and they are among the few animals that produce an intrinsic gut cellulase.

The Lepidotrichidae should be mentioned in view of this family's important phylogenetic position. Prior to 1961 the family was known only as fossils. However, in 1961 Wygodzinsky discovered a living representative, *Tricholepidon gertschi,* in California, that possesses a large number of primitive characters. This species is clearly the most archaic living apterygote insect discovered to date and is likely to be similar to the common ancestor of other modern groups.

Literature

Accounts of the biology of Microcoryphia and Zygentoma are provided by Delaney (1957) and Sharov (1966). Keys for identification are given in the papers by Delaney (1954) [British species], Slabaugh (1940) [North American domiciliary Lepismatidae], and Wygodzinsky (1972) [North and Central American Lepismatidae].

Delaney, M. J., 1954, Thysanura and Diplura, R. Entomol. Soc. Handb. Ident. Br. Insects 1(2):1–7.
Delaney, M. J., 1957, Life histories in the Thysanura, Acta Zool. Cracov. 2:61–90.
Sharov, A. G., 1966, Basic Arthropodan Stock, Pergamon Press, Oxford.
Slabaugh, R. E., 1940, A new thysanuran and a key to the domestic species of Lepismatidae (Thysanura) found in the United States, Entomol. News 51:95–98.
Wygodzinsky, P., 1961, On a surviving representative of the Lepidotrichidae (Thysanura), Ann. Entomol. Soc. Am. 54:621–627.
Wygodzinsky, P., 1972, A review of the silverfish (Lepismatidae, Thysanura) of the United States and the Caribbean Area, Am. Mus. Novit. 2481:26 pp.

6

Paleoptera

1. Introduction

In the infraclass Paleoptera are the orders Ephemeroptera (mayflies) and Odonata (dragonflies and damselflies), the living species of which represent the few remains of two formerly very extensive groups. Although both are placed in the Paleoptera, it should be realized that the Ephemeroptera and Odonata are two very different groups that must have diverged at a very early stage in the evolution of winged insects. They possess the following common features which unite them as Paleoptera: wings that cannot be folded back against the body when not in use, retention of the anterior median wing vein, netlike arrangement of wing veins (many crossveins), aquatic juvenile stage, and considerable change from juvenile to adult form. In members of both orders, wing development is external, though this feature is not, of course, restricted to Paleoptera.

2. Ephemeroptera

SYNONYMS: Plectoptera, Ephemerida   COMMON NAMES: mayflies, shadflies

Adults small to medium-sized elongate fragile insects; antennae short and setaceous, mouthparts vestigial, compound eyes large, three ocelli present; generally two pairs of membranous wings (though hind pair greatly reduced) held vertically over body when at rest, with many crossveins; abdomen terminated with two very long cerci and frequently a median caudal filament.

Larvae aquatic; body campodeiform; antennae short, compound eyes well developed, biting mouthparts; abdomen usually with long cerci and a median caudal filament, and four to seven pairs of segmental tracheal gills.

Approximately 2000 species of this widely distributed order have been described. Nearly 600 species occur in the United States, and approximately 50 species have been reported from Great Britain.

Structure

Adult. The head is triangular in shape when viewed from above. The compound eyes are large (especially in the male where they almost meet in the midline) and typically are divided horizontally into an upper region with large facets and a lower region with smaller facets. The antennae are small, multisegmented, setaceous structures. The mouthparts are vestigial. The thoracic region is dominated by the large mesothoracic segment. Pleural sulci are poorly developed or absent even on the pterothorax. Two pairs of fragile wings are generally present, though the hind pair is always reduced or absent entirely. The wing venation is primitive, the median vein being divided into anterior and posterior branches. The legs are sometimes reduced, associated with the habit of passing the entire adult life on the wing. However, the forelegs of males are usually enlarged and used to grip a female during mating. Primitively there are five tarsal segments, but the basal one or two segments may fuse with the tibia in the higher families. In females paired gonopores open behind the seventh abdominal sternum. A typical ovipositor is absent. In males a pair of claspers occurs on the ninth sternum. Between these claspers lies a pair of penes.

The most noteworthy internal feature is the modification of the gut as an aerostatic organ to reduce the specific gravity of the insect. The esophagus is a narrow tube equipped with muscles that regulate the amount of air in the gut. Swallowed air is held in the midgut, which no longer has a digestive function and is lined with pavement rather than columnar epithelium. The hindgut also has a valve to prevent loss of air. The reproductive organs are very primitive; accessory glands are absent, and the gonoducts are paired in both sexes.

Larva. Mayfly larvae exhibit a wide range of body form associated with the diverse habitats in which they are found. The body is of variable shape but is often flattened dorsoventrally. The antennae, compound eyes, and ocelli differ little from those of adults. Larvae possess well-developed biting mouthparts. The structure of the legs varies according to whether a larva is a swimming, burrowing, or clinging form. The abdomen is terminated with a pair of long cerci and usually a median caudal filament. Between four and seven pairs of tracheal gills occur on the abdomen. In open-water forms the gills are usually lamellate; in burrowing species they tend to be plumose. In some species gills may not be directly important in gaseous exchange. They are capable of coordinated flapping movements and may serve simply to create a current of water flowing over the body. In some species accessory gill-like respiratory structures develop on the thorax and head.

Life History and Habits

Adult mayflies are commonly found in the vicinity of water, often in huge mating swarms. They are short-lived creatures, existing for only a few hours (generally nocturnal species) or a few days. A swarm consists generally only of males. Females enter the swarm to find a mate, though in some species copulation occurs away from the swarm. The egg-laying habits are quite variable, as is the number of eggs laid. In some short-lived species eggs are laid en masse on the water surface. The clutch breaks up and the eggs sink, becoming scattered over the substrate. In species which survive for several days the eggs may be laid in small batches; Baetis spp. females descend below the water surface to secure the eggs on the substrate. Eggs often have special structures which serve to anchor them in position. They usually hatch within 10–20 days, but in a few species the eggs enter a diapause which is broken only by low winter temperatures. Consequently they do not hatch until the following spring. In some species that have a relatively long adult life (up to three weeks) ovoviviparity occurs, females retaining fertilized eggs in the genital tract for several days prior to oviposition. Embryos then hatch from the eggs within a few minutes of deposition.

In most mayfly species the larval life-span is 2–4 months; however, some mayfly larvae are long-lived, with a development time of at least a year and, in some instances, of two or three years' duration. During this period they molt many times (more than 20 is not uncommon). Larvae occupy a wide range of habitats, though each one is characteristic for a particular species. They may burrow into the substrate, hide beneath stones and logs, clamber about among water plants, or cling to the upper surface of rocks and stones in fast-flowing streams. With the exception of a few carnivorous forms, larvae feed on algae, or plant detritus. By their sheer weight of numbers, larvae frequently form the major constituent in the diet of trout and other game fish.

Mayflies are unique among the Pterygota in that they molt in the adult stage. A mature larva, on emergence from its aquatic environment, molts into a subimago, a winged adult form, often capable of flight. A subimago can be distinguished from the imago into which it molts by its duller coloration and by the translucent wings, which are often fringed with hair. A subimago exists usually for only a short time before molting to the imago. Under adverse climatic conditions, however, a subimago may survive for many days. In a few exceptional species the subimago never molts but is the reproductive stage. In subtropical and tropical mayflies the emergence of the larvae tends to occur over a considerable period of time. In species from cooler climates emergence is highly synchronized, and this leads to the production of enormous swarms for short periods of the year.

Phylogeny and Classification

Although the basic groups within the Ephemeroptera have been recognized since the work of Eaton (1883–1888, cited in Edmunds, 1962), considerable difference of opinion exists with regard to the taxonomic rank that should be assigned to these groups, the limits of the groups, and the relationships between the groups (see discussion in Peters and Peters, 1973). The primary obstacle to determining these relationships is the high degree of parallel evolution that has occurred between members of different groups. In many insect groups this problem can be overcome usually by comparing a number of different characters from all stages of the life history (Edmunds, 1972). Unfortunately, many mayflies are known only from the juvenile or adult form.

Edmunds (1962) has proposed that the order be subdivided into 19 families shared among 5 superfamilies whose relationships are depicted in Figure 6.1. It is in the Heptagenioidea that the mayflies with the greatest number of primitive features are found, belonging to the family Siphlonuridae. It is likely that the other major groups evolved from some siphlonuridlike ancestor. The first line to split off led to the modern Caenoidea and Prosopistomatoidea. The four

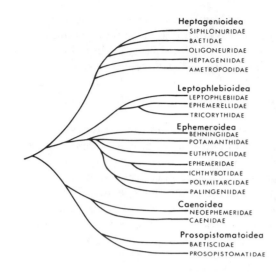

Heptagenioidea
SIPHLONURIDAE
BAETIDAE
OLIGONEURIDAE
HEPTAGENIIDAE
AMETROPODIDAE

Leptophlebioidea
LEPTOPHLEBIIDAE
EPHEMERELLIDAE
TRICORYTHIDAE

Ephemeroidea
BEHNINGIIDAE
POTAMANTHIDAE
EUTHYPLOCIIDAE
EPHEMERIDAE
ICHTHYBOTIDAE
POLYMITARCIDAE
PALINGENIIDAE

Caenoidea
NEOEPHEMERIDAE
CAENIDAE

Prosopistomatoidea
BAETISCIDAE
PROSOPISTOMATIDAE

FIGURE 6.1. Proposed phylogenetic relationships within the Ephemeroptera. [After G. F. Edmunds, Jr., 1962, The principles applied in identifying the hierarchic levels of the higher categories of Ephemeroptera, *Syst. Zool.* **11**:22–31. By permission of the American Museum of Natural History.]

families within these two superfamilies exhibit much parallelism, and their relationships are controversial. Edmunds (1962), however, believes that it is most reasonable to assume that the Baetiscidae and Prosopistomatidae are sister groups. The leptophlebioid and ephemeroid lines appear to have had a common ancestry with the heptagenioid group. In the evolution of the Leptophlebioidea, larvae retained the bottom-crawling habit of their ancestors. The trend in the Ephemeroidea was toward a burrowing habit, although in the primitive families Potamanthidae and Euthyplociidae larvae have remained crawlers.

Superfamily Heptagenioidea

The family SIPHLONURIDAE is a fairly large family of mayflies with a worldwide distribution but especially diverse in the holarctic region. The streamlined, active larvae are found on the bottom of fast-flowing streams or among vegetation in still-water habitats. Some are predaceous. Adults are medium- to large-sized mayflies, and the sexes are similar in coloration. In both sexes the compound eyes are large and have a transverse band dividing the upper and lower regions. In males the eyes are usually contiguous.

The BAETIDAE (Figure 6.2) form a very large family with a worldwide distribution. The torpedo-shaped larvae are found in a variety of habitats but commonly on the bottom of fast-flowing streams where they may be well camouflaged. Adults are generally small and sexually dimorphic. The hind wings are greatly reduced or absent. The compound eyes of males are large and divided horizontally into distinct parts; in females the eyes are small and simple.

The HEPTAGENIIDAE (Figure 6.3) rank next to the Baetidae in terms of number of species. They are an almost entirely holarctic and oriental group and are not represented in the Australasian region. Heptageniid larvae are the typical mayflies found clinging to the underside (occasionally the exposed face) of stones in fast-flowing streams and on wave-washed shores of large lakes. They

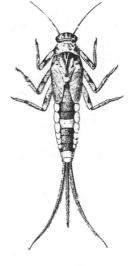

FIGURE 6.2. Larva of *Baetis vagans* (Baetidae). [From B. D. Burks, 1953, The mayflies, or Ephemeroptera, of Illinois, *Bull. Ill. Nat. Hist. Survey* **26**(1). By permission of the Illinois Natural History Survey.]

are remarkably well adapted for this life. Their body is extremely flattened dorsoventrally; the femora are broad and flat; the tarsal claws have denticles on the lower side; the gills are strengthened on their anterior margin; in some species the entire body takes on the shape (and function) of a sucking disc. Larvae are generally dark in color. Adults are of variable size and color. The eyes of males are large but not contiguous.

Superfamily Leptophlebioidea

The family Leptophlebiidae is another large and diverse group of world-wide distribution but especially common in the Southern Hemisphere. A good deal of parallel evolution of habits and morphology appears to have taken

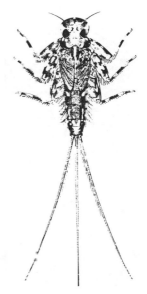

FIGURE 6.3. Larva of *Heptagenia flavescens* (Heptageniidae). [From B. D. Burks, 1953, The mayflies, or Ephemeroptera, of Illinois, *Bull. Ill. Nat. Hist Survey* **26**(1). By permission of the Illinois Natural History Survey.]

place between the Leptophlebiidae in the Australian region and the Baetidae and Heptageniidae in the holarctic region. Thus, many leptophlebiid species are found as larvae in still or slow-moving water, and, in some instances, the adults closely resemble baetids. Larvae of other species are found clinging to rocks in fast-flowing waters and resemble heptageniid larvae.

Although not large in terms of the number of species, the EPHEMERELLIDAE are a very widely distributed group with representatives in all the major regions of the world. Larvae are found in a wide variety of still- and moving-water habitats, especially cold, fast-flowing streams. Adults are small to medium-sized mayflies.

Superfamily Ephemeroidea

The family EPHEMERIDAE (Figure 6.4) is widely distributed, though it has not been found in Australia. Larvae are burrowers in the mud or sand of large lakes, rivers, and streams. The mandibles ("tusks") are long and used for lifting the roof of the burrow. The forelegs are flattened for digging. Most of the body and the appendages are covered with fine hairs. These become coated with silt, and the insect is thereby well camouflaged. Adults are generally moderately-sized to large insects. Their wings are hyaline, though they may be spotted in some species.

Superfamily Caenoidea

The CAENIDAE (Figure 6.5) constitute a relatively small but widely distributed family of generally small mayflies. The hairy larvae are burrowers found in still or slow-moving water. The second pair of gills is enlarged and strengthened, forming a plate that overlaps and protects the remaining four pairs of gills. The plate is alternately raised and lowered to effect water circula-

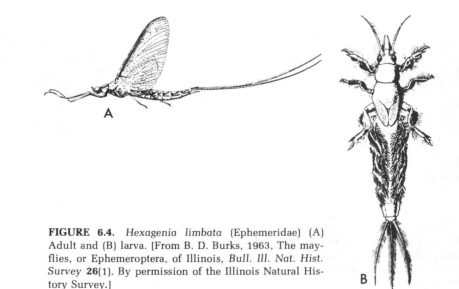

FIGURE 6.4. *Hexagenia limbata* (Ephemeridae) (A) Adult and (B) larva. [From B. D. Burks, 1963, The mayflies, or Ephemeroptera, of Illinois, *Bull. Ill. Nat. Hist. Survey* **26**(1). By permission of the Illinois Natural History Survey.]

FIGURE 6.5. Larva of *Caenis simulans* (Caenidae). [From B. D. Burks, 1953, The mayflies, or Ephemeroptera, of Illinois, *Bull. Ill. Nat. Hist. Survey* **26**(1). By permission of the Illinois Natural History Survey.]

tion. Male and female adults appear almost identical. The compound eyes are not especially large, but the lateral ocelli are about half the size of the compound eyes. The hind wings are absent.

Superfamily *Prosopistomatoidea*

The family BAETISCIDAE (Figure 6.6) contains a single North American genus, *Baetisca*. Larvae are remarkable in having an enormous, posteriorly projecting mesonotal shield that protects the gills. They live in sand at the bottom of lakes and rivers. Adults are medium-sized, brownish insects. The compound eyes of males are large and almost contiguous, but not divided horizontally. The mesothorax is unusually large in both sexes.

Literature

The work of Needham, Traver and Hsu (1935) remains the best general account of the Ephemeroptera and also contains detailed descriptions of most North American species. For an appreciation of the controversy regarding the phylogeny and classification of Ephemeroptera the reader should consult the

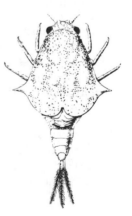

FIGURE 6.6. Larva of *Baetisca bajkovi* (Baetiscidae). [From B. D. Burks, 1953, The mayflies, or Ephemeroptera, of Illinois, *Bull. Ill. Nat. Hist. Survey* **26**(1). By permission of the Illinois Natural History Survey.]

papers given by Edmunds, Koss, Landa, and Riek at the First International Conference on Ephemeroptera (see Peters and Peters, 1973) and the reviews by Edmunds (1962, 1972). Edmunds, Allen, and Peters (1963) provide a key and information on distribution for the world families and subfamilies of Ephemeroptera larvae. Edmunds (1978) includes a key to North American families, as well as a list of references to the taxonomy of mayflies in different North American regions. More specific keys include those of Day (1956), Edmunds (1959), and Edmunds *et al.* (1976) [North American genera]; and Kimmins (1972) and Macan (1970) [British species].

Day, W. C. 1956, Ephemeroptera, in: *Aquatic Insects of California* (R. L. Usinger, ed.), University of California Press, Berkeley.

Edmunds, G. F., Jr., 1959, Ephemeroptera, in: *Freshwater Biology* (W. T. Edmondson, ed.), Wiley, New York.

Edmunds, G. F., Jr., 1962, The principles applied in determining the hierarchic level of the higher categories of Ephemeroptera, *Syst. Zool.* **11:**22–31.

Edmunds G. F., Jr., 1972, Biogeography and evolution of Ephemeroptera, *Annu. Rev. Entomol.* **17:**21–42.

Edmunds, G. F., Jr., 1978, Ephemeroptera, in: *An Introduction to the Aquatic Insects of North America* (R. W. Merritt and K. W. Cummins, eds.), Kendall/Hunt, Dubuque, Iowa.

Edmunds, G. F., Jr., Allen, R. K., and Peters, W. L., 1963, An annotated key to the nymphs of the families and subfamilies of mayflies (Ephemeroptera), *Univ. Utah Biol. Ser.* **13**(1):55 pp.

Edmunds, G. F., Jr., Jensen, S. L., and Berner, L., 1976, *The Mayflies of North and Central America,* University of Minnesota Press, Minneapolis.

Kimmins, D. E., 1972, A revised key to the adults of the British species of Ephemeroptera with notes on their ecology (second revised edition), *Sci. Publ. F.W. Biol. Assoc.* **15:**75 pp.

Macan, T. T., 1970, A key to the nymphs of British species of Ephemeroptera (second revised edition), *Sci. Publ. F.W. Biol. Assoc.* **20:**63 pp.

Needham, J. G., Traver, J. R., and Hsu, Y.-C., 1935, *The Biology of Mayflies,* Comstock, New York.

Peters, W. L., and Peters, J. G. (ed.), 1973, *Proceedings of the First International Conference on Ephemeroptera,* Brill, Leiden.

3. Odonata

SYNONYMS: none   COMMON NAMES: dragonflies and damselflies

Adults medium-sized to large elongate insects, frequently strikingly marked; head with antennae short and setaceous, compound eyes prominent, biting mouthparts; thorax with two pairs of membranous wings of approximately equal size and with netlike venation, pterostigma usually present; abdomen of male with copulatory organs on second and third sterna.

Larvae aquatic; body campodeiform; head equipped with extensible "mask" (modified labium) for catching prey, antennae small, compound eyes large; abdomen terminated with three processes, either short and stocky or extended into large lamellate structures.

Almost 5000 species of Odonata have been identified from different areas of the world. Nearly 10% of these are from North America. Approximately 50 species are found in the British fauna, and 250 species have been described from Australia.

Adult. The body of adult Odonata is remarkable for its colors, both pigmentary and structural, that frequently form a characteristic pattern over the dorsal region (Figure 6.11A). Most adults range from 30 to 90 mm in length and are sturdy, actively flying insects. The head is freely articulated with the thorax, and a large part of its surface, especially in Anisoptera (dragonflies) is occupied by the well-developed compound eyes. Three ocelli form a triangle on the vertex. The antennae are short, hairlike structures which, apparently, carry few sense organs. The mouthparts are powerful structures of the biting and chewing type. The thorax is somewhat parallelogram-shaped, with the legs placed anteroventrally and the wings situated posteroventrally. The prothorax is distinct but small, and in female Zygoptera (damselflies) is sculptured so as to articulate with the claspers of the male during mating. The mesothorax and metathorax are large and fused together. The pleura of these segments are very large and possess prominent sulci. The legs are weak and unsuitable for walking. They serve to grasp the prey (usually caught in flight) and hold it to the mouth during feeding. In Zygoptera the fore and hind wings are identical; in Anisoptera the hind wing is somewhat broader near the base. A prominent pterostigma is present on each wing in all except a few species. The wing venation is a primitive netlike arrangement. Only the anterior branch of the media and the posterior branch of the cubitus are present. Ten abdominal segments are visible with segments 1–8 bearing spiracles. In male Odonata the sterna of segments 2 and 3 bear secondary copulatory structures for transfer of sperm to the female genital tract. The true genital opening is located behind the ninth sternum. In female Zygoptera and many Anisoptera that are endophytic a well-developed ovipositor is present. In other, exophytic, Anisoptera it is reduced or absent.

Most internal organs are greatly elongated because of the narrow body. The testes extend from abdominal segment 4 to 8 and the ovaries occupy the whole length of the abdomen. Between 50 and 70 Malpighian tubules, united in groups of 5 or 6, enter the alimentary canal at the junction of the midgut and hindgut. The respiratory system is well developed and in many species includes a large number of air sacs in the thoracic region. The nervous system is generally primitive, although the brain is enlarged transversely due to the presence of large optic lobes.

Larva. Odonate larvae are usually shorter and stockier than adults. In general the larval head resembles that of the adult, though it differs in possession of the "mask," the elongated labium (Figure 6.7), used to capture prey. At rest the mask (so-called because it covers the other mouthparts) is folded at the junction of the postmentum and prementum and held between the bases of the legs. It is extended extremely rapidly by means of localized blood pressure changes, and the prey is grasped by the hooklike labial palps. In contrast to those of adults, the legs are normally positioned on the thorax, well developed and quite long. At the tip of the abdomen there are three appendages, one mediodorsal and two lateral (the cerci). These are small in Anisoptera but enlarged to form caudal lamellae in Zygoptera.

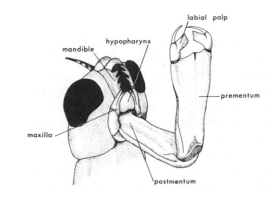

FIGURE 6.7. Lateroventral view of head of dragonfly larva showing mask. [After A. D. Imms, 1957, *A General Textbook of Entomology*, 9th ed. (revised by O. W. Richards and R. G. Davies). By permission of Chapman and Hall Ltd.]

Internally larvae differ from adults in several features. In the foregut there is a well-developed gizzard for breaking up food. There are initially only a few Malpighian tubules, though the number increases in each instar. Although undoubtedly a proportion of gaseous exchange takes place directly across the body wall *sensu stricto*, odonate larvae do possess special respiratory structures. In Anisoptera the wall of the rectum is greatly folded and well supplied with tracheae, forming "rectal gills." Water is continually pumped in and out of the rectum. In Zygoptera the caudal lamellae appear to supplement the surface area available for gaseous exchange (although in highly oxygenated water the larvae appear to survive perfectly well when the lamellae are removed). Finally, in some Zygoptera paired gills occur on most abdominal segments.

Life History and Habits

After emergence, immature adult Odonata spend some time away from water, usually among trees or tall grass where they hunt for prey and become sexually mature. It is during this maturation phase that some Odonata migrate over long distances. The maturation period, which lasts anywhere from a few days in smaller species to a month in large dragonflies, is followed by the reproductive phase in which mating and oviposition occur. Usually these two processes occur at the same site, although this is not necessarily so. A male frequently occupies and defends a "territory" much as do the males of many species of birds and mammals. Should a female of the species enter the territory and be recognized by the male (probably using visual cues), he will attempt to mate with her. Using his legs the male grasps the female on the pterothorax. He curls his abdomen around, and, using the terminal claspers, grasps the female's prothorax (Zygoptera) or head (most Anisoptera), which, as noted earlier, is specially sculptured to provide a close fit. The male's legs then release their grip. The female bends her abdomen forward until its tip contacts the accessory genitalia on the male's second and third abdominal segments, which have been previously charged with sperm (Figure 6.8). Copulation lasts generally between 5 and 30 min. Oviposition occurs shortly after copulation with the male remaining close (most Anisoptera) or still attached (Zygoptera) to the female.

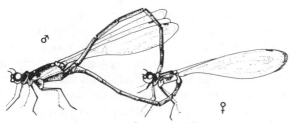

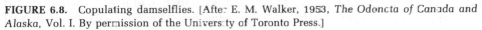

FIGURE 6.8. Copulating damselflies. [After E. M. Walker, 1953, *The Odonata of Canada and Alaska*, Vol. I. By permission of the University of Toronto Press.]

Oviposition is either endophytic (in Zygoptera and a few Anisoptera) or exophytic (most Anisoptera). Endophytic species usually have elongate eggs, which are deposited in leaves and stems of aquatic plants. Frequently a female climbs a considerable distance below the water surface before laying the eggs. In the exophytic dragonflies the eggs are simply released into the water or loosely attached to vegetation.

Eggs usually hatch within 2 to 5 weeks, although in some species eggs enter diapause, which takes them through adverse conditions. Larvae are facultative predators, feeding on whatever animals of appropriate size are available. Detection of prey is achieved primitively through the use of a variety of sense organs. In advanced species the eyes become of primary importance. Odonate larvae are preyed upon by larger members of the same order as well as aquatic vertebrates. For this reason they tend to be rather cryptic in habit. Many forms, which as adults oviposit exophytically, burrow in the substrate or hide among detritus. Larvae of species that lay their eggs endophytically (mainly Zygoptera) are generally weed-dwellers and escape detection through the use of camouflage. When detected, larvae can escape rapidly be either expelling water from the rectal cavity—a form of jet propulsion (Anisoptera)—or using rapid undulating movements of the abdomen and caudal lamellae (Zygoptera). The duration of larval development is variable. In temperate species which overwinter in the egg stage, the time from hatching to emergence may be only one to two months. In the large, temperate Anisoptera, larval development may take two or more years. In many temperate species that overwinter as larvae there is a diapause in the final instar. This feature ensures synchronous emergence the following spring and thus greatly improves the chances of successful reproduction (see also Chapter 23, Section 3.2.1). In tropical regions, the larvae usually emerge at dusk to avoid predation. In higher latitudes or altitudes, where the low nightly temperatures are unsuitable for emergence, the larvae must emerge during the day. At this time they are particularly vulnerable as prey, especially for birds.

Phylogeny and Classification

In contrast to that of the other Paleopteran order, the Ephemeroptera, the fossil record of the Odonata is remarkably extensive. The record indicates that the earliest true odonates (from the Upper Permian period) belonged to the suborder Zygoptera. These early forms underwent a rapid evolution during the

Triassic period (when members of the suborder Anisozygoptera first appeared) so that by the Jurassic period representatives of recent families of Zygoptera and Anisoptera were already in existence. The proposed relationships of the extant groups of Odonata are shown in Figure 6.9. According to Fraser (1954, 1957) an early dichotomy in the ancient zygopteran stock led, on the one hand, to the modern families of Coenagrioidea and, on the other, to the lestinoids, from which the remaining groups of Odonata evolved. One line led to the modern families of Lestinoidea. The other line branched almost immediately to give rise to the Agrioidea and the Anisozygoptera. The Anisozygoptera are an almost entirely fossil group, with only two recent species described from Japan and India. Members of this group show a combination of zygopteran and anisopteran features. The relic family Hemiphlebiidae, containing only a single species from Victoria, Australia, is sufficiently aberrant that Fraser places it in a separate superfamily originating close to the base of the agrioid line. The remaining suborder, Anisoptera, evolved from anisozygopteran stock, and the modern families fall into two major groups, the more primitive Aeshnoidea and the Libelluloidea.

Suborder Zygoptera (Damselflies)

Damselflies are characterized by the following structural features: fore and hind wings identical in shape and venation, quadrangular discoidal cell never

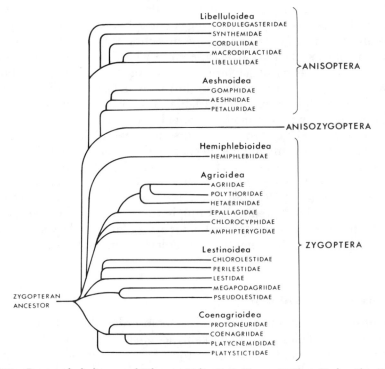

FIGURE 6.9. Proposed phylogeny of Odonata. [After F. C. Fraser, 1957, *A Reclassification of the Odonata*. By permission of the Royal Zoological Society of New South Wales.]

longitudinally divided, eyes far apart, and larvae with three (rarely two) caudal lamellae.

Superfamily Coenagrioidea

The family COENAGEIIDAE (=COENAGRIONIDAE) is the most successful zygopteran group, containing about 60 genera and a very large number of species. The family as a whole is cosmopolitan, and certain genera, for example, *Coenagrion* and *Ischnura* (Figure 6.10), are found throughout the world. Larvae are found among vegetation in still or slowly moving water. The generally small adults are weak fliers and rest with their narrow wings closely apposed over the body. The sexes are differently colored, with males usually much brighter. Commonly the dorsal surface of males is a complex pattern of pale blue and black markings. Females are usually uniformly dark in color, though this may change with age.

Superfamily Agrioidea

Members of the cosmopolitan family AGRIIDAE (=CALOPTERYGIDAE=AGRIONIDAE) are medium to large, broad-winged damselflies characterized by the brilliant metallic coloring of their bodies and, in males, the wings also. Larval Agriidae are found at the margins of fast-flowing water; they have long spidery legs and elongate caudal lamellae.

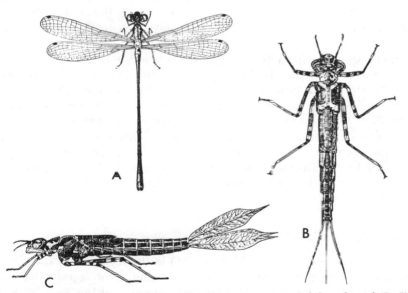

FIGURE 6.10. A damselfly, *Ischnura cervula* (Coenagriidae). (A) Adult male and (B, C) larva, dorsal and lateral views. [Reproduced by permission of the Smithsonian Institution Press from *Smithsonian Institution United States National Museum Proceedings*, Volume 49, "Notes on the life history and ecology of the dragonflies (Odonata) of Washington and Oregon," July 28, 1915, by C. H. Kennedy: Figures 77, 120, and 121. Washington, D.C., U.S. Government Printing Office, 1916.]

Superfamily Lestinoidea

The LESTIDAE constitute another very large and cosmopolitan family of damselflies. Its members are, as adults, medium-sized, metallically colored insects that rest with their wings partially or completely outspread. They are found in the vicinity of still water. Eggs are laid in emergent vegetation and almost always show delayed development, an adaptation to overcome adverse climatic conditions such as drought or cold. Larvae are elongate, streamlined creatures that clamber about among the underwater vegetation. They are usually well camouflaged.

Suborder Anisoptera (Dragonflies)

Distinguishing features of dragonflies include fore and hind wings dissimilar in venation and, usually, shape; discoidal cell divided horizontally into two triangular areas; eyes contiguous or nearly so; and larvae stout and without caudal lamellae.

Superfamily Aeshnoidea

The AESHNIDAE form one of the most primitive of recent dragonfly families. This is indicated by the retention in females of a primitive zygopteran ovipositor, which is found among other Anisoptera only in the small and equally archaic family PETALURIDAE. The family is large and cosmopolitan. It contains the largest and strongest-flying of the modern dragonflies, characterized by the enormous eyes that meet broadly in the midline of the head. Larvae are stout, elongate insects found among vegetation in still or slow-moving water.

The GOMPHIDAE form another primitive family whose members have eyes which are widely separated. The ovipositor is, however, rudimentary, and eggs are laid by simply dipping the tip of the abdomen into the water. Because they are laid in fast-flowing waters, the eggs have special adhesive properties to prevent their being washed away. Gomphid larvae are burrowers or sprawlers in the substrate, and some burrowing species have a greatly elongated tenth abdominal segment in order to retain respiratory contact with the water. Frequently the forelegs are fossorial. Adults are generally black and yellow, with one or the other color predominating according to the habitat in which the insects are found.

Superfamily Libelluloidea

The family LIBELLULIDAE (Figure 6.11) is the largest and most heterogeneous group of Anisoptera, and its members are found throughout the world in a variety of still-water habitats. Larvae hide among rotting vegetation on the bottom of lakes or ponds, though a few have become secondarily adapted for a more active existence among growing vegetation. Adults vary greatly in size and coloration, though the pale wings commonly have a number of characteristic spots and bands of dark pigment on them.

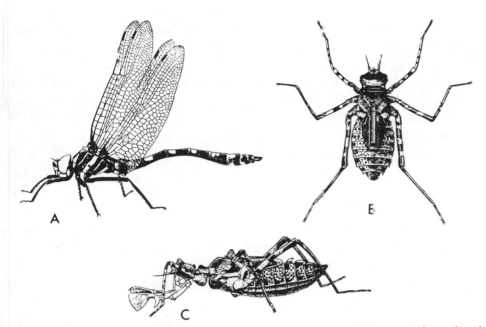

FIGURE 6.11. A dragonfly, *Macromia magnifica* (Libellulidae). (A) Adult male; (B) larva, dorsal view; and (C) larva, lateral view with labium extended. [Reproduced by permission of the Smithsonian Institution Press from *Smithsonian Institution United States National Museum Proceedings*, Volume 49, "Notes on the life history and ecology of the dragonflies (Odonata) of Washington and Oregon," July 28, 1915, by C. H. Kennedy Figures 134, 146, and 147. Washington, D.C., U.S. Government Printing Office, 1916.]

Literature

The Odonata have been one of the most popular insect groups for study, and the literature on them is abundant. Good general accounts of their biology are given by Walker (1953), Corbet, Longfield, and Moore (1960), and Corbet (1962). Fraser (1954, 1957) has discussed the phylogeny and classification of the order. Species may be identified through the keys of Walker (1953, 1958), Walker and Corbet (1975), Needham and Westfall (1955), and Smith and Pritchard (1956) [North American forms]; and Fraser (1949) and Corbet et al. (1960) [British species].

The international journal *Odonatologica*, first published in 1972, is devoted entirely to publication of papers on Odonata.

Corbet, P. S., 1962, *A Biology of Dragonflies*, Witherby, London.
Corbet, P. S., Longfield, C., and Moore, N. W., 1960, *Dragonflies*, Collins, London.
Fraser, F. C., 1949, Oconata, *R. Entomol. Soc. Handb. Ident. Br. Insects* **1** (10); 48 pp.
Fraser, F. C., 1954, The origin and descent of the order Odonata based on the evidence of persistent archaic characters, *Proc. R. Entomol. Soc. London Ser. B* **23**:89–94.
Fraser, F. C., 1957, *A Reclassification of the Order Odonata*, Royal Zoological Society of New South Wales, Sydney.
Needham, J. G., and Westfall, M. J., Jr., 1955, *A Manual of the Dragonflies of North America (Anisoptera)*, University of California Press, Berkeley.
Smith, R. F., and Pritchard, A. E., 1956, Odonata, in: *Aquatic Insects of California* (R. L. Usinger, ed.), University of California Press, Berkeley.

Walker, E. M., 1953, *The Odonata of Canada and Alaska*, Vol. 1: *General and Zygoptera*, University of Toronto Press, Toronto.

Walker, E. M., 1958, *The Odonata of Canada and Alaska*, Vol. 2: *Anisoptera—4 Families*, University of Toronto Press, Toronto.

Walker, E. M., and Corbet, P. S., 1975, *The Odonata of Canada and Alaska*, Vol. 3: *Anisoptera—3 Families*, University of Toronto Press, Toronto.

7

The Orthopteroid Orders

1. Introduction

In this chapter we shall deal with the following nine orders: Plecoptera, Embioptera, Dictyoptera, Isoptera, Grylloblattodea, Dermaptera, Phasmida, Orthoptera, and Zoraptera. Members of these orders can be distinguished from other exopterygotes [the hemipteroid orders (Chapter 8)] by the following features: generalized biting mouthparts, wing venation usually well developed with numerous crossveins (though less netlike than that of Paleoptera), cerci present, terminalia of male may be asymmetrical and reduced, many Malpighian tubules, and generalized nervous system with several discrete abdominal ganglia.

2. Plecoptera

Synonym: Perlaria   Common name: stoneflies

Moderate-sized to fairly large soft-bodied insects; head with long setaceous antennae, weak mandibulate mouthparts, well-developed compound eyes and 2 or 3 ocelli; thorax almost always with 2 pairs of membranous wings (sometimes reduced), hind pair in most species with a large anal lobe, venation frequently specialized, legs identical and with a 3-segmented tarsus; abdomen of most species terminated by long multisegmented cerci, females lacking an ovipositor.

Larvae aquatic, generally resembling adults except for presence of a variable number of tracheal gills.

More than 1200 species of this very ancient order have been described. Though the order has representatives throughout the world, most families have a rather restricted distribution (Illies, 1965).

Structure

Adult. The plecopteran head is prognathous and bears a pair of elongate, multisegmented antennae, well-developed compound eyes, three (rarely two) ocelli, and weak, biting mouthparts. Usually all the mouthparts are present, but in members of a few families the mandibles are vestigial. The thorax is primitive. Its segments are free and the prothorax is large. Two pairs of membranous wings are nearly always present, though brachypterous and apterous species occur at high altitudes and latitudes. The hind wing typically has a large anal fan, but this is reduced in the more advanced families. The wing venation is generally primitive, but considerable variation is seen within the order. In members of primitive families a typical archedictyon is developed to a greater or lesser degree; in those of advanced groups the number of branches of the longitudinal veins and the number of crossveins are greatly reduced. The abdomen contains 10 complete segments, with the 11th represented by the epiproct, paraprocts, and long cerci. In Nemouridae, however, the latter are reduced to an unsegmented structure used in copulation.

The esophagus is very long, the gizzard rudimentary, and midgut and hindgut short. There are between 20 and 60 Malpighian tubules. In primitive families the central nervous system includes three thoracic and eight abdominal ganglia, but in advanced groups the sixth to eighth abdominal ganglia fuse. In males the testes meet in the midline, but their products are carried by separate vasa deferentia to a pair of seminal vesicles. Usually there is a median ejaculatory duct, but in some species the vasa deferentia remain separate until they reach the median gonopore. In females the panoistic ovarioles arise from a common duct that joins the oviducts of each side. A spermatheca is usually present.

Larva. In general form larvae resemble adults, except for the absence of wings and the presence, in most species, of several pairs of gills. Primitively there are five or six pairs of abdominal gills, but in members of more advanced groups these are reduced in number and secondary respiratory structures may appear on more anterior regions of the body (submentum, neck, thorax, and coxae) or may encircle the anus. In many species the legs are fringed with hairs that assist swimming.

Life History and Habits

Adult stoneflies are weak flyers and seldom found far from the banks of streams or edges of lakes where they rest on vegetation, rocks, logs, etc. Nocturnal species usually hide in crevices or among vegetation during the day. Many stoneflies do not feed as adults. Others feed on lichens, acellular algae, and pollen.

Mating occurs on the ground. Large numbers of eggs are laid, usually in batches of 100 or more. In flying species females hover over the water and dip the abdomen beneath the surface. Brachypterous and apterous forms crawl to the water's edge in order to oviposit. Eggs frequently develop adhesive properties on contact with water. A few species are ovoviviparous. Larvae are typically found in streams or lakes whose bottom is covered with stones under which they can hide. Development is slow, frequently taking more than a year in the larger species. Many molts occur, 33 having been recorded over a period of 3 years for one species. Most stonefly larvae (of the suborders Archiperlaria and Filipalpia) are phytophagous, feeding on lichens, algae, moss, and diatoms. Typically these are the species which also feed in the adult stage.

Juveniles of other species, belonging to the suborder Setipalpia, are carnivorous, living on other insects. These species do not feed as adults. Like that of Odonata and mayflies, emergence of stoneflies is frequently highly sychronized.

Phylogeny and Classification

The Plecoptera constitute a very ancient order of insects that probably had its origins from a closely similar group, the Paraplecoptera, in the Upper Carboniferous period. Indeed, by the Permian period there existed stoneflies assignable to the still extant families Eustheniidae and Taeniopterygidae. Other specimens from the fossil family Paleoperlidae, allied to modern Perlidae, indicate that all major stonefly groups were in existence more than 200 million years ago.

Stoneflies usually have been separated into two suborders, Filipalpia and Setipalpia. However, Illies (1965), whose system is used here, considers that the extremely primitive families Eustheniidae and Diamphipnoidae are sufficiently distinct from the remaining Filipalpia that they should be grouped in a separate suborder, the Archiperlaria.

Suborder Archiperlaria

Members of the suborder Archiperlaria exhibit several extremely primitive features, for example, large size, rather brightly colored body and wings (usually red, yellow, or green), many crossveins in all wings, anal fan in the hind wing with eight or nine anal veins, and larvae with four to six pairs of abdominal gills. Members of the small family EUSTHENIIDAE are considered by Illies to represent the prototype of plecopteran organization. The family is entirely restricted to eastern Australasia and southern South America.

Suborder Setipalpia

The suborder Setipalpia contains stoneflies of medium and large size, with well-developed cerci. In larvae the labial paraglossae are much larger than the glossae, and there is a reduction in the number of abdominal gills, accessory repiratory structures being developed on the thorax. Larvae are usually carnivorous (except Pteronarcidae) and found in warmer waters. Adults generally do not feed. Eggs often possess specialized adhesive structures. Members of this suborder are restricted mainly to the Northern Hemisphere, though representatives of the Perlidae are found in South America as a result of migration from the north. The suborder includes two families, Pteronarcidae and Perlidae, each probably meriting separate superfamily status.

The family PTERONARCIDAE (Figure 7.1A) is the most primitive of the Setipalpia. Primitive features include large body size and many crossveins in the wings (though not as many as in those of Eustheniidae). It is primarily a North American family that has invaded eastern Asia in relatively recent times. Its members are found in medium- to large-sized rivers. Larvae are phytophagous.

The family PERLIDAE (Figure 7.1B,C) is the largest family in the order.

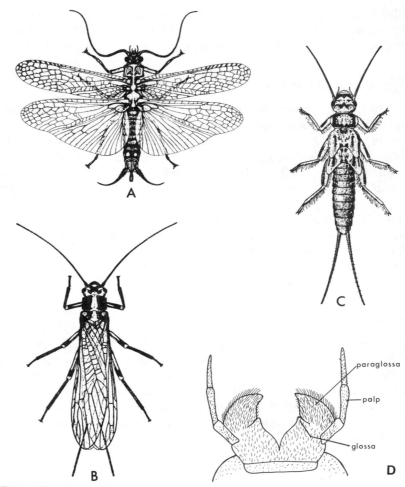

FIGURE 7.1. Plecoptera (suborder Setipalpia). (A) *Pteronarcys californica* (Pteronarcidae) adult, (B) *Isoperla confusa* (Perlidae) adult, (C) *I. confusa* larva, and (D) ventral view of labium of *I. confusa*, showing unequal size of paraglossae and glossae. [A, from R. A. Gaufin, W. E. Ricker, M. Miner, P. Milam, and R. A. Hayes, 1972, The stoneflies (Plecoptera) of Montana, *Trans. Am. Entomol. Soc.* **98**:1–161. By permission of the American Entomological Society. B, C, D, from T. H. Frison, 1935, The stoneflies, or Plecoptera, of Illinois, *Bull. Ill. Nat. Hist. Surv.* **20**(4). By permission of the Illinois Natural History Survey.]

Though primarily a tropical group, the family has many representatives in north temperate regions in both the Old and New Worlds, and a few are found in southern latitudes of South America. That this is a rather advanced group is indicated by the reduced glossae, reduced first abdominal sternite, the fusion of the first two abdominal ganglia with that of the metathorax, and the absence of abdominal gills in larvae. Larvae are generally carnivorous.

Suborder Filipalpia

Most members of the suborder Filipalpia may be recognized by their generally small size, paraglossae and glossae of about equal size, typically reduced cerci (modified as copulatory organs in males), and reduced wing venation (especially the number of crossveins). Larvae (and in many species adults also) are phytophagous. Larvae respire by means of gills encircling the anus, at the base of the legs, or on the neck. In some species gills are lacking. Within the suborder two groups of families can be separated: (1) the more primitive Austroperloidea from Australasia and east Asia, in which the wings still contain many crossveins and cerci are long, and whose larvae respire by means of simple, tubular anal gills; and (2) the Taeniopterygoidea, a primarily holarctic group whose members possess all the features listed above as characteristic of the suborder. Larvae either have specialized anal gills or gills on the legs and neck, or lack gills entirely. Representatives of all four families of Taeniopterygoidea are found both in Western Europe (including the British Isles) and in North America.

Superfamily Taeniopterygoidea

That the TAENIOPTERYGIDAE constitute the most primitive family of the Taeniopterygoidea is indicated by the comparatively rich wing venation, large anal lobe on the hind wing, and 5- or 6-segmented cerci. Specializations include gills on the leg bases and the form of the male genitalia. These stoneflies are commonly found in large streams and rivers. The adults of many species are known as winter stoneflies because of their habit of emerging, in the Northern Hemisphere, between January and April. In some species adults are pollen feeders.

In CAPNIIDAE (Figure 7.2A) the wing venation is usually reduced, as is the size of the anal fan in the hind wing. The cerci, however, remain long. Like members of the previous family, Capniidae usually emerge during the winter. Males are frequently brachypterous.

LEUCTRIDAE are recognized by their ability to fold the wings down over the sides of the abdomen. Larval Leuctridae are typically found in small streams in mountainous regions. This is an advanced family in which the anal area and cerci are reduced, and the male genitalia are specialized.

Although the wing venation is primitive, other advances make the NEMOURIDAE (Figure 7.2B–D) the most specialized family in this group. The cerci and external and internal genitalia of the male are highly modified. The abdominal nerve cord contains only five or six ganglia. Nemourid larvae are found in a variety of habitats from which they emerge in spring or early summer.

Literature

Claassen (1931), Hynes (1941, 1942, 1976), Brinck (1949), and Hitchcock (1974) provide much information on the general biology of stoneflies. The phylogeny of the order is discussed by Ricker (1950, 1952) and Illies (1965). Keys for identification of the North American Plecoptera are provided by Jewett (1956) and Ricker (1959), and for the British species by Kimmins (1950) and Hynes (1967).

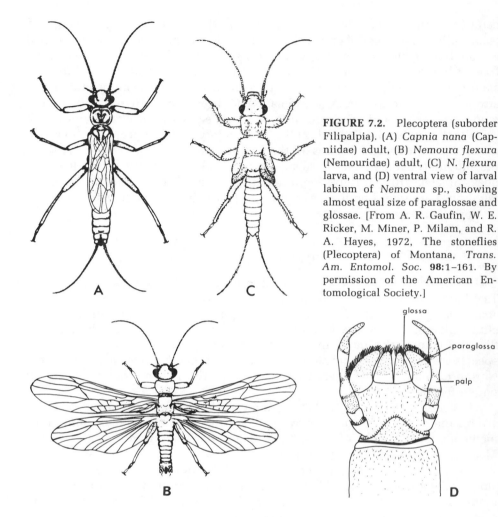

FIGURE 7.2. Plecoptera (suborder Filipalpia). (A) *Capnia nana* (Capniidae) adult, (B) *Nemoura flexura* (Nemouridae) adult, (C) *N. flexura* larva, and (D) ventral view of larval labium of *Nemoura* sp., showing almost equal size of paraglossae and glossae. [From A. R. Gaufin, W. E. Ricker, M. Miner, P. Milam, and R. A. Hayes, 1972, The stoneflies (Plecoptera) of Montana, *Trans. Am. Entomol. Soc.* **98**:1–161. By permission of the American Entomological Society.]

Brinck, P., 1949, Studies on Swedish stoneflies, *Opusc. Entomol. Suppl.* **11**:250 pp.

Claassen, P. W., 1931, Plecoptera nymphs of America (north of Mexico), *Thomas Say Foundation, Publ.* **3**:199 pp.

Hitchcock, S. W., 1974, Guide to the insects of Connecticut. Part VII. The Plecoptera or stoneflies of Connecticut, *Conn. State Geol. Nat. Hist. Surv., Bull.* **107**:262 pp.

Hynes, H. B. N., 1941, The taxonomy and ecology of the nymphs of British Plecoptera with notes on the adults and eggs, *Trans. R. Entomol. Soc. London* **91**:459–557.

Hynes, H. B. N., 1942, A study of the feeding of adult stoneflies (Plecoptera), *Proc. R. Entomol. Soc. London Ser. A.* **17**:81–82.

Hynes, H. B. N., 1967, A key to the adults and nymphs of British stoneflies (Plecoptera) (2nd ed.), *F.W. Biol. Ass. Sci. Publ.* **17**:86 pp.

Hynes, H. B. N., 1976, Biology of Plecoptera, *Annu. Rev. Entomol.* **21**:135–154.

Illies, J., 1965, Phylogeny and zoogeography of the Plecoptera, *Annu. Rev. Entomol.* **10**:117–140.

Jewett, S. G., Jr., 1956, Plecoptera, in: *Aquatic Insects of California* (R. L. Usinger, ed.), University of California Press, Berkeley.

Kimmins, D. E., 1950, Plecoptera, *R. Entomol. Soc. Handb. Ident. Br. Insects* **1**(6):18 pp.

Ricker, W. E., 1950, Some evolutionary trends in Plecoptera, *Proc. Indiana Acad. Sci.* **59**:197–209.

Ricker, W. E., 1952, Systematic studies in Plecoptera, *Indiana Univ. Publ. Sci. Ser.* **18**:1–200.

Ricker, W. E., 1959, Plecoptera, in: *Freshwater Biology* (W. T. Edmondson, ed.), Wiley, New York.

3. Embioptera

SYNONYMS: Embiodea, Embiidina   COMMON NAMES: webspinners, embiids

Elongate, small or moderately sized insects that live gregariously in silk tunnels; head with filiform antennae, compound eyes, and mandibulate mouthparts but lacking ocelli; males of almost all species with 2 pairs of nearly identical wings in which radial vein is thickened, females apterous, tarsi 3-segmented and basal segment of fore tarsus greatly enlarged; cerci 2-segmened and usually asymmetrical in males.

The Embioptera (Figure 7.3) are mostly confined to the larger land masses in tropical or subtropical areas of the world, though a few have found their way even to oceanic islands. Although fewer than 200 species have been described, Ross (1970a,b) has suggested that this figure may represent only 10% of the world total.

Structure

As a group, the Embioptera are of remarkably uniform structure, a feature related to the widespread similarity of the tunnels in which they live. Webspinners are soft-bodied insects that fly only weakly or not at all. The prognathous head bears filiform antennae, compound eyes (often large and kidney-shaped in males, small in females) and mandibulate mouthparts. Ocelli are absent. No trace of wings can be seen in females, but in males the fore and hind wings are very similar. The radius is thickened, the other veins are reduced. The wings are flexible and able to fold at any point. This facilitates

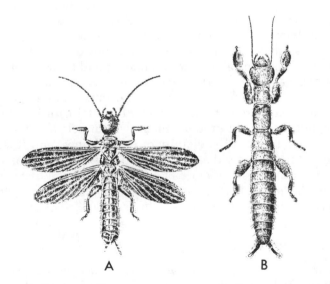

A   B

FIGURE 7.3. *Embia major* (Embioptera). (A) Male and (B) female. [From A. D. Imms 1957, *A General Textbook of Entomology*, 9th ed. (revised by O. W. Richards and R. G. Davies). By permission of Chapman and Hall Ltd.]

backward movement along the tunnels. For flight the wings are made more rigid by pumping blood into the radius. The forelegs are stout, and the basal tarsal segment is swollen to accommodate the silk glands, which number about 200. Ducts from the glands carry the product to the exterior via hairlike ejectors. The hind femur is also enlarged to contain a large tibial depressor muscle. This is correlated with the ability to run backward with great speed. There are 10 obvious abdominal segments. The cerci are 2-segmented and, in males, usually asymmetrical.

The internal structure is generalized. The gut is straight, and a large number of Malpighian tubules open into it. The ventral nerve cord is paired and includes three thoracic and seven abdominal ganglia. Each ovary consists of five ovarioles that are connected at intervals with the oviduct. A spermatheca is present. The five testes on each side are also arranged serially along the vas deferens, which swells proximally into a seminal vesicle. Two pairs of accessory glands occur in males.

Life History and Habits

Both adult and juvenile Embioptera can produce silken tunnels which are just wide enough to permit the animals to move forward or backward along them. Generally, many embiids are found associated together in a "nest" of interconnected tubes. It must be emphasized, however, that this gregarious behavior is in no way social. In humid regions an entire nest may be exposed, but in drier parts of the world it is usually partially subterranean as a protection against desiccation and fire. Nests are constructed in the immediate vicinity of a food source, and tunnels often extend directly into this. Embiids are phytophagous, with dead grass and leaves, lichens, moss, and bark constituting the main food. Early workers believed that males might be carnivorous on account of the rather distinct mandibles. It is now known, however, that the structure of the latter is correlated with their use as grasping organs during copulation, and, in many species, the mature males do not feed.

A typical nest contains a few mature females and their developing young. Mature males are generally short-lived and, in some species, are eaten by the female after mating. Parthenogenesis probably occurs in some species. Eggs are laid in a tunnel and guarded by the female. Parental care is extended to the young larvae, but these soon produce their own tunnels in which to develop. New colonies are formed in the vicinity of the old ones, and it is during this short migration to new sites that embiids are especially vulnerable. The absence of wings in females has more or less restricted the distribution of the Embioptera to the major land masses, though some species, perhaps transported by man, are found on remote Pacific islands.

Phylogeny and Classification

Embioptera constitute a very ancient insect order with a fossil record that extends to the Permian period. Its members are clearly orthopteroid, and their nearest living relatives appear to be the Plecoptera (Illies, 1965; Ross, 1970b). According to Illies, they probably evolved from Paraplecoptera.

Because of the neotenous nature of females, identification and classification can be carried out with certainty only by examining mature males.

Ross (1970a,b) recognizes eight families of living Embioptera, but it is not yet possible to draw many conclusions regarding their phylogenetic relationships because of the general structural uniformity of the order and the amount of parallel evolution that has taken place between families. The northern South American and West Indian family CLOTHODIDAE is the most primitive group. In this family, to which certain Miocene fossils are assigned, the cerci of the male are symmetrical and comprise two smooth segments. The largest family, EMBIIDAE, is a rather heterogeneous group of Old and New World forms. It seems probable that this group will be subdivided when more knowledge is available. Another large family is the OLIGOTOMIDAE, a rather primitive group with representatives in Asia, Australia, southern Europe, and possibly East Africa. Three species of *Oligotoma* have been introduced accidentally into the United States. Other families are the AUSTRALEMBIIDAE (restricted to eastern Australia and Tasmania), NOTOLIGOTOMIDAE (Southeast Asia and Australia), EMBONYCHIDAE (East Asia), TERATEMBIIDAE (South America and southern United States), and ANISEMBIIDAE (Central America and southern United States).

Literature

The biology of the Embioptera is dealt with by Ross (1940, 1970a, 1970b), who also (1944) has described and provided keys for identification of the North and South American species. Ross (1970b) has discussed the evolution and classification of the order.

Illies, J., 1965. Phylogeny and zoogeography of the Plecoptera, *Annu. Rev. Entomol.* **10**:117–140.
Ross, E. S., 1940, A revision of the Embioptera of North America, *Ann. Entomol. Soc. Am.* **33**:629–676.
Ross, E. S., 1944, A revision of the Embioptera, or web-spinners, of the New World, *Proc. U.S. Nat. Mus.* **94**:401–504.
Ross, E. S., 1970a, Embioptera, in: *The Insects of Australia* (I. M. Mackerras, ed.), Melbourne University Press, Carlton, Victoria.
Ross, E. S., 1970b, Biosystematics of the Embioptera, *Annu. Rev. Entomol.* **15**:157–172.

4. Dictyoptera*

SYNONYMS: Oothecaria Blattiformia, Blattopteriformia

COMMON NAMES: cockroaches and mantids

Small to very large terrestrial insects of varied form; head hypognathous with filiform, multisegmented antennae, mandibulate mouthparts and well-developed compound eyes, ocelli present (Mantodea) or usually absent (Blattodea); pronotum large and disclike (Blattodea) or elongate (most Mantodea), legs with 5-segmented

*Prof. D. K. McE. Kevan (personal communication) has pointed out that this ordinal name should be spelled Dictuoptera and that the name Dictyoptera was given originally to a miscellany of "neuropteroid" insects. However, the well-established use of Dictyoptera for the order under discussion will be continued here.

tarsi, fore wings modified as tegmina, brachyptery, and aptery common; ovipositor reduced and hidden, male genitalia complex and concealed, cerci fairly short but multisegmented.

As defined above the order Dictyoptera contains some 5500 described species that fall into two clearly defined suborders, Blattodea (cockroaches), with at least 3500 species, and Mantodea (mantids), with 2000 or more. The order is mainly tropical or subtropical. Several species of cockroaches are important domestic pests.

Structure

Cockroaches are typically flattened, oval-shaped insects whose head is covered by the large disclike pronotum. In contrast, mantids are elongate and easily recognized by their raptorial forelegs, prominent, movable head, and usually elongate pronotum. Almost all mantids are procryptically colored, though it is not known whether such camouflage is more important in concealing them from prey or from would-be predators.

The head is hypognathous. Compound eyes are well developed in most forms but may be reduced or absent in cockroaches that live in caves, ants' nests, etc. Three ocelli are present in mantids, but in most cockroaches the ocelli have degenerated, being represented by a pair of transparent areas on the cuticle, the fenestrae. The antennae, which in some species are very long, are filiform and multisegmented. Well-developed mandibulate mouthparts are present. The legs are essentially similar in cockroaches, but in most mantids the forelegs are greatly enlarged and bear spines for catching prey. Wings may be fully developed, shortened, or absent. The fore wings are moderately sclerotized and form tegmina. The hind wings have large anal areas. The venation is primitive, with the longitudinal veins much branched and large numbers of crossveins present. Ten obvious segments are present in the abdomen, with the eleventh represented in both sexes by the paraprocts and short, multisegmented cerci. In males the ninth sternum forms the subgenital plate, which bears a pair of styli. The genitalia, which are partially hidden by the subgenital and supra-anal (tenth tergal) plates, are membranous and asymmetrical. In females the subgenital plate is formed from the seventh sternum, which envelops the small ovipositor. Sterna 8 to 10 are reduced and internal.

The gut, which is long and coiled in cockroaches, short and straight in mantids, contains a large crop, well-developed gizzard, and a short midgut attached to which are eight ceca. Up to 100 or more Malpighian tubules originate at the anterior end of the hindgut. The nervous system is generalized, and three thoracic and six or seven abdominal ganglia are usually present. In some cockroaches only four or five abdominal ganglia can be seen, as a result of coalescence of the anterior ones with the metathoracic ganglion. The testes comprise four or more follicles enclosed in a peritoneal sheath. The vasa deferentia enter the ejaculatory duct, at the anterior end of which are the seminal vesicles and various accessory glands. A large conglobate gland of uncertain function opens separately to the exterior in male cockroaches. There are several panoistic ovarioles in each ovary The lateral oviducts lead to the common oviduct, which opens into a large genital chamber. The spermatheca also enters this chamber on its dorsal side. Various accessory glands whose secretions form the ootheca also open into the genital chamber. Various subcutaneous glands, whose secretions may be either repugnatory or important in courtship, are found in cockroaches.

Life History and Habits

As the differences are so great the life history and habits of Blattodea and Mantodea are described separately.

Blattodea. Cockroaches are mostly secretive, primarily nocturnal, typically ground-dwelling insects that hide by day in cracks and crevices, under stones, in rotting logs, among decaying vegetation, etc. Some, however, live on foliage, etc. well above the ground. Most species prefer a rather humid environment, though some are found in semidesert or even desert conditions and others in semiaquatic situations. A few live in caves, ants' nests, and similar places. Some species may be gregarious, insects at the same stage of development occupying the same hiding places and feeding together. A few species exhibit subsocial behavior. Generally cockroaches are omnivorous but are rarely active predators. A few species feed on rotting wood, which is digested by symbiotic bacteria or protozoans in the cockroaches' gut. These microorganisms are very similar to those found in termites.

Usually simple courtship behavior precedes mating, which may take a considerable time. Cockroaches may be (1) *oviparous*, the eggs being enclosed in a leathery or horny ootheca which frequently may be seen projecting from a female's genital chamber prior to deposition; (2) *ovoviviparous*, where the eggs are enclosed in a membranous ootheca which is retained in the genital tract throughout embryonic development; or (3) *viviparous*, in which case the ootheca is incompletely formed within the genital chamber, and the embryos receive nourishment from the mother. Parthenogenesis occurs in some species. Development is often slow, taking nearly a year and involving up to 12 molts. Adults are frequently long-lived.

Mantodea. The life-style of mantids is in marked contrast with that of cockroaches. Mantids live a solitary existence, mostly in shrubs, trees, and other vegetation, where they wait motionless for the arrival of suitable prey, usually other insects, though anything of appropriate size is fair game. Occasionally, mantids will stalk their prey until they are within grasping distance. This is normally the situation with ground-living species (which are mostly found in arid regions).

Mating in mantids is sometimes risky for a male, since his partner, almost always larger, may regard him as being more desirable as a meal than as a lover! Eggs are laid in a mass of frothy material that hardens to form an ootheca. Usually this is attached to an object some distance from the ground, though a few species deposit the ootheca in the soil. Parental care is shown by females in a few species. Obligate parthenogenesis occurs, rarely, for example, in *Brunneria borealis* from the southern United States. As in cockroaches, the development time is rather long and there may be many molts.

Phylogeny and Classification

Cockroaches with forms similar to those of modern species existed in the Carboniferous period, some 300 million years ago. The point at which the mantid and cockroach lines diverged is not clear, however, since the earliest fossil record for undoubted mantids comes from the early Tertiary, approximately 60 million years ago, by which time the group was already well established.

There have been several attempts to interpret the phylogeny of the suborder Blattodea, all of which were hampered because of the high degree of parallel evolution that has occurred within the group. McKittrick (1964) made a comparative examination of the external genitalia of both sexes, oviposition behavior, and crop structure in a wide variety of species. As a result, she recognizes five families whose relationships are expressed in Figure 7.4. It can be seen from this figure that cockroach evolution has proceeded along two lines, one leading to the families Cryptocercidae and Blattidae, the other to the families Polyphagidae, Blattellidae (=Ectobiidae), and Blaberidae. Durden (1969), on the other hand, made an extensive study of Carboniferous cockroaches, and while, in most respects, his conclusions do not differ fundamentally from those of McKittrick, he recognizes several additional superfamilies. Of those superfamilies that contain recent species, two should be noted, the Cryptocercoidea and Polyphagoidea, which contain the Cryptocercidae and Polyphagidae, respectively.

Classification of the suborder Mantodea is also difficult because of parallel evolutionary trends among the constituent groups. Beier (1964) divides the suborder into eight families, contained within the single superfamily Mantoidea, of which the Amorphoscelidae and Mantidae are the largest. The six remaining families are small, tropical groups of restricted distribution.

Suborder Blattodea

In members of the suborder Blattodea the head is covered with a large, shield-shaped pronotum; the legs are identical; and the gizzard is strongly dentate.

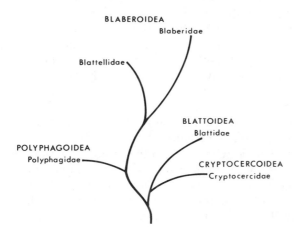

FIGURE 7.4. Proposed phylogeny of Blattodea. [From F. A. McKittrick, 1964, Evolutionary studies of cockroaches, *Mem. Cornell Univ. Agric. Expt. Station #389.* By permission of the Cornell University Agriculture Experimental Station.]

Superfamily Cryptocercoidea

The single family CRYPTOCERCIDAE is represented by the one genus, *Cryptocercus*. *C. punctulatus*, from mountainous regions in eastern and western United States, has been particularly important in establishing evolutionary links between the Blattodea and termites. *C. punctulatus* lives in colonies containing individuals of all ages beneath rotting logs and exhibits subsocial behavior. It feeds on wood, which, as in the "lower" termites, is digested by flagellate protozoans present in the hindgut. Since the lining and contents of the hindgut are lost at each molt, insects must obtain a fresh supply of protozoans. This they do by eating fecal pellets.

Superfamily Blattoidea

The cockroaches in the superfamily Blattoidea are included in a single large, cosmopolitan family, the BLATTIDAE. They are generally fairly large (2–5 cm in length) and may be recognized by the numerous spines on the ventroposterior margin of the femora. The family contains several species that are closely associated with man and do considerable damage to his property, as well as cause health hazards through contamination of food. *Blatta orientalis* (the Oriental cockroach) (Figure 7.5A) appears to be a native of the Mediterranean region but has been distributed through commerce to many parts of the world. It is the major cockroach pest in Britain and is widely distributed throughout North America. It prefers generally cool situations and is typically found in cellars, basements, toilets, bathrooms, and kitchens. It can tolerate warmer conditions provided that water is available. Four species of *Periplaneta*, *P. americana* (the American cockroach) (Figure 7.5B), *P. australasiae* (the Australian cockroach), *P. fuliginosa* (the smokey-brown cockroach), and *P. brunnea* (the brown cockroach), which are of African origin, are also found in and around human habitations. All four species prefer warmer, moister habitats than those enjoyed by *B. orientalis*, and are frequently found in outdoor habitats in subtropical regions.

Superfamily Polyphagoidea

The POLYPHAGIDAE constitute a rather small but widely distributed family that includes the most primitive living cockroaches. They are generally small (2 cm or less in length) and often have a hairy pronotum. Some inhabit arid regions, and a few species are inquilines in ants' nests.

Superfamily Blaberoidea

BLATTELLIDAE are generally small cockroaches (not usually more than about 1 cm in length), with relatively long, slender legs. The family is large, widely distributed, and contains two major pest species, *Blattella germanica* (the German cockroach) (Figure 7.5C) and *Supella longipalpa* (formerly *supellictilium*) (the brown-banded cockroach). The German cockroach ranks second to the Oriental cockroach in economic importance. It prefers warm, humid

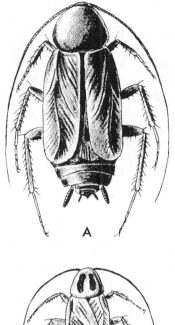

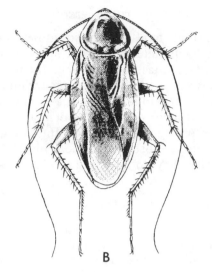

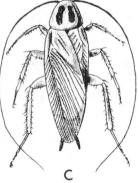

FIGURE 7.5. Cockroaches. (A) The Oriental cockroach, *Blatta orientalis* (Blattidae); (B) the American cockroach, *Periplaneta americana* (Blattidae); and (C) the German cockroach, *Blattella germanica* (Blattellidae). [From L. A. Swan and C. S. Papp, 1972, Copyright 1972 by L. A. Swan and C. S. Papp. Reprinted by permission of Harper & Row, Publishers, Inc.]

surroundings, such as are found in bakeries, restaurants, and domestic kitchens. Like *B. germanica, S. longipalpa* is probably of African origin. It became established in Florida at the beginning of the twentieth century and has now been reported from all states. It is also common in some areas of Canada. There are few records from Great Britain.

The family BLABERIDAE is the most recently evolved cockroach family and the one which has undergone the most extensive adaptive radiation. The group is primarily tropical and contains the largest cockroach species. Its members are generally found under logs, in humus, etc., though some species are arboreal. A few species may occasionally become associated with man, for example, *Pycnoscelus surinamensis* (the Surinam cockroach), *Leucophaea maderae* (the Madeira cockroach), and *Nauphoeta cinerea* (the lobster cockroach).

Suborder Mantodea

In members of the suborder Mantodea the head is not covered with a pronotum; three ocelli are present; the forelegs are raptorial; and the gizzard is not well developed.

The family AMORPHOSCELIDAE is best represented in the Australasian region, though species are also found in Asia, Africa, and southern Europe. Two morphological features distinguish members of this family from other mantids. The pronotum is short, and the tibiae and femora of the raptorial forelegs lack spines. Many species are procryptically colored and have various spines and prominences on the head and pronotum.

The family MANTIDAE (Figure 7.6) is easily the largest of the Mantodea and has a wide distribution throughout the tropical and warmer temperate regions of the world. Mantidae are almost always elongate and vary in length from about 3 to 10 cm when fully grown. They are frequently well camouflaged. Most of the species of mantids found in North America belong to this family, including several that have been introduced, for example, *Mantis religiosa*, the "soothsayer" or "praying mantis" of southern Europe.

Literature

Good general accounts of cockroaches are given by Guthrie and Tindall (1968) and Cornwell (1968, 1976). These authors also deal with their economic importance and provide extensive bibliographies. The biotic associations of cockroaches are discussed by Roth and Willis (1960), and their medical and veterinary importance by the same authors (1957). The phylogeny and classification of Blattodea is dealt with by McKittrick (1964) and Durden (1969). Probably the best accounts of mantid biology are provided by Chopard (1938, 1949), though in his description of the North American species Gurney (1950) includes much general information. Beier (1964) should be consulted for a recent version of mantid classification.

A key to the British cockroaches is provided by Hincks (1949). The North American Dictyoptera may be identified from Rehn (1950), Gurney (1950), and Helfer (1963).

Beier, M., 1964, Blattopteroidea. Ordnung Mantodea Burmeister 1838 (Raptoriae Latreille 1802; Mantoidea Handlirsch 1903; Mantidea auct.), *Bronn's Kl. Ordn. Tierreichs* **6**:849–870.
Chopard, L., 1938, *La Biologie des Orthoptères*, Lechevalier, Paris.
Chopard, L., 1949, Ordre des Dictyoptères, in: *Traité de Zoologie* (P.-P Grassé, ed.), Vol. IX, Masson, Paris.
Cornwell, P. B., 1968, 1976, *The Cockroach*, Vols. I and II, Hutchinson, London.
Durden, C. J., 1969, Pennsylvanian correlation using blattoid insects, *Can. J. Earth Sci.* **6**:1159–1177.

FIGURE 7.6. The Carolina mantid, *Stagmomantis carolina* (Mantodea). [From M. Hebard, 1934, The Dermaptera and Orthoptera of Illinois, *Bull. Ill. Nat. Hist. Surv.* **20**(3). By permission of the Illinois Natural History Survey.]

Gurney, A. B., 1950, Praying mantids of the United States, native and introduced, *Annu. Rep. Smithson. Inst.* **1950**:339–362.

Guthrie, D. M., and Tindall, A. R., 1968, *The Biology of the Cockroach*, Edward Arnold, London.

Helfer, J. R., 1963, *How to Know the Grasshoppers, Cockroaches, and Their Allies*, Brown, Dubuque, Iowa.

Hincks, W. D., 1949, Dermaptera and Orthoptera, *R. Entomol. Soc. Handb. Ident. Br. Insects* **1**(5):20 pp.

McKittrick, F. A., 1964, Evolutionary studies of cockroaches, *Mem. Cornell Univ. Agric Expt. Station* **389**:177 pp.

Rehn, J. W. H., 1950, A key to the genera of North American Blattaria, including established adventives, *Entomol. News* **61**:64–67.

Roth, L. M., and Willis, E. R., 1957, The medical and veterinary importance of cockroaches, *Smithson. Misc. Collect.* **134**(10):147 pp.

Roth, L. M., and Willis, E. R., 1960, The biotic associations of cockroaches, *Smithson. Misc. Collect.* **141**(4422):470 pp.

5. Isoptera

SYNONYMS: Termitina, Termitida, Socialia

COMMON NAMES: termites, white ants

Polymorphic social insects living in colonies that comprise reproductives, soldiers, and workers; head with moniliform multisegmented antennae and mandibulate mouthparts, compound eyes present but frequently degenerate, ocelli often absent; wings when present almost identical (except *Mastotermes*) and membranous, lying horizontally over abdomen at rest, capable of being shed by a predetermined basal fracture (except *Mastotermes*), legs identical and with a large coxa, tarsi almost always 4-segmented (5-segmented in *Mastotermes*); cerci short and with few segments, external genitalia lacking in both sexes of most species.

Nearly 2000 species of termites have been described, mainly from tropical and subtropical areas, though a few species are found in moderate temperate climates such as those of southern Europe and southern and western North America as far north as southern Canada. Several species have been transported to new areas by commerce, and some of these have become established in heated buildings (for example, in Hamburg and Toronto) well outside their normal range of climatic tolerance.

Structure

In almost all species the mature termite colony contains individuals of remarkably different form and function. Each group of individuals that perform the same function is known as a caste. In most species three castes occur, reproductive (primary and secondary), soldier (often of more than one kind), and worker, though in the more primitive termites (Mastotermitidae, Kalotermitidae, and many Hodotermitidae) one or both (Mastotermitidae) castes are absent, and the functions of workers are taken over by the larger juvenile stages (called pseudergates). Since the castes are of different form, it is convenient to describe them separately.

Reproductive. The body of PRIMARY REPRODUCTIVES is normally quite dark in color and fairly well sclerotized. However, in physogastric queens, that is, females whose abdomen becomes enormously swollen through the stretching of the intersegmental membrane (Figure 7.10C), the abdomen is pale and the original tergal and sternal plates are the only areas of sclerotization. The head is round or oval and carries well-developed compound eyes, moniliform antennae with a variable number of segments (generally fewer in more advanced termites), and mandibulate mouthparts. In some termite families a small opening the fontanelle, occurs in the midline between or behind the compound eyes. This marks the opening of the frontal gland, a structure that appears to be functionless in reproductives. In the thorax the pronotum is usually of a distinctive shape and is of use in taxonomic studies. The thoracic sterna are membranous. Except in Mastotermitidae, the two pairs of wings are very similar in appearance, with strongly sclerotized veins in the anterior portion and a basal (humeral) suture along which fracture of the wing occurs. In *Mastotermes* the wings are much less specialized, the hind wings having a large anal lobe as in cockroaches. The legs are all quite similar, having large coxae and, almost always, 4-segmented tarsi; in *Mastotermes* the tarsi are 5-segmented. On the coxae of the mid- and hindlegs a well-developed meron is present. There are 10 obvious abdominal segments with the eleventh tergum having fused with the tenth, and the eleventh sternum being represented by the paraprocts. Short cerci are present which are 3- to 8-segmented in lower termites but are reduced to an unsegmented or 2-segmented tubercle in higher forms. External genitalia are absent in almost all species.

The esophagus is a long, narrow tube and is followed by the scarcely differentiated crop. The gizzard wall is greatly folded longitudinally, each fold having cuticular thickenings and, often, teeth. The gizzard has a remarkably uniform structure in all termites. The midgut is a tube of uniform diameter that contains a peritrophic membrane. In physogastric queens the midgut is enormously enlarged, a development presumably associated with absorption of the large quantities of saliva fed to them by the workers. The hindgut is well developed and differentiated into a number of regions, the most prominent of which is the large paunch in which are contained bacterial or protozoan symbionts. The posterior wall of the paunch contains columnar epithelium and is probably a region of absorption. In lower termites (except Mastotermitidae) eight Malpighian tubules enter the gut at the junction of the midgut and hindgut. In Termitidae there are only four tubules. The central nervous system is orthopteroid, though the brain is relatively large. There are three thoracic and six abdominal ganglia. The testes each comprise up to 10 fingerlike follicles that enter the paired vasa deferentia. At the junction of the vasa deferentia and ejaculatory duct there is a pair of seminal vesicles. In females each ovary contains an extremely variable number of panoistic ovarioles (8–10 in some lower termites, to several hundred or thousands in the physogastric queens of higher termite species). The paired oviducts enter the short common oviduct, which leads into the genital chamber. A spermatheca and accessory glands also enter this chamber.

In SECONDARY REPRODUCTIVES (=SUPPLEMENTARY REPRODUCTIVES of lower termites, REPLACEMENT REPRODUCTIVES of higher forms) the body is less sclerotized and paler than that of the primaries. The compound eyes are usually reduced. Another important difference between the primary and secondary reproductives is that in the latter fully developed wings never occur. The supplementary reproductives may have wing pads or they may be apterous, according to the form of termite from which they differentiate. Female secondary reproductives are never physogastric. In other aspects of their anatomy and morphology they resemble the primary reproductives.

Soldier. This caste is easily recognized by the large, well-sclerotized head of its members, which may be of either sex. Primitively the mandibles too are very large, sometimes enormous, and suited for biting. In other species in which the mandibles are large they may serve as pincers, or they may be asymmetrical and hinged so as to snap open at great speed, thus delivering a powerful blow to an adversary. In the soldiers of many higher termites the mandibles are, however, reduced in size or vestigial. Correlated with this is an enlargement of the frons to form a more or less pointed rostrum, at the tip of which opens the frontal gland. The latter is well developed, often projecting posteriorly into the abdomen, and produces a noxious secretion that may be ejected for a considerable distance. Generally soldiers are apterous, though in exceptional circumstances they may develop from juveniles with wing buds. They are never fully winged. In most other aspects they resemble the reproductives, though their reproductive organs are atrophied.

Worker. In most species the body of workers is generally pale and weakly sclerotized. The head resembles that of a primary reproductive, except that the compound eyes are usually reduced (except in species that forage by day above ground), and the mandibles are typically more powerful. Workers are wingless and may be of either sex. The general form of the abdomen resembles that of a reproductive.

Internally workers resemble reproductives, except for the following points: the salivary glands may be greatly enlarged in accord with their product being used as food by primary reproductives, the paunch is typically much larger than that of reproductives and soldiers, the brain is relatively much smaller than that of reproductives, and the reproductive organs are atrophied.

Life History and Habits

New colonies may be formed in various ways. By far the commonest method is swarming, in which large numbers of winged individuals (alates) leave the parent colony. These individuals separate into pairs, lose their wings, and search for a suitable site in which to mate and rear their young. The onset of swarming is closely correlated with climatic conditions. In tropical species it occurs typically at the onset of the rainy season, an adaptation which facilitates nest formation in the damp, soft earth for subterranean species. In species from temperate climates swarming occurs during the summer. Flights may occur at any time of the day, but for a given species there are frequently specific hours during which swarming takes place. Swarming may be temporarily postponed, however, if environmental conditions are unsuitable. The distance traveled by the alates is usually only a few hundred yards unless they are assisted by wind. It is at this time that termites are most susceptible to predators.

On landing individuals lose their wings, and a male is attracted (probably chemically) to a female, which he follows until she locates a suitable nesting site. After closing the entrance to the nest, the royal pair, as the founding pair are called, mate within a few hours or days. (Mating is, however, periodically repeated throughout the life of the pair.) Egg laying begins soon after the royal pair have become established, but the rate of egg production is low during the first year or two. Initially only workers are produced, but as the number of individuals increases soldiers, and, in the lower termites, supplementary reproductives differentiate. The ratio of soldiers to nonsoldiers varies, according to species, between 1 : 12 and 1 : 100. Comparable ratios for secondary reproductives are not available. As their name suggests supplementary reproductives are generally produced in colonies of lower termites to enhance the total egg production, though this is probably no more than one or two eggs per day. This compares unfavorably with the figure for higher termite colonies in which the single primary female may lay many thousands of eggs per day. It is likely, however, that such a high rate is not sustained on a year-round basis but is seasonal. The differentiation of the various castes and their maintenance in a fixed ratio to each other are complex phenomena, controlled by the interaction of pheromonal, nutritional, hormonal, and perhaps other factors (see Chapter 21, Section 7). A colony matures (that is, becomes capable of producing all castes, including winged reproductives) after several years, but it continues to increase in size after this time. It is obviously difficult to estimate the number of individuals in fully grown colonies, but in the lower termites the figure is usually several hundred or thousands, while in the higher species it may be several million.

Two other methods of colony foundation are known. In some species, in which the nest is rather a diffuse structure, groups of individuals may become more or less isolated from the rest of the colony. In these groups supplementary reproductives differentiate, and the group becomes independent of the parent colony. This is described as budding. The foundation of new colonies by deliberate social fragmentation (sociotomy) has been reported for a few species. In this situation a large number of individuals of all castes (often including the original royal pair) emerge from the parent colony and march to a new location. The original colony them becomes headed by replacement reproductives.

Termite nests exhibit a wide range of form, the complexity of which parallels approximately the phylogeny of the order. In the primitive Kalotermitidae and some Hodotermitidae (subfamily Termopsinae) the nest is simply a series of cavities and tunnels excavated in wood. Few partitions are constructed by these termites, and there is no differentiation of the nest into specific regions. In other lower termites the nest may be in wood or subterranean, but even in the former situation contact with the ground is almost always maintained by a series of tunnels. This ensures that the humidity of the nest remains high. Most Hodotermitidae build completely subterranean nests, in which the beginnings of specialization are seen Food is stored in chambers immediately below the surface of the ground. The main chamber, which is considerably subdivided by both horizontal and vertical walls, is several feet below the surface. However, in nests of this family there is no chamber specifically for the royal pair. Nests of Rhinotermitidae may be entirely in soil or in wood or in both these media. In only a few Rhinotermitidae of the genus *Coptotermes* and in most Termitidae are epigeous (above ground) nests constructed (Figure 7.7), though it should be emphasized that even in these species a considerable portion of the nest may be subterranean. In the simplest epigeous nests little differentiation occurs between the peripheral and internal parts, which comprise a mass of interconnecting, uniform chambers; the royal chamber is either absent or located in the subterranean part of the nest. In more complex nests the above-ground component comprises a thick peripheral wall enclosed within which is the habitacle (nursery) and surrounding food chambers. The royal chamber is usually located near the base of the structure.

A major problem for all social insects, which increases in magnitude in proportion to the size of the colony, is maintenance of a suitable nest climate. Regulation of relative humidity, temperature, and carbon dioxide concentration must occur. For termites that live in wetter regions humidity regulation is not a serious problem, and the relative humidity within the nest is generally 96–99%. In termites from regions with long dry spells various behavioral adaptations ensure the well-being of a colony. The commonest of these is for the termites to move more deeply into the ground where the moisture content is greater. Other species behave like honeybees and regurgitate saliva or crop contents on to the walls of the nest, especially in the nursery region. A species that lives in the Sahara Desert burrows deeply into the ground to the level of the water table and brings moisture-laden particles up into the nest area. Other desert species probably do likewise.

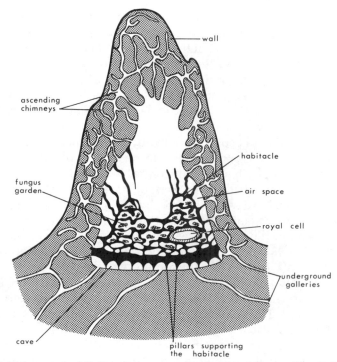

FIGURE 7.7. Mature nest of *Bellicositermes natalensis* (Termitidae). [After P.-P. Grassé (ed.), 1949, *Traité de Zoologie*, Vol. IX. By permission of Masson, Paris].

Temperature is also regulated in some termite nests to a remarkable degree. To some extent this is facilitated by the location of the nests in wood or soil, which serve as excellent buffers against sudden changes in external temperature. In cold weather termites behave much like bees, clustering together in the center of the nest and effectively reducing the "operating space," whose temperature must be maintained by metabolic heat. In mound-building termites, whose nest may be fully exposed to the sun, the temperature in the center of the nest is held steady as a result of the extremely thick outer walls which provide excellent insulation. However, as the degree of insulation from external temperature fluctuations increases, so does the problem of gaseous exchange. Although it has been shown experimentally that termites can withstand very high carbon dioxide concentrations, field studies have indicated that under natural conditions they do not face this problem because of the nest's air-conditioning system. Convection currents, created by the different temperatures at the center and periphery of the nest are the basis of the system. In *Bellicositermes natalensis* the heat created in the central (nursery) area causes the air in this region to rise to the upper chamber (Figure 7.7). The air then moves along the radial ducts to the peripheral region of the nest, which comprises a system of thin-walled tubes. Carbon dioxide and oxygen can diffuse easily across these walls. As the "fresh" air in the peripheral tubes cools, it sinks into the "cellar" of the nest, eventually to be drawn by convection back into the central area.

Termites are primitively wood-eating insects, and this habit is retained in most lower termites and many of the higher forms. Since, however, termites do not themselves have the ability to digest wood, complex relationships have evolved with other organisms (bacteria, protozoa, and fungi) that are able to degrade the food. In its simplest form, seen in Mastotermitidae, Kalotermitidae, and some Hodotermitidae, the relationship is one in which the wood is digested by flagellate protozoans in the paunch of the host termite. A variation on this is the replacement of wood by other vegetable matter such as seeds, leaves, and grass (as in the harvester termites of the family Hodotermitidae). Accompanying this modification is the evolution of foraging habits and storage of food in special chambers within the nest. In the most advanced family, Termitidae, a wide range of modifications of the primitive feeding habit is seen. However, a feature common to all members of the family is the replacement of the symbiotic gut protozoa by bacteria. Many Termitidae feed on wood like the lower termites. Others feed on humus. Perhaps the most interesting group, however, is the subfamily Macrotermitinae, most species of which culture a basidiomycete fungus of the genus *Termitomyces* in special "fungus gardens." Although the occurrence of these structures has been known since 1779, it is only recently that the precise relationship between the termite and fungus has been established. In a typical fungus garden the fungus grows on sheets of reddish-brown "comb" (decaying vegetable material) and is visible as a whitish mycelium containing conidia and conidiophores. This latter observation led early authors to suggest that the young termites were fed on the fungus, though it soon became apparent that the small amount of fungus would not satisfy even their requirements. It was some time before the authors realized that the comb was a dynamic structure, being removed from below and built up on its upper surface or in the space beneath. In other words, the comb forms the food of the termites. Using staining techniques, it has been shown that the primary role of the fungus is digestion of the lignin component of the comb, releasing cellulose that is then broken down by bacteria in the termites' gut. Secondarily, however, the fungus also provides vitamins and a source of nitrogen. Another point of contention was the method of comb construction. It was believed originally that the termites regurgitated chewed-up food to produce comb, but recent work has shown that the comb is derived from feces. Thus, the vegetable material passes twice through the gut of Termitidae, a situation that is comparable with that in other termite families in which proctodeal feeding is an important method of extracting the maximum nutrition from the food (see below).

Only workers and/or pseudergates are able to feed themselves. Members of other castes and very young stages must be fed, and, furthermore, their diet, as in other social insects, is different to a greater or lesser degree from that of workers. Exchange of food material is effected either by anus-to-mouth transfer (proctodeal feeding) or by mouth-to-mouth transfer (stomodeal feeding). The former method takes place in all families except the Termitidae, and normally it occurs only between workers or larger juveniles, although occasionally soldiers may act as donors. Proctodeal food is a liquid containing protozoans, products of digestion, and undigested food. Stomodeal food is either a semisolid material comprising the regurgitated contents of the crop, which are fed to soldiers in the lower termite families, or saliva, which appears to be the only food received by reproductives of all families, very young stages of lower termites, and all juvenile stages and soldiers of Termitidae.

Phylogeny and Classification

There is little doubt that termites are derived from cockroachlike ancestors that were similar in many ways to *Cryptocercus punctulatus*, a subsocial, wood-eating cockroach. Indeed, some authorities, for example, McKittrick (1964), consider the similarities between termites and cockroaches to be sufficiently great as to include the former as a suborder of the Dictyoptera.

The proposed relationships of the modern termite families are expressed diagrammatically in Figure 7.8. There is general agreement regarding the major subdivisions of the order, although some authorities feel that certain subfamilial groups are sufficiently distinct as to warrant familial status. The Mastotermitidae are in most respects the most primitive termite family. It includes a single living species, *Mastotermes darwiniensis*, which bears many similarities to the cockroaches (see below). The Kalotermitidae are believed to have evolved from the Mastotermitidae, with which they share a number of features (two teeth on left mandible, presence of ocelli, and an arolium on the tarsus). According to Krishna (1970), early in termite evolution a split occurred, with one line giving rise to the mastotermitid–kalotermitid branch and the other line leading to the Hodotermitidae and other families. In some respects the Hodotermitidae retain more primitive features than do the Mastotermitidae; for example, some genera have three teeth on the left mandible, a feature found in cockroaches. Some earlier authors divided the Hodotermitidae into two familial groups, Termopsidae (containing the more primitive, damp-wood termites

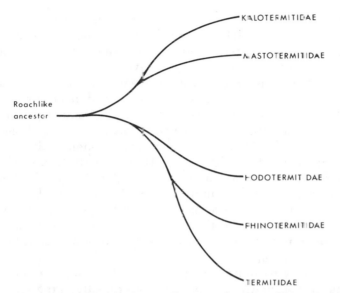

KALOTERMITIDAE

MASTOTERMITIDAE

Roachlike
ancestor

HODOTERMITIDAE

RHINOTERMITIDAE

TERMITIDAE

FIGURE 7.8. Proposed phylogeny of Isoptera. [After K. Krishna, 1970, Taxonomy, phylogeny and distribution, in: *Biology of Termites,* Vol. II (K. Krishna and F. M. Weesner, eds.). By permission of Academic Press Inc. and the author.]

of temperate regions) and the Hodotermitidae (*sensu stricto*) (in which are placed the harvester termites of warm, dry regions). Additional work has shown, however, that the distinction between the two is not clear-cut. The Rhinotermitidae evolved from an early hodotermitid ancestor, in which the three-toothed left mandible and ocelli were retained. The position of the Serritermitidae is questionable. The family, which contains the single genus *Serritermes* from Brazil, has been included previously in the Rhinotermitidae and Termitidae on account of its mixture of characters. It is probably best to consider the group as a distinct family that has evolved from early rhinotermitid stock. The families Mastotermitidae, Kalotermitidae, Hodotermitidae, Rhinotermitidae, and Serritermitidae are collectively known as the lower termites. Common to them all is a mutualistic relationship between the termite host and certain flagellate protozoans found in the hindgut. In the remaining termite family, Termitidae, regarded as higher termites, there are generally few or no protozoans in the hindgut, and the relationship between them and the host is never mutualistic. Where such a relationship exists it is between the termite and the bacteria of the hindgut. The Termitidae evolved from a rhinotermitidlike ancestor.

Superfamily Mastotermitoidea

The single family MASTOTERMITIDAE contains a number of fossil, and one living, species. *Mastotermes darwiniensis,* from tropical areas of northern Australia and New Guinea, has a large number of features which indicate the close relationship of the cockroaches and termites. These include the 5-segmented tarsi, long multisegmented antennae, well-developed compound eyes and ocelli, netlike wing venation, distinct anal lobe in the hind wing, absence of a basal suture in both wings, certain structural similarities in the gizzard and genitalia of both groups, and the laying of eggs in an ootheca, a feature found in no other termite family. There is no caste system in *M. darwiniensis*, whose social organization is therefore very primitive.

Members of the family KALOTERMITIDAE are commonly known as dry-wood termites from their habit of living in sound, dry wood that is not in contact with the ground. No distinct worker caste exists in this family whose soldiers are characterized by their truncate (phragmotic) head, used for plugging holes in the wood in which they live. A few species are economically important.

The primitive members of the family HODOTERMITIDAE are the damp-wood termites found in rotting logs in temperate regions. As in Kalotermitidae, no worker caste occurs in these primitive forms (Figure 7.9). The more advanced Hodotermitidae are the harvester termites that forage above ground for grass, leaves, etc., typically in drier regions of the world. A distinct soldier caste exists in these groups. A few species of both wood-eating and foraging forms may be of economic importance.

Most RHINOTERMITIDAE are subterranean forms that live in buried, rotting wood. Some species, however, construct nests directly in the soil, or in rotting logs above ground, and yet others build a mound nest. All species are wood

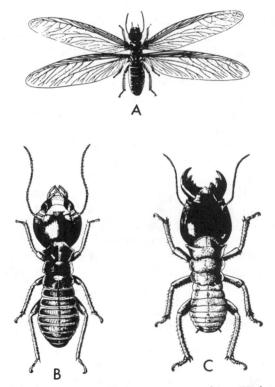

FIGURE 7.9. Castes of the lower termite, *Hodotermes mossambicus* (Hodotermitidae). (A) Alate, (B) pseudergate, and (C) soldier. [From W. H. G. Coaton, 1958, The Hodotermitid harvester termites of South Africa, *Union of South Africa, Department of Agricultural Science Bulletin*, Vol. 375. By permission of the South African Department of Agricultural Technical Services.]

eaters, and many are extremely important economically. An interesting feature of many species is the occurrence of dimorphism in the soldiers. The larger form retains the large mandibles; the smaller form has reduced mandibles, but the labrum is elongated and used for the dispersal of a noxious fluid produced by the frontal gland.

It is within the family TERMITIDAE (Figure 7.10), which contains about three-quarters of the living termite species, that the greatest range of social development and specialization exists. Four subfamilies can be distinguished: (1) Amitermitinae, containing the most primitive Termitidae and showing close affinities with the Rhinotermitidae; (2) Termitinae, in which the major evolutionary trend is toward the development in soldiers of snapping mandibles; (3) Macrotermitinae, which contain the Old World fungus-growing termites; and (4) Nasutitermitinae, the largest subfamily, with nearly 500 species, characterized by the evolutionary development in soldiers of a rostrum (nasus) at the tip of which opens the frontal gland (Figure 7.11). They are consequently known as "nasute soldiers" The contents of the gland are forcefully ejected in the form of a sticky irritant fluid.

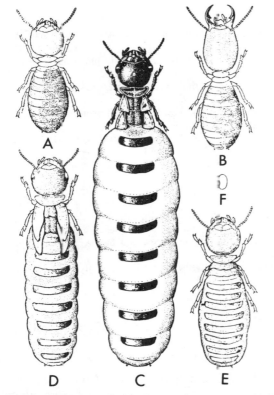

FIGURE 7.10. Castes of the higher termite, *Amitermes hastatus* (Termiticae). (A) Worker, (B) soldier, (C) physogastric queen, (D) secondary queen, (E) tertiary queen, and (F) egg. All figures are to same scale. The worker is about 5 mm. long. [From S. H. Skaife 1954, The black-mound termite of the Cape, *Amitermes atlanticus* Fuller, *Trans. R. Soc. S. Afr* **34**:251–271. By permission of the Royal Society of South Africa.]

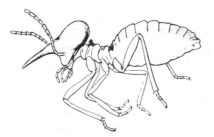

FIGURE 7.11. Nasute soldier of *Trinervitermes* sp. (Termitidae). [From P.-P. Grassé (ed.), 1949, *Traité de Zoologie*, Vol. IX. By permission of Masson, Paris.]

Literature

Krishna and Weesner (1969, 1970) have edited two recent volumes entitled "Biology of Termites" in which all aspects of termite biology are discussed by specialists. Volume I covers anatomy, physiology, and behavior; Volume II deals with taxonomy and zoogeography. Other useful introductions to the biology of the order are given by Wilson (1971), Howse (1970), Weesner (1960), Skaife (1961), and Harris (1971). Their economic importance is reviewed by Harris (1969, 1971) and Hickin (1971). North American termites are discussed by Weesner (1965) (also 1970, in "Biology of Termites," Vol. II) and may be identified in Helfer (1963).

Harris, W. V., 1969, *Termites as Pests of Crops and Trees*, Commonwealth Institute of Entomology, London.
Harris, W. V., 1971, *Termites—Their Recognition and Control* (2nd ed.), Longmans-Green, London.
Helfer, J. R., 1963, *How to Know the Grasshoppers, Cockroaches, and Their Allies*, Brown, Dubuque, Iowa.
Hickin, N. E., 1971, *Termites: A World Problem*, Hutchinson, London.
Howse, P. E., 1970, *Termites*, Hutchinson, London.
Krishna, K., 1970, Taxonomy, phylogeny and distribution, in: *Biology of Termites*, Vol. II (K. Krishna and F. M. Weesner, eds.), Academic Press, New York.
Krishna, K., and Weesner, F. M., 1969, 1970, *Biology of Termites*, Vols. I and II, Academic Press, New York.
McKittrick, F. A., 1964, Evolutionary studies of cockroaches, *Mem. Cornell Univ. Agric. Expt. Station* **389:**177 pp.
Skaife, S. H., 1961, *Dwellers in Darkness*, Doubleday, New York.
Weesner, F. M., 1960, Evolution and biology of termites, *Annu. Rev. Entomol.* **5:**153–170.
Weesner, F. M., 1965, *The Termites of the United States*, National Pest Control Association, Elisabeth, N.J.
Wilson, E. O., 1971, *The Insect Societies*, Harvard University Press, Cambridge.

6. Grylloblattodea

SYNONYMS: Notoptera, Grylloblattaria   COMMON NAME: rock crawlers

Elongate insects; head prognathous, mouthparts mandibulate, compound eyes reduced or absent, ocelli absent, antennae long and filiform; thoracic segments similar and well distinguished, legs virtually identical with large coxae and 5-segmented tarsi, wings and auditory organs absent; abdomen with long, segmented cerci, females with well-developed ovipositor, males with articulated coxites on 9th sternum and asymmetrical genitalia.

This is an extremely small order, comprising only about a dozen species, whose members are found in western North America eastern Siberia, and Japan. The first representative of the order was described by Walker in 1914, though specimens taken in 1906 are known (D. K. McE. Kevan, personal communication).

Structure

Grylloblattodea (Figure 7.12) are elongate, wingless insects which reach a length of about 3 cm. Their head is flattened and prognathous. It carries well-developed mandibulate mouthparts and long, filiform antennae. The compound eyes are reduced to a few ommatidia or are entirely absent. There are no ocelli. The thoracic segments are more or less identical, though the prothorax is slightly larger than the other two. The six legs are similar in structure, each with a large coxa and a 5-segmented tarsus. The abdomen has 10 obvious segments and the eleventh is represented by the epiproct and paraprocts. The ovipositor of females comprises three pairs of valves. The ninth sternite in males carries a pair of asymmetric coxites, each bearing a small style. The cerci are long, 8-segmented structures.

The internal structure is orthopteroid in nature.

Life History and Habits

Grylloblattodea are secretive, nocturnal insects, generally found in cold, wet locations, often at relatively high altitudes, under stones, in rotting logs, etc. They favor low temperatures (optimally ca. 4°C) and go underground during warmer months. They are omnivorous or, more commonly, predaceous. The black eggs are deposited singly, and embryonic development is said to take about one year. There are eight juvenile stages that together occupy about five years.

Phylogeny and Classification

The relationship of the Grylloblattodea to other orthopteroid groups has been the subject of considerable discussion. The insects possess a combination

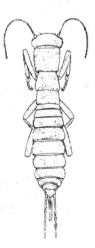

FIGURE 7.12.   *Grylloblatta campodeiformis* [Grylloblattodea]. [From E. M. Walker, 1914, A new species of Orthoptera, forming a new genus and family, *Can. Entomol.* **46**:93–99. By permission of the Entomological Society of Canada.]

of primitive "blattoid" and "orthopteroid" features, together with specialized features of their own. The multisegmented cerci, 5-segmented tarsi, large coxae, and asymmetric male genitalia suggest a relationship with cockroaches, whereas the well-developed ovipositor and the structure of the tentorium are more orthopteroid in nature. According to some authors, for example, Zeuner (1939), they appear to be specialized survivors of a primitive protorthopteran stock from which the Dictyoptera and Orthoptera evolved. However, the studies of Giles (1963) and Kamp (1973) indicate that their closest relatives are Dermaptera and that the two groups may have had a common ancestry.

The living Grylloblattodea discovered to date are placed in a single family, GRYLLOBLATTIDAE, containing three genera: *Grylloblatta* (North American forms); *Grylloblattina* (from Siberia); and *Galloisiana* (from Japan).

Literature

The biology of Grylloblattodea is described by Walker (1937) and Mills and Pepper (1937). Gurney (1948, 1961) and Kamp (1979) discuss the taxonomy and distribution of the group. Gurney (1961) also provides a bibliography of other work on the order. Kamp (1973) discusses their affinities with other orthopteroid insects.

Giles, E. T., 1963, The comparative external morphology and affinities of the Dermaptera, *Trans. R. Entomol. Soc. London* **115**:95–164.

Gurney, A. B., 1948, The taxonomy and distribution of the Grylloblattidae (Orthoptera), *Proc. Entomol. Soc. Wash.* **50**:86–102.

Gurney, A. B., 1961, Further advances in the taxonomy and distribution of the Grylloblattidae (Orthoptera), *Proc. Biol. Soc. Wash.* **74**:67–76.

Kamp, J. W., 1973, Numerical classification of the orthopteroids, with special reference to the Grylloblattodea, *Can. Entomol.* **105**:1235–1249.

Kamp, J. W., 1979, Taxonomy, distribution, and zoogeographic evolution of *Grylloblatta* in Canada (Insecta: Notoptera), *Can. Entomol.* **111**:27–38.

Mills, H. B., and Pepper, J. H., 1937, Observations on *Grylloblatta campodeiformis* Walker, *Ann. Entomol. Soc. Am.* **30**:269–274.

Walker, E. M., 1914, A new species of Orthoptera forming a new genus and family, *Can. Entomol.* **46**:93–99.

Walker, E. M., 1937, *Grylloblatta*, a living fossil, *Trans. R. Soc. Canada* **31**:1–10.

Zeuner, F. E., 1939, *Fossil Orthoptera Ensifera*, 2 vols., British Museum of Natural History, London.

7. Dermaptera

SYNONYMS: Euplexoptera, Euplecoptera,   COMMON NAME: earwigs
Dermoptera, Labiduroida,
Forficulida, etc.

Generally elongate insects; head prognathous with biting mouthparts and multisegmented antennae, compound eyes present (reduced or absent in epizoic forms), ocelli absent; wings generally present, fore wings modified into short smooth veinless tegmina, hind wings semicircular and membranous with veins arranged radially, legs subequal and with 3-segmented tarsi; abdomen with unsegmented forcepslike cerci, ovipositor of females reduced or absent.

Dermaptera are found in all except the frigid regions of the world, though they are most common in the warmer parts. Approximately 1200 species have been described, almost all of which are free-living forms. About a dozen species are epizoic* on South East Asian bats (two species of *Arixenia*) or African rodents (genus *Hemimerus*).

Structure

The prognathous head carries a pair of multisegmented antennae, a pair of large compound eyes (except in epizoic species where they are reduced or absent), and mandibulate mouthparts. Ocelli are absent. The pronotum is enlarged somewhat and has a quadrangular shape. The metanotum is also large and has two lines of spines which lock the tegmina in the resting position. Both the tegmina and hind wings are absent in epizoic forms and some free-living species. In other species the degree of wing development is variable. The tegmina are smooth, veinless structures that do not extend beyond the metathorax. The hind wings are membranous, semicircular structures composed mainly of a very large anal area supported by a number of raciating branches of the first anal vein. The preanal area is reduced to a small anterior area and contains only the radius and cubitus veins. The wings fold both longitudinally and transversely and are stored beneath the tegmina. The legs are more or less similar in form, increasing in size posteriorly. They bear a 3-segmented tarsus. The 10-segmented abdomen is flattened dorsoventrally and telescopic (due to the overlapping of the tergal plates). At its posterior tip is a smooth pygidium (called the eleventh segment by some authors). The form of the cerci is variable. In the Forficulina they are unsegmented, forcepslike structures, more curved in the male than in the female. They are used offensively and defensively, and for assisting in the opening and closing of the wings, and during copulation. In epizoic species they are hairy, stylelike structures. In females of some primitive Forficulina a reduced ovipositor is found, but other Dermaptera lack this structure. Paired penes are found in males of some species, but usually one of these organs is reduced or absent.

The gut resembles that of orthopteroids, except that it lacks mesenteric ceca. Between 8 and 20 Malpighian tubules occur, usually in groups of 4 or 5. The nervous system is generalized, the ventral nerve cord containing three thoracic and six abdominal ganglia. The tracheal system, which lacks air sacs, opens to the exterior via two thoracic and eight abdominal spiracles. In both sexes the reproductive system shows considerable variability. In males there is a pair of testes, paired vasa deferentia, single or paired seminal vesicles, and one or two ejaculatory ducts. In females, the ovaries are paired and contain polytrophic ovarioles. The ovarioles are arranged in two ways, either in three rows along the length of the lateral oviduct, or as a group at the anterior end of the oviduct.

*Species of *Arixenia* and *Hemimerus* are usually described as ectoparasitic. However, their parasitic nature has not been established.

Life History and Habits

The free-living earwigs are secretive, nocturnal creatures that hide in crevices, under stones, in logs, etc. They are fast runners and, although many have wings, they use them only rarely. Most species are omnivorous, and a few may damage young shoots and buds of plants. When animal food is available, however, it seems to be preferred.

In species from warmer regions reproduction occurs through the year; in temperate species egg laying is probably restricted to the summer. Female earwigs of some species exhibit parental care. The eggs are laid in a short tunnel constructed by a female who remains in attendance until the young hatch and for the first week or two of larval life. In warmer regions young earwigs develop rapidly, and there are several generations per year; in temperate species larval development is arrested with the onset of cold weather, to be completed the following spring. The epizoic Dermaptera are viviparous, eggs developing within the follicle of the ovariole (see Figure 20.15). Approximately six embryos develop at a time, apparently being nourished via a placentalike structure attached to the head of each embryo.

Phylogeny and Classification

According to Giles (1963), the nearest relatives of the Dermaptera are the Grylloblattodea, the two orders having evolved from some common protorthopteran stock. This conclusion is supported by Kamp's (1973) numerical analysis of selected orthopteroid insects. The Dermaptera have been arranged traditionally in three suborders: the Forficulina (free-living forms), Arixeniina (two species of *Arixenia* (Figure 7.13B) that live in close association with East Indian cave bats), and Hemimerina [about 10 species of *Hemimerus* (Figure 7.13C), all of which are epizoic on African giant rats of the genus *Cricetomys*]. As a result of his comparative study of the external morphology of Dermaptera, Giles concludes that this arrangement is the most satisfactory.

Popham (1965a), however, has devised a rather different scheme for the classification of Dermaptera based on internal and external anatomical features (especially the genitalia) and on the geographical distribution of the group. Popham considers that the features by which *Arixenia* differs from free-living forms are simply adaptations to living on bats and their feces. In many basic features *Arixenia* resembles the Forficulina, and Popham therefore places the Arixeniidae in the superfamily Labioidea. In contrast, Popham argues that the differences between the Forficulina and Hemimerina are of sufficient magnitude that the latter group should be given ordinal status.

Suborder Forficulina

Superfamily Pygidicranoidea

Members of the superfamily Pygidicranoidea possess many primitive features and appear to be the closest to the ancestral dermapteran stock. Species are found in Asia, Australia, South Africa, and South America.

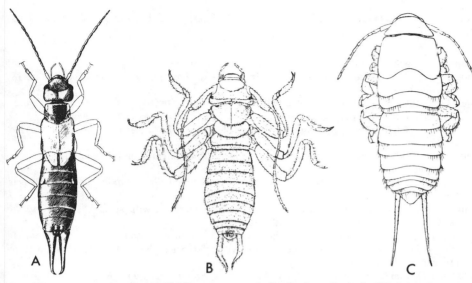

FIGURE 7.13. Dermaptera. (A) The European earwig *Forficula auricularia*, (B) *Arixenia* sp., and (C) *Hemimerus* sp. [A from D. J Borror and D. M. Delong, 1971 *An Introduction to the Study of Insects*, 3rd ed. By permission of Holt, Rinehart and Winston, Inc. B, C, from P.-P. Grassé (ed.), 1949, *Traité de Zoologie*, Vol. IX. By permission of Masson, Paris.]

Superfamily Karschielloidea

Members of the superfamily Karschielloidea, which is restricted to South Africa, are large, carnivorous Dermaptera that are specialized for feeding on ants.

Superfamily Labioidea

As defined by Popham the superfamily Labioidea includes three families, LABIIDAE, CARCINOPHORIDAE, and ARIXENIIDAE, only the first of which is likely to be commonly encountered. Common genera include *Labia, Prolabia*, and *Spongiphora*.

Superfamily Forficuloidea

Three families are included in the superfamily Forficuloidea, the CHELISOCHIDAE, LABIDURIDAE, and FORFICULIDAE. Only members of the last two are at all common. The Labiduridae are a widely distributed family that includes the seaside earwig, *Anisolabis maritima*, found on both Atlantic and Pacific coastlines. Other common genera are *Labidura* and *Euborellia*. The commonest species of Forficulidae is the European earwig, *Forficula auricularia* (Figure 7.13A), which through commerce has become distributed throughout the cooler regions of the world and is often considered a pest.

Literature

Chopard (1938, 1949) deals with the biology of earwigs. The evolution and classification of the order are discussed by Giles (1963) and Popham (1965a).

Hebard (1934), Hincks (1949), Popham (1965b), and Helfer (1963) have provided keys for identification.

Chopard, L., 1938, *La Biologie des Orthoptères*, Lechevalier, Paris.

Chopard, L., 1949, Ordre des Dermaptères, in: *Traité de Zoologie* (P.-P. Grassé, ed.), Vol. IX, Masson, Paris.

Giles, E. T., 1963, The comparative external morphology and affinities of the Dermaptera, *Trans. R. Entomol. Soc.* London 115:95–164.

Hebard, M., 1934, The Dermaptera and Orthoptera of Illinois, *Ill. Nat. Hist. Survey, Bull.* 20:125–279.

Helfer, J. R., 1963, *How to Know the Grasshoppers, Cockroaches, and Their Allies*, Brown, Dubuque, Iowa.

Hincks, W. D., 1949, Dermaptera and Orthoptera, *R. Entomol. Soc. Handb. Ident. Br. Insects* 1(5):20 pp.

Kamp, J. W., 1973, Numerical classification of the orthopteroids, with special reference to the Grylloblattodea, *Can. Entomol.* 105:1235–1249.

Popham, E. J., 1965a, The functional morphology of the reproductive organs of the common earwig (*Forficula auricularia*) and other Dermaptera with reference to the natural classification of the order, *J. Zool.* 146:1–43.

Popham, E. J., 1965b, A key to Dermaptera subfamilies, *Entomologist* 98:126–136.

8. Phasmida

SYNONYMS: Phasmodea, Cheleutoptera, Phasmatodea, Gressoria, etc.

COMMON NAMES: stick insects, leaf insects, walking sticks, phasmids

Moderate-sized to very large insects, usually of elongate cylindrical form, occasionally leaflike; head with well-developed compound eyes and mandibulate mouthparts, ocelli often absent; prothorax short, mesothorax and metathorax long, with or without wings, all legs very similar with small widely separated coxae and in most species 5-segmented tarsi; ovipositor small and concealed, male genitalia asymmetrical and hidden, cerci short and unsegmented; specialized auditory and stridulatory structures absent in most species.*

More than 2500 species of this predominantly tropical group have been described.

Structure

Most members of this order are remarkable for their close resemblance to the plants on which they are normally found (Figure 7.14). The body, which in some species may exceed 30 cm in length, is commonly elongate, wingless, or brachypterous and resembles a twig. In a few species which retain wings the body is dorsoventrally flattened and sculptured so as to resemble a leaf or a group of leaves.

The prognathous head bears a pair of antennae which may be short to very long. Compound eyes are always present, but ocelli are found only in some of

---

Phyllium and related genera have a stridulatory apparatus on the antennae.

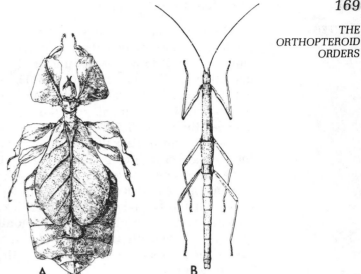

FIGURE 7.14. Phasmida. (A) A leaf insect, *Phyllium* sp. and (B) a stick insect, *Carausius morosus*. [A, from P.-P. Grassé (ed.), 1949, *Traité de Zoologie*, Vol. IX. By permission of Masson, Paris. B, from A. D. Imms, 1957, *A General Textbook of Entomology*, 9th ed. (revised by O. W. Richards and R. G. Davies). By permission of Chapman and Hall Ltd.]

the winged species, when they may be confined to males. In accord with the phytophagous habit, the mouthparts are of a generalized mandibulate form. The prothorax is small, while the mesothorax and metathorax are elongate with the latter firmly connected to the first abdominal segment. Wings may be present or absent, and all intermediate conditions of brachyptery are known. When wings are present they are typically fully developed in males but reduced in females. The fore wings take the form of tegmina, which in many species are much shorter than the hind wings. In leaflike species the venation is much modified to mimic the veins of a leaf. The legs are all similar, with small, widely separated coxae and, in the leaflike forms, broadly flattened tibiae and femora. Eleven abdominal segments are present, though the eleventh is represented only by the epiproct, paired paraprocts, and unsegmented cerci. In males the terminal abdominal segments and the aedeagus are of variable form. In some species the cerci are secondarily modified as claspers. The ovipositor, which comprises three pairs of small valves, is covered by the operculum, a keel-shaped structure formed from the eighth sternum.

The gut is straight and comprises a large crop whose posterior part functions as a gizzard, a long midgut that bears numerous external papillae over its posterior part, and a short hindgut. Numerous Malpighian tubules, arranged in two groups, enter the gut via a common duct. The central nervous system is primitive and includes three thoracic and seven abdominal ganglia. The male reproductive system consists of a pair of tubular testes and short vasa deferentia that open together with various accessory glands into the ejaculatory duct. In females several panoistic ovarioles lie alongside the lateral oviducts. One or two spermathecae and a pair of accessory glands open into the dorsally placed bursa copulatrix near its junction with the common oviduct. Paired repugnatory glands are found in the prothorax of many species. Their ducts open to the exterior in front of the fore coxae.

Life History and Habits

Phasmids are sedentary insects whose ability to escape from would-be predators is largely dependent on their close resemblance to twigs or leaves. However, they do possess a number of other devices that can be brought into operation should they be detected. These include falling to the ground and entering a cataleptic state, in which they may avoid further detection, secretion of an obnoxious material from the repugnatory glands, and reflex autotomy between the femur and trochanter should the leg be seized. Many phasmids are nocturnal, and during the day they remain completely immobile. Typically they are green or brown insects but the color of different populations of the same species may vary considerably. A few species exhibit phase polymorphism analogous to that which occurs in locusts. It should be emphasized, however, that the behavioral differences found between the different phases of locusts are not seen in phasmids. In addition, a few species can change their color physiologically in a matter of hours. This color change results from the aggregation or dispersion of pigment in the epidermal cells and is controlled by the endocrine system.

Phasmids in general are rather uncommon, and this is partly associated, in many species, with the development of facultative parthenogenesis. In a few species, for example, *Carausius* (=*Dixippus*) *morosus*, a common experimental insect, parthenogenesis is virtually obligate, males being extremely rare. Eggs, which frequently resemble seeds, are laid singly and allowed to fall to the ground. Development time is variable, even among eggs laid by the same female. Eggs may remain viable for two or more years. Postembryonic development takes several months, and there are on average, six molts in females, one or two less in males.

Phylogeny and Classification

Phasmids for long remained almost entirely unknown as fossils and elucidation of their relationships with the other orthopteroid groups was based on comparative morphology. Sharov (1968), however, brings together much fossil evidence to suggest that their affinities are closest to the Orthoptera (suborder Caelifera). In contrast, Kamp's (1973) phenetic analysis indicates that the Phasmida are most closely related to the Dermaptera and Grylloblattodea, and that perhaps the three form a natural group. Günther (1953) divides the order into two families, PHYLLIIDAE (leaf insects and related sticklike forms) and PHASMATIDAE (the remaining stick insects), though the two groups show much parallelism and the distinction between them is not usually marked.

Literature

Phasmid biology is reviewed by Chopard (1938, 1949), Clark (1974), and Bedford (1978). The taxonomy and geographical distribution of the order is discussed by Günther (1953). Helfer (1963) provides a key to the North American forms.

Bedford, G. E., 1978, Biology and ecology of the Phasmatodea, Annu. Rev. Entomol. 23:125–149.

Chopard, L., 1938, La Biologie des Orthoptères, Lechevalier, Paris.

Chopard, L., 1949, Ordre des Chéleutoptères, in: Traité de Zoologie (P.-P. Grassé, ed.), Vol. IX, Masson, Paris.

Clark, J. T., 1974, Stick and Leaf Insects, Sherlock, Winchester, Hants.

Günther, K., 1953, Über die taxonomische Gliederung und die geographische Verbreitung der Insektenordnung der Phasmatodea, Beitr. Entomol. 3:541–563.

Helfer, J. R., 1963, How to Know the Grasshoppers, Cockroaches, and Their Allies, Brown, Dubuque, Iowa.

Kamp, J. W., 1973, Numerical classification of the orthopteroids, with special reference to the Grylloblattodea, Can. Entomol. 105:1235–1249.

Sharov, A. G., 1968, The phylogeny of the Orthopteroidea, Tr. Paleontol. Inst. Akad. Nauk SSSR 118:1–213 (Engl. transl., 1971).

9. Orthoptera

SYNONYMS: Saltatoria, Saltatoptera, Orthopteroida, etc.

COMMON NAMES grasshoppers, locusts, katydids, crickets

Medium-sized to large, winged, brachypterous or apterous insects; head with mandibulate mouthparts; well-developed compound eyes, either long or relatively short antennae; prothorax large, hindlegs in almost all species enlarged for jumping, coxae small and widely spaced, tarsi usually 3- or 4-segmented, when present fore wings usually forming thickened tegmina; females usually with well-developed exposed ovipositor, males with concealed copulatory structures, cerci short to moderately long and unsegmented; auditory and stridulatory organs very often present.

The Orthoptera constitute a large order of insects with more than 20,000 species having a worldwide, though mainly tropical, distribution.

Structure

Taken as a group the Orthoptera, in keeping with their varied biology, exhibit a wide range of anatomical and morphological features. The head is usually hypognathous, but may be prognathous in some burrowing species and tree crickets. The compound eyes are typically large, and the antennae vary from comparatively short (suborder Caelifera) to very long (suborder Ensifera). Eyes and antennae are reduced in size in burrowing or cave-dwelling forms. The mouthparts are mandibulate but show some modifications according to the diet of the insect and occasionally to other considerations. The pronotum is large and extends lateroventrally to cover the pleural region. The mesothoracic and metathoracic nota are closely associated, though the basic components can still be readily distinguished (see Chapter 3). The legs are unequally developed. The forelegs of primarily digging forms are short but much enlarged. In some predaceous species the fore femora and/or tibiae are equipped with rows of long spines. In almost all Ensifera the fore tibiae bear auditory organs (see Chapter 12). The hind femora of most species are greatly enlarged to accommodate the muscles used in jumping, but they may also be modified for production of sound. Typically, the two pairs of wings are well developed, but varying degrees of reduction of the fore and hind wings, or the latter alone, may occur, especially in females. In several groups, the wings may be modified to resemble leaves, grass blades, or stems, so as to camouflage the insect. Whether fully developed or not, the fore wings are thickened and known as tegmina. The hind wings, unless reduced, are broad, due to the development of a large anal area. The wing venation, particularly of the tegmina, is variable and frequently modified in conjunction with stridulation. Paired auditory (tympanal) organs (see Chapter 12), if present, occur in both sexes and are found either on the first abdominal segment (in Acridoidea) or on the fore tibiae (in Tettigonioidea and Grylloidea). Eleven abdominal segments are distinguishable, though the sternum of the first is closely associated with the metathorax, and the most posterior tagmata are modified in conjunction with reproduction. In females a well-developed ovipositor is usually found, except in primarily burrowing forms. It is made up of three pairs of valves, though the inner pair may be reduced (see Chapter 3, Section 5.2.1). In Ensifera the ovipositor is long and used to place eggs in crevices, soft ground, or plant tissues. In Caelifera the ovipositor valves are short and stout in accordance with the digging function that they perform. In most Orthoptera the "external" copulatory structures of males are enclosed within a pouch formed by the greatly enlarged ninth abdominal sternum. The cerci are unsegmented and, in most species, short and rigid. In many Ensifera (especially Grylloidea) they are rather long and flexible, and in males of many species (especially Tettigonioidea) they are modified to form clasping structures used during copulation.

The internal structure of Orthoptera is usually rather generalized. In the gut the crop is large and proventriculus fairly to very well developed. The anterior midgut possesses mesenteric ceca, two in Ensifera, usually six in Caelifera. There are many Malpighian tubules which enter the gut directly or via a common duct. In the central nervous system there are three thoracic and six or seven abdominal ganglia. The tracheal system, which communicates with the exterior by means of two thoracic and eight abdominal pairs of spiracles, is often modified, particularly in migratory forms, through the development of large, segmentally arranged air sacs. These presumably serve to increase the volume of air exchanged during breathing movements (see Chapter 17). The paired testes of males are typically united in a middorsal position. A variety of accessory glands are found in the reproductive system. A pair of lateral ovaries, comprising a number of panoistic ovarioles, is found in females.

Life History and Habits

Most species are active, diurnal (most Caelifera), or nocturnal (many Ensifera) insects, capable of jumping in order to escape from would-be predators, or to launch themselves into flight. Flying ability is mostly rather limited, but a number of species (mostly members of the Acrididae) are capable of sustained flight for many hours. A few groups are typically cryptozoic, with their members living in humus or beneath logs and stones. A few species are true soil dwellers and some live in caves. Although normally "solitary" insects, a number of species (especially of Caelifera) may become gregarious under certain conditions to form swarms, which may reach enormous proportions (see Superfamily Acridoidea). Most Orthoptera are phytophagous and consequently are extremely important because of the damage they may do to crops. Many species of Ensifera are omnivorous and a few are carnivorous, feeding on other insects, which in some instances they catch with their specialized forelegs.

Stridulation is practiced by a large number of Orthoptera and serves to bring sexes together and to elicit certain behavioral responses leading to copulation. Stridulation is achieved in a great variety of ways, but, in most species, the principal method of sound production is either tegminal (crickets and katydids) or femoroalary (in many grasshoppers). In the former, the sound is typically produced when one tegmen is drawn across the other. The especially strengthened, toothed vein ("file") on one tegmen makes contact with the edge of the other tegmen, and an adjacent area of wing, the "mirror," is caused to resonate. In femoroalary stridulation it is normally the drawing of the hind femora across the tegmina that creates the sound. Another common method of stridulation, particularly among more primitive forms, is femoroabdominal, whereby the inside of the femur is rubbed over teeth or ridges on the side of the abdomen. Some grasshoppers make a clattering sound in flight either by striking the hind wing on the tegmen or by the rapid opening and closing of the fanlike hind wings. And there are numerous other less common means of sound production (see Kevan, 1955). It is generally the male that stridulates, though in some species the female also may perform this act. Stridulation is almost always species-specific, that is, each species has its own "song," and this probably serves as an isolating mechanism between closely related species. Exceptions to this rule may sometimes occur, however, if there are other isolating mechanisms. For example, the songs of *Gryllus pennsylvanicus* and *G. veletis* are almost identical, but the former occurs as adults only in late summer and autumn and the latter in late spring and early summer.

Reproduction is almost always sexual, though a few facultatively parthenogenetic species are known. Eggs are laid singly (in many Ensifera) or in batches (in Caelifera) in a variety of locations. The Ensifera typically lay them in or on stems or leaves, or in loose soil or humus. The Caelifera mostly dig into the soil, making alternate opening and closing movements of the ovipositor, and typically deposit the eggs in a mass of frothy material. This soon hardens to form a cylindrical mass known as the egg pod. Usually several batches of eggs are laid. The duration of embryonic development is variable, and many temperate species pass the winter in a diapausing egg stage, though some overwinter as juveniles. The pronymph (first-instar larva) is covered with a loose cuticle which serves to protect the insect as it emerges from the substrate. This cuticle is shed usually within a few minutes of hatching. In most species there are four to six additional larval instars in which the young form increasingly resembles the adult.

Phylogeny and Classification

The evolutionary relationships of the families of Orthoptera are indicated in Figure 7.15. Paleontology and comparative morphology make it clear that there was a very early separation, in the Upper Carboniferous period, of the two

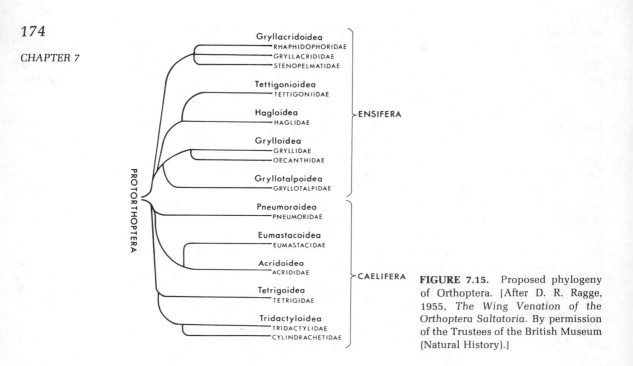

FIGURE 7.15. Proposed phylogeny of Orthoptera. [After D. R. Ragge, 1955, *The Wing Venation of the Orthoptera Saltatoria*. By permission of the Trustees of the British Museum (Natural History).]

major evolutionary lines, one of which led to the crickets, long-horned grasshoppers, etc., and the other to the short-horned grasshoppers and their allies. Indeed, Ragge (1955) suggested that the two groups probably evolved independently from different protorthopteran ancestors. This is in accord with the findings of Sharov (1968) and also with Hennig (1969). Thus, some authorities, for example, Vickery *et al.* (1974) and Kevan (1976), accord each group ordinal status, the crickets, long-horned grasshoppers, etc. being placed in the order Grylloptera, with the short-horned grasshoppers and relatives in the order Orthoptera *sensu stricto*. Other workers do not consider the differences between each group to warrant their separation into two orders, and so classify them as the suborders Ensifera and Caelifera, respectively, within the order Orthoptera. The latter is the system used here.

Within the Ensifera, the central family of those extant appears to be the Haglidae (=Prophalangopsidae), an almost entirely fossil group containing only three living species, which Chopard (1949) places in its own superfamily, Hagloidea. In agreement with Zeuner (1939), Ragge (1955) considers that it was from early members of this group that the remaining Ensifera arose.

Judging by the wing venation and structure of the male external genitalia, the Pneumoridae, a small South and East African family comprising approximately 20 species, appear to be among the most primitive groups of living Caelifera. However, the modern representatives of this family differ in so many morphological features from other grasshoppers that they must have left the main evolutionary path at an early date (Dirsh, 1961). They are included, along with a few peculiar American species of quite different appearance, in a separate superfamily, Pneumoroidea. The family Tetrigidae has been placed by many authors in the Acridoidea, but its members differ from acridoids in

several ways (summarized in Dirsh), and it appears best to segregate them in
their own superfamily, Terrigoidea. In yet another superfamily, Tridactyloidea,
are placed the Tridactylidae (pygmy mole crickets) and Cylindrachetidae (false
mole crickets). These two families were formerly classified in the Grylloidea
(using this name in its old, wide, sense) because of their superficial similarities
in structure and habits with the Gryllotalpidae. It is now realized that these
similarities are the result of parallel evolution, and the true affinities of the two
families lie with the Caelifera rather than the Ensifera.

Suborder Ensifera

Members of the suborder Ensifera are characterized by their multisegmented antennae, which are as long or longer than the body. Tympanal organs,
when present, are located on the fore tibiae. The principal superfamilies are the
Gryllacridoidea, Tettigonioidea, and Grylloidea.

Superfamily Gryllacridoidea

The superfamily Gryllacridoidea is a primitive superfamily that contains
more than 1000 species. Its members are usually somewhat cricketlike in
appearance, with longish cerci. However, they may be at once distinguished
from true crickets because males lack a stridulatory mirror on the tegmen, and females have a laterally compressed, not needlelike, ovipositor. They may be
winged, brachypterous, or apterous and may or may not have tympana on the fore
tibiae. They are secretive, nocturnal creatures. The group is mainly restricted to
tropical or subtropical regions or to temperate parts of the Southern Hemisphere. Included in the superfamily are (1) cave or camel crickets (family
RHAPHIDOPHORIDAE), often darkly colored, humpbacked insects that live in
caves, hollow logs, under stones, and in other humid situations; the genus
Ceuthophilus (Figure 7.16A) is well known in the basements of North American houses, and *Tachycines asynamorus* is an oriental species commonly
found in greenhouses; (2) Jerusalem crickets, stone crickets, true wetas, and

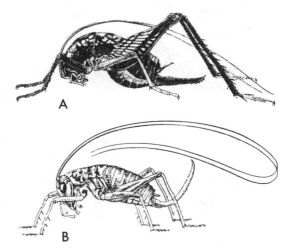

FIGURE 7.16. Gryllacridoidea (A) A
camel cricket, *Ceuthophilus macula-
tus*; and (B) a leaf roller, *Camptonotus
carolinensis*, [A, from M. Hebard, 1934,
The Dermaptera and Orthoptera of
Ilinois, *Bull. Ill. Nat. Hist. Surv.*
20(3). By permission of the Illinois
Natural History Survey. B, from
W. S. Blatchley, 1920, *Orthoptera of
Northeastern America with Especial
Reference to the Faunas of Indiana and
Florida*, The Nature Publishing Co., In-
dianapolis, Ind.]

king crickets (family STENOPELMATIDAE), large, omnivorous or carnivorous forms, found under stones, in rotting logs, etc.; some are subterranean and, as such, may have vestigial eyes and antennae, and enlarged forelegs; and (3) the leaf-rolling "grasshoppers" (family GRYLLACRIDIDAE) (Figure 7.16B), so-called because many build shelters by rolling up a leaf and tying it with a silklike secretion from glands that open near the mouth. This is the largest family, with about 600 species. Its members are mainly predaceous.

Superfamily Tettigonioidea

The superfamily Tettigonioidea, which includes the common long-horned grasshoppers and katydids (bush crickets) (Figure 7.17), is the largest ensiferan superfamily, with more than 5000 described species. The superfamily is primarily tropical, though very many species occur in temperate regions of the world. They generally live among herbage or in trees and are commonly green. Males almost always possess a stridulatory apparatus, and in many species females also can stridulate. The species that live above the ground are usually fully winged, and their wings often resemble leaves. In ground-dwelling forms, which are less numerous, the wings are often reduced so that only the sound-producing area remains. The ovipositor is a laterally flattened, bladelike structure, either curved or straight, that is sometimes as long as the body itself or even longer. Tettigonioids in general are phytophagous or omnivorous, though a few are apparently entirely carnivorous. Occasionally its population density is sufficiently high that a species becomes economically important; for example, the Mormon cricket, *Anabrus simplex*, is frequently a serious pest of field crops and may reach "plague" proportions in western and midwestern North America. At least one African species of the genus *Ruspolia* (=*Homorocor*-

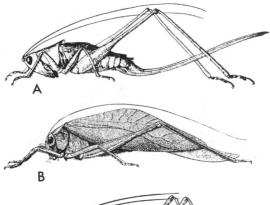

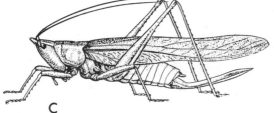

FIGURE 7.17. Tettigonioidea. (A) The meadow katydid, *Conocephalus strictus;* (B) a bush katydid, *Microcentrum rhombifolium;* and (C) a long-horned grasshopper, *Neoconocephalus palustris.* [A, B, from M. Hebard, 1934, The Dermaptera and Orthoptera of Illinois, *Bull. Ill. Nat. Hist. Surv.* **20**(3). By permission of the Illinois Natural History Survey. C, from W. S. Blatchley, 1920, *Orthoptera of Northeastern America, with Especial Reference to the Faunas of Indiana and Florida,* The Nature Publishing Co., Indianapolis, Ind.]

yphus) may form migratory flying swarms. It is widely collected as human food.

Superfamily Grylloidea

The Grylloidea (true crickets) form another very large superfamily, containing well in excess of 2000 described species. They bear a slight resemblance to the Tettigonioidea but differ from the latter in having only three tarsal segments and, in females, an ovipositor that is generally needlelike or cylindrical, usually straight or only slightly curved and comprising only two pairs of valves. They have rather long, flexible cerci, and a few species are green. Most species are typically nocturnal insects that hide in humid microhabitats during the day. Some species are commonly found adjacent to ponds and streams, and a number are cave dwellers. Like the tettigonioids, the grylloids exhibit the full range of wing development. They are generally omnivorous and occasionally of economic importance through their feeding on crops (field crickets, family GRYLLIDAE) or the damage they cause by ovipositing in twigs (tree crickets, family OECANTHIDAE). A few species may be common in buildings, for example, the house cricket (*Acheta domesticus*) (Figure 7.18) and tropical house cricket (*Gryllodes sigillatus*) (Gryllidae).

Superfamily Gryllotalpoidea

The Gryllotalpoidea (mole crickets) (Figure 7.19) form a very distinct group, comprising about 50 species. As their common name indicates, they are subterranean in habit, having greatly enlarged forelegs, reduced eyes, and vestigial ovipositor. Most species are fully winged, though brachypterous and apterous forms are known. Mole crickets are omnivorous and often cause significant damage to field crops, other cultivated plants, and turf, particularly in warmer climates.

Suborder Caelifera

In members of the suborder Caelifera, the antennae are comparatively short with, in most species, less than 30 segments. Tympanal organs, when present, are on the first abdominal segment. The principal superfamilies are the Eumastacoidea, Acridoidea, Tetrigoidea, and Tridactyloidea.

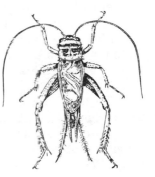

FIGURE 7.18. The house cricket, *Acheta domesticus* (Gryllidae). [From A. D. Imms, 1957, *A General Textbook of Entomology*, 9th ed. (revised by O. W. Richards and R. G. Davies). By permission of Chapman and Hall Ltd.]

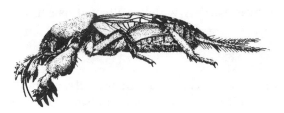

FIGURE 7.19. A mole cricket, *Gryllotalpa hexadactyla* (Gryllotalpidae). [From M. Hebard, 1934, The Dermaptera and Orthoptera of Illinois, *Bull. Ill. Nat. Hist. Surv.* **20**(3). By permission of the Illinois Natural History Survey.]

Superfamily Eumastacoidea

The curious grasshopperlike insects comprising the Eumastacoidea are sometimes known as monkey grasshoppers because of the fancifully simian appearance of their faces. They almost all possess extremely short antennae and rest with their legs splayed. Some are winged, some not. The superfamily is quite extensive but mainly tropical. Eumastacoids are diurnal and phytophagous, mostly living in bushes and trees. Few are of economic importance.

Superfamily Acridoidea

The Acridoidea (short-horned grasshoppers and locusts) are the largest orthopteran superfamily, with about 12,000 described species, most of them in the family ACRIDIDAE (Figure 7.20). Although most species are found in warmer areas of the world, a large number occur in temperate climates, where usually they overwinter in the egg stage. Most acridoids inhabit grassland and other low vegetation, but a number live in trees. They are predominantly diurnally active insects. They are virtually exclusively phytophagous, eating mainly living plants. Most species are winged and some are strong fliers. There are several important subfamilies of Acrididae. The CYRTACANTHACRIDINAE are a subfamily of rather large forms confined mainly to tropical and subtropical regions of the world. The subfamily contains a number of economically very important species, including several locusts, for example, *Schistocerca gregaria gregaria* (the desert locust of Africa and southwest Asia), *S. americana* (the South American locust, comprising at least two major subspecies), and *Nomadacris septemfasciata* (the red locust). The OEDIPODINAE are another subfamily of worldwide distribution, but whose members are found mainly in warmer, drier areas. It contains many important pest species, including *Locusta migratoria* (the migratory locust of Africa, northern Australia, Asia, and southern Europe, with several distinct subspecies) and *Chortoicetes terminifera* (the Australian plague grasshopper). The CATANTOPINAE constitute another large, widely distributed subfamily. As presently constituted, it is a rather heterogeneous assemblage and is in gradual process of revision. Usually included in this subfamily are the genera *Melanoplus* and *Dichropus*, several species of which are major pests, particularly in the grassland areas of North and South America, respectively.

The term "locust" is applied to about 20 species of Acrididae that are capable, under certain conditions, of aggregating in immense swarms that may migrate for considerable distances and cause massive damage to vegetation. It is now established that locusts (and, for that matter, some Tettigonioidea, Phasmida, and Lepidoptera) can exist in more than one form or "phase." These

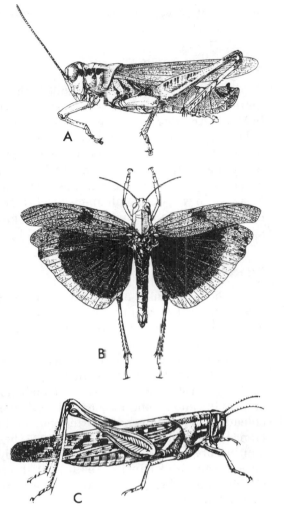

FIGURE 7.20. Acrididae. (A) The differential grasshopper, *Melanoplus differentialis;* (B) the Carolina grasshopper, *Dissosteira carolina;* and (C) the American locust, *Schistocerca americana.* [A, B, from L. Chopard, 1938, *Encyclopédie Entomologique. XX. La Biologie des Orthoptères.* By permission of Lechevalier, Paris. C, from M. Hebard, 1934, The Dermaptera and Orthoptera of Illinois, *Bull. Ill. Nat. Hist. Surv.* **20**(3). By permission of the Illinois Natural History Survey.]

phases differ not only in color and morphology, but in their physiology, ecology, and behavior. Phase polymorphism is largely a density-dependent phenomenon. Under extended conditions of low population density, the locusts exist in the "solitary phase." If conditions change so that the population density increases, the locusts enter a "transition phase," and under high density conditions change to the "gregarious phase." It is the latter phase that is of importance with reference to swarming. Gregarious larvae (hoppers) are highly active and "march" in vast groups from place to place. As adults they may be carried for very considerable distances by wind currents (see Chapter 22, Section 5.1). New swarms in most species originate in more or less permanent breeding areas (outbreak areas), though desert locusts seem to be more strictly nomadic, even in the solitary phase. Though adults may breed in the regions to which they are distributed for a period of several years, the population gradually decreases because of unsuitable conditions. The effects of population density on phase development are mediated via the endocrine system (see Chapter 21).

Superfamily Tetrigoidea

The more than 1000 species of Orthoptera that make up the superfamily Tetrigoidea, comprising the single family TETRIGIDAE (Figure 7.21), are known in North America as grouse locusts. They are distinguished by an enormously enlarged pronotum that projects posteriorly to cover most of the abdomen, 2-segmented tarsi, and, in most species, short antennae with 12 or fewer segments. In many species the wings (especially the tegmina) are reduced. Auditory and stridulatory structures are absent. The grouse locusts are most common in warmer areas of the world, though many species are found in temperate regions, where they overwinter in the adult stage. They prefer rather moist habitats and some species are semiaquatic. They feed on algae and moss.

Literature

General accounts of the Orthoptera are provided by Chopard (1938, 1949). The periodical *Acrida* is devoted entirely to all aspects of Orthopteran biology. Shotwell (1941) and Uvarov (1966, 1977) deal specifically with Acridoidea. The *Bulletins* and *Memoirs* of the Centre for Overseas Pest Research (formerly the Anti-Locust Research Centre) deal with aspects of the biology and control of locusts. The evolutionary relationships of the winged Orthoptera are discussed by Ragge (1955), who bases his conclusions on a study of wing venation, amplifying and extending the work of Zeuner (1939) on Ensifera. Sharov (1968) considers the phylogeny of Orthoptera and related orders, especially with reference to fossil forms. Dirsh (1961, 1975) has discussed the relationships of the families of Acridoidea and also provided keys for their separation. Other useful keys are those of Hincks (1949) [British species]; Froeschner (1954), Brooks (1958), and Helfer (1963) [North American forms]; and Vickery *et al.* (1974) [Eastern Canada].

Brooks, A. R., 1958, Acridoidea of southern Alberta, Saskatchewan and Manitoba (Orthoptera), *Can. Entomol., Suppl.* **9**:92 pp.

Chopard, L., 1938, *La Biologie des Orthoptères*, Lechevalier, Paris.

Chopard, L., 1949, Ordre des Orthoptères, in: *Traité de Zoologie* (P.-P. Grassé, ed.), Vol. IX, Masson, Paris.

Dirsh, V. M., 1961, A preliminary revision of the families and subfamilies of Acridoidea (Orthoptera, Insecta), *Bull. Br. Mus. Nat. Hist. Entomol.* **10**:351–419.

Dirsh, V. M., 1975, *Classification of the Acridomorphoid Insects*, Classey, Faringdon, Oxon.

Froeschner, R. C., 1954, The grasshoppers and other Orthoptera of Iowa, *Iowa State Coll. J. Sci.* **29**:163–354.

Helfer, J. R., 1963, *How to Know the Grasshoppers, Cockroaches and Their Allies*, Brown, Dubuque, Iowa.

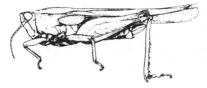

FIGURE 7.21. A grouse locust, *Tetrix subulata* (Tetrigidae). [From J. A. G. Rehn and H. J., Grant, Jr., 1961, *A Monograph of the Orthoptera of North America (North of Mexico)*, Vol. I. Monographs of the Academy of Natural Sciences of Philadelphia. By permission of the Academy of Natural Sciences of Philadelphia.]

Hennig, W., 1969, *Die Stamesgeschichte der Insecten*, Kramer, Frankfurt-am-Main.

Hincks, W. D., 1949, Dermaptera and Orthoptera, *R. Entomol. Soc. Handb. Ident. Br. Insects* **1**(5):20 pp.

Kevan, D. K. McE., 1955, Méthodes inhabituelles de production de son chez les Orthoptères, *Ann. Epiphyt. Fasc. Hors. Série* **1954**:103–142.

Kevan, D. K. McE., 1976, Suprafamilial classification of orthopteroid and related insects, applying the principles of symbolic logic—A draft scheme for discussion and consideration, XV International Congress of Entomology, Washington, D.C., 1976 (unpublished manuscript).

Ragge, D. R., 1955, *The Wing-Venation of the Orthoptera Saltatoria*, British Museum of Natural History, London.

Sharov, A. G., 1968, The phylogeny of the Orthopteroidea. *Tr. Paleontol. Inst. Akad. Nauk SSSR* **118**:1–213 (Engl. transl., 1971).

Shotwell, R. L., 1941, Life-histories and habits of some grasshoppers of economic importance on the Great Plains, *U.S. Dep. Agric., Tech. Bull.* **774**:47 pp.

Uvarov, B. P., 1966, 1977, *Grasshoppers and Locusts. A Handbook of General Acridology*, Vols. 1 and 2, Cambridge University Press, Cambridge.

Vickery, V. R., Johnstone, D. E., and Kevan, D. K. McE., 1974, The orthopteroid insects of Quebec and the Atlantic provinces of Canada, *Mem. Lyman Entomol. Mus. Res. Lab.* **1**:204 pp.

Zeuner, F. E., 1939, *Fossil Orthoptera Ensifera*, 2 vols., British Museum of Natural History, London.

10. Zoraptera

SYNONYMS: none   COMMON NAMES: none

Minute gregarious insects; head with 9-segmented moniliform antennae and biting mouthparts; commonly apterous but occasionally winged, with greatly reduced venation, tarsi 2-segmented; cerci short and unsegmented, ovipositor absent, male genitalia specialized and frequently asymmetrical.

The Zoraptera constitute a very small order of insects, containing about 20 species in a single genus *Zorotypus* and family. They are found in all geographical regions except continental Australia.

Structure

Zoraptera are 3 mm or less in length and usually pale and wingless (Figure 7.22). In most species a darker, winged form has also been found. The hypognathous head bears a pair of moniliform antennae and biting mouthparts.

FIGURE 7.22. *Zorotypus guineensis* (Zoraptera). [From A. D. Imms, 1957, *A General Textbook of Entomology*, 9th ed. (revised by O. W. Richards and R. G. Davies). By permission of Chapman and Hall Ltd.]

Compound eyes and three ocelli are found only in the winged form. A prominent Y-shaped epicranial suture can be seen on the head. The prothorax is prominent. The mesonotum and metanotum are simple in apterous individuals. When present, the wings are membranous and have a much-reduced venation. They are frequently shed at the base, though there is no basal suture as occurs in termites. The legs have a large coxa and a 2-segmented tarsus. The 11-segmented abdomen carries a pair of unsegmented cerci. There is no ovipositor, and, in males, the external genitalia are frequently asymmetrical. The homologies of the genitalia are uncertain.

The gut contains a very large crop, a short midregion, and a convoluted hind part. Six Malpighian tubules occur. The nervous system is specialized with only two abdominal ganglia. The reproductive organs are typically orthopteroid.

Life History and Habits

The life history and habits of Zoraptera are poorly understood. Species are found in rotting logs, sawdust piles, humus, etc., where they appear to feed on fungal spores and minute arthropods. Although they are gregarious and dimorphic, there appears to be no social organization. Both apterous and alate forms are sexually functional. The young are of two forms, one with and the other without wing buds. The number of juvenile instars is not known.

Phylogeny and Classification

The relationship of the Zoraptera with other insect orders is unclear. The general form of the head, mandibulate mouthparts, structure of the thorax, presence of cerci, and nature of the male genitalia are orthopteroid, especially blattoid. The reduced wing venation, small number of Malpighian tubules, and concentrated abdominal ganglia are, however, hemipteroid. It may be speculated that they left the orthopteroid branch at a point close to that at which the hemipteroid line originated.

Literature

The biology of this little known order is summarized by Gurney (1938, 1974) and Riegel (1963).

Gurney, A. B., 1938, A synopsis of the order Zoraptera, with notes on the biology of *Zorotypus hubbardi* Caudell, *Proc. Entomol. Soc. Wash.* **40:**57–87.
Gurney, A. B., 1974, Class Insecta. Order Zoraptera, *Dep. Agric. Tech. Serv. Rep. South Afr.* **38:**32–34.
Riegel, G. T., 1963, Distribution of *Zorotypus hubbardi* (Zoraptera), *Ann. Entomol. Soc. Am.* **56:**744–747.

8

The Hemipteroid Orders

1. Introduction

The four orders (Psocoptera, Phthiraptera, Hemiptera, and Thysanoptera) that constitute the hemipteroid group are united by the following features: specialized, usually suctorial, mouthparts; small anal lobe in hind wing; wing venation reduced; cerci absent; few Malpighian tubules; and ventral nerve cord with few discrete ganglia. On the whole, the hemipteroid group is more homogeneous than the orthopteroid group, although two evolutionary lines have developed, leading to the Psocoptera–Phthiraptera, on the one hand, and the Hemiptera–Thysanoptera, on the other.

2. Psocoptera

SYNONYMS: Corrodentia, Copeognatha   COMMON NAMES barklice, booklice, psocids

Small or minute soft-bodied insects head with long filiform antennae and specialized chewing mouthparts, compound eyes usually prominent but reduced in some species; prothorax small, wings present or absent, legs with 2- or 3-segmented tarsi; external genitalia of both sexes concealed, cerci absent.

This order, containing about 1700 described species, has a worldwide distribution. Some 150 species occur in North America, and about 80 species are found in Great Britain.

Structure

Psocoptera are stocky, soft-bodied insects whose length is usually less than 10 mm. The large, mobile head bears a swollen postclypeus, long filiform antennae, and, usually, prominent compound eyes, though the latter are reduced in some wingless species. Three ocelli are present in winged forms but absent in apterous species. The Y-shaped epicranial suture is prominent. The mouthparts, though retaining a chewing function, are specialized. The mandibles are dissimilar, though each has both grinding and cutting edges. In the maxillae the cardo and stipes are not always distinct. The galea is a large, fleshy lobe, whereas the lacinia is a narrow, sclerotized rod (the "pick"), which may be used to scrape food from the substrate. The hypopharynx is well developed and has a characteristic structure. The lingua bears a pair of ventral sclerites which are connected to the median sitophore sclerite by five ligaments. The sitophore sclerite is situated on the ventral surface of the base of the cibarium. Opposite to it, on the dorsal surface of the cibarium wall, is a knoblike process that is believed to move against the sclerite in the manner of a mortar and pestle and facilitate the grinding up of food.

In winged forms the small prothorax is largely concealed by the pterothorax. The wings are membranous and have a prominent but reduced venation. The anterior pair are larger than the hind pair. At rest they are held rooflike over the body. The fore and hind wings are coupled both during flight and at rest. Varying degrees of brachyptery occur, even within the same species, and aptery is common, especially in females. In some species the legs carry what is believed to be a stridulatory organ. The abdomen is 9-segmented and terminates in a dorsal epiproct and a pair of lateral paraprocts. The external genitalia of males are weakly developed, and their homologies are uncertain. A small ovipositor is usually present in females, though it may be considerably reduced or absent in some forms.

Four Malpighian tubules originate at the posterior end of the midgut, which is long and convoluted. The nervous system is highly modified and comprises only five ganglionic centers: brain, subesophageal ganglion, prothoracic ganglion, a composite pterothoracic ganglion, and a composite abdominal ganglion. Each ovary contains three to five polytrophic ovarioles. The lateral oviducts are short and open into a larger median duct. A spermatheca is present. The testes are roundish or three-lobed. The vasa deferentia lead into large seminal vesicles which appear to produce the material of the spermatophore.

Life History and Habits

Most Psocoptera are found on vegetation or under bark, though some live among leaf litter or under stones. A few species are associated with man and may be encountered in houses or buildings in which food materials are stored. Though they may occur in vast numbers, they are seldom of economic importance. They are primarily phytophagous, feeding on algae, lichens, fungi, pollen, and decaying fragments of higher plants; occasionally they eat dead animal matter. Species associated with man live on cereal products or, in the case of the common booklice, molds that develop on old books. Many species are gregarious, with individuals of all ages living together, often beneath a silken web produced from secretions of the modified labial glands. Most outdoor species are fully winged, though brachyptery and aptery are common under certain environmental conditions.

Males are unknown in some species, while in others facultative parthenogenesis may occur. In dioecious species a male typically "courts" a female prior to mating. Eggs, between 20 and 100 at a time, are laid singly or in groups. A few species are viviparous. Larvae usually pass through six instars prior to metamorphosis, but this figure is variable in polymorphic species.

Modern Psocoptera are but the remnants of an order that had already undergone an extensive evolution by the end of the Permian period. Of the four hemipteroid orders, the Psocoptera are generally considered to be closest to the ancestral stock. Living Psocoptera are divisible into three well-defined suborders: Trogiomorpha, Troctomorpha, and Psocomorpha (Eupsocida). The Trogiomorpha contains the most primitive and the Psocomorpha the most advanced Psocoptera. Unfortunately, insufficient systematic work has been done to permit firm conclusions to be reached with respect to the relationships of the many families or, in many cases, to which family certain genera belong. This is especially true for the Psocomorpha, to which at least 75% of the recent species belong.

Suborder Trogiomorpha

Distinguishing characters of Trogiomorpha include antennae with more than 20 segments, never secondarily annulated; tarsi 3-segmented; labial palps 2-segmented; pterostigma not thickened, or absent; and paraprocts with strong posterior spine.

Only members of the cosmopolitan family TROGIIDAE are common in North America. *Trogium pulsatorium* (Figure 8.1A) is a common booklouse which feeds on paper, vegetable matter and cereal products. Another common species is *Lepinotus inquilinus*, which is found especially in granaries and warehouses.

Suborder Troctomorpha

Features of Troctomorpha are 12- to 17-segmented antennae, with some flagellar segments secondarily annulated; 2- or 3-segmented tarsi; 2-segmented

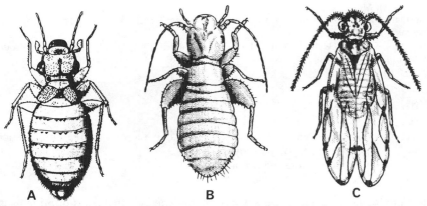

FIGURE 8.1. Psocoptera. (A) *Trogium pulsatorium* (Trogiidae) (distal antennal segments omitted), (B) *Liposcelis divinatorius* (Liposcelidae), and (C) *Ectopsocus californicus* (Peripsocidae). [A, from P.-P. Grassé (ed.), 1951, *Traité de Zoologie*, Vol. X. By permission of Masson, Paris. B, C, from L. A. Swan and C. S. Papp 1972, *The Common Insects of North America.* Copyright 1972 by L. A. Swan and C. S. Papp. Reprinted by permission of Harper & Row, Publishers, Inc.]

labial palps; pterostigma not thickened; and paraprocts without a strong posterior spine.

The cosmopolitan family LIPOSCELIDAE, whose members are recognized by their greatly enlarged hind femora, includes a number of common North American species. *Liposcelis diviniatorius* (Figure 8.1B) is perhaps the most common booklouse. It is frequently found associated with *T. pulsatorium*.

Suborder Psocomorpha

Features of Psocomorpha include antennae almost always 13-segmented; tarsi 2- or 3-segmented, if latter, then flagellar segments of antennae not secondarily annulated; labial palps unsegmented; pterostigma not thickened; and paraprocts without a strong posterior spine.

Judging by their common features, this suborder is probably derived from or had a common ancestry with the Troctomorpha. Most members of the suborder are found outdoors, on growing vegetation, in leaf litter, or on bark. A few species do occur in buildings of various kinds. The 10-spotted psocid, *Ectopsocus* (=*Peripsocus*) *californicus* (PERIPSOCIDAE) (Figure 8.1C), is a common booklouse. Its relative *E. pumilis* is a widespread indoor species, found especially in warehouses and granaries. *Lachesilla pedicularia* (LACHESILLIDAE), found in both Europe and North America, is frequently encountered in granaries and the like where vegetable matter is being stored.

Literature

General introductions to the morphology and biology of Psocoptera are given by Badonnel (1951) and Imms (1957). More specific information is presented by Sommerman (1943, 1944).

Badonnel, A., 1951, Ordre des Psocoptères, in: *Traité de Zoologie* (P.-P. Grassé, ed.), Vol. X, Masson, Paris.
Imms, A. D., 1957, *A General Textbook of Entomology,* 9th ed. (revised by O. W. Richards and R. G. Davies), Methuen, London.
Sommerman, K. M., 1943, Bionomics of *Ectopsocus pumilis* (Banks) (Corrodentia, Caeciliidae), *Psyche* **50**:53–63.
Sommerman, K. M., 1944, Bionomics of *Amapsocus amabilis* (Walsh) (Corrodentia, Psocidae), *Ann. Entomol. Soc. Am.* **37**:359–364.

3. Phthiraptera

SYNONYMS: Pseudorhynchota, Mallophaga (in part), Lipoptera (in part), Anopula (in part), Siphunculata (in part)

COMMON NAMES: lice, sucking lice (in part), chewing lice or bird lice (in part)

Minute to small, apterous, dorsoventrally flattened ectoparasites of birds or mammals; head prognathous or hypognathous, compound eyes reduced or absent, ocelli absent, antennae 3- to 5-segmented, mouthparts of chewing or piercing–sucking type with palps of maxillae and labium reduced or absent; prothorax free or fused with pterothorax, legs with unsegmented or 2-segmented tarsi, more or less modified for clinging to hair or feathers, one, two, or no tarsal claws; abdomen 7- to 10-segmented, cerci absent.

The order includes about 4000 species, distributed about evenly between the chewing lice (including the suborders Amblycera, Ischnocera, and Rhyncophthirina), most of which parasitize birds, and the sucking lice (suborder Anoplura), which are parasites of placental mammals. Many species are important pests, in their role as disease vectors, of domesticated animals and man.

Structure

The minute to small (0.35–20 mm long) body is dorsoventrally flattened and shows a variable degree of sclerotization. In chewing lice the head is relatively large and bears chewing mouthparts which show certain resemblances to those of Psocoptera. In sucking lice the relatively small head has partially retracted, suctorial mouthparts whose homologies have been difficult to interpret. The labrum forms a short eversible proboscis. Three stylets are contained within a pouch that runs ventrally off the cibarium. The ventral stylet represents the modified labium, the middle stylet probably is an extension of the opening from the salivary duct, and the dorsal stylet is either the modified maxillae or the hypopharynx. The mandibles disappear during embryogenesis. The head bears 3- to 5-segmented antennae which may be filiform (in Anoplura), capitate and in grooves (Amblycera), or modified for grasping (Ischnocera). Compound eyes are reduced or absent and ocelli are never present. In Anoplura the thoracic segments are fused, but in other lice the prothorax is distinct from the pterothorax. The legs include a 2-segmented or unsegmented tarsus and one or two tarsal claws. Eight to 10 visible abdominal segments occur.

In sucking lice the cibarium and pharynx form a strong sucking pump, but the crop and gizzard are poorly differentiated. In chewing lice the crop is large. In all Phthiraptera the midgut is large and has two mesenteric ceca. Four Malpighian tubules enter the midgut posteriorly. The nervous system is highly modified and includes a composite metathoracoabdominal or thoracoabdominal ganglion. The internal reproductive organs resemble those of Psocoptera.

Life History and Habits

Lice are host-specific, and, though several species may be found on the same host, they may occupy characteristic areas of the body. They are highly sensitive to changes in temperature and humidity and survive for only a few days should the host die.

Most species of chewing lice live among the feathers of birds where they feed on fragments of feathers and skin. A few species, for example, the chicken body louse, *Menacanthus stramineus*, feed on blood in addition to epidermal products and are able to pierce the skin or developing quills. The members of two small families of chewing lice are parasitic on mammals, though their general habits are the same as those found on birds.

The sucking lice feed exclusively on the blood of the host, always a placental mammal. It has been suggested that a possible reason for the high host specificity of sucking lice is the lethal effect an unsuitable host's blood might have on the symbiotic bacteria present in certain gut cells or in the mycetome, a structure closely associated with the gut.

Heavy infestations of lice may render a host more susceptible to disease and cause economic loss due to reduction in quality. In addition, some sucking lice are important disease vectors (see Suborder Anoplura).

In Anoplura, at least, mating occurs frequently, presumably due to the absence in females of a spermatheca. In many species of chewing lice, males are less common than females and facultative parthenogenesis may occur. Eggs are usually cemented to hairs or feathers by means of a secretion from the female's accessory gland. Postembryonic development is rapid, juveniles passing through three molts, and adults become sexually mature within a few days of the final molt. Transfer to new hosts is by physical contact and occurs during mating, communal roosting, and brooding and feeding the young. Some blood-sucking Diptera may act as intermediaries in the transfer process.

Phylogeny and Classification

Phthiraptera are evidently derived from free-living Psocoptera-like ancestors, though the order is not known as fossils, and the geological period during which the separation took place is uncertain. Attempts to determine evolutionary relationships in the order are complicated by the presence of both primitive and specialized characters in the group or species under consideration.

Traditionally, the chewing (bird) lice and sucking lice were arranged in separate orders, the Mallophaga and Anoplura, respectively, the great difference in the mouthparts being taken as sufficient justification for this separation. According to Clay (1970), however, all lice should be included in the same order, Phthiraptera, subdivisible into four suborders, Amblycera, Ischnocera, Rhyncophthirina, and Anoplura. (The first three comprised the old order Mallophaga.)

It is generally accepted that the suborder Amblycera contains the more primitive species, and it was from this group that the suborder Ischnocera evolved. The third suborder, Rhyncophthirina, containing the single genus *Haematomyzus*, is of particular interest because its living representatives probably resemble the ancestors of the sucking lice, suborder Anoplura (Hopkins, 1949). According to Hopkins, this suborder was probably derived from a trichodectoid ancestor. Evidence from host–parasite associations indicates that Amblycera existed in the Jurassic period, though Hopkins questions whether this is early enough for the evolution of a new suborder. The supposed trichodectoid ancestor arose in the late Jurassic or early Cretaceous, and Anoplura evolved during the middle Cretaceous.

Suborder *Amblycera*

Members of the suborder Amblycera have the following characteristics: antennae capitate, 4-segmented, lying in grooves; mandibles horizontal; maxillary palpi present; and mesothorax and metathorax usually separate.

Hopkins (1949) and Clay (1949) recognize six families in this suborder. Of these, three are restricted to avian hosts, two to marsupials, and one to placental mammals. The MENOPONIDAE are the largest family and have a cosmopolitan distribution. Its members infest birds, and several species are important pests of poultry, for example, *Menacanthus stramineus* (the chicken body louse) and *Menopon gallinae* (the shaft louse) (Figure 8.2A). The LAEMOBOTHRIIDAE are a small family containing the single genus *Laemobothrion*, species of which are parasites of water birds and hawks. The RICINIDAE, containing only two genera, are found on hummingbirds and several families of passerines. The Australian BOOPIIDAE and closely related South American TRIMENOPONIDAE are with few exceptions found on marsupials and histricomorph rodents, respectively. The remaining family, GYROPIDAE, is endemic to South and Central America, though two species, *Gyropus ovalis* and *Gliricola por-*

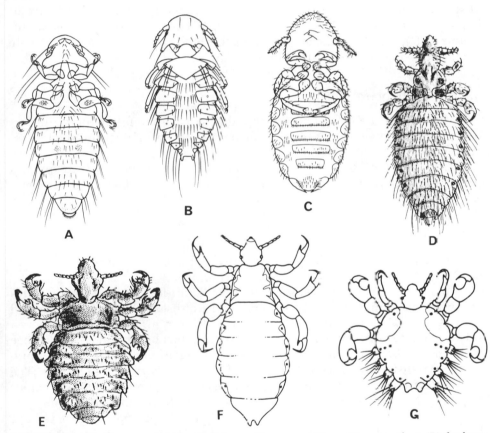

FIGURE 8.2. Phthiraptera. (A) The shaft louse, *Menopon gallinae* (Menoponidae); (B) the large turkey louse, *Chelopistes meleagridis* (Philopteridae); (C) the red cattle louse, *Bovicola bovis* (Trichodectidae); (D) the long-nosed cattle louse, *Linognathus vituli* (Linognathidae); (E) the short-nosed cattle louse, *Haematopinus eurysternus* (Haematopinidae); (F) the human body louse, *Pediculus humanus corporis* (Pediculidae); and (G) the crab louse, *Phthirus pubis* (Pediculidae). [A–C, F, G, from D. J. Borror, D. M. Delong, and C. A. Triplehorn, 1976, *An Introduction to the Study of Insects*, 4th ed. By permission of Holt, Rinehart and Winston, Inc. D, E. from L. A. Swan and C. S. Papp 1972, *The Common Insects of North America*. Copyright 1972 by L. A. Swan and C. S. Papp. Reprinted by permission of Harper & Row, Publishers, Inc.]

celli, which are found on guinea pigs, have been spread by commerce to other parts of the world. Most species of Gyropidae live on histrocomorph rodents.

Suborder Ischnocera

Characteristics of Ischnocera are antennae filiform, 3- to 5-segmented, exposed; mandibles vertical, maxillary palps absent; and mesothorax and metathorax usually fused.

Included in this suborder are two important families. The PHILOPTERIDAE are the largest family in the order and have a worldwide distribution. Its members are parasitic on birds and include a number of pest species found on poultry, for example, *Cuclutogaster heterographus* (the chicken head louse) and *Chelopistes meleagridis* (the large turkey louse) (Figure 8.2B). The family TRICHODECTIDAE, which is restricted to placental mammals, contains a number of species found on domesticated animals, for example, *Damalinia (Bovicola) bovis* (cattle) (Figure 8.2C). *D. ovis* (sheep), *D. equi* (horses), *Felicola subrostratus* (cats), and *Trichodectes canis* (dogs), which can serve as an intermediate host for the dog tapeworm, *Dipylidium caninum*.

Suborder Rhyncophthirina

Members of the suborder Rhyncophthirina have the following characteristics: head prolonged into a rostrum; mandibles at apex of rostrum; and labium and maxillae vestigial.

This suborder contains only two species, *Haematomyzus elephantis*, a parasite of both Indian and African elephants, and *H. hopkinsi*, which infests warthogs.

Suborder Anoplura

Members of the suborder Anoplura are recognized by their relatively small head, styliform mouthparts, and lack of mandibles.

Ferris (1951) recognizes six families. The ECHINOPHTHIRIIDAE are parasites of pinnipede carnivores (seals, sea lions, and walruses). Their body is covered with strong spines or scales which retain a film of air over the body when submerged. It is evident that many of these lice must be very long lived, since their hosts spend most of their life at sea, coming ashore to breed (when presumably transfer of the parasites occurs) for only a short time each year. The LINOGNATHIDAE parasitize dogs (*Linognathus setosus*) and ruminants, including sheep (*L. ovillus* and *L. pedalis*), cattle (*L. vituli*) (Figure 8.2D), and goats (*L. stenopsis*). Members of the closely related family NEOLINOGNATHIDAE have been found on elephant shrews in Africa. The HAEMATOPINIDAE are a large family whose members parasitize ungulates, including several domesticated forms on which they may become serious pests. *Haematopinus eurysternus* (Figure 8.2E) occurs on cattle, *H. suis* on pigs, and *H. asini* on horses. The HOPLOPLEURIDAE are also a large group whose hosts are mainly rodents and rabbits, but also include insectivores and primates. The PEDICULIDAE parasitize primates. The genus *Pediculus* is found on man (as *P. humanus*) and other hominoids as well as New World monkeys. *P. humanus* exists as two sub-species,* *P. h. capitis*, the head louse, and *P. h. corporis* (=*vestimenti*), the body louse (Figure 8.2F), which differ in size and habits. The smaller head louse attaches its eggs to hair, whereas the body louse or "cootie" lays its eggs on clothing to which it usually remains attached. The body louse is an important vector of diseases such as endemic typhus and trench fever, caused by blood-borne rickettsias. Relapsing fever, caused by a spirochaete, is also spread by the louse. *Phthirus pubis*, the pubic or crab louse (Figure 8.2G), is another parasite of man, though this species does not appear to be a disease vector.

Literature

A good deal of general information on Phthiraptera is given by Hopkins (1949), who also discusses the phylogeny of the order, by Rothschild and Clay (1952), and by Askew (1971). Clay (1949, 1970) discusses the phylogeny of the order. Ferris (1951) deals specifically with the Anoplura, providing a systematic treatment of the world species.

Askew, R. R., 1971, *Parasitic Insects*, American Elsevier, New York.
Clay, T., 1949, Some problems in the evolution of a group of ectoparasites, *Evolution* 3:279–299.
Clay, T., 1970, The Amblycera (Phthiraptera: Insecta), *Bull. Br. Mus. Nat. Hist.* **25**:75–98.
Ferris, G. F., 1951, The sucking lice, *Mem. Pac. Coast Entomol. Soc.* **1**:320 pp.
Hopkins, G. H. E., 1949, The host associations of the lice of mammals, *Proc. Zool. Soc. London* **119**:387–604.
Rothschild, M., and Clay, T., 1952, *Fleas, Flukes and Cuckoos. A Study of Bird Parasites*, Collins, London.

4. Hemiptera

SYNONYM: Rhynchota   COMMON NAME: true bugs

Minute to large insects, head opisthognathous (Homoptera) or prognathous (Heteroptera), compound eyes usually well developed but rarely absent, two or three ocelli usually present, antennae with few segments, mouthparts suctorial with mandibles and maxillae in form of stylets enclosed within a labial sheath; two pairs of wings usually present with fore wings of harder consistency than hind pair; abdomen with 9–11 segments, external genitalia variable in both sexes, cerci absent.

This group is the largest and most heterogeneous group of exopterygotes, containing approximately 55,000 described species from all regions of the world. The order contains two distinct suborders, the Homoptera and Heteroptera.

Structure

Hemiptera range in size from about 1 mm to 9 cm. They are frequently procryptically colored in shades of green or brown, or aposematically colored

*J. R. Busvine (1978) *Systematic Entomology* 3:1–8) argues that the two should be regarded as separate species.

in striking patterns, often in red and white. Yet others mimic ants and other insects with which they live. The head is either opisthognathous and lacking a "gula" in Homoptera or prognathous and with a sclerotized "gula" in most Heteroptera. The antennae comprise only four or five segments in most Hemiptera. The compound eyes are typically well developed and of variable shape and size. Two or three ocelli are usually present. The unifying feature of the order are the suctorial mouthparts, which are remarkably similar in all except a very few members of the group. The mandibles and maxillae form two pairs of piercing stylets that are contained within the flexible, segmented labium (see Chapter 3, Section 3.2.2 and Figure 3.17). Maxillary and labial palps are absent, though the labium, which never enters the tissue that is pierced, has a number of sensory hairs at its tip.

In Heteroptera the pronotum is large, the mesonotum and metanotum small. In most Homoptera the pronotum is small and collarlike, the mesonotum is well developed, the metanotum somewhat less so. Usually two pairs of wings are present, though brachyptery and aptery are common, sometimes occurring in the same species. In Homoptera the fore wings are of uniform texture and are often of a harder consistency than the hind pair. Among Heteroptera there is typically a marked difference in the consistency of the fore and hind wings. The fore wings ("hemelytra") usually are well sclerotized basally, with only a small distal portion remaining membranous. The hind pair remain membranous and, at rest, are folded beneath the hemelytra. The legs are generally identical and suited for a cursorial habit. However, in certain groups they are modified for a variety of functions, such as catching prey, swimming, jumping, moving over the water surface, and production of sound, or they may be vestigial or absent in females that are sedentary.

In its least specialized condition the abdomen has 11 segments of which the first two may be modified in connection with sound production, the eighth and ninth possess the external genitalia, and the last two are extremely reduced. Frequently reduction of up to three anterior segments occurs. The ovipositor is complete in many Homoptera and those Heteroptera that lay their eggs in plant tissues, but reduced or absent in other groups. The homologies of the male genitalia are complex. Primitively, the enlarged ninth sternum bears a pair of lateral claspers, while the cavity formed by invagination of the membrane between the ninth and tenth sterna contains the aedeagus and a pair of parameres. However, either the claspers or both claspers and parameres are absent in some groups.

The alimentary canal presents a wide variety of structural modifications associated with the liquid diet of the order. Posterior to the cibarial sucking pump, which is not part of the alimentary canal, is a short foregut. The midgut is large, frequently occupying a major part of the abdomen and usually differentiated into several regions, of which the most anterior is a croplike structure. The posterior region is tubular and from it, in many Heteroptera, arise a large number of ceca that contain symbiotic bacteria. In many plantsucking Hemiptera the hindgut is long and convoluted. The anterior part of the hindgut and the Malpighian tubules may come to lie alongside the swollen midgut, thus providing a "short cut" to the hindgut for the large volumes of liquid feces (see Figure 16.6). In a few Heteroptera the midgut ends blindly or is entirely separated from the hindgut. Four Malpighian tubules are the rule, but rarely there are only three or two. In Aphidoidea there are no Malpighian tubules, and the function of excretion is taken over by the midgut. The central nervous system is highly specialized, and discrete abdominal ganglia never occur. At most, there are three ventral ganglia (subesophageal, first thoracic, and a composite ganglion containing the meso- and metathoracic and all the abdominal centers). Frequently, only the subesophageal and a composite thoracoabdominal ganglion are found. In some Hemiptera there is a single composite ventral ganglion Usually each ovary comprises between four and eight acrotrophic ovarioles arranged in a group at the distal end of the lateral oviduct. In a few Homoptera, for example, aphids, there may, however, be up to 100 ovarioles. Paired accessory glands and zero to three spermathecae are also present. Each testis contains from one to seven follicles, which may or may not be enclosed in a follicular sheath. Well-developed accessory glands also occur in males. Various types of subcutaneous glands are common in Hemiptera. Many Heteroptera possess repugnatorial glands that open via a pair of pores situated near the hind coxae. Unicellular wax-secreting glands occur in many Homoptera, usually on the dorsal surface of the abdomen. The function of these glands is presumably protective.

Life History and Habits

Homoptera and the majority of Heteroptera are terrestrial insects. Other Heteroptera show varying degrees of adaptation to an aquatic existence. Some occur in the littoral or intertidal zone of the seashore, on marshy ground, or in damp moss. Others live on the surface of water. Among the truly aquatic forms there is a range of adaptation from those species that must periodically visit the surface to respire (and which periodically fly from one location to another) to those that remain submerged permanently and respire by means of a plastron (see Chapter 15).

All Homoptera and many Heteroptera feed on fluids from plants. All parts of the plant are attacked: roots, stem, leaves, flowers, and seeds (often when these have fallen to the ground). The remaining Heteroptera are predaceous, living on the body fluids of other arthropods and vertebrates. It is primarily because of these feeding habits, assisted in many cases by extraordinarily high rates of reproduction (see Superfamily Aphidoidea), that the order is considered by many people to be the most economically important insect group. The damaging effect on plants may be direct or indirect. When insect populations are large, the loss of sap results in stunted growth and poor yield and quality. Many species, when feeding, inject saliva that causes necrosis of plant tissue. Indirectly, the effects are to weaken the plant, making it more susceptible to attack by other pathogens, especially fungi and viruses. Most important of all, however, is the role of Hemiptera (especially aphids) as vectors of viruses that cause major plant diseases, for example, mosaic, leaf roll yellows. Among the predaceous Hemiptera, a few species may act as vectors for the transmission of disease, for example, certain South American Reduviidae are carriers of *Trypanosoma cruzi*, a flagellate protozoan that causes trypanosomiasis (Chagas' disease), a form of sleeping sickness among humans. On the beneficial side, many Homoptera play an important part in weed control; for example, *Dactylopius* species have been successfully used in the control of the prickly pear cactus (*Opuntia*), and many predaceous Heteroptera exert a major controlling effect on some arthropod pests.

The majority of Hemiptera are bisexual and oviparous. There are, however, species which are facultatively parthenogenetic, and some which are ovoviviparous or viviparous. In many aphids all these reproductive conditions may be met within the same species in the course of a year. In aphids parthenogenesis and viviparity commonly occur together in the spring and early summer, thus enabling the insect to exploit fully the increased food available at this time. Insemination is typically of the usual intragenital type. However, in Cimicoidea various forms of hemocoelic insemination occur (see Chapter 19). The eggs of Homoptera are generally simple, ovoid structures; those of Heteroptera are very diverse in form and coloring. They are glued to plant surfaces, inserted in crevices or between adjacent parts of a plant, laid in litter or soil, or in the case of predaceous species, in the vicinity of a host if possible. In a few Heteroptera the eggs are stuck on the dorsal surface of males. Parental care is exhibited by a few species. Usually, there are between three and seven instars (almost always six in Heteroptera). The juvenile stages tend to feed on the same part of the plant as the adult, though in some species juveniles are found on the roots, while adults occur on the upper parts. Although most Hemiptera are typically exopterygote in their postembryonic development, there occurs in a few Homoptera (some Aleurodidae and winged male Coccoidea) one or two resting instars, the "pupal" and "prepupal" stages. These instars generally do not feed but undergo some degree of metamorphosis. A similar phenomenon occurs in Thysanoptera.

Phylogeny and Classification

Like the Psocoptera, the Hemiptera are a very ancient order whose fossil record extends into the Lower Permian period. The earliest Hemiptera were undoubtedly Homoptera. This suborder evidently underwent a rapid radiation during the Permian period, by the end of which representatives of at least three extant superfamilies (Cicadelloidea, Cercopoidea, and Psylloidea) occurred. Within the Homoptera there seem to be three main evolutionary lines that led to the modern Coleorrhyncha, Auchenorrhyncha, and Sternorrhyncha (Figure 8.3). The Coleorrhyncha contains a single superfamily, Peloridioidea, species of which exhibit many primitive features and are considered by some authori-

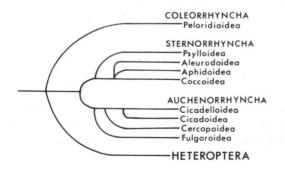

FIGURE 8.3. Proposed phylogeny of Homoptera. Based on G. Heslop-Harrison, 1956, The age and origin of the Hemiptera, with special reference to the sub-order Homoptera, [*Proc. Univ. Durham Philos. Soc. Ser. A* **12**:150–169.]

ties to be the remnants of the ancestral stock from which the suborder Heteroptera evolved. Along the auchenorrhynchan line, the earliest group to split off was the superfamily Fulgoroidea. The main line of evolution then divided giving rise to the Cercopoidea, on the one hand, and the Cicadoidea and Cicadelloidea, on the other. The Psylloidea appear to have left the main line of evolution in the Sternorrhyncha at an early date. The main line then split, giving rise to the other three superfamilies, Coccoidea, Aphidoidea, and Aleurodoidea. It should be noted that use of the terms Coleorrhyncha, Auchenorrhyncha, and Sternorrhyncha, while expressing natural relationships in the sense that there have been three distinct lines of evolution within the Homoptera, is somewhat misleading because it tends to overemphasize certain relationships. Thus, the Peloridioidea, Fulgoroidea, and Psylloidea, all early derivatives from the ancestral homopteran stock, yet placed in three different groups, are probably more closely related than, say, the Fulgoroidea and Cicadelloidea, both of which are included in the Auchenorrhyncha.

Fossil Heteroptera are not known with certainty prior to the Lower Jurassic period, although, in view of the diversity of form that these fossils exhibit, it is evident that the group had been in existence for a much longer time, probably since the Lower Permian or even earlier. The prognathous head of most Heteroptera is taken as evidence that an early trend in the evolution of the group was toward a predaceous habit. This feature is retained in many recent families, though a number have reverted to the phytophagous habits of their homopteran ancestors. It would only be after the development of predaceous habits that groups became truly aquatic. Traditionally, Heteroptera were divided into the Cryptocerata (truly aquatic groups, generally with short, concealed antennae) and Gymnocerata (terrestrial and amphibious forms, with long, prominent antennae). More recently the tendency has been to arrange the Heteroptera in three series, Hydrocorisae (=Cryptocerata), Amphibicorisae (amphibious forms), and Geocorisae (terrestrial groups). Although these terms are useful in that they describe easily recognizable groups of Heteroptera, they are misleading in the sense that they do not indicate the varying degree of diversity that has occurred in each group. Thus, in the Hydrocorisae and Amphibicorisae there appears to have been a single major evolutionary line from the basic heteropteran stock. In contrast, in the Geocorisae three major and several minor evolutionary lines appear to have arisen from the ancestral stock. In other words, the Cimicoidea and Pentatomoidea, for example, which are both superfamilies of the Geocorisae, are probably no more closely related than, say, the Cimicoidea and the Gerroidea (Amphibicorisae). On the basis of his extensive study of the structure and biology of the eggs of Heteroptera, and of other unpublished data, Cobben (1968) has concluded that modern Heteroptera have evolved from extinct amphibicorisous stock, and that the Geocorisae is a polyphyletic assemblage. A suggested phylogeny of Heteroptera is indicated in Figure 8.4.

Suborder Homoptera

Characteristics of Homoptera include head opisthognathous and without a "gula"; pronotum small; and fore wings, when present, in form of tegmina with uniform texture and held rooflike over body at rest

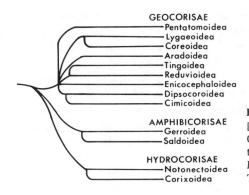

FIGURE 8.4. Evolution within the Heteroptera. [After W. E. China and N. C. E. Miller, 1959, Check-list and keys to the families and subfamilies to the Hemiptera–Heteroptera, *Bull. Br. Mus. Nat. Hist. Entomol.* **8**(1):1–45. By permission of the Trustees of the British Museum (Natural History).]

Series I Coleorrhyncha

Superfamily Peloridioidea

Contained in the superfamily Peloridioidea is a single, Southern Hemisphere family, PELORIDIIDAE, whose 20 species are small, flattened, cryptically colored Hemiptera found among moss and liverworts or in caves. They show many primitive features, which suggests that they are remnants of the ancestral stock from which other Hemiptera arose. They have an interesting disjunct distribution, occurring in South America and the Australian region.

Series II Auchenorrhyncha

Superfamily Fulgoroidea

In the large, heterogeneous superfamily Fulgoroidea, whose members are commonly known as plant hoppers, there are about 9000 species arranged in some 20 families. Plant hoppers are mainly phloem feeders on higher plants, but a few feed on fungi. The majority of species are 1 cm or less in length, though some tropical forms may reach a length of 5 cm or more. Some of the larger and commonly encountered families are as follows. The CIXIIDAE (Figure 8.5A), with more than 1000 species distributed throughout the world, are regarded as the most primitive fulgoroid family. Little is known of their biology, but the young stages are subterranean and typically feed on grass roots. Larvae of a few species have been encountered in ants' nests. DELPHACIDAE (Figure 8.5B) are small (less than 1 cm in length) fulgoroids that frequently have reduced wings. This is the largest family of fulgoroids, and its more than 1100 species are recognized by the large spur on the tibia of the hindlegs. A few members of this family are serious pests through their acting as vectors of virus diseases. The DERBIDAE constitute a mainly tropical family, whose more than 800 species typically have very long wings and feed on fungi or higher plants. DICTYOPHARIDAE (Figure 8.5C) are medium-sized Fulgoroidea whose head bears a distinct anterior process. The family, which contains more than 500 species, is widely distributed through arid or semiarid areas of the world. The FULGORIDAE, with about 650 species, are a widely distributed group, many of whose members are known as lantern flies because the inflated anterior part of

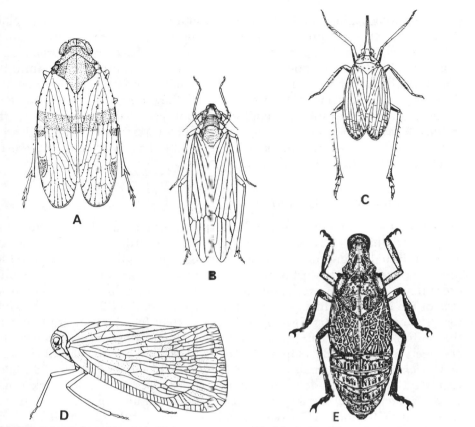

FIGURE 8.5. Fulgoroidea (A) *Cixius angustatus* (Cixiidae), (B) *Stenocranus dorsalis* (Delphacidae), (C) *Scolops perdix* (Dictyopharidae), (D) *Anormenis septentrionalis* (Flatidae), and (E) *Fitchiella robertsoni* (Issidae). [From H. Osborn, 1938, The Fulgoridae of Ohio, *Ohio Biol. Surv. Bull.* **6**(6):283–349 (Bulletin #35). By permission of the Ohio Biological Survey.]

the head was believed originally to be luminous. The largest plant hoppers are members of this family. The ACHILIDAE form a widely distributed, though primarily tropical family containing about 350 species. The larvae live beneath bark or among rotting wood. More than 1000 species are included in the family FLATIDAE (Figure 8.5D), a highly specialized, mainly tropical, group. Members of this family resemble moths by virtue of their triangular, opaque tegmina, which are folded to form a steep roof over the body. The ISSIDAE (Figure 8.5E) are another large family, with more than 1000 species. Its members are mostly dull colored and frequently have a squat, beetlelike facies.

Superfamily Cercopoidea

The members of this small, rather homogeneous superfamily are arranged in three families, CERCOPIDAE, APHROPHORIDAE, and MACHAEROTIDAE. Species are seldom more than 15 mm in length and frequently strikingly colored as adults. The larvae of a few Cercopidae are subterranean, but mostly they live either in a mass of froth (Cercopidae and Aphrophoridae), when they are known as "cuckoo-spit insects" or "spittlebugs," or in a calcareous tube

(Machaerotidae). The function of these structures, which are affixed to plant stems, is to provide protection from predators and to prevent desiccation. Adults are active, hopping insects that in many cases bear a crude resemblance to a frog, hence the common name of "froghopper." Aphrophoridae form the largest family with a wide distribution. Its members are usually found on herbaceous plants, but a few live on trees to which they sometimes do considerable damage. Cercopidae are also widespread but particularly common in the tropics. A few species are pests of grasses and clovers, for example, *Philaenus spumarius*, the meadow spittlebug (Figure 8.6). The Machaerotidae are restricted to Asia, tropical Africa, and Australia.

Superfamily Cicadoidea

Cicadas are common insects in all the warmer regions of the world. They are generally between 2 and 5 cm in length and are particularly well known on account of their sound-producing abilities and the length of time required for juvenile development. Most of the 1500 or so species are placed in the family CICADIDAE. The sound-producing organs (TYMBALS) are located on the dorsal side of the first abdominal segment of males only in the Cicadidae. The auditory TYMPANA are better developed in males than females and are found on the ventral side of the anterior abdominal segments. Most cicadas require several years for juvenile development. Probably the best known "periodic cicada" is *Magicicada septendecim* (Figure 8.7), which, in the eastern United States, requires 17 years for its development; in contrast, the southern form takes only 13 years to mature. A larva spends this entire period underground, feeding on roots, especially those of perennial plants. Prior to the final molt the larva leaves the soil to complete its metamorphosis on a tree or other object. Eggs are laid in twigs, a process that may cause considerable dieback of the tree.

Superfamily Cicadelloidea

Most of the Cicadelloidea are placed in two very large families, CICADELLIDAE (JASSIDAE) (leafhoppers) and MEMBRACIDAE (treehoppers). The Cicadellidae form a large, cosmopolitan group, and are found on almost all types of plants. They rank second only to the Aphididae in the enormity of their numbers and, in consequence, constitute major pests. They cause a wide variety of

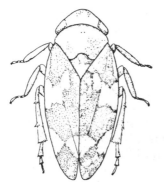

FIGURE 8.6. Cercopoidea. A froghopper, *Philaenus spumarius* (Cercopidae). [From D. J. Borror, D. M. Delong, and C. A. Triplehorn, 1976, *An Introduction to the Study of Insects*, 4th ed. By permission of Holt, Rinehart and Winston, Inc.]

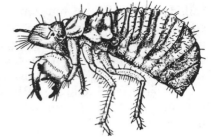

FIGURE 8.7. Cicadcidea. Larva of the periodical cicada, *Magicicada septendecim* (Cicadidae). [From A. D. Imms, 1957, *A General Textbook of Entomology*, 9th ed. (revised by O. W. Richards and R. G. Davies). By permission of Chapman and Hall Ltd.]

injuries to plants. They may remove large quantities of sap, block the phloem tubes, or destroy the chlorophyll so that growth is stunted. Many are vectors of viruses that causes disease. A few damage the plants by their oviposition habits. Two of the best-known pests are *Circulifer tenellus*, the beet leafhopper [Figure 8.8A), and *Empoasca fabae*, the potato leafhopper (Figure 8.8B), which feeds on solanaceous plants, beans, celery, alfalfa, and various flowers. Several other species of *Empoasca* are also important pests (see Swan and Papp, 1972}. Membracidae are easily recognized by their enormous pronotum that projects backward over the abdomen and often assumes bizarre shapes. The family is primarily neotropical. They are generally gregarious and attended by ants for the honeydew they produce. They are seldom of economic importance; however, *Stictocephala bubalus*, the buffalo treehopper (Figure 8.8C), may damage young fruit trees and nursery stock as a result of its egg-laying activity.

Series III Sternorrhyncha

Superfamily Psylloidea

The 1250 described species of Psylloidea are placed in a single family, PSYLLIDAE. The jumping plant lice, as members of the family are commonly

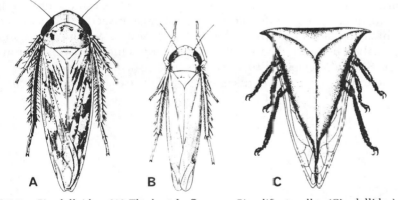

FIGURE 8.8. Cicadelloidea. (A) The beet leafhopper, *Circulifer tenellus* (Cicadellidae); (B) the potato leafhopper, *Empoasca fabae* (Cicadellidae); and (C) the buffalo treehopper, *Stictocephala bubalus* (Membracidae). [A, B, from D. M. Delong, 1948, The leafhoppers, or Cicadellidae, of Illinois, *Bull. Ill. Nat. Hist. Surv.* **24**(2):97–376. By permission of the Illinois Natural History Survey. C, from D. J. Borror, D. M. Delong, and C. A. Triplehorn, 1976, *An Introduction to the Study of Insects*, 4th ed. By permission of Holt, Rinehart and Winston, Inc.]

called, are small (2–5 mm long) and resemble miniature cicadas. They have strong hindlegs, and wings are present in both sexes. The family contains many pest species, which may cause galls or a general stunting of host plants. The commonly held view that this stunting is caused by viruses for which psyllids serve as vectors is now believed to be wrong; it is the toxic saliva injected during feeding which causes the damage. Examples of pest species are *Psylla pyricola* and *P. mali* (Figure 8.9), two species introduced into North America from Europe, which feed on pear and apple, respectively.

Superfamily Coccoidea

Many families are included in the superfamily Coccoidea, a large, heterogeneous group of more than 6000 species. Despite this heterogeneity a unifying characteristic of the group is the more or less degenerate females. These are apterous, and may be scalelike, gall-like, or covered with a waxy or powdery coating. For this reason they are commonly known as "scale insects" or "mealybugs." Adult males are either apterous or have fore wings only, and have nonfunctional mouthparts. Females are oviparous (in which case the eggs are usually retained within the scaley covering of a female), ovoviviparous, or viviparous. Parthenogenesis is common and hermaphroditism is known to occur in one species. Many species have become cosmopolitan as a result of distribution by Man. Notes on some of the commoner families are given below. The DIASPIDIDAE (armored scales) form the largest family of coccoids. Females are covered with a hard, waxy layer that is separate from the body. Included in the family are many pests of trees and shrubs, for example, *Lepidosaphes ulmi*, the oystershell scale (Figure 8.10A,B), and *Quadraspidiotus perniciosus*, the San Jose scale. Another large family is the COCCIDAE (soft scales, wax scales), the female members of which show a wide range of form. The family contains several pests that are now widespread through commerce; for example, *Pulvinaria innumerabilis*, the cottony maple scale, introduced from Europe, is now found throughout North America feeding on forest, shade, and fruit trees. The LACCIFERIDAE, a mainly tropical and subtropical group, have females that are extremely degenerate and live in a resinous cell. *Laccifer lacca*, the Indian lac insect, produces a secretion from which shellac is prepared. The PSEUDOCOCCIDAE are the common mealybugs, so-called because females are covered with a mealy or filamentous waxy secretion. Several species of *Pseudococcus* (Figure 8.10C) are major pests through their functioning as vectors of disease-causing viruses. The ERIOCOCCIDAE were

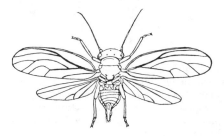

FIGURE 8.9. Psylloidea. The apple sucker, *Psylla mali* (Psyllidae). [From A. D. Imms, 1957, *A General Textbook of Entomology*, 9th ed. (revised by O. W. Richards and R. G. Davies). By permission of Chapman and Hall Ltd.]

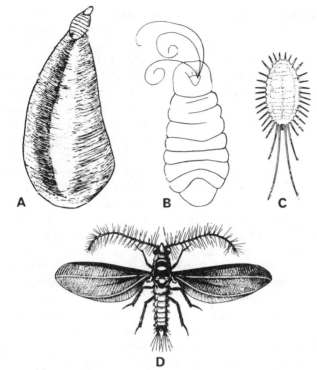

FIGURE 8.10. Coccoidea. (A) The oystershell scale, *Lepidosaphes ulmi* (Diaspididae); (B) female *L. ulmi*; (C) female long-tailed mealybug, *Pseudococcus longispinus* (Pseudococcidae) and (D) male cottony-cushion scale, *Icerya purchasi* (Margarodidae). [A, B, from D. J. Borror, D. M. Delong, and C. A. Triplehorn, 1976, *An Introduction to the Study of Insects*, 4th ed. By permission of Holt, Rinehart and Winston, Inc. C,D, from P.-P. Grassé (ed.), 1951, *Traité de Zoologie*, Vol. X. By permission of Masson, Paris.]

formerly included in the Pseudococcidae because of their close resemblance to mealybugs. The family contains a number of pest species, including some gall-formers, as well as some potentially beneficial ones. Various *Dactylopius* species, for example, have been introduced into Australia in an attempt to control the prickly pear weed. The MARGARODIDAE form a small but widely distributed family, included in which is *Icerya purchasi*, the cottony-cushion scale (Figure 8.10D), an Australian species that was transplanted through commerce to many regions of the world where it became an important pest of citrus fruit. In California this pest has been controlled successfully following the introduction of the predaceous beetle *Vedalia cardinalis* (see Chapter 24, Section 2.3). The ORTHEZIIDAE are another small family mainly from the Nearctic and Palaearctic. Several of its members, which are found on roots, have been transported to all parts of the world and are important pests.

Superfamily Aphidoidea

More than 3600 species of Aphidoidea have been described, including some of the world's most important insect pests. Members of the family are characterized by their complex life cycles, in which the species takes on a variety of forms and frequently alternates between two taxonomically distinct host plants. By far the largest of the four families of Aphidoidea is the APHIDIDAE (plant lice, greenfly, and blackfly). Most aphids are found on leaves, shoots, or buds, though a few species live in rather specialized situations, for example, in unfolded leaves or in earth shelters especially constructed for them by ants with which they are associated. Most species are polymorphic in different generations and reproduce in a variety of ways. They may also show host alternation. A typical life cycle, that of *Dysaphis plantaginea*, the rosy apple aphid, is shown in Figure 8.11. Usually aphids overwinter in the egg stage. In spring the eggs hatch and give rise to wingless, viviparous parthenogenetic females. A variable number of such generations occur, and these are followed by the production of winged individuals that migrate to the alternate host on which reproduction continues. Later in the season sexual males and females are produced, and the aphids return to the original host. They mate and females lay the overwintering eggs. Parthenogenesis provides the means by which aphids can increase their population extremely rapidly. Fortunately, the occurrence of a large number of predators and adverse weather conditions usually keep their numbers in check.

Aphids in sufficient numbers may have a direct effect on their hosts, causing wilting and stunted growth. They are, however, economically more important through their role as vectors of disease-producing viruses. Migratory species which are not particularly host-specific are especially important, since these can transmit diseases among a wide variety of plants. *Myzus persicae*, the green peach aphid (Figure 8.12A), is the classic example, being known as a vector for more than 100 virus diseases.

The PEMPHIGIDAE (ERIOSOMATIDAE) are closely related to the Aphididae. The family is small, widely distributed (often by man's activities), and includes both above- and below-ground feeders and many gall-making species. Many of the "woolly aphids," so-called because of the woollike waxy filaments that they produce, are included in this family. Most species undergo host alternation.

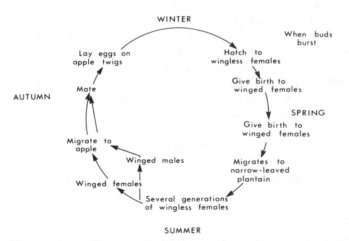

FIGURE 8.11. Life cycle of the rosy apple aphid, *Dysaphis plantaginea*. [After D. J. Borror, D. M. Delong, and C. A. Triplehorn, 1976, *An Introduction to the Study of Insects*, 4th ed. By permission of Holt, Rinehart and Winston, Inc.]

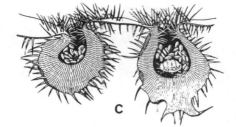

FIGURE 8.12. Aphidoidea. (A) The green peach aphid, *Myzus persicae* (Aphididae); (3) winged female of the grape phylloxera, *Phylloxera vitifoliae* (Phylloxeridae); and (C) section through galls on grape leaf showing wingless female and eggs. [A, reproduced by permission of the Smithsonian Institution Press from Smithsonian Scientific Series, Volume Five (*Insects: Their Ways and Means of Living*) by Robert Evans Snodgrass: Fig. 101B, page 171. Copyright © 1930, Smithsonian Institution, New York. B, C, from P.-P. Grassé (ed.), 1951, *Traité de Zoologie*, Vol. X. By permission of Masson, Paris.]

The ADELGIDAE (CHERMIDAE) are a small family whose members are confined to conifers. They feed on needles, twigs, or within galls. Most species alternate hosts, the primary one always being spruce. The PHYLLOXERIDAE are a small but widely distributed family that includes a number of pests, for example, *Viteus* (= *Phylloxera*) *vitifoliae*, the vine or grape phylloxera that attacks the leaves of the grape, forming small galls in which it lives and reproduces (Figure 8.12B,C). The life cycle is complex, with a high degree of polymorphism, though host alternation does not occur.

Superfamily Aleurodoidea

Included in the superfamily Aleurodoidea is a single family, ALEURODIDAE (ALEYRODIDAE), with approximately 1100 species, commonly known as whiteflies (Figure 8.13). They are small insects, usually 3 mm or less in length, generally covered with a whitish waxy secretion. They are commonly found on the underside of leaves. The group is mainly tropical or subtropical, though a few species are pests of greenhouse crops in temperate regions. The life cycle is complex, and parthenogenesis is commonly involved. Larvae are sedentary, and the final stage ("pupa") does not feed but undergoes a marked metamorphosis to the adult.

Suborder Heteroptera

In Heteroptera the head is usually prognathous, almost always with a "gula"; the pronotum is well developed; and fore wings when present are in the form of hemelytra, with wings held flat over body when at rest.

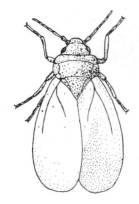

FIGURE 8.13. Aleurodoidea. The greenhouse whitefly, *Trialeurodes vaporariorum* (Aleurodidae). [From L. Lloyd, 1922, The control of the greenhouse whitefly (*Asterochiton vaporariorum*) with notes on its biology, *Ann. Appl. Biol.* **9**:1–32. By permission of the Association of Applied Biologists.]

Series I Geocorisae

Superfamily Dipsocoroidea

The superfamily Dipsocoroidea is a small but widely distributed super-family whose members are minute, predaceous insects found in damp habitats such as moss, leaf litter, and ants' nests. Usually all the species are placed in a single family, DIPSOCORIDAE (CRYPTOSTEMMATIDAE). The affinities of the group are uncertain; probably it represents an early offshoot of the cimicoid line.

Superfamily Enicocephaloidea

The 50 or so species are arranged in a single family, ENICOCEPHALIDAE, which has a wide distribution. They are small insects (3 or 4 mm in length) that often appear in swarms, hence the common name of gnat bugs. The affinities of the group are uncertain but appear to lie with the Cimicoidea and Reduvioidea.

Superfamily Cimicoidea

Approximately 6000 species of Cimicoidea are known, arranged in some eight families of which only four are common. There is a general tendency toward a bloodsucking habit, though most members of the largest family, MIRIDAE (CAPSIDAE), are phytophagous. Some species are, however, facultative bloodsuckers, should the occasion present itself. Some 5000 species of Miridae have already been described and the group includes some important pest species, for example, *Lygus lineolaris,* the tarnished plant bug (Figure 8.14A), which feeds on cotton, alfalfa, hay, and various vegetables and fruits. Its relative, *Poecilocapsus lineatus,* the four-lined plant bug (Figure 8.14B), feeds on gooseberry, currant, rose, and various annual flowers. The NABIDAE, containing about 300 species, were formerly included in the Reduvioidea, but the demonstration of hemocoelic insemination in certain species has led to their transfer to the Cimicoidea. The CIMICIDAE are a small but widely distributed family whose members are bloodsucking ectoparasites of birds and mammals. Included in the family are the bedbugs, *Cimex lectularius* (Figure 8.14C) (cosmopolitan) and *C. hemipterus* (mainly southern Asia and Africa). These are particularly

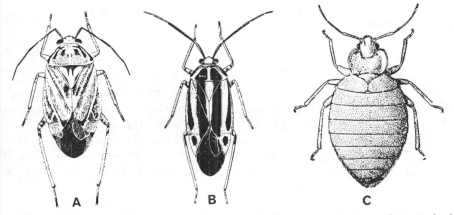

FIGURE 8.14. Cimicoidea. (A) The tarnished plant bug, *Lygus lineolaris* (Miridae); (B) the four-lined plant bug, *Poecilocapsus lineatus* (Miridae); and (C) the bedbug, *Cimex lectularius* (Cimicidae). [A, B, from H. H. Knight, 1941, The plant bugs, or Miridae, of Illinois, *Bull. Ill. Nat. Hist. Surv.* **22**(1):234 pp. By permission of the Illinois Natural History Survey.]

common in unhygienic and/or overcrowded conditions. They hide during the day, and also lay their eggs, in crevices, coming out to feed at night. So far as is known, they are not responsible for transmission of any disease. The AN-THOCORIDAE, a group of some 300 species that feed on blood or eggs of other arthropods, are often included in the Cimicidae to which they are closely related. Anthocorids are found on flowers, under bark, or in leaf litter.

Superfamily Reduvioidea

The more than 4000 described species of Reduvioidea are placed in one family, the REDUVIIDAE, commonly known as "assassin bugs" [Figure 8.15]. All species are predaceous, particularly on other arthropods, though a number feed on the blood of vertebrates, including man. Most species inject a saliva that paralyzes the tissue as well as assists in its digestion, thereby causing severe and painful "bites." Species of *Triatoma* and *Rhodnius prolixus*, a "famous" experimental insect, carry *Trypanosoma cruzi*, which causes a fatal form of sleep-

FIGURE 8.15. Reduvioidea. The bloodsucking conenose, *Triatoma sanguisuga* (Reduviidae). [From R. C. Froeschner, 1944, Contributions to a synopsis of the Hemiptera of Missouri, *Am. Midl. Nat.* **31**(3):638–683. By permission of the American Midland Naturalist.]

ing sickness (Chagas' disease) in humans. In many species the forelegs are raptorial.

Superfamily Tingoidea

About 700 species of Tingoidea are known and included in the single family TINGIDAE (lacebugs). They are easily recognized by the lacelike pattern on the dorsal surface of the head and body, including the wings (Figure 8.16). They are phytophagous and occasionally cause damage when present in large numbers. The group appears to have a blend of both reduvioid and cimicoid features and is probably an early offshoot of the ancestral heteropteran stock.

Superfamily Aradoidea

Slightly more than 400 species of Aradoidea are known, almost all of which are placed in the family ARADIDAE (Figure 8.17), the remaining few, which are found only in termites' nests, being included in the TERMITAPHIDIDAE. The Aradidae are widely distributed and found typically beneath bark or rotting wood. They feed on the sap of fungi. Their mouthparts are extremely long and at rest are coiled within the head.

Superfamily Pentatomoidea

Included among the 4000 species of Pentatomoidea are the familiar shield and stink bugs. The group is subdivided into a variable number of families of which by far the largest, with more than 3400 species, are the PENTATOMIDAE. Most pentatomids are phytophagous, but a few are predaceous on other insects, especially caterpillars. They are typically brightly colored or conspicuously marked insects, capable of emitting an obnoxious fluid from thoracic repugnatorial glands. A few species are economically important, for example, *Murgantia histrionica*, the harlequin cabbage bug (Figure 8.18) found on cabbage and other Cruciferae. The CYDNIDAE (negro bugs) resemble the Pentatomidae in shape but are darkly colored and live under stones or dead leaves, or occasion-

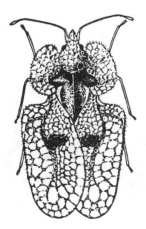

FIGURE 8.16. Tingoidea. The sycamore lacebug, *Corythuca ciliata* (Tingidae). [From R. C. Froeschner, 1944, Contributions to a synopsis of the Hemiptera of Missouri, *Am. Midl. Nat.* **31**(3):638–683. By permission of the American Midland Naturalist.]

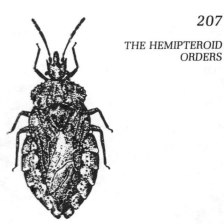

FIGURE 8.17. Aradoidea. *Aradus acutus* (Aradidae). [From R. C. Froeschner, 1942, Contributions to a synopsis of the Hemiptera of Missouri, *Am. Midl. Nat.* **27**(3):591–609. By permission of the American Midland Naturalist.]

ally in ants' nests. The SCUTERELLIDAE are easily recognized by their enormously enlarged mesoscutellum, which extends posteriorly to cover the entire abdomen and wings.

Superfamily Lygaeoidea

Most of the more than 2000 species of Lygaeoidea belong to the family LYGAEIDAE, a widely distributed group whose members are usually found on the ground, among vegetation, or under stones or low plants, where they feed usually on mature seeds that have fallen to the ground. Some species live off the ground, feeding on stems or immature seeds of a variety of plants, especially grasses. A few others are predaceous. Several species are pests, the best-known example being *Blissus leucopterus*, the chinch bug (Figure 8.19A), which attacks maturing cereal crops in the United States. A small but economically important family is the PYRRHOCORIDAE, whose members are widely distributed, commonly black and red bugs. Several species of *Dysdercus* (Figure 8.19B) are major pests of cotton and other Malvaceae on whose seeds (bolls) they feed. During this activity the bolls become contaminated with a fungus, which later stains the cotton fibers, hence the common name of "cotton stain-

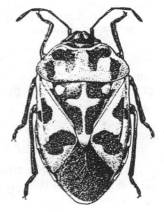

FIGURE 8.18. Pentatomoidea. The harlequin cabbage bug, *Murgantia histrionica* (Pentatomidae). [From R. C. Froeschner, 1941, Contributions to a synopsis of the Hemiptera of Missouri, *Am. Midl. Nat.* **26**(1):122–146. By permission of the American Midland Naturalist.]

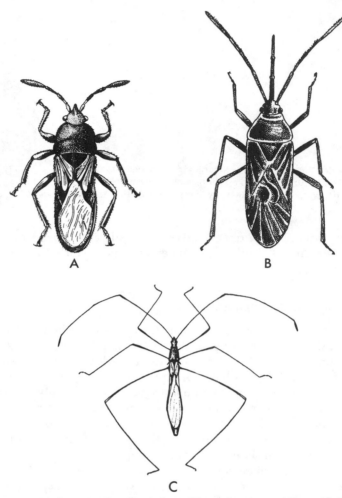

FIGURE 8.19. Lygaeoidea. (A) The chinch bug, *Blissus leucopterus* (Lygaeidae); (B) a cotton stainer, *Dysdercus suturellus* (Pyrrhocoridae); and (C) a stilt bug, *Jalysus wickhami* (Berytidae). [A, from P.-P. Grassé, 1951, *Traité de Zoologie*, Vol. X. By permission of Masson, Paris. B, from L. A. Swan and C. S. Papp, 1972, *The Common Insects of North America*. Copyright 1972 by L. A. Swan and C. S. Papp. Reprinted by permission of Harper & Row, Publishers, Inc. C, from R. C. Froeschner, 1942, Contributions to a synopsis of the Hemiptera of Missouri, *Am. Midl. Nat.* **27**(3):591–609. By permission of the American Midland Naturalist.]

ers" for these insects. The BERYTIDAE (stilt bugs) (Figure 8.19C) are secretive, slow-moving bugs found among dense undergrowth. The family is small but widely distributed. Its members have narrow, elongated bodies and long slender legs and antennae.

Superfamily Coreoidea

The Coreoidea are closely related to the Lygaeoidea, and the demarcating line between the two groups is often difficult to define. There is also disagreement over the higher classification of the superfamily, some authorities recognizing up to five families, others placing all species in the single family COREIDAE. The more than 2000 species are generally dark colored bugs, most common in Asia, Africa, and South America. They are all phytophagous, and some are pests, for example, the squash bug, *Anasa tristis* (Figure 8.20), on Cucurbitaceae in North America. Many species rival pentatomids in their abilities to produce obnoxious smells.

Series II Amphibicorisae

Superfamily Gerroidea

Although only a small superfamily with just over 500 species, the Gerroidea contain some of the best known Hemiptera. All species are predaceous and semiaquatic, living on the surface of still or slow-running water or in moist habitats close to water. The commonest family is the GERRIDAE (GERRIDIDAE) whose 200 species are known as "pond skaters" or "water striders" (Figure 8.21A). They are mainly found on fresh water, but a few species are marine. They feed on insects that fall onto the water surface. Another large family is the VELIIDAE whose 200 species are mainly from neotropical and oriental regions. They resemble gerrids, but their body is widened in the region of the mesothorax and metathorax. The HYDROMETRIDAE ("water measurers") form a small, mainly tropical, family, although the genus *Hydrometra* (Figure 8.21B) is cosmopolitan. These insects are superficially like stick insects. The MESOVELIIDAE are usually found at the margins of ponds, etc., crawling among the vegetation and debris. They are small bugs, 5 mm or less in length. The HEBRIDAE are another small family of minute bugs found among moss or on ponds in which there is an abundance of floating or emergent vegetation.

Superfamily Saldoidea

There is still controversy as to whether the Saldoidea are Amphibicorisae or Geocorisae. Some authors consider that they are an early derivation from the gerroid line, while others believe that they are an independent line that arose from the basic heteropteran stock between the cimicoid and pentatomoid lines of development. The group comprises three families: LEPTOPODIDAE (mainly

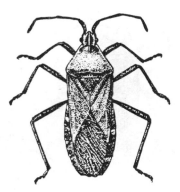

FIGURE 8.20. Coreoidea. The squash bug, *Anasa tristis* (Coreidae). [From R. C. Froeschner, 1942, Contributions to a synopsis of the Hemiptera of Missouri, *Am. Midl. Nat.* **27**(3):591–609. By permission of the American Midland Naturalist.]

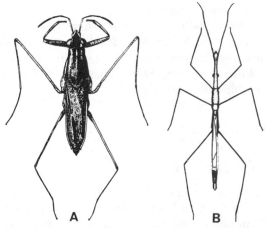

FIGURE 8.21. Gerroidea. (A) A pond skater, *Gerris marginatus* (Gerridae); and (B) a water measurer, *Hydrometra martini* (Hydrometridae). [A, from R. C. Froeschner, 1962, Contributions to a synopsis of the Hemiptera of Missouri, *Am. Midl. Nat.* **67**(1):208–240. By permission of the American Midland Naturalist. B, from A. R. Brooks and L. A. Kelton, 1967, Aquatic and semiaquatic Heteroptera of Alberta, Saskatchewan and Manitoba (Hemiptera), *Mem. Entomol. Soc. Can.* **51**:92 pp. By permission of the Entomological Society of Canada.]

oriental), LEOTICHIIDAE (two species, from southeast Asia), and SALDIDAE (Figure 8.22), a predominantly holarctic group of about 150 species that frequent the edges of ponds, streams, muddy estuaries, and the seashore. They are predaceous and can move rapidly, using a combination of running and flying.

Series III Hydrocorisae

Superfamily Notonectoidea

Most of the more than 500 species of Notonectoidea are included in four families: NOTONECTIDAE, NAUCORIDAE, BELOSTOMATIDAE, and NEPIDAE. About 40 species are divided equally among the widely distributed but rare families PLEIDAE and HELOTREPHIDAE. The Notonectoidea, along with the Corixoidea,

FIGURE 8.22. Saldoidea. *Teloleuca pellucens* (Saldidae). [From A. R. Brooks, and L. A. Kelton, 1967, Aquatic and semiaquatic Heteroptera of Alberta, Saskatchewan and Manitoba (Hemiptera), *Mem. Entomol. Soc. Can.* **51**:92 pp. By permission of the Entomological Society of Canada.]

are the most completely adapted of the Heteroptera for an aquatic existence. Juvenile stages spend their entire time submerged; almost all adults, however, require atmospheric oxygen, which they obtain by periodically surfacing or, in Nepidae, through the development of a posterior, respiratory tube (siphon). All species are predaceous. The Notonectidae (back swimmers) (Figure 8.23A), as their common name suggests, swim with the true ventral surface uppermost. The hindlegs are long and oarlike, with a fringe of bristles on the posterior margin. Back swimmers usually rest at the water surface and, when disturbed, swim actively downward and grasp onto a submerged object. Cosmopolitan genera are *Notonecta* and *Anisops*. The mainly tropical Naucoridae are slow-moving bugs found among submerged vegetation. In the family Belostomatidae (giant water bugs) are found the largest members of the Heteroptera, some South American species reaching a length of 11 cm. *Lethocerus* has a worldwide distribution. Its species are brown, oval-shaped bugs with raptorial forelegs (Figure 8.23B). The Nepidae are called "water scorpions" because of their respiratory siphon. The group is mainly tropical or subtropical, though the genera *Ranatra* (Figure 8.23C) and *Nepa* (Figure 8.23D) are cosmopolitan.

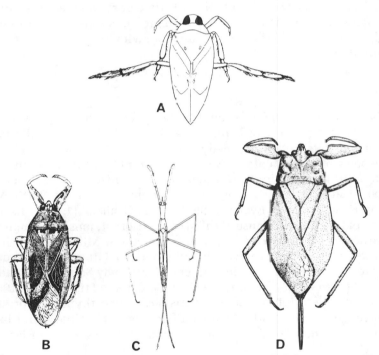

FIGURE 8.23. Notonectoidea. (A) A back swimmer, *Notonecta undulata* (Notonectidae); (B) the giant water bug, *Lethocerus americanus* (Belostomatidae); (C) a water scorpion, *Ranatra fusca* (Nepidae); and (D) a water scorpion, *Nepa apiculata* (Nepidae). [A, from D. J. Borror, D. M. Delong, and C. A. Triplehorn, 1976, *An Introduction to the Study of Insects*, 4th ed. By permission of Holt, Rinehart and Winston, Inc. B, C, from A. R. Brooks and L. A. Kelton, 1967, Aquatic and semiaquatic Heteroptera of Alberta, Saskatchewan and Manitoba (Hemiptera) *Mem. Entomol. Soc. Can.* **51:**92 pp. By permission of the Entomological Society of Canada. D, from R. C. Froeschner, 1962, Contributions to a synopsis of the Hemiptera of Missouri, *Am. Midl. Nat.* **67**(1):208–240. By permission of the American Midland Naturalist.]

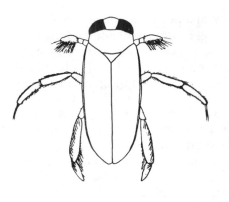

FIGURE 8.24. Corixoidea. A water boatman, *Sigara atropodonta* (Corixidae). [From D. J. Borror, D. M. De-long, and C. A. Triplehorn, 1976, *An Introduction to the Study of Insects* 4th ed. By permission of Holt, Rinehart and Winston, Inc.]

Superfamily Corixoidea

The 200 or so species of Corixoidea are placed in a single family, CORIXIDAE (water boatmen) (Figure 8.24). These are usually regarded as microphagous bugs that feed on detritus, algae, etc., which they scoop up with their flattened, hairy fore tarsi. However, many species are known to be predaceous. They generally cling onto the substrate or submerged vegetation with their midlegs, only surfacing to renew their air supply. The hindlegs are enlarged, flattened, and fringed with hairs for swimming purposes. *Corixa* is a cosmopolitan genus.

Literature

Probably the best general account of the order are the chapters by Poisson (1951) and Pesson (1951) in "Traité de Zoologie." Miller (1971) discusses the biology of the Heteroptera. Kennedy and Stroyan (1959) and Auclair (1963) review aspects of aphid biology. Swan and Papp (1972) include much useful information on geographic distribution, food, life history, and (where appropriate) pest status, in their survey of the common insects of North America. Heslop-Harrison (1956), Evans (1963), and Cobben (1968) discuss the phylogeny of the order. Because of the size of the order, most systematic papers deal with a specific group of Hemiptera. China and Miller (1959), however, provide a key to the families and subfamilies of the heteropteran suborder. Keys and descriptions of the British Heteroptera are given by Southwood and Leston (1959) and Macan (1965) [aquatic species]. Le Quesne (1960, 1965, 1969) provides keys and descriptions of the British Auchenorrhyncha. Borror *et al.* (1976) have keys to the North American families of Hemiptera and include a selection of references for those requiring greater access to the order.

Auclair, J. L., 1963, Aphid feeding and nutrition, *Annu. Rev. Entomol.* **8:**439–490.

Borror, D. J., Delong, D. M., and Triplehorn, C. A., 1976, *An Introduction to the Study of Insects*, 4th ed., Holt, Rinehart and Winston, New York.

China, W. E., and Miller, N. C. E., 1959, Check-list and keys to the families and subfamilies of the Hemiptera–Heteroptera, *Bull. Br. Mus. Nat. Hist. Entomol.* **8**(1):1–45.

Cobben, R. H., 1968, *Evolutionary Trends in Heteroptera*, Part I, Center for Agricultural Publishing and Documentation, Wageningen, Netherlands.

Evans, J. W., 1963, The phylogeny of the Homoptera, *Annu. Rev. Entomol.* **8:**77–94.

Heslop-Harrison, G., 1956, The age and origin of the Hemiptera, with special reference to the sub-order Homoptera, *Proc. Univ. Durham Philos. Soc. Ser. A* **12**:150–169.

Kennedy, J. S., and Stroyan, H. L. G., 1959, Biology of aphids, *Annu. Rev. Entomol.* **4**:139–160.

Le Quesne, W. J., 1960, 1965, 1969, Hemiptera–Fulgoromorpha and Hemiptera–Cicadomorpha, *R. Entomol. Soc. Handb. Ident. Br. Insects* **2**(3):1–68; **2**(2):1–64 65–148.

Macan, T. T., 1965, A revised key to the British water bugs (Hemiptera–Heteroptera), (2nd revised ed.), *Sci. Publ. F.W. Biol. Assoc.* **16**:78 pp.

Miller, N. C. E., 1971, *The Biology of the Heteroptera*, 2nd ed., Hill, London.

Pesson, P., 1951, Ordre des Homoptères, in: *Traité de Zoologie* (F.-P. Grassé, ed.), Vol. X, Masson, Paris.

Poisson, R., 1951, Ordre des Heteroptères, in: *Traité de Zoologie* (P.-P. Grassé, ed.), Vol. X, Masson, Paris.

Southwood, T. R. E., and Leston, D., 1959, *Land and Water Bugs of the British Isles*, Warne, London.

Swan, L. A., and Papp, C. S., 1972, *The Common Insects of North America*, Harper & Row, New York.

5. Thysanoptera

SYNONYM: Physapoda   COMMON NAME: thrips

Small to minute slender-bodied insects; head with short 6- to 9-segmented antennae, asymmetrical suctorial mouthparts, small but prominent compound eyes, ocelli present or absent; prothorax large and free, legs with unsegmented or 2-segmented tarsi and terminal eversible vesicles, wings when present are narrow and fringed with long setae; external genitalia variable, cerci always absent.

Some 4500 species belong to this widely distributed order.

Structure

Thrips are generally yellowish, brown, or black elongate insects that range in length from about 0.5 to 10 mm. The squarish head is devoid of sutures and sulci, prognathous or hypognathous, and bears a pair of prominent compound eyes, 6- to 9-segmented antennae inserted close together on the head, and asymmetrical suctorial mouthparts. The labrum and labium form a short cone-shaped rostrum that encloses the styletlike left mandible and paired maxillae. The right mandible disappears during embryogenesis. Labial and maxillary palps are present. Ocelli (three) are found only in winged adults. The prothorax is large and free, the mesothorax and metathorax fused. The legs are all similar and each bears an unsegmented or 2-segmented tarsus at the tip of which is a bladderlike structure that can be everted by means of hemolymph pressure to enable the insect to walk on a variety of surfaces. Wings may be fully developed, shortened, or absent. When present they are membranous, narrow structures with few or no veins. Each possesses a fringe of long setae. The fore and hind wings are coupled during flight. The 11-segmented abdomen tapers posteriorly Females of the suborder Tubulifera have a delicate, reversible, chute-like ovipositor, but in Terebrantia this is a strong, external structure with four saw-like valves. Both tubuliferan and terebrantian males have well-developed external genitalia.

The internal structure of Thysanoptera is generally similar to that of Hemiptera. There is a large cibarial pump anterior to the alimentary canal. The foregut is short and leads in to a large midgut, which is differentiated into an anterior croplike region and tubular hind portion. There are four Malpighian tubules. The nervous system is highly specialized. The subesophageal and prothoracic ganglia are fused, those of the mesothorax and metathorax remain separate, and those of the abdomen have coalesced to form a single center. Each ovary contains four panoistic ovarioles. A spermatheca occurs, but accessory glands may or may not be present. In males the testes are fusiform and connect via short vasa deferentia to the ejaculatory duct. The latter also receives ducts from one or two pairs of large accessory glands.

Life History and Habits

Most species of thrips are found on growing vegetation, particularly among flowers. A few species live on the ground among rotting vegetation. In general they feed on sap, but a few live on fungi or are predaceous on other small arthropods. Not surprisingly, a large number of thrips are important pests because of the damage they cause, either directly by weakening plants and perhaps preventing fruit formation, or indirectly by acting as transmitters of disease-causing viruses. Partially offsetting their importance as pests is the benefit they produce by assisting in pollination and, in predaceous species, killing harmful insects. Most thrips seem to be rather inactive insects, though a number can run rapidly on occasion. Many can fly but only rarely resort to this activity.

Most species are bisexual and oviparous; a few, however, practice facultative or obligate parthenogenesis and some are viviparous. The eggs of Terebrantia are somewhat kidney-shaped and laid in plant tissues; those of Tubulifera are oval and simply deposited on the surface of a plant. The postembryonic development of thrips is of interest because it parallels in some respects that found in endopterygote orders. The first two juvenile stages are typically exopterygote in that the insect generally resembles the adult except for the lack of wings and in the possession of fewer antennal segments. However, the remaining instars (two in Terebrantia, three in Tubulifera) are resting stages that do not feed and in which some degree of metamorphosis occurs. The first of these resting stages is called the prepupa (propupa), the remaining one or two, the pupal stage(s). Whether the latter is homologous with the pupa of endopterygotes is open to discussion (see Chapter 2, Section 3.3).

Phylogeny and Classification

The Thysanoptera appear to be most closely related to the Hemiptera, with which they may have had a common ancestor in the Upper Carboniferous or Lower Permian period. It has been suggested (Grinfel'd, 1959, cited in Stannard, 1968) that the order arose as a pollen-feeding group in which the mouthparts, already evolving toward the suctorial type, became asymmetrical owing to loss of the right mandible. It appears that a single mandible would be more effective than a pair of these structures in piercing pollen grains. Unfortunately, the fossil record is extremely poor for this order. The earliest fossil, *Permothrips*, from the Upper Permian, already has the narrow, straplike wings typical of the order and appears to be close to the modern Aeolothripidae. Of the two suborders, the Terebrantia, whose members retain wing veins and, in females, a well-developed ovipositor, are undoubtedly more primitive than the Tubulifera, whose species have veinless wings and no ovipositor. Within the Terebrantia, the Aeolothripidae are believed to form the most primitive group, as indicated by the slightly broadened wings with several veins, the 9-segmented antennae, and 3-segmented maxillary palps. From some aeolothripidlike ancestor the three remaining terebrantian families evolved. These are the Merothripidae, Heterothripidae, and Thripidae. Members of these three families still retained their ancestral association with vascular plants, as pollen or sap feeders. The single family that constitutes the Tubulifera, the Phlaeothripidae, probably arose from an early thripid ancestor (Stannard, 1968). Many of the Phlaeothripidae exploited new feeding niches such as fungal sap, or the moisture in humus or beneath bark.

Suborder Terebrantia

Though the AEOLOTHRIPIDAE are considered the most primitive family of Terebrantia, most of its members prey on other insects, and only a few suck plant juices. The HETEROTHRIPIDAE and MEROTHRIPIDAE are small families confined to the neotropical region. The THRIPIDAE are the second largest family of Thysanoptera and contain most of the pest species. They are primarily sap feeders and cause damage either directly by generally weakening their hosts or indirectly by serving as vectors of disease-causing viruses. Several pest species have been transported by man to many parts of the world. *Taeniothrips inconsequens*, the pear thrips (Figure 8.25A), is a European species now widespread throughout the United States. *T. simplex*, the gladiolus thrips, is another wide-

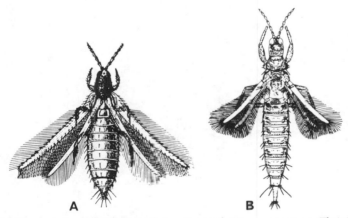

A   **B**

FIGURE 8.25.   Thysanoptera. (A) The pear thrips, *Taeniothrips inconsequens* (Thripidae); and (B) *Liothrips citricornis* (Phlaeothripidae). [A, from P.-P. Grassé (ed.), 1951, *Traité de Zoologie*, Vol. X. By permission of Masson, Paris. B, from L. J. Stannard, Jr., 1968, The thrips, or Thysanoptera, of Illinois, *Bull. Ill. Nat. Hist. Surv.* **29**(4):215–552. By permission of the Illinois Natural History Survey.]

spread species. *Thrips tabaci*, the onion thrips, is a cosmopolitan species, which, although preferring onions, is known to feed also on many other plants, including tomatoes, tobacco, cotton, and beans. It is known to transmit the virus that causes spotted wilt of tomatoes. Species of *Limothrips* (grain thrips) are important pests of cereal crops in various parts of the world.

Suborder Tubulifera

Members of the single family PHLAEOTHRIPIDAE (Figure 8.25B) that comprise the suborder Tubulifera are readily distinguished from other Thysanoptera by many ecological as well as morphological features. They are generally large thrips, occasionally reaching more than 1 cm in length, often gregarious, living mostly beneath leaf litter where they feed on fungi or small insects. A few species are sap feeders and may become pests. This is the largest family of thrips, some 2700 species having been recognized.

Literature

Stannard (1968) and Lewis (1973) have good accounts of the biology of thrips, the former also providing keys for and descriptions of many North American species as well as discussing the phylogeny of the order. Morison (1949) provides a key to most of the British species.

Lewis, T., 1973, *Thrips, Their Biology, Ecology and Economic Importance*, Academic Press, New York.

Morison, G. D., 1949, Thysanoptera of the London area, *London Naturalist Reprint* **59**:1–131.

Stannard, L. J., Jr., 1968, The thrips, or Thysanoptera, of Illinois, *Bull. Ill. Nat. Hist. Surv.* **29**(4):215–552.

9

The Panorpoid Orders

1. Introduction

In this and the following chapter we shall deal with the endopterygote insects—those that have a distinct pupal instar in which the insect undergoes a drastic metamorphosis from the larval to the adult form. As noted in Chapter 2, considerable difficulty has arisen in deciding whether endopterygote orders are monophyletic or polyphyletic. Five orders, Mecoptera, Trichoptera, Lepidoptera, Diptera, and Siphonaptera, show clear affinities which enable them to be grouped together as the panorpoid complex. The remaining orders, Megaloptera, Raphidioptera, Neuroptera, Coleoptera, Hymenoptera, and Strepsiptera show few affinities with the panorpoid group.

2. Mecoptera

SYNONYMS: Panorpatae, Panorpina   COMMON NAME: scorpion flies

Slender medium-sized insects; head prolonged ventrally into a broad rostrum with long filiform antennae, well-developed compound eyes, and biting mouthparts; usually with two pairs of identical membranous wings with primitive venation and carried horizontally at rest; abdomen with short cerci and, in males, prominent genitalia.

Larvae usually eruciform with compound eyes, biting mouthparts, and thoracic legs; abdominal legs present or absent. Pupae decticous and exarate.

This is a small order containing about 400 described species, nearly half of which belong to two genera, *Panorpa* and *Bittacus*. The order is particularly common in the Northern Hemisphere.

Structure

Adult.   A characteristic feature of Mecoptera is the ventral prolongation of the head into a broad rostrum. Incorporated into this structure are the clypeus, labrum, and maxillae. Compound eyes are well developed, and in most species there are three ocelli. The antennae are multisegmented and filiform. The

217

mouthparts are mandibulate, except in *Nannochorista*, where they are specialized and appear to foreshadow the suctorial type seen in lower Diptera. The prothorax is small, the pterothorax well developed. The legs are long and thin and adapted for walking. They have a 5-segmented tarsus. In Bittacidae the fifth tarsal segment folds back on the fourth and is used for catching prey. Two pairs of fully developed, identical, membranous wings are present in most species; the venation is primitive. A wing-coupling apparatus is present. The abdomen of females is clearly 11-segmented and bears a pair of 2-segmented cerci. Only rarely (in Boreidae) is an ovipositor present. In males segment 9 is bifurcate, and bears a pair of bulbous claspers. Segment 10 is inconspicuous and bears unsegmented cerci. The aedeagus lies at the base of the claspers. In Panorpidae the terminal segments are turned upward and resemble somewhat a scorpion's sting, hence the common name for the order.

The foregut has two interesting features. The esophagus contains two dilations which appear to form a sucking apparatus, and the crop is provided with long setae which may act as a filter. Six Malpighian tubules occur. The nervous system is generalized, with three thoracic and between five and eight abdominal ganglia. Each testis comprises three or four follicles. The paired vasa deferentia open separately into a median seminal vesicle which also receives paired accessory glands. In females each ovary contains 7–19 polytrophic ovarioles (panoistic ovarioles in Boreidae). The paired oviducts unite before entering a genital pouch. The ducts from the spermatheca and accessory glands also lead into the pouch.

Larva and Pupa. Larvae are caterpillarlike, with a distinct head capsule which bears compound eyes. Prolegs occur on the first eight abdominal segments, and the apex of the abdomen bears either a suction disc or a pair of hooks. In Boreidae larvae are grublike, lacking prolegs and a terminal suction disc. Pupae are decticous and exarate.

Life History and Habits

Adult scorpion flies are most frequently encountered in cool, shaded locations, especially among low vegetation. They can fly actively when disturbed, though they normally rest on grass, under leaves, etc. Males prey on soft-bodied insects, which they capture, both on the wing and on plants, with their legs. It is not known for certain whether females capture prey or feed only on nectar plus food received from a male during courtship. In some species a male captures prey and then attracts females by secreting a volatile chemical from the posterior abdominal segments. After copulation the male offers the prey to the female. When she has eaten the pair separates. Only males with food are attractive to females, and a male may use the same prey in order to woo several females. In other species prey is not offered to the female by the male, but fluids are exchanged orally. Eggs are either dropped randomly or, more commonly, deposited in batches in moist depressions in the ground. Larvae are carnivorous (herbivorous in Boreidae) and pass through four to seven instars before constructing an earthen cell in which to pupate. Development time is variable. There are probably two generations per year in some species from warmer regions, one per year in those from temperate climates.

The Mecoptera and Neuroptera were flourishing orders at the beginning of the Permian period, though the nature of their ancestors remains unsolved. Within the early Mecoptera three suborders are distinguishable. These are (1) Protomecoptera, a small group containing a few fossil forms and three archaic recent species; (2) Paramecoptera, containing two Permian genera and from which Trichoptera were possibly derived; and (3) Eumecoptera, in which the majority of Mecoptera, both fossil and recent, are placed. Almost all Permian Eumecoptera were placed by Tillyard (1935) in the family Permochoristidae, though this is now usually subdivided into eight distinct families. It is from this heterogeneous group of fossils that the origins of recent Eumecoptera and other panorpoid orders can be traced. Of the eight families five were extinct by the middle of the Mesozoic. The sixth, Nannochoristidae, has survived to the present time. The Mesochoristidae gradually merged into the modern Choristidae and were perhaps ancestral to two other recent families, the Panorpidae and Bittacidae. The final family, Permotipulidae, is now known to be a composite group that includes fossils that are properly placed in the Nannochoristidae or the Diptera. The remaining Permotipulidae, belonging to the genus *Permotipula*, although not the direct ancestors of the Diptera, must be very close to the point at which the two orders diverged. The origin and affinities of the recent family Boreidae are unclear. There are many differences between this group and other Mecoptera, and it is clear that the boreids must have left the main mecopteran group at an early date, perhaps from some permochoristid family (Tillyard, 1935). Unfortunately, there is no fossil evidence to support this suggestion. Hinton (1958) considers the Boreidae sufficiently different as to warrant being placed in a separate order, Neomecoptera.

Suborder Protomecoptera

Only three species arranged in two families are included in the suborder Protomecoptera. *Notiothauma reedi* (NOTIOTHAUMIDAE) occurs in Chile; a Western Australian species (*Austromerope poultoni*) and one species (*Merope tuber*) from the eastern United States are placed in the MEROPIDAE. Practically nothing is known of their life history.

Suborder Eumecoptera

Five families are included in the suborder Eumecoptera. BOREIDAE (Figure 9.1A) have a holarctic distribution. They are small (2–5 mm in length), and the wings are bristlelike in males and scalelike or vestigial in females. The largest mecopteran family, the PANORPIDAE (Figure 9.1B), is another essentially holarctic family, though it also has representatives in Asia and Australia. The BITTACIDAE (Figure 9.1C) are another large family with a cosmopolitan distribution. Its members are recognized by their raptorial tarsi. They are found typically hanging from vegetation by their forelegs, waiting for suitable prey, which they grasp with their hind tarsi. NANNOCHORISTIDAE form a small, primitive Southern Hemisphere family. Adults live in the vicinity of streams or lakes,

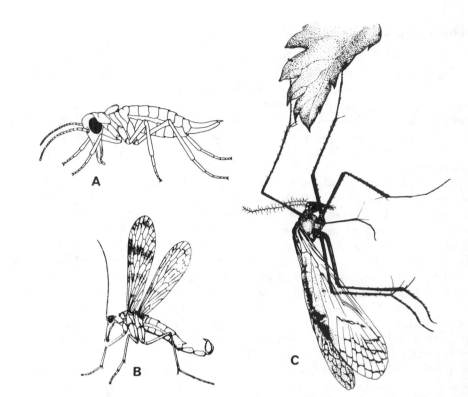

FIGURE 9.1. Mecoptera. (A) *Boreus brumalis* (Boreidae), (B) *Panorpa helena* (Panorpidae), and (C) *Bittacus pilicornis* (Bittacidae). [A, B, from D. J. Borror, D. M. Delong, and C. A. Triplehorn, 1976, *An Introduction to the Study of Insects*, 4th ed. By permission of Holt, Rinehart and Winston, Inc. C, from D. W. Webb, N. D. Penny, and J. C. Marlin, 1975, The Mecoptera, or scorpionflies, of Illinois, *Bull. Ill. Nat. Hist. Surv.* **31**:251–316. By permission of the Illinois Natural History Survey.]

and larvae are aquatic. CHORISTIDAE, with only four species, are restricted to Australia.

Literature

Considering the central position that this order occupies in any discussion of the evolution of most endopterygote orders, surprisingly little work has been done on Mecoptera. Information on the biology of Mecoptera is provided by Carpenter (1931), Setty (1940), Byers (1963), and Webb *et al.* (1975). The evolution of the order and its relationship to the other panorpoid groups are discussed by Tillyard (1935) and Hinton (1958). Byers (1954) provides a list of the North American species, with notes on their distribution.

Byers, G. W., 1954, Notes on North American Mecoptera, *Ann. Entomol. Soc. Am.* **47**:484–510.
Byers, G. W., 1963, The life history of *Panorpa nuptialis* (Mecoptera: Panorpidae), *Ann. Entomol. Soc. Am.* **56**:142–149.
Carpenter, F. M., 1931, The biology of the Mecoptera, *Psyche* **38**:41–55.
Hinton, H. E., 1958, The phylogeny of the panorpoid orders, *Annu. Rev. Entomol.* **3**:181–206.
Setty, L. R., 1940, Biology and morphology of some North American Bittacidae, *Am. Midl. Nat.* **23**:257–353.

Tillyard, R. J., 1935, The evolution of the scorpion-flies and their derivatives, *Ann. Entomol. Soc. Am.* **28**:1–45.

Webb, D. W., Penny, N. D., and Marlin, J. C., 1975, The Mecoptera, or scorpionflies, of Illinois, *Bull. Ill. Nat. Hist. Surv.* **31**(7):251–316.

3. Trichoptera

SYNONYM: Phryganoidea   COMMON NAME: caddis flies

Small to medium-sized mothlike insects; head with setaceous antennae, reduced mandibulate mouthparts, and usually small compound eyes (large in some males); prothorax small, two pairs of wings almost always present and covered with hairs, held rooflike over body at rest, legs identical with 5-segmented tarsi

Larvae aquatic usually eruciform and case dwellers. Pupae decticous and exarate.

Nearly 7000 species of Trichoptera have been described, including some 1200 species from North America. The order has a worldwide distribution.

Structure

Adult. Caddis flies are between 1.5 and 40 mm in length. They are mothlike and usually drab in color. The setaceous antennae are always long, sometimes several times the length of the wings. The compound eyes are usually small, but in some males they are very large and almost meet at the vertex. Ocelli (three in number) are present in some species, absent in others. The mouthparts are reduced; the mandibles are vestigial, the maxillae are small and closely associated with the labium. The hypopharynx is well developed and in several groups modified for sucking up liquid food. The prothorax is small and ringlike; the mesothorax and metathorax are well developed. The legs are long and slender and have 5-segmented tarsi. Two pairs of membranous wings covered with long hairs are almost always present, though one or both pairs may be greatly reduced in a few species. The fore and hind wings are coupled during flight. The wing venation is generalized and resembles that of some primitive Lepidoptera. There is in some species a whitish spot, the thyridium, devoid of hairs near the center of each wing. The wings are held rooflike over the body when not in use. Ten abdominal segments can be distinguished. In males the genitalia comprise a pair of claspers and bilobed aedeagus, or an aedeagus alone. In females of some species the terminal segments are retractile and function as an ovipositor.

From what little is known, the internal structure is quite generalized. The gut is short and straight, and there are six Malpighian tubules. In the ventral nerve cord three thoracic and seven abdominal ganglia are found, the metathoracic and first abdominal ganglia having fused together. The testes are saclike; the ovaries contain numerous polytrophic ovarioles.

Larva and Pupa. Larvae are generally campodeiform (non-case-making forms) (Figure 9.3A) or eruciform (case-building forms) (Figure 9.5A). The head is well sclerotized and carries a pair of very short antennae, mandibulate mouthparts, and two lateral clusters of ocelli. The thorax is variably sclerotized and bears well-developed legs which have an unsegmented tarsus. The forelegs are fairly short and used more for holding food and constructing the case than for walking. The abdomen is 10-segmented. The first abdominal segment of most species of Limnephiloidea has three prominent, retractile papillae which bear sensory hairs; these may enable the insect to maintain its position in the case. Prolegs are absent on all but the last abdominal segment. These have a pair of strong hooks for anchoring the insect to the case or substrate. In a few species respiration is entirely cutaneous; however, in most species gills are developed. In case bearers these are usually simple filamentous structures developed on the abdomen, occasionally also on the thorax. Sometimes they are arranged in groups, or they may be branched basally. Blood gills (nontracheated) occur in some species; they are usually eversible and perhaps have an osmoregulatory function.

Pupae are decticous and exarate: they respire using the larval gills or cutaneously.

Life History and Habits

Caddis flies, which are usually found close to fresh water, are mainly crepuscular or nocturnal, hiding by day among vegetation. They do not take solid food but are capable of sucking up nectar or water, which enables them to survive for several weeks. Coupling takes place during flight, but the insects come to rest for the actual process of insemination. Eggs are laid in ribbons or masses either directly into the water or on some object immediately above it, the young larvae dropping into the water on hatching. Development may be rapid and in most species includes five larval instars. Most temperate region caddis flies have only one generation per year (univoltine), though some species are bivoltine or semivoltine. Diapause in the final larval or adult instar, rarely the egg stage, is characteristic of many species. In more primitive families larvae do not build cases. They either live under stones or construct a nonportable silken web. In addition to providing them with protection, the web also serves to trap food. Many larvae, however, construct a portable case whose shape and the nature of the materials incorporated into it are variable but characteristic for a particular family or even genus. The case has a silken lining to which other materials both organic and inorganic are stuck. Members of most case-bearing species are omnivorous, with a preference for vegetable matter.

Prior to pupation, a case-bearing larva attaches the case to the substrate and closes off each end with a perforated silken wall. The pupa either lies freely in the case or spins a very flimsy cocoon. Members of non-case-bearing species spin a silken cocoon, which is often strengthened by the addition of foreign materials. The pharate adult (often incorrectly referred to as the "pupa") (see Chapter 21) bites its way out of the case or cocoon and crawls or swims to a suitable emergence site. Molting takes place at the water surface and adults can fly immediately.

Phylogeny and Classification

The fossil record of Trichoptera is poor. Tillyard (1935), mainly on the basis of a comparison between the wing venation of primitive Trichoptera and fossil Eumecoptera, concluded that the ancestor of the order was a member of the Permian eumecopteran family Permochoristidae. However, *Belmontia*, a Permian fossil belonging to the Paramecoptera, has a venation which is intermediate between that of primitive Mecoptera and that of primitive Trichoptera, suggesting that the latter are derived from paramecopteran rather than eumecopteran stock (Ross, 1967). The earliest truly trichopteran fossil is generally believed to be *Necrotaulius* from the early Jurassic period. During the Mesozoic, the order seems to have undergone a wide radiation, and by the end of this era most recent families had been established. Within the order the evolutionary development of the larval stage has far outstripped that of the adult. The order is presumed to have evolved in a forested habitat, through which small streams ran. Such a habitat would not be conducive to a wide adaptive radiation of adults, but larvae, evolving in a "new," relatively unexplored habitat, were able to diversify in a remarkable manner. The net result of this is that, whereas larval Trichoptera can be identified relatively easily from a variety of ecological and morphological characters, many adults (especially those from families which still inhabit the "primitive" environment) are only distinguished with the greatest difficulty. Their evolution has been concerned for the most part with changes in the genitalia and development of specific behavior patterns to prevent interspecific mating (Ross, 1967).

The many modern families fall into several natural groups (Figure 9.2). The seven families of retreat (or web) makers are included in the suborder Annulipalpia, containing the single superfamily Hydropsychoidea. The remaining 27 families that together make up the other suborder, Integripalpia,

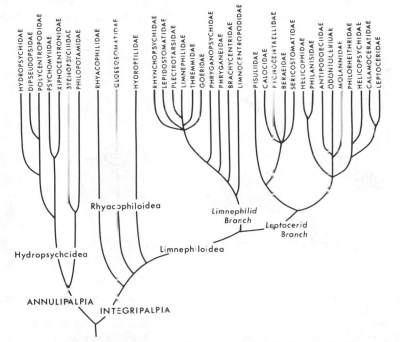

FIGURE 9.2. A suggested phylogeny of Trichoptera. [After H. H. Ross, 1967, The evolution and past dispersal of the Trichoptera. Reproduced, with permission, from the *Annual Review of Entomology*, Volume 12 © 1967 by Annual Reviews Inc.]

fall into three groups, the Rhyacophiloidea, and the limnephilid and leptocerid branches of the Limnephiloidea. The Rhyacophiloidea include three primitive families that have been separated for a considerable time, yet retain a number of common features. The case-building Limnephiloidea appear to have evolved from a hydroptilid ancestor.

Suborder Annulipalpia

Members of the suborder Annulipalpia are web-making Trichoptera with the following features: terminal segment of maxillary palp annulated; adults without a supratentorium; females with distinct cerci; larval hindlegs projecting downward; and larval anal hooks large, slender, and curved.

Superfamily Hydropsychoidea

Four of the seven families recognized by Ross (1967) are large and common. The three remaining groups are small and frequently included by other authorities in the larger families. The PHILOPOTAMIDAE (Figure 9.3A) are among the most primitive members of the superfamily Hydropsychoidea. Adults are small (6 to 9 mm long) and generally have brownish bodies and grey or blackish wings. Sometimes females are apterous. Larvae live in fast-flowing streams and build tubular webs whose entrance is larger than the exit. A larva stays in its web and feeds on material caught in it. PSYCHOMYIIDAE and POLY-CENTROPODIDAE (Figure 9.3B) tend to be larger than the members of the previous family, with a length of up to 2 cm. They are usually dark in color. Larvae are found in a variety of situations, from rapidly flowing streams to large lakes. They

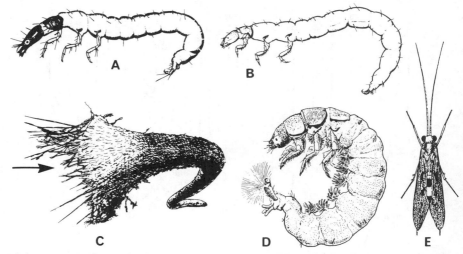

FIGURE 9.3. Hydropsychoidea. (A) *Chimarra* sp. (Philopotamidae) larva; (B) *Neureclipsis bimaculata* (Polycentropodidae) larva; (C) capturing net of *N. bimaculata*, arrow indicating direction of current; (D) *Hydropsyche simulans* (Hydropsychidae) larva; and (E) *H. simulans* adult. [A–C, from G. B. Wiggins, 1977, *Larvae of the North American Caddisfly Genera.* By permission of the Royal Ontario Museum. D, E, from H. H. Ross, 1944, The caddisflies, or Trichoptera, of Illinois, *Bull. Ill. Nat. Hist. Surv.* **23**:1–326. By permission of the Illinois Natural History Survey.]

construct tubular or conical webs (Figure 9.3C), which are sometimes buried in the substrate. The HYDROPSYCHIDAE (Figure 9.3D,E) form a large family whose members breed in small, fast-flowing streams, especially where there are rapids. Larvae live in a tubular retreat, frequently covered with foreign matter in front of which is a wide, cup-shaped net in which food material is trapped.

Suborder Integripalpia

The Integripalpia are free-living or case-building forms with the following characteristics: apical segment of maxillary palps not annulated; adults with a supratentorium; females without cerci; larval hindlegs projecting sideways; larval anal hooks short and stout.

Superfamily Rhyacophiloidea

Members of the superfamily Rhyacophiloidea are either free-living or builders of relatively simple cases. The family RHYACOPHILIDAE (Figure 9.4A,B) contains the most archaic members of the order. The family is a large one, with a worldwide distribution. Adults are generally brownish or mottled and vary in length from 3 to 13 mm. Larvae are predaceous and found only in clear, cold, fast-flowing streams. They do not construct a case but move freely over the bottom of the stream, producing a continuous strand of silk. Mature larvae spin a silken cell, often incorporating a layer of sand grains, in which pupation occurs. GLOSSOSOMATIDAE are little different as adults from Rhyacophilidae, in which they are often included as a distinct subfamily. The different habits of

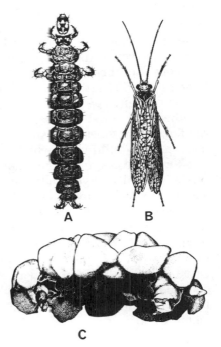

FIGURE 9.4. Rhyacophiloidea. (A) *Rhyacophila fuscula* (Rhyacophilidae) larva, (B) *Rhyacophila fenestra* adult, and (C) *Glossosoma* sp. (Glossosomatidae) case with larva. [A, C, from G. B. Wiggins, 1977, *Larvae of the North American Caddisfly Genera*. By permission of the Royal Ontario Museum. B, from H. H. Ross, 1944, The caddisflies, or Trichoptera, of Illinois, *Bull. Ill. Nat. Hist. Surv.* **23**:1–326. By permission of the Illinois Natural History Survey.]

the larvae are, however, considered sufficient justification for their separation. Larvae, found in swiftly flowing water, are phytophagous and live in saddle-like or turtle-shaped cases (Figure 9.4C); that is, the ventral side of the case is flat and composed of fine sand grains, and the dorsal surface is strongly convex and built of coarser material. When larvae are about to pupate they cut away the ventral part of the case and fix the dorsal component to the substrate. Adult HYDROPTILIDAE are hairy, minute Trichoptera (1.5–6 mm in length), often called micro-caddis flies, whose larvae are found in a variety of permanent waters. The larvae are of particular interest in that they exhibit hypermetamorphosis (see Chapter 21). The first four instars are active, free-living larvae which feed on algae and show very little growth. Nearly all growth occurs in the last, case-making instar, which is very different in appearance from its predecessors. In many species the case has a purselike shape, from which members of the family are called purse-case makers.

Superfamily Limnephiloidea

The most specialized Trichoptera are included in the superfamily Limnephiloidea. In adults the mouthparts are considerably modified with the development of the hypopharynx as a haustellumlike structure. Larvae are always tubular case builders. The case often has a very specific design and is enlarged by the larva as it grows. Pupation takes place in the larval case. Within the Limnephiloidea two distinct evolutionary lines, the limnephilid branch and the leptocerid branch, can be distinguished on morphological grounds, though much parallelism occurs between the biology of the two groups. Notes on the commonly encountered families are given below. The LIMNEPHILIDAE (Figure 9.5A,B) form the largest family of the order with more than 200 North American species. It is a primarily holarctic group of small to large caddis flies (7–25 mm in length), which as adults are brownish with patterned wings. Larvae are found in a variety of habitats, principally ponds and slow-moving streams, and construct their cases from a variety of materials. The PHRYGANEIDAE are large caddis flies (14–25 mm in length) that breed mostly in marshes or lakes. Larvae of many species live in cases built of strips of vegetable matter glued together in a spiral (Figure 9.5C). Although it includes only a few species, the family HELICOPSYCHIDAE contains some of the best-known caddis larvae because they build cases which are shaped like a snail's shell (Figure 9.5D). The SERICOSTOMATIDAE form another large, widely distributed (perhaps polyphyletic) family of generally small caddis flies, whose larvae are found in a variety of slowly moving or still water habitats. The LEPTOCERIDAE constitute a large family of small caddis flies (5–15 mm in length) that, as adults, may be recognized by their antennae, whose length may be twice that of the body (Figure 9.5E). Larvae are found in many locations ranging from cold, swiftly flowing streams to relatively warm, marshy ponds. Cases are frequently constructed of fine stones or sand and are very elongate. In some species there is a changeover to the use of vegetable matter for case construction in the later instars (Figure 9.5F) when spiral cases similar to those of Phryganeidae may be built.

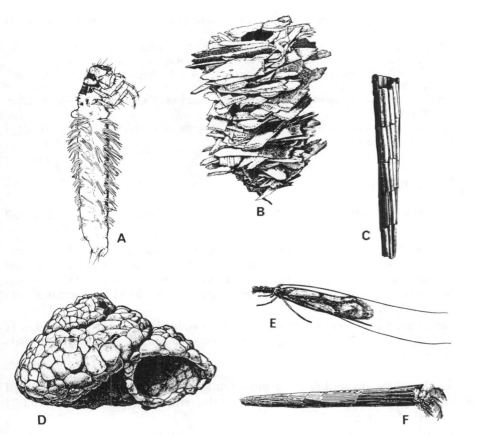

FIGURE 9.5. Limnephiloidea. (A) *Limnephilus indivisus* (Limnephilidae) larva, (B) *L. indivisus* case, (C) *Phryganea cinerea* (Phryganeidae) case, (D) *Helicopsyche borealis* (Helicopsychidae) case, (E) *Triaenodes tarda* (Leptoceridae) adult, and (F) *T. tarda* larva in case. [A–D, from G. B. Wiggins, 1977, *Larvae of the North American Caddisfly Genera.* By permission of the Royal Ontario Museum. E, F, from H. H. Ross, 1944, The caddisflies, or Trichoptera, of Illinois, *Bull. Ill. Nat. Hist. Surv.* **23**:1–326. By permission of the Illinois Natural History Survey.]

Literature

Betten (1934), Hickin (1967), and Wiggins (1977) provide accounts of the biology of Trichoptera. The evolution of the order is dealt with by Ross (1956, 1964, 1967). Keys for the identification of Trichoptera are given by Denning (1956), Ross (1959), and Wiggins (1977) [North American forms]; and Hickin (1946, 1949, 1967) [larvae and pupae of British species].

Betten, C., 1934, The caddisflies or Trichoptera of New York State, *Bull. N.Y. State Mus.* **292**: 576 pp.

Denning, D. G., 1956, Trichoptera, in: *Aquatic insects of California* (R. L. Usinger, ed.), University of California Press, Berkeley.

Hickin, N. E., 1946, Larvae of the British Trichoptera, *Trans. R. Entomol. Soc. London* **97**:187–212.

Hickin, N. E., 1949, Pupae of British Trichoptera, *Trans. R. Entomol. Soc. London* **100**:275–289.

Hickin, N. E., 1967, *Caddis Larvae*, Hutchinson, London.

Ross, H. H., 1956, *Evolution and Classification of the Mountain Caddisflies*, University of Illinois Press, Urbana.

Ross, H. H., 1959, Trichoptera, in: *Freshwater Biology* (W. T. Edmondson, ed.), Wiley, New York.

Ross, H. H., 1964, Evolution of caddisworm cases and nets, *Am. Zool.* **4**:209–220.

Ross, H. H., 1967, The evolution and past dispersal of the Trichoptera, *Annu. Rev. Entomol.* **12**:169–206.

Tillyard, R. J., 1935, The evolution of the scorpion-flies and their derivatives, *Ann. Entomol. Soc. Am.* **28**:1–45.

Wiggins, G. B., 1977, *Larvae of the North American Caddisfly Genera (Trichoptera)*, University of Toronto Press, Toronto.

4. Lepidoptera

SYNONYM: Glossata   COMMON NAMES: butterflies and moths

Insects whose body and appendages are covered with scales; head with large compound eyes and mouthparts in form of suctorial proboscis; prothorax in most species reduced, two pairs of membranous wings present in almost all species with few crossveins; posterior abdominal segments much modified in connection with reproduction, cerci absent.

Larvae eruciform with well-developed head, mandibulate mouthparts, and 0 to 11 (usually 8) pairs of legs. Pupae in most species adecticous and obtect, in others decticous.

Lepidoptera are perhaps the most familiar and easily recognized of all insects. About 112,000 species have been described, including about 10,000 from North America.

Structure

Adult. Adult Lepidoptera have a very constant fundamental structure, a feature which led to great difficulties in determining the phylogeny of the order. They range in length from a few millimeters up to 10 cm or more. The entire body and appendages are covered with scales (modified hairs). The compound eyes are large and cover a major portion of the head capsule. Two ocelli may be present in most species concealed by scales. The antennae are of variable size and structure. In most Lepidoptera mandibles are absent and the maxillae are modified as a suctorial proboscis (see Chapter 3, Section 3.2.2 and Figure 3.12). When not in use the proboscis is coiled away beneath the thorax. In most species the prothorax is reduced and collarlike. The mesothorax is the larger of the pterothoracic segments that bear large tegulae, a characteristic feature of the order. Auditory organs are present on the metathorax of some moths. Both the fore and hind wings are generally large, membranous, and covered with scales. The latter are flattened macrotrichia supported on a short, thin stalk. The surface of the scale may be striated, which leads to the production of iridescent colors. Pigments may be deposited within the scales. In males of some species certain scales (androconia) are modified so as to facilitate the volatization of material produced in the underlying scent glands. In primitive

Lepidoptera the wing venation is identical in the fore and hind wings (homoneurous) and resembles that of primitive Trichoptera. In advanced forms there is considerable divergence between the venation of the fore and hind wings (the heteroneurous condition). The wing-coupling apparatus of primitive Lepidoptera comprises simply the small jugum of the fore wing, which lies on top of the hind wing. Occasionally a few short spines are present on the anterior part of the hind wing, which assist in coupling. In higher Lepidoptera the coupling apparatus is usually made up of the retinaculum of the fore wing and frenulum of the hind wing (see Chapter 3, Section 4.3.2 and Figure 3.28). However, in certain families the frenulum has been lost, and wing coupling is achieved simply by overlapping. The humeral area of the hind wing is greatly enlarged and strengthened, and lies beneath the fore wing. This is the amplexiform system. In females of a few species wings are reduced or lost. The abdomen comprises 10 segments, though the sternum of segment 1 is missing and the ninth and tenth (sometimes also the seventh and eighth) segments are modified in relation to the genitalia. The male genitalia are complex and their homologies unclear. In females there are two basic types of genitalia (Imms, 1957; but see Mutuura, 1972 for a different interpretation). In the zeuglopteran and monotrysian type there is usually a single cloacal opening on fused sterna 9 and 10 which serves both for insemination and oviposition, as well as for defecation. In Hepialoidea, however, there are separate openings, but they are both on fused segment 9/10. In the ditrysian type there are separate openings for insemination (on sternum 8) and egg laying (sterna 9/10) (Figure 19.3). In most species the genital aperture is flanked by a pair of soft lobes, but these may be strongly sclerotized and function as an ovipositor in species which lay their eggs in crevices or plant tissue. A pair of auditory organs is found on segment 1 in some moths. Cerci are absent.

The anterior region of the foregut is modified as a pharyngeal sucking pump from which a narrow esophagus leads posteriorly and, in primitive forms, expands to form a crop. In higher Lepidoptera, however, the crop is a large lateral dilation. The midgut is short and straight, the hindgut longer and coiled. In most species there are six Malpighian tubules which enter the gut via two lateral ducts. Only two tubules occur in some species. The nervous system is somewhat concentrated. In the most primitive condition three thoracic and five abdominal ganglia are found. In the majority of Lepidoptera, however, there are two thoracic and four abdominal centers. In most male Lepidoptera the four testis follicles on each side are intimately fused to form a single median gonad, though the two vasa deferentia remain distinct and open posteriorly into seminal vesicles. The latter each receive an elongate accessory gland. In most species four polytrophic ovarioles occur in each ovary that enters the common oviduct via a short lateral tube. Various accessory glands open into the oviduct. In the monotrysian arrangement sperm are deposited directly in the spermatheca. In the ditrysian structure sperm are deposited initially in a large chamber, the bulla seminalis, from which they eventually migrate to the spermatheca via the seminal duct.

Larva and Pupa. Larvae (caterpillars) are eruciform, and the three primary body divisions are easily recognizable. The head is heavily sclerotized and bears strong, biting mouthparts, six ocelli in most species, and short, 3-segmented antennae. The prementum carries a median spinneret which receives the ducts of the silk (modified salivary) glands. Three pairs of thoracic legs and, in most species, five pairs of abdominal prolegs occur; the latter are located on segments 3 to 6 and on segment 10. Micropterygidae have eight pairs of prolegs. At the opposite extreme, many leaf-mining larvae are apodous. Larvae are usually equipped with some method for protecting themselves from would-be predators. They may be colored either procryptically or aposematically; they may build shelters in which to hide, or they may possess a variety of repugnatory devices, such as glands, that produce an obnoxious fluid or long, irritating hairs that invest the body. The pupae of most Lepidoptera are adecticous and obtect. In Zeugloptera and Dachnonypha, however, they are decticous and exarate.

Life History and Habits

Lepidoptera are found in a wide variety of habitats but are almost always associated with higher plants, especially angiosperms. Almost all adult Lepidoptera feed on nectar, the juice of overripe fruit, or other liquids. Members of a few species do not feed as adults and their mouthparts are correspondingly reduced. Most Lepidoptera are strong fliers, and some may migrate for considerable distances. They are usually strikingly colored, and balanced polymorphism occurs in some species. To facilitate reproduction, females may secrete pheromones that attract males, sometimes from a considerable distance. Likewise, pheromones secreted by males serve to initiate the mating response of a female. Lepidoptera almost always reproduce sexually and are oviparous, though facultative parthenogenesis and ovoviviparity occur in a few species. Egg-laying habits and the number of eggs laid are extremely variable. Some species simply drop their eggs in flight. Others are attracted to the larva's host plant by its odor and lay their eggs in a characteristic pattern either on or in it. Some species lay only a few eggs, others lay many thousands. Larvae of almost all species are phytophagous, and no parts of plants remain unexploited. Because of their phytophagous habits and high reproductive rate, many species are important pests. Members of a few species are carnivorous or feed on various animal products. Larval development is usually rapid, and there may be several generations each year. However, in some species larval development requires 2 or 3 years for completion. In most Lepidoptera pupation takes place within a cocoon constructed in a variety of ways. It may be made entirely of silk or, more frequently, a mixture of silk and foreign materials. In Zeugloptera and Dachnonypha the pharate adult bites open the cocoon and wriggles out of it shortly before eclosion. However, eclosion frequently takes place within the cocoon, in which case the adult has various devices that facilitate escape. These include temporary spines for cutting open the cocoon and the secretion from the mouth of a special softening fluid. In many instances the cocoon itself is specially constructed to allow easy egress. Pupae of butterflies are usually naked and suspended from a small pad of silk attached to a support. This pad may represent the remains of the ancestral cocoon.

The earliest fossil Lepidoptera are from the Eocene period, although it is clear from their specialized nature that the order must have been in existence for a considerable period before this. It becomes clear from a comparison of the structure of their primitive members that Trichoptera are the nearest living relatives of Lepidoptera, with which they perhaps had a common ancestor in the Triassic period. The Lepidoptera remained a small group until the Cretaceous period, when they underwent a remarkable radiation in conjunction with the evolution of flowering plants. Of particular interest in a discussion of lepidopteran evolution is the position of the Micropterygidae, very small mothlike insects with a worldwide distribution. They possess a combination of lepidopteran, trichopteran, mecopteran, and specialized features. In their wing venation, the absence of both a thyridium and an anal plate, presence of scales, and other features they resemble typical Lepidoptera. However, their mandibles are functional, the maxillae possess a lacinia, and the galeae are not fused. The larvae have eight pairs of abdominal prolegs, and the pupae are decticous and exarate. According to the relative significance with which different authors view these features, the Micropterygidae have been placed in the Trichoptera, Lepidoptera, or in a separate order, Zeugloptera. The modern consensus of opinion is that they are archaic Lepidoptera, sufficiently distinct from other members of the order as to warrant separate subordinal status.

Several systems have been proposed for classifying the recent Lepidoptera, the most familiar being the division of the group into Rhopalocera (butterflies) and Heterocera (moths). Hinton (1946) has described this arrangement as absurd, since it implies that the two divisions are of equal taxonomic rank, which is certainly not the case. Another arrangement is the separation into Macrolepidoptera and Microlepidoptera based mainly on size, but this is an artificial rather than a natural distinction. Division of the order into Jugatae (primitive Lepidoptera with a jugate coupling apparatus) and Frenatae (those possessing a frenulate type of coupling) is an unnatural system because even within the same family there are species which have a frenulum and species which do not. Other systems based on the nature of the mouthparts and form of the pupa have also been proposed. The system adopted here, which appears to be a natural arrangement, is based on the female genital opening. Four suborders are recognized: Zeugloptera, Dacnonypha, Monotrysia (in which the Dacnonypha were formerly included), and Ditrysia. The first three suborders together include only about 2% of the world's species. There is still much confusion regarding classification beyond the subordinal level and the rank that groups should be given. The arrangement used here is that of Common (1970) who recognizes 18 superfamilies of Ditrysia, of which two (Hesperioidea and Papilionoidea) comprise the butterflies. A tentative phylogeny of the group is shown in Figure 9.6.

Suborder Zeugloptera

The very small, diurnal moths of the suborder Zeugloptera have the following features: functional dentate mandibles, maxilla with lacinia and free galeae,

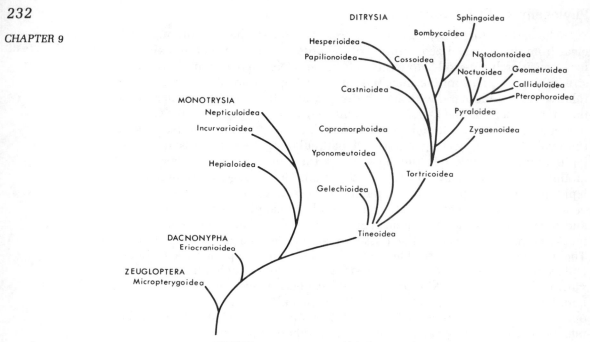

FIGURE 9.6. A suggested phylogeny of Lepidoptera.

hypopharynx modified for grinding pollen grains; homoneurous wing venation, jugate wing-coupling; larvae with eight pairs of prolegs; and pupae decticous and exarate.

The suborder contains a single superfamily, Micropterygoidea, and one small family, the MICROPTERYGIDAE, which has a worldwide distribution, though only three species occur in North America. Adults are metallic-colored moths with a body length of only a few millimeters. They feed on pollen. Larvae were believed until recently to feed on moss and liverworts, but it now seems that they are detritus feeders.

Suborder Dacnonypha

The suborder Dacnonypha includes very small, generally diurnal moths with the following features: mandibles either absent or present but not dentate, lacinia reduced or absent, galeae unspecialized or fused to form a short proboscis; homoneurous wing venation, jugate wing-coupling; larvae apodous; and pupae decticous and exarate.

All members of this suborder are included in the superfamily Eriocranioidea. Almost all the species belong to the family ERIOCRANIIDAE, a primarily holarctic group. Larvae are known as leaf miners* through their habit of living within the leaf tissue of chestnut, birch, oak, and hazel. Pupation takes place in a cocoon buried in soil.

---

*Note: Larvae of many other Lepidoptera, Diptera, and Hymenoptera of similar habit are also called leaf miners.

Included in the suborder Monotrysia are moths of variable size with the following features: mandibles and laciniae absent; galeae usually fused to form a proboscis; wing venation homoneurous or heteroneurous, wing-coupling in most species frenate, in a few jugate; females with one (rarely two) genital openings on fused segment 9/10; larvae either with prolegs or apodous; and pupae adecticous and obtect.

The suborder is somewhat heterogeneous; a clear-cut division exists between the superfamily Hepialoidea, whose members have a homoneurous wing venation, and the superfamilies Nepticuloidea and Incurvarioidea, which probably had a common origin. Their members have a heteroneurous venation. The three groups are combined in one suborder on the basis of the large number of primitive features they exhibit.

Superfamily Hepialoidea

Most members belong to the family HEPIALIDAE (ghost or swift moths), a widely distributed, though primarily Australian, group. The group appears to be an early offshoot from the main monotrysian line, judging by the homoneurous venation, the absence of ocelli, vestigial maxillary palps, and reduced labial palps of adults. These moths are generally medium-sized or large and are rapid fliers. Larvae feed either on underground roots or within wood. Pupae are active and equipped with spines and ridges that enable them to make their way to the surface prior to eclosion.

Superfamily Nepticuloidea (Stigmelloidea)

Included in the superfamily Nepticuloidea are two small but widespread families, the NEPTICULIDAE (STIGMELLIDAE) and OPOSTEGIDAE. Adults are some of the smallest Lepidoptera, with a wingspan of only a few millimeters. Larvae are leaf, stem, or bark miners, rarely gall formers.

Superfamily Incurvarioidea

Four families are usually included in the superfamily Incurvarioidea: the INCURVARIIDAE, HELIOZELIDAE, TISCHERIIDAE, and PRODOXIDAE, although the Tischeriidae are sometimes treated under the Hepialoidea, and the Prodoxidae are often considered as a subfamily of the Incurvariidae. Adults are small and equipped with a sclerotized ovipositor capable of piercing plant tissue. Larvae are apodous or have short prolegs and in early instars are leaf miners. Later they construct a portable case in which they eventually pupate.

Suborder Ditrysia

The suborder Ditrysia includes variably sized moths and butterflies with the following features: mandibles absent, galeae almost always fused to form proboscis; wing venation heteroneurous and reduced, wing-coupling frenate or amplexiform; females with two genital openings, each on a different segment; larvae variable; and pupae adecticous and obtect.

Superfamily Cossoidea

Cossoidea form one of the most primitive ditrysian superfamilies. They show affinities with both the Tineoidea and Tortricoidea and have often been included in the latter group. They constitute a single, widely distributed family, COSSIDAE, whose members are commonly known as goat, carpenter, or leopard moths (Figure 9.7). They are small to large, swift-flying, nocturnal moths, generally greyish in color, often with dark wing spots. Larvae ("carpenter worms") are mainly wood borers, though a few live in soil and feed externally on plant roots.

Superfamily Tortricoidea

The Tortricoidea are small moths that are more common in temperate than in tropical regions. Adults are greyish or brownish, frequently mottled, moths. Larvae are concealed feeders, living in shelters formed by rolling leaves or tying several together, or as miners of fruits, seeds, bark, etc. Two families are generally recognized, PHALONIIDAE and TORTRICIDAE, the latter containing several pest species, for example, *Enarmonia* (*Carpocapsa*) *pomonella*, the codling moth (Figure 9.8), whose larvae mine in apples, and *Choristoneura fumiferana*, the spruce budworm, which defoliates a wide variety of evergreen trees.

Superfamily Tineoidea

Tineoidea are the most primitive of the tineoid group of superfamilies (see Figure 9.6). Adult moths are small or medium-sized and most have narrow wings bordered with long hairs. Larvae are concealed feeders that live in portable cases, silken tubes, or mines within the food. There are several common families. The PSYCHIDAE (Figure 9.9A,B) are commonly known as bagworm

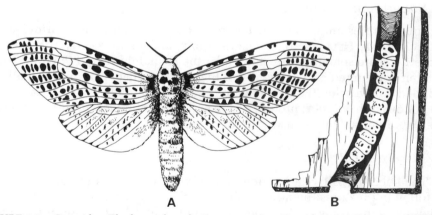

A   B

FIGURE 9.7. Cossoidea. The leopard moth, *Zeuzera pyrina* (Cossidae). (A) Female and (B) larva in burrow. [After A. D. Imms, 1957, *A General Textbook of Entomology*, 9th ed. (revised by O. W. Richards and R. G. Davies). By permission of Chapman and Hall Ltd.]

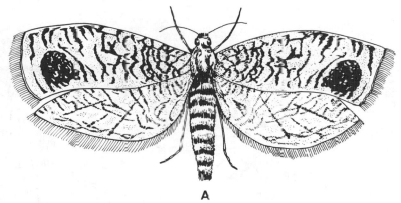

A

B

FIGURE 9.8. Tortricoidea The codling moth, *Enarmonia pomonella* (Tortricidae). (A) Adult and (B) mature larva on apple (cut away to show damage). [A, from L. A. Swan and C. S. Papp, 1972, *The Common Insects of North America*. Copyright 1972 by L. A. Swan and C. S. Papp. Reprinted by permission of Harper & Row, Publishers, Inc. B, after W. J. Holland, 1920, *The Moth Book*, Doubleday and Co., Inc.]

moths because of the portable case composed of silk and vegetable matter that a larva inhabits. Pupation occurs in the case. Females of many species are wingless and apodous; they do not leave the case but lay eggs directly in it. The affinities of the Psychidae are uncertain. Although they are included here in the Tineoidea, other authors believe that they are closer to the Lymantriidae (Noctuoidea) or Heterogynidae (Zygaenoidea). The family TINEIDAE includes the clothes moths *Tineola bisselliella* and *Tinea pellionella* (Figure 9.9C,D), a case-bearing species, and the carpet moth *Trichophaga tapetzella*. Larvae of this family may or may not live in a portable case and feed usually on dried vegetable or animal matter. LYONETIIDAE are very small moths whose larvae are either miners in leaves, stems or bark, or web builders. GRACILLARIIDAE form a cosmopolitan family of small moths recognized by their narrow, fringed wings. The larvae, which are leaf miners, undergo hypermetamorphosis. When young they are flattened and have bladelike mandibles which they use to lacerate the cells on whose sap they feed. Later they metamorphose, develop normal mouthparts, and feed in the typical manner on parenchyma.

Superfamily Yponomeutoidea

The constituent families of this small group are frequently considered as part of the Tineoidea. There are two common families. GLYPHIPTERYGIDAE are

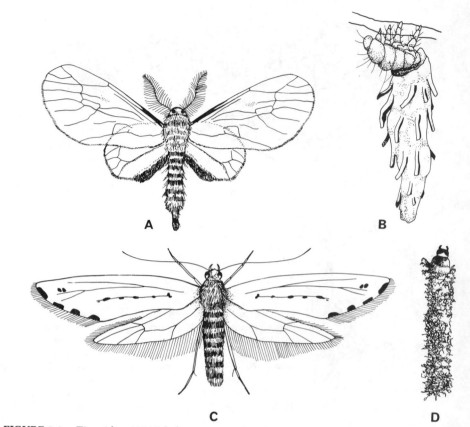

FIGURE 9.9. Tineoidea. (A) Male bagworm moth, *Thyridopteryx ephemeraeformis* (Psychidae); (B) *T. ephemeraeformis* larva with bag; (C) the case-making clothes moth, *Tinea pellionella* (Tineidae); and (D) *T. pellionella* larva in case. [A, C, D, from L. A. Swan and C. S. Papp, 1972, *The Common Insects of North America*. Copyright 1972 by L. A. Swan and C. S. Papp. Reprinted by permission of Harper & Row, Publishers, Inc. B, after W. J. Holland, 1920, *The Moth Book*, Double-day and Co., Inc.]

especially common in the Southern Hemisphere. They are small, diurnal, frequently metallic-colored moths whose larvae construct webs and, in some species, are gregarious. YPONOMEUTIDAE are very small moths whose larvae are miners, web builders, or exposed feeders. Included in this family are the diamondback moths, of which *Plutella maculipennis* (Figure 9.10) is a major pest of cabbage and other Cruciferae.

Superfamily Gelechioidea

The families of Gelechioidea are often placed in the Tineoidea. The superfamily is a large one, containing about 10,000 species, more than 4000 of which are Australian. Common (1970) includes 18 families in the group, of which four are very large. The family OECOPHORIDAE contains at least 3000 species, about 80% of which are Australian. Larvae build a portable case, mine in wood or

FIGURE 9.10. Yponomeutoidea. The diamondback moth, *Plutella maculipennis* (Yponomeutidae).

stems, join leaves, or construct subterranean tunnels. The GELECHIIDAE form a widely distributed family containing about 4000 species. Larvae of most species are web builders, but a few are miners or case bearers. The family includes many cosmopolitan pests, for example, *Pectinophora gossypiella*, the pink bollworm, damages cotton, and *Sitotroga cerealella*, the Angoumois grain moth, causes much damage to stored grains. COSMOPTERYGIDAE are miners, web builders or leaf-tying species. A few prey on scale insects. The family XYLORICTIDAE includes some of the largest tineoid Lepidoptera, species with wingspans up to 7.5 cm. Adults are typically procryptically colored, nocturnal insects. Larvae are miners or leaf-tying forms.

Superfamily Copromorphoidea

Copromorphoidea form a small, probably heterogeneous group whose three families are of doubtful systematic position. The CARPOSINIDAE, found in Hawaii and Australasia, are sometimes placed in the Tortricoidea or Tineoidea. The larvae are bark or fruit miners. The COPROMORPHIDAE are a primarily Australian family, probably most closely related to the gelechioid group. The larvae are miners. The ALUCITIDAE are the many-plume moths, so-called because the wings are cleft into six or more plumelike divisions. The larvae are miners. The family is tineoid in some respects, but more like the Pyralidae (Pyraloidea) in others.

Superfamily Castnioidea

The superfamily Castnioidea contains only about 160 species in one family, CASTNIIDAE, distributed in tropical America, India, Malaysia, and Australia. The adults are brightly colored and in some respects resemble nymphalid or hesperid butterflies, a feature that has led some authors to suggest that they are the ancestral group from which butterflies are derived.

Superfamily Zygaenoidea

The affinities of the superfamily Zygaenoidea are uncertain. Its members have certain features in common with Cossoidea, and members of the small family HETEROGYNIDAE are very similar in structure and habits to the Psychidae (Tineoidea). Zygaenoid larvae are stout, sluglike creatures that are exposed feeders on plants, or ectoparasitic on Homoptera or in ants' nests. ZYGAENIDAE are diurnal, often brightly colored moths that resemble butterflies. LIMACODIDAE form a mainly tropical family in which larvae are aposematically colored and possess stinging hairs.

Superfamily Pyraloidea

The Pyraloidea may be a heterogeneous assembly of families whose common features are the result of convergence rather than true affinity. The THYRIIDAE form a small, mainly tropical family, which is of particular interest in view of its possible relationship with other large groups of Lepidoptera. It is considered by some authors to be the ancestral group from which butterflies have evolved. The vast majority of species belong to the family PYRALIDAE, a group whose members can be arranged in several distinct subfamilies. Adults have tympanal organs on the first abdominal segment, a feature not found in other Pyraloidea. Larvae are extremely variable in their habits. They are miners, web or tube builders. They are adapted to terrestrial or aquatic habitats and a large number are pests. The grass moths (CRAMBINAE), whose larvae bore into stems, include *Chilo suppressalis*, the rice-stem borer, and *Diatraea saccharalis* (Figure 9.11), the sugarcane borer. The GALLERIINAE include the wax moth, *Galleria mel-*

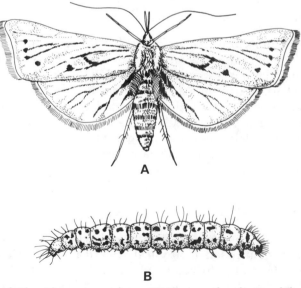

A

B

FIGURE 9.11. Pyraloidea. The sugarcane borer, *Diatraea saccharalis* (Pyralidae). (A) Adult and (B) larva. [From L. A. Swan and C. S. Papp, 1972, *The Common Insects of North America.* Copyright 1972 by L. A. Swan and C. S. Papp. Reprinted by permission of Harper & Row, Publishers, Inc.]

lonella, which lives in beehives. In the large subfamily PYRAUSTINAE, the larvae build a web among leaves and fruits. Important pests belonging to this group are *Pyrausta (Ostrinia) nubilalis*, the European corn borer, which attacks maize, and various webworms belonging to the genera *Loxostege* and *Diaphania*.

Superfamily Pterophoroidea

The single family PTEROPHORIDAE that constitutes the superfamily Pterophoroidea is frequently included in the Pyraloidea, with which it has strong affinities. Adult Pterophoridae are moths whose wings are split up into distinct plumes. The larvae are mainly exposed feeders, though a few are initially leaf or stem miners.

Superfamily Hesperioidea

Hesperioidea and members of the next superfamily constitute the butterflies. The group includes two families, the MEGATHYMIDAE, a small, mainly tropical group and the HESPERIIDAE (skippers) (Figure 9.12) a very large and widely distributed family. Adults receive their common name from their jerky erratic flight. Larvae construct shelters by joining or rolling leaves. The larvae of primitive species feed on dicotyledons, those of advanced forms on monocotyledons. Of particular phylogenetic interest is the subfamily EUSCHEMONINAE, whose members are very mothlike. *Euschemon* has often been placed in the Castniidae because of this similarity. Tillyard (1918, cited in Imms, 1957) has in fact pointed out that if the wing-coupling apparatus alone is considered, then male *Euschemon* are moths and females are butterflies!

Superfamily Papilionoidea

Papilionoidea form a very large group that contains five families. NYMPHALIDAE (Figure 9.13A,B) form the dominant butterfly family with more than

FIGURE 9.12. Hesperioidea. The common sooty wing, *Pholisora catullus* [Hesperiidae]. [From L. A. Swan and C. S. Papp, 1972, *The Common Insects of North America*. Copyright 1972 by L. A. Swan and C. S. Papp. Reprinted by permission of Harper & Row, Publishers, Inc.]

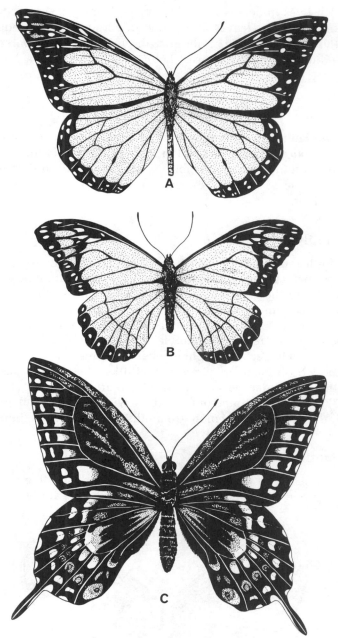

FIGURE 9.13. Papilionoidea. (A) The monarch butterfly, *Danaus plexippus* (Nymphalidae); (B) the viceroy, *Limenitis archippus* (Nymphalidae), a mimic of *D. plexippus*; and (C) the black swallowtail, *Papilio polyxenes asterius* (Papilionidae). [From L. A. Swan and C. S. Papp, 1972, *The Common Insects of North America.* Copyright 1972 by L. A. Swan and C. S. Papp. Reprinted by permission of Harper & Row, Publishers, Inc.]

5000 described species. It is frequently split up into a number of familial groups. Members of the family are recognized by their short, functionless, hairy forelegs. Many are distasteful or mimic distasteful species. The major subgroups are the DANAINAE (milkweed butterflies and monarchs), SATYRINAE (satyrs, wood nymphs, meadow browns, heaths, etc.), NYMPHALINAE (fritillaries, peacocks, admirals, tortoiseshells), HELICONIINAE (heliconians), and MORPHINAE (morphos).

PAPILIONIDAE (swallowtails) (Figure 9.13C) are large, mainly tropical or subtropical butterflies. Adults are strikingly colored and in many species mimic Danainae. The sexes are frequently dimorphic, and among females polymorphism is common. Larvae are usually procryptically colored. They may be smooth, with a row of raised tubercles on the dorsal surface. Situated dorsally on the prothorax is an eversible osmaterium which emits a pungent odor. The family PIERIDAE (whites, sulfurs, orange tips) includes some very common butterflies. The family is primarily tropical, though well represented in temperate regions. Included in the genus *Pieris* are several species that are pests, especially of Cruciferae. The commonest of these is *Pieris rapae*, probably the most economically important of all butterflies. The LYCAENIDAE (blues, coppers, hairstreaks) form a widespread family of small to medium-sized butterflies. The upper surfaces of the wings are metallic blue or coppery in color, the undersides are somber and often with eye spots or streaks. Larvae are onisciform (shaped like a wood louse). Many species are carnivorous on Homoptera, and a few live in ants' nests, feeding on eggs and larvae. Many of the phytophagous species are nocturnal feeders, hiding in holes by day. LIBYTHEIDAE are mainly neotropical butterflies of medium size. Adults are commonly known as snout butterflies from the way their elongated labial palps are held close together and project forward from the head. As in Nymphalidae, to which they are closely related, the forelegs are reduced, especially in males. Libytheid larvae resemble those of Pieridae.

Superfamily Geometroidea

Geometroidea form an extremely large group with more than 12,000 species, about 10% of which are found in North America. Adults (with the exception of those in one small family of doubtful affinity) have abdominal tympanal organs. In most species adults and larvae are extremely well camouflaged in their natural habitat. About 90% of the geometroids belong to the family GEOMETRIDAE. Adults are generally small slender-bodied, and with large wings that are held horizontally when the moths are resting. In some species females are apterous. Larvae, which frequently resemble twigs, have the anterior two or three pairs of prolegs reduced or absent. They are often known as geometers or inchworms from the way they move in looping fashion. The family includes a number of important defoliators of fruit and shade trees, for example, *Paleacrita vernata*, the spring cankerworm (Figure 9.14), and *Alsophila pometaria*, the fall cankerworm.

Superfamily Calliduloidea

Calliduloidea form a small group of two families, CALLIDULIDAE and PTEROTHYSANIDAE, whose members are diurnal moths that resemble geometrids. The group is primarily oriental.

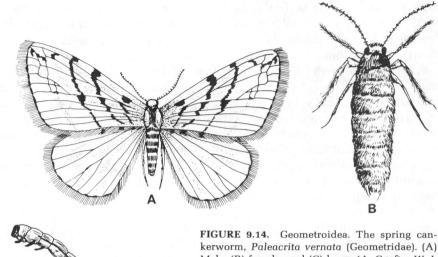

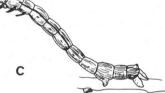

FIGURE 9.14. Geometroidea. The spring cankerworm, *Paleacrita vernata* (Geometridae). (A) Male, (B) female, and (C) larva. [A, C, after W. J. Holland, 1920, *The Moth Book*, Doubleday and Co., Inc. B, from L. A. Swan and C. S. Papp, 1972, *The Common Insects of North America*. Copyright 1972 by L. A. Swan and C. S. Papp. Reprinted by permission of Harper & Row, Publishers, Inc.]

Superfamily Bombycoidea

The outstanding feature of Bombycoidea is the gradual reduction or loss of characters that its members display. Ocelli and tympanal organs are never present. Reduction or loss of the frenulum occurs and the amplexiform method of wing-coupling is developed. The proboscis is rudimentary or absent in the more specialized families. The superfamily is split up into many families, but only a few of these are common. The family LASIOCAMPIDAE (eggars, lappet moths) is a widely distributed group of medium-sized or large, sexually dimorphic moths with stout bodies. Larvae are usually hairy and in many species are gregarious, living in a communal silk nest. These "tent caterpillars" (Figure 9.15A) leave the nest to feed during the day. Some species are important defoliators. The family SATURNIIDAE (giant silkworm moths) (Figure 9.15B,C) includes some of the largest Lepidoptera with a wingspan of up to 25 cm. The group is mainly tropical, with few representatives in temperate regions. Larvae are characterized by the scoli (spiny protruberances) on their dorsal surface. Several oriental species build cocoons from which the silk is of commercial value. The BOMBYCIDAE (including the EUPTEROTIDAE of some authors) form a mainly Old World family. Larvae are usually covered with tufts of hair, although *Bombyx mori*, the silkworm, is naked and has a short anal horn.

Superfamily Sphingoidea

The superfamily Sphingoidea contains one family, the SPHINGIDAE (hawk moths) (Figure 9.16), whose affinities lie with the Bombycoidea. The family is mainly tropical and contains medium-sized to large moths that generally possess a very long proboscis. Larvae are smooth and characterized by a large

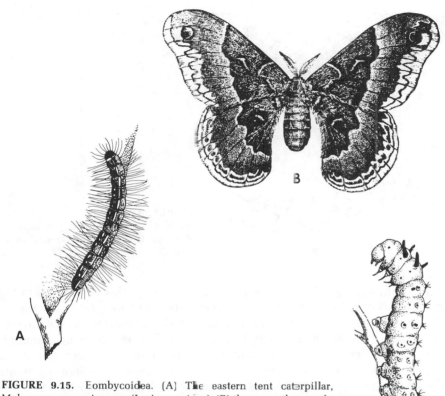

FIGURE 9.15. Bombycoidea. (A) The eastern tent caterpillar, *Malacosoma americanum* (Lasiocampidae); (B) the promethea moth, *Callosamia promethea* (Saturniidae); and (C) *C. promethea* larva. [A, C, after W. J. Holland, 1920, *The Moth Book*, Doubleday and Co. Inc. B, reprinted from Comstock: *An Introduction to Entomology*. Copyright © 1940 by Comstock Publishing Co., Inc. Used by permission of the publisher, Cornell University Press.]

dorsal horn on the eighth abdominal segment. This gives them their common name of hornworms. Some species are pests of solanaceous plants such as tobacco, tomato, and potato.

Superfamily Notodontoidea

Almost all the members of the superfamily Notodontoidea are contained in the family NOTODONTIDAE (prominents). They are closely related to the Noctuoidea, in which they are sometimes included. Adults are exclusively nocturnal moths with stout bodies. Larvae are exposed feeders found on trees and shrubs. When disturbed they raise the anterior and posterior ends of the body into the air and become motionless. In this attitude they vaguely resemble a twig or dead leaf.

Superfamily Noctuoidea

Noctuoidea form the largest lepidopteran superfamily. They are a remarkably homogeneous group, and the constituent families are difficult to define.

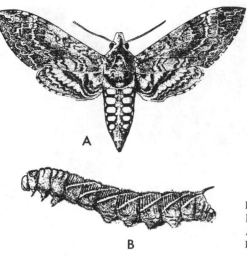

A

B

FIGURE 9.16. Sphingoidea. The tobacco hornworm, *Manduca sexta* (Sphingidae). (A) Adult and (B) larva. [By permission of U.S. Department of Agriculture.]

Four families are particularly large. About 2700 species of NOCTUIDAE (AG-ROTIDAE) occur in North America. These mainly nocturnal moths are procryptically colored with dark fore wings; usually they rest on tree trunks during the day. Larvae are typically phytophagous, though some prey on Homoptera. They usually have four pairs of prolegs, but in some species one or more anterior pairs are reduced, and the caterpillar moves in a looping manner. Pupation in most species takes place in the ground. The family contains a large number of major pests. These include *Pseudaletia (Leucania) unipuncta*, the armyworm, so-called because of its habit of appearing in massive numbers and marching gregariously to feed on cereal crops; *Heliothis zea*, the corn earworm, which feeds on maize cobs and other plants; *Trichoplusia ni*, the cabbage looper; and several other species belonging to different genera (for example, *Agrotis, Prodenia, Feltia*) that are commonly known as cutworms (Figure 9.17) from their habit of cutting off the plant stem level with the ground. LYMANTRIIDAE (LIPARIDAE) (tussock moths) are medium-sized moths, very similar to the noctuids, whose larvae are often densely hairy and may have osmateria on the sixth and seventh abdominal segments. A few species, including *Lymantria (Porthetria) dispar*, the gypsy moth, a European import, are important defoliators of shade trees. ARCTIIDAE (tiger moths and footman moths) are nocturnal, stout-bodied moths, whose wings are typically conspicuously spotted or striped. Their larvae are more or less hairy and feed on low herbaceous plants or on lichens on tree trunks. The AMATIDAE (CTENUCHIDAE) constitute an almost entirely tropical family of small to medium-sized, diurnal moths. Many are brilliantly colored and resemble aculeate Hymenoptera in having the anterior part of the abdomen constricted. Larvae are hairy and generally resemble those of Arctiidae.

Literature

There is an enormous volume of literature on Lepidoptera. This deals especially with the identification and distribution of butterflies and moths, a

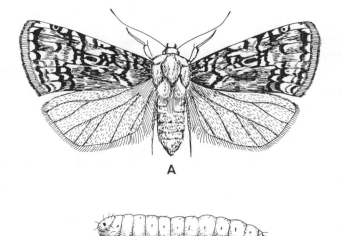

FIGURE 9.17. Noctuoidea. The pale western cutworm, *Agrotis orthogonia* (Noctuidae). (A) Adult and (B) larva. [After C. J. Sorenson and H. F. Thornby, 1941, The pale western cut worm, *Bulletin #297*. By permission of the College of Agriculture and Agricultural Experiment Station, Utah State University.]

feature which is not surprising when the popularity of this group with insect collectors is considered. The biology of Lepidoptera is described by Clark (1932) and Ford (1945, 1972). Discussions on the phylogeny of the order include those of Hinton (1946, 1952), Turner (1947), Ehrlich (1958), Ross (1965), Brock (1971), Mutuura (1972), and Common (1975). The North American species may be identified through the works of Klots (1951), Ehrlich and Ehrlich (1961), Holland (1968), Dominick *et al.* (1971–), and Howe (1975). South (1943a,b) and Mansell and Newman (1968) deal with the British Lepidoptera.

Brock, J. P., 1971, A contribution towards an understanding of the morphology and phylogeny of the Ditrysian Lepidoptera, *J. Nat. Hist.* **5:**29–102.

Clark, A. H., 1932, The butterflies of the District of Columbia and vicinity. *Bull. U.S. Nat. Mus.* **137:**337 pp.

Common, I. F. B., 1970, Lepidoptera, in: *The Insects of Australia* (I. M. Mackerras, ed.), Melbourne University Press, Carlton, Victoria.

Common, I. F. B., 1975, Evolution and classification of the Lepidoptera, *Annu. Rev. Entomol.* **20:**183–203.

Dominick. R. B., Edwards, C. R., Ferguson, D. C., Franclemont, J. G., Hodges, R. W., and Munroe, E. G. (eds.), 1971–, *The Moths of America North of Mexico*, Classey Hampton, Middlesex.

Ehrlich, P. R., 1958, The comparative morphology, phylogeny and higher classification of the butterflies (Lepidoptera: Papilionoidea), *Univ. Kans. Sci. Bull.* **39:**305–370.

Ehrlich, P. R., and Ehrlich, A. H., 1961, *How to Know the Butterflies*, Brown, Dubuque, Iowa.

Ford, E. B., 1945, *Butterflies*, Collins, London.

Ford, E. B., 1972, *Moths*, 3rd ed., Collins, London.

Hinton, H. E., 1946, On the homology and nomenclature of the setae of lepidopterous larvae, with some notes on the phylogeny of the Lepidoptera, *Trans. R. Entomol. Soc. London* **97:**1–37.

Hinton, H. E., 1952, The structure of the larval prolegs of the Lepidoptera and their value in the classification of the major groups, *Lepid. News* **6:**1–6.

Holland, W. J., 1968, *The Moth Book*, Dover, New York.

Howe, W. H. (ed.), 1975, *The Butterflies of North America*, Doubleday, Garden City, New York.

Imms, A. D., 1957, *A General Textbook of Entomology*, 9th ed. (revised by O. W. Richards and R. G. Davies), Methuen, London.

Klots, A. B., 1951, *A Field Guide to the Butterflies*, Houghton Mifflin, Boston.

Mansell, E., and Newman, H. L., 1968, *The Complete British Butterflies in Colour*, Rainbird, London.

Mutuura, A., 1972, Morphology of the female terminalia in Lepidoptera, and its taxonomic significance, *Can. Entomol.* **104**:1055–1071.

Ross, H. H., 1965, *A Textbook of Entomology*, 3rd ed., Wiley, New York.

South, R., 1943a, *The Moths of the British Isles*, 2 vols., Warne, London.

South, R., 1943b, *The Butterflies of the British Isles*, Warne, London.

Turner, A. J., 1947, A review of the phylogeny and classification of the Lepidoptera, *Proc. Linn. Soc. N.S.W.* **71**:303–338.

5. Diptera

SYNONYMS: none   COMMON NAMES: true flies; includes mosquitoes, midges, deer- and horseflies, houseflies

Generally minute to small soft-bodied insects; head highly mobile with large compound eyes, antennae of variable size and structure, and suctorial mouthparts; prothorax and metathorax small and fused with large mesothorax, wings present only on mesothorax, halteres present on metathorax; legs with 5-segmented tarsi; abdomen with variable number of visible segments, female genitalia simple in most species, male genitalia complex, cerci present.

Larvae eruciform and in most species apodous; head in many species reduced and retracted. Pupae adecticous and obtect or exarate, the latter enclosed in a puparium.

The more than 80,000 species of Diptera described to date probably represent about one half of the world total. The order has a worldwide distribution. It includes some of the commonest insects and a large number of species of veterinary and medical importance.

Structure

The great structural diversity found in the Diptera is but a reflection of the variety of niches that the true flies have exploited.

Adult.   Adults range in size from about 0.5 mm to several centimeters and they are generally soft-bodied. The head is relatively large and highly mobile. It carries well-developed compound eyes, which in males are frequently holoptic. The antennae are of variable size and structure and are important taxonomically. In most Diptera (Cyclorrhapha) there is a ∩-shaped ptilinal (frontal) suture that runs transversely above the antennae and extends downward on each side of them. This suture indicates the position of the ptilinum, a membranous sac that is exserted and distended at eclosion in order to rupture the puparium and assist a fly in tunneling through soil, etc. The mouthparts are adapted for sucking and are described in Chapter 3 (see Section 3.2.2 and Figures 3.14 to 3.16). The prothoracic and metathoracic segments are narrow and fused intimately with the very large mesothorax, which bears the single pair of membranous wings. The hind wings are extremely modified, forming halteres, small, clublike structures important as organs of balance (see Chapter 14). In a few species wings are reduced or absent. The legs almost always have 5-segmented tarsi, and one or more pairs may be modified for grasping prey in some species. Primitively, there are 11 abdominal segments, but in most Diptera this number is reduced and rarely more than 4 or 5 are readily visible. Frequently the more posterior segments (postabdomen) are telescoped into the anterior part of the abdomen (preabdomen) (see Figure 3.29). The postabdomen thus formed is used as an extensible ovipositor. The male genitalia are complex and their homologies uncertain due to the rotation of the abdomen and asymmetric growth of the individual components during the pupal stage.

In most Diptera the cibarium is strongly muscular and serves as a pump for sucking up liquids into the gut. In the bloodsucking Tabanidae and Culicidae a large pharyngeal pump is also present. The alimentary canal is, in most primitive forms, relatively unconvoluted. In Cyclorrhapha, however, it is much more coiled due to the increase in length of the midgut. The esophagus divides posteriorly into the gizzard and, usually, one diverticulum, the food reservoir (misleadingly called the 'crop"). In Culicidae three diverticula are found. In Nematocera the midgut is a short, saclike structure; in Cyclorrhapha it is long and convoluted. Generally there are four Malpighian tubules that arise in pairs from a common duct on either side of the gut. In the nervous system a complete range of specialization is seen. In primitive Nematocera three thoracic and seven abdominal ganglia occur, but all intermediate conditions are found between this arrangement and the situation in the more advanced Cyclorrhapha where a composite thoracoabdominal ganglion exists. In females the paired ovaries comprise a variable number of polytrophic ovarioles. In viviparous species there may be only one or two, but in the majority of oviparous flies there may be more than 100. In viviparous forms the common oviduct is dilated to form a uterus, and the accessory glands produce a nutritive secretion. One to three spermathecae are always present. In males the testes are generally small, ovoid, and pigmented. The short, paired vasa deferentia lead into a muscular ejaculatory sac. Paired accessory glands may occur.

Larva and Pupa. Larvae are usually elongate and cylindrical. Body segmentation is usually distinct, but in a few groups the true number of segments is masked as a result of secondary division or fusion of the original segments. Most larvae are apodous, though prolegs may occasionally be present on the thorax and/or abdomen. In primitive Diptera the head is a distinct, sclerotized capsule (the eucephalous condition). In most species, however, it is much reduced (hemicephalous) or entirely vestigial (acephalous) (see Figure 3.13). The antennae and mouthparts are well developed in Nematocera. In higher Diptera a variable degree of reduction or modification takes place, culminating in the situation in Cyclorrhapha, where the antennae are in the form of minute papillae, and the mouthparts are reduced to a pair of curved hooks (probably the original maxillae). The internal structure of larvae generally resembles that of adults. Dipteran pupae are always adecticous. In Nematocera and Brachycera pupae are obtect, whereas those of Cyclorrhapha are secondarily exarate and coarctate, being enclosed in a puparium, the hardened cuticle of the third larval instar.

Life History and Habits

Adult Diptera are active, mostly free-living insects that are found in all major habitats. They are predominantly diurnal and usually associated with flowers or with decaying organic matter. With the exception of a few species that do not feed as adults, flies feed entirely on liquids. The majority feed on nectar or the juices from decaying organic matter, but a few groups are adapted for feeding on the body fluids of other animals especially arthropods and vertebrates. This is achieved in some species by simply cutting the skin or squeezing prey with the labella and sucking up the exuded fluid. In the majority of body-fluid feeders, however, a fine proboscis is used to pierce the skin and penetrate directly to the fluid, usually blood. The habit is usually confined to females. It is through the bloodsucking habit and the subsequent importance of these insects as vectors of disease-causing microorganisms that the order is generally considered the most important of the entire class from the medical and veterinary point of view. Although parthenogenesis is known to occur in a few species, most Diptera reproduce bisexually. Copulation is preceded in some species by an elaborate courtship. Usually females actively search for, and lay eggs directly on, the larval food source. Members of a few groups are ovoviviparous or viviparous. Egg development is normally rapid and hatching occurs in a few days. Larvae are usually found in moist locations such as soil, mud, decaying organic matter, and plant or animal tissues, though a few are truly aquatic. The majority are liquid feeders or microphagous. Some aquatic forms trap their food in specially developed mouth brushes. Larvae of many species are of agricultural or medical importance. Usually four larval instars occur, but up to eight are found in some species, and in Cyclorrhapha the fourth is suppressed. Prior to pupation, larvae generally crawl to a drier location. Pupae may be naked, but those of many Nematocera and Brachycera are enclosed in a cocoon, and those of Cyclorrhapha are ensheathed by the puparium.

Phylogeny and Classification

It is clear from a structural comparison of primitive living Mecoptera and Diptera and from the relatively scarce fossil evidence that Diptera evolved from an ancient mecopteran group in the Permian period. The Australian fossil *Permotipula* was originally assigned to the suborder Protodiptera of the order Diptera on the basis of its wing venation (described from a single fossil imprint). However, with the discovery of a more intact specimen with *four* wings, *Permotipula* was transferred to the Mecoptera. Other features indicate that, although it was not on the main line of evolution leading to the Diptera, *Permotipula* must have been close to the point of divergence of this order from the order Mecoptera. The only true Permian Diptera so far described are two other Australian genera, *Permotanyderus* and *Choristotanyderus*, belonging to the Permotanyderidae. However, these are known only from single wing imprints, and if new discoveries show that they were four-winged, a reappraisal of their position would be necessary. Although very little fossil evidence is available from the Triassic, it is evident that the most primitive suborder, Nematocera, underwent a considerable radiation during this period. By the Lower Jurassic, well-developed Nematocera (some looking remarkably like modern Tipulidae) and primitive Brachycera were present. The remaining suborder, Cyclorrhapha, evolved from brachyceran stock, probably in the late Jurassic. The great radiation of the order, and the establishment of many of the structures and habits of modern flies, took place in the Cretaceous period. This was, of course, correlated with the evolution of the flowering plants and mammals. By the Eocene period, the dipteran fauna was similar in many respects to that which survives today. Indeed, many Eocene fossils are assigned to modern genera. A possible phylogeny of the order is shown in Figure 9.18.

Classification of the modern Diptera presents problems, particularly concerning the rank assignable to different groups. The problem arises because the order is extremely old, it contains many extinct groups and others which are in decline, and yet there are also groups which are still evolving at an extremely rapid rate. Thus, ancient families have well-established differences and are easily separated. Relatively recent groups (for example, the so-called "families" of Muscoidea) are really little more than convenient divisions of the superfamily because of the vast number of species that it contains The differences between these families, therefore are relatively minor. In the following classification, which is slightly modified from that of Colless and McAlpine

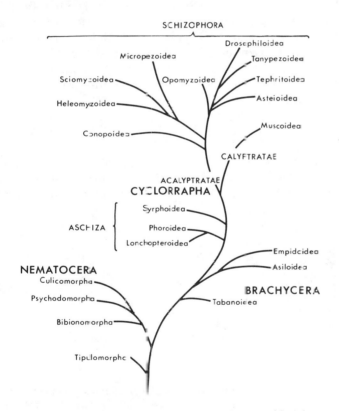

FIGURE 9.18. A suggested phylogeny of the Diptera. [After H. Oldroyd, 1964, *The Natural History of Flies*, Weidenfeld and Nicolson. By permission of Mrs. J. M. Oldroyd.]

(1970), the nematoceran families are arranged in "divisions" rather than "superfamilies." The purpose of this is to indicate that the constituent families are much more widely separated than are the families grouped together in the Brachycera or Cyclorrhapha.

Suborder Nematocera

Most Nematocera are small, delicate, flies, with long antennae of simple structure, and 3- to 5-segmented maxillary palps. Larvae have a well-developed head and chewing mandibles that move in the horizontal plane.

The suborder contains the oldest families of Diptera, most of which are now on the decline. Some, however, like the Culicidae, have undergone relatively recent radiations and are among the most successful modern groups.

Division Tipulomorpha

Members of the division Tipulomorpha are generally found in cool, moist habitats and have long legs and wings. Their larvae are generally semiaquatic or aquatic. Three families are included in the group. The TIPULIDAE (Figure 9.19) (crane flies, daddy longlegs) form one of the largest and commonest families of Diptera, with over 11,000 world species. The family contains some large insects whose wingspan may reach 7.5 cm. PTYCHOPTERIDAE are moderate-sized flies that resemble crane flies both morphologically and in many of their habits. TRICHOCERIDAE (winter crane flies) are easily confused with the true crane flies. They carry the preference for cool, moist habitats to an extreme, and many species are common in caves and mines. Adults are often encountered in large swarms during winter.

Division Psychodomorpha

Psychodomorpha are small to large flies that are among the most primitive of living Diptera. The two families that comprise the division are the TANYDERIDAE, a mainly Australasian group whose members resemble crane

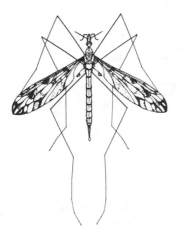

FIGURE 9.19. Tipulomorpha. A crane fly, *Tipula trivittata* (Tipulidae). [From F. R. Cole and E. I. Schlinger, 1969, *The Flies of Western North America.* By permission of the University of California Press.]

flies and are found in the same general habitat, and the PSYCHODIDAE (moth flies) (Figure 9.20), a widely distributed group of small flies recognizable by the hairy wings held rooflike over the body when at rest. Although most species do not feed as adults, some females feed on blood, including those of the genus *Phlebotomus* (sand flies), species of which are vectors of various virus- and leishmania-induced diseases.

Division Culicomorpha

Included in Culicomorpha a group of generally small and delicate flies, are some extremely well-known Diptera. The group is divided into five families in the present arrangement. The CULICIDAE form a large, widespread family that includes mosquitoes (subfamily CULICINAE) (Figure 9.21) and phantom midges (CHAOBORINAE) (Figure 9.22). The general habits of members of these two subfamilies present an interesting contrast. Larval mosquitoes are filter feeders that strain microorganisms from the water in which they live. As adults males do not feed, but females are voracious bloodsuckers and as such are responsible for the spread of some human diseases, for example, malaria, yellow fever, and filariasis. (Some incidently, also spread the myxomatosis virus of rabbits and are, therefore, of some positive economic value.) Larval Chaoborinae, on the other hand, are predators (particularly of mosquito larvae!). The adults, however, are nectar feeders. The DIXIDAE form a small but widely distributed group that is frequently considered a subfamily of the Culicidae, mainly on the basis of the adult wing venation and the similarity between the larval and pupal stages of the two groups. The CHIRONOMIDAE (TENDIPEDIDAE) (Figure 9.23) constitute a large, widely distributed family of more than 2000 species. Adults are small, mosquitolike flies, though they do not feed. They often form massive swarms in the vicinity of water. Larvae are aquatic and either are free-living or lie buried in the substrate, members of many species constructing a special tube. The CERATOPOGONIDAE (biting midges, punkies, no-see-ums) (Figure 9.24) form a widespread family of minute or small flies, some of which suck the blood of vertebrates and other arthropods, or prey on other insects. Members of most species, however, feed on nectar and/or pollen and render considerable benefit through cross-fertilization of the plants. The SIMULIIDAE (blackflies, buffalo gnats) (Figure 9.25) form a widespread family of biting flies that attack birds, mammals, and other insects. The larvae are found in swiftly flowing water, attached to the substrate by an anal sucker. The THAUMELIDAE constitute a small family of minute midges

FIGURE 9.20. Psychodomorpha. A moth fly, *Psychoda* sp. (Psychodidae). [From F. R. Cole and E. I. Schlinger, 1969, *The Flies of Western North America.* By permission of the University of California Press.]

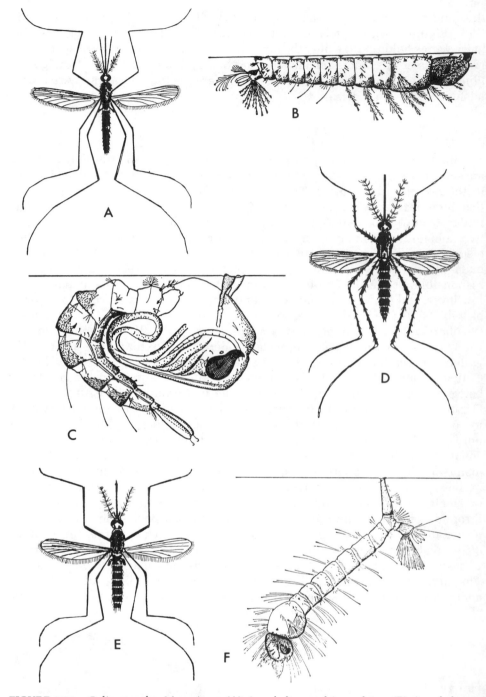

FIGURE 9.21. Culicomorpha. Mosquitoes. (A) *Anopheles quadrimaculatus*, (B) *Anopheles* sp. larva, (C) *Anopheles* sp. pupa, (D) *Aedes canadiensis*, (E) *Culex pipiens*, and (F) *Culex* sp. larva. [A, D, E, from S. J. Carpenter and W. J. LaCasse, 1955, *Mosquitoes of North America*. By permission of the University of California Press. A, E, drawn by Saburo Shibata. D, drawn by Kei Daishoji. B, C, F, from J. D. Gillett, 1971, *Mosquitos*, Weidenfeld and Nicolson. By permission of the author.]

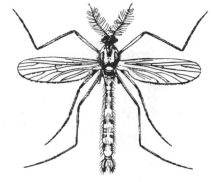

FIGURE 9.22. Culicomorpha. The clear lake gnat, *Chaoborus astictopus* (Culicidae). [From F. R. Cole and E. I. Schlinger, 1969, *The Flies of Western North America*. By permission of the University of California Press.]

restricted to higher altitudes and latitudes. The systematic position of the family is uncertain. In some features its members resemble the other Culicomorpha, in others the Bibionomorpha.

Division Bibionomorpha

Bibionomorpha form a large and diverse group of Nematocera that is subdivided into more than a dozen families. Most of these are small and will not be mentioned here. The MYCETOPHILIDAE and SCIARIDAE, which together comprise about 2000 species, are sometimes included in a single family of the former name. They are commonly known as fungus gnats from the observation that the larvae feed mainly on fungi and decaying plant material. Adults are commonly encountered in cool, damp situations. The BIBIONIDAE (March flies) (Figure 9.26A) are robust, hairy flies of medium to small size. They frequent grassy places where the larvae feed gregariously on roots or decaying vegetable matter. The SCATOPSIDAE, which are sometimes included as a subfamily of the Bibionidae, are very small flies whose larvae feed on decaying organic matter, especially dung. The CECIDOMYIIDAE (gall midges) form a very large family of minute flies, most of which feed, in the larval stage, on plant tissues, frequently

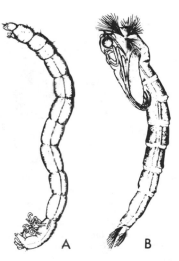

FIGURE 9.23. Culicomorpha. *Chironomus tentans* (Chironomidae). (A) Larva and (B) pupa. [From O. A. Johannsen, 1937, Aquatic Diptera. Part IV. Chironomidae: Subfamily Chironominae, *Mem. Cornell Univ. Agric. Expt. Station* **210:**52 pp. By permission of Cornell University Agriculture Experimental Station.]

A   B

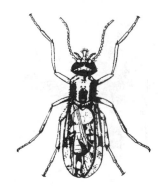

FIGURE 9.24. Culicomorpha. A punkie, *Culicoides dovei* (Ceratopogonidae). [From F. R. Cole and E. I. Schlinger, 1969, *The Flies of Western North America*. By permission of the University of California Press.]

causing the formation of galls. There are, however, other saprophagous or predaceous species. Within the family are several economically important species, for example, the Hessian fly, *Phytophaga* (*Mayetiola*) *destructor* (Figure 9.26B), whose larvae feed on wheat shoots. Many species are paedogenetic, the full-grown larvae becoming sexually mature and reproducing parthenogenetically. As the young larvae grow, they devour their parent from within. Several generations of paedogenetic larvae may develop in a season, and the larval population can thus increase enormously. Eventually, the larvae pupate normally and sexual reproduction follows.

Suborder Brachycera

Most Brachycera are rather stout flies with short, generally 3-segmented antennae of variable structure, and unsegmented or 2-segmented maxillary

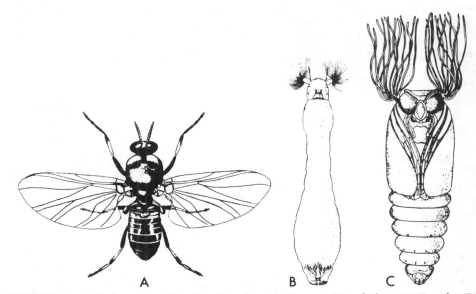

FIGURE 9.25. Culicomorpha. A blackfly, *Simulium nigricoxum* (Simuliidae). (A) Female, (B) mature larva, and (C) pupa. [From A. E. Cameron, 1922, The morphology and biology of a Canadian cattle-infesting blackfly, *Simulium simile* Mall. (Diptera, Simuliidae), *Bulletin #5—New Series (Technical)*. By permission of Agriculture Canada.]

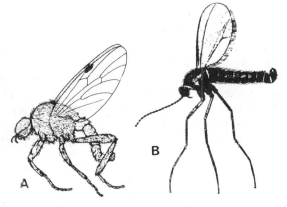

FIGURE 9.26. Bibionomorpha. (A) A March fly, *Bibio albipennis* (Bibionidae); and (B) the Hessian fly, *Phytophaga destructor* (Cecidomyiidae). [A from D. J. Borror and D. M. Delong, 1971, *An Introduction to the Study of Insects*, 3rd ed. By permission of Holt, Rinehart and Winston, Inc. B, from L. A. Swan and C. S. Papp, 1972. *The Common Insects of North America.* Copyright 1972 by L. A. Swan and C. S. Papp. Reprinted by permission of Harper & Row, Publishers, Inc.]

palps. Larvae are hemicephalous, with sickle-shaped mandibles that move in the vertical plane.

Superfamily Tabanoidea

Tabanoidea form the most diverse, though not the largest, of the three brachyceran superfamilies. As many as 10 distinct families are recognized, though many of these are small and of restricted distribution. The largest family is the TABANIDAE (Figure 9.27A), with more than 3000 species, which includes those bloodsucking insects commonly known as horse- and deerflies, clegs, and probably many other, less polite names! The bloodsuckers belong in fact to only three genera, *Tabanus*, *Chrysops*, and *Haematopota*, whose evolution has closely followed that of the hoofed mammals. Although they are known to be capable of transmitting various diseases both human and of livestock, tabanids cause far greater economic losses by their disturbance and irritation of livestock, resulting in lower yields of milk and meat. Only female horseflies suck blood, in the absence of which they feed, like males and like members of most tabanid species, on nectar and pollen. Another large and

FIGURE 9.27. Tabanoidea. (A) A horsefly, *Tabanus opacus* (Tabanidae); and (B) a soldier fly, *Odontomyia hoodiana* (Stratiomyidae). [A, from J. F. McAlpine, 1961, Variation, distribution and evolution of the *Tabanus (Hybomitra) frontalis* complex of horse flies (Diptera: Tabanidae), *Can. Entomol.* **93:**894–924. By permission of the Entomological Society of Canada. B, from F. R. Cole and E. I. Schlinger, 1969, *The Flies of Western North America.* By permission of the University of California Press.]

well-distributed family is the STRATIOMYIDAE (Figure 9.27B), containing some 1500 species, commonly known as soldier flies. The weakly flying adults are encountered among low-growing herbage and are most probably nectar feeders. Members of many species are conspicuously striped, and some species are excellent wasp mimics. The RHAGIONIDAE (snipe flies) are perhaps the most primitive of the Brachycera. The family, which contains more than 400 species, is widely distributed, though seldom encountered, because of the secretive, solitary habits of adult flies. Members of many species are nectar feeders, others are predaceous, and a few are probably facultative blood suckers.

Superfamily Asiloidea

The ASILIDAE (robber flies) (Figure 9.28A), numbering more than 4000 species, form the largest family of Brachycera. Adults prey on a variety of other insects, whose body fluids they suck. They are powerful fliers and catch their prey on the wing, have well-developed eyes and some degree of stereoscopic vision, possess strong legs for grasping the prey and are usually hairy, especially around the face, a feature which perhaps protects them during the struggle. In contrast, adult BOMBYLIIDAE (bee flies) (Figure 9.28B), of which there are about 3000 described species, are nectar feeders. The common name has double significance. First, the flies resemble bumblebees, and second, in many species, female flies deposit eggs at the mouth of a solitary bee's or wasp's nest so that the larvae may feed on the pollen, honey, and even young Hymenoptera. In a few species eggs are laid directly onto larvae of Lepidoptera or Hymenoptera. The THEREVIDAE form a widely distributed group of about 500 species that generally resemble robber flies, but there is doubt as to the adult feeding habits.

Superfamily Empidoidea

The Empidoidea contain two large families of advanced Brachycera, the EMPIDIDAE (dance flies) and DOLICHOPODIDAE (long-legged flies). There are about 2000 species of Empididae, which are predaceous in both adult and juvenile stages. The family is largely restricted to the temperate regions of both hemi-

FIGURE 9.28. Asiloidea. (A) A robber fly, *Mallophorina pulchra* (Asilidae); and (B) a bee fly, *Poecilanthrax autumnalis* (Bombyliidae). [From F. R. Cole and E. I. Schlinger, 1969, *The Flies of Western North America*. By permission of the University of California Press.]

spheres, and its members gain their common name from the swarming behavior of most, though by no means all, species. In many species males have an elaborate courtship display in which a female is offered a gift of food (real or imitation). Presumably this served originally as an insurance policy against the male's life. The Dolichopodidae, which constitute a family of about 2000 species, appear to be a specialized offshoot of the empid line. They are generally found in cool, moist habitats, including the seashore and salt marshes. Like Empididae, they are predaceous on other insects, especially Diptera.

Suborder Cyclorrhapha

Cyclorrhapha are flies with 3-segmented antennae with an arista usually dorsal in position, unsegmented maxillary palps, and, in most species, a ptilinum. Larvae are acephalic (maggotlike). Pupae are coarctate, enclosed within a puparium.

The Cyclorrhapha are arranged in two series. The first, Aschiza, is perhaps a polyphyletic group, containing the more primitive members, which lack a ptilinal suture. The second, Schizophora, includes those flies in which a ptilinal suture is present. In older classifications a third series, Pupiparia, is often found. This is now known to be an unnatural (i.e., polyphyletic) group whose families should be placed in the Schizophora. The majority of Schizophora (i.e., those that are winged as adults) can be arranged in two subdivisions, the Calyptratae, which contain flies that possess a calypter, or lobe, at the base of the fore wing that covers the halteres, and the Acalyptratae, whose members have no such lobe. Though this division is a natural one, that is, it represents a true evolutionary divergence, it should be remembered that (1) some groups have secondarily gained or lost the calypter, and (2) the adults of some parasitic families are wingless, though their affinities are clearly either calyptrate or acalyptrate.

Series Aschiza

Superfamily Lonchopteroidea

This group contains a single monogeneric family LONCHOPTERIDAE whose 20 or so species show features in common with the Dolichopodidae and the Phoridae (see below). Parthenogenesis seems to be the usual mode of reproduction.

Superfamily Phoroidea

Two small families are included in this group, the PHORIDAE and the PLATYPEZIDAE. Although many phorids are free-living, fully winged flies found among low vegetation, they seem to prefer to run rather than fly, a feature which foreshadows the brachypterous or apterous condition of the many species that live underground, frequently in ants' or termites' nests. The Platypezidae form a small family containing flies that are usually found swarming above or running over low vegetation. The larvae are usually found in fungi.

Superfamily Syrphoidea

One small and one very large family comprise the Syrphoidea. The PIPUN-CULIDAE are small, humpbacked flies with large heads that tend to be found hovering over flowers. Their larvae are parasites of Homoptera. The SYRPHIDAE (Figure 9.29) are the well-known hover flies. They form one of the largest and most easily recognized groups of Diptera. They are generally brightly colored, often striped, and many mimic bees or wasps. In some species there are obvious reasons for the preciseness of this mimicry, for the hover fly lays its eggs in the nests of Hymenoptera and, because of its similarity, presumably avoids detection. For other species the reason is less obvious, and no relationship is apparent between the mimic and its model. In contrast to the rather uniform, nectar-feeding habits of adult hover flies, those of larvae are extremely varied, phytophagous, zoophagous, and saprophagous species being known.

Series Schizophora

Superfamily Conopoidea

The superfamily Conopoidea, the most primitive of the Schizophora, contains the single family CONOPIDAE, whose members closely mimic bees, wasps, and hover flies. Despite the closeness of the resemblance, however, the relationship between the conopid and its model is not always specific; that is, the conopid parasitizes a range of morphologically different Hymenoptera, including both wasps and bumblebees.

Superfamily Tephritoidea

The superfamily Tephritoidea includes six families, of which three are relatively large and widely distributed. The OTITIDAE are quite common in dense vegetation, and the larvae feed on decaying plant material. The PLATY-STOMATIDAE are of generally similar habits, though the larvae utilize a wider range of food materials, being found in both living and dead plants as well as on dead animal tissues. The largest and best-known family is the TRYPETIDAE (TEPHRITIDAE), the fruit flies, a group that includes some major agricultural pests. Although they are phytophagous, the larvae feed on a variety of plant

FIGURE 9.29. Syrphoidea. A hover fly, *Eupeodes volucris* (Syrphidae). [From a drawing by Charles S. Papp. By permission of the artist.]

materials. They may be leaf or stem miners, gall formers, flower-inhabiting species, or fruit and seed eaters. In the latter category are the Mediterranean fruit fly, *Ceratitis capitata* (Figure 9.30), which attacks citrus and other fruits, and *Rhagoletis pomonella*, the apple maggot fly, whose larvae tunnel into apples, pears, etc.

Superfamily Micropezoidea

Micropezoidea form a small group of some five families whose members are few and mainly tropical. The adults are found in close association with the larval food source, which is usually decaying vegetable matter or animal excreta.

Superfamily Tanypezoidea

Tanypezoidea form another small superfamily whose members have a variety of phytophagous habits. The PSILIDAE from a mainly holarctic family that includes *Psila rosae*, the carrot fly, whose larvae often cause much damage to carrots, celery, and other root crops.

Superfamily Sciomyzoidea

Colless and McAlpine (1970) suggest that the Sciomyzoidea may be polyphyletic. Most of the eight families that are included in the group are small and restricted to a particular world region. The SCIOMYZIDAE are common insects in marshes and other wet habitats. The larvae may be phytophagous, but more commonly parasitize slugs and snails. The LAUXANIIDAE (SAPROMYZIDAE) form a large family whose members are found in a wide variety of locations, both wet and dry. The larvae feed on decaying plant material.

Superfamily Heleomyzoidea

The superfamily Heleomyzoidea is another relatively small group containing small families whose members feed on decaying organic material. Particularly well known, though not a large group, are the COELOPIDAE (kelp flies) that in the larval stage feed on rotting seaweed lying just above the high-water mark. Members of other families feed on animal feces.

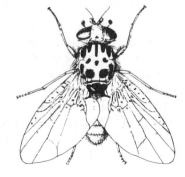

FIGURE 9.30. Tephritoidea. The Mediterranean fruit fly, *Ceratitis capitata* (Tephritidae). [From F. R. Cole and E. I. Schlinger, 1969, *The Flies of Western North America*. By permission of the University of California Press.]

Superfamily Opomyzoidea

The superfamily Opomyzoidea is another, possibly polyphyletic, group that includes several small families of restricted distribution. Mostly the larvae feed on rotting plant and animal tissue. However, larvae of AGROMYZIDAE, a relatively large and common group, are leaf or stem miners, rarely gall formers, in living plants.

Superfamily Asteioidea

Asteioidea are small to minute flies that usually occur among grasses on which the larvae feed. They are arranged in five small families, mostly of restricted distribution.

Superfamily Drosophiloidea

The superfamily Drosophiloidea contains several families of flies. The family EPHYDRIDAE (shore flies) (Figure 9.31A), with more than 1000 described species, is one of the best known because of the remarkable variety of habitats that its members occupy. They are typically found near water, both fresh and salt, and in many species the larvae are truly aquatic. Larvae of some species feed on algae, but generally they and the adults are carnivorous or carrion feeders, sometimes almost to the point of being parasitic. Two examples, illustrating the extreme habitats in which Ephydridae are found, are *Ephydra riparia*, which is found in the Great Salt Lake of Utah, and *Psilopa (Helaeomyia) petrolei*, whose larvae live in pools of crude petroleum in California. The closely related DROSOPHILIDAE (pomace or fruit flies) are small flies generally seen in the vicinity of decaying vegetation or fruit, or near breweries and vinegar factories. The larvae are mostly fungivorous, though a few are leaf miners or prey on other insects. Various species of *Drosophila* have, of course, been extensively used in genetic studies. Another common group of flies are the CHLOROPIDAE, which are generally found in meadows, where the larvae feed on leaves, stems, or grasses. Some species, including the frit fly, *Oscinella frit* (Figure 9.31B), are important pests of cereals. Members of a few species which feed on decaying organic matter are sometimes attracted to animal secretions. They seem particularly attracted to the eyes and are responsible for transmission of certain eye diseases in some parts of the world.

Superfamily Muscoidea

Muscoidea form a very large group of Diptera that comprise the calyptrate families and a number of parasitic groups. Within the superfamily there is a distinct division of the families into a muscid group (houseflies and relatives) and a calliphorid group (blowflies, flesh flies, etc.), which are sometimes placed in a separate superfamily, the Calliphoroidea. The two groups almost certainly had a common ancestor whose larva was a compost feeder. During the evolution of Muscoidea emphasis has been placed on feeding in the adult stage. Thus, in the muscid group adults feed on living plant tissue, animal excreta, carrion, living insects, or vertebrate blood. In contrast, members of the

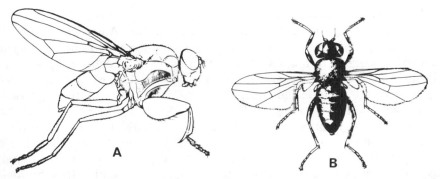

Figure 9.31. Drosophiloidea. (A) An ephydrid with raptorial forelegs, *Ochthera mantis* (Ephydridae); and (B) the European frit fly, *Oscinella frit* (Chloropidae). [A, from F. R. Cole and E. I. Schlinger, 1969, *The Flies of Western North America*. By permission of the University of California Press. B, from L. A. Swan and C. S. Papp, 1972, *The Common Insects of North America*. Copyright 1972 by L. A. Swan and C. S. Papp. Reprinted by permission of Harper & Row, Publishers, Inc.]

calliphorid group are strictly flesh eaters in the larval stage (including some true parasites) and mostly carrion feeders or bloodsuckers as adults.

Most muscid flies are placed in the family MUSCIDAE, which includes many common flies, for example, the various species of housefly such as *Musca domestica* (Figure 9.32A) and *Fannia canicularis*, and blood-sucking species such as the stable fly (*Stomoxys calcitrans*), face fly (*Musca autumnalis*), and tsetse flies (*Glossina* spp.). These are well known, of course, for the harm they do to man, either directly or indirectly, but it should be emphasized that the majority of species are perfectly innocucus in this regard. The ANTHOMYIIDAE form another large family of flies that are phytophagous in the larval stage. Many species are economically important, for example, the cabbage root fly, *Hylemya brassicae*, and the wheat bulb fly, *H. coarctata* (Figure 9.32B). A closely related group is the family SCATOPHAGIDAE, which contains many familiar dung flies (Figure 9.32C). The larvae feed on animal excreta, the adults prey on small insects.

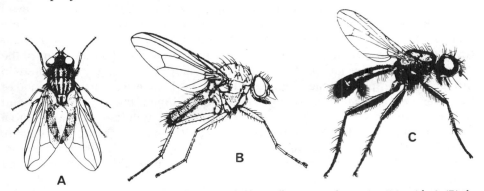

FIGURE 9.32. Muscoidea. Muscid flies. (A) The housefly, *Musca domestica* (Muscidae); (B) the wheat bulb fly, *Hylemya coarctata* (Anthomyiidae); and (C) a dung fly, *Cordilura criddlei* (Scatophagidae). (A, from V. B. Wigglesworth, 1959, Metamorphosis, polymorphism, differentiation. Copyright © February 1959 by Scientific American, Inc. All rights reserved. B, C, from F. R. Cole and E. I. Schlinger, 1969, *The Flies of Western North America*. By permission of the University of California Press.]

The CALLIPHORIDAE are a large, cosmopolitan family that includes blowflies, green- and bluebottles, and screwworm flies. Among members of the family a complete spectrum of larval feeding habits can be seen, ranging from true carrion feeders, through species which feed on exudates or open wounds of living animals, to truly parasitic forms. Calliphoridae of medical or veterinary importance include the sheep blowflies (*Lucilia* spp.), the screwworms [*Cochliomyia* (*Callitroga*) spp.] (Figure 9.33A), and bluebottles (*Calliphora* spp.), which are vectors of human diseases. Closely related to the calliphorids are the SARCOPHAGIDAE (flesh flies) (Figure 9.33B), whose larvae feed on decaying animal tissue or are true parasites of arthropods, molluscs, or annelids. Most species are viviparous, depositing first-instar larvae directly into the food source. The TACHINIDAE (Figure 9.33C,D) form one of the largest families of Diptera. The larvae, with one doubtful exception, are parasitic on other arthropods, mainly insects. An egg, or in the many viviparous species, a larva, is frequently deposited directly on the body of the host. Alternatively, the egg is laid on the host's food plant. The host usually dies as a result of the parasitism, and there is little doubt that tachinids play a role equal to that of many parasitic Hymenoptera in controlling the population level of certain species. Not surprisingly, some have been employed by Man in an effort to reduce the numbers of some economically important pests in certain areas. Having definite affinities with the tachinids are the OESTRIDAE and the GASTEROPHILIDAE, commonly known as bot- and warble flies. The Oestridae (Figure 9.33E,F) are hairy, beetlelike flies with vestigial mouthparts, whose larvae are internal parasites of mammals. The larvae are found either in the nasal and pharyngeal cavities (bots) or beneath the skin (warbles), and feed on fluid exudates from the damaged tissue. The Gasterophilidae (Figure 9.33G) are a family of flies, most of whose larvae live attached to the lining of the stomach of members of the horse family, rhinoceroses, and elephants. Some species lay their eggs on vegetation, while others lay them on the hair or around the mouth of the host.

We complete this survey of the Diptera by noting three remarkable families of flies, the HIPPOBOSCIDAE, STREBLIDAE, and NYCTERIBIIDAE, that were formerly included in a special group, the Pupiparia, though this is now believed to be an unnatural arrangement. In all the families, however, the adults are bloodsucking ectoparasites of birds or mammals, and the larvae develop entirely within the genital tract of the female, being nourished by secretions of the "uterine" wall. The larvae do not leave the mother until they are about to pupate. The Hippoboscidae (Figure 9.34A) (louse flies) are the least changed of the three groups. Members of most species are winged, though rarely fly, and they resemble in several ways the tsetse flies, with which they may have had a common ancestor. Most Streblidae (Figure 9.34B) have wings, though they are folded in pleats to permit easy movement through the host's fur. The Nycteribiidae (Figure 9.34C) are wingless and look remarkably like "six-legged spiders."

Literature

There exists a massive volume of literature on Diptera, including many books, too numerous to mention specifically, on particular groups of flies or

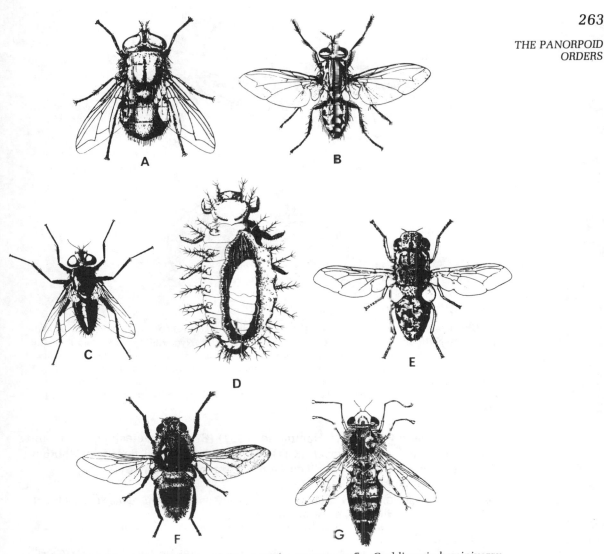

FIGURE 9.33. Muscoidea. Calliphorid flies. (A) The screwworm fly, *Cochliomyia hominivorax* (Calliphoridae); (B) *Sarcophaga kellyi* (Sarcophagidae), a parasite of grasshoppers; (C) the bean beetle tachinid, *Aplomyiopsis epilachnae* (Tachinidae); (D) *A. epilachnae* larva inside bean beetle grub; (E) the sheep botfly, *Oestrus ovis* (Oestridae); (F) the cattle warble fly, *Hypoderma bovis* (Oestridae); and (G) the horse botfly, *Gasterophilus intestinalis* (Gasterophilidae). [A, G, from M. T. James, 1948, The flies that cause myiasis in Man, *U.S. Dep. Agric., Misc. Publ. #631.* By permission of the U.S. Department of Agriculture. B, from L. A. Swan and C. S. Papp, 1972, *The Common Insects of North America.* Copyright 1972 by L. A. Swan and C. S. Papp. Reprinted by permission of Harper & Row, Publishers, Inc. C, D, by permission of U.S. Department of Agriculture. E, F, from A. Castellani and A. J. Chambers, 1910, *Manual of Tropical Medicine.* By permission of Bailliere and Tindall.]

aspects of their biology. Oldroyd's (1964) book is probably the best overall treatment of the Diptera from the point of view of the biology and phylogeny of the group. In addition, the book contains an extensive bibliography. Keys for identification of Diptera are given by Oldroyd (1949) (and other authors in the

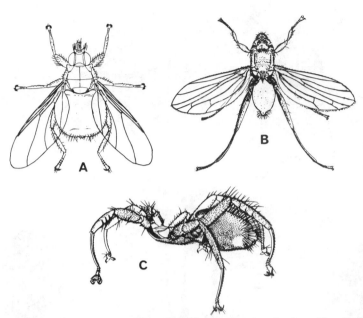

FIGURE 9.34. (A) *Lynchia americana* (Hippoboscidae), a parasite of owls and hawks; (B) a bat fly, *Strebla vespertilionis* (Streblidae); and (C) *Cyclopodia greefi* (Nycteribiidae). [A, from F. R. Cole and E. I. Schlinger, 1969, *The Flies of Western North America*. By permission of the University of California Press. B, from Q. C. Kessel, 1925, A synopsis of the Streblidae of the world, *J. N.Y. Entomol. Soc.* **33**:11–33. By permission of the New York Entomological Society. C, from H. Oldroyd, 1964. *The Natural History of Flies*, Weidenfeld and Nicolson. By permission of Mrs. J. M. Oldroyd.]

same series) and Colyer and Hammond (1951) [British species]; and Cole and Schlinger (1969) and Borror *et al.* (1976) [North American families]. Rohdendorf (1974) has an extensive discussion of dipteran phylogeny.

Borror, D. J., Delong, D. M., and Triplehorn, C. A., 1976, *An Introduction to the Study of Insects*, 4th ed., Holt, Rinehart and Winston, New York.

Cole, F. R., and Schlinger, E. I., 1969, *The Flies of Western North America*, University of California Press, Berkeley.

Colless, D. H., and McAlpine, J. F., 1970, Diptera, in: *The Insects of Australia* (I. M. Mackerras, ed.), Melbourne University Press, Carlton, Victoria.

Colyer, C. N., and Hammond, C. O., 1951, *Flies of the British Isles*, Warne, London.

Oldroyd, H., 1949, Diptera: Introduction and key to families, *R. Entomol. Soc. Handb. Ident. Br. Insects* **9**(1):49 pp.

Oldroyd, H., 1964, *The Natural History of Flies*, Weidenfeld and Nicolson, London.

Rohdendorf, B., 1974, *The Historical Development of Diptera*, University of Alberta Press, Edmonton.

6. Siphonaptera

SYNONYMS: Aphaniptera, Suctoria   COMMON NAME: fleas

Small, wingless, laterally compressed, jumping ectoparasites of birds and mammals; head sessile without compound eyes, antennae short and lying in grooves, mouthparts of piercing and sucking type; coxae large, tarsi 5-segmented; abdomen 10-segmented and bearing unsegmented cerci.

Larvae eruciform and apodous. Pupae adecticous and exarate, enclosed in a cocoon.

More than 1300 species of fleas have been described, about 90% of which are parasitic on mammals. More than 200 species occur in North America.

Structure

Adult. Fleas are highly modified for their ectoparasitic life, a feature which has made determination of their relationships with other Insecta extremely difficult. The adults, which are between 1 and 6 mm in length, are highly compressed laterally and heavily sclerotized. The many hairs and spines on the body are directed posteriorly to facilitate forward movement. The head is broadly attached to the body and carries the short 3-segmented antennae in grooves. Compound eyes are absent, but there may be two lateral ocelli. The mouthparts are modified for piercing the skin and sucking up blood. The lacinia are elongate and together with the epipharynx form a piercing organ which rests in the grooved prementum. The thoracic segments are freely mobile and increase in size posteriorly. The legs are adapted for jumping and clinging to the host. The coxae are very large, and the tarsi terminate in a pair of strong claws. Ten abdominal segments occur, the last three of which are modified for reproductive purposes, especially in males, where the sternum and tergum of the ninth segment form clasping organs.

Both the cibarium and pharynx are strongly muscular for sucking up blood. The small proventriculus is fitted with cuticular rods that may serve to break up blood corpuscles. The midgut (stomach) is large and fills most of the abdomen. Four Malpighian tubules arise at the anterior end of the short hindgut. The nervous system is primitive and includes three thoracic and seven or eight abdominal ganglia. The testes are fusiform and are connected to a small seminal vesicle by means of fine vasa deferentia. The ovaries contain from four to eight panoistic ovarioles.

Larva and Pupa. Larvae (see Figure 9.35D) are white and vermiform. They have a well-developed head that in some respects resembles that of nematocerous Diptera. The mouthparts, though modified, are of the biting type. There are 13 body segments, but the distinction between thoracic and abdominal regions is poor. Pupae are adecticous and exarate and enclosed in a cocoon. Traces of wings can be seen on the pupae of some species.

Life History and Habits

Adult fleas are exclusively bloodsucking ectoparasites, though their association with a host is a rather loose one, that is, they spend a considerable time off the host's body. Host–parasite specificity is variable. A few species of fleas are monoxenous (restricted to a single host species) but most are polyxenous, with more than 20 potential hosts recorded for some species. Conversely, an animal may host one or many flea species. It is this lack of host specificity which makes fleas important vectors of certain diseases. The cosmopolitan rat

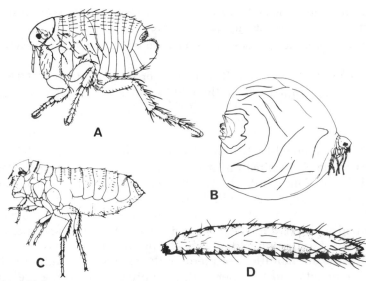

FIGURE 9.35. Siphonaptera. (A) The human flea, *Pulex irritans* (Pulicidae); (B) the female chigoe flea, *Tunga penetrans* (Tungidae); (C) the sand-martin flea, *Ceratophyllus styx* (Ceratophyllidae); and (D) larva of *Spilopsyllus cuniculi* (Pulicidae). [A, from L. A. Swan and C. S. Papp, 1972, *The Common Insects of North America.* Copyright 1972 by L. A. Swan and C. S. Papp. Reprinted by permission of Harper & Row, Publishers, Inc. B–D, from R. R. Askew, 1971, *Parasitic Insects.* By permission of Heinemann Educational Books Ltd.]

flea, *Xenopsylla cheopis*, for example, is responsible for transmitting bubonic plague and typhus, normally diseases of rodents, to humans. Fleas also are the intermediate hosts of the dog and rodent tapeworms which can infect man. Adult fleas may survive for several months in the absence of a host and may live for more than a year when food is available.

Mating and maturation of the oocytes in at least some species follows feeding. In *Spilopsyllus cuniculi*, the rabbit flea, the entire reproductive cycle is controlled by the gonadal hormone level in the host's blood. Eggs are almost never attached to the host but fall to the floor. Larvae of most species are not parasitic but feed on organic debris. After two molts the larvae pupate in a cocoon. Newly emerged adults may survive for some time without food.

Phylogeny and Classification

The origin and classification of fleas are highly controversial subjects. The major difficulty lies in the almost complete lack of fossil evidence, there being only a few descriptions of fossils fleas, based on a single specimen (now lost!) from Baltic Amber and a few records from the Lower Cretaceous period. Conclusions regarding the evolution of fleas must therefore be based entirely on comparative studies of living material, an approach which is not helped by the high degree of convergence that has occurred within the order, in accord with the rather uniform habits of the group. Almost entirely on the basis of larval characters, Hinton (1958) suggests that the order is derived from a *Boreus*-like ancestor. Holland (1964) appears to concur with this view. Modern fleas are arranged in three superfamilies, as indicated below.

Two families are contained in the Pulicoidea. The PULICIDAE are associated with a range of birds and mammals. The family contains many cosmopolitan fleas which attack man and domestic animals, including *X. cheopis*, *Pulex irritans* (the so-called "human flea") (Figure 9.35A) whose normal host is the pig, *Ctenocephalides canis* and *C. felis* (dog and cat fleas), and *Echidnophaga gallinacea* (the sticktight flea). Most Pulicidae have a rather loose association with their hosts, but sticktight fleas become permanently attached in the manner of ticks. The TUNGIDAE (Figure 9.35B) form a mainly tropical group of fleas, females of which burrow under the skin of the host, especially under the toenails or between the toes. Hosts include birds, various rodents, and occasionally man.

Superfamily Malacopsylloidea

The two families that comprise the Malacopsylloidea were originally included in the Ceratophylloidea. The family RHOPALOPSYLLIDAE includes fleas of both sea birds and mammals (mainly rodents). The group is mainly neotropical, but representatives are found also in Australia and southern North America. The MALACOPSYLLIDAE are found solely on armadilloes in South America.

Superfamily Ceratophylloidea

Holland (1964) subdivided the very large group, Ceratophylloidea, into 12 families, the largest of which are mentioned below. The HYSTRICHOPSYLLIDAE constitute the largest family of fleas, containing 45 genera. Representatives occur throughout the world, though the group is mainly a holarctic one. Most species are parasites of rodents and shrews, though a few are found on carnivores and marsupials. Another large group of fleas is the family CERATOPHYLLIDAE (Figure 9.35C), which includes several cosmopolitan species. Ceratophyllids are found mainly on rodents, though some occur on birds. Several species are believed to be capable of transmitting plague from rodents to man, and others can serve as the intermediate host for the tapeworm, *Hymenolepis diminuta*. Related to the previous family are the LEPTOPSYLLIDAE, found usually on small rodents, but also known from rabbits and lynx in North America. The ISCHNOPSYLLIDAE are restricted entirely to bats. The PYGIOPSYLLIDAE form a primarily Australian family of fleas found on a wide range of marsupials, monotremes, rodents, and passerine and sea birds.

Literature

Rothschild and Clay (1952), Askew (1971), and Rothschild (1975) include a good deal of general information on fleas, especially concerning host-parasite relationships. Flea classification and phylogeny is dealt with by Holland (1964). Keys for identification are given by Ewing and Fox (1943) [North American forms] and Smit (1957) [British species].

Askew, R. R., 1971, *Parasitic Insects*, American Elsevier, New York.
Ewing, H. E., and Fox, ., 1943, The fleas of North America, *U.S. Dep. Agric., Misc. Publ.* **500:**142 pp.

Hinton, H. E., 1958, The phylogeny of the panorpoid orders, *Annu. Rev. Entomol.* **3**:181–206.

Holland, G. P., 1964, Evolution, classification, and host relationships of Siphonaptera, *Annu. Rev. Entomol.* **9**:123–146.

Rothschild, M., 1975, Recent advances in our knowledge of the order Siphonaptera, *Annu. Rev. Entomol.* **20**:241–259.

Rothschild, M., and Clay, T., 1952, *Fleas, Flukes and Cuckoos: A Study of Bird Parasites*, Collins, London.

Smit, F. G. A. M. 1957, Siphonaptera, *R. Entomol. Soc. Handb. Ident. Br. Insects* **1**(16):94 pp.

10

The Remaining Endopterygote Orders

1. Introduction

The six remaining endopterygote orders dealt with in this chapter are quite distinct from those that form the panorpoid complex. Of the six, the order Hymenoptera appears most isolated phylogenetically and is sometimes considered in a distinct superorder, the Hymenopteroidea. The other orders are then united in the Neuropteroidea.

2. Megaloptera

SYNONYM: Sialoidea (in order Neuroptera *sensu lato*)

COMMON NAMES: alderflies and dobsonflies

Large, soft-bodied insects; head with chewing mouthparts, elongate antennae, and large compound eyes, three ocelli present (Sialidae) or absent (Corydalidae); two pairs of identical wings with primitive venation and large number of crossveins, abdomen 10-segmented without cerci; ovipositor absent.

Larvae aquatic with chewing mouthparts and paired abdominal gills. Pupae decticous and exarate.

Representatives of this small order are found in all world regions, though their distribution is discontinuous. About 50 species have been described in North America, with about the same number occurring in the United Kingdom.

Structure

Adult. Adult Megaloptera are generally large insects, with members of some species having a wingspan of about 12 cm. Their head carries well-developed compound eyes, long multisegmented antennae, and chewing mouthparts. Three ocelli are present in Sialidae but absent in Corydalidae. The thoracic segments are well developed and freely movable; the pronotum is broad. The legs are all similar. Four membranous wings occur, with all the major veins and a large number of crossveins present. The wings lack a pterostigma. The wing-coupling apparatus is of the jugofrenate type. The abdomen is 10-segmented and lacks cerci. Females have no ovipositor.

The structure of the internal organs is poorly known. The alimentary canal has a mediodorsal food reservoir; in most species eight Malpighian tubules are present; the nervous system is primitive with three thoracic and generally seven abdominal ganglia; females have a variable number of polytrophic ovarioles.

Larva and Pupa. Larvae are elongate and in some species may reach a length of 8 cm. The prognathous head is well developed and carries chewing mouthparts. The thorax bears three pairs of strong legs, the abdomen seven or eight pairs of gills. Pupae are decticous, exarate, and not enclosed in a cocoon.

Life History and Habits

Adult Megaloptera are generally found in the vicinity of streams or in other cool, moist habitats where they feed on nectar and soft-bodied insects. Reproduction appears to be entirely sexual and eggs are attached, in batches of several hundred to several thousand, to stones, vegetation, etc., usually near water. Larvae are aquatic and predaceous. Development in most species is completed in a season, but in some large forms it may take 2 or 3 years. Prior to pupation larvae leave the water and burrow into soil or moss or under stones where pupation occurs. Before eclosion pharate adults wriggle to the surface of the pupation medium.

Phylogeny and Classification

Fossil Megaloptera are recognizable from the Lower Permian and may have had a common ancestry with Mecoptera and other neuropteroids (Raphidioptera and Neuroptera) in the Upper Carboniferous. Recent Megaloptera are placed in a single superfamily Sialoidea, which includes two families, SIALIDAE (alderflies) (Figure 10.1A,B) and CORYDALIDAE (dobsonflies) (Figure 10.1C,D).

Literature

The biology of Megaloptera is treated in more detail by Berland and Grassé (1951). Hinton (1958) discusses the phylogenetic position of the order. Ross (1937) provides information on the life history and habits of the Sialidae.

Berland, L., and Grassé, P.-P., 1951, Super-ordre des Néuroptéroides, in Traité de Zoologie (P.-P. Grassé, ed.), Vol. X, Fasc. I, Masson, Paris.
Hinton, H. E., 1958, The phylogeny of the panorpoid orders, *Annu. Rev. Entomol.* **3**:181–206.
Ross, H. H., 1937, Nearctic alderflies of the genus *Sialis*, *Bull. Ill. Nat. Hist. Surv.* **21**(3):57–78.

3. Raphidioptera

SYNONYMS: Raphidiodea, Raphidioidea
(in order Neuroptera *sensu lato*)   COMMON NAME: snakeflies

Large insects similar to Megaloptera but distinguished by elongate "neck"; head with chewing mouthparts, bulging compound eyes, and elongate antennae; thorax

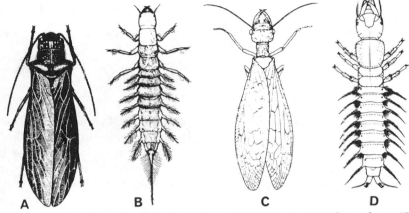

FIGURE 10.1. Sialoidea. (A) An alderfly, *Sialis mohri* (Sialidae); (B) *Sialis* sp. larva; (C) a dobsonfly, *Corydalus cornutus* (Corydalidae); and (D) *Corydalus* sp. larva. [A, B, from H. H. Ross, 1937, Studies of nearctic aquatic insects. I. Nearctic alderflies of the genus *Sialis* (Megaloptera, Sialidae), *Bull. Ill. Nat. Hist. Surv.* **21**(3). By permission of the Illinois Natural History Survey. C, from D. J. Borror, D. M Delong, and C. A. Triplehorn, 1976, *An Introduction to the Study of Insects*, 4th ed. By permission of Holt, Rinehart and Winston, Inc. D, from A. Peterson, 1951, *Larvae of Insects*. By permission of Mrs. Helen Peterson.]

with two pairs of identical wings; abdomen 10-segmented, cerci absent, females with elongate ovipositor.

Larvae terrestrial with chewing mouthparts. Pupae decticous and exarate.

Members of the order, which comprises about 100 species, are found on all continents except Australasia. About 20 species have been described from western North American.

Structure

Adult. Snakeflies essentially resemble Megaloptera but may be distinguished by the elongate "neck" formed from the prothorax and narrow, posterior part of the head. The head carries a pair of bulging compound eyes, chewing mouthparts, and long, multisegmented antennae. Ocelli are present (Raphidiidae) or absent (Inocelliidae). The wings are identical, and have a primitive venation of many crossveins, and a pterostigma. Females have a long hairlike ovipositor.

Larva and Pupa. Larvae are elongate and have a prognathous head with chewing mouthparts. The thoracic legs are all identical. The abdomen lacks appendages. Pupae are decticous, exarate, and closely resemble adults.

Life History and Habits

Adult Raphidioptera may be encountered on flowers, foliage, tree trunks, etc. where they feed especially on aphids and caterpillars. Females lay their eggs in cracks in the bark of trees. Larvae are found under loose bark, especially of conifers, where they prey on other insects. Mature larvae form a cell in which to pupate. Pupae may actively move about until eclosion.

FIGURE 10.2   Raphidioidea. A snakefly, *Agulla adnixa* (Raphidiidae). [From D. J. Borror, D. M. Delong, and C. A. Triplehorn, 1976, *An Introduction to the Study of Insects*, 4th ed. By permission of Holt, Rinehart and Winston, Inc.]

Phylogeny and Classification

Raphidioptera, like Megaloptera, are among the most primitive endopterygotes and probably had their origin in the Upper Carboniferous period. Recent members of the order are included in a single superfamily Raphidioidea, containing two families, RAPHIDIIDAE (Figure 10.2) and INOCELLIIDAE. North American species are restricted to two genera, *Agulla* (Raphidiidae) and *Inocellia*.

Literature

Aspöck and Aspöck (1975) summarize the biology of Raphidioptera. The phylogenetic relationships of the order are discussed by Achtelig and Kristensen (1973). Carpenter (1936) deals with the systematics of North American species.

Achtelig, M., and Kristensen, N. P., 1973, A re-examination of the relationship of the Raphidioptera (Insecta), *Z. Zool. Syst. Evoutionsforsch.* **11**:268–274.
Aspöck, H., and Aspöck, A., 1975, The present state of knowledge on the Raphidioptera of America (Insecta, Neuropteroidea), *Pol. Pismo Entomol.* **45**:537–546.
Carpenter, F. M., 1936, Revision of the nearctic Raphidiodea (recent and fossil), *Proc. Am. Acad. Arts Sci.* **71**:89–157.

4. Neuroptera

SYNONYM: Planipennia (in order Neuroptera *sensu lato*)

COMMON NAMES: lacewings, mantispids, antlions

Minute to large soft-bodied insects; head with chewing mouthparts, long multisegmented antennae, and well-developed compound eyes; two pairs of identical wings present, most species with primitive venation and veins bifurcated at wing margins; abdomen 10-segmented, cerci absent.

Larvae of most species terrestrial with suctorial mouthparts. Pupae decticous and exarate, enclosed in a cocoon.

As constituted here, the order contains about 4000 species and is represented in all world regions. Representatives are especially common in warmer climates.

Structure

Adult.   Adults range in size from a few millimeters to several centimeters. Members of most species are soft-bodied, weakly flying insects whose head carries a pair of well-developed compound eyes, long, multisegmented antennae, and chewing mouthparts. In most species the legs are identical, though in Mantispidae the forelegs are large and raptorial. The four wings are membranous, more or less equal in size, and generally have a primitive venation. In Coniopterygidae, however, the number of longitudinal and crossveins is much reduced. The abdomen is 10-segmented and lacks cerci.

Larva and Pupa. Larvae have a well-developed, prognathous head with suctorial mouthparts. Each mandible is sickle-shaped and grooved on the inner side. The lacinia is closely apposed to the groove forming a tube up which soluble food can be drawn. The alimentary canal is occluded posterior to the midgut. The Malpighian tubules are secondarily attached at their tips to the rectum and secrete silk with which mature larvae spin a pupal cocoon. Pupae are decticous and exarate, with well-developed, functional mandibles.

Life History and Habits

Adult Neuroptera feed on soft-bodied insects and nectar. They appear to reproduce solely by sexual means, and eggs are laid singly or in small batches. Larvae are predaceous and, with the exception of Sisyridae, terrestrial. Development in most species is completed in a season. Mature larvae spin a cocoon in which to pupate. Prior to eclosion, the pharate adult chews its way out of the cocoon using the pupal mandibles.

Phylogeny and Classification

Neuroptera, like members of the previous two orders, probably evolved in the Upper Carboniferous period from a common ancestor with the Mecoptera. The extant Neuroptera can be arranged in several, clearly defined superfamilies. Of these, the Ithonoidea contain the most generalized members of the order and appear to link the Neuroptera with the Megaloptera.

Superfamily Ithonoidea

Ithonoidea are large, mothlike insects. Less than 10 species have been described, mainly from Australia, and all are included in the family ITHONIDAE. The family is placed by some authors in the Osmyloidea.

Superfamily Coniopterygoidea

The 100 or so species of the small superfamily Coniopterygoidea are contained in the single family CONIOPTERYGIDAE. Adults are minute, aphidlike Neuroptera, frequently only 1 to 2 mm long, whose bodies are covered with a white powdery exudate.

Superfamily Osmyloidea

Osmyloidea form another small group, most of whose approximately 100 species are included in the OSMYLIDAE. Adults are large insects which live close to water. Larvae are terrestrial or semiaquatic. The group occurs mainly in the Southern Hemisphere.

Superfamily Mantispoidea

Mantispoidea appear to be closely related to Osmyloidea. The superfamily contains three small families, BEROTHIDAE, SISYRIDAE, and MANTISPIDAE. The first two are widely but discontinuously distributed throughout the world. Sisyrid larvae are aquatic and feed on freshwater sponges. Their abdomen bears seven pairs of gills. Adult Mantispidae (Figure 10.3), as their name suggests, are mantislike in that their forelegs are large and raptorial. They are restricted to the warmer regions of the world, including the southern United States.

Superfamily Hemerobioidea

The superfamily Hemerobioidea contains three families. HEMEROBIIDAE (brown lacewings) are generally nocturnal insects found in wooded areas. The group is widely distributed, mainly in temperate regions. The CHRYSOPIDAE (green lacewings) (Figure 10.4) constitute a familiar group of insects, found in long grass. Adults are called "stink flies" by some people through their ability to produce a disagreeable odor when caught. PSYCHOPSIDAE are rather like Hemerobiidae in both structure and general habits. The family is a small one, with representatives in Australia, South Africa, and Asia.

Superfamily Myrmeleontoidea

Included in the group is the largest neuropteran family, the MYRMELEONTIDAE (Figure 10.5), whose larvae are commonly called "antlions." These generally remain concealed under stones or debris, or cover themselves with particles of sand, etc., and await their prey (passing insects). Larvae of a few species construct pits at the bottom of which they conceal themselves and wait for some unfortunate insect to fall in. Adults resemble damselflies in general appearance. The closely related ASCALAPHIDAE are actively flying insects that catch their prey on the wing. Their larvae are similar in habits to antlions. Other families in this group are the NYMPHIDAE (Australian), STILBOPTERYGIDAE (Australian and South American), and NEMOPTERIDAE (Old World and Australian).

FIGURE 10.3. Mantispoidea. *Mantispa cincticornis* (Mantispidae). [From D. J. Borror, D. M. Delong, and C. A. Triplehorn, 1976, *An Introduction to the Study of Insects,* 4th ed. By permission of Holt, Rinehart and Winston, Inc.]

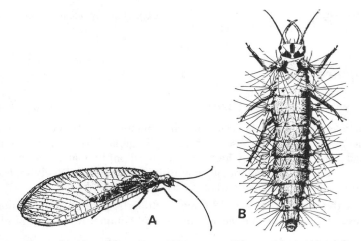

FIGURE 10.4   Hemerobioidea. A lacewing, *Chrysopa* sp. (Chrysopidae). [A] Adult and (B) larva. [By permission of the Illinois Natural History Survey.]

Literature

The biology of Neuroptera is dealt with by Withycombe (1925), Berland and Grassé (1951), and Throne (1971). Hinton (1958) discusses the phylogenetic relationships of the order.

Berland, L., and Grassé, P.-P., 1951, Super-ordre des Néuroptéroides, in: *Traité de Zoologie* (P.-P. Grassé, ed.), Vol. X, Fasc. I, Masson, Paris.

Hinton, H. E., 1958, The phylogeny of the panorpoid orders, *Annu. Rev. Entomol.* **3:**181–206.

Throne, A. L., 1971, The Neuroptera-suborder Planipennia of Wisconsin. Parts I and II, *Mich. Entomol.* **4:**65–78, 79–87.

Withycombe, C. L., 1925, Some aspects of the biology and morphology of the Neuroptera with

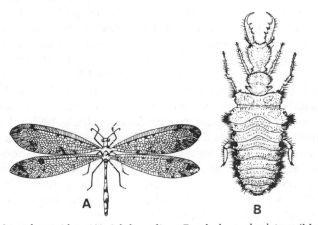

FIGURE 10.5.   Myrmeleontoidea. (A) Adult antlion, *Dendroleon obsoletum* (Myrmeleontidae); and (B) *Myrmeleon* sp. (Myrmeleontidae) larva. [A, from D. J. Borror, D. M. Delong, and C. A. Triplehorn, 1976, *An Introduction to the Study of Insects*, 4th ed. By permission of Holt, Rinehart and Winston, Inc. B, from A. Peterson, 1951, *Larvae of Insects*. By permission of Mrs. Helen Peterson.]

special reference to the immature stages and their possible phylogenetic significance, *Trans. Entomol. Soc. London* **1924**:303–411.

5. Coleoptera

SYNONYMS: Eleutherata, Elytroptera   COMMON NAME: beetles

Minute to very large insects; head with chewing mouthparts and extremely variable antennae, eyes present or absent; prothorax large and freely movable, fore wings modified into rigid elytra which usually meet middorsally and cover most of abdomen, hind wings membranous and usually folded beneath elytra, occasionally reduced or absent; abdomen varied.

Larvae with a distinct head and biting mouthparts, with or without thoracic legs, only rarely with prolegs. Pupae adecticous and commonly exarate, rarely obtect.

The Coleoptera are the most diverse order of insects and, indeed, of animals, with approximately 300,000 described species. Examination of their basic structure does little to suggest why the group should be so successful, yet they have come to occupy an amazing variety of habitats, with the single exception of the sea, though many littoral species occur.

Structure

Adult.   Among the living Coleoptera are some of both the largest and smallest of recent Insecta. The scarabaeid *Dynastes hercules* reaches 16 cm in length, in contrast to many Ptiliidae which are 0.5 mm or less. The head is usually heavily sclerotized and of variable shape. Compound eyes are present or absent; occasionally they are so large as to meet both dorsally and ventrally. Ocelli (not more than two) are usually absent. The antennae are typically 11-segmented, but their length and form are extremely varied. The mouthparts typically are of the chewing type, but their precise structure is varied. In many species the mandibles are sexually dimorphic, being enormously enlarged and frequently branched in the male. The prothorax is the largest of the thoracic segments and is usually quite mobile. The mesothorax is small and the meta-thorax relatively large, except in species in which the hind wings are reduced or absent. The fore wings are modified as hard elytra which meet in the midline but are not fused except in species in which the hind wings are lacking. The metathoracic wings are membranous and typically are folded beneath the elytra when not in use. The legs are usually all similar, though one or more pairs may be modified for the performance of particular functions. The number of visible abdominal segments is varied. Basically there are 10 segments, though the first is much reduced, and the last two or three are reduced and/or telescoped within the more anterior ones.

In accord with their varied diet, the alimentary canal shows a range of structure. Basically, however, it comprises a short, narrow pharynx, a widened expansion, the crop, followed by a poorly developed gizzard, a midgut which is highly variable, though usually possessing a large number of ceca, and a hindgut of variable length. Typically four or six Malpighian tubules occur, and in some species a cryptonephridial arrangement exists (see Chapter 18). The entire range of concentration of the central nervous system is found in Coleoptera, from the primitive condition in which three thoracic and seven or eight abdominal ganglia can be distinguished to that in which all the thoracic and abdominal ganglia are fused to form a composite structure. In males the paired testes may be simple, coiled tubular structures (Adephaga) or subdivided into a number of discrete follicles (Polyphaga). Paired vasa deferentia lead to the median ejaculatory duct, into which also open accessory glands of variable number and structure. The ovarioles of females are polytrophic (Adephaga) or acrotrophic (Polyphaga), and varied in number. A single spermatheca and its associated accessory gland enter the vagina by means of a long duct.

Larva and Pupa. The general form of beetle larvae is widely varied, though in all species the head is well developed and sclerotized, and the thoracic and abdominal segments (usually 10, rarely 9 or 8) are readily distinguishable. Thoracic legs are present or absent; abdominal prolegs are absent. Four basic larval types are found, campodeiform, eruciform, scarabaeiform, and apodous (see Chapter 21 for further details of these types). In a few families, for example, the Meloidae, hypermetamorphosis occurs when a larva passes through all four forms during its development. Pupae are always adecticous and in most species exarate, though in a few Staphylinidae they are obtect.

Life History and Habits

Although they are found in large numbers in most major habitats, beetles are among the least frequently observed insects by virtue of their secretive habits. The vast majority (ca. 95%) of the world's species are terrestrial, approximately 5000 species live in fresh water, and a few species have managed to invade the littoral zone of the shore where they are submerged twice daily in seawater. The terrestrial species are most common in soil and rotting plant and animal remains, though many live on or in living plants (including fungi). Many species are associated with man and live in his clothes, carpets, furniture, and food. Some groups are very well adapted for survival in extremely dry situations. A wide range of adaptation for an aquatic existence is found. In some species only the larvae are aquatic, though the adults live in the vicinity of water and may be able to survive short periods of submersion. Other species have both aquatic larvae and adults, the latter having various devices for retaining air in a layer around their body, since their spiracles are still functional (see Chapter 15). Perhaps surprisingly, there are very few truly aquatic pupae; larvae almost always leave the water to pupate or construct a submerged but air-filled cocoon. The only exception to this is in a few species of Psephenidae (Dryopoidea), where pupae respire by means of spiracular gills.

The majority of beetles are phytophagous in both the larval and adult stages, living in or on plants, wood, fungi, and a variety of stored products, including cereals, tobacco, and dried fruits. Some are voracious carnivores, capturing almost anything of suitable size. Others feed on dead animal materials such as wool and leather. Very few Coleoptera are parasitic. It will be obvious from the above that many beetles will be of economic importance to man. The majority of these are pests, though many are beneficial through their feeding on weeds or other insects such as aphids.

Most beetles reproduce sexually and are oviparous. Parthenogenesis, paedogenesis, viviparity, etc., are extremely rare phenomena in this order. The number of eggs laid varies among species, as does the subsequent pattern of development. There may be one or several generations per annum, or larval development may take several years. Parental care is practiced in some species. Pupation usually takes place in the soil or in the food plant. In some groups a cocoon is spun.

Phylogeny and Classification

The modern consensus of opinion is that the Coleoptera were derived from some Megaloptera-like ancestor, probably in the early part of the Permian period. Evidence in support of this opinion was produced with the discovery of the Lower Permian fossil *Tshekardocoleus* whose features are intermediate between those of primitive existing Coleoptera and Megaloptera. The earliest undoubted beetle fossils are from the Upper Permian and belong to the suborder Archostemata, a few species of which survive to the present day. Fossils of the suborder Adephaga are known from the Lower Jurassic, though Crowson (1960) suggests that the group is as old as the Archostemata. It appears certain that the earliest members of the suborder Polyphaga also evolved in the late Permian, though convincing fossil evidence is not available in support of this. Crowson speculates that the early radiation of the Coleoptera led to the formation of three stocks; carnivorous Adephaga, wood-boring Archostemata, and Polyphaga (including Myxophaga). The latter was initially a small group of wood or fungus eaters which only later (in the Cretaceous) underwent a massive radiation correlated with the advent of the flowering plants. More than 80% of the world's species are included in this suborder. The phylogeny of Coleoptera is shown diagrammatically in Figure 10.6.

Suborder Archostemata

Adult members of the suborder Archostemata have the following characters: wings with distal part coiled when at rest; hind wings with oblongum cell; notopleural sulcus present on prothorax; and hind coxae not immovably fixed to metasternum. Larvae have 5-segmented legs with one or two claws and mandibles with a molar area. Urogomphi are absent.

Less than 30 species of beetles are contained in this suborder, all of them in the superfamily Cupedoidea. All except one species of cupedoid belong to the family CUPEDIDAE, members of which are found under bark or in rotting logs in North America, Asia, and southeast Africa. The MICROMALTHIDAE, containing only one species, *Micromalthus debilis*, which has been reported from North and South America, South Africa, and Gibralter, are a family of uncertain affinities. Some authors consider that it is a polyphagan group, related to the Lymexyloidea or Cantharoidea.

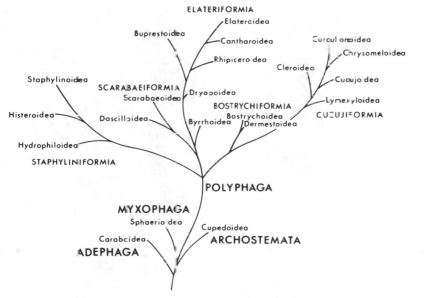

FIGURE 10.6 A suggested phylogeny of Coleoptera.

Suborder Adephaga

The suborder Adephaga includes beetles with the following features: hind wings with an oblongum; notopleural sulcus on prothorax; hind coxae immovably attached to metasternum; testes tubular and coiled; and ovarioles polytrophic. Larvae have 5-segmented legs with one or two claws. Their manibles lack a molar area. Segmented urogomphi are present in most species.

The suborder includes a single superfamily Caraboidea, most of whose members are predaceous beetles, with only a few species secondarily phytophagous. Members of the superfamily are traditionally arranged in two sections, Geadephaga (terrestrial forms) and Hydradephaga (aquatic forms). The former section contains the families RHYSODIDAE and CARABIDAE (Carabidae are split into many families by some coleopterists). Rhysodidae are small, black beetles which live in rotting wood both in the adult and larval stages. About 125 species are known, mostly from warmer areas. In contrast, about 35,000 species of Carabidae have been described from all parts of the world. About 80% of these belong to the subfamily CARABINAE (ground beetles), which, as their common name suggests, live in the soil, under stones or bark, or in logs. The elytra are frequently fused together and the wings atrophied. Both adults and larvae are carnivorous, and some species are of considerable benefit through their destruction of pest Lepidoptera. *Calosoma sycophanta* (Figure 10.7A,B) was introduced into the United States from Europe at the turn of the century for the control of the gypsy and browntail moths. Members of the subfamily CICINDELINAE (tiger beetles) (Figure 10.7C) are voracious predators, especially in the larval stage. The subfamily comprises about 2000 species and is mainly tropical or subtropical. Among the several families of Hydradephaga are the HALIPLIDAE, DYTISCIDAE, and GYRINIDAE. The Haliplidae (Figure 10.7D)

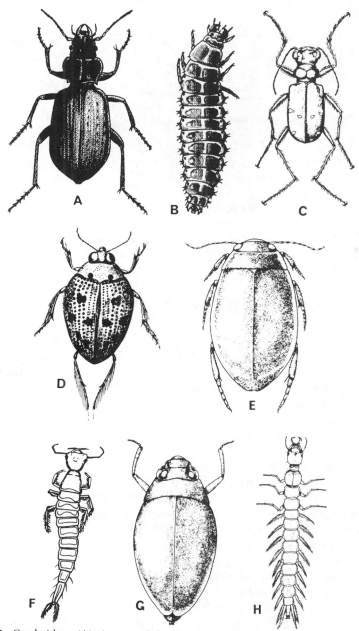

FIGURE 10.7 Caraboidea. (A) A ground beetle, *Carabus sycophanta* (Carabidae); (B) *C. sycophanta* larva; (C) a tiger beetle, *Cicindela sexguttata* (Carabidae); (D) *Peltodytes edentulus* (Haliplidae); (E) a diving beetle, *Dytiscus verticalis* (Dytiscidae); (F) *Dytiscus* sp. larva; (G) a whirligig beetle, *Dineutes americanus* (Gyrinidae); and (H) *D. americanus* larva. [A, B, from L. A. Swan, and C. S. Papp, 1972, *The Common Insects of North America.* Copyright 1972 by L. A. Swan and C. S. Papp. Reprinted by permission of Harper & Row, Publishers, Inc. C, D, from R. H. Arnett, Jr., 1968, *The Beetles of the United States (A Manual for Identification).* By permission of the author. E, G, from D. J. Borror, D. M. Delong, and C. A. Triplehorn, 1976, *An Introduction to the Study of Insects,* 4th ed. By permission of Holt, Rinehart and Winston, Inc. F, from E. S. Dillon and L. S. Dillon, 1972, *A Manual of Common Beetles of Eastern N. America.* By permission of Dover

constitute a small but widely distributed and common family of water beetles. Adults generally crawl about among the algae on which they feed, though they can swim. Larvae are predaceous. The family Dytiscidae (Figure 10.7E,F) contains about 4000 species and is especially common in the palearctic region. Both adults and larvae are predaceous. The family Gyrinidae (Figure 10.7G,H), with about 400 species, includes the familiar "whirligig" beetles which swim in vast numbers on the surface of ponds.

Suborder Myxophaga

Myxophaga are minute beetles with clubbed antennae, a prothorax with a notopleural sulcus, and hind wings that have an oblongum and fringe of long hairs and are coiled apically. Larvae are aquatic and have mandibles with a molar area.

Crowson (1955) proposed erection of this suborder to contain four families, totalling about 20 species, that were previously included in the Polyphaga. The four families (LEPICERIDAE, SPHAERIIDAE, HYDROSCAPHIDAE, and CALYPTOMER-IDAE) are included in the single superfamily Sphaeriodea. The suborder appears to be very ancient with an origin close to that of the Archostemata and Adephaga.

Suborder Polyphaga

The beetles of the suborder Polyphaga have hind wings that lack an oblongum and are never coiled distally. A notopleural sulcus is absent from the prothorax, the hind coxae are movable, the testes are not tubular and coiled, and the ovarioles are acrotrophic. The legs of larvae are either 4-segmented (without tarsus) plus single claw, vestigial, or absent. The larval mandibles have a molar area.

Within the Polyphaga five major series (evolutionary lines) can be recognized (Figure 10.6). Crowson (1960) suggests that the first adaptive radiation of the Polyphaga occurred in the Triassic when three ancestral stocks had their origins: the staphyliniform, eucinetoid, and dermestoid. The former gave rise to the modern Staphyliniformia. The eucinetoid group evolved into the Scarabaeiformia and Elateriformia, and the dermestoid group was ancestral to the Bostrychiformia and Cucujiformia. The latter series includes more than one half of the total beetle species and can therefore be considered as the most highly evolved group within the order.

Series Staphyliniformia

Three superfamilies are included in the series Staphyliniformia: the Histeroidea, Hydrophiloidea and Staphylinoidea. The Hydrophiloidea appear to be the most primitive group.

Publications, New York. H, from A. G. Böving, and F. C. Craighead, 1930, An illustrated synopsis of the principal larval forms of the order Coleoptera, Entomol. Am. XI:1–351. Published by the Brooklyn Entomological Society. By permission of the New York Entomological Society.]

Superfamily Hydrophiloidea

More than 80% of the 2400 species included in the superfamily Hydrophiloidea belong to the family HYDROPHILIDAE (Figure 10.8), whose members are mainly aquatic beetles somewhat similar to Dytiscidae. However, they are scavengers rather than predators, adults do not usually rest head downward at the surface of the pond, and, when swimming, they move their legs alternately rather than synchronously. The larvae are predaceous. Some hydrophilids are terrestrial, but restricted to damp places, for example, among decaying plant material or in dung.

Superfamily Histeroidea

Almost all of the approximately 2500 species of Histeroidea are placed in the family HISTERIDAE (Figure 10.9), both the adults and larvae of which feed on other insects. These small, usually shiny black beetles with short elytra (which leave the last one or two abdominal segments exposed), are found under bark, in rotting animal or vegetable material, or in ants' nests.

Superfamily Staphylinoidea

The very large superfamily Staphylinoidea contains more than 35,000 species, adults of which are characterized by unusually short elytra that leave about half of the abdomen visible. Some 27,000 of these species belong to the family STAPHYLINIDAE (rove beetles) (Figure 10.10A), a group of very diverse habits. The majority appear to be carnivorous or saprophagous, though precise details of their feeding habits are not known. They occur in many places, for example, in decaying animal or vegetable matter, under stones or bark, on flowers, under seaweed, in moss or fungi, in the nests of birds, mammals, and social insects. The PSELAPHIDAE (about 5000 species) closely resemble rove beetles both morphologically and in habits. They are found, moreover, in similar habitats. Adult SCYDMAENIDAE (1200 species), in contrast to those of the two families above, have fully developed elytra. They are hairy and somewhat antlike in appearance (Figure 10.10B), being found under stones, in humus, or in ants' nests. The ANISOTOMIDAE (LEIODIDAE) form a family of about 1100

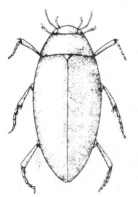

FIGURE 10.8. Hydrophiloidea. *Hydrophilus triangularis* (Hydrophilidae). [From D. J. Borror, D. M. Delong, and C. A. Triplehorn, 1976, *An Introduction to the Study of Insects*, 4th ed. By permission of Holt, Rinehart and Winston, Inc.]

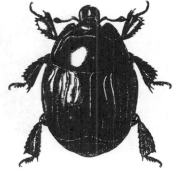

FIGURE 10.9. Histeroidea. *Margarinotus immunis* (Histeridae). [From R. H. Arnett, Jr., 1968, *The Beetles of the United States (A Manual for Identification)*. By permission of the author.]

species of beetles that are found in decaying organic matter, under bark, etc. A number of species have become adapted for a cave-dwelling existence.

Series Scarabaeiformia

Superfamily Scarabaeoidea

Nearly 19,000 species are placed in superfamily Scarabaeoidea, about 17,000 of which are included in the family SCARABAEIDAE. Adults may be recognized by the lamellate, terminal segments of the antennae. Many common beetles belong to the family, including dung beetles (SCARABAEINAE=COPRINAE), cockchafers (May or June bugs) (MELOLONTHINAE) (Figure 10.11B), shining leaf chafers (RUTELINAE), and the large and striking elephant (rhinoceros) beetles (DYNASTINAE) (Figure 10.11A). Most scarabaeids feed on decaying organic matter, especially dung, both in the adult and juvenile stages, though there are many variations of this theme. Larvae of some species feed underground on plant roots and a few live in termites' nests. Adults frequently feed on nectar, foliage, or fruit, or they do not feed at all. Two other common scarabaeid families are the LUCANIDAE and PASSALIDAE. The family Lucanidae (stag beetles) (Figure 10.11C) includes about 750 species in which the adults are sexually dimorphic. The mandibles of males are enormously enlarged,

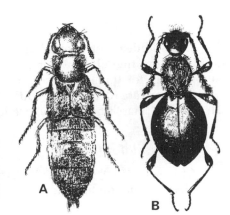

FIGURE 10.10. Staphylinoidea. (A) The hairy rove beetle, *Staphylinus maxillosus villosus* (Staphylinidae); and (B) an antlike stone beetle, *Euconnus clavipes* (Scydmaenidae). [From R. H. Arnett, Jr., 1968, *The Beetles of the United States (A Manual for Identification)*. By permission of the author.]

A

B

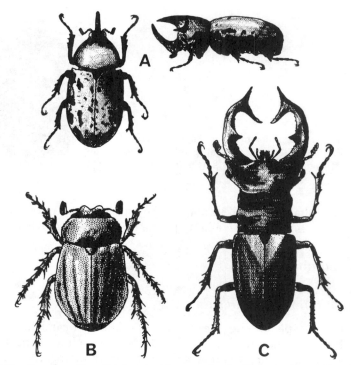

FIGURE 10.11.   Scarabaeoidea. (A) A male rhinoceros beetle, *Dynastes tityus* (Scarabaeidae); (B) a May beetle, *Phyllophaga rugosa* (Scarabaeidae); and (C) the giant stag beetle, *Lucanus elaphus* *(Lucanidae).* [From L. A. Swan, and C. S. Papp, 1972, *The Common Insects of North America.* Copyright 1972 by L. A. Swan and C. S. Papp. Reprinted by permission of Harper & Row, Publishers, Inc.]

though the significance of this is not yet understood. Larvae generally feed on rotting wood. Adults are mainly nectar, occasionally foliage, feeders. The Passalidae form a mainly tropical, forest-dwelling family of about 500 species. Large numbers of these beetles are often found in the same log, and it appears that the adults assist in feeding the larvae by partially chewing the rotting wood beforehand.

Superfamily Dascilloidea

Dascilloidea form a small group of about 500 species, usually arranged in four families, though Crowson (1960) has suggested the erection of a separate superfamily for three of these. The two largest families are the DASCILLIDAE, whose larvae are root feeders, while the adults feed on nectar or foliage, and the HELODIDAE (marsh beetles), whose larvae are aquatic.

Series Elateriformia

Crowson (1960) suggests that the Elateriformia and Scarabaeiformia may have had a common ancestor which perhaps resembled the recent Dascillidae in general form but which was semiaquatic in habit and fed on algae. Crowson points out that in contrast to the Adephaga, Staphyliniformia, and Cucujiformia, in which the larval life is short and adult life long, in the Elateriformia larvae are long-lived and adults short-lived, frequently not taking any protein food. Thus, the adaptations of larvae are usually more important than those of adults. Within the series six superfamilies are recognized.

Superfamily Byrrhoidea

About 90% of the 300 species of Byrrhoidea are in the family BYRRHIDAE (pill beetles). These are ground-dwelling beetles that feed mainly on moss in both the larval and adult stage.

Superfamily Dryopoidea

Most of the slightly more than 1000 species in the superfamily Dryopoidea are subaquatic, the adults generally living in mud or on vegetation at the margin of ponds or streams, and the larvae being truly aquatic and showing a number of adaptations for this mode of life. The largest families are the ELMIDAE and DRYOPIDAE, each with about 300 species. In these families both the adults and larvae are usually aquatic, though species in which the adults are terrestrial are known.

Superfamily Buprestoidea

The 11,500 species of Buprestoidea are placed in a single family BUPRES-TIDAE (Figure 10.12). They are commonly known as "jewel beetles" on account of their usually brilliant coloration. The family is particularly common in forests where the adults are found on flowers, while the larvae, which may sometimes become serious pests, bore in wood (living or dead) or herbaceous plants where they cause galls.

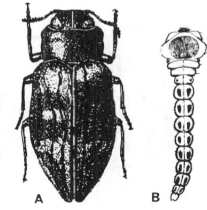

FIGURE 10.12. Buprestoidea. (A) *Chrysobothris femorata* (Buprestidae) and (B) *Chrysobothris* sp. larva. [A, from E. S. Dillon and L. S. Dillon, 1972, *A Manual of Common Beetles of Eastern N. America.* By permission of Dover Publications, New York. B, from A. D. Imms, 1957, *A General Textbook of Entomology,* 9th ed. (revised by O. W. Richards and R. G. Davies). By permission of Chapman and Hall Ltd.]

A   B

Superfamily Rhipiceroidea

Rhipiceroidea form a small superfamily (180 species), found mainly in tropical areas, whose adults are recognized by their large flabellate antennae. Almost nothing is known of the habits of this group. One species of RHIPICERIDAE is parasitic in the larval stage on immature cicadas. Larval CALLIRHIPIDAE appear to be wood borers.

Superfamily Elateroidea

About 7000 of the 8000 species in the superfamily Elateroidea belong to the family ELATERIDAE (Figure 10.13), the adults and larvae of which are commonly known as "click beetles" and "wireworms," respectively. Adults are phytophagous and are found in flowers, or under bark. Larvae are subterranean, feed on young roots, and frequently cause extensive damage to cereal crops, beans, cotton, and potatoes. Most of the remaining Elateroidea belong to the EUCNEMIDAE, a group of elateridlike beetles found mainly in tropical areas, where they feed on rotting wood.

Superfamily Cantharoidea

Nearly 12,500 species of Cantharoidea are known, most of these being assigned to four families. In almost all species both adults and larvae are predaceous. The MELYDRIDAE (Figure 10.14A) (4000 species) form the largest family. Its members are found on flowers and are frequently brightly colored. Some authors consider that the similarities between adults of this family and of the CANTHARIDAE are due to convergence, and the true position of the Melydridae is in the Cleroidea. The family Cantharidae (Figure 10.14B) contains about 3500 species of soft-bodied, generally hairy beetles, commonly known as "soldier beetles." Like melydrids, they are usually found on flowers. The approximately 1700 species of LAMPYRIDAE (fireflies) (Figure 10.14C) are renowned for their ability to produce light. Both larvae and adults show this feature, and, although it has obvious sexual significance in mature insects, its purpose is by no means clear in the juvenile stage. In adults, light is produced by special cells on the

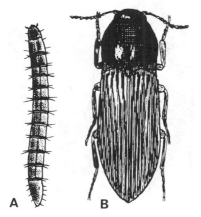

FIGURE 10.13. Elateroidea. The wheat wireworm, *Agriotes mancus* (Elateridae). (A) Larva and (B) adult. [From L. A. Swan and C. S. Papp, 1972, *The Common Insects of North America.* Copyright 1972 by L. A. Swan and C. S. Papp. Reprinted by permission of Harper & Row, Publishers, Inc.]

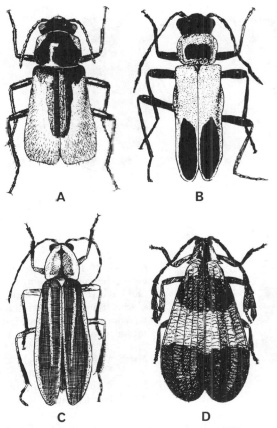

FIGURE 10.14. Cantharoidea. (A) A soft-winged flower beetle, *Malachius aeneus* (Melydridae); (B) the soldier beetle, *Chauliognathus pennsylvanicus* (Cantharidae); (C) a firefly, *Photuris pennsylvanica* (Lampyridae); and (D) a net-winged beetle, *Calopteron reticulatum* (Lycidae). [From E. S. Dillon and L. S. Dillon, 1972, *A Manual of Common Beetles of Eastern N. America.* By permission of Dover Publications, New York.]

abdomen; in larvae the entire body is faintly luminescent. The LYCIDAE (net-winged beetles) (Figure 10.14D) (3000 species) are cantharidlike beetles whose larvae live in soil or beneath bark. Adults, which occur on tree trunks or foliage, are brightly colored, apparently distasteful insects which are mimicked by a number of other insects.

Series Bostrychiformia

Superfamily Dermestoidea

Most of the 800 or so species in the superfamily Dermestoidea belong to the family DERMESTIDAE (skin and carpet beetles) (Figure 10.15), a group that contains a number of economically important species. Both adults and larvae are scavengers on a variety of plant and animal materials including furs, hides, wool, museum specimens, clothing, carpets, and various foods such as bacon

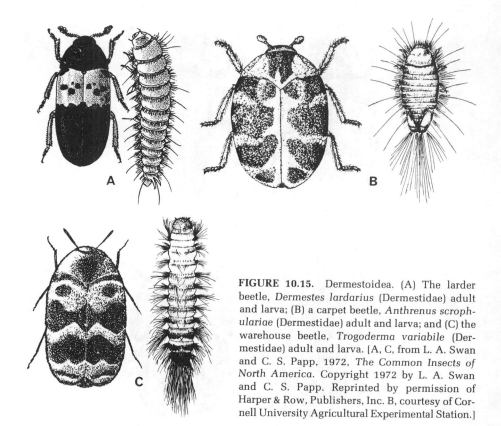

FIGURE 10.15. Dermestoidea. (A) The larder beetle, *Dermestes lardarius* (Dermestidae) adult and larva; (B) a carpet beetle, *Anthrenus scrophulariae* (Dermestidae) adult and larva; and (C) the warehouse beetle, *Trogoderma variabile* (Dermestidae) adult and larva. [A, C, from L. A. Swan and C. S. Papp, 1972, *The Common Insects of North America.* Copyright 1972 by L. A. Swan and C. S. Papp. Reprinted by permission of Harper & Row, Publishers, Inc. B, courtesy of Cornell University Agricultural Experimental Station.]

and cheese. Most damage is done by the hairy larvae. *Dermestes, Anthrenus, Attagenus,* and *Trogoderma* are genera that contain economically important species.

Superfamily Bostrychoidea

About one-half of the 2300 species of Bostrychoidea belong to the family ANOBIIDAE, which includes the "deathwatch" beetles, so-called because the tapping noise they make as they bore was supposed to be a sign of a future death in the house. Most species are wood borers in the larval stage, but adults leave the wood for mating purposes. A few species do not live in wood but attack stored products such as cereal products and tobacco. Economically important species include the cosmopolitan furniture beetle, *Anobium punctatum,* and the drugstore beetle or biscuit weevil, *Stegobium paniceum* (Figure 10.16A). The related family PTINIDAE (spider beetles) (Figure 10.16B), with about 700 species, includes no wood-boring forms, but only beetles associated with stored food, and dried animal or vegetable matter. The BOSTRYCHIDAE (about 450 species) (Figure 10.16C) mainly bore in felled timber or dried wood. Adults resemble scolytids but may be distinguished by their strongly deflexed head and rasplike pronotum. The closely related LYCTIDAE form a small but widely distributed family whose members are commonly known as powder-post beetles, since they usually reduce the dry wood in which they burrow to a powder.

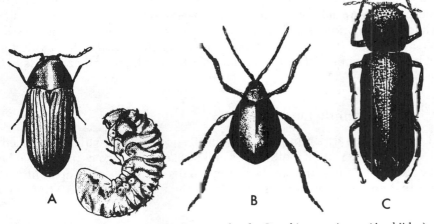

FIGURE 10.16. Bostrychoidea. (A) The drugstore beetle, *Stegobium paniceum* (Anobiidae) adult and larva; (B) the humpbacked spider beetle, *Gibbium psylloides* (Ptinidae); and (C) the apple twig borer, *Amphicerus hamatus* (Bostrychidae). [A, B, from L. A. Swan and C. S. Papp, 1972, *The Common Insects of North America*. Copyright 1972 by L. A. Swan and C. S. Papp. Reprinted by permission of Harper & Row, Publishers, Inc. C, from E. S. Dillon and L. S. Dillon, 1972, *A Manual of Common Beetles of Eastern N. America*. By permission of Dover Publications, New York.]

Series Cucujiformia

Within the immense group Cucujiformia are five superfamilies. Three of these, Cleroidea, Lymexyloidea, and Cucujoidea are considered to be "primitive" or "lower" groups when compared with the other two, the Chrysomeloidea and Curculionoidea.

Superfamily Cleroidea

The superfamily Cleroidea contains about 4000 species of beetles, about 3400 of which belong to the mainly tropical family CLERIDAE (Figure 10.17). Checkered beetles, as clerids are commonly known on account of their striking color patterns, are mainly found on or in tree trunks where they prey on other insects, especially scolytid beetle larvae. A few species are found in stored products with a high fat content, for example, ham, cheese, and copra. The family TROGO-

FIGURE 10.17. Cleroidea. A checkered beetle, *Enoclerus nigripes* (Cleridae). From E. S. Dillon and L. S. Dillon, 1972, *A Manual of Common Beetles of Eastern N. America*. By permission of Dover Publications, New York.]

STIDAE (OSTOMIDAE). containing nearly 600 species, is a mainly tropical group whose members are mainly predaceous on other insects. Included in this family is the cadelle, *Tenebrioides mauritanicus*, a cosmopolitan species found in stored cereals. Although it may cause some damage to the cereals, this tends to be offset by the beneficial effect this species has by preying on other insects in the grain.

Superfamily Lymexyloidea

Lymexyloidea are a group of less than 40 species, all of which are placed in one family, LYMEXYLIDAE. Both adults and larvae are elongate insects that bore in tree stumps and logs that are beginning to decay. They feed apparently on a fungus that grows in the tunnels, transfer of the fungus to new locations being effected by the beetles.

Superfamily Cucujoidea

Though outnumbered by the Chrysomeloidea and Curculionoidea in terms of species, the Cucujoidea are the most diverse superfamily of beetles as is indicated by the large number of families (more than 50) that it contains. Most of these are relatively small, that is, contain less than 500 species, and the majority of the 41,000 plus species in the superfamily are contained in a few very large families. The largest families are noted below.

The NITIDULIDAE (sap beetles) (Figure 10.18A) (2200 species) constitute a diverse family of beetles. Adults of some species live in flowers, where they feed on pollen and nectar; others feed on fungi or ripe fruit, or are leaf miners, or prey on other insects. A few species live in the nests of social Hymenoptera. Most CRYPTOPHAGIDAE (800 species) are very small, hairy beetles that feed on fungi. They are most common in moldy materials, including stored products. A few species live in the nests of birds, mammals, bees, and wasps. The EROTYLIDAE (1500 species) constitute a mainly South American group of beetles whose larvae feed on the fruiting bodies of fungi. The frequently brightly colored adults are usually found beneath the bark of rotting logs. The family COCCINELLIDAE (ladybugs, ladybird beetles) (Figure 10.18B), comprising about 5000 species, is a very important group of mainly brightly colored. spotted, or striped beetles. The majority of species are carnivorous as both adults and larvae on Homoptera, other soft-bodied insects, and mites. Some species have been used successfully in biological control programs, the best example being the introduction of the Australian *Vedalia cardinalis* to control the cottony-cushion scale, *Icerya purchasi*, in the citrus fruit orchards of California. Other coccinellids are phytophagous, however, and may cause much damage to solanaceous or bean crops. The ENDOMYCHIDAE (1100 species) are brightly colored, mainly tropical beetles found under bark or in rotting wood where they feed on fungi. The COLYDIIDAE, with more than 1400 species, form a mainly tropical family of beetles that are generally associated with rotting wood, where it is thought that they prey on the larvae of wood-boring beetles. A few species occur in ants' or bees' nests.

Containing more than 15,000 species, the TENEBRIONIDAE (darkling bee-

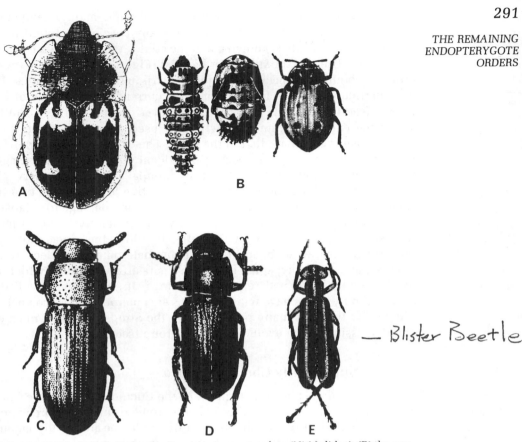

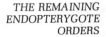
— Blister Beetle

FIGURE 10.18. Cucujoidea. (A) A sap beetle, *Prometopia sexmaculata* (Nitidulidae); (B) the convergent ladybug, *Hippodamia convergens* (Coccinellidae); (C) the confused flour beetle, *Tribolium confusum* (Tenebrionidae); (D) adult of the yellow mealworm, *Tenebrio molitor* (Tenebrionidae); and (E) the striped blister beetle, *Epicauta vittata* (Meloidae). [A, D, E, from E. S. Dillon and L. S. Dillon, 1972, *A Manual of Common Beetles of Eastern N. America.* By permission of Dover Publications, New York. B, C, from L. A. Swan and C. S. Papp, 1972, *The Common Insects of North America.* Copyright 1972 by L. A. Swan and C. S. Papp. Reprinted by permission of Harper & Row, Publishers, Inc.]

tles) (Figure 10.18C,D) are one of the largest beetle families. Adults, though generally dark in color, show a remarkable divergence of form. In contrast, larvae are extremely uniform and generally resemble wireworms. Most species are ground-dwelling insects found beneath stones or logs, but they are also found in rotting wood, fungi, nests of birds and social insects, and stored products where they may occur in enormous numbers and do considerable damage. It is to the latter group that the familiar flour beetles (*Tribolium* spp.) and mealworms (larvae of *Tenebrio* spp.) belong. Members of some species are capable of withstanding considerable periods of desiccation, and are able to absorb moisture from the air should the need arise. The ALLECULIDAE (1100 species) are small, generally dark beetles that are found as adults on leaves or flowers where they are believed to feed on pollen and nectar. The larvae occur in rotting wood, etc., and generally resemble wireworms. The OEDEMERIDAE

(1500 species) form a widely distributed, though mainly temperate family whose larvae prefer moist rotting wood. Many species are found in driftwood on the seashore. Little is known about the adults, though they are usually found on flowers. The MELOIDAE (blister beetles) (Figure 10.18E) form a well-known group of about 2000 species of frequently strikingly colored beetles. The family is of particular interest for the general occurrence in its larvae of heteromorphosis (see Chapter 21) and the production in the adults of cantharidin, a substance extracted from the elytra and used in certain drugs (and which causes blisters when applied to the skin). The life history is somewhat varied, but generally the early instars are parasitic on grasshopper or bee eggs; a non-feeding, overwintering stage, the "pseudopupa" or "coarctate larva," then follows, after which develops either another active feeding stage or a true pupal stage. Adults are phytophagous and may do much damage to solanaceous and leguminous crops. A smaller but closely related family is the RHIPIPHORIDAE (250 species), a unique family of beetles in which the larvae are at least temporarily endoparasitic, in addition to being heteromorphic. Bees, wasps, cockroaches, and, possibly, some beetles are parasitized. Adults, which are short-lived, usually have both elytra and wings. In the subfamily RHIPIDIINAE, however, only males have wings; females are apterous and somewhat larviform. The family thus has many parallels with the Strepsiptera and must enter into any discussion of the evolution of this group (see Section 6).

Superfamily Chrysomeloidea

Though almost as large a group as the Cucujoidea, the superfamily Chrysomeloidea is much more uniform, being divided into only three families, two very large and the third small. Both adults and larvae are phytophagous or xylophagous. The CERAMBYCIDAE (20,000 species) are commonly known as long-horned beetles (Figure 10.19A). The family, which is mainly tropical but has representatives throughout the world, contains some of the largest insects, with adults of some species more than 7 cm in length (excluding the antennae). Adults are brightly colored, cryptically colored, or resemble other insects. Larvae generally bore into wood, though a few are restricted to the softer roots and stems of herbaceous plants. In most species the larval stage lasts for several years. The BRUCHIDAE (1200 species) are commonly known as "pea and bean weevils," though they are not weevils in the true sense. Larvae feed almost exclusively in the seeds of Leguminosae, though some species attack Palmaceae (coconuts, etc.). The group includes two cosmopolitan pests, the pea weevil, *Bruchus pisorum,* which attacks growing, though not stored (i.e., dried) peas, and the bean weevil, *Acanthoscelides obtectus* (Figure 10.19B), which attacks both beans and peas, either in the field or in storage. Some 20,000 species of CHRYSOMELIDAE (leaf beetles) have been described. Adults are frequently brightly colored beetles which feed on foliage and flowers. The larvae are varied in their habits. They may feed exposed on leaves or stems, or mine into them. Some live below ground and feed on roots. Others live in ants' nests and a few are aquatic. The family includes a large number of pest species, perhaps the most infamous of which is the Colorado potato beetle, *Leptinotarsa decemlineata* (Figure 10.19C), belonging to the subfamily CHRYSOMELINAE. Most pest

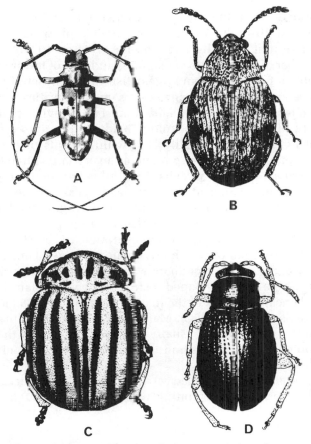

FIGURE 10.19. Chrysomeloidea. (A) The spotted pine sawyer, *Monochamus maculosus* (Cerambycidae); (B) the bean weevil, *Acanthoscelides obtectus* (Bruchidae); (C) the Colorado potato beetle, *Leptinotarsa decemlineata* (Chrysomelidae); and (D) the potato flea beetle, *Epitrix cucumeris* (Chrysomelidae). [A, from L. A. Swan and C. S. Papp, 1972, *The Common Insects of North America.* Copyright 1972 by L. A. Swan and C. S. Papp. Reprinted by permission of Harper & Row, Publishers, Inc. B, from R. H. Arnett, Jr., 1968, *The Beetles of the United States (A Manual for Identification).* By permission of the author. C, D, from E. S. Dillon and L. S. Dillon, 1972, *A Manual of Common Beetles of Eastern N. America.* By permission of Dover Publications, New York.]

species, however, are in the subfamilies GALERUCINAE and HALTICINAE. For example, *Acalymma vittata*, the striped cucumber beetle, feeds in the adult stage on a variety of cucurbits and, as a larva, on the roots of various plants. This species is known to act as a vector of certain diseases which cause wilting. Flea beetles (Halticinae), so-called because of their jumping ability, attack a variety of crops, for example, turnips (*Phyllotreta* spp.), and tobacco, potatoes, tomatoes, and eggplants (*Epitrix* spp.) (Figure 10.19D).

Superfamily Curculionoidea

Most adult Curculionoidea are easily recognized by the prolongation of the head to form a rostrum at the tip of which are located the mouthparts. It is estimated that the superfamily already includes more than 65,000 described species (Britton, 1970), about 60,000 of which belong to the family CURCULIONIDAE (weevils, snout beetles) (Figure 10.20A). Arrangement of this vast number of species into subfamilies and taxa of lower rank is perhaps the major problem for coleopteran systematists at the present time. More than 100 subfamilies are already recognized by some authorities, though some of these are given familial status by others. Almost all weevils are phytophagous, and no parts of plants are left unexplored. Adults bore into seeds, fruit, and other parts of plants. Larvae usually feed within plants or externally below ground. Not surprisingly, the group includes a very large number of pests, for example, the granary and rice weevils, *Sitophilus* (=*Calandra*) *granarius* and *S. oryzae*, respectively, which attack cereal seeds, peas, and beans, and various *Anthonomus* species, of which the best known is *A. grandis*, the boll weevil, which attacks cotton. Two subfamilies that require specific mention are the PLATYPODINAE (pinhole borers) and SCOLYTINAE (bark beetles, ambrosia beetles) (Figure 10.20B), each of which are frequently given familial rank. Members of these two groups are woodborers and, unlike the majority of weevils, adults do not have a well-developed rostrum. In both subfamilies the beetles generally live beneath the bark of the tree where they construct a characteristic pattern of tunnels (Figure 10.20C). Many species attack healthy trees, which they girdle and kill. The beetles usually feed on fungi, which they cultivate in the tunnels. Dutch elm disease is transmitted by the European elm bark beetle, *Scolytus multistriatus*.

The ANTHRIBIIDAE (2400 species) form a mainly tropical family, most species of which are found in rotting wood or fungi, though a few feed on

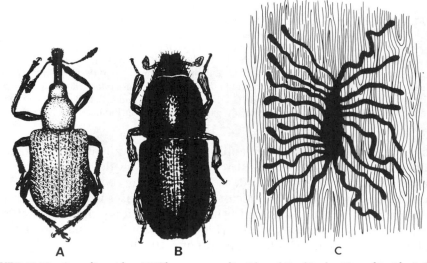

A   **B**   **C**

FIGURE 10.20 Curculionoidea. (A) The rose curculio, *Rhynchites bicolor* (Curculionidae); (B) the shot-hole borer, *Scolytus rugulosus* (Curculionidae); and (C) boring pattern of S. rugulosus. [A, B, from E. S. Dillon and L. S. Dillon, 1972, *A Manual of Common Beetles of Eastern N. America*. By permission of Dover Publications, New York. C, from L. A. Swan and C. S. Papp, 1972, *The Common Insects of North America*. Copyright 1972 by L. A. Swan and C. S. Papp. Reprinted by permission of Harper & Row, Publishers, Inc.]

seeds, etc., including the coffee bean weevil, *Araecerus fasciculatus*, an important pest of coffee and cocoa beans, dried fruit, and nutmeg. The BRENTIDAE (1200 species) are mainly confined to the tropical forests where the larvae feed on rotting wood. The APIONIDAE (1000 species) constitute a widely distributed family, whose larvae mainly burrow in seeds, stems, or roots of plants. Some species are occasionally pests of Leguminosae.

Literature

Much general information on Coleoptera is given by Dillon and Dillon (1972) and Arnett (1968). Reviews of the ecology of particular families are those of Richter (1958) and Linsley (1959). Imms (1957) and Britton (1970) provide comprehensive reviews of the order, with emphasis on taxonomy, the latter author dealing with the Australian forms. The classification and phylogeny of Coleoptera are discussed by Crowson (1955, 1960). The British species may be identified using Linnsen (1959), or the Royal Entomological Society handbooks (Crowson, 1956, and other authors in the same series). Böving and Craighead (1930), Jacques (1951), Dillon and Dillon (1972), Arnett (1968), and Borror *et al.* (1976) deal with the American species.

Arnett, R. H., Jr., 1968. *The Beetles of the United States (A Manual for Identification)*, American Entomological Institute, Ann Arbor, Mich.

Borror, D. J., Delong, D. M., and Triplehorn. C. A., 1976, *An Introduction to the Study of Insects*, 4th ed., Holt, Rinehart and Winston, New York.

Böving, A. G. and Craighead, F. C., 1930, An Illustrated synopsis of the principal larval forms of the order Coleoptera, *Entomol. Am.* **11**:1–351.

Britton, E. B., 1970, Coleoptera, in: *The Insects of Australia* (I. M. Mackerras, ed.), Melbourne University Press, Carlton, Victoria.

Crowson, R. A., 1955, *The Natural Classification of the Families of Coleoptera*, Nathanial Lloyd, London.

Crowson, R. A., 1956, Coleoptera. Introduction and keys to families, *R. Entomol. Soc. Handb. Ident. Br. Insects* **4**(1):59pp.

Crowson, R. A., 1960, The phylogeny of Coleoptera, *Annu. Rev. Entomol.* **5**:111–134.

Dillon, E. S., and Dillon, L. S., 1972, *A Manual of Common Beetles of Eastern N. America*, Dover, New York.

Imms, A. D., 1957, *A General Textbook of Entomology*, 9th ed. (revised by O. W. Richards and R. G. Davies), Methuen, London.

Jacques, H. E., 1951, *How to Know the Beetles*, Brown, Dubuque, Iowa.

Linnsen, E. F., 1959, *Beetles of the British Isles*, 2 vols., Warne, London.

Linsley, E. G., 1959, Ecology of Cerambycidae, *Annu. Rev. Entomol.* **4**:99–138.

Richter, P. O., 1958, Biology of Scarabaeidae, *Annu. Rev. Entomol.* **3**:311–334.

6. Strepsiptera

SYNONYMS: none   COMMON NAME: stylopoids

Males free living; mouthparts degenerate, antennae conspicuous and flabellate; fore wings reduced, metathorax and hind wings well developed, legs lacking a trochanter. Females larviform and viviparous, usually parasitoid and enclosed in puparium; with 3 to 5 secondarily segmental genital openings.

Larvae heteromorphic; pupae adecticous and exarate in males, suppressed in females.

The Strepsiptera are an order containing about 300 species of most highly specialized insects, the males of which are free living, the females usually parasitoid, and larvae always so in other insects. The order has been recorded from all the major zoogeographical areas.

Structure

Adult Male.   Males (Figure 10.21A) are small (1.5–4.0 mm in length) and usually black or brownish. The head is very distinct with its flabellate antennae and protruding, berrylike compound eyes. The mouthparts are mandibulate but invariably reduced. The prothorax and mesothorax are small, the metathorax is very large and usually accounts for about half the body length. The fore wings are reduced and without veins. They may function like the halteres of Diptera. The hind wings are large but have few veins. The legs are weak and lack a trochanter. They are used for holding onto the female during copulation. The abdomen is 10-segmented, has a usually hooked aedeagus, but lacks cerci.

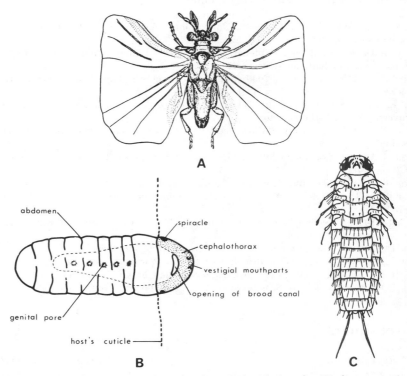

FIGURE 10.21.   Strepsiptera. (A) *Stylops pacifica* (Stylopidae) male, (B) diagrammatic representation of a female strepsipteran in host (ventral view), and (C) triungulin larva of *S. pacifica* (ventral view). [A, C, from R. M. Bohart, 1941, A revision of the Strepsiptera with special reference to the species of N. America, *Calif. Univ. Publ., Entomol.* **7**(6):91–160. By permission of University of California Press. B, from R. R. Askew, 1971, *Parasitic Insects.* By permission of Heinemann Educational Books Ltd.]

Adult Female. In only one family, the Mengeidae, are females free-living and larviform; all other female Strepsiptera are parasitoid. Only the greatly reduced head and thorax protrude from the host's body, the abdomen remaining enclosed within the last larval cuticle, the puparium, which is itself in the abdomen of the host (see Figure 10 21B). Between the puparium and the true ventral surface of the female lies a flattened cavity, the brood passage. It is along this passage that insemination occurs and the first-instar larvae emerge. Secondary genital openings (median invaginations of the ventral integument) occur on abdominal segments 2 to 5.

In both sexes the gut is reduced and ends blindly; Malpighian tubules are absent; the nervous system is highly concentrated, with an anterior ganglionic mass which includes the original thoracic centers and the first two abdominal ganglia, and a posterior mass comprising the posterior abdominal ganglia. The reproductive system of males includes paired testes and tubes which lead to a common duct. In females the reproductive organs apparently disintegrate and release the eggs into the hemocoel.

Larva and Pupa. The first-instar larva is an active creature known as a triungulin (Figure 10.21C). This lacks antennae and has reduced mouthparts, but has well-formed legs. Should this reach a suitable host it metamorphoses into an apodous, grublike form. Pupae are adecticous and exarate in males but suppressed in females except in Mengeidae.

Life History and Habits

The principal hosts of Strepsiptera are Homoptera–Auchenorrhyncha, and Sphecoidea, Vespoidea, and Apoidea within the Hymenoptera. Other hosts include Thysanura, Orthoptera, Dictyoptera, Heteroptera, Diptera, and other aculeate Hymenoptera. Members of a given strepsipteran family usually parasitize only one or a very few major insect groups, for example, Mengeidae are known only from Thysanura. However, at the generic and specific levels the degree of host specificity is variable. It is generally high when a hymenopteran is the host, lower if the host is a homopteran. Both sexes of the host are parasitized, and both sexes of the parasite may occur on the same host, in variable numbers. The location at which the parasite protrudes from the host varies but is generally specific for a particular genus.

After emergence, males, which survive for only a few hours, are quite active in their search for a female. Copulation occurs on the host, except in Mengeidae. In some species it appears that parthenogenesis is probable. The embryos develop within the female's body, which presumably receives nourishment by direct absorption through the very thin puparial cuticle. Upon hatching, the triungulins (usually several thousand are produced by a female) escape via the brood passage. They usually remain on the host in large numbers until an opportunity presents itself for entering a new (immature) host. The precise details of this process have rarely been observed, but it is assumed to occur in the nest in species parasitizing Hymenoptera or on plants in species parasitizing other groups. The triungulins are active and, in many species, capable of jumping for distances of 2 or 3 cm. They appear to enter the host through a combination of enzymatic and physical activity. Fluid is secreted from the mouth and appears to partially dissolve the host's cuticle. Within the host a larva soon molts to the second stage, a more grublike form. It is not known whether further molts occur beyond this stage or whether the insect grows by simple expansion of the cuticle. Early development takes place entirely within the host, but later the larva works its way to the integument where it forces its anterior end outward. This usually coincides with the final larval or pupal stage of the host. The presence of the parasite is not without effect on the host, which is said to be "stylopized." Most noticeable are changes in the structure of the external genitalia and other external sexual characters, and atrophy of the gonads and other internal structures. Whether these changes are directly caused by the parasite or are merely the result of inadequate nutrition, is not known. Parasitized female hosts are not fertile, but males may be.

Phylogeny and Classification

Fossil Strepsiptera are known only from Baltic amber (Oligocene period), and these are clearly assignable to the recent family Mengeidae. Thus, conclusions about the evolution of the group must be based entirely on comparative studies of recent forms. On different occasions, the stylopoids have been considered as a separate order, as a superfamily within the Coleoptera–Cucujiformia, and even as a family, Stylopidae, of aberrant beetles. Some authors consider that the group's nearest relatives are Coleoptera, whereas others have suggested they are derived from the same ancestral stock as Hymenoptera. The many similarities in form and habits between the Rhipiphoridae and the most primitive Strepsiptera is the traditional evidence taken to support the inclusion of the stylopoids in the Coleoptera. However, the modifications for parasitism have been taken much further in the stylopoids than in the rhipiphorids, implying that the former is a much older group (Crowson, 1960). On the basis of these similarities, Crowson suggests that the stylopoids merit the rank of a superfamily—Stylopoidea—within the Coleoptera, and were probably derived from a lymexyloid ancestor, probably in the Cretaceous period. Other authors such as Pierce (1964) and Ulrich (1966), who emphasize the differences between Strepsiptera and Coleoptera, conclude that the similarities are due to convergence and argue that the stylopoids deserve ordinal status.

Recent stylopoids can be arranged in five families, two of which, the STYLOPIDAE and HALICTOPHAGIDAE, are cosmopolitan, the other three having a sporadic distribution. The family MENGEIDAE includes the most primitive Strepsiptera, which are free-living in the adult stage. Species have been found in the Mediterranean region, China, Australia, and North America, where the larvae are parasitic in Thysanura. The family Stylopidae is the largest in the order. Its members parasitize bees and wasps. The CORIOXENIDAE (CALLI-PHARIXENIDAE), found in Ethiopia, Japan, New Guinea, and Australia, are parasites of Heteroptera. The Halictophagidae (including the Elenchinae) are mainly parasitic on Homoptera but have also been recorded from Heteroptera, Orthoptera, Dictyoptera, and Diptera. Members of the MYRMECOLACIDAE, a family found in eastern Asia, Australia, Africa, and South America, are unusual in that males and females apparently develop in widely different hosts. Males have been taken only from ants, whereas females parasitize Orthoptera and Mantodea.

Literature

The ecology of Strepsiptera is described by Bohart (1941), Pierce (1964), and Askew (1971). Crowson (1955, 1960), Pierce (1964), and Ulrich (1966) have discussed the evolutionary position of the group.

Askew, R. R., 1971, *Parasitic Insects*, American Elsevier, New York.
Bohart, R. M., 1941, A revision of the Strepsiptera with special reference to the species of North America, *Univ. Calif. Publ. Entomol.* 7:91–160.
Crowson, R. A., 1955, *The Natural Classification of the Families of Coleoptera*, Nathaniel Lloyd, London.
Crowson, R. A., 1960, The phylogeny of Coleoptera, *Annu. Rev. Entomol.* 5:111–134.
Pierce, W. D., 1964, The Strepsiptera are a true order, unrelated to Coleoptera, *Ann. Entomol. Soc. Am.* 57:603–605.
Ulrich, W., 1966, Evolution and classification of the Strepsiptera, *Proc. 1st Int. Congr. Parasitol.* 1:609–611.

7. Hymenoptera

SYNONYMS: none

COMMON NAMES: bees, wasps, ants, ichneumon flies

Minute to medium-sized insects; head usually with well-developed compound eyes, mandibulate mouthparts (though usually adapted for sucking also); two pairs of transparent wings present in most species, fore and hind wings coupled, venation reduced; abdomen in most species markedly constricted between segments 1 and 2, with former intimately fused with metathorax; females with an ovipositor which is modified in some species for purposes in addition to egg laying.

Larvae caterpillarlike (Symphyta) or maggotlike (Apocrita). Pupae adecticous and in most species exarate, often in a cocoon.

The order Hymenoptera, which includes more than 100,000 described species, contains some of the most advanced and highly specialized insects. In the evolution of the order, emphasis has been laid not so much on structural and physiological modifications as has occurred in other orders, but on the development of complex behavior patterns. These are particularly related to provision of food for the progeny and have led ultimately to the evolution of sociality in several groups.

Structure

Adult. The hypognathous head usually is very mobile and bears very large, in some species holoptic, compound eyes. Three ocelli are usually present. The antennae contain between 9 and 70 segments and are sometimes sexually dimorphic. The mouthparts show a wide range in form from the generalized mandibulate type found in the suborder Symphyta to a highly specialized, sucking type found in the most advanced Apocrita, such as the bees (see Chapter 3, Section 3.2.2). The prothorax is small; its tergum is collar-like and fused with the large mesonotum. The prosternum is very small and usually can be seen only with difficulty. Two pairs of wings are generally present, with the fore wings larger than the hind pair. The venation is much reduced, and, rarely, veins are completely absent. The fore and hind wings are coupled by means of hamuli. Brachyptery or aptery occurs, for example, in ant workers and some Chalcidoidea. Hymenoptera are unique among Insecta in that a trochantellus is present on at least some legs. This is actually part of the femur, though it appears as a second segment of the trochanter. The legs are frequently specialized for particular functions, for example, digging, grasping and carrying prey, and collection of pollen (see Chapter 3, Section 4.3.1 and Figure 3.25). In Symphyta the first abdominal segment is clearly recognizable as a part of the abdomen. In Apocrita, however, the tergum has become intimately fused with the metathorax and is distinguishable only by the presence of spiracles. In this condition it is known as the propodeum. The first abdominal sternite has disappeared entirely. In Apocrita, a marked constriction, the petiole, separates the first from the remaining abdominal segments; the latter constitute the "gaster" or "metasoma," which normally bears distally a pair of unsegmented cercilike structures, the pygostyles. The male terminalia are usually large and comprise the lateral parameres, the aedeagus, and a pair of ventral lobes, the volsellae. The ovipositor is well developed in females and is frequently modified for sawing, boring, piercing, or stinging, although only in the latter does it no longer participate in egg laying (see Chapter 3, Section 5.2.1 and Figure 3.32).

The gut is of quite uniform structure throughout the order. The esophagus is narrow and long, especially when the petiole is elongate, and leads to a thin-walled crop (honey stomach) in the anterior part of the abdomen. Behind the crop is the proventriculus, which serves apparently to regulate the entry of food into the usually large stomach (ventriculus). This is followed by the ileum and rectum. The number of Malpighian tubules varies from two, in most parasitic forms, to more than 100 in some nectar feeders. The nervous system exhibits various degrees of specialization; in primitive forms, three thoracic and nine abdominal ganglia occur in the ventral nerve cord, whereas in most Apocrita three thoracic and between two and six abdominal ganglia can be found. The paired testes are separate in Symphyta and a few Apocrita, but fused in other Hymenoptera. The vasa deferentia are swollen basally into vesicula seminales which lead into paired ejaculatory canals. The latter also receive the ducts of the two large accessory glands prior to forming a common tube. In females the number of polytrophic ovarioles varies from one to more than 100 per ovary. The two oviducts fuse to form the vagina, which also receives the duct of the median spermatheca. The vagina is swollen posteriorly to form the bursa copulatrix.

Larva and Pupa. Larvae of Symphyta have a distinct, well-sclerotized head, and 3 thoracic and 9 or 10 abdominal segments. The mandibulate mouthparts are well developed. If a larva is a surface feeder, as in the sawflies, it usually has well-developed thoracic legs and six to eight pairs of prolegs. In boring or mining species the thoracic legs are reduced and prolegs absent. Larvae of Apocrita are apodous and resemble maggots. The head is only weakly sclerotized or much reduced and, in parasitic forms sunk into the prothorax. Heteromorphosis is common in parasitic species. The first-instar larva is extremely variable in form, but the final instar is always maggotlike. Pupae are adecticous and in most species exarate. With the exception of certain groups (Cynipidae, Chalcidoidea, most Apoidea, and many Formicoidea), a cocoon is spun in which pupation occurs.

Life History and Habits

Most adult Hymenoptera are nectar- or honeydew-feeding insects and are thus found on or near flowers. A few are predaceous, while others feed on plant tissue or fungi. Through their search for nectar and pollen for feeding the larvae, the social Hymenoptera play an extremely important role in the cross-pollination of flowering plants. In addition, some parasitoid and predatory forms are important agents in the control of insects that could otherwise become pests. For these reasons the Hymenoptera are considered to be the insect order of greatest benefit to Man.

In the complexity of its members' behavior the Hymenoptera surpass all other insect orders. The development of this behavior has accompanied the evolution of parental care and, ultimately, social life in the order. Within the order the importance of the male sex shows a gradual diminution, and facultative or cyclic diploid parthenogenesis (Chapter 20, Section 8.1) becomes the rule rather than the exception in some groups of social Hymenoptera. The majority of individuals produced are females, which, however, do not mature sexually but merely "serve" the colony as a whole. Males are produced only for the purpose of founding new colonies; they develop from unfertilized eggs and thus are haploid.

Four broad categories of life history can be recognized in Hymenoptera, and these are paralleled by an increasing complexity of behavior patterns exhibited by females. These life histories are based on the feeding habits of the larvae. The simplest life history is found in Symphyta, whose larvae are generally foliage eaters or wood borers and females simply deposit their eggs on a host plant. The second stage is seen in the primitive Apocrita, whose larvae are parasitoids, that is, they feed on the tissues of a host which they eventually kill. This necessitated the evolution, in females, of the ability to search out a suitable (sometimes highly specific) host on which to oviposit. In more advanced Apocrita such as the solitary wasps and bees, we see the beginning of parental care. A female builds a special cell in which to deposit an egg. She then practices either "mass provisioning," that is, the supplying of sufficient food at one time for the whole of larval development, or "progressive provisioning" in which food is supplied at intervals throughout larval life. Associated with this development is the evolution of the ovipositor as a sting, for paralyzing, though not killing, the prey. Thus, the stage has been reached where a female comes into contact with her progeny. The fourth type of life history is found in the social Hymenoptera where a larva is fed throughout its development on food provided by the parent or another adult (which is usually sterile). The food is animal material (in wasps), or plant material such as pollen and nectar (bees), or seed, tissues, or fungi (ants). In primitive social species mass provisioning is practiced. At a more advanced stage progressive provisioning occurs, and there may be several egg-producing females in the colony. Gradually, division of labor evolves, and the colony then contains a single egg-laying female, the queen, and large numbers of structurally distinguishable females, the workers, which perform various duties in the colony. In some species there is a temporal separation of these duties; that is, an individual performs different duties at different ages.

Phylogeny and Classification

Hymenoptera appear to be relatively isolated from other endopterygotes, and speculations on the origin of the group have been widely divergent. In certain of their features, for example, the large number of Malpighian tubules, they resemble orthopteroid insects, and more than one author has suggested that the order evolved from Protorthoptera, implying a diphyletic origin of the endopterygotes. More frequently, on the basis of many common morphological features and habits possessed by Mecoptera and primitive Hymenoptera, it has been concluded that Hymenoptera are a very early offshoot of the panorpoid complex. This separation probably took place in the Upper Carboniferous period because the earliest fossil Hymenoptera, from the Lower Triassic, are already well developed and clearly assignable to the recent family Xyelidae. Fossils belonging to other symphytan families were present in the Upper Triassic and Jurassic periods. The earliest fossil Apocrita have been found in the Jurassic period and by the Cretaceous representatives of many recent superfamilies were present.

It is generally accepted that Apocrita evolved from the Symphyta. However, the latter are primarily phytophagous, and the evolution of the primitive, parasitoid Apocrita from such a group has been the subject of much speculation (reviewed by Malyshev, 1968). On different occasions all the major symphytan groups (superfamilies) have been suggested as being ancestral to Apocrita. Malyshev considers that, of these groups, the Cephoidea are the closest relatives, though not the *direct* ancestors, of Apocrita. Malyshev suggests that from some "procephoid" ancestor two distinct lines developed. In one, the larva was a mining form which actively searched for suitable plant tissues on which to feed. It was from this line that the modern Cephoidea arose. In the other, the female, during oviposition, secreted fluid which caused the plant tissue surrounding the egg to proliferate and eventually form a gall, inside which the larvae fed. Malyshev argues that galls, being a rich food source, would attract other endophytic Hymenoptera whose larvae were initially phytophagous. Obviously, this would present excellent opportunities for contact between the original and "guest" larvae and for facultative zoophagy. Gradually this relationship became more definitive, and obligate zoophagy resulted. However, as soon as the factor that attracted an ovipositing female to a gall was not the gall itself but the egg or larva within, the importance of the gall as an oviposition site was lost. Thus, the primitive parasitoid females would now be able to seek out eggs or larvae in quite different situations. This stage in hymenopteran evolution is seen in several modern superfamilies, namely the Megalyroidea, Trigonaloidea, Ichneumonoidea, Evanioidea, Proctotrupoidea,

Cynipoidea, and Chalcidoidea. In the latter two groups a large number of species have reverted to the gall-forming habit. These seven superfamilies collectively form the "Parasitica" of earlier authors. The majority of Hymenoptera, however, have proceeded beyond this stage. In these forms (the Aculeata) the ovipositor took on a new function as a sting for paralyzing (though not killing) the prey. The advantage of this was twofold. First, the prey remained "fresh" while the larva consumed it, and second, the immobile prey could not carry the parasitoid into a different, perhaps inhospitable habitat.

From this ancestral group of paralyzers, the recent groups of higher Apocrita evolved. Three major evolutionary lines can be distinguished. One led to the Formicoidea (ants), members of which are social. Primitive ants are carnivorous or feed on animal products, especially honeydew, but higher forms are secondarily phytophagous and live on plant tissue, seeds, or fungi. This dietary change was perhaps a response to the difficulty of obtaining sufficient animal food as the size of a colony grew. A second line of evolution led to the solitary and social wasps (Scolioidea, Sphecoidea, and Vespoidea). This line differed from the "ant-line," in that its members began to construct special cells in which to place the prey and lay their eggs. In the third line, probably derived from some sphecoid ancestor, arose the bees (Apoidea). In this group, which contains both solitary and social forms, the diet changed from a carnivorous one to one in which plant products, namely, pollen and nectar, were stored in the cells to provide food for developing larvae.

The probable phylogeny of the Hymenoptera is summarized in Figure 10.22.

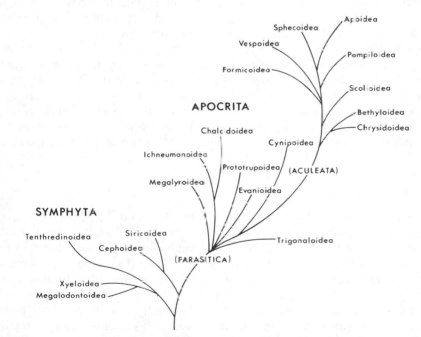

FIGURE 10.22   A possible phylogeny of Hymenoptera. [After H. E. Evans and M. J. W. Eberhard, 1970, The Wasps. By permission of the University of Michigan Press.]

Suborder Symphyta (Chalastogastra)

In adults of the suborder Symphyta the abdomen is broadly attached to the thorax, and there is no marked constriction between the first and second abdominal segments. Larvae have a well-developed head, thoracic, and, in most species, abdominal legs.

Superfamily Xyeloidea

Xyeloidea form a small and extremely primitive superfamily containing the single family XYELIDAE. Adults feed on flowers and have a generalized wing venation. Larvae are found in flowers and have prolegs on all abdominal segments.

Superfamily Megalodontoidea

The primitive superfamily Megalodontoidea includes two families, the PAMPHILIIDAE and MEGALODONTIDAE. Adults feed on flowers, while larvae, which lack prolegs, live gregariously in webs or rolled leaves.

Superfamily Tenthredinoidea

Tenthredinoidea form a very large group with diverse habits. Females have a sawlike rather than a boring ovipositor and are commonly called sawflies. Most species belong to the family TENTHREDINIDAE. Adults are often carnivorous. Parthenogenesis is common, and larvae are usually caterpillarlike in form and habits. A few are leaf miners and apodous. Some species are economically important, for example, *Nematus ribesii*, the imported currant worm (Figure 10.23A), and *Pristiphora erichsonii*, the larch sawfly. Other families, which are not large but contain economically important species, are the CIMBICIDAE (including *Cimbex americana*) (Figure 10.23B), which defoliate various broadleaved trees) and DIPRIONIDAE (conifer sawflies) (Figure 10.23C), which include several of North America's most important forest pests (*Neodiprion* spp. and *Diprion* spp.). The PERGIDAE form a large family of sawflies which are found mainly in Australia and South America. Larvae are gregarious and lack prolegs.

Superfamily Siricoidea

Female Siricoidea have a boring ovipositor, and larvae have reduced thoracic legs, no prolegs, and live in wood. The largest family is the SIRICIDAE (horntails) (Figure 10.24), whose adults are large, often brightly colored insects. Larvae burrow extensively in the heart wood of both deciduous and evergreen trees and may cause much damage, especially those species which live symbiotically with rot-producing fungi. Although only a small family, the ORUSSIDAE (parasitic wood wasps) should be mentioned, since its members possess some unusual features, for example, the ovipositor is extremely long and coiled within the body; larvae are apodous and probably parasitic on buprestid beetle larvae.

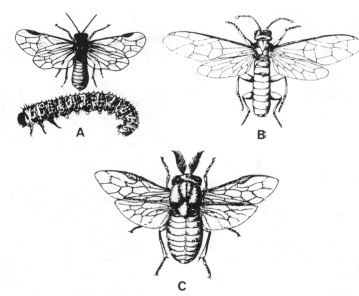

FIGURE 10.23. Tenthredinoidea. (A) The imported currant worm, *Nematus ribesii* (Tenthredinidae) adult and larva; (B) the elm sawfly, *Cimbex americana* (Cimbicidae); and (C) the redheaded pine sawfly, *Neodiprion lecontei* (Diprionidae). [A, from L. A. Swan and C. S. Papp, 1972, *The Common Insects of North America.* Copyright by L. A. Swan and C. S. Papp. Reprinted by permission of Harper & Row, Publishers, Inc. B, C, from U. S. Department of Agriculture.]

Superfamily Cephoidea

All Cephoidea are included in the family CEPHIDAE (stem sawflies). Larvae bore into the stems of grasses and berries. Some species are of considerable economic importance, for example, *Cephus cinctus*, the wheat stem sawfly (Figure 10.25), and *Janus integer*, which bores in the stems of currants.

Suborder Apocrita

Almost all adult Apocrita have the first abdominal segment (propodeum) intimately fused with the thorax and a constriction between the first and second abdominal segments. Larvae are apodous and have a reduced head capsule.

FIGURE 10.24 Siricoidea. The pigeon tremex, *Tremex columba* (Siricidae). [From H. E. Evans and M. J. W. Eberhard, 1970, *The Wasps.* By permission of the University of Michigan Press.]

FIGURE 10.25. Cephoidea. The wheat stem sawfly, *Cephus cinctus* (Cephidae). [From L. A. Swan and C. S. Papp, 1972, *The Common Insects of North America.* Copyright 1972 by L. A. Swan and C. S. Papp. Reprinted by permission of Harper & Row, Publishers, Inc.]

Superfamily Megalyroidea

Megalyroidea form a primitive apocritan group containing two small families, MEGALYRIDAE and STEPHANIDAE. These were formerly included in the Ichneumonoidea, but their members differ from true ichneumons in a number of structural features and habits. Indeed, it is possible that the Megalyroidea are derived from an orussid rather than a cephoid ancestor as were the remaining Apocrita. Like those of Orussidae, larvae of Megalyroidea are parasitic on wood-boring beetles.

Superfamily Trigonaloidea

Trigonaloidea form a small group of archaic but highly specialized Apocrita whose members are all contained in the family TRIGONALIDAE. Structurally they possess a combination of features of Parasitica and Aculeata. The trigonalids are hyperparasites of sawfly larvae or lepidopteran caterpillars. The eggs are laid on plant tissue, which is then eaten by the sawfly larvae or caterpillars. These are then parasitized by other Hymenoptera (ichneumons) or Diptera (tachinids), which, in turn, become the primary host for the trigonalids.

Superfamily Ichneumonoidea

Ichneumonoidea form probably the second largest superfamily within the Hymenoptera, with some 16,000 described species. Most of them are assigned to two families, the ICHNEUMONIDAE (ichneumon flies) (Figure 10.26A) and BRACONIDAE (Figure 10.26B). Members of both these groups are parasitic on other insects, especially larvae or pupae of Lepidoptera, but including those of Diptera, Coleoptera, Neuroptera, and other Hymenoptera, and arachnids. The degree of host specificity varies among species. Many species are beneficial to man because they exert a major control on populations of pest insects. As examples may be cited *Bracon cephi*, which parasitizes larvae of the wheat stem sawfly, and several *Apanteles* spp. that attack cabbageworms, the tobacco hornworm, clothes moth larvae, and caterpillars of the gypsy moth, among others. Several ichneumons and braconids have been used in the biological control of pests.

Superfamily Evanioidea

The constituent families EVANIIDAE, AULACIDAE, and GASTERUPTIIDAE, are included by some authors in the previous superfamily with whose members

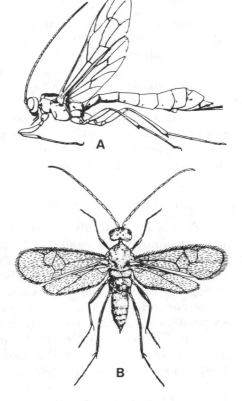

FIGURE 10.26. Ichneumonoidea. (A) An ichneu-
mon fly, *Rhyssa persuasoria* (Ichneumonidae).
Note that the entire ovipositor is not drawn. (B)
A braconid, *Apanteles cajae* (Braconidae). A, reproduced by permission of the Smithsonian Institution Press from *United States National Museum
Bulletin 216, Part 2, (Ichneumon-Flies of America
North of Mexico: 2. Subfamilies Ephialtinae Xori-
dinae Acaenitinae)* by Henry and Marjorie Townes:
Figure 302b, page 598. Washington, D.C., 1960.
Smithsonian Institution. B, from R. R. Askew,
1971, *Parasitic Insects.* By permission of Heine-
mann Educational Books Ltd.]

they show many affinities Evaniidae are widespread and appear to be exclusively parasites of cockroach oothecae. Aulacidae, which generally resemble
ichneumons, are parasitic on wood-boring Coleoptera and Hymenoptera. Gasteruptiidae parasitize solitary bees and wasps.

Superfamily Proctotrupoidea (Serphoidea)

Proctotrupoidea are usually small, black, chalcidlike insects. They are
parasitoids on the larvae, pupae, and, especially, eggs of a wide range of
insects and spiders. A few are hyperparasites and some are inquilines. The
group is subdivided into as many as nine families, though most species belong
to either the PLATYGASTERIDAE or the SCELIONIDAE. The former are parasites of
Diptera, especially cecidomyiids, including a number of pest species, for
example, the Hessian fly, over which they exert a major population control.
Platygasteridae lay their eggs among those of the host, but the eggs do not
hatch until after those of the host. Thus, they are parasitoid on the larvae.
Scelionidae, on the other hand, are true egg parasitoids, usually of Lepidoptera,
Hemiptera or orthopteroid insects, and occasionally of spiders. Phoresy is
common; that is, a female scelionid locates and is carried about by a suitable
host until the latter begins to oviposit. The scelionid then leaves the host and
lays her own eggs among those of the host.

Superfamily Cynipoidea

Most of the 1600 species so far ascribed to the Cynipoidea belong to the family CYNIPIDAE, subfamily CYNIPINAE (gall wasps) (Figure 10.27), and are either gall makers or inquilines in already-formed galls. The host plant and the form of the gall are usually specific for a particular species of cynipid, though in some primitive species a range of host plants is chosen. In some species the life history is complex with two generations per year. The first generation comprises entirely females which reproduce parthenogenetically. The second generation, which develops in a different gall on a different plant, includes both males and females. Members of other cynipid subfamilies and other families in the group are parasitoids or hyperparasites of other insects.

Superfamily Chalcidoidea

This is a highly diverse group of mostly small to minute, parasitic or phytophagous (usually gall-forming) insects, most of which are arranged in a number of large and important families. The family AGAONTIDAE (fig insects) contains perhaps 500 species of chalcidoids, which show some remarkable biological features. The species are restricted to living in the receptacles of species or varieties of fig, whose flowers a female pollinates in her search for an oviposition site. The Smyrna fig, a cultivated variety, requires to be cross-pollinated with the wild fig before fruit can be formed. To facilitate this, fig growers place branches of the wild fig, from whose receptacles female fig wasps emerge, among those of the Smyrna fig. As a female searches for a suitable egg-laying site, she accidentally visits the Smyrna fig flowers, though, because they are the wrong shape, she does not oviposit in them. The TORYMIDAE exhibit a wide range of life histories. Most species are parasitoids of gall-forming insects, though a few are themselves gall formers or inquilines of galls. Some are parasites of mantid oothecae, others develop in bees' or wasps' nests, and a few feed on seeds. The CHALCIDIDAE are a mainly South American family, though they have representatives in all parts of the world, whose members are parasitoids of lepidopteran, dipteran, and coleopteran larvae or pupae, or hyperparasites of tachinids or ichneumon flies. The EURYTOMIDAE (Figure 10.28A) constitute a family of very diverse habits. Commonly its members produce galls on the stems of grasses, including cereals. Other species feed on

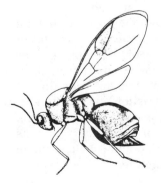

FIGURE 10.27. Cynipoidea. A gall wasp, *Diplolepis rosae* (Cynipidae). [From D. J. Borror, D. M. Delong, and C. A. Triplehorn, 1976, *An Introduction to the Study of Insects*, 4th ed. By permission of Holt, Rinehart and Winston, Inc.]

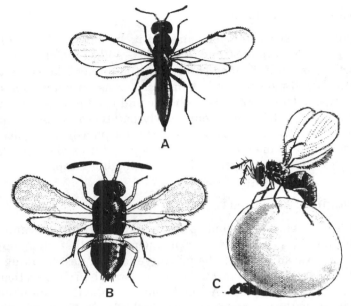

FIGURE 10.28. Chalcidoidea. (A) The wheat jointworm, *Harmolita (Tetramesa) tritici* (Eurytomidae); (B) *Aphelinus mali* (Eulophidae), a parasitoid of the woolly apple aphid (*Eriosoma lanigerum*); and (C) ovipositing *Trichogramma minutum* (Trichogrammatidae), a parasitoid of the eggs of more than 200 species of Lepidoptera. [From L. A. Swan and C. S. Papp, 1972, *The Common Insects of North America*. Copyright 1972 by L. A. Swan and C. S. Papp. Reprinted by permission of Harper & Row, Publishers, Inc.]

seeds, are inquilines in the nests of bees and wasps, are parasitoids on gall-forming insects, or are egg parasites of Orthoptera. Another very large family is the EULOPHIDAE (Figure 10.28B), a group whose members are very small parasitoids or hyperparasites. The parasitoids are important control agents of many insect pests, especially scale insects and aphids. The TRICHOGRAM-MATIDAE (Figure 10.28C) are minute egg parasites, especially of Lepidoptera. Some species have been reared in large numbers in efforts to control certain pests. The MYMARIDAE are some of the smallest insects, with some species having a length of 0.21 mm! They are egg parasites of Coleoptera and Hemiptera, including some aquatic species. The ENCYRTIDAE are mainly parasitoids of eggs, larvae, and pupae, especially of Lepidoptera and Homoptera, though representatives of most other orders are attacked. Like members of the previous group, the PTEROMALIDAE attack a wide variety of insect hosts, at all stages in their life history.

Superfamily Chrysidoidea

This group, which includes the single family CHRYSIDIDAE, has similarities with both Chalcidoidea and Bethyloidea and is frequently included in the latter. The species are commonly known as cuckoo wasps from their habit of laying an egg in the cells of solitary wasps or bees. Usually, the egg does not hatch until the host larva has consumed its own food supply.

Superfamily Bethyloidea

This group is rather heterogeneous, and its members show affinities with the Chrysidoidea, Pompiloidea, and Scolioidea. Within the group two families are noteworthy for the links they show with the higher aculeate groups. The Dryinidae are small parasites of homopteran larvae on whose abdomen they form large cysts. A female's tarsus is chelate and used to grasp the host while she deposits an egg. The host is temporarily paralyzed (stung by the female) but soon recovers. The Bethylidae are external parasites of coleopteran or lepidopteran larvae. Some species drag their prey into a sheltered location, a habit which foreshadows the situation in the digging wasps.

Superfamily Scolioidea

Scolioidea are closely related to Bethyloidea. Within the group, whose members are mostly parasitoids on Coleoptera and other Hymenoptera, are some species whose adults exhibit a primitive form of nest-building behavior. Scoliidae (digging wasps) (Figure 10.29A) are large, hairy wasps that are especially common in tropical areas. A female burrows into the ground and locates a beetle larva, usually a scarabaeid, on which she lays an egg, having first paralyzed it. In some species females build a special cell around the beetle

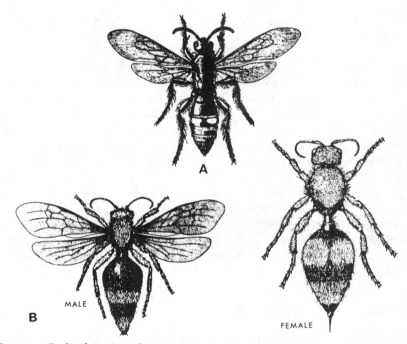

A

MALE

B

FEMALE

FIGURE 10.29. Scolioidea. (A) A digging wasp, *Scolia dubia* (Scoliidae); and (B) a velvet ant, *Dasymutilla occidentalis* (Mutillidae), male and female. [A, from L. A. Swan and C. S. Papp, 1972, *The Common Insects of North America.* Copyright 1972 by L. A. Swan and C. S. Papp. Reprinted by permission of Harper & Row, Publishers, Inc. B, from D. J. Borror, D. M. Delong, and C. A. Triplehorn, 1976, *An Introduction to the Study of Insects,* 4th ed. By permission of Holt, Rinehart and Winston, Inc.]

larva. Female Tiphiidae also attack scarabaeid larvae, in addition to ground-dwelling solitary and social bees and wasps. Females of many species are wingless and are either transported to flowers by males, or fed by males on nectar or honeydew. Mutillidae (velvet ants) (Figure 10.29B) are hairy, antlike scolioids whose females are apterous. They parasitize both social and solitary bees and wasps, and some Diptera.

Superfamily Formicoidea

The several thousand species of ants that comprise the Formicoidea are included in the single family Formicidae. They are social, with the exception of a few secondarily "solitary" parasitic forms. In the latter there is no worker caste, and a female deposits her eggs in the nest of a closely related species whose workers then rear the resulting larvae. Polymorphism reaches the extreme in ants, where, in some species, several different forms of worker occur, each performing a particular function. Workers form the vast bulk of individuals in the nest. There may be one to several queens; males are few in number and probably produced seasonally. Except for a short period prior to swarming, all individuals are apterous. However, winged queens and males are produced in most species (not in Dorylinae) in order to found new nests. After a short mating flight, a queen alone finds a suitable nesting site, divests her wings, and eventually begins egg laying. Except in a few primitive species, the queen does not forage for food during this initial phase of nest building but lives entirely on fat body reserves and the products of wing muscle degeneration.

Several distinct subfamilies are recognized. The most primitive ants, which are carnivorous and form only small colonies, are the Ponerinae (Figure 10.30), which nest in the ground or rotting logs. In the ponerine ants there is little structural difference between the queen and workers which are monomorphic. Dorylinae are nomadic ants (army ants) of tropical regions. Like members of the previous subfamily, they are carnivorous, and in their

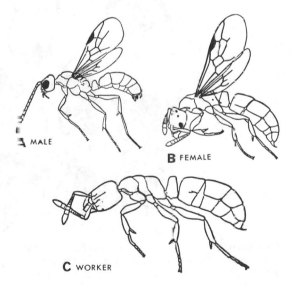

A MALE

B FEMALE

C WORKER

FIGURE 10.30. Formicoidea. Castes of the primitive ant, *Ponera pennsylvanica* (Formicidae: Ponerinae). (A) Male, (B) female, and (C) worker. [From W. M. Wheeler, 1910, *Ants. Their Structure, Development and Behaviour.* By permission of Columbia University Press.]

search for food may form massive columns whose length may cover 100 m or more. The remaining ant subfamilies have solved the problem of obtaining sufficient food by changing from a carnivorous to a generally phytophagous diet. The MYRMICINAE (Figure 10.31) form the largest and most common subfamily. Many species are harvester ants, so-called because they collect seeds which they store in the nest. Others grow fungi on decaying leaf fragments and ant excreta in special subterranean chambers. It is to this subfamily that many of the inquiline and parasitic species belong. DOLICHODERINAE and FORMICINAE, which form the second largest subfamily, are generally nectar or honeydew feeders. Workers usually have a flexible integument that stretches remarkably as food is imbibed. In the honey ants there is a distinct form of worker, the replete (Figure 10.32), which spends its life in the nest and serves as a living bottle in which food can be stored. Many species have established a symbiotic relationship with honeydew-secreting insects, mainly Homoptera. In return for a copious supply of honeydew the ants move the insects to new "pasture," carry them off if there is a disturbance, build shelters for them, and store their eggs during the winter.

In addition, ants have a close relationship with many other insects that actually live in their nest. The relationship ranges from one in which the inquilines are scavengers or predators and are treated with hostility by the ants,

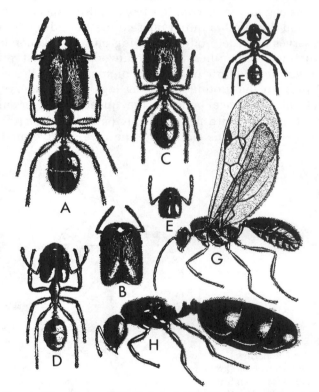

FIGURE 10.31. Formicoidea. Castes of *Pheidole instabilis* (Formicidae: Myrmicinae). (A) Soldier, (B–E) intermediate workers, (F) typical worker, (G) male, and (H) dealated female. [From W. M. Wheeler, 1910, *Ants. Their Structure, Development and Behaviour.* By permission of Columbia University Press.]

FIGURE 10.32. Formicoidea. Replete of the honey ant, *Myrmecocystus hortideorum* (Formicidae: Formicinae). [From W. M. Wheeler, 1910, *Ants. Their Structure, Development, and Behaviour*. By permission of Columbia University Press.]

through one where the ants behave indifferently toward the visitors, to a situation in which the ants "welcome" their guests, to the extent of feeding and rearing them. In return for this hospitality, the guests appear to exude substances which are highly attractive to the ants. Frequently, two species of ants may live quite amicably in close proximity. From this it is only a short step to social parasitism in which a queen of one species becomes adopted by the workers of another species. They tend the eggs laid by her and feed the larvae which develop. The host queen is killed, and eventually the nest is taken over entirely by the intruders. A relationship of a different form is shown by the slave-making ants, which capture workers of another species; these then perform the domestic duties within the colony.

Superfamily Vespoidea

Vespoidea comprise the true wasps. Typically, in both solitary and social forms, larvae are reared in specially constructed cells and fed animal material. Adults feed on nectar, honeydew, or ripe fruit. In some species larvae are also fed pollen and nectar. The solitary forms can be arranged in two families, MASARIDAE and EUMENIDAE. The former is a small group of highly specialized wasps which feed their larvae on pollen and nectar. The Eumenidae are varied in their nest-building habits. Many construct cells in wood or stems; others, the common mason or potter wasps (Figure 10.33A), build cells of mud fastened to twigs and other objects. Once a cell is provisioned [with larvae of Lepidoptera or Coleoptera, usually) and an egg laid, the entrance is sealed.

The social wasps (yellow jackets and hornets] (Figure 10.33B–D] are included in the family VESPIDAE. They construct nests built of "wasp paper," a substance made from fragments of chewed wood mixed with saliva. Differentiation of queens, workers, and males is not always distinct. In contrast to the solitary forms, vespids feed their young by progressive provisioning. Colonies of some species are very large and may contain several queens.

Superfamily Pompiloidea

Pompiloidea form a large and fairly homogeneous group whose phylogenetic position appears to be close to the Bethyloidea and Scolioidea. Almost all species are included in the family POMPILIDAE [Figure 10.34), a group of

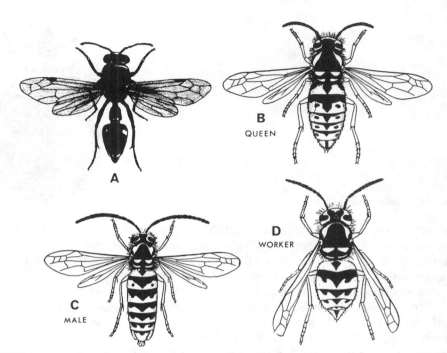

FIGURE 10.33. Vespoidea. (A) A potter wasp, *Eumenes fraterna* (Eumenidae); and (B–D) a hornet, *Vespula pennsylvanica* (Vespidae) queen, male, and worker. [A, from L. A. Swan and C. S. Papp, 1972, *The Common Insects of North America.* Coypright 1972 by L. A. Swan and C. S. Papp. Reprinted by permission of Harper & Row, Publishers, Inc. B–D, from E. O. Essig, 1954, *Insects of Western North America.* Reprinted with permission of the Macmillan Publishing Co., Inc. Copyright 1926 by Macmillan Publishing Co., Inc., renewed 1954 by E. O. Essig.]

predatory, digging wasps, often large in size, that typically parasitize spiders. A spider is stung by a wasp and may be either paralyzed or killed outright. The wasp usually constructs a nest (which may contain one or several cells) in the ground and deposits the prey in it. Females of other species simply use the spider's own burrow.

Superfamily Sphecoidea

Sphecoidea are solitary wasps, females of which construct cells either below or above ground and are known as mud daubers (Figure 10.35), sand

FIGURE 10.34. Pompiloidea. A spider wasp, *Episyron quinquenotatus* (Pompilidae). [From D. J. Borror, D. M. Delong, and C. A. Triplehorn, 1976, *An Introduction to the Study of Insects,* 4th ed. By permission of Holt, Rinehart and Winston, Inc.]

FIGURE 10.35. Sphecoidea. The black and yellow mud dauber, *Sceliphron caementarium* (Sphecidae). [From L. A. Swan and C. S. Papp, 1972, *The Common Insects of North America.* Copyright 1972 by L. A. Swan and C. S. Papp. Reprinted by permission of Harper & Row, Publishers, Inc.]

wasps, or digger wasps. Females of most species practice mass provisioning, but in a few species progressive provisioning occurs. Prey, which is always paralyzed or killed outright, comprises larvae of a wide variety of insects and Arachnida. In some species a high degree of specificity exists between the wasp and its prey. Opinions differ as to whether the group should be subdivided into a number of families, or whether all species should be included in a single family, SPHECIDAE.

Superfamily Apoidea

Apoidea, the solitary and social bees, are clearly closely related to and probably evolved from an early ancestral form of the Sphecoidea. Bees differ from sphecoid wasps, in that they use pollen and nectar rather than animal matter for feeding the larvae. The most obvious structural difference between the two groups is the presence on bees of branched or plumose body hairs. The vast majority of the 20,000 or so described species are solitary forms, the social species being restricted to three families, Halictidae, Anthophoridae, and Apidae. Females of a few species deposit their eggs in the nests of other bees.

The family COLLETIDAE is particularly common in the Southern Hemisphere, especially Australia. It contains the most primitive bees whose females construct simple nests in soil, hollow stems, or holes in wood. They are sometimes called plasterer bees from their habit of lining the nest with a salivary secretion that dries to form a thin transparent sheet. The HALICTIDAE form a large cosmopolitan family, most species of which are solitary bees. In some species, however, large numbers of egg-laying females occupy the same nest site, usually a hole in the ground, though there is no division of labor within the colony. A few species are truly social; each season a single female constructs a nest and rears a brood of young, all of which develop into workers, though these may not be structurally much different from the queen. The workers care for the eggs laid subsequently, which develop into both males and females. Only fertilized females overwinter. The ANDRENIDAE (Figure 10.36A) are the common solitary bees of the holarctic region, with about 1100 species described from North America alone. Andrenids typically nest in burrows in the ground. Often large numbers nest together in the same "apartment," each bee with her own "suite." The MEGACHILIDAE form another very large family, which includes the familiar leaf-cutter bees (Figure 10.36B,C), females of which build nests from leaf fragments. Others, however, build nests from mud or live in burrows, under stones, and in other suitable holes. Some species are

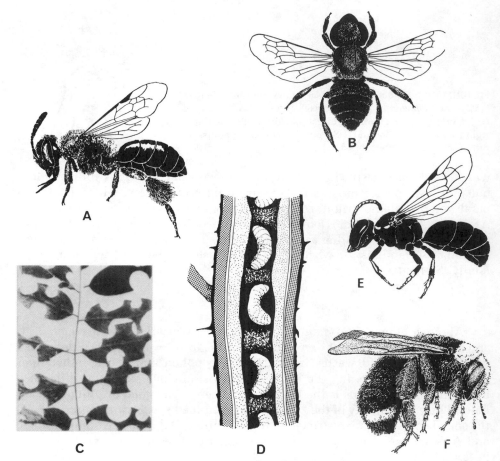

FIGURE 10.36. Apoidea. (A) *Andrena* sp. (Andrenidae); (B) a leaf-cutter bee, *Megachile latimanus* (Megachilidae); (C) the work of leaf-cutter bees; the removed portions of the leaves were used in nest building; (D) nest of the small carpenter bee, *Ceratina dupla* (Anthophoridae); (E) *Ceratina acantha* (Anthophoridae); and (F) the yellow-faced bumblebee, *Bombus vosnesenskii* (Apidae). [A–C, E, F, from E. O. Essig, 1954, *Insects of Western North America.* Reprinted with permission of the Macmillan Publishing Co., Inc. Copyright 1926 by Macmillan Publishing Co., Inc.; renewed 1954 by E. O. Essig. D, after D. J. Borror, D. M. Delong, and C. A. Triplehorn, 1976, *An Introduction to the Study of Insects*, 4th ed. By permission of Holt, Rinehart and Winston, Inc.]

parasitic on other bees. *Megachile rotundata* is now cultured on a large scale in North America and used as an alfalfa pollinator. The weaker honeybee can pollinate this plant but experiences difficulty in forcing its way into the flower and soon learns that there are easier sources of food. Use of leaf-cutter bees can increase the yield of seed severalfold. The ANTHOPHORIDAE form a large group of mainly solitary or parasitic bees that are often included in the next family. Females of solitary forms frequently nest in large numbers close together, usually in holes in the ground (ANTHOPHORINAE) (digger bees) or burrow into wood or plant stems (XYLOCOPINAE) (carpenter bees) (Figure 10.36D,E). In some Xylocopinae there is a primitive social organization in which the queen, morphologically identical with her offspring (the workers), lays eggs and is long lived, whereas the workers do not usually lay eggs and live for only a relatively short time. The family APIDAE includes all the highly social bees and a few neotropical solitary species. The family includes the BOMBINAE (bumblebees) (Figure 10.36F) and APINAE (honeybees and relatives). The former are common, large, hairy bees, found mainly in the holarctic region. The social organization of bumblebees is primitive, and workers frequently differ from the queen only in size. They do not construct a true comb but rear larvae in "pots." Often, these are sealed off after egg laying, and only older larvae are fed regularly. Only the queen overwinters, and new nests are produced annually. Some female Bombinae (genus *Psithyrus*) lay their eggs in the nests of other bumblebees, occasionally killing the host queen but more often living side by side. Bumblebee workers then attend to and rear the *Psithyrus* larvae in preference to those of their own species. Among the Apinae, varying degrees of social organization occur. In most species, the queen is well differentiated from the workers. However, in nests of *Melipona* and *Trigona* (mainly neotropical forms) mass provisioning of cells occurs, and there is no contact between the parent and the young. Young queens of *Trigona* are reared in specially built cells, as are those of *Apis*. However, in contrast to *Apis*, a young queen and attendants form a new nest; the old queen, being too large and heavy to fly, cannot move to a new site as occurs in *Apis*. *Apis mellifera*, the honeybee (Figure 10.37), is a European species, though it has been introduced to many other parts of the world by man.

Literature

Because of their fascinating behavior, Hymenoptera, especially the social forms, have been the subject of much literary effort, and only selected examples

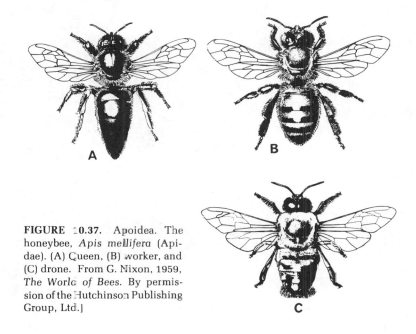

FIGURE 10.37. Apoidea. The honeybee, *Apis mellifera* (Apidae). (A) Queen, (B) worker, and (C) drone. From G. Nixon, 1959, *The World of Bees*. By permission of the Hutchinson Publishing Group, Ltd.]

of this can be given here. Malyshev (1968), in dealing with the phylogeny of Hymenoptera, provides much information on their ecology. The book also contains a large bibliography. Doutt (1959) and Askew (1971) have reviewed the ecology of the parasitic Hymenoptera. Authors dealing with the biology of specific groups include Wheeler (1928), Michener and Michener (1951), and Richards (1953) [social forms]; Nixon (1954) and Stephen *et al.* (1969) [bees]; Free and Butler (1959) [bumblebees]; Evans and Eberhard (1970) [wasps]; and Sudd (1967) [ants]. The British Hymenoptera may be identified from Step (1932) or from the Royal Entomological Society's handbooks (Richards, 1956, and other authors). A key to the North American families is given by Borror *et al.* (1976).

Askew, R. R., 1971, *Parasitic Insects*, American Elsevier, New York.

Borror, D. J., Delong, D. M., and Triplehorn, C. A., 1976, *An Introduction to the Study of Insects*, 4th ed., Holt, Rinehart and Winston, New York.

Doutt, R. L., 1959, The biology of parasitic Hymenoptera, *Annu. Rev. Entomol.* **4:**161–182.

Evans, H. E., and Eberhard, M. J. W., 1970, *The Wasps*, University of Michigan Press, Ann Arbor.

Free, J. B., and Butler, C. G., 1959, *Bumblebees*, Collins, London.

Malyshev, S. M., 1968, *Genesis of the Hymenoptera and the Phases of Their Evolution* (O. W. Richards and B. Uvarov, eds.), Methuen, London.

Michener, C. D., and Michener, M. H., 1951, *American Social Insects*, Van Nostrand, New York.

Nixon, G., 1954, *The World of Bees*, Hutchinson, London.

Richards, O. W., 1953, *The Social Insects*, MacDonald, London.

Richards, O. W., 1956, Hymenoptera. Introduction and keys to families, *R. Entomol. Soc. Handb. Ident. Br. Insects* **6**(1)**:**1–94.

Step, E., 1932, *Bees, Wasps, Ants and Allied Insects of the British Isles*, Warne, London.

Stephen, W. P., Bohart, G. E., and Torchio, P. F., 1969, *The Biology and External Morphology of Bees with a Synopsis of the Genera of Northwestern America*, Agricultural Experiment Station, Oregon State University, Corvallis.

Sudd, J. H., 1967, *An Introduction to the Behaviour of Ants*, Arnold, London.

Wheeler, W. M., 1928, *The Social Insects, Their Origin and Evolution*, Harcourt, Brace, and World, New York.

II

Anatomy and Physiology

11

The Integument

1. Introduction

The integument of insects (and other arthropods) comprises the basement membrane, epidermis, and cuticle. It is often thought of as the "skin" of an insect but, functionally speaking, it is far more than just that (Locke, 1974). Not only does it provide physical protection for internal organs but, because of its rigidity, it serves as a skeleton to which muscles can be attached. It also reduces water loss to a very low level in most Insecta, a feature which has been of great significance in the evolution of this predominantly terrestrial class. In addition to these primary functions, the cuticular component of the integument performs a number of secondary duties. It acts as a metabolic reserve, to be used cyclically to construct the next stage, or during periods of great metabolic activity or starvation. It prevents entry of foreign material, both living and nonliving, into a insect. The color of insects is also a function of the integument, especially the cuticular component.

The integument is not a uniform structure. On the contrary, both its cellular and acellular components may be differentiated in a variety of ways to suit an insect's needs. Epidermal cells may form specialized glands which produce components of the cuticle or may develop into particular parts of sense organs. The cuticle itself is variously differentiated according to the function it is required to perform. Where muscles are attached or where abrasion may occur it is thick and rigid; at points of articulation it is flexible and elastic; over some sensory structures it may be extremely thin.

2. Structure

The innermost component of the integument (Figure 11.1) is the basement membrane, an amorphous, acellular layer of neutral mucopolysaccharide on the outer side of which rests the epidermis. It is up to 0.5 μm thick and may be produced by hemocytes, though there are indications that epidermal cells also participate in its synthesis.

The epidermis (hypodermis) is a more or less continuous sheet of tissue, one-cell thick. During periods of inactivity, its cells are flattened and intercel-

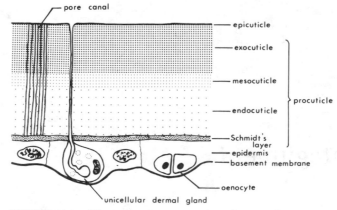

FIGURE 11.1.   Diagrammatic cross section of mature integument.

lular boundaries are indistinct. When active, the cells are more or less cuboidal, and their plasma membranes are readily apparent; one to several nucleoli are evident, during ribonucleic acid synthesis. The density of cells in a particular area varies, following a sequence which can be correlated with the molting cycle. The cells often contain granules of a reddish-brown pigment, insectorubin, which in some insects contributes significantly to their color. However, in most insects color is produced by the cuticle—see Section 4.3.

Epidermal cells may be differentiated into sense organs or specialized glandular cells. Oenocytes are large, ductless, often polyploid cells, up to 100 μm in diameter. They occur in pairs or small groups and the cells of each group may be derived from one original epidermal cell. Usually they move to the hemocoelic face of the epidermis, though in some insects they form clusters in the hemocoel or migrate and reassemble within the fat body. Oenocytes show signs of secretory activity which can be correlated with the molting cycle, and, on the basis of certain staining reactions, it has been suggested that they produce the lipoprotein component of epicuticle. Dermal glands of various types are also differentiated. In their simplest form the glands may be unicellular and have a long duct that penetrates the cuticle to the exterior. More commonly, they are composed of several cells. The gland cells again exhibit cyclical activity associated with new cuticle production, and it has been proposed that they secrete the cement layer of epicuticle.

The cuticle is produced by epidermal cells. In addition to covering the entire body, it also lines ectodermal invaginations such as the foregut and hindgut, tracheae and tracheoles, and certain glands. The cuticle includes two primary layers, the inner procuticle and outer epicuticle. The procuticle forms the bulk of the cuticle and in most species is differentiated into two zones, endocuticle and exocuticle, which differ markedly in their physical properties but only slightly in their chemical composition. In some cuticles the border between the two is not clear and an intermediate area, the mesocuticle is visible. Adjacent to the epidermal cells a narrow amorphous layer, Schmidt's layer, may be seen. It was suggested originally that this was an adhesive layer holding the cuticle to the cells. It is now believed to represent newly secreted, less well stabilized endocuticle.

The endocuticle is composed of lamellae (Figure 11.2). Electron microscopy reveals that each lamella is made up of a mass of microfibers arranged in a succession of planes, all fibers in a plane being parallel to each other. The orientation changes slightly from plane to plane making cuticle like plywood with hundreds of layers. The exocuticle may be defined as the region of procuticle "adjacent to the epicuticle which is so stabilized that it is not attacked by the molting fluid and is left behind with the exuvium at molting" (Locke, 1974, pp. 139–140). Not only is the exocuticle chemically inert, it is hard and extremely strong. It is, in fact, procuticle that has been "tanned" (see Section 3.3). Exocuticle is absent from areas of the integument where flexibility is required, for example, at joints and intersegmental membranes, and along the ecdysial line. In many soft-bodied endopterygote larvae the exocuticle is extremely thin and frequently cannot be distinguished from the epicuticle.

Procuticle is composed almost entirely of protein and chitin. The latter is a nitrogenous polysaccharide consisting primarily of N-acetyl-D-glucosamine residues together with a small amount of glucosamine linked in a β1,4 configuration (Figure 11.3). In other words, chitin is very similar to cellulose, another polysaccharide of great structural significance, except that the hydroxyl group of carbon atom 2 of each residue is replaced by an acetamide group. Because of this configuration, extensive hydrogen bonding is possible between adjacent chitin molecules which link together (like cellulose) to form microfibers. Chitin makes up between 25 and 60% of the dry weight of procuticle but is not found in the epicuticle. It is associated with the protein component, being linked to protein molecules by covalent bonds, forming a glycoprotein complex. The protein of the procuticle can be resolved into water-soluble and water-insoluble fractions. The former is called "arthropodin" but this term now has little meaning, since it implies a single protein species. This is not the case; the water-soluble fraction contains several different proteins. The water-insoluble fraction includes proteins bound to chitin and/or to one another.

In the exocuticle, adjacent protein molecules are linked together by a quinone molecule, and the cuticle is said to be tanned (see Section 3.3). The tanned protein, which is known as "sclerotin," comprises several different molecules. Resilin, discovered in 1960, is a rubberlike material found in areas of the cuticle that undergo springlike movements, for example, wing hinges. Like rubber, resilin, when stretched, is able to store the energy involved. When the tension is released, the stored energy is used to return the protein to its original length.

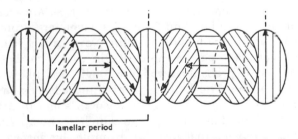

lamellar period

FIGURE 11.2. Diagram showing orientation of microfibers in lamellae of endocuticle. [From A. C. Neville and S. Caveney, 1969, Scarabaeid beetle exocuticle as an optical analogue of cholesteric liquid crystals, Biol. Rev. 44:531–562. By permission of Cambridge University Press, New York.]

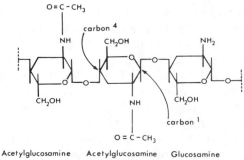

FIGURE 11.3. The chemical structure of chitin.

In addition to these structural proteins, enzymes also exist in the cuticle, including diphenol oxidase, which catalyzes the oxidation of dihydric phenols used in the tanning process. This enzyme appears to be located in or just beneath the epicuticle.

A variety of pigments have been found in the cuticle (or in the epidermis) which may give an insect its characteristic color (see Section 4.3).

Certain processes occur at the surface of the cuticle after it has been formed, for example, secretion and repair of the wax layer and tanning of the outer procuticle. Thus, a route of communication must remain open between the epidermis and cuticular surface. This route takes the form of pore canals which are formed as the new procuticle is deposited (see Section 3.1), and which may or may not contain a cytoplasmic process. Most often, the canals do not contain an extension of the epidermal cell but have at least one "filament" produced by the cell. Locke (1974) suggests that the filament(s) might keep a channel open in the newly formed cuticle until the latter hardens, and anchor the cells to the cuticle. In some insects the pore canals become filled with cuticular material once epicuticle formation (including tanning) is complete. The pore canals terminate immediately below the epicuticle. Running from the tips of the pore canals to the outer surface of the epicuticle are lipid-filled channels known as wax canals.

The epicuticle is a composite structure produced partly by epidermal cells and partly by specialized glands. It ranges in thickness from a fraction of a micrometer to several micrometers. Mainly on the basis of its staining reactions earlier authors concluded that the epicuticle was a four-layered structure, and electron microscopy has tended to confirm this. However, one or more of these layers may be absent in some species and even over certain areas of cuticle in an individual under particular conditions. The layers are, from outside to inside, cement, wax, polyphenol, and cuticulin. Nowadays it is considered that the polyphenol layer is the part of the cuticulin which has been tanned. The nature of the cement is probably variable, though it is likely to be approximately similar to shellac. The latter is a mixture of laccose and lipids. The cement is undoubtedly a hard, protective layer in some insects. In others it appears to be more important as a sponge which soaks up excess wax. The latter could quickly replace that lost, for example, by surface abrasion. The waxes are long-chain fatty acid esters. Within the wax layer three regions can be distinguished. Adjacent to the cuticulin is a monolayer of tightly packed molecules in liquid form. The molecules are oriented with their nonpolar (water-repellent) groups facing outward, giving the epicuticular surface its high contact angle with water and its resistance to water loss. Most wax is in the middle layer, which is less ordered and permeates the cement. The outer layer, which comprises crystalline wax blooms, is not present in all insects. The cuticulin *sensu stricto* (that is, the untanned inner portion) extends over the entire body surface and ectodermal invaginations, including the most minute tracheoles, but is absent from specific areas of sense organs and from the tips of certain gland cells. It may be considered the most important layer of the epicuticle for the following reasons (Locke, 1974). (1) It is a selectively permeable barrier. During breakdown of the old cuticle, it allows the "activating factor" for the molting gel to move out and the products of cuticular hydrolysis to enter, yet it is impermeable to the enzymes in the molting fluid. It is permeable to waxes (since these are deposited only after the cuticulin layer has formed) and, in some insects, it permits the entry of water. (2) It is inelastic and, therefore, serves as a limiter of growth. (3) It provides the base on which the wax monolayer sits. The nature of the cuticulin will therefore determine whether the wax molecules are oriented with their polar or nonpolar groups facing outward and, therefore, the surface properties of the cuticle. (4) It may play a role in determining the surface pattern of the cuticle. Despite its importance the composition of the cuticulin is unknown, though its staining properties indicate that it contains lipoprotein, which becomes tanned in the outer polyphenol layer.

3. Cuticle Formation

Formation of new cuticle (Figure 11.4) may be viewed largely as a succession of syntheses by epidermal cells, with dermal glands and oenocytes adding their products at the appropriate moment (Locke, 1974). It must be realized, however, that other, related processes such as dissolution of old cuticle are going on concurrently and that cuticle formation is partly a preecdysial and partly a postecdysial event; that is, much endocuticle formation, tanning of the outer procuticle, wax secretion, and other processes occur after the remains of the old cuticle are shed.

3.1. Preecdysis

In most species the onset of a molting cycle is marked by an increase in the volume of the epidermal cells and/or by epidermal mitoses. These events are soon followed by apolysis, the detachment of the epidermis from the old cuticle. The epidermal cells, at this time, show signs of preparation for future synthetic activity. One or more nucleoli become prominent, the number of ribosomes increases, and the ribonucleic acid content of the cells is elevated. As apolysis occurs, the space between the epidermis and old cuticle becomes filled with molting gel. Initially the gel is inactive, and it is only after formation of the cuticulin layer of the epicuticle that the epidermal cells release an "activation factor" so that digestion of the old endocuticle can begin. Molting fluid

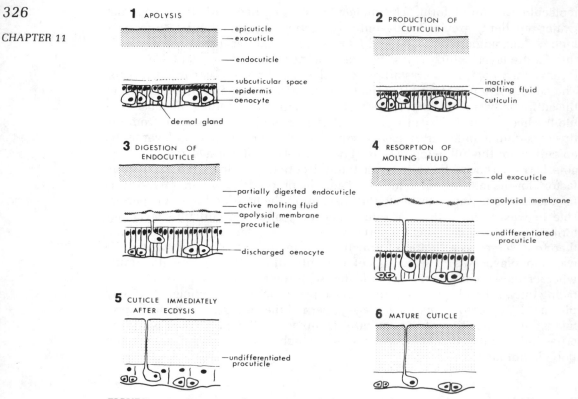

FIGURE 11.4. Summary of events in cuticle formation. [After R. F. Chapman, 1971, *The Insects: Structure and Function.* By permission of Elsevier North-Holland, Inc., and the author.]

contains proteinases and a chitinase. Between 80 and 90 percent of the old cuticle is digested, resorbed, and reused in the production of new cuticle. However, molting fluid does not affect the exocuticle, or muscle insertions and sensory structures in the integument. Thus, an insect is able to move and receive information from the environment more or less to the point of ecdysis.

The first layer of new cuticle deposited is the cuticulin component of the epicuticle. Minute papillae form on the distal side of the epidermal cells and fine droplets of material are secreted from them. Eventually the droplets coalesce to form a continuous layer. Oenocytes are maximally active at this time, and it is possible that they are involved in cuticulin formation, perhaps by synthesizing a precursor. Initially the cuticulin layer is smooth but soon becomes wrinkled. This permits expansion of the cuticle after molting. Tanning of the outer cuticulin layer then occurs, following secretion of polyphenols.

At about the same time, production and deposition of the new procuticle begin. In contrast to the epicuticle, whose layers are produced sequentially from inside to outside, the new procuticle is produced with the newest layers on the inside. As the procuticle increases in thickness, the original papillae elongate to become the cytoplasmic processes within the newly formed pore canals.

Deposition of the wax layer of the epicuticle begins a few hours prior to ecdysis. The wax is secreted by the epidermal cells, probably as lipid–water liquid crystals, and passes along the pore canals to the outside. Wax production continues after ecdysis and, in some insects, throughout the entire intermolt period.

3.2. Ecdysis

At the time of ecdysis the old cuticle comprises only the original exocuticle and epicuticle In many insects it is separated from the new cuticle by an air space and a thin ecdysial membrane that is formed from undigested inner lamellae of the endocuticle. Shortly before molting an insect begins to swallow air (or water, if aquatic), thereby increasing the hemolymph pressure by as much as 90 mm Hg. Hemolymph is then localized in the head and thorax following contraction of intersegmental abdominal muscles. In many insects these muscles become functional only at the time of ecdysis and histolyze after each molt. The local increase in pressure in the anterior part of the body causes the old cuticle to split, along the ecdysial line when this is present. An insect continues to swallow air or water after the molt in order to stretch the new cuticle prior to tanning.

3.3. Postecdysis

Several processes are continued or initiated after ecdysis. Wax secretion continues, and the major portion of the endocuticle is deposited at this time. Indeed, endocuticle production in some insects appears to be a more or less continuous process throughout the intermolt period. It is also at this time that the cement layer of the epicuticle is laid down.

The most striking postecdysial event, however, is the hardening* of the outer procuticle (i.e., the differentiation of the exocuticle). Hardening is usually accompanied by darkening (melanization), though the two may be distinct processes; that is, some species have pale but very hard cuticles. Much information on the mechanism of hardening has come from study of tanning of the fly puparium and cockroach ootheca, though both are "special cases" and may not indicate precisely the nature of tanning in normal procuticle. In fly larvae the level of the amino acid tyrosine in the hemolymph reaches a maximum prior to puparium formation, then declines as tanning proceeds. The tyrosine is apparently accumulated by hemocytes where it is oxidized to "dopa" (dihyroxyphenylalanine) which is then decarboxylated forming dopamine. It has been suggested that dopamine is accumulated by the epidermal cells and converted to N-acetyldopamine which moves, presumably via pore canals, to the epicuticle where it is oxidized to a quinone under the influence of diphenol oxidase. The quinone diffuses back into the procuticle and links protein molecules together. The process is summarized in Figure 11.5. According to Hackman (1974), the quinone may polymerize, a process which makes the molecule a larger, more effective tanning agent. The polymer will have more reactive sites and be capable of bridging larger distances, the result being that

*As Hackman (1974) points out, the hardness of a cuticle is not related to chitin content, though the term "heavily chitinized" is sometimes used to describe a hard cuticle.

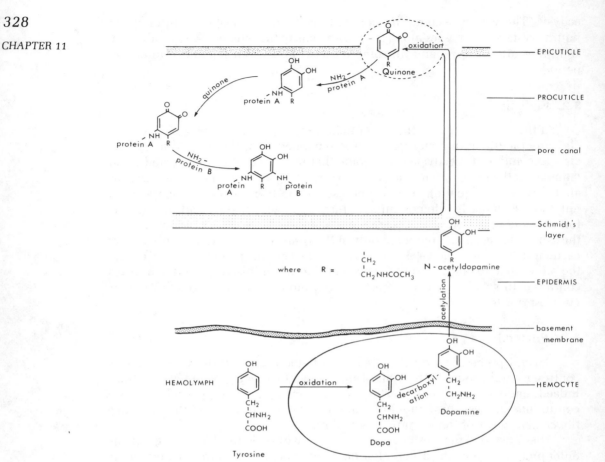

FIGURE 11.5. Summary of the tanning process.

more protein molecules can be linked. The quinone combines covalently with the terminal amino group or a sulfhydryl group of the protein to form an N-catechol protein, which in the presence of excess quinone is oxidized to an N-quinonoid protein. The latter is able to react with the terminal amino group or sulfhydryl group of another protein to link the two molecules. In addition, the quinone can react with the ε-amino groups of lysine residues in the protein molecule to effect further binding. The possibility of autotanning also exists, though the process has never been demonstrated. This involves the oxidation of the constituent aromatic amino acids (for example, tyrosine) in a polypeptide chain to form quinone compounds which can subsequently link with amino groups in adjacent chains. This process could account for tanning that occurs in cuticles known to contain little free diphenol.

The color of tanned cuticle depends on the amount of o-quinone that is present. When this molecule is present in small quantities the cuticle is pale; if it is in excess, and especially if it polymerizes, the cuticle when tanned is dark. In some cuticles production of the characteristic dark brown or black color (melanization) is entirely separate from tanning, although it involves tyrosine and its oxidation product dopa. Oxidation of dopa yields dopaquinone which can form an indole ring derivative, dopachrome. Decarboxylation, oxidation, and polymerization then occur to produce the pigment melanin.

As tanning occurs, the cuticle undergoes a number of physical rearrangments. Its water content decreases due to a decline in the number of hydrophilic groups such as —NH$_2$ because of their involvement in tanning and to a tighter packing of the chitin and protein molecules. This leads to an overall decrease in the thickness of the cuticle.

3.4. Coordination of Events

It is essential that the complex series of events comprising a molt cycle be coordinated. Central to this coordination are hormones (see Chapter 13), though the expression of their effects can be modified by other factors, for example, nutrition and injury. Many events within a molt cycle are influenced by hormones, but still unsolved is the problem of whether this influence is direct or indirect; that is, do hormones directly regulate all these events or merely trigger the start of the series? (See Locke, 1974.) Further complicating the picture is the observation that more than one hormone may influence the same event.

Ecdysone affects many events throughout a molt cycle. It initiates apolysis, stimulates epidermal cell mitosis, increases chitinase activity, and induces the synthesis of several enzymes involved in tanning. The site of action of ecdysone has been studied by Karlson et al. (see references in Neville, 1975) in relation to tanning. Ecdysone stimulates synthesis of the messenger RNA specific for dopa decarboxylase. Karlson's group suggested that ecdysone acted on the epidermal cells; however, it now seems likely that the hemocytes are a target organ. Though the modifying influence of juvenile hormone on ecdysone effects has been known for a considerable time (see Chapter 21), it has been shown only recently that it, too, acts on the genome. Its effect is to program the epidermis to secrete a cuticle characteristic of the juvenile stage.

Bursicon, like ecdysone, is known to affect many processes, though all are related to tanning and melanization of the cuticle. Bursicon's primary effect appears to be to increase the permeability of the hemocyte wall to tyrosine and the epidermal cell wall to dopamine. The hormone may exert this effect via a cyclic AMP-mediated system. In addition, bursicon may activate tyrosinase in hemocytes, which catalyzes the oxidation of tyrosine to dopa.

Finally, ecdysis is regulated by an eclosion hormone secreted by the brain in silk moths (see Chapter 21, Section 6.2) and wax synthesis and endocuticle deposition require the presence of the corpus allatum/corpus cardiacum complex in the lepidopteran Calpodes ethlius.

4. Functions of the Integument

Most functions of the integument relate to the physical structure of the cuticle though the latter may serve as a source of metabolites during periods of starvation. These functions may be discussed under three headings, strength and hardness, permeability, and production of color.

4.1. Strength and Hardness

The few studies that have been carried out on the mechanical properties of insect cuticle indicate that it is of medium rigidity and low tensile strength (Locke, 1974). There is, however, wide variation from this general statement, for example, the cuticles of most endopterygote larvae are extremely plastic, whereas the mandibular cuticle of many biting insects may be extremely hard, enabling them to bite through metal. Further, there is an obvious difference in properties between sclerites and intersegmental membranes, and between typical nonelastic cuticle and that which contains a high proportion of resilin.

Though the above properties indicate that the cuticle is satisfactory as a "skin" preventing physical damage to internal organs, discussion of the suitability of the cuticle as a skeletal component must include an appreciation of overall body structure (Locke, 1974). Most components of insect (and other arthropod) bodies may be considered as cuticular cylinders or spheres. Such a tubular shell (used here in the engineering sense to mean a surface-supporting structure which is thin in relation to total size) is about three times as strong as a solid rod of the same material having the same cross-sectional area as the shell (i.e., they both contain the same amount of skeletal material). The force required to distort the shell is proportional to the thickness of the shell and inversely proportional to the cross-sectional area of the whole body. Thus, in small organisms where the thickness of the shell is great relative to the cross-sectional area of the body, the use of a shell as an exoskeletal structure is quite feasible. In larger organisms the advantage of the extra strength due to a shell type of skeleton is greatly outweighed (in a literal sense!) by the massive increase in thickness of the shell that would be necessitated and, perhaps, the physiological problems of producing the large amounts of material required for its construction.

4.2. Permeability

For different insects there exists a wide range of materials that are potential permeants of the integument, and of factors that affect their rate of permeation. Sometimes specific regions of the integument are constructed to facilitate entry or exit of certain materials; more often the integument is structured to prevent entry or loss. As this time we shall consider only the permeability of the cuticle to water and insecticides, of which the latter may now be considered a "natural" hazard for most insects. The passage of gases through the integument is considered in Chapter 15.

Water. Water may be either lost or gained through the integument. In terrestrial insects, which exist in humidities that are almost always less than saturation, the problem is to prevent loss through evaporation. In freshwater forms the problem is to prevent entry due to osmosis.

In many terrestrial insects the rate of evaporative water loss is probably less than 1% per hour of the total water content of the body (i.e., of the order of 1–3 mg/cm² per hr for most species). Most of this loss occurs via the respiratory system, despite the evolution of mechanical and physiological features to reduce such loss (see Chapter 15). Water loss through the integument (*sensu stricto*) is extremely slight, due mainly to the highly impermeable epicuticle and in particular the wax components. Early experiments demonstrated that permeability of the integument is relatively independent of temperature up to a certain point (the transition temperature), above which it increases markedly. As a result of his studies on both artificial and natural systems, Beament (1961) has concluded that the initial impermeability is due to the highly ordered wax monolayer whose molecules sit on the tanned cuticulin at an angle of about 25° to the perpendicular axis, with their polar ends facing inward and nonpolar ends outward. In this arrangement the molecules are closely packed and held tightly together by Van der Waals forces. As temperature increases, the molecules gain in kinetic energy, and eventually the bonds between them rupture. Spaces appear and water loss increases significantly. The nature of the wax and its transition temperature can be correlated with the normal niche of the insect. Insects from humid environments or that have access to moisture in their diet, for example, aphids, caterpillars, and bloodsucking insects, have "soft" waxes, with low transition temperatures. Forms from dry environments or stages with water-conservation problems, for example, eggs and pupae, are covered with "hard" wax, whose transition temperature is high (in most species above the thermal death point of the insect)

Some insects which are normally found in extremely dry habitats and may go for long periods without access to free water, for example, *Tenebrio molitor* and prepupae of fleas are able to take up water from an atmosphere in which the humidity is relatively high. Until recently it was believed that uptake occurred across the body surface perhaps via the pore canals. However, it has now been demonstrated that uptake occurs across the wall of the rectum (see Chapter 18).

In many freshwater insects, for example, adult Heteroptera and Coleoptera, the cuticle is highly impermeable because of its wax monolayer and water gain is probably 4% or less of the body weight per day In most aquatic insects, however, the wax layer is absent. Thus, gains of up to 30% of the body weight per day are experienced, the excess water being removed via the excretory system (see Chapter 18).

Insecticides. Economic motives have stimulated an enormous interest in the permeability of the integument to chemicals, especially insecticides and their solvents (Ebeling, 1974). Though, for the most part, the cuticle acts as a physical barrier to decrease the rate of entry of such materials, there is evidence that in some insects it may bring about metabolic degradation of certain compounds, and consequently reduction of their potency. It follows that increased resistance to a particular compound may result from changes in either the structure or the metabolic properties of the integument (see also Chapter 16). The primary barrier to the entrance of insecticides is for most insects the wax layer of the epicuticle. In addition, the cement layer probably provides some protection against penetration. The procuticle offers both lipid and aqueous pathways along which an insecticide may travel, but the precise rate at which a compound moves depends on many variables, especially thickness of the cuticle, presence or absence of pore canals, and whether the latter are filled with cytoplasmic extensions or other material. It follows that the rate of penetration will vary according to the location of an insecticide on the integument. Thin, membranous cuticle such as occurs in intersegmental regions or covers tactile or chemosensory hairs generally provides little resistance to penetration. The tracheal system is another site of entry. The extent to which tanning of the procuticle occurs is also related to penetration rate. As the chitin–protein micelles become more tightly packed and the cuticle partially dehydrated, permeability decreases.

In addition, but obviously related to the physical features of the cuticle, the physicochemical nature of an insecticide is an important factor in determining the rate of entry. Especially significant is the partition coefficient (the relative solubility in oil and in water) of an insecticide or its solvent. In order to penetrate the epicuticular wax the material must be relatively lipid soluble. However, in order to penetrate the relatively polar material of the procuticle and, eventually, to leave the integument to move toward its site of action the material must be partially water-soluble. Thus, correct formulation of an insecticidal solution is an important consideration.

It should be apparent from the above discussion that few generalizations can be made. At the present time, therefore, the suitability of an insecticide must be considered separately for each species. Because of the factors which affect the entry of insecticides, a great difference usually exists between "real toxicity," that is, toxicity at the site of action, and "apparent toxicity," the amount of material that must be applied topically to bring about death of the insect. The chief feature which relates the two is obviously the "penetration velocity," that is, the rate at which material passes through the cuticle. When the rate is high, the real and apparent toxicity values will be nearly identical.

4.3. Color

As in other animals, the color of insects serves to conceal them from predators (sometimes through mimicry), frighten or "warn" predators that potential prey is distasteful, or facilitate intraspecific and/or sexual recognition. It may be used also in thermoregulation. The color of an insect generally depends on the integument. Rarely, an insect's color may be due to pigments in tissues or hemolymph below the integument. For example, the red color of *Chironomus* larvae is due to hemoglobin in solution in the hemolymph. Integumental colors may be produced in two ways. Pigmentary colors are produced when pigments in the integument (usually the cuticle) absorb certain wavelengths of light and reflect others (see Fuzeau-Bresch, 1972). Physical (structural) colors result when light waves of a certain length are reflected as a result of the physical features of the surface of the integument.

Pigmentary colors result from the presence in molecules of particular bonds between atoms. Especially important are double bonds such as C=C, C=O, C=N, and N=N which absorb particular wavelengths of light (see Hackman, 1974). The integument may contain a variety of pigment molecules which produce characteristic colors. Usually the molecule, known as a chromophore, is conjugated with a protein to form a chromoprotein. The brown or black color of many insects results usually from melanin pigment. Melanin is a molecule composed of polymerized indole or quinone rings. Typically, it is located in the cuticle, but in *Carausius* it occurs in the epidermis, where it is capable of movement and may be concerned with thermoregulation as well as concealment. Carotenoids are common pigments of phytophagous insects. They are acquired through feeding as insects are unable to synthesize them. Carotenoids generally produce yellow, orange, and red colors, and, in combination with a blue pigment, mesobiliverdin, produce green. Examples of the use of carotenoids include the yellow color of mature *Schistocerca* and the red color of *Pyrrhocoris* and *Coccinella*. Pterines, which are purine derivatives, are common pigments of Lepidoptera and Hymenoptera and produce yellow, white, and red colors. Ommochromes, which are derivatives of tryptophan, an amino acid, are an important group of pigments that produce yellow, red, and brown colors. Examples of colors resulting from ommochromes are the pink of immature adult *Schistocerca*, the red of Odonata, and the reds and browns of nymphalid butterflies. In some insects the characteristic red or yellow body color is due to flavones originally present in the foodplant.

Physical colors are produced by scattering, interference, or diffraction of light though the latter is extremely rare. Most white, blue, and iridescent colors are produced using the first two methods. White results from the scattering of light by an uneven surface or by granules that occur below the surface. When the irregularities are large relative to the wavelength of light, all colors are reflected equally, and white light results. An interference color is produced by laminated structures when the distance between successive laminae is similar to the wavelength of light that produces that particular color. As light strikes the laminae light waves of the "correct" length will be reflected by successive surfaces, and the color they produce will therefore be reinforced. Light waves of different lengths will be out of phase. Changing the angle at which light strikes the surface (or equally the angle at which the surface is viewed) is equivalent to altering the distance between laminae. In turn, this will alter the wavelength that is reinforced and color that is produced. This change of color in relation to the angle of viewing is termed iridescence. Iridescent colors are common in many Coleoptera and Lepidoptera.

5. Summary

The integument not only provides physical protection for internal organs but serves as a skeleton, reduces water loss (entry) in terrestrial (aquatic) insects, acts as a metabolic reserve, and usually is responsible for the color of insects. It comprises a basement membrane, epidermis, procuticle (including untanned endocuticle and tanned exocuticle), epicuticle, specialized glands, and sensory structures. Cuticle formation is a complex sequence of syntheses coordinated by hormones. Cuticle is highly impermeable to water, mainly due to the epicuticular wax. The wax is also the primary barrier to the entry of insecticides, though protection is also provided by the cement and exocuticle. Integumentary colors may be either pigmentary or physical.

6. Literature

Reviews on aspects of the integument are given by Ebeling (1974) [permeability of cuticle], Hackman (1974) [chemistry of cuticle], Locke (1974) [structure and formation of integument], and Neville (1975) and Hepburn (1976) [general]. The latter two texts contain very large reference lists.

Beament, J. W. L., 1961, The water relations of insect cuticle, *Biol. Rev.* **36**:281–320.

Ebeling, W., 1974, Permeability of insect cuticle, in: *The Physiology of Insecta*, 2nd ed., Vol. VI (M. Rockstein, ed.), Academic Press, New York.

Fuzeau-Bresch, S., 1972, Pigments and color changes, *Annu. Rev. Entomol.* **17**:403–424.

Hackman, R. H., 1974, Chemistry of the insect cuticle, in: *The Physiology of Insecta*, 2nd ed., Vol. VI (M. Rockstein, ed.), Academic Press, New York.

Hepburn, H. R. (ed.), 1976, *The Insect Integument*, American Elsevier, New York.

Locke, M., 1974, The structure and formation of the integument in insects, in: *The Physiology of Insecta*, 2nd ed., Vol. VI (M. Rockstein, ed.), Academic Press, New York.

Neville, A. C., 1975, *Biology of the Arthropod Cuticle*, Springer-Verlag, New York.

12

Sensory Systems

1. Introduction

Organisms constantly monitor and respond to changes in their environment (both external and internal) so as to maintain themselves under the most favorable conditions for growth and reproduction. The structures which receive these environmental cues are sense cells, and the cues are always forms of energy, for example, light, heat, kinetic (as in mechanoreception and sound reception), and potential (as in chemoreception, the sense of smell and taste) (Dethier, 1963). The sensory structures use the energy to do work, namely, to generate a message that can be conducted to a decoding area, the central nervous system, so that an appropriate response can be initiated. The message is, of course, in the form of a nerve impulse. Sensory structures are generally specialized so as to respond to only one energy form and are usually surrounded by accessory structures that modify the incident energy.

As Dethier (1963) points out, the small size and exoskeleton of insects have had marked influence on their sensory and nervous systems. Smallness and, presumably, short neural pathways would provide for a very rapid response to stimuli. However, it would also mean relatively few axons and, therefore, a limitation in the number of responses possible to a given stimulus. This has led to a situation in insects where stimulation of a single sense cell may trigger a series of responses. Further, almost all insect sense cells are PRIMARY sense cells, that is, they not only receive the stimulus but initiate and transmit information to the central nervous system; in other words, they are true neurons. In contrast, in vertebrates, almost all sensory systems include both a specialized (SECONDARY) sense cell and a sensory neuron that transmits information to the central nervous system. The cuticle provides protection and support by virtue of its rigid, inert nature, yet sense cells must be able to respond to very subtle (minute) energy changes in the environment. Thus, only where the cuticle is sufficiently "weakened" (thinner and more flexible) will the energy change be sufficient to stimulate the cell. An insect, therefore, must strike a balance between safety and sensitivity. In contrast to mammalian skin, which has millions of generally distributed sensory structures, the surface of an insect has only a few thousand such structures, and most of these are restricted to particular regions of the body.

Two broad morphological types of sense cells are recognizable (Dethier, 1963), those associated with cuticle (and therefore including invaginations of the body wall) (Type I neurons) and those which are never associated with cuticle and lie on the inner side of the integument, on the wall of the gut, or alongside muscles or connective tissue where they function as proprioceptors (Type II neurons) (see Section 2.2). A Type I neuron and its associated cells are derived embryonically from the same epidermal cell. They and the associated cuticle form the sensillum (sense organ). All types of sensilla, with the possible exception of the ommatidia of the compound eye, are homologous and derived from cuticular hairs.

2. Mechanoreception

Insects receive and respond to a wide variety of mechanical stimuli: they are sensitive to physical contact with solid surfaces (touching and being touched); they detect air movements, including sound waves; and they have gravitational sense, that is, through particular mechanosensilla they gain information about their body position in relation to gravity. This is especially important in flying or swimming insects which are in a homogeneous medium; they receive information about their body posture and the relationship of different body components to each other, and they obtain information on physical events occurring within the body, for example, the extension of muscles in movement, the filling of the gut by food, and the stretching of the oviduct when mature eggs are present.

Information on the above is gathered by a spectrum of mechanosensilla associated with which, in most cases, are accessory structures that transform the energy of the stimulus into usable form, namely, a mechanical deformation of the sense cell's protoplasm.

2.1. Sensory Hairs

The simplest form of mechanosensillum is seen in sensory hairs (sensilla trichodea) (Figure 12.1), which occur on all parts of the body but are in greatest concentration on those which frequently come into contact with the substrate, the tarsal segments of the legs, antennae, and mouthparts. In its simplest form a sensillum comprises a rigid hair set in a membranous socket and four associated cells; these are the inner sheath cell (also known as the trichogen or generative hair cell), outer sheath cell (tormogen or membrane-producing cell), neurilemma cell, which ensheathes the cell body and axon of the sensory neuron, and the sensory neuron whose dendrite often is cuticularized and includes a terminal cuticular filament (scolopale). (Some hairs include several neurons, but only one of these is mechanosensory, the remainder are chemosensory.)

Hairs differ in their sensitivity; long, delicate hairs respond to the slightest force, even air-pressure changes, whereas shorter, thicker spines require considerable force for stimulation. Associated with this variable sensitivity are differences in the electrophysiology of the hairs. Delicate hairs typically adapt

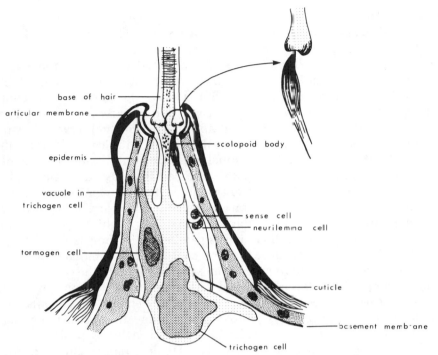

base of hair

articular membrane

epidermis

vacuole in
trichogen cell

tormogen cell

scolopoid body

sense cell

neurilemma cell

cuticle

basement membrane

trichogen cell

FIGURE 12.1. Section through base of tactile hair. [After F. Hsü, 1938, Étude cytologique et comparée sur les sensilla des insectes, *La Cellule* **47**:5–60. By permission of *La Cellule*.]

rapidly, that is, rapidly lose their sensitivity to a continuously applied stimulus. More strongly built hairs, however, adapt only very slowly. Most hairs respond to a stimulus only while moving and are said to be "velocity-sensitive" and the response is "phasic." Such hairs are found on structures that "explore" the environment. The remainder respond continuously to a static deformation ("pressure-sensitive" forms with a "tonic" response) and are found usually in groups (hair plates) (Figure 12.2) where information on position with respect to gravity or posture can be obtained, for example, at joints or on genitalia. In such situations they are serving as proprioceptors.

2.2. Proprioceptors

Proprioceptors are sense organs able to respond continuously to deformations (changes in length) and stresses (tensions and compressions) in the body. They provide an organism with information on posture and position. Five

FIGURE 12.2. Hair plate at joint of coxa with pleuron. [After J. W. S. Pringle, 1938, Proprioceptors in insects. III. The function of the hair sensilla at the joints, *J. Exp. Biol.* **15**:467–473. By permission of Cambridge University Press, New York.]

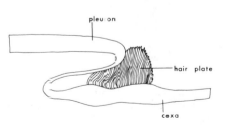

pleuron

hair plate

coxa

types of proprioceptors occur in insects: hair plates, campaniform sensilla, chordotonal organs, stretch receptors, and nerve nets. In common, they respond tonically and adapt very slowly to a stimulus.

A campaniform sensillum (Figure 12.3A) includes all the components of a tactile hair with which it is homologous except for the hair shaft, which is replaced by a dome-shaped plate of thin cuticle. The plate may be slightly raised above the surrounding cuticle, flush with it, or recessed, but in all cases it is contacted at its center by the distal tip of the neuron and serves as a stretch or compression sensor. In many species the plate is elliptical and has a stiffening rod of cuticle running longitudinally on the ventral side, to which the neuron tip is attached (Figure 12.3B). The sensillum is stimulated only by forces which cause the rod to bend or straighten, that is, which cause the dome to become more or less arched. Typically, sensilla are arranged in groups with their longitudinal axes parallel to the normal axis along which stress is applied. In *Periplaneta,* for example, their axis is parallel to the long axis of the leg, so that when the insect is standing they are continuously stimulated because of the stress in the cuticle. Information from sensilla passes to the central nervous system where it inhibits the so-called "righting reflex." When the insect is turned on its back there are no longer stresses in the cuticle, the sensilla are not stimulated, the righting reflex is not inhibited, and the insect undertakes a series of kicking movements in order to regain the standing position.

Chordotonal (scolophorous) sensilla (=scolopidia) (Figure 12.4A) are another widely distributed form of proprioceptor in insects. Unlike the sensilla discussed earlier, chordotonal sensilla lack a specialized exocuticular component, though, it should be emphasized, they are believed to be homologous with the sensillum trichodeum and other types of sensilla. They are associated with the body wall, internal skeletal structures, tracheae, and structures in which pressure changes occur. Though they are found singly, more commonly

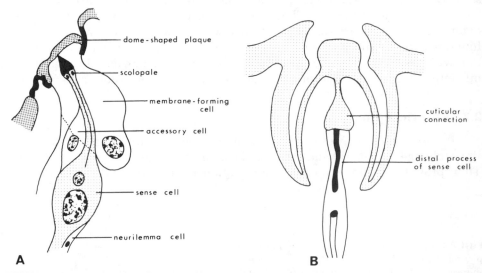

A   B

— dome-shaped plaque
— scolopale
— membrane-forming cell
— accessory cell
— sense cell
— neurilemma cell

cuticular connection
distal process of sense cell

FIGURE 12.3. (A) Campaniform sensillum and (B) section through tip of campaniform sensillum to show stiffening rod of cuticle running along cuticular plate present in some species. [After V. G. Dethier, 1963, *The Physiology of Insect Senses,* John Wiley and Sons, Inc. By permission of the author.]

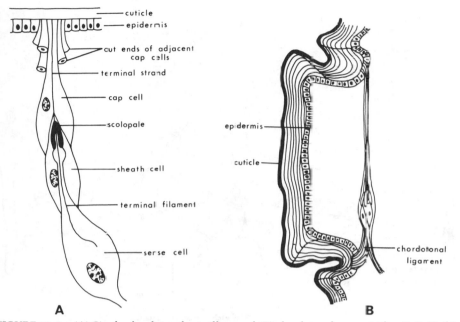

cuticle
epidermis
cut ends of adjacent cap cells
terminal strand
cap cell
scolopale
sheath cell
terminal filament
sense cell

epidermis
cuticle
chordotonal ligament

A   **B**

FIGURE 12.4. (A) Single chordotonal sensillum and (B) chordotonal organ [After V. G. Dethier, 1953, Mechanoreception, in: *Insect Physiology* (K. D. Roeder, ed.). Copyright © 1953 John Wiley and Sons, Inc. Reprinted by permission of John Wiley and Sons, Inc.]

they occur in groups. Chordotonal organs exist as strands of tissue which stretch between two points. The proximal end of the sensory neuron is attached to one point by means of a ligament and the distal end is covered by a cap cell which is attached to the second point (Figure 12.4B). A change in the relative position of the points will cause the strand's length to be altered and the sense cell to be stimulated. Frequently the alteration of position of the points is brought about as a result of pressure changes, for example, in the air within the tracheal system, in the hemolymph within the body cavity, or in aquatic insects in the water in which they are swimming. In relatively few insects chordotonal sensilla are aggregated in large numbers and capable of being stimulated by changes in external air pressure, that is, sound waves (see Section 3).

Stretch receptors (Figure 12.5) comprise a multipolar neuron (Type II) whose dendrites terminate in a strand of connective tissue or a modified muscle cell, the ends of which are attached to the body wall, intersegmental membranes, and/or muscles. As the points to which the ends are attached move with respect to each other, the receptor is stimulated. Stretch receptors are probably most important in providing information to the central nervous system on rhythmically occurring events within the insect, for example, breathing movements, waves of peristalsis along the gut, and locomotion.

A peripheral nerve plexus (net), containing bipolar and multipolar sensory neurons is located beneath the body wall in many larvae whose cuticle is thin and flexible, or beneath the intersegmental and arthrodial membranes of insects with a rigid integument. The nerve endings are presumably stimulated by tension in the body wall or movements of joints. A similar arrangement is present in the wall of the alimentary canal, though the sensory neurons in this case pass their information on to the visceral nervous system (see Chapter 13).

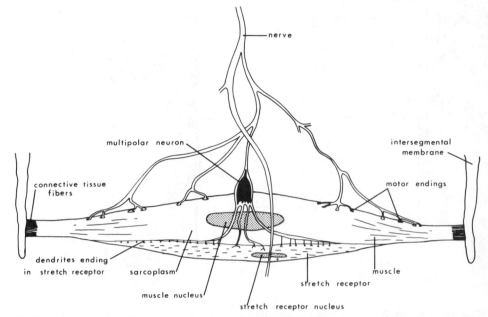

FIGURE 12.5. Stretch receptor of male mosquito. [After R. F. Chapman, 1971, *The Insects: Structure and Function.* By permission of Elsevier/North-Holland, Inc., and the author.]

3. Sound Reception

Sounds are waves of pressure detected by organs of hearing. A sound wave is produced when particles are made to vibrate, the vibration causing displacement of adjacent particles. Usually sound is thought of as an airborne phenomenon; however, it should be appreciated that sounds can pass also through liquids and solids. It will be apparent, therefore, that the distinction between sound reception and mechanoreception is not clear-cut. Indeed, many insects which lack specialized auditory organs can clearly "hear," in that they respond in a characteristic manner to particular sounds. For example, caterpillars stop all movements and contract their bodies in response to sound. If, however, their bodies are coated with water or powder, or the hairs removed, the response is abolished. Insects with specialized sound sensors continue to respond to sounds of low frequency even after the specialized organs have been damaged or removed. The structures which respond to these low-frequency sound waves (see Figure 12.8) are the most delicate mechanosensilla, namely, the sensilla trichodea and, probably, chordotonal sensilla distributed over the body surface. In some species, hairs sensitive to sound may be restricted to particular areas, for example, antennae or cerci.

The specialized auditory organs possessed by many species comprise groups of chordotonal sensilla and associated accessory structures that enhance the sensitivity of the organ. They include Johnston's organ of male mosquitoes and chironomids, tympanal organs, and subgenual organs. The first two are sensitive to airborne vibrations, the latter to vibrations in solids.

3.1. Johnston's Organ

Johnston's organ, that is, one or more groups cf chordotonal sensilla located in the pedicel of the antenna, is present in all adult and many larval insects and generally serves a proprioceptive function, providing information on the position of the antenna with respect to the head, the direction and strength of air or water currents, or, in back swimmers, the orientation of the insect in the water. In male mosquitoes and chironcmids, however, the structure has become specialized to perceive sounds, including those produced by the wings of females in flight. Experimentally, it has been shown that male *Aedes aegypti* are attracted especially to sounds in the frequency range 500–550 cps, which compare with the flight tone of females, 449–603 cps. At these frequencies, males show a characteristic mating response. Sounds of other frequencies (including a male's flight-tone, which is somewhat higher than that of a female) are also perceived by males, though these do not attract them. A male's antennae are extremely "bushy," covered with long, fine hairs which vibrate in unison at certain frequencies, causing the flagellum to move in its "socket," the pedicel (Figure 12.6). The bulbous pedicel accommodates a large number of sensilla which are arranged in two primary groups, the inner and

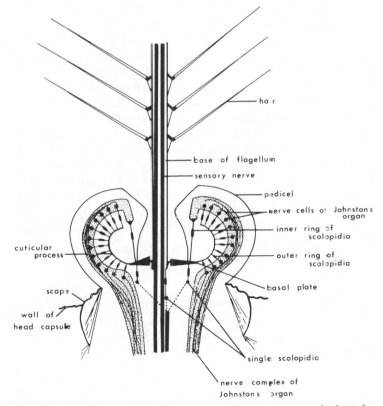

FIGURE 12.6. Johnston's organ. [After H. Autrum, 1963, Anatomy and physiology of sound receptors in invertebrates, in: *Acoustic Behaviour of Animals* (R G. Busnel, ed.). By permission of Elsevier/North-Holland Biomedical Press, Amsterdam.]

outer rings. It is suggested that Johnston's organ resolves the sound into two components, a component running parallel to the flagellum, to which the inner sensilla are most sensitive, and a component perpendicular to the flagellum, which stimulates the outer sensilla. By "estimating" the relative strength of the stimulus from each component, a male is able to determine a female's position.

3.2. Tympanal Organs

Tympanal organs are present in the adults of many species of Orthoptera (on the fore tibiae of Tettigoniidae and Gryllidae, on the first abdominal segment of Acrididae), Lepidoptera (abdominal in Geometroidea and Pyraloidea, metathoracic in Noctuoidea), and Hemiptera (abdominal in Cicadidae, thoracic in others). Though their detailed structure is variable, all tympanal organs have three common features: a thin cuticular membrane; a large tracheal air sac appressed to the membrane, the two structures forming a "drum"; and a group of chordotonal sensilla (Figure 12.7).

Sound waves which strike the drum cause it to vibrate and, therefore, the sensilla to be stimulated. The range of frequency of the waves which stimulate tympanal organs is high. For example, in Acrididae, it extends from less than 1 kcps to about 50 kcps. Over this range the sensitivity of the organ varies greatly, with a maximum in the 2- to 15-kcps range (Figure 12.8). In contrast, the human ear is most sensitive to a frequency of 1 to 3 kcps. An important feature of tympanal organs, as displacement receivers,* is their directional sensitivity; that is, they are most sensitive to sounds which strike the drum perpendicularly. Thus, insects with these organs can locate the source of a sound.

The functional significance of tympanal organs varies. In Orthoptera and Hemiptera, the ability to hear is complemented by the ability to produce sounds (see Chapter 3), and in these orders the organs are important in species aggregation and/or mate location. In Lepidoptera, tympanal organs are used to detect the approach of insectivorous bats. Interestingly, the tympanal organs are most sensitive to frequencies in the range of 15 to 60 kcps, which comes within the frequency range of the sounds uttered by the bats as they echolocate. A moth's response varies according to the intensity of the sound. At low intensity (i.e., when the bat is 100 feet or more distant) a moth moves away from the sound. At high intensity, a moth takes more striking action, flying an erratic course, or dropping to the ground.

3.3. Subgenual Organs

These are chordotonal organs present in the tibiae of most insects, excluding Thysanura, Coleoptera, and Diptera. The organ, which comprises between 10 and 40 sensilla, appears to detect vibrations in the substrate though the mechanism by which the organ is stimulated is not known.

---

*Sound may be perceived either by pressure receivers, devices sensitive to the pressure changes caused by the vibrating particles, for example, the mammalian ear and most microphones, or by displacement receivers. Such receivers have a sensitive diaphragm whose displacement, induced by vibrating particles, can be measured.

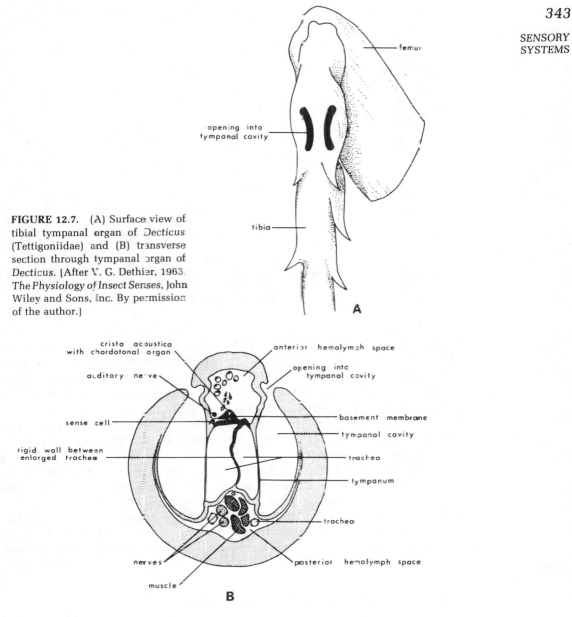

FIGURE 12.7. (A) Surface view of tibial tympanal organ of Decticus (Tettigoniidae) and (B) transverse section through tympanal organ of Decticus. [After V. G. Dethier, 1963. *The Physiology of Insect Senses*, John Wiley and Sons, Inc. By permission of the author.]

4. Chemoreception

Chemoreception, essentially taste and smell, is an extremely significant process in the Insecta, as it initiates some of their most important behavior patterns, for example, feeding behavior, selection of an oviposition site, host or mate location, behavior integrating caste functions in social insects, and responses to man-made attractants and repellents (Hodgson, 1974).

Though taste and smell are distinguished traditionally, such a distinction has no firm morphological or physiological basis. The sensilla for the two

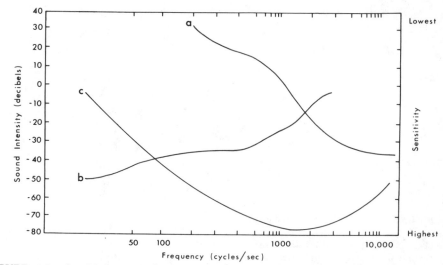

FIGURE 12.8. Sensitivity curves for (a) tympanal organ of *Locusta,* (b) cercal hairs of *Gryllus,* and (c) human ear. [After V. G. Dethier, 1953, Mechanoreception, in: *Insect Physiology* (K. D. Roeder, ed.). Copyright © 1953 John Wiley and Sons, Inc. Reprinted by permission of John Wiley and Sons, Inc.]

senses are structurally identical; indeed, in some species the same structure is used for both olfaction (smell) and gustation (taste). Further, stimulation of a sensillum by either tastes or odors probably entails the same subcellular or molecular interactions. Any difference between smell and taste is, then, a matter of degree. Smell may be defined as chemostimulation by compounds in very low concentration but volatile at physiological temperatures, and taste as chemostimulation by higher concentrations of liquids that are not volatile at physiological temperatures.

In addition to taste and smell, insects have a third method of detecting chemical stimuli, the common chemical sense. This is the response of an insect (always an avoiding reaction) to high concentrations of noxious chemicals. It is not a response due to stimulation of normal chemosensilla, because the response is not abolished after surgical removal of the structures bearing these sensilla. It would seem to be a nonspecific response of other types of sensory neurons.

4.1. Location and Structure of Sensilla

Behavioral and, more recently, electrophysiological experiments have been used to establish the location and nature of chemosensilla. Organs of taste are common on the mouthparts, especially the palps, though they have been identified also on the antennae (Hymenoptera), tarsi (many Lepidoptera, Diptera, and the honeybee), ovipositor (parasitic Hymenoptera and some Diptera), and on the general body surface. The antennae are the primary site of olfactory organs and often bear many thousands of these structures. The mouthparts also carry olfactory structures in many species.

Earlier authors classified chemosensilla according to their morphology, but this system has proved inadequate because structures which look similar at the light microscope level have been shown, using electron microscopy combined with electrophysiological studies, to have quite different structure and function. Slifer (1970) groups chemosensilla in two categories, "thick-walled" and "thin-walled," which, though a seemingly simple structural criterion for separation, is valid in all except a few cases, and broadly correlates with their functions as organs of taste and smell, respectively.

Thick-walled (uniporous) chemosensilla (Figure 12.9A) take the form of hairs (including some of the sensilla basiconica of earlier authors), pegs, or papillae. Generally, they serve as taste sensilla, though some are also sensitive to strong odors. The chemosensitive hairs and pegs broadly resemble tactile hairs, though they can be distinguished with the electron microscope. Whereas tactile hairs have a sharply pointed top and are innervated by a single neuron whose dendrite terminates at the base of the hair, thick-walled chemosensory hairs have a rounded tip with a terminal pore, multineuronal innervation, and dendrites which extend along the length of the hair to terminate at the pore. The dendrites are usually enclosed in a cuticular sheath. Occasionally, the thick-walled pegs may be set in pits, when they are known as sensilla coeloconica and perhaps have an olfactory function. Papillae having the same general features as chemosensory hairs have been observed in the food canal of aphids, on the labellum of flies, and on the cockroach hypopharynx. Some thick-walled hairs are both mechano- and chemosensory.

Thin-walled (multiporous) chemosensilla of various types (Figure 12.9B) have been described from the antennae of all species sufficiently well studied. They have in common a thin cuticular covering (0.1 μm or less in width) perforated by many (up to several thousand) pores. In most cases multineuronal innervation of the sensillum occurs, though occasionally only one neuron is observed. The dendrites from each neuron branch and the branches terminate at the pores. The tip of each dendrite branch is covered with about two dozen "pore filaments." Whether these filaments represent the true endings of the dendrites or an extracellular secretion is controversial. Thin-walled chemosensilla exist as surface hairs and pegs, pegs in pits, and plate organs (sensilla placodea). The latter are roundish, flat, or domed areas of cuticle, which occur in great density on the antennae of many Hymenoptera, some Coleoptera, and some Homoptera. For some time their function was controversial; however, recent electrophysiological and electron-microscope studies have supported the view that they are olfactory.

4.2. Physiology of Chemoreception

Early studies mainly behavioral, established broad parameters for the senses of smell and taste in insects and revealed some interesting comparisons between the senses in insects and those in humans. Like humans, insects appear to "recognize" the four basic taste qualities, sweet (acceptable), salty, acidic, and bitter [all nonacceptable]. The response of an insect, however, varies according to its "physiological state," for example, age, sex, normal diet, and immediate history. The sensitivity of an insect to taste stimuli is, broadly speaking, equal to or greater than that of man. Sucrose, for example, can be

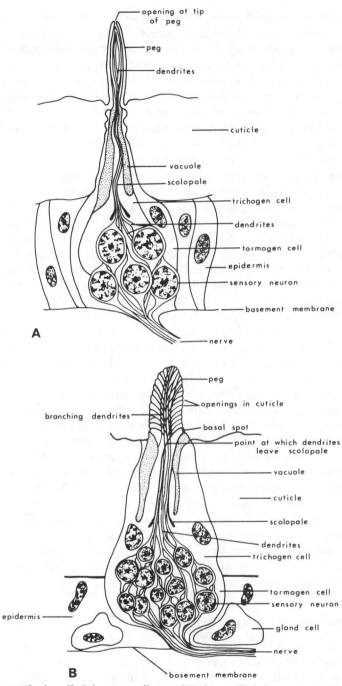

FIGURE 12.9. (A) Thick-walled chemosensillum and (B) thin-walled chemosensillum. [A, after E. H. Slifer, J. J. Prestage, and H. W. Beams, 1957, The fine structure of the long basiconic sensory pegs of the grasshopper (Orthoptera; Acrididae) with special reference to those on the antenna, *J. Morphol.* **101:**359–397. By permission of the Wistar Press. B, after E. H. Slifer, J. J. Prestage, and H. W. Beams, 1959, The chemoreceptors and other sense organs on the antennal flagellum of the grasshopper (Orthoptera; Acrididae), *J. Morphol.* **105:**145–191. By permission of the Wistar Press.]

detected by man at a concentration of 2×10^{-2} M, by the honeybee at 6×10^{-2} M; hydrochloric acid stimulates at 1.25×10^{-3} M in man, at 10^{-3} M (in 1 M sucrose) in the honeybee. However, as in man, sensitivity to a substance increases if that substance has not been experienced for some time. The red admiral butterfly, *Pyrameis atalanta*, for example, fed regularly on sucrose has a tarsal threshold sensitivity of $10^{-} - 10^{-2}$ M. If, however, the sugar is withheld for some time sensitivity increases so that a concentration of 8×10^{-5} M now elicits a response. Chemosensilla on different parts of the body have differing sensitivity to particular chemicals, for example, in the fly, *Calliphora vomitaria*, those on the tarsi are 16 times more sensitive to sucrose than those on the labellum.

Disaccharides are more stimulating than monosaccharides. Trisaccharides are generally nonstimulating and polysaccharides never so. Of the inorganic ions, cations exhibit increasing stimulation in parallel with their partition coefficient and ionic mobility; that is, H^+ is more stimulating than $NH_4^+ > K^- > Ca^{2+} > Mg^{2+} > Na^+$. For anions, the situation appears more complex, and the relationship between ability to stimulate and physical properties of the ions is unclear. Organic acids stimulate in proportion to their degree of dissociation, indicating that the H^+ ion is the principal factor in stimulation. The stimulating power of nonelectrolytes is proportional to the oil : water partition coefficient, though contradictions to this generalization occur.

Some taste sensilla are capable of being stimulated by various substances. Electrophysiological work has shown that this is possible because sensory neurons in the sensillum respond differentially to the substances. For example, in the labellar hairs of *Phormia* one neuron is sugar-sensitive, one is salt-sensitive, one is mechanosensory (and terminates at the base of the hair), and a fourth neuron is water-sensitive.

Species vary in their sensitivity and response to odors, as do individuals in different physiological states. As with taste, insects are especially sensitive to odors of significance to them, and insects which are exposed to a wide range of odors have a better sense of smell (are more sensitive) than insects whose "olfactory environment" is relatively uniform. Sensitivity depends on several factors, including the number of sensilla, the number of sensory neurons, the number of branches from each dendrite and the number of receptor sites. The last two mentioned enable summation of subthreshold responses to occur, so that a sensory cell is stimulated.

For both taste and smell, a stimulant causes depolarization of the plasma membrane of the sensory neuron. Organic molecules, whose effect is highly specific, that is, even closely related molecules have much less effect than the original stimulant, probably bind to a membrane receptor site in a highly stereospecific manner. This binding causes a structural change in the membrane to effect depolarization. Cations are assumed to bind at particular anionic sites, and vice versa.

5. Humidity Perception

Many observations on their behavior indicate that insects are able to monitor the amount of water vapor in the surrounding air. Insects actively seek out (see Chapter 13) a "preferred" humidity in which to rest, or orient themselves toward a source of liquid water. The value of the preferred humidity varies with the physiological state of the insect, especially its state of desiccation. Normally, for example, the flour beetle, *Tribolium castaneum*, prefers dry conditions; however, after a few days without food and water, it develops a preference for more humid conditions. Ablation experiments have established that humidity detectors are typically located on the antennae, though they occur on the anterior sternites in *Drosophila* larvae, and surround the spiracles in *Glossina*, where they monitor the air leaving the tracheal system. The nature and physiology of the sensilla is poorly understood, though it seems likely that they exist as thin-walled hairs or pegs. It is possible that the sensory structures respond to the number of water molecules attaching to specific sites on the sensillum, in a manner analogous to chemoreception. Alternatively, the sensory structure may be affected by the rate at which it loses water to the environment. Evaporation might cause the osmotic pressure of the fluid within the sensory cell to increase, thereby effecting stimulation. Another hypothesis suggests that the cooling effect produced by evaporation might be the stimulus.

6. Temperature Perception

This is the least understood of insect senses. Insects clearly respond to temperature in a behavioral sense, by seeking out a "preferred" temperature. For example, outside its preferred range, the desert locust, *Schistocerca gregaria*, becomes active. This locomotor activity is random but may take the insect away from the unfavorable conditions. Within the preferred range, the insect remains relatively inactive. Under field conditions, the locust will alter its orientation to the sun, raise or lower its body relative to the ground, or climb vegetation in order to keep its body temperature in the preferred range.

Parasitic insects such as *Rhodnius*, *Cimex*, and mosquitoes, which feed on mammalian blood, are able to orient to a heat source. Though the ability to sense heat is present over the entire body surface, it appears that in bloodsucking species, the antennae and/or legs are especially sensitive and probably carry specialized sensilla in the form of thick-walled hairs (see Davis and Sokolove, 1975; Reinouts van Haga and Mitchell, 1975). In some insects, specialized structures have not been identified and it is possible that temperature sensitivity is a response of sensory structures with other functions.

7. Photoreception

Almost all insects are able to detect light energy by means of specialized photosensory structures—compound eyes, ocelli, or stemmata. In the few species which lack these structures, for example, some cave-dwelling forms, there is commonly sensitivity to light over the general body surface.

The use which insects make of light varies from a situation in which it serves as a general stimulant of activity, through one of simple orientation (positive or negative phototropism), to a state where it enables an insect to carry out complex navigation, and/or to perceive form, patterns, and colors. At all levels of complexity, however, the basic mechanism of stimulation is very likely the same; that is, the solar energy striking the photosensory cell is absorbed by pigment in the cell. The pigment undergoes a slight conformational change which causes, in an as yet unknown way, a momentary increase in permeability of the receptor cell membrane and, thereby, the initiation of a nerve impulse which travels to the central nervous system.

7.1. Compound Eyes

Paired compound eyes, the main photosensory system, are well developed in most adult insects, but may be reduced or absent in parasitic or sedentary forms, such as lice, fleas, and female scale insects. Typically, the eyes occupy a relatively large proportion of head surface, from which they bulge out to provide a wide visual field. In dragonflies, male tabanids, and horseflies, the eyes meet in the midcorsal line, the holoptic condition In some other species, the eye is divided into readily distinguishable dorsal and ventral regions. Occasionally, these regions are physically separate, as in the mayfly, *Cloeon*, and water beetle, *Gyrinus*.

A compound eye comprises a variable number of photosensilla, the ommatidia. The number ranges from 1 in the ant *Ponera punctatissima* to more than 10,000 in dragonflies. Each ommatidium (Figure 12.1C) includes a light-gathering component (lens plus crystalline cone); the primary sense cells (retinular cells), which collect and transduce light energy; and various enveloping (pigment) cells. The lens, a region of transparent cuticle, is produced by the primary pigment cells. Its surface is usually smooth, but in some nocturnal species it is covered with numerous minute pimples, about 0.2μm high, which are believed to improve the amount of light transmitted. The crystalline cone is a clear, hard material produced by four cells (Semper's cells). The material is typically intracellular, and the nuclei are situated around it (eucone type). In some species the material is extracellular (pseudocone type); in others there is no crystalline cone and the cells, which are transparent, occupy the area (acone type). Primitively, eight retinular cells occur beneath the crystalline cone, though one of these is usually degenerate or eccentrically located. The seven remaining cells are arranged around a central axis in most species; occasionally, they exist in two tiers of three and four cells, respectively. The mature sensory cells are unipolar, that is, lack dendrites. Instead, their inner surface is modified to form a receptive area, the rhabdomere. Collectively, the rhabdomeres form a rhabdom. By means of electron microscopy, the rhabdomeres can be seen to comprise closely packed microvilli, of diameter about 500 Å, which extend from the cell surface. In cross section the microvilli are hexagonal. The details of rhabdom construction are variable though, in general, "open" and "closed" types can be distinguished (Figure 12.10C,D). In the open type, found in Diptera and Hemiptera, individual rhabdomeres are physically separated; in the closed type, common to most insects, the rhabdomeres are wedge-shaped and closely packed around the central axis. However, even in the most compact rhabdoms, there is extracellular space between microvilli to permit the ionic movements which are the basis of impulse transmission (Goldsmith and Ber-

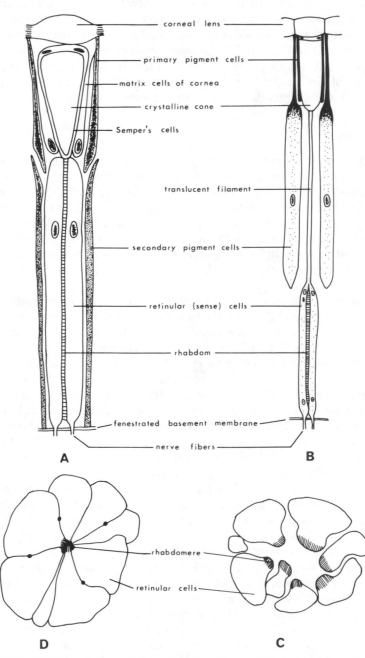

FIGURE 12.10 (A) Photopic ommatidium, (B) scotopic ommatidium, (C) open rhabdom, and (D) closed rhabdom. [A, B, after V. B. Wigglesworth, 1965, *The Principles of Insect Physiology*, 6th ed. By permission of the author. C, D, after T. H. Goldsmith, and G. D. Bernard, 1974, The visual systems of insects, in: *The Physiology of Insecta*, Vol. II, 2nd ed. (M. Rockstein, ed.). By permission of Academic Press Inc., and the authors.]

nard, 1974). The rhabdom contains visual pigments which resemble those of the vertebrate eye; that is, they are conjugated proteins called rhodopsins. The photosensitive component of the molecule, the chromophore, is retinaldehyde (retinal), the aldehyde of vitamin A, or closely related derivatives (see Section 7.1.3). Surrounding the photosensitive cells are the secondary pigment cells which contain granules of red, yellow, and brown pigments (mainly ommochromes). Proximally, the retinular cells narrow to form discrete axons which enter the optic lobe. Between the axons, outside the basement membrane, tracheae may occur which, in addition to their respiratory function, may serve to reflect light onto the rhabdom.

Two ommatidial types can be distinguished according to the arrangement of retinular and pigment cells (Figure 12.10A,B). In photopic (apposition) ommatidia characteristic of diurnal insects, the retinular cells span the distance between the crystalline cone and basement membrane. The secondary pigment cells lie alongside the retinular cells, and the pigment within the former does not migrate longitudinally. Scotopic (superposition) ommatidia, found in nocturnal or crepuscular species, have short retinular cells whose rhabdom is often connected to the crystalline cone by a translucent filament which serves to conduct light to the rhabdom. The secondary pigment cells do not envelop the retinular cells and their pigment granules are capable of marked longitudinal migration.

7.1.1. Form and Movement Perception

It seems not unlikely that early students of insect vision would assume that insects "see" in the same manner as humans; that is, a reasonably sharply defined image would form in the compound eye. Thus, early ideas on compound eye function were attempts to reconcile the observable structure of the eye with this assumed function. The classic "mosaic theory" of insect vision, introduced by Müller (1829) and expanded by Exner (1891) (cited from Goldsmith and Bernard, 1974), proposed that each ommatidium is sensitive only to light which enters at a small angle to its longitudinal axis. More oblique light rays are absorbed by pigment in the cells surrounding an ommatidium. It was assumed that little overlap existed between the visual fields of adjacent ommatidia, and thus, it was suggested, there formed in an eye an erect, mosaic image, being a composite of a large number of point source images, each formed in a separate ommatidium. The image would focus at the level of the rhabdom and its "sharpness," that is, visual acuity, would depend on the number of ommatidia per unit surface area of the eye.

Exner noted that the corneal lens and crystalline cone (which together function as a lens cylinder) appeared laminated (Figure 12.11) and suggested that this was due to a gradient of refractive index (decreasing from the axis outward). The lens cylinder would serve to bend light rays diverging from a point source back toward the axis, where they would form a point image. Exner proposed that the two types of ommatidium distinguishable morphologically and ecologically, that is, the photopic ommatidia of diurnal insects and the scotopic ommatidia of nocturnal forms, would form images in different ways. He suggested that in photopic eyes the lens cylinder had a focal length equal to its absolute length which would cause the light rays to focus at the base of the

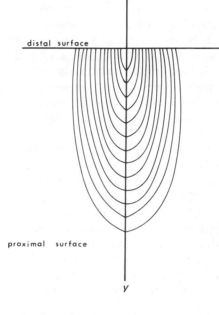

x

distal surface

proximal surface

y

FIGURE 12.11 Lens cylinder, comprising a series of concentric lamellae of different refractive index. Refractive index is greatest along the axis xy. [After V. G. Dethier, 1963, *The Physiology of Insect Senses*, John Wiley and Sons, Inc. By permission of the author.]

cylinder, that is, on the upper end of the rhabdom (Figure 12.12A). The mosaic image formed in the eye would be of the "apposition" type. In scotopic eyes the focal length of the lens cylinder was supposedly one-half the absolute length of the lens cylinder with the result that the light rays are brought to a focus, that is, the image is formed, some distance behind the cylinder, specifically halfway between the corneal surface and the center of curvature of the eye. According to Exner, this coincided with the tip of the rhabdom (Figure 12.12B). Further, in the dark-adapted position pigment granules in the enveloping pigment cells aggregate in the distal end of the cell, thereby permitting light rays at a somewhat greater angle to the axis to be bent back toward the axis as they pass through adjacent lens cylinders. In other words, the point image formed on each rhabdom would be derived not from a single pencil of rays, as in the apposition type, but from a group of such pencils, that is, it would be formed by the superposition of light from a number of adjacent facets and is described therefore as a superposition image. In support of his ideas Exner managed to photograph objects, using the eye of the firefly *Lampyris* as a lens system.

During the early part of the twentieth century, Exner's proposals gained acceptance and were supported by most observations that were made on insect compound eyes. However, within the last two decades, coincident with the development of more refined techniques, his ideas have been severely criticized for the following reasons: (1) images either may be entirely absent or are formed some distance below a rhabdom (the situation in *Lampyris*, though real, is exceptional); (2) the optics of image formation in cleaned corneas (used by Exner) are different from those in a living eye; (3) the region between the lens cylinder and rhabdom in scotopic ommatidia is not of uniform refractive index; this would therefore alter the plane of focus of light rays entering an ommatidium from adjacent facets; (4) in many scotopic ommatidia a translu-

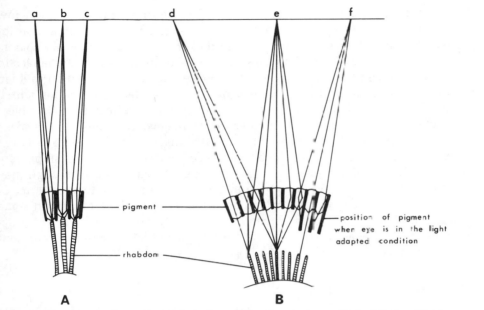

FIGURE 12.12. Image formation in photopic eye (A) and scotopic eye (B). (a–f) Paths of light rays. [After V. G. Dethier, 1963, *The Physiology of Insect Senses*, John Wiley and Sons, Inc. By permission of the author.]

cent filament occurs which guides light from the cone to the rhabdom; (5) it is doubtful whether the crystalline cone has a gradient of refractive index in many insects (although recent work indicates the existence of such a gradient in *Lampyris*); (6) extensive overlap of visual fields occurs, even in photopic systems, for example, in *Locusta* the angle between adjacent ommatidia is 1–2°, yet an ommatidium is stimulated by light rays at an angle of 20° to its longitudinal axis; and (7) good evidence now exists that in many open rhabdoms the functional unit is the individual rhabdomere, not the rhabdom, as is implicit in Exner's proposals.

Despite the above criticisms of Exner's ideas, it is apparent that in many species an image of sorts is formed at the level of the rhabdom. However, it is now accepted that for the majority of insects an image per se has no physiological significance. The real function of the compound eye appears to be that of movement perception. The eye's structure is ideally suited for this function, with its large number of ommatidia, each of which is sensitive only to light shining parallel with or at very small angles to the longitudinal axis of the ommatidium. It can be shown that an angular displacement of the light source of 0.1° or less can be detected by the eye. Many behavioral observations indicate a preference by insects for moving objects or objects with complex shapes. Bees prefer moving to stationary flowers, and they are attracted more by multistriped than by solid patterns. Dragonfly larvae will attempt to capture prey only when it is moving. Stimulation of the eye by a series of changes of light intensity is known as the "flicker effect." The number of stimuli per unit time to which the eye is sensitive, which depends on the rate at which the eye recovers from a previous stimulus, is known as the flicker fusion frequency, and provides the basis for grouping eyes into "slow" and "fast" categories. Slow eyes have a low flicker fusion frequency and are found in more slowly moving, nocturnal insects, whereas fast eyes, with a very high flicker fusion frequency, are characteristic of fast-flying, diurnal species. As the position of the insect changes in relation to an object, the eye will receive a succession of light stimuli. Provided that the rate of change of position does not lead to a rate of stimulation which exceeds the flicker fusion frequency, the insect will, in effect, scan the object and obtain a sense of its shape. In other words, the eye translates form in space into a sequence of events in time. It follows that insects whose eyes have a high flicker fusion frequency have the best form perception. Bees, for example, are readily trained to distinguish solid shapes from striped patterns, though they cannot distinguish between two solid shapes or between two patterns of stripes. Many insects which hunt on the wing, such as dragonflies and some wasps, have excellent form perception especially species which are prey-specific. Some solitary wasps find their nest by recognizing landmarks adjacent to the entrance.

7.1.2. Distance Perception

An ability to judge distance is especially important for quick-moving and/or predaceous insects. This ability is dependent on binocular vision, which, in insects, is achieved when ommatidia in each eye are stimulated by the same light source, that is, when the ommatidial axes cross. For this to occur, the surface of the eyes is highly curved, which ensures a considerable overlap of their visual fields. Many predaceous species will not attack prey unless it is within catchable distance. The ability to differentiate distance appears to derive from the fact that only when certain ommatidia are stimulated is the catching reflex induced (Figure 12.13). The axes of these ommatidia cross at a point within the range of the capturing device, for example, the labium in odonate larvae and the mandibles of tiger beetles.

7.1.3. Spectral Sensitivity and Color Vision

Light-sensitive cells do not respond equally to all wavelengths of light, rather they are particularly sensitive to certain parts of the spectrum. This differential sensitivity may be due to the presence in the cells of either a single visual pigment that has peak absorption at two or more wavelengths, or two or more visual pigments, each with a characteristic peak of absorption. Though the range of the spectrum to which insects are sensitive is about the same as in man, it is shifted toward the shorter wavelengths. Many species, representative of most orders, have been shown to be very sensitive to ultraviolet light (300–400 nm) (UV) and some very significant phototaxes (behavioral responses to light) are initiated by it. For example, ants are negatively phototactic to UV and when given a choice will always congregate in a region not illuminated by UV. Many other insects are attracted by UV. Bees are attracted to yellow flowers by the pattern of UV that the latter reflect. In many species of butterflies the members of one sex have a characteristic pattern of UV-reflecting scales on the wings, invisible to would-be vertebrate predators, which facilitates intraspecific recognition and mating.

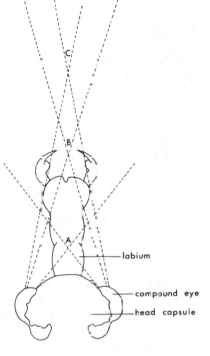

FIGURE 12.13. Distance perception in *Aeshna*. The insect can perceive the distance of any point that simultaneously stimulates ommatidia in both compound eyes (e.g., points A, B, and C). However, the insect extends its labium only when ommatidia, whose visual axes fall between A and B, are stimulated. [After V. B. Wigglesworth, 1965, *The Principles of Insect Physiology*, 6th ed. By permission of the author.]

In addition to UV, the compound eye is sensitive to other wavelengths of light, though the peaks of sensitivity differ among species and even among regions of the same eye. The fly *Calliphora*, for example, shows maximum sensitivity at 470 nm (blue), 490 nm (blue-green), and 520 nm (yellow). The honeybee drone has peaks at 447 nm and 530 nm. Worker bees are also attracted to certain red flowers, but this is because of the UV which the petals reflect.

Although, as noted above, an insect's eye may show peaks of sensitivity to different wavelengths of light, this in itself does not constitute color vision. The latter requires that the insect has the ability to discriminate between wavelengths to which it is sensitive. Field observations and, more recently, behavioral experiments have demonstrated that representatives of many orders are able to distinguish between colors. Clearly the ability to discriminate requires at least two types of receptors (retinular cells) each of which responds to light of a particular wavelength. Further, it seems that in the insect eye, as in the human eye, the variable response of the color receptors is achieved through the possession of different visual pigments. However, as yet, little direct evidence, for example, studies on individual retinular cells, exists for this supposition.

7.1.4. Sensitivity to Polarized Light

As light waves travel from their source, they oscillate sinusoidally about their longitudinal axis. The planes in which they oscillate are usually scattered randomly, through 360°, around the longitudinal axis, but under certain circumstances more waves may travel in a specific plane and the light is said to be polarized. Sunlight entering the earth's atmosphere is "filtered," that is, light rays in certain planes are reflected by dust particles, etc., and the light striking the earth's surface is therefore partially polarized. Some insects can detect and make use of the plane of polarization for navigation and orientation. The classic studies of von Frisch (1949, 1950) (see von Frisch, 1967) showed that foraging honeybees "measure" the angle between a food source, the hive, and the sun and, on returning to the hive, communicate this information to their fellows through the performance of a "waggle dance" on the honeycomb. Actual sight of the sun is not essential for the "light compass reaction," provided that a bee can see a patch of blue sky and thus determine the plane of polarized light. The site of detection of polarized light appears to be the microtubules of the retinular cells. Individual cells within an ommatidium appear to be differentially sensitive to light in different planes, possibly because of the differing orientation of microtubules (or pigment within them) among the cells. For additional information on the detection of polarized light by insects, see Wehner (1976).

7.2. Simple Eyes

Many adult insects and juvenile exopterygotes possess in addition to compound eyes, simple eyes, dorsofrontal in position, known as ocelli. The larvae of endopterygotes have, as their sole photosensory structure, stemmata (lateral ocelli).

Ocelli. The structure of an ocellus is shown in Figure 12.14A. It comprises usually about 500 to 1000 photosensitive cells beneath a common cuticular lens (Goodman, 1970). The cells are arranged in groups of two to five cells, and, distally, each differentiates into a rhabdomere to form a central rhabdom. In contrast to the retinular cells of ommatidia, the photosensitive cells of dorsal ocelli are second-order sense cells; that is, their axons do not themselves conduct information to the central nervous system, but synapse within the ocellar nerve with axons of cells originating in the brain. Ocellar function is poorly understood. Though an ocellus is able to form an image, it does so below the level of the rhabdom and, therefore, the image has no physiological significance. Ocelli respond to the same wavelengths as compound eyes but are much more sensitive than compound eyes; that is, they are stimulated by very low light intensities. They may measure light intensity, and the information derived from them may be used to modify an insect's response to stimuli received by the compound eye. Painting ocelli may cause temporary reversal or inhibition of light-directed behavior, or reduce the rapidity with which an insect responds to light stimuli. Such observations suggest that ocelli act as "stimulators" of the nervous system, so that an insect detects and responds more rapidly to light entering the compound eyes. In addition, in some species ocelli appear essential for the maintenance of diurnal locomotor rhythms.

Stemmata. These are of three types: (1) In sawfly and beetle larvae each stemma, one on either side of the head, resembles a dorsal ocellus, that is, comprises a single cuticular lens lying beneath which are groups of photosensitive cells with a central rhabdom. (2) In larvae of Neuroptera, Trichoptera, and Lepidoptera, there are usually several laterally placed stemmata on each side of the head. Each resembles an ommatidium in that it includes both a corneal and

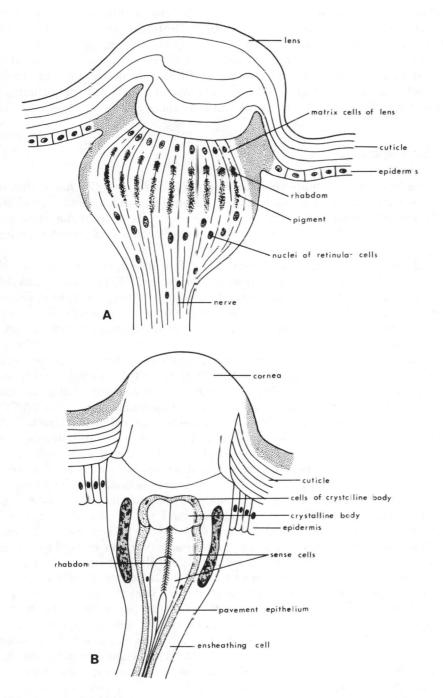

FIGURE 12.14. (A) Ocellus of *Aphrophora* (Cercopidae: Hemiptera) and (B) stemma of *Gastropacha* (Lepidoptera). [After V. B. Wigglesworth, 1965, *The Principles of Insect Physiology*, 6th ed. By permission of the author.]

crystalline lens and a group of retinal cells with a central rhabdom (Figure 12.14B). (3) Finally, in larvae of cyclorrhaph Diptera there are no external signs of stemmata, but a pocket of photosensitive cells occurs on each side of the pharyngeal skeleton. The stemmata of caterpillars have been best studied with regard to function. They are capable of limited form recognition, perhaps through a flicker effect as noted for the compound eye, and some caterpillars are apparently able to discriminate color via the stemmata.

8. Summary

Insect sensory cells are almost always primary sense cells; that is, they both receive a stimulus and conduct it to the central nervous system. Type I sense cells are associated with cuticle and accessory cells to form the sense organ (sensillum); Type II cells are not associated with cuticle and always function as proprioceptors.

Mechanosensilla include sensory hairs, hair plates, campaniform sensilla, chordotonal organs, stretch receptors, and nerve nets. Some insects can detect sound by means of fine sensory hairs (sound of low frequency) or specialized chordotonal structures, such as Johnston's organ and tympanal organs. In both mechanoreception and sound reception the stimulus leads to mechanical deformation of a sense cell's cytoplasm.

Chemosensilla occur as thick-walled (uniporous) hairs, pegs, or papillae which generally have a gustatory function, or as thin-walled (multiporous) hairs, pegs, or plate organs whose usual function is olfactory. Each chemosensillum includes several sensory cells, each of which may respond to a different stimulus. The stimulant probably binds to the sensory cell membrane to induce depolarization. For organic molecules binding is probably highly stereospecific.

Little is known about the nature of humidity sensilla, though in some species they appear to have the form of thin-walled hairs or pegs. Equally, the structure and mode of operation of heat detectors is poorly understood.

Insects detect light energy via compound eyes, ocelli, or stemmata, rarely a dermal light sense. Compound eyes, the chief photosensory structures, are composed of ommatidia, each of which includes a light-focusing system, photosensitive (retinular) cells, and enveloping pigment cells. Two types of ommatidia occur, the photopic characteristic of diurnal insects and the scotopic found in crepuscular and nocturnal species. Retinular cells are differentiated along their inner longitudinal axis into a rhabdomere, a series of closely packed microvilli in which are contained visual pigments. Even when an image is formed at the level of the retinular cells it has no functional significance. The primary function of the compound eyes appears to be movement perception, though some appreciation of form is gained as a result of the flicker effect. The ability to perceive distance is present in some species and is based on the considerable overlap of the visual fields of the two compound eyes. Color vision occurs in some insects and probably results from the presence of retinular cells with different visual pigments. Ocellar function is poorly understood, though it may include general stimulation of the nervous system and provision of information on sudden changes in light intensity. Stemmata appear capable of limited form perception through a flicker effect and, in some species, of color discrimination.

9. Literature

Dethier's (1963) monograph remains the best general review of insect senses. Reviews dealing with particular aspects of sensory perception include those of Schwartzkopff (1974), McIver (1975), and Rice (1975) [mechanoreception including sound reception]; Haskell (1961) [sound reception]; Slifer (1970) and Hodgson (1974) [chemoreception]; Burtt and Catton (1966), Mazokhin-Porshnyakov (1969), Goldsmith and Bernard (1974), and Horridge (1975) [vision—mainly compound eyes]; and Goodman (1970) [dorsal ocelli].

Burtt, E. T., and Catton, W. T., 1966, Image formation and sensory transmission in the compound eye, *Adv. Insect Physiol.* **3:**1–52.

Davis, E. E., and Sokolove, P. G., 1975, Temperature responses of antennal receptors of the mosquito, *Aedes aegypti, J. Comp. Physiol.* **96:**223–236.

Dethier, V. G., 1963, *The Physiology of Insect Senses*, Wiley, New York.

Frisch, K. von, 1967, *The Dance Language and Orientation of Bees*, Springer-Verlag, New York.

Goldsmith, T. H., and Bernard, G. D., 1974, The visual system of insects, in: *The Physiology of Insecta*, 2nd ed., Vol. II (M. Rockstein, ed.), Academic Press New York

Goodman, L. J., 1970, The structure and function of the insect dorsal ocellus *Adv. Insect Physiol.* **7:**97–195.

Haskell, P. T., 1961, *Insect Sounds*, Witherby, London.

Hodgson, E. S., 1974, Chemoreception, in: *The Physiology of Insecta*, 2nd ed., Vol. II (M. Rockstein, ed.), Academic Press, New York.

Horridge, G. A. (ed.), 1975, *The Compound Eye and Vision of Insects*, Clarendon, Oxford.

Mazokhin-Porshnyakov, G. A., 1969, *Insect Vision*, Plenum Press, New York.

McIver, S. B., 1975, Structure of cuticular mechanoreceptors of arthropods, *Annu. Rev. Entomol.* **20:**381–397.

Reinouts van Haga, H. A., and Mitchell, B. K., 1975, Temperature receptors on tarsi of the tsetse fly, *Glossina morsitans* West, *Nature (London)* **255:**225–226.

Rice, M. J., 1975, Insect mechanoreceptor mechanisms, in: *Sensory Physiology and Behavior* (R. Galun, ed.), Plenum Press, New York.

Schwartzkopff, J., 1974, Mechanoreception, in: *The Physiology of Insecta*, 2nd ed., Vol. II (M. Rockstein, ed.), Academic Press, New York.

Slifer, E. H., 1970, The structure of arthropod chemoreceptors, *Annu. Rev. Entomol.* **15:**121–142.

Wehner, R., 1976, Polarized-light navigation by insects, *Sci. Am.* **235**(July):106–115.

13

Nervous and Chemical Integration

1. Introduction

Animals constantly monitor both their internal and their external environment and make the necessary adjustments in order to maintain themselves optimally and thus to develop and reproduce at the maximum rate. The adjustments they make may be immediate and obvious, for example, flight from predators, or longer-term, for example, entry into diapause to avoid impending adverse conditions. The nature of the response depends, obviously, on the nature of the stimulus. Only very rarely does a stimulus act directly on the effector system; almost always a stimulus is received by an appropriate sensory structure and taken to the central nervous system, which "determines" an appropriate response under the circumstances. When a response is immediate, that is, achieved in a matter of seconds or less, it is the nervous system that transfers the message to the effector system. Such responses are usually temporary in nature. Delayed responses are achieved through the use of chemical messages (viz., hormones) and are generally longer-lasting. The nervous and endocrine systems of an individual are, then, the systems that coordinate the response with the stimulus. Pheromones, which constitute another chemical regulating system, coordinate the behavior and development of a group of individuals of the same species. This system is intimately related to nervous and endocrine regulation and is appropriately discussed in the same chapter.

2. Nervous System

The structural unit of nerve impulse conduction is the nerve cell or neuron, which comprises a cell body (perikaryon) where a nucleus, many mitochondria, and other organelles are located; an axon, an extremely elongate cytoplasmic process, often branched distally; and dendrites, which are short, branched outgrowths of the perikaryon. Mitochondria are distributed along the axon and provide the energy required for active transport of ions necessary to restore the correct ionic balance following impulse transmission (see Section 2.3). The structural relationship of these components is variable (Figure 13.1). Motor (efferent) neurons, which carry impulses from the central nervous system,

A MONOPOLAR

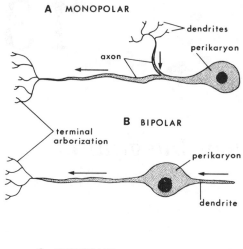

dendrites

perikaryon

axon

terminal
arborization

B BIPOLAR

perikaryon

dendrite

C MULTIPOLAR

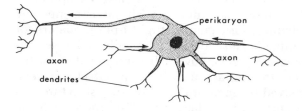

perikaryon

axon

axon

dendrites

FIGURE 13.1. Neurons found in the insect nervous system. Arrows indicate direction of impulse conduction. [After R. F. Chapman, 1971, *The Insects: Structure and Function*. By permission of Elsevier North-Holland, Inc., and the author.]

are monopolar, and their perikarya are located within a ganglion. Sensory (afferent) neurons are usually bipolar but may be multipolar, and their cell bodies are adjacent to the sense organ; internuncial (association) neurons (= interneurons), which transmit information from sensory to motor neurons, are bipolar and their cell bodies occur in a ganglion. Interneurons may be intersegmental and branched, so that the variety of pathways along which information can travel and, therefore, the variety of responses are increased. Neurons are not directly connected to each other but are separated by a minute fluid-filled cavity, the synapse. The normal diameter of axons is 5 μm or less; however, some interneurons within the ventral nerve cord, the so-called "giant fibers," have a diameter up to 60 μm. These giant fibers may run the length of the nerve cord without synapsing and are unbranched except at their termini. They are well suited, therefore, for very rapid transmission of information from sense organ to effector organ; that is, they facilitate a very rapid but stereotyped response to a stimulus and may be important in escape reactions, though the latter has recently been disputed (see Hoyle, 1974).

Neurons are aggregated into nerves and ganglia. Nerves include only the axonal component of neurons, whereas ganglia include axons, perikarya, and dendrites. The typical structures of a ganglion and interganglionic connective are shown in Figure 13.2. In a ganglion there is a central neuropile which comprises a mass of efferent, afferent, and association axons. Frequently visible within the neuropile are groups of axons running parallel, known as fiber tracts. The perikarya of motor and association neurons are normally found in clusters adjacent to the neuropile. Neurons are surrounded by glial cells which

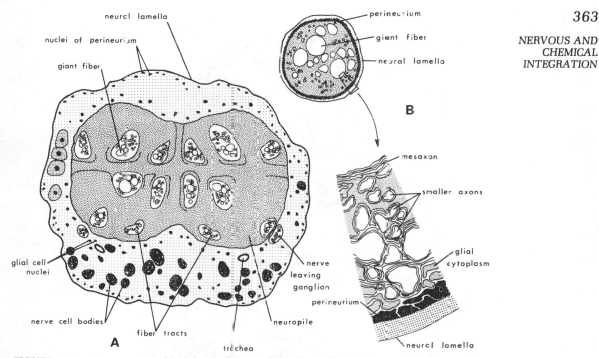

FIGURE 13.2. Cross sections through (A) abdominal ganglion and (B) interganglionic connective to show general structure. [A, after K. D. Roeder, 1963, *Nerve Cells and Insect Behaviour*. By permission of Harvard University Press. B, after J. E. Treherne and Y. Pichon, 1972, The insect blood–brain barrier, *Acv. Insect Physiol.* **9:**257–313. By permission of Academic Press Ltd., London, and the authors.]

are differentiated according to their position and function. The peripheral glial cells, which form the perineurium, are loaded with mitochondria and glycogen and serve as the energy store for the neurons. In addition, they secrete the neural lamella, a protective sheath which contains collagen fibrils and mucopolysaccharide. The lamella is freely permeable, enabling the perineural cells to accumulate nutrients from the hemolymph. The inner glial cells occur among the perikarya into which they extend fingerlike extensions of their cytoplasm, the trophospongium (Figure 13.3A). The function of these cells is to transport nutrients from perineural cells to the perikarya. Once in the perikarya, nutrients are transported to their site of use by cytoplasmic streaming. Wrapped tightly around each axon or groups of smaller axons are other glial (Schwann) cells (Figure 13.3B) These cells effectively isolate axons from the hemolymph in which they are bathed. This is essential in view of the ionic basis of impulse transmission and is especially important in insects whose hemolymph composition is highly variable.

Structurally, the nervous system may be divided into (1) the central nervous system and its peripheral nerves, and (2) the visceral nervous system.

2.1. Central Nervous System

The central nervous system arises during embryonic development as an ectodermal delamination on the ventral side. Each embryonic segment in-

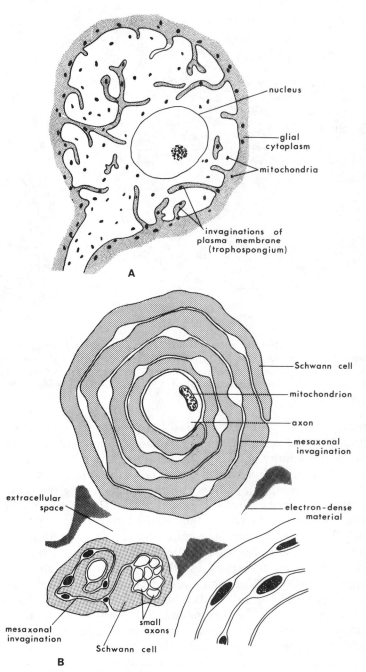

FIGURE 13.3. (A) Cell body of motor neuron showing trophospongium and (B) cross section through axons and surrounding Schwann cells. [A, after V. B. Wigglesworth, 1965, *The Principles of Insect Physiology*, 6th ed., Methuen and Co., London. By permission of the author. B, after J. E. Treherne and Y. Pichon, 1972, The insect blood–brain barrier, *Adv. Insect Physiol.* **9**:257–313. By permission of Academic Press Ltd., London, and the authors.]

cludes initially a pair of ganglia, though these soon fuse. In addition, varying degrees of anteroposterior fusion occur so that composite ganglia result. Thus, in a mature insect the central nervous system comprises the brain, subesophageal ganglion, and a variable number of ventral ganglia.

The brain (Figure 13.4A) is probably derived from the ganglia of three segments and forms the major association center of the nervous system. It includes the protocerebrum, deutocerebrum, and tritocerebrum. The protocerebrum, the largest and most complex region of the brain, contains both neural and endocrine (neurosecretory) elements. Anteriorly it forms the proximal part of the ocellar nerves (the only occasion on which the cell bodies of sensory neurons are located other than adjacent to the sense organ), and laterally is fused with the optic lobes. Within the protocerebrum is a pair of corpora pedunculata, the mushroom-shaped bodies, so-called because of their outline in cross section. Their size can be broadly correlated with the development of complex behavior patterns, and they are most highly developed in the social Hymenoptera. In worker ants, for example, they make up about one-fifth the volume of the brain. The median central body is another important associa-

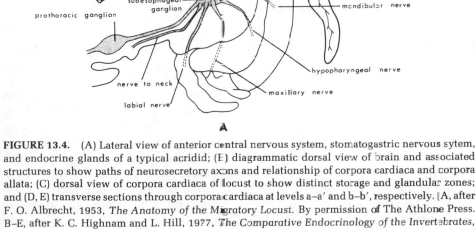

A

FIGURE 13.4. (A) Lateral view of anterior central nervous system, stomatogastric nervous sytem, and endocrine glands of a typical acridid; (B) diagrammatic dorsal view of brain and associated structures to show paths of neurosecretory axons and relationship of corpora cardiaca and corpora allata; (C) dorsal view of corpora cardiaca of locust to show distinct storage and glandular zones; and (D, E) transverse sections through corpora cardiaca at levels a–a' and b–b', respectively. [A, after F. O. Albrecht, 1953, *The Anatomy of the Migratory Locust.* By permission of The Athlone Press. B–E, after K. C. Highnam and L. Hill, 1977, *The Comparative Endocrinology of the Invertebrates,* 2nd ed. By permission of Edward Arnold Publishers Ltd.]

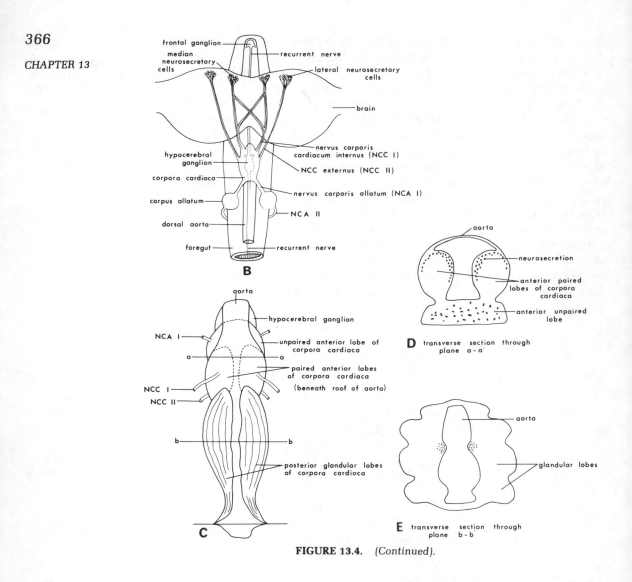

FIGURE 13.4. (Continued).

tion center, one function of which appears to be the coordination of segmental motor activities, for example, respiratory movements. Each optic lobe contains three neuropilar masses in which light stimuli are interpreted. The deutocerebrum is largely composed of the paired antennal lobes. These two neuropiles include both sensory and motor neurons and are responsible for initiating both responses to antennal stimuli and movements of the antennae. The tritocerebrum is a small region of the brain located beneath the deutocerebrum and comprises a pair of neuropiles which contain axons, both sensory and motor, leading to/from the frontal ganglion and labrum. As indicated in Figure 13.5, almost all paired neuropiles send tracts of axons to the opposite partner.

The subesophageal ganglion is also composite and includes the elements of the embryonic ganglia of the mandibular, maxillary, and labial segments. From this ganglion, nerves containing both sensory and motor axons run to the

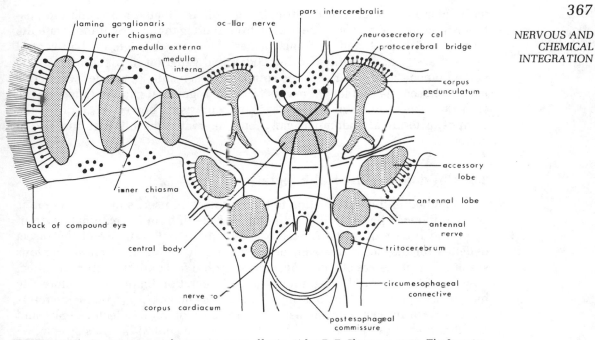

FIGURE 13.5. Cross section to show major areas of brain. [After R. F. Chapman, 1971, *The Insects: Structure and Function.* By permission of Elsevier North-Holland, Inc., and the author.]

mouthparts, salivary glands, and neck. The ganglion also appears to be the center for maintaining (though not initiating) locomotor activity.

In most insects the three segmental thoracic ganglia remain separate. Though details vary from species to species, each ganglion innervates the leg and flight muscles (direct and indirect), spiracles, and sense organs of the segment in which it is located.

The maximum number of abdominal ganglia is eight, seen in the adult thysanuran *Machilis* and larvae of many species, though even in these insects the terminal ganglion is composite, including the last four segmental ganglia of the embryonic stage. Varying degrees of fusion of the abdominal ganglia occur in different orders and sometimes there is fusion of the composite abdominal ganglion with the ganglia of the thorax to form a single thoracoabdominal ganglion. (Consult Chapters 5 to 10 for details of individual orders.)

2.2. Visceral Nervous System

According to Wigglesworth (1965) the visceral (sympathetic) nervous system includes three parts: the stomatogastric system, the unpaired ventral nerves, and the caudal sympathetic system. The stomatogastric system, shown partially in Figure 13.4, arises during embryogenesis as an invagination of the dorsal wall of the stomodeum. It includes the frontal ganglion, recurrent nerve which lies mediodorsally above the gut, hypocerebral ganglion, a pair of inner esophageal nerves, a pair of outer esophageal (gastric) nerves, each of which terminates in an ingluvial (ventricular) ganglion situated alongside the posterior foregut, and various fine nerves from these ganglia which innervate the foregut and midgut, and, in some species, the heart. A single median ventral nerve arises from each thoracic and abdominal ganglion in some insects. The nerve branches and innervates the spiracle on each side. In species where this nerve is absent, paired lateral nerves from the segmental ganglia innervate the spiracles. The caudal sympathetic system, comprising nerves arising from the composite terminal abdominal ganglion, innervates the hindgut and sexual organs.

2.3. Physiology of Neural Integration

As noted in the Introduction to this chapter, an insect's nervous system is constantly receiving stimuli of different kinds both from the external environment and from within its own body. The subsequent response of the insect depends on the net assessment of these stimuli within the central nervous system. The processes of receiving, assessing, and responding to stimuli collectively constitute neural integration. Neural integration includes, therefore, the biophysics of impulse transmission along axons and across synapses, the reflex pathways (in insects, intrasegmental) from sense organ to effector organ, and coordination of these segmental events within the central nervous system.

Impulse transmission along axons and across synapses appears to be essentially the same in insects as in other animals and will not be discussed here in detail. It is based on the initially dissimilar ionic concentrations on each side of the axonal membrane. In the "resting" (unstimulated) state there is an excess of sodium ions external to the membrane and an excess of potassium ions in the axonal cytoplasm. Because of the different rates at which sodium and potassium ions are able to diffuse across the membrane and the inability of large organic anions within the axon to move through the membrane, the inside is negative with respect to the outside to the extent of 60–80 mV. This is termed the resting potential. A stimulus of sufficient magnitude received by a neuron results in a momentary increase in the permeability of the membrane at the site of stimulus so that a massive influx of sodium ions occurs, and the inside becomes positive with respect to the outside. This is depolarization. Almost immediately, potassium ions flow out and, once again, the inside becomes negative with respect to the outside. This is the reverse polarization phase. Active transport of sodium and potassium ions then occurs during the refractory period, to restore the original ionic concentrations so that the neuron can be restimulated. The momentary change in polarization at one point on the axonal membrane induces changes in permeability in adjacent sections of the membrane. Thus, an impulse is, in effect, a wave of depolarization traveling along the membrane.

Transmission across a synapse, depending as it does on diffusion of molecules through fluid, is relatively slow and may take up about 25% of the total time for conduction of an impulse through a reflex arc. Rarely, when a synaptic gap is narrow (i.e., pre- and postsynaptic membranes are closely apposed), the ionic movements across the presynaptic membrane are sufficient to directly induce depolarization of the postsynaptic membrane (Huber, 1974).

Mostly, however, when an impulse reaches a synapse, it causes release of a chemical (a neurotransmitter) from membrane-bound vesicles. The chemical diffuses across the synapse and, in excitatory neurons, brings about depolarization of the postsynaptic membrane. According to Pichon (1974), the nature of the neurotransmitter substance(s) liberated at excitatory synapses has not been conclusively demonstrated, though the most likely candidate appears to be acetylcholine. Histochemical and biochemical studies indicate that levels of acetylcholine and cholinesterase are high in the insect nervous system. Pharmacologically, it has been shown that insect axons are highly sensitive to acetylcholine, which causes depolarization of the axonal membrane. Other substances present in the nervous system, whose possible neurotransmitter role cannot be discounted, include catecholamines (especially dopamine, but not epinephrine which is present in very small amounts), and indolamines such as 5-hydroxytryptamine (serotonin). Low concentrations of these amines applied to the heart, gut, reproductive tract, etc., have an excitatory effect, and it may be that these substances serve as neurotransmitters in the visceral nervous system.

Sometimes a single nerve impulse arriving at the presynaptic membrane does not stimulate the release of a sufficient amount of neurotransmitter. Thus, the magnitude of depolarization of the postsynaptic membrane is not large enough to initiate an impulse in the postsynaptic axon. If additional impulses reach the presynaptic membrane before the first depolarization has decayed, sufficient additional neurotransmitter may be released so that the minimum level for continued passage of the impulse (the "threshold" level) is exceeded. This additive effect of the presynaptic impulses is known as temporal summation. A second form of summation is spatial, which occurs at convergent synapses. Here, several sensory axons synapse with one internuncial neuron. A postsynaptic impulse is initiated only when impulses from a sufficient number of sensory axons arrive at the synapse simultaneously. Divergent synapses are also found where the presynaptic axon synapses with several postsynaptic neurons. In this arrangement the arrival of a single impulse at a synapse may be sufficient to initiate impulse transmission in, say one of the postsynaptic neurons. The arrival of additional impulses in quick succession will lead to the initiation of impulses in other postsynaptic neurons whose threshold levels are higher. Thus, synapses play an important role in selection of an appropriate response for a given stimulus.

Eventually, an impulse reaches the effector organ, most commonly muscle. Between the tip of the motor axon and the muscle cell membrane is a fluid-filled space, comparable to a synapse, called a neuromuscular junction. Again, to achieve depolarization of the muscle cell membrane and, ultimately, muscle contraction, a chemical released from the tip of the axon diffuses across the neuromuscular junction. In insects, this chemical is probably glutamate (Pichon, 1974).

Thus far, we have discussed only the physiology of stimulatory (excitatory) neurons. However, also important in neural integration are inhibitory neurons; that is, neurons whose synaptic neurotransmitter causes hyperpolarization of the postsynaptic or effector cell membrane. When inhibition occurs at a synapse, that is, within the central nervous system, it is known as central inhibition. Central inhibition is the prevention of the normal stimulatory output from the central nervous system and may arise spontaneously within the system or result from sensory input. For example, copulatory movements of the abdomen in the male mantis, which are regulated by a segmental reflex pathway located within the terminal abdominal ganglion, are normally inhibited by spontaneous impulses arising within the brain and passing down the ventral nerve cord. In the fly *Phormia* the stimulation of stretch receptors during feeding results in decreased sensitivity to taste due to central inhibition of the positive stimuli received by the brain from the tarsal chemoreceptors.

When inhibition of an effector organ occurs it is known as peripheral inhibition. One method of achieving such inhibition is the release, at the motor axon tip, of a chemical which causes hyperpolarization of the effector cell membrane. At the neuromuscular junction of insects, this chemical is probably γ-aminobutyric acid (Pichon, 1974), a substance which increases the resting potential of the muscle cell membrane.

The simplest unit of impulse conduction from a sense organ to an effector organ is the reflex arc, which involves the sequential transmission of information along a sensory neuron, at least one interneuron, and a motor neuron (Figure 13.6). As Wigglesworth (1965) points out, however, the situation, that is, what happens after a sense organ is stimulated, is far more complex, involving many pathways and responses by several effectors whose activities are coordinated by higher centers.

In insects reflex responses are segmental; that is, a stimulus received by a sense organ in a particular segment initiates a response which travels via an interneuron located in that segment's ganglion to an effector organ in the same segment. This is easily demonstrated by isolating individual segments. For example, in an isolated thoracic segment preparation of a grasshopper, touching the tarsus causes the leg to make a stepping movement. Of course, in an intact insect such a stimulus also leads to compensatory movements of other legs to maintain balance or to initiate walking, activities which are cooordinated via association centers in the subesophageal ganglion. Touching the tip of the isolated ovipositor in *Bombyx*, for example, initiates typical egg-laying

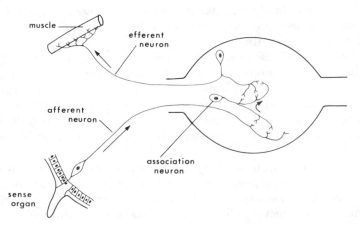

FIGURE 13.6. Diagrammatic representation of reflex arc. (After V. B. Wigglesworth, 1965, *The Principles of Insect Physiology*, 6th ed., Methuen and Co. By permission of the author.)

movements, provided that the terminal ganglion and its nerves are intact. In other words, each segmental ganglion possesses a good deal of reflex autonomy.

Nervous activity of the type described above, which occurs only after an appropriate stimulus is given, is said to be exogenous. However, an important component of nervous activity in insects is endogenous, that is, does not require sensory input but is based on neurons with intrinsic pacemakers. Such neurons possess specialized membrane regions which undergo periodic, spontaneous changes in excitability (permeability?) and where impulses are thereby initiated. A wide variety of motor responses are organized, in part, by endogenous activity. For example, ventilation movements of the abdomen are initiated by endogenous activity in individual ganglia. Even walking and stridulation are motor responses under partially endogenous control (Huber, 1974). An obvious question to ask, therefore, is "Why don't insects walk or stridulate continuously?" The answer to this question is that these and all other motor responses are "controlled" by higher centers, specifically the brain and/or subesophageal ganglion. By "control" we mean that the association centers assess all the information coming in via sensory neurons and, on this total assessment, determine the nature of the response. In addition, the centers coordinate and modify identical segmental activities, such as ventilation movements, so that they operate most efficiently under a given set of conditions.

Early evidence for the role of the brain and subesophageal ganglion as coordinating centers came from fairly crude experiments in which one or both centers were removed and the resultant behavior of an insect observed. More recent experiments involving localized destruction or stimulation of parts of these centers has confirmed and added to the general picture obtained by earlier authors. To illustrate the complexity of coordination and control of motor activity, let us take walking as an example. As noted above, some of the motor neurons to the leg muscles originating in each thoracic ganglion show endogenous activity. However, this activity may not be of sufficient magnitude that contraction of the muscles occurs. The activity of these neurons can be increased, however, by impulses emanating from the subesophageal ganglion, and the net result would be the initiation of leg movements. Walking requires the coordination of such movements and this is achieved through the involvement of sensory neurons associated with the proprioceptors on each leg. Information from these sense organs is sent to each thoracic ganglion and the movement of each leg is modified so that stepping becomes coordinated. The corpora pedunculata and central body are also involved in the regulation of walking. Impulses originating in the corpora pedunculata inhibit locomotor activity, presumably by decreasing the excitability of the subesophageal ganglion. Moreover, reciprocal inhibition may occur between the corpus pedunculatum on each side of the brain, and this is the basis of the turning response. In contrast, the central body appears to be an important excitatory system in locomotion, because its stimulation evokes fast running, jumping, and flying in some species. As yet, however, the interaction between these two cerebral association centers is not understood. Ultimately, of course, whether or not the insect walks and, if so, at what speed and in what direction, depends on the summed effects of the sensory stimuli it receives.

At the outset, insect behavior is dependent on the environmental stimuli received, though, as noted earlier, not all behavior patterns originate exogenously; many common patterns have a spontaneous, endogenous origin. Axons may be branched; synapses may be convergent or divergent; temporal or spatial summation of impulses may occur at synapses; neurons may be excitatory or inhibitory in their effects. Thus, an enormous number of potential routes are open to impulses generated by a given set of stimuli. The eventual routes taken and, therefore, the motor responses which follow, depend on the size, nature, and frequency of these stimuli.

2.4. Learning and Memory

The translation of sensory input in the motor response takes place within a matter of milliseconds and thereby fits well into the broad definition of "nervous control." However, another important aspect of neural physiology is learning, which, along with the related event, memory, may occupy time intervals measured in hours, days, or even years. Learning is the ability to associate one environmental condition with another; memory is the ability to store information gathered by sense organs. Within this broad definition of learning, several phenomena can be included. Habituation, perhaps the simplest form of learning, is adaptation (eventual failure to respond) of an organism to stimuli which are not significant to its well-being. For example, a cockroach normally shows a striking escape reaction when air is blown over the cerci. If, however, this treatment is continued for a period of time, the insect eventually no longer responds to it. Conditioning is learning to respond to a stimulus which initially has no effect. Related to this is trial-and-error learning where an animal learns to respond in a particular way to a stimulus, having initially attempted to respond in other ways for which acts it received a negative reaction. Latent learning is the ability to relate two or more environmental stimuli, though this does not confer an immediate benefit on the organism, for example, recognition of landmarks in relation to a nesting site. The digger wasp, *Philanthus*, makes a short orientation flight in the immediate vicinity of its nest prior to hunting and, on this basis, can recognize the landmarks and remember the nest location on its return an hour or more later. Circadian rhythms are also a form of learning. Many organisms perform particular activities at set times of the day, and these activities are initially triggered by a certain environmental stimulus, for example, the onset of darkness. Even if this stimulus is removed, say, by keeping an animal in constant light, the activity continues to be initiated at the normal time.

Though there is no doubt that insects are able both to learn and to memorize, the physiological/molecular basis for these events is not known. However, some generalized statements can be made. The mushroom-shaped bodies appear to be the center where complex behavior is learned, and, as noted earlier, these structures occupy a relatively greater proportion of the brain volume in insects such as the social Hymenoptera (particularly the worker caste), which exhibit the greatest learning capacity. However, there is evidence that simpler forms of learning can occur in other ganglia, for example, those of the thorax. Headless insects can learn to keep a leg in a certain position so as to

avoid repeated electric shocks. Associated with this is a decline in cholinesterase activity in the segmental ganglion, though whether this observation has
any significance in the learning process is yet to be determined.

3. Endocrine System

Study of the role of hormones in the control of physiological processes
(including behavior) is one of the most popular areas of research in entomology
at the present time, though our knowledge of insect hormones still lags far
behind that of vertebrate hormones, especially those of mammals. Insects, like
vertebrates, possess both epithelial endocrine glands (the corpora allata and
molt glands, derived during embryogenesis from groups of ectodermal cells
in the region of the maxillary pouches) and glandular nerve cells (neurosecretory cells), which are found in all the ganglia of the central nervous system and
release their products into the hemolymph via neurohemal organs.

The functions of hormones are many, and discussion of these is best
treated in conjunction with specific physiological systems. In this chapter,
therefore, only the structure of the glands, the nature of their products, and the
principles of neuroendocrine integration will be examined.

3.1. Neurosecretory Cells and Corpora Cardiaca

The best studied neurosecretory cells are the median neurosecretory cells
(mNSC) of the protocerebrum. They occur in two groups, one on each side of
the midline, and their axons pass down through the brain, crossing over on
route, and normally terminate in a pair of neurohemal organs, the corpora
cardiaca, where neurosecretion is stored (Figure 13.4). In some species, for
example, *Musca domestica*, some neurosecretory axons do not terminate in the
corpora cardiaca but pass through the glands to the corpora allata. In many
Hemiptera–Heteroptera, the axons bypass the corpora cardiaca and, instead,
terminate in the adjacent aorta wall (Dogra, 1967). In aphids, some neurosecretory axons transport their product directly to the target organ (Johnson,
1963). And the axons of the mNSC which produce bursicon terminate in the
fused thoracoabdominal ganglion of higher Diptera and in the last abdominal
ganglion of cockroaches and locusts (Highnam and Hill, 1977). The corpora
cardiaca are closely apposed to the dorsal aorta into which neurosecretion and
intrinsic products of the corpora cardiaca are released. Many of the mNSC
contain discrete granules which take up characteristic stains, for example,
paraldehyde fuchsin and Victoria blue and are known as "A-type" cells. Observation that the amount of stainable material in these cells is variable under
different physiological conditions, and other experimentation leave no doubt
as to the endocrine nature of the cells. It should be noted, however, that these
stains almost certainly do not stain hormones *per se* but their carrier proteins.
Thus, it is likely that the term "A-type cell" includes a mixture of cells producing different hormones. The observation that destruction of these cells affects a
wide range of physiological processes (see later chapters) supports this proposal. Other cells in the group (the so-called B, C, and D cells), which do not

contain stainable droplets, are also considered by many authors to be neurose-cretory, though it must be emphasized that there is no definite (experimental) proof for this. Yet other workers consider these cells to be A-type cells in nonstorage phases of their secretory cycle. Also in the protocerebrum are two groups of lateral neurosecretory cells (lNSC) whose axons do not decussate but travel to the corpus cardiacum of the same side. Though the lNSC contain stainable droplets, there is almost no experimental evidence for their presumed neurosecretory function.

The corpora cardiaca arise as invaginations of the foregut during em-bryogenesis at the same time as the stomatogastric nervous system and are, in fact, modified nerve ganglia. Indeed, they are referred to in older publications as postcerebral or pharyngeal ganglia. Though their main function is to store neurosecretion, many of their intrinsic cells also produce hormones. In some species, for example, the desert locust, the neurosecretory storage zone and glandular zone (zone of intrinsic cells) are distinct (Figure 13.4C–E); in others, the neurosecretory axons terminate among the intrinsic cells.

Neurosecretory cells (that is, cells which resemble histologically the A-type mNSC) are found in all the ventral ganglia, and their axons which contain stainable droplets can be traced to a series of segmental neurohemal organs, the perisympathetic organs adjacent to the unpaired ventral nerve. It should be appreciated that in only a very few cases has experimental evidence for their neurosecretory function been obtained.

Many functions have been ascribed to neurosecretory hormones and the intrinsic hormones produced by the corpora cardiaca, but for relatively few of these functions is there good experimental evidence. Products of the mNSC include thoracotropic hormone, which activates the molt glands (see Chapter 21); allatotropic hormone, which regulates the activity of the corpora allata (Chapters 19 and 21); bursicon, which is important in cuticular tanning (Chapter 11); and diuretic hormone, which affects osmoregulation (Chapter 18). In addi-tion, neurosecretion from the mNSC affects behavior, though in many cases this is certainly an indirect action, and is important in protein synthesis. The intrinsic cells of the corpora cardiaca produce hyperglycemic and adipokinetic hormones important in carbohydrate and lipid metabolism (Chapter 16) and hormones that stimulate heartbeat rate (Chapter 17), gut peristalsis, and with-ing movements of Malpighian tubules. It appears, however, that the mNSC may be involved in the elaboration of these materials because extracts of these cells exert similar, though less strong, effects on these processes. Neurosecre-tion from the subesophageal ganglion is, in cockroaches, synthesized and re-leased regularly and controls the circadian rhythm of locomotor activity. In female pupae* of *Bombyx*, two large neurosecretory cells in the subesophageal ganglion produce a diapause hormone which promotes the development of eggs which enter diapause (see Chapter 22). In *Rhodnius*, diuretic hormone is produced not by the cerebral neurosecretory cells but by the hindmost group of neurosecretory cells in the fused ganglion of the thoracic and first abdominal segments.

The biochemical nature of insect neurosecretions is poorly understood and, in some cases, conflicting reports have appeared. Japanese workers in the

---

*In many Lepidoptera, including *Bombyx*, egg development begins in the pupa.

late 1950s first reported on the nature of the thoracotropic hormone from pupae of the silkmoth, *Bombyx mori*. These workers identified the hormone as cholesterol, though several subsequent attempts by other scientists to confirm this failed. In 1961, two other Japanese scientists, working again with B. mori, reported that the thoracotropic hormone was proteinaceous in nature, an observation which is more in keeping with the nature of both other insect neurosecretory materials and neurosecretions from other animals. Bursicon, for example, appears to be a protein with a molecular weight of about 40,000. The diuretic, hyperglycemic, adipokinetic, and heart-accelerating factors are also polypeptidic in nature.

3.2. Corpora Allata

Typically the corpora allata are seen as a pair of spherical bodies lying one on each side of the gut, behind the brain (Figure 13.4A,B). However, in some species the glands may be fused in a middorsal position above the aorta, or each gland may fuse with the corpus cardiacum on the same side. In larvae of cyclorrhaph Diptera the corpora allata, corpora cardiaca, and molt glands fuse to form a composite structure, Weismann's ring, which surrounds the aorta. Each gland receives a nerve from the corpus cardiacum on its own side, though the axons which form this nerve are probably those of mNSC, and also a nerve from the subesophageal ganglion.

The corpora allata produce a hormone known variously as juvenile hormone, metamorphosis-inhibiting hormone, or neotenin, with reference to its function in juvenile insects (see Chapter 21), and gonadotropic hormone to indicate its function in adults (Chapter 19). Though it seems probable that, for a given species, the chemical nature of the hormone is identical in both juvenile and adult stages, there are some differences in its chemical structure between species. Juvenile hormone is a terpenoid compound (Figure 13.7A) and, to date, three naturally occurring forms, JH-I, JH-II, and JH-III, containing 18, 17, and 16 carbon atoms, respectively, have been identified. In addition, a large number of related compounds have been shown to exert juvenilizing and/or gonadotropic effects.

3.3. Molt Glands

The paired molt glands generally comprise two strips of tissue, frequently branched, which are interwoven among the tracheae, fat body, muscles, and

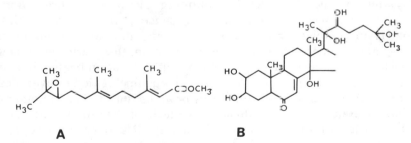

FIGURE 13.7. (A) Locust juvenile hormone (= C_{16}JH = JHIII) and (B) β-ecdysone.

connective tissue of the head and anterior thorax. In accord with their variable position, they have been described as prothoracic glands, ventral head glands, tentorial glands, pericardial glands, and peritracheal glands, though these structures are homologous. Except in primitive apterygotes, "solitary" locusts, and, apparently, worker and soldier termites, the glands are found only in juvenile insects and degenerate shortly after the molt to the adult. The molt glands show distinct cycles of activity correlated with new cuticle formation and ecdysis. Their product, ecdysone, initiates several important events in this regard (see Chapters 11 and 21). Like juvenile hormone, ecdysone exists in several slightly different forms. Basically, however, it is a steroid (Figure 13.7B) and has the same carbon skeleton as cholesterol, which is probably its precursor.

3.4. Other Supposed Endocrine Structures

A variety of structures in insects have been proposed as endocrine glands at one time or another.

The oenocytes, which become active early in the molt cycle and again at the onset of sexual maturity in adult females, were considered by some authors to produce hormones. However, it now seems certain that they are concerned with lipoprotein metabolism in conjunction with epicuticle production in the molt cycle (see Chapter 11) and eggshell formation during reproduction.

In contrast to those of vertebrates, the gonads of insects do not produce sex hormones which influence the development of secondary sexual characters. A large number of experiments in which insects were castrated could be cited to support this statement. The absence of secondary sexual characters in some parasitized insects was at one time considered to be evidence in support of sex hormones. However, differences between normal and parasitized forms are now known to be due to differences in the nutritional state of the insects and not to a specific effect on the gonads as was originally suggested. An apparent exception to the above generalization is found in the firefly *Lampyris noctiluca* (Coleoptera) where an androgenic hormone produced by the testes induces the development of male sexual characters. Thus, implantation of these organs into a female larva causes sex reversal. Ovarian tissue, however, does not produce a hormone for induction of femaleness.

A few recent reports indicate that the ovaries produce hormones which affect reproductive development. Adams and co-workers (see Adams *et al.*, 1975 for earlier references) have demonstrated that the maturing ovaries of the housefly, *Musca domestica*, produce an oostatic hormone which regulates the cyclic pattern of egg maturation by inhibiting release of neurosecretion from the median neurosecretory cells. In the absence of neurosecretion, juvenile hormone is not produced by the corpora allata and, therefore, development of a further egg batch does not occur. After oviposition, the ovaries no longer produce oostatic hormone, and a new cycle of egg maturation begins. In contrast, the antigonadotropin produced by cells in the pedicel region of ovarioles containing mature eggs in the bug *Rhodnius prolixus* does not act on other endocrine centers. Rather, it appears to act at the level of the follicle cells, blocking the action of juvenile hormone (see Chapter 19) (Huebner and Davey, 1973).

The ovary is also a source of ecdysone in the mosquito, *Aedes aegypti* (Hagedorn *et al.*, 1975) where it stimulates the synthesis of specific yolk proteins (vitellogenins) in the fat body (see Chapter 19). It is not known whether the ovary actually synthesizes the ecdysone or perhaps accumulates it from the hemolymph where it has been stored during the larval stage when it is being produced by the molt glands. Ecdysone has also been detected in adults of several other insect species, a significant observation in view of the known degeneration of the molt glands immediately after metamorphosis. Elucidation of the source and function of the hormone in these species will be especially interesting.

4. Pheromones

Pheromones are chemicals produced by one individual which induce a particular behavioral or developmental response in other individuals of the same species. Like hormones, they are produced in small quantities and serve as chemical messengers. Indeed, they are referred to in older literature as "ectohormones," a term which may take on renewed significance following the discovery that in termites juvenile hormone apparently serves also as a pheromone regulating caste differentiation (see Section 4.2). Pheromones may be volatile and therefore capable of being detected as an odor over considerable distances, or they may be nonvolatile, requiring actual physical contact among individuals for their dissemination. They may be highly specific, even to the extent that only a particular isomer of a substance induces the typical effect in a given species. (As a corollary, closely related species often utilize different isomeric forms of a given chemical.) Pheromones are released only under appropriate conditions, that is, in response to appropriate environmental stimuli, and examples of this are given below. Thus, whereas the neural and endocrine systems coordinate the behavior and/or development of an individual, pheromones regulate these processes within populations.

Pheromones may be arranged in rather broad, sometimes overlapping, categories based on their functions. There are sex pheromones, caste-regulating pheromones, aggregation pheromones, alarm pheromones, and trail-marking pheromones.

4.1. Sex Pheromones

In the term "sex pheromones" are included chemicals which (1) excite and/or attract members of the opposite sex (sex attractants), (2) accelerate or retard sexual maturation (in either the opposite and/or same sex), or (3) enhance fecundity and/or reduce receptivity in the female following their transfer during copulation. Sex attractants are typically volatile chemicals produced by either male or female members of a species, whose release and detection by the partner are essential prerequisites to successful courtship and mating. Male-attracting substances are produced by virgin females of species representing many insect orders, but especially Lepidoptera and Coleoptera (for lists, see Jacobson, 1965, 1974). Females release their pheromones only in response to a specific stimulus. Many species of moths begin "calling" (everting their pheromone-secreting glands and exuding the chemical) 1.5 to 2.0 hours before dawn. Other Lepidoptera are stimulated to release pheromone by the scent of the larval food plant. *Rhodnius prolixus* females release pheromone only after a blood meal, and pheromone production in *Periplaneta americana* is arrested when the ootheca is formed. It is likely, in species such as *R. prolixus* and *P. americana*, which have repeated cycles of oocyte development over a period of time, that pheromone production is under the control of the endocrine system, especially the corpora allata, and that the effects of stimuli such as food intake or presence of an ootheca are mediated via the neuroendocrine system.

Typically, the pheromone-producing glands of female Lepidoptera are eversible sacs located in the intersegmental membrane behind the seventh or eighth abdominal sternites. In the homopteran *Schizaphis borealis* the glands are probably on the hind tibiae; in *Periplaneta* the pheromone seems to be produced in the gut and is released from fecal pellets. In Coleoptera the glands are abdominal.

The components of some male attractants have been identified and, broadly speaking, in Lepidoptera appear to be long-chain, unsaturated alcohols (Figure 13.8A) or the corresponding acetate. Usually, only one isomeric form of a component is attractive in a given species. When an attractant contains two or more components, these may occur in species-specific proportions. As a result, under natural conditions males respond only to the pheromone produced by females of their species. Other factors which serve to prevent interspecific attraction between males and females of species producing similar phero-mones include differences in the time of day at which males are sensitive to pheromone, differences in geographic location of the species, differences in the time of year when the species are sexually mature, and the need for additional stimuli, perhaps auditory, visual, or chemical, before a male is attracted to a female. For example, the "initial" separation of two species may occur on the basis of their attraction to different host plants. Thus, in the vicinity of a host plant, the chances of being attracted to a conspecific female will be greatly increased. Nevertheless, other stimuli will be necessary to "confirm" the con-specific nature of the partner.

In many species (for lists, see Jacobson, 1965, 1974) it is males which produce a sex attractant. For example, the male cockroach, *Nauphoeta cinerea*, produces "seducin," which both attracts and pacifies the unmated female so

$$CH_3-(CH_2)_2-CH=CH-CH=CH-(CH_2)_8-CH_2OH$$

A Bombykol

$$CH_3-\overset{\overset{\displaystyle O}{\|}}{C}-(CH_2)_5-CH=CH-COOH$$

B 9-Oxodecenoic acid

$$CH_3-\overset{\overset{\displaystyle OH}{|}}{CH}-(CH_2)_5-CH=CH-COOH$$

C 9-Hydroxydecenoic acid

$$CH_3-(CH_2)_9-CH_3$$

D Undecane

$$CH_3-(CH_2)_4-COOH$$

E Caproic acid

FIGURE 13.8. Pheromones. (A) Bombykol, the sex attractant of the silk moth, *Bombyx mori*; (B, C) honeybee queen pheromones; (D) undecane, an alarm pheromone produced by many formicine ants; and (E) caproic acid, a major component of the trail-marking secretion of the termite *Zooter-mopsis nevadensis*.

that a connection can be established. The pheromone produced by the male boll weevil, *Anthonomus grandis*, attracts only females in summer; however, in the fall it serves as an aggregation pheromone, attracting both males and females. Male *Tenebrio molitor* produce both a female attractant and an antiaphrodisiac which inhibits the response of other males to the female's pheromone. The pheromone-producing glands of males are much more variable in their location than those of females. For example, pheromone is secreted from mandibular glands in ants, from glands in the thorax and abdomen of some beetles, from glands at the base of the fore wings in some Lepidoptera, from abdominal glands in other Lepidoptera and cockroaches, and from rectal glands in certain Diptera. As in females, the attractants produced by males are usually long-chain alcohols or their aldehydic derivatives.

Because of their specificity and effectiveness in very low concentrations, the use of sex attractants in pest control has great potential an aspect which will be more fully discussed in Chapter 24.

Sexual maturation-accelerating or -inhibiting pheromones are produced by many insects which tend to live in groups, including social species. These substances serve either to synchronize reproductive development so that both sexes mature together or, in social species, to inhibit the reproductive capability of almost all individuals so that their energy can be redirected to other functions. Mature male desert locusts, for example, produce a volatile pheromone which speeds up maturation of younger males and females. Conversely, immature migratory locusts of both sexes appear to secrete a pheromone which retards the maturation of slightly older individuals. Production of pheromones occurs in the epidermal cells over the entire body of the locust and is under the control of the corpora allata. Interestingly, the pheromones appear to operate by regulating the activity of the corpora allata in other individuals. The chemical nature of the pheromones is unknown.

Each colony of social insects has only one or very few reproducing individuals of each sex. Most members of the colony, the "workers," devote their effort to maintaining the colony and never mature sexually. Their failure to mature results from the production by reproductives of inhibitory pheromones. In a colony of honeybees the queen produces, in the mandibular glands, a material called "queen substance," 9-oxo-2-decenoic acid (Figure 13.8B), which is spread over the body during grooming to be later licked off by attendant workers. Mutual feeding among workers results in the dispersal of queen substance through the colony. The glands also produce a volatile pheromone, 9-hydroxy-2-decenoic acid (Figure 13.8C), which, together with queen substance, inhibits ovarian development in workers. As the queen ages and/or the number of individuals in a colony increases, the amount of pheromone available to each worker declines, and the latter's behavior changes. The workers construct queen cells in which the larvae are fed a special diet so that they develop into new queens. The first new queen to emerge kills the others and proceeds on a nuptial flight accompanied by drones. Apparently, queen substance produced by the new queen now serves as a sex attractant for the drones. When the new queen returns to the hive from the flight, part of the colony swarms, that is, the old queen, accompanied by workers, leaves the hive to found a new colony. Again, it is queen substance, along with 9-hydroxy-

2-decenoic acid, which enables workers to locate and congregate around the queen. This example shows how the function of a pheromone may vary according to the particular environmental circumstances in which the pheromone is released.

In lower termites, in contrast to the honeybee, there is a pair of equally important primary reproductives, the king and queen. When a colony is small, the development of additional reproductives is inhibited by means of sex-specific pheromones secreted by the royal pair. The pheromones apparently are released in the feces and transferred from termite to termite by trophallaxis. Lüscher (1972) has suggested that juvenile hormone may be the pheromone which inhibits reproductive development, though how such an arrangement could act in a sex-specific manner has not been satisfactorily explained. In addition, there are indications that the king secretes a pheromone which enhances the development of female supplementary reproductives in the absence of the queen.

In some species of mosquitoes and other Diptera, Orthoptera, Hemiptera, and Lepidoptera, the male, during mating, transfers to the female via the seminal fluid, chemicals which inhibit receptivity (willingness to mate subsequently) or enhance fecundity (by increasing the rate at which eggs mature and are laid) (see Gillott and Friedel, 1977). In contrast to other pheromones, the substances which are produced in the accessory reproductive glands or their analogue are proteinaceous in nature. Both receptivity-inhibiting substances and fecundity-enhancing substances signal that insemination has occurred. The former ensure that a larger number of virgin females will be inseminated; the latter, acting as they do by triggering oviposition, increase the probability of fertilized (viable) eggs being laid (in most species unfertilized eggs are inviable). Both types of pheromones serve, therefore, to increase the reproductive economy of the species.

4.2. Caste-Regulating Pheromones

As noted in the previous section, in social insects very few members of a colony ever mature sexually, their reproductive development being inhibited by pheromones so that these individuals (forming the worker caste) can perform other activities for the benefit of the colony as a whole. In addition to the worker caste, there exists in termites and ants a soldier caste. The number of soldiers present is proportional to the size of the colony, a feature which suggests that soldiers regulate the numbers in their ranks by production of a soldier-inhibiting pheromone. However, the situation is made more complicated by a positive influence on soldier production (presumably pheromonal) on the part of the reproductives.

The regulation of caste differentiation in social insects is a morphogenetic phenomenon, just as are the changes from larva to larva, larva to pupa, and pupa to adult. Such changes depend on the activity of the corpora allata (level of juvenile hormone in the hemolymph) for their manifestation. It is not surprising, therefore, to learn that the pheromones which regulate caste differentiation (including the development of reproductives) exert their effect, ultimately, via the corpora allata (see Chapter 21). In lower termites, for example, soldier formation can be induced in experimental colonies by administration of juvenile hormone (through feeding, topical application, or as vapor). Thus, the soldier-inhibiting pheromone may act by inhibiting the corpora allata or by competing with juvenile hormone at its site of action (Lüscher, 1972).

4.3. Aggregation Pheromones

Aggregation pheromones are produced by either one or both sexes in some species (especially Coleoptera) and serve to attract other individuals to a feeding or mating site. They have been best studied in the bark and ambrosia beetles (Scolytinae). Males of *Ips confusus*, for example, begin producing and releasing pheromone within a few hours of locating a new host tree. However, males do not become attractive until feces production has begun, indicating that the pheromone is produced by glandular cells in the gut. Work on other species indicates that the cells are located at the anterior end of the hindgut and begin production of pheromone in the presence of digested food particles. From what little work has been done in identifying the active principle, it appears that the "pheromone" is, in fact, a mixture of compounds, mostly cyclic alcohols or aldehydes, which act synergistically; that is, each compound alone exerts little or no effect, but combinations of two or more of the compounds induce a strong response.

4.4. Alarm Pheromones

As their name indicates, alarm pheromones warn members of a species of impending danger. They are produced by insects which live in groups, including social forms. In social insects only workers and female reproductives produce alarm pheromones. The response to a pheromone varies with its concentration. At low concentrations, members of a colony will orient themselves toward the source of the pheromone; at higher levels the response is one of greater activity, either rapid movement away from or an attack on the source. The site of production and chemical nature of alarm pheromones are variable (see Jacobson, 1974). In ants, for example, mandibular glands, Dufour's glands, supraanal glands, or venom glands secrete alarm pheromones which may be long-chain hydrocarbons, alcohols, aldehydes, ketones, or other organic molecules (Figure 13.8D). In many species a mixture of several of these compounds may be secreted. In contrast to sex pheromones, alarm substances (1) are typically cross-specific, that is, will induce an alarm reaction in several (related) species (2) must·be released in relatively large amounts to elicit a response, and (3) are effective over relatively short distances. Further, different (but related) compounds may elicit an alarm response in a species, whereas a species reacts usually only to one isomeric form of sex attractant (Jacobson, 1974).

4.5. Trail-Marking Pheromones

Trail-marking pheromones are an extremely important means of communicating information re the location and quantity of food available to a colony of social insects. In termites the sternal glands on the abdomen release pheromone each time the abdomen is pressed to the substrate. In ants trail-marking pheromones may be produced in the hindgut, Dufour's gland, ventral glands, or on the metathoracic legs, and are released as the abdomen or limbs make contact with the substrate. The pheromone-producing glands of the honeybee worker are distributed over the body and the pheromone is transferred during grooming to the feet. It is thereby deposited at the opening to the hive where it both attracts returning workers and stimulates them to enter.

Generally, trail-marking pheromones are laid by foraging workers as they return to the nest, though they may be used to direct other individuals during swarming or to sites where nest repairs are necessary. The trail may be a continuous line of secretion or a series of spots at intervals. Most trails are relatively "short-lived" and fade within a matter of minutes unless continously reinforced. Some ants, however, make trails which last for several days. When a source of food is good, more returning workers secrete pheromone, thereby establishing a "strong" trail to which more workers will be attracted.

The chemical nature of only a few trail pheromones is known. In termites they appear to be long-chain acids, alcohols or hydrocarbons (Figure 13.8E), and are cross-specific.

5. Environmental, Neural, and Endocrine Interaction

Only very rarely is a physiological event directly influenced by environmental stimuli. Temperature changes, through their effect on reaction rate, can, in poikilotherms at least, alter the rate at which an event is occurring. However, in almost all situations stimuli are first received by sensory structures which send information to the central nervous system to be dealt with. Some of these stimuli require an immediate response which, as noted earlier in this chapter, is achieved via the motor neuron-effector organ system. The information received by the central nervous system as a result of other stimuli, however, initiates longer-term responses mediated via the endocrine system. This information, then, must be first "translated" within the brain into hormonal language. The center for translation is the neurosecretory system. Depending on the stimulus and, presumably, the site of termination of the internuncial neurons, different neurosecretory cells will be stimulated to synthesize and/or release their product. This material may then act directly on target organs ("one-step" or "first-order" neurosecretory control) or exert a tropic effect on other endocrine glands ("two-step" or "second order" control).* A number of examples may be cited to illustrate these two levels of control. In *Rhodnius* feeding leads to stimulation of stretch receptors in the abdominal wall. Information from these receptors passes along the ventral nerve cord to the composite thoracic ganglion where the posterior neurosecretory cells are stimulated to

---

*In vertebrates "three-step" ("third-order") neurosecretory control is also found where hypothalamic neurosecretion exerts a tropic effect on specific cells of the anterior pituitary gland. The products of these cells are themselves "tropic" hormones which act on other epithelial endocrine glands.

release diuretic hormone. The latter facilitates rapid excretion of the excess water present in the blood meal. In juvenile *Rhodnius* similar information is also received by the brain whose neurosecretory cells liberate thoracotropic hormone to trigger a new molting cycle. A variety of stimuli may enhance the rate of egg production in the female insect. Copulation, oviposition, pheromones, photoperiod, and feeding are all stimuli which cause neurosecretory cells to release an allatotropic hormone. The latter activates the corpora allata whose secretion promotes egg development.

Figure 13.9 summarizes the variety of factors, environmental and experimental. which operate via the neurosecretory system to affect physiological events in insects.

6. Summary

The nervous system of insects, like that of other animals, comprises neurons (sensory, internuncial, and motor), synapses (minute fluid-filled cavities separating adjacent neurons), and protective cells which wrap around neurons to effectively isolate them from the hemolymph. Neurons are aggregated to form nerves and a series of segmental ganglia. In some cases the ganglia fuse to form a composite structure. The central nervous system includes the brain, subesophageal ganglion, and a variable number of thoracic and abdominal ganglia. The visceral nervous system includes the stomatogastric system, unpaired ventral nerves, and caudal sympathetic system. Impulse transmission along an axon and across a synapse is essentially like that of other animals. Individual reflex responses are segmental. though the overall coordination of motor responses is achieved largely within the brain and/or sub-

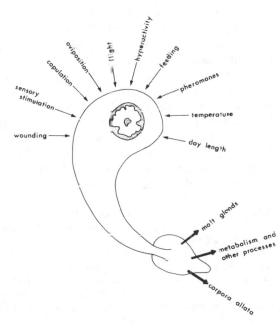

FIGURE 13.9. Diagrammatic summary of the variety of environmental and other factors which influence neurosecretory activity in insects. [After K. C. Highnam, 1965, Some aspects of neurosecretion in arthropods, *Zool. Jahrb., Abt. Allg. Zool. Physiol. Tiere* **71**:558–582. By permission of VEB Gustav Fischer Verlag.]

esophageal ganglion. Insects are able both to learn and to memorize, and, though the physiological/molecular bases for these abilities are not known, it is likely that they occur within the corpora pedunculata of the brain.

The endocrine system comprises neurosecretory cells, found in most ganglia, corpora cardiaca, which both store cerebral neurosecretion and synthesize intrinsic hormones, corpora allata, molt glands, and, possibly, ovaries. Neurosecretory hormones, which are probably polypeptidic, either act directly on target tissues or exert a tropic effect on other endocrine glands. In other words, the neurosecretory system "translates" neural information derived from environmental stimuli into hormonal messages which regulate a variety of physiological processes. Juvenile (=gonadotropic) hormone, produced by the corpora allata is a terpenoid compound containing 16, 17, or 18 carbon atoms. Molting hormone (ecdysone) is a steroid resembling cholesterol, from which it probably is biosynthesized.

Pheromones induce behavioral or developmental responses in other individuals of the same species. They may be volatile, exerting their influence over considerable distance, or nonvolatile when they are spread by physical contact (especially trophallaxis) among individuals. They include (1) sex pheromones, which modify reproductive behavior or development; (2) caste-regulating pheromones, which determine the proportion of different castes in colonies of social insects; (3) aggregation pheromones, which attract insects to a common feeding and/or mating site; (4) alarm pheromones, which warn of impending danger; and (5) trail-marking pheromones, which provide information on the location and quantity of food available to a colony of social insects.

7. Literature

Reviews on various aspects of nervous and chemical integration are numerous, and the following list represents but a small selection of relatively recent publications. General treatments of the structure and function of insect nervous systems (including neurosecretion) are given by Bullock and Horridge (1965) and Treherne (1974). The neural and endocrine bases of behavior are dealt with by Roeder (1967), Markl (1974), Lindauer and Stockhammer (1974), Barton Browne (1974), and Howse (1975). Learning and memory are examined by Alloway (1972). Neurosecretion is reviewed by Gabe (1966) and Goldsworthy and Mordue (1974), while general reviews of insect hormones are those of Novak (1975) and Highnam and Hill (1977). Jacobson (1965, 1974) has summarized existing knowledge of pheromones and provided a list of reviews which deal with specific types of pheromone.

Adams, T. S., Grugel, S., Ittycheriah, P. I., Olstad, G., and Caldwell, J. M., 1975, Interactions of the ring gland, ovaries and juvenile hormone with brain neurosecretory cells in *Musca domestica*, *J. Insect Physiol.* **21**:1027–1043.

Alloway, T. M., 1972, Learning and memory in insects, *Annu. Rev. Entomol.* **17**:43–56.

Barton Browne, L., 1974, *Experimental Analysis of Insect Behaviour*, Springer-Verlag, New York.

Bullock, T. H., and Horridge, G. A., 1965, *Structure and Function in the Nervous Systems of Invertebrates*, Freeman Publishing Co., San Francisco, Calif.

Dogra, G. S., 1967, Neurosecretory system of Heteroptera (Hemiptera) and role of the aorta as a neurohaemal organ, *Nature (London)* **215**:199–201.

Gabe, M., 1966, *Neurosecretion*, Pergamon Press, Oxford.

Gillott, C., and Friedel, T., 1977, Fecundity-enhancing and receptivity-inhibiting substances produced by male insects: A review, *Adv. Invert. Reprod.* **1**:199–218.

Goldsworthy, G. J., and Mordue, W., 1974, Neurosecretory hormones in insects, *J. Endocrinol.* **60**:529–558.

Hagedorn, H. H., O'Connor, J. D., Fuchs, M. S., Sage, B., Schlaeger, D. A., and Bohm, M. K., 1975, The ovary as a source of α-ecdysone in an adult mosquito, *Proc. Natl. Acad. Sci. USA* **72**:3255–3259.

Highnam, K. C., and Hill, L, 1977, *The Comparative Endocrinology of the Invertebrates*, 2nd ed., Edward Arnold, London.

Howse, P. E., 1975, Brain structure and behavior in insects, *Annu. Rev. Entomol.* **20**:359–379.

Hoyle, G., 1974, Neural control of skeletal muscle, in: *The Physiology of Insecta*, 2nd ed., Vol IV (M. Rockstein, ed.), Academic Press, New York.

Huber, F., 1974, Neural integration (central nervous system), in: *The Physiology of Insecta*, 2nd ed., Vol. IV (M. Rockstein, ed.), Academic Press, New York.

Huebner, E., and Davey, K. G., 1973, An antigonadotropin from the ovaries of the insect *Rhodnius prolixus Stal*, *Can. J. Zool.* **51**:113–120.

Jacobson, M., 1965, *Insect Sex Attractants*, Wiley, New York.

Jacobson, M., 1974, Insect pheromones, in: *The Physiology of Insecta*, 2nd ed., Vol. III (M. Rockstein, ed.), Academic Press, New York.

Johnson, B., 1963, A histological study of neurosecretion in aphids, *J. Insect Physiol.* **9**:727–740.

Lindauer, M., and Stockhammer, K. A., 1974, Social behavior and mutual communication, in: *The Physiology of Insecta*, 2nd ed., Vol. III (M. Rockstein, ed.), Academic Press, New York.

Lüscher, M., 1972, Environmental control of juvenile hormone (JH) secretion and caste differentiation in termites, *Gen. Comp. Endocrinol.*, *Suppl.* **3**:509–514

Markl, H., 1974, Insect behavior: Functions and mechanisms, in: *The Physiology of Insecta*, 2nd ed., Vol. III (M. Rockstein, ed.), Academic Press, New York.

Novak, V. J. A., 1975, *Insect Hormones*, 2nd Engl. ed., Chapman and Hall, London.

Pichon, Y., 1974, The pharmacology of the insect nervous system, in: *The Physiology of Insecta*, 2nd ed., Vol. IV (M. Rockstein, ed.), Academic Press, New York.

Roeder, K. D., 1967, *Nerve Cells and Insect Behavior*, 2nd ed., Harvard University Press, Cambridge, Mass.

Treherne, J. E. (ed.), 1974, *Insect Neurobiology*, Elsevier North-Holland Publishing Co., New York.

Wigglesworth, V. B., 1965, *The Principles of Insect Physiology*, 6th ed., Methuen, London.

14

Muscles and Locomotion

1. Introduction

The ability to move is a characteristic of living animals and facilitates distribution, food procurement, location of a mate or egg-laying site, and avoidance of unsuitable conditions. Insects, largely through their ability to fly when adult, are among the most mobile and widely distributed of animals. Development of this ability early in the evolution of the class has made the Insecta the most diverse and successful animal group (see Chapter 2). However, flight is only one method of locomotion employed by insects. Terrestrial species may walk, jump, or crawl over the substrate, or burrow within it. Aquatic forms can swim in a variety of ways or run on the water surface.

In their locomotory movements, insects conform to normal dynamic and mechanical principles. However, their generally small size and light weight have led to the development of some unique structural, physiological, and biochemical features in their locomotory systems.

2. Muscles

Essentially, the structure and contractile mechanism of insect muscle are comparable to those of vertebrate skeletal (cross-striated) muscle; that is, there are no muscles in insects of the smooth (nonstriated) type. Within muscle cells, the contractile elements actin and myosin have been identified, and Huxley's sliding filament theory of muscle contraction is assumed to apply. Though insect muscles are always cross-striated, there is considerable variation in their structure, biochemistry, and neural control, in accord with specific functions.

Because of their small size and the variable composition of the hemolymph of insects the neuromuscular system has some unique features (Hoyle, 1974). Being small means that an insect has a limited space for muscles which are, accordingly, reduced in size. Though this is achieved to some extent by a decrease in the size of individual cells (fibers), the principle change has been a decline in the number of fibers per muscle such that some insect muscles comprise only one or two cells. Thus, to achieve a graded muscle contraction, each fiber must be capable of a variable response, in contrast to the vertebrate situation where graded muscle responses result in part from stimulation of a variable number of fibers. Equally, the volume of nervous tissue is strictly limited, so that there are few motor neurons for the control of muscle contraction. The hemolymph surrounding muscles may contain high concentrations of ions (especially divalent ions such as Mg^{2+}) (see also Chapter 17) that could inhibit impulse transmission at the synapse and neuromuscular junction. That this does not occur is due to the evolution of a myelin sheath which covers ganglia, nerves, and neuromuscular junctions.

2.1. Structure

Insect muscles can be arranged in two categories: (1) skeletal muscles whose function is to move one part of the skeleton in relation to another, the two parts being separated by a joint of some kind, and (2) visceral muscles, which form layers of tissue enveloping internal organs such as the heart, gut, and reproductive tract.

Attachment of a muscle to the integument must take into account the fact that periodically the remains of the old cuticle are shed; therefore, an insertion must be able to break and reform easily. As Figure 14.1 indicates, a muscle terminates at the basement membrane lying beneath the epidermis. The muscle cells and epidermal cells interdigitate, increasing the surface area for attachment by about 10 times, and desmosomes occur at intervals, replacing the basement membrane. Attachment of a muscle cell to the rigid cuticle is effected

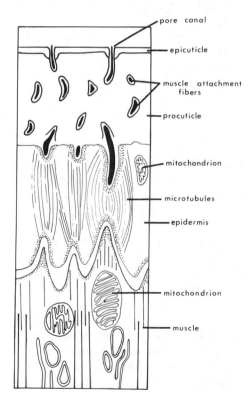

FIGURE 14.1. Muscle insertion. [After A. C. Neville, 1975, *Biology of the Arthropod Cuticle*. By permission of Springer-Verlag New York, Inc.]

by large numbers of parallel microtubules (called "tonofibrillae" by earlier authors). Distally, the epidermal cell membrane is invaginated, forming numbers of conical hemidesmosomes on which the microtubules terminate. Running distad from each hemidesmosome is one, rarely two, muscle attachment fibers (=tonofibrils of earlier authors). Each fiber passes along a pore canal to the cuticulin layer of the epicuticle to which they are attached by a special cement. As the cuticulin layer is the first one formed during production of a new cuticle (see Chapter 11), attachment of newly formed fibers can readily occur. Until the actual molt, however, these are continuous with the old fibers and, therefore, normal muscle contraction is possible (Neville, 1975).

Muscles comprise a variable number of elongate, multinucleated cells (fibers) (not to be confused with the muscle attachment fibers mentioned above) which may extend along the length of a muscle. A muscle is arranged usually in units of 10 to 20 fibers, each unit being separated from the others by a tracheolated membrane. Each unit has a separate nerve supply. The cytoplasm (sarcoplasm) of each fiber contains a variable number of mitochondria (sarcosomes). Even at the light microscope level, the transversely striated nature of muscles is visible. Higher magnification reveals that each fiber contains a large number of myofibrils (=fibrillae=sarcostyles) lying parallel in the sarcoplasm and extending the length of the cell. Each myofibril comprises the contractile filaments, made up primarily of two proteins, actin and myosin. The thicker myosin filaments are surrounded by the thinner but more numerous actin filaments. Filaments of each myofibril within a cell tend to be aligned, and it is this which creates the striated appearance (alternating light and dark bands) of the cell. The dark bands (A bands) correspond to regions where the actin and myosin overlap, whereas the lightly stained bands indicate regions where there is only actin (I bands) or myosin (H bands) (Figure 14.2). Electron microscopy has revealed in addition to these bands a number of thin transverse structures in the muscle fiber. Each of these Z lines (discs) runs across the fiber in the center of the I bands, separating individual contractile segments called sarcomeres. Attached to each side of the Z line are the actin filaments, which in contracted muscle are connected to the myosin filaments by means of cross bridges present at each end of the myosin. Periodically, the plasma membrane (sarcolemma) of the muscle fiber is deeply invaginated and forms the so-called T system (transverse system).

Though the above description is applicable to all the muscles of insects, different types of muscles can be distinguished, primarily on the basis of the arrangement of myofibrils, mitochondria, and nuclei; the degree of separation of the myofibrils; the degree of development of the sarcoplasmic reticulum; and the number of actins surrounding each myosin (Figure 14.3). These include tubular (lamellar) close-packed, and fibrillar muscles, all of which are skeletal, and visceral muscles. Leg and segmental muscles of many adult insects and the flight muscles of primitive fliers, such as Odonata and Dictyoptera, are of the tubular type, in which the flattened (lamellate) myofibrils are arranged radially around the central sarcoplasm. The nuclei are distributed within the core of sarcoplasm and the slablike mitochondria are interspersed between the myofibrils. The body musculature of apterygotes and some larval pterygotes, the leg muscles of some adult pterygotes, and the flight muscles of Orthoptera and

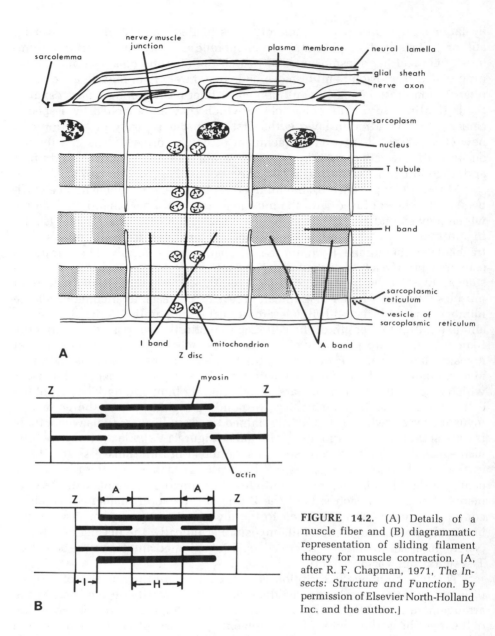

FIGURE 14.2. (A) Details of a muscle fiber and (B) diagrammatic representation of sliding filament theory for muscle contraction. [A, after R. F. Chapman, 1971, *The Insects: Structure and Function*. By permission of Elsevier North-Holland Inc. and the author.]

Lepidoptera are of the close-packed type. Here the myofibrils and mitochondria are concentrated in the center of the fiber and the nuclei are arranged peripherally. In close-packed flight muscles, the fibers are considerably larger than those of tubular flight muscles. In addition, tracheoles deeply indent the fiber, whereas in tubular muscles tracheoles simply lie alongside each fiber. It should be appreciated that the tracheoles do not actually penetrate the muscle cell membrane; that is, they are extracellular. In most insects the indirect muscles, which provide the power for flight, are nearly always fibrillar, so-called because individual fibrils are characteristically very large and, together with the

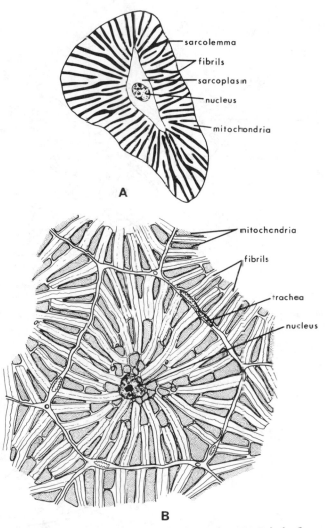

FIGURE 14.3. Transverse sections of insect skeletal muscles. (A) Tubular leg muscle of *Vespa* (Hymenoptera), (B) tubular flight muscle of *Enallagma* [Odonata], (C) close-packed flight muscle of a butterfly, and (D) fibrillar flight muscle of *Tenebrio* (Coleoptera . (Not to same scale.) [A, after H. E. Jordan, 1920, Studies on striped muscle structure. VI, *Am. J. Anat.* **27**:1–66. By permission of The Wistar Press. B,C, redrawn from electron micrographs in D. S Smith, 1965, The flight muscles of insects, *Scientific American*, June 1965, W. H. Freeman and Co. By permission of the author. D, redrawn from an electron micrograph in D. S. Smith. 1961, The structure of insect fibrillar muscles. A study made with special reference to the membrane systems of the fiber, *J. Biophys. Biochem. Cytol.* **10**:123–158. By permission of the Rockefeller Institute Press and the author.]

massive mitochondria, occupy almost all the volume of the fiber. Very little sarcoplasm is present, and the nuclei are squeezed randomly between the fibrils. Because of their size, there are often only a few fibrils per cell, and these are frequently quite isolated from each other by the massively indented and intertwining system of tracheoles. The presence of large quantities of cytochromes in the mitochondria gives these muscles a characteristic pink or yel-

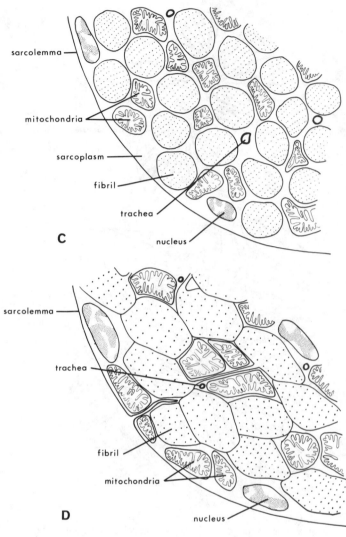

sarcolemma

mitochondria

sarcoplasm

fibril

trachea

C

nucleus

sarcolemma

trachea

fibril

mitochondria

D

nucleus

FIGURE 14.3. (*Continued*).

low color. It should be apparent even from this brief description that fibrillar muscles are designed to facilitate a high rate of aerobic respiration in connection with the energetics of flight. Visceral muscles differ from skeletal muscles in several ways. The cells comprising them are uninucleate, may branch, and are joined to adjacent cells by septate desmosomes. Their contractile elements are not arranged in fibrils and contain a larger proportion of actin to myosin. Nevertheless, the visceral muscles are striated (sometimes only weakly), and their method of contraction is apparently identical to that of skeletal muscles.

All skeletal muscles and many visceral muscles are innervated. The skeletal muscles always receive nerves from the central nervous system, whereas the visceral muscles are innervated from either the stomatogastric or the central nervous system. Within a particular muscle unit, each fiber may be innervated by one, two, or three functionally distinct axons. One of these is always excitatory; where two occur (the commonest arrangement), they are usually both excitatory ["fast" and "slow" axons (see Chapter 13)] but may be a "slow" excitatory axon plus an inhibitory axon; in some cases all three types of axon occur. This arrangment, known as polyneuronal innervation, facilitates a variable response on the part of a muscle (see Section 2.2). Each axon, regardless of its function, is much branched and, in contrast to the situation in vertebrate muscle, there are several motor neuron endings from each axon on each muscle fiber (multiterminal innervation) (Figure 14.4).

2.2. Physiology

Like those of vertebrates, insect muscles appear to contract according to the "sliding filament theory." The arrival of a suitable excitatory nerve impulse at a neuromuscular junction causes depolarization of the adjacent sarcolemma. A wave of depolarization spreads over the fiber and into the interior of the cell via the T system. Depolarization of the T system membranes induces a momentary increase in the permeability of the adjacent sarcoplasmic reticulum, so that calcium ions, stored in vesicles of the reticulum, are released into the sarcoplasm surrounding the myofibrils. The precise role of the calcium ions is not yet known but they activate cross-bridge formation between the actin and myosin, enabling the filaments to slide over each other so that the distance between adjacent Z lines is decreased (Figure 14.2B). The net effect is for the muscle to contract. Energy derived from the hydrolysis of adenosine triphosphate (ATP) is required for contraction, though its precise function is unknown. It may be used in breaking the cross-bridges, or for the active transport of the calcium ions back into the vesicles, or for both these processes. In addition to sliding over each other, both the actin and myosin filaments may

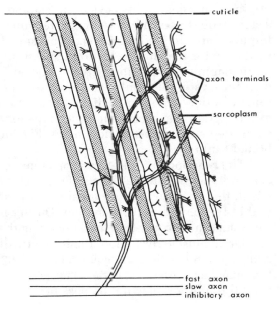

FIGURE 14.4. Polyneuronal and multiterminal innervation of an insect muscle. [After G. Hoyle, 1974, Neural control of skeletal muscle, in: *The Physiology of Insecta*, 2nd ed., Vol. IV (M. Rockstein, ed.). By permission of Academic Press, Inc., and the author.]

shorten (by coiling), thus enabling an even greater degree of contraction to occur.

Extension (relaxation) of a muscle may result simply from the opposing elasticity of the cuticle to which the muscle is attached. More commonly, muscles occur in pairs, each member of the pair working antagonistically to the other; that is, as one muscle is stimulated to contract, its partner (unstimulated) is stretched. Normally, the previously unstimulated muscle is stimulated to begin contraction while active contraction of the partner is still occurring (cocontraction). This is thought to bring about dampening of contraction, perhaps thereby preventing damage to a vigorously contracting muscle. Also, in slow movements, it provides an insect with a means of precisely controlling such movements (Hoyle, 1974). Muscle antagonism is achieved by central inhibition, that is, at the level of interneurons within the central nervous system (see Chapter 13). Thus, for a given stimulus, the passage of impulses along an axon to one muscle of the pair will be permitted, and hence that muscle will contract. However, passage of impulses to the partner is inhibited and the muscle will be passively stretched. It should be emphasized that in this arrangement the axon to each muscle is excitatory. In slow walking movements, for example, alternating stimulation of each muscle is quite distinct. At higher speeds, however, this reciprocal inhibition breaks down, and one of the muscles remains permanently in a mildly contracted state, serving as an "elastic restoring element" (Hoyle, 1974). The other muscle continues to be alternately stimulated and thus provides the driving power for the activity.

As noted earlier, commonly muscles receive two excitatory axons, one "slow," the other "fast." These terms are somewhat misleading for they do not indicate the speed at which impulses travel along the axons, but rather the speed at which a significant contraction can be observed in the muscle. Thus, an impulse traveling along a fast axon induces a strong contraction of the "all or nothing" type; that is, a further contraction cannot be initiated until the original ionic conditions have been restored. In contrast, a single impulse from a slow axon causes only a weak contraction in the muscle. However, additional impulses arriving in quick succession are additive in their effect (summation) so that, with the slow axon arrangement a graded response is possible for a particular muscle, despite the relatively few fibers it may contain. Muscles with dual innervation use only the slow axon for most requirements; the fast axon functions only when immediate and/or massive contraction is necessary. For example, the extensor tibia muscle of the hindleg of a grasshopper is ordinarily controlled solely via the slow axon. For jumping, however, the fast axon is brought into play.

The function of inhibitory axons remains questionable. Electrophysiological work has shown that in normal activity the inhibitory axon is electrically silent, that is, shows no electrical activity, and is clearly being inhibited from within the central nervous system. During periods of great activity, impulses can sometimes be observed passing along the axon, and here it may be used to accelerate muscle relaxation, though normally the use of antagonistic muscles and central inhibition is adequate. Hoyle (1974) suggests that peripheral inhibition may be necessary at certain stages in the life cycle, such as molting, when central inhibition may not be possible.

3.1. Movement on or through a Substrate

3.2.1. Walking

Insects can walk at almost imperceptibly low speed (witness a mantis stalking its prey) or run at seemingly very high rates (try to catch a cockroach). The latter is, however, a wrong impression created by the smallness of the organism, the rate at which its legs move, and the rate at which it can change direction. Ants, "scurrying" about on a hot summer's day are traveling only about 1.5 km/hr, and the elusive cockroach has a top speed of just under 5 km/hr (Hughes and Mill, 1974).

Nevertheless, an insect leg is structurally well-adapted for locomotion. Like the limbs of other actively moving animals, it tapers toward the distal end, which is light and easily lifted. Its tarsal segments are equipped with claws or pulvilli that provide the necessary friction between the limb and substrate. A leg comprises four main segments (see Chapter 3) which articulate with each other and with the body. The coxa articulates proximally with the thorax, usually by means of a dicondylic joint and distally with the fused trochanter and femur, also via a dicondylic joint. Dicondylic joints permit movement in a single plane. However, the two joints are set at right angles to each other and, therefore, the tip of a leg can move in three dimensions.

The muscles which move a leg are both extrinsic (having one end inserted on the wall of the thorax) and intrinsic (having both ends inserted within the leg) (Figure 14.5). The majority of extrinsic muscles move the coxa, rarely the fused trochantofemoral segment, whereas the paired intrinsic muscles move leg segments in relation to each other. Some of the extrinsic muscles have a dual function, serving to bring about both leg and wing movements. Typically, the leg muscles include (1) the coxal promotor and its antagonist, the coxal remotor, which run from the tergum to the anterior and posterior edges, respectively, of the coxa; contraction of the coxal promotor causes the coxa to twist forward, thereby effecting protraction (a forward swing) of the entire leg; (2) the coxal adductor and abductor (attached to the sternum and pleuron, respectively), which move the coxa toward or away from the body; (3) anterior and posterior coxal rotators, which arise on the sternum and assist in raising and moving the leg forward or backward; and (4) an extensor (levator) and flexor (depressor) muscle in each leg segment, which serve to increase and decrease, respectively, the angle between adjacent segments. It should be noted that the muscles which move a particular segment are actually located in the next more proximal segment. For example, the tibial extensor and flexor muscles, which alter the angle between the femur and tibia, are located within the femur and are attached by short tendons inserted at the head of the tibia.

It is the coordinated actions of the extrinsic and intrinsic muscles which move a leg and propel an insect forward. In considering how propulsion is achieved, it must be remembered also that another important function of a leg is to support the body, that is, to keep it off the ground. In the latter situation, a leg may be considered as a single-segmented structure—a rigid strut. If the strut

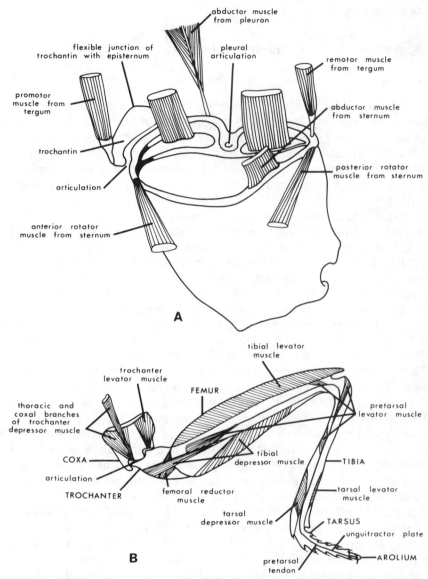

FIGURE 14.5. (A) Musculature of coxa, (B) segmental musculature of leg, and (C) musculature of hindleg of grasshopper. [A, C, from R. E. Snodgrass, *Principles of Insect Morphology*. Copyright 1935 by McGraw-Hill, Inc. Used with permission of McGraw-Hill Book Company. B, reproduced by permission of the Smithsonian Institution Press from *Smithsonian Miscellaneous Collections*, Volume 80, Morphology and mechanism of the insect thorax, Number 1, June 25, 1927, 108 pages, by R. E. Snodgrass: Figure 39, page 89. Washington, D.C., 1928, Smithsonian Institution.]

is vertical, the force along its length (axial force) will be solely supporting and will have no propulsive component. If the strut is inclined, the axial force can be resolved into two components, a vertical supportive force and a horizontal propulsive force. Because the leg protrudes laterally from the body, the horizontal force can be further resolved into a transverse force pushing the insect

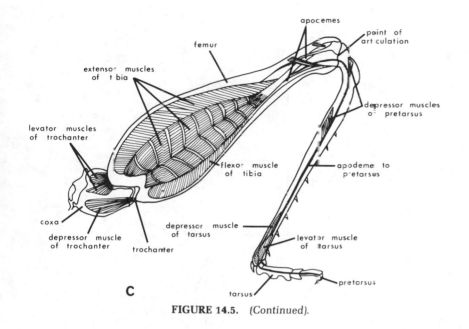

FIGURE 14.5. (Continued).

sideways and a longitudinal force which causes backward or forward motion. The relative sizes of these horizontal forces depends on (1) which leg is being considered and (2) the position of that leg. Figure 14.6 indicates the size of these forces for each leg at its two extreme positions. It will be apparent that in almost all its positions the foreleg will inhibit forward movement, whereas the mid- and hindlegs always promote forward movement. In equilibrium, that is, when an insect is standing still, the forces will be equal and opposite. Movement of an insect's body will occur only if the center of gravity of the body falls. This occurs when the forces become imbalanced, for example, by changing the position of a foreleg so that its retarding effect no longer is equal to the promoting effect of the other legs, whereupon the insect topples forward (Hughes and Mill, 1974).

Also important from the point of view of locomotion is the leg's ability to function as a lever, that is, a solid bar which rotates about a fulcrum and on which work can be done. The fulcrum is the coxothoracic joint and the work is done by the large, extrinsic muscles. Because of the large angle through which it can rotate and because of its angle to the body, the foreleg is most important as a lever. In contrast, the mid- and hindlegs, which each rotate through only a small angle, exert only a slight lever effect and serve primarily as struts (in the fully extended, rigid position). For the foreleg in its fully protracted position, contraction of the retractor muscle (i.e., the lever effect) will be sufficient to overcome the opposing retarding (strut) effect and, provided that the frictional forces between the ground and tarsi are sufficient, the body will be moved forward.

However, the largest component of the propulsive force is derived as a result of the legs' ability to flex and extend by virtue of their jointed nature. Flexure (a decrease in the angle between adjacent leg segments) will raise the

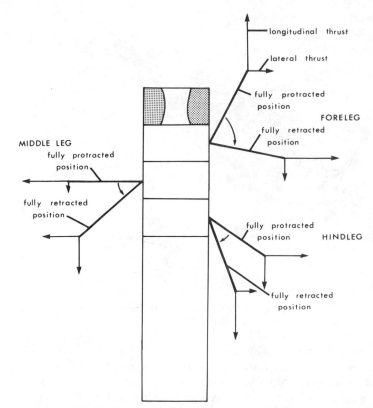

FIGURE 14.6. Magnitude of the longitudinal and lateral forces due to the strut effect for each leg in its extreme position. [After G. M. Hughes, 1952, The coordination of insect movements. I. The walking movements of insects, *J. Exp Biol.* **29**:267–284. By permission of Cambridge University Press, New York.]

leg off the ground so that it can be moved forward without the need to over-come frictional forces between it and the ground. In the case of the foreleg, flexure first will remove, by lifting the leg from the ground, the retarding effect due to its action as a strut and, second, when the leg is replaced on the sub-strate, will cause the body to be pulled forward. Flexure of the foreleg con-tinues until the leg is perpendicular to the body, at which point extension begins so that now the body is pushed forward. For the mid- and hindlegs, flexure serves to bring the legs into a new forward position. Extension, as in the case of the foreleg, will push the body forward. Because the hindleg is usually the largest of the three, it exerts the greatest propulsive force.

As noted above, the horizontal axial force along each leg has a transverse as well as a longitudinal component. Thus as an insect moves, its body zigzags slightly from side to side, the transverse forces exerted by the fore- and hindlegs of one side being balanced by an opposite force exerted by the middle leg of the opposite side in the normal rhythm of leg movements.

Rhythms of Leg Movements.   Most insects use all six legs during normal walking. Other species habitually employ only the two anterior or the two posterior pairs of legs but may use all six legs at higher speeds. In all instances, however, the legs are lifted in an orderly sequence (though this may vary with the speed of the insect), and there are always at least three points of contact with the substrate forming a "triangle of support" for the body. (In some species which employ two pairs of legs, the tip of the abdomen may serve as a point of support.) Two other generalizations that may be made are (1) no leg is lifted until the leg behind has taken up a supporting position, and (2) the legs of a segment alternate in their movements.

In the typical hexapodal gait at low speed, only one leg at a time is raised off the gound, so that the stepping sequence is R3, R2, R1, L3, L2, L1 (where R and L are right and left legs, respectively, and 1, 2, and 3 indicate the fore-, mid-, and hindlegs, respectively). With increase in speed, overlap occurs between both sides so that the sequence first becomes R3 L1, R2, R1 L3, L2, etc., and, then, R3 R1 L2, R2 L3 L1, etc., that is, a true alternating tripod gait.

The orthopteran Rhipipteryx has a quadrupedal gait, using only the anterior two pairs of legs and using the tip of the abdomen as a support. Its stepping sequence is R1 L2, R2 L1, etc. The mantis is likewise quadrupedal at low speed, using the posterior two pairs of legs (sequence R3 L2, R2 L3, etc.). At high speed, the forelegs are brought into action though the insect remains effectively quadrupedal (sequence L1 R3, L3 R2, L2 R1, etc.).

Coordination of the movements both among segments of the same leg and among different legs requires a high level of neural activity, both sensory and motor. Unfortunately, this is an area which, to date, has been little studied. From experiments on insects turned upside down, that is, resting on their backs, when for a short time at least the stepping sequence is continued, it is clear that the stepping movements are coordinated by endogenous activity within the central nervous system. However, for continued coordinated responses to occur and for adjustment of the basic rhythm, the legs must receive sensory input from proprioceptors located on and in them.

3.1.2. Jumping

Jumping is especially well developed in grasshoppers, fleas, flea beetles, click beetles, and Collembola. In the first three mentioned groups, jumping involves the hindlegs, which, like those of other jumping animals, are elongate and capable of great extension. Their length ensures that the limbs are in contact with the substrate for a long time during takeoff. Extension is achieved as the initially acute angle between the femur and tibia is increased to more than 90° by the time the tarsi leave the substrate. The length and extension together enable sufficient thrust to be developed that the insect can jump heights and distances many times its body length. For example, a fifth-instar locust (length about 4 cm) may "high jump" 30 cm and, concurrently, "long jump" 70 cm.

In Orthoptera the power for jumping is developed by the large extensor tibiae muscle in the femur (Figure 14.5C). The muscle is arranged in two masses of tissue which arise on the femur wall and are inserted obliquely on a long flat apodeme attached to the upper end of the tibia. The resultant herringbone arrangement increases the effective cross-sectional area of muscle attached to the apodeme, thereby increasing the power which the muscle develops. Because the apodeme is attached to the upper end of the tibia, slight contraction of the muscle will cause a relatively enormous movement at the tarsus (ratio of movements 60 : 1 when the tibiofemoral joint is tightly flexed).

It may be calculated that for the locust to achieve the maximum thrust for takeoff, the body must be accelerated at about 1.5×10^4 cm/sec^2 over a time span of 20 msec. The force exerted by each extensor muscle is about 5×10^5 dyne (=500 g wt) for an insect weighing 3 g (Alexander, 1968, cited from Hughes and Mill, 1974). To withstand this force, the apodeme must have a strength which approaches that of moderate steel. The extremely short time period over which this acceleration is developed makes it unlikely that jumping occurs as a direct result of muscle contraction. Indeed, Heitler (1974) has shown that the initial energy of muscle contraction is stored as elastic energy as a result of a cuticular locking device which holds the flexor tendon in opposition to the force developed within the extensor muscle. At a critical level, the tendon is released, allowing the tibia to rotate rapidly backward.

Likewise, in fleas, the energy of muscle contraction is first stored as elastic energy. Prior to jumping, the flea contracts various extrinsic muscles of the metathorax which are inserted via a tendon on the fused trochantofemoral segment. This serves to draw the leg closer to the body, compressing a pad of resilin and causing the pleural and coxal walls to bend. At a certain point of contraction, the thoracic catches (pegs of cuticle) slip into notches on the sternum, thereby "cocking the system." The jump is initiated when other, laterally inserted muscles contract to pull the catches out of the notches, thus allowing the stored energy to be rapidly released (Rothschild *et al.*, 1972).

3.1.3. Crawling and Burrowing

Many endopterygote larvae employ the thoracic legs for locomotion in the manner typical of adult insects; that is, they step with the legs in a specified sequence, the legs of a given segment alternating with each other. Usually, however, changes in body shape, achieved by synchronized contraction/relaxation of specific body muscles, are used for locomotion in soft-bodied larvae. In this method the legs, together with various accessory locomotory appendages, for example, the abdominal prolegs of caterpillars, are used solely as friction points between the body and substrate. Apodous larvae depend solely on peristalsis of the body wall for locomotion.

Where changes in body shape are used for locomotion the body fluids act as a hydrostatic skeleton. In other words, the insect employs the principle of incompressibility of liquids, so that contraction of muscles in one part of the body, leading to a decrease in volume, will require a relaxation of muscles and a concomittant increase in volume in another region of the body. Special muscles keep the body turgid, enabling the locomotor muscles to effect these volume changes.

Crawling in lepidopteran caterpillars is probably the best studied method of locomotion in endopterygote larvae and comprises anteriorly directed waves of contraction of the longitudinal muscles, each wave causing the body to be pushed upward and forward. Three main phases can be recognized in each wave of contraction (Figure 14.7). First, contraction of the dorsal longitudinal muscles and transverse muscles causes a segment to shorten dorsally and its

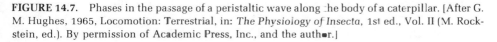

dorsal longitudinal
muscle

dorsoventral
muscle

ventral longitudinal
muscle

proleg

retractor muscle
of proleg

FIGURE 14.7. Phases in the passage of a peristaltic wave along the body of a caterpillar. [After G. M. Hughes, 1965, Locomotion: Terrestrial, in: *The Physiology of Insecta*, 1st ed., Vol. II (M. Rockstein, ed.). By permission of Academic Press, Inc., and the author.]

posterior end to be raised so that the segment behind is lifted from the substrate. The dorsoventral muscles and leg retractor muscles then contract, lifting both feet of the segment from the substrate. Finally contraction of the ventral longitudinal muscles, combined with the relaxation of the dorsoventral and leg retractor muscles, moves the segment forward and down to the substrate. Little work has been done on the neural coordination of crawling, though it seems probable that endogenous activity within the central nervous system is responsible. However, proprioceptive stimuli undoubtedly influence the process.

Crawling or burrowing in apodous larvae is comparable to peristaltic locomotory movements found in other invertebrates, for example, molluscs and annelids. Larvae, which crawl over the surface of the substrate, grip the substrate with, for example, protrusible prolegs or creeping welts (transversely arranged thickenings equipped with stiff hairs) situated at the posterior end of the body. A peristaltic wave of contraction then moves anteriorly, lengthening and narrowing the body. The anterior end is attached to the substrate while the posterior is released and pulled forward as the anterior longitudinal muscles contract. In many burrowing forms peristalsis proceeds in the opposite direction to movement, so that the narrowing and elongation begins at the anterior end and runs posteriorly. As the anterior end relaxes behind the peristaltic wave, it expands. This expansion serves both to anchor the anterior end and to enlarge the diameter of the burrow (Hughes and Mill, 1974).

3.2. Movement on or through Water

Progression on or through water presents very different problems to movement on a solid substrate. For small organisms, such as insects that live on the water surface, surface tension is a hindrance in production of propulsive leg movements. For submerged insects, the liquid medium offers considerable resistance to movement, especially for actively swimming forms.

Insects which move slowly over the surface of the water, for example,

Hydrometra (Hemiptera), or crawl along the bottom, for example, larval Odonata and Trichoptera, normally employ the hexapodal gait described above for terrestrial species. More rapidly moving species typically operate the legs in a rowing motion; that is, both legs of the segment move synchronously. Some species do not use legs but have evolved special mechanisms to facilitate rapid locomotion.

3.2.1. Surface Running

The ability to move rapidly over the surface of water has been developed by most Gerroidea (Hemiptera), whose common names include pondskaters and waterstriders. To stay on the surface, that is, to avoid becoming water-logged, these insects have developed various waterproofing features, especially hydrophobic (waxy) secretions, on the distal parts of the legs. However, these features considerably reduce the frictional force between the legs and water surface which is necessary for locomotion. This problem is overcome in many species by having certain parts of the tarsus, particularly the claws, penetrate the surface film and/or by having special structures, for example, an expandable fan which opens when the leg is pushed backward. In the Gerridae, however, the backward push of the legs is sufficiently strong that a wave of water is produced which acts as a "starting block" against which the tarsi can push (Nachtigall, 1974).

The functional morphology and mechanics of movement have been examined in detail in *Gerris* (Brinkhurst, 1959; Darnhofer-Demar, 1969). This insect has greatly elongated middle legs through which most of the power for movement is supplied. Some power is derived from the hindlegs, though these function primarily as direction stabilizers. The articulation of the coxa with the pleuron is such that the power derived from contraction of the large trochanteral retractor muscles is used exclusively to move the legs in the horizontal plane, that is, to effect the rowing motion. Equally, the coxal muscles serve only to lift the legs from the water surface during protraction. At the beginning of a stroke, the forelegs are lifted off the surface. The middle legs are rapidly retracted so that a wave of water forms behind the tarsi. As the legs are accelerated backward, the tarsi then push against this wave, causing the insect to move forward. After each acceleration stroke, the insect glides over the surface for distances up to 15 cm. The power developed in each leg of a segment is identical and the insect glides, therefore, in a straight line. Turning can occur only between strokes and is achieved by the independent backward or forward movement of the middle legs over the water surface.

3.2.2. Swimming by Means of Legs

Both larval and adult aquatic Coleoptera (Dytiscidae, Hydrophilidae, Gyrinidae, Haliplidae) and Hemiptera (Corixidae, Belostomatidae, Nepidae, Notonectidae) swim by means of their legs. Normally, only the hindlegs or the mid- and hindlegs are used and these are variously modified so that their surface area can be increased during the propulsive stroke and reduced when the limbs are moving anteriorly. Modifications include (1) an increase in the relative length and a flattening of the tarsus; (2) arrangement of the leg articulation, so that during the active stroke the flattened surface is presented perpendicularly to the direction of the movement, whereas during recovery the limb is pulled with the flattened surface parallel to the direction of movement; the leg is also flexed and drawn back close to the body during recovery; (3) development of articulated hairs on the tarsus and tibia which spread perpendicularly to the direction of movement during the power stroke, yet lie flat against the leg during recovery; such hairs may increase the effective area by up to five times; and (4) in *Gyrinus*, development of swimming blades on the tibia and tarsus (Figure 3.24)—these are articulated plates which normally lie flattened against each other. During the power stroke, the water resistance causes them to rotate so that their edges overlap and their flattened surface is perpendicular to the direction of movement of the leg.

In addition to the surface area presented, the speed at which the leg moves is proportional to the force developed. Thus, it is important for the propulsive stroke to be rapid, whereas the recovery stroke is relatively slow. Accordingly, the retractor muscles are well developed compared with the protractor muscles.

In the best swimmers other important structural changes can be seen, such as streamlining of the body and restriction of movement and/or change in position of the coxa. In adult *Dytiscus*, for example, the coxa is inserted more posteriorly than in terrestrial beetles and is fused to the thorax. Thus, the fulcrum for the rowing action is the dicondylic coxotrochanteral joint which operates like a hinge so that the leg moves only in one plane. Because of this arrangement, all the muscle power can be used to effect motion in this plane.

Several variations are found in the rhythms of leg movements. Where a single pair of legs is used in swimming, both legs retract together. When both the midlegs and hindlegs are used, both members of the same body segment usually move simultaneously, but are in opposite phase with the legs of the other segment; that is, when one pair is being retracted, the other pair is being protracted. In adult Haliplidae and Hydrophilidae and many larval beetles all three pairs of legs are used, in a manner comparable with the tripod gait of terrestrial insects.

Steering in the horizontal plane (control of yawing) is achieved by varying the power exerted by the legs on each side. For vertical steering (movement up or down) the nonpropulsive legs become involved. These may be held out from the body in the manner of a rudder or may act as weakly beating oars. By varying the angle to the body at which the legs are placed the insect will either dive, surface, or move horizontally through the water. Most aquatic insects are quite stable in the rolling and pitching planes because of their dorsoventrally flattened body. (See Figure 14.14 for explanation of the terms yawing, pitching, and rolling.)

3.2.3. Swimming by Other Means

A variety of other methods for moving through water can be found in insects, including body curling found in many larval and pupal Diptera, body undulation (larval Ephemeroptera and Zygoptera), jet propulsion (larval Anisoptera), and flying (a few adult Lepidoptera and Hymenoptera) (Nachtigall, 1974).

Many midge and mosquito larvae rapidly coil the body, first in one direction, then the other, to achieve a relatively inefficient form of locomotion. Chironomids, for example, lose 92% of the energy expended in the power stroke during recovery. Consequently, a 5.5-mm larva oscillating its body 10 times per second moves at only 1.7 mm/sec through the water. Mosquito larvae possess flattened groups of hairs (swimming fans) or solid "paddles" at the tip of the abdomen and are consequently more efficient and active swimmers than chironomids.

Larvae of Zygoptera and some Ephemeroptera undulate the abdomen, which is equipped at its tip with three flattened lamellae (Zygoptera, Figure 6.10) or swimming fans (Ephemeroptera, Figure 6.2). Some ephemeropteran larvae supplement the action of the fans by rapidly folding their abdominal gills against the body.

Dragonfly larvae (Anisoptera) normally take in and expel water from the rectum during gaseous exchange (Chapter 15). In emergencies this arrangement can be converted into a jet propulsion system for moving an insect forward at high speed (up to 50 cm/sec). Rapid contraction of longitudinal muscles causes the abdomen to shorten by up to 10%. Simultaneous contraction of the dorsoventral muscles leads to an increase in hemolymph pressure which forces water out of the rectum via the narrow anus at speeds approaching 250 cm/sec.

Female *Hydrocampa nympheata* (Lepidoptera) and adult *Dacunsa* (Hymenoptera) use their wings in addition to legs for swimming underwater. Other Hymenoptera (*Polynema* and *Limnodites*) swim solely by the use of their wings.

3.3. Flight

As noted earlier, wings probably arose as paranotal lobes which functioned as gliding planes. In time, these flattened expansions became articulated, perhaps in response to the need for attitudinal control (Chapter 2). Subsequent evolution led to the wings taking on a new function—propulsion of an insect through the air—partly as a result of which insects were able to move into new environments to become the diverse group we know today. Despite this diversity, there is sufficient similarity of skeletal and neuromuscular structure and function to suggest that wings had a monophyletic origin (Pringle, 1974).

Examination of the form and mode of operation of the pterothorax reveals certain trends, all of which lead to an improvement in flying ability. Primitively, the power for wing movement was derived from various "direct" muscles, that is, those which are directly connected with the wing articulations. These muscles serve also to determine the nature of the wing beat. Even today, the direct muscles remain important power suppliers in the Odonata, Orthoptera, Dictyoptera (Blattodea), and Coleoptera. In other insect groups, efficiency is increased by separating the control of wing beat (by the direct muscles) from power production, which becomes the job of large "indirect" muscles located in the thorax.

Other important differences between poor and good flyers are seen in the fine structure and neuromuscular physiology of the flight muscles. Generally (though there are exceptions), poor fliers flap their wings relatively slowly (up to 100 beats/sec, each beat of the wings being initiated by a burst of impulses to the power-producing muscles, which are of the tubular or close-packed type. In contrast, in the best fliers whose wing-beat frequency is high (up to 1000 beats/sec), muscle contraction is not in synchrony with the arrival of nerve impulses at the neuromuscular junction. Rather, the rhythm of contraction is generated within the muscles themselves, which are fibrillar. Accordingly, the two forms of rhythm are described as synchronous (neurogenic) and asynchronous (myogenic), respectively.

3.3.1. Structural Basis

Each wing-bearing segment is essentially an elastic box whose shape can be changed by contractions of the muscles within, the changes in shape causing the wings to move up and down. The skeletal components of a generalized wing-bearing segment are shown in Figure 3.18. The essential features are as follows. Each segment contains two large intersegmental invaginations, the prephragma and postphragma, between which the dorsal longitudinal muscles stretch. The alinotum bears on each side an anterior and a posterior notal process, to which the wing is attached via the first and third axillary sclerites. The pleuron is largely sclerotized and articulates with the wing by means of the pleural wing process, above which sits the second axillary sclerite. The hinge so formed is especially important in wing movement because of the large amount of resilin which it contains (see Section 3.3.3) Two other important articulating sclerites, which are usually quite separate from the sclerotized portion of the pleuron, are the anterior basalar and posterior subalar. Internally, the pleuron and sternum are thickened, forming the pleural and sternal apophyses, respectively, which brace the pterothorax. In some insects these apophyses are fused, but generally they are joined by a short but powerful pleurosternal muscle (Figures 14.8B and 14.13A'—C').

The muscles used in flight may be separated into three categories according to their anatomical arrangement (Figure 14.8A,B). The indirect flight muscles include the dorsal longitudinal muscles, dorsoventral muscles, oblique dorsal muscles, and oblique intersegmental muscles The direct muscles are the basalar and subalar muscles, and the axillary muscles attached to the axillary sclerites (including the wing flexor muscle, which runs from the pleuron to the third axillary sclerite). In the third category are the accessory indirect muscles which comprise the pleurosternal, tergopleural, and intersegmental muscles. Their function is to brace the pterothorax or to change the position of its components relative to each other. In addition, certain extrinsic leg muscles may also be important in wing movements. For example, in Coleoptera, whose coxae are fused to the thorax, the coxotergal muscles can assist the dorsoventral muscles in supplying power to raise the wings. In other species with articulated coxae the upper point of insertion of the extrinsic coxal muscles may change to the basalar or subalar. Thus the muscles can alter both leg and wing positions, that is, they may have a dual locomotory function. For example, in *Schistocerca gregaria* the anterior and posterior tergocoxal (indirect) muscles act synergistically during flight (1) to provide power for the upstroke of the wings and (2) to draw the legs up close to the body. This is achieved

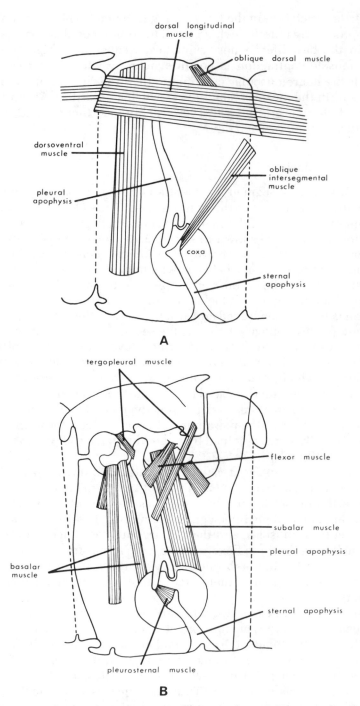

FIGURE 14.8. (A) Indirect and (B) accessory indirect muscles and direct muscles of right side of wing-bearing segment, seen from within. [After J. W. S. Pringle, 1957, *Insect Flight*. By permission of Cambridge University Press, New York.]

through polyneuronal innervation, whereby the parts of the tergocoxal muscles
which receive slow and inhibitory motor axons are responsible for the drawing
up of the legs, and the muscle fibers innervated by the fast axon move the
wings. In contrast, when the insect is running, the anterior and posterior tergocoxal muscles function antagonistically, effecting promotion and remotion,
respectively, of the legs. In the same species, the (direct) second basalar and
subalar muscles, while acting synergistically to aid the indirect muscles in the
production of power for the downstroke of the wings, act antagonistically to
bring about wing twisting (pronation and supination, respectively) or, when
running, promotion and remotion, respectively, of the legs (Figure 14.9) (Wilson, 1962).

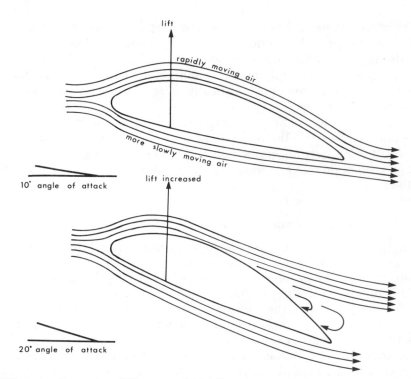

FIGURE 14.10. Generation of lift as a result of different air speeds above and below an aerofoil.

supplied by engines. In insects and other flying animals, the wings are movable and supply horizontal propulsive force (thrust) as well as lift. The thrust must be sufficient to overcome the opposing drag forces due to air resistance. In order to develop both lift and thrust during its stroke, the wing both moves up and down and changes its angle of attack (Figure 14.11A). The relative values of lift and thrust will change throughout the wing beat. In the middle of a downstroke, the rapid downward movement of the wing operating in conjunction with the already moving horizontal stream of air over the wings (assuming the insect is in flight) will result in a positive angle of attack (i.e., the air will strike the underside of the wing) and give rise to a strongly positive lift. Concurrently, because the wing is pronated (its leading edge is pulled down), there will be a slightly positive thrust (Figure 14.11B). During an upstroke, the angle of attack becomes slightly negative (pressure of air above is greater than pressure of air below the wing) causing slightly negative lift, yet increasing the positive thrust (Figure 14.11C). The angle at which an insect holds the body in flight also results in positive lift, though this amounts to less than 5% of the total lift in the desert locust and about 20% in some Diptera.

3.3.3. Mechanics of Wing Movements

All insects use the indirect tergosternal or tergocoxal muscles to raise the wing. Contraction of these muscles pulls the tergum down so that its points of articulation with the wing fall below the articulation of the wing with the

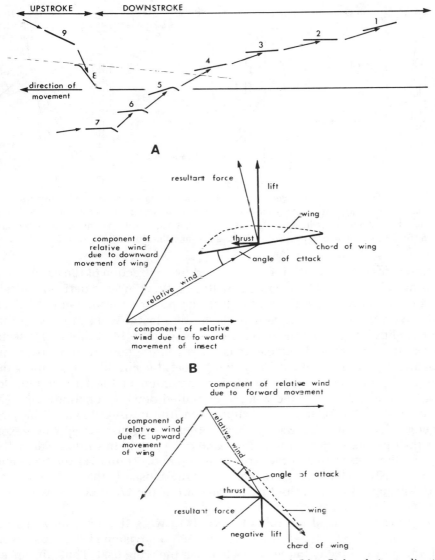

FIGURE 14.11. (A) Changes in the angle at which a wing is held in flight relative to direction of movement. Arrows indicate angle at which air strikes the wing. Numbers indicate chronological sequence of wing positions during a stroke. (B,C) Magnitude of lift and thrust approximately midway through downstroke and upstroke, respectively. [A, after M Jensen, 1956, Biology and physics of locust flight. III. The aerodynamics of locust flight, *Philos. Trans. R. Soc. London, Ser. B* **239**:511–552. By permission of The Royal Society, London, and the author. B, C, after R. F. Chapman, 1971, *The Insects: Structure and Function.* By permission of Elsevier North-Holland Inc. and the author.]

pleural wing process which serves as a fulcrum (Figure 14.12A). In most insects indirect muscles are also used to lower the wing. Shortening of the dorsal longitudinal muscles causes the tergum to bow upward, raising the anterior and posterior notal processes above the tip of the pleural wing process and, therefore, the wing to be lowered (Figure 14.12B). In Coleoptera and Orthoptera

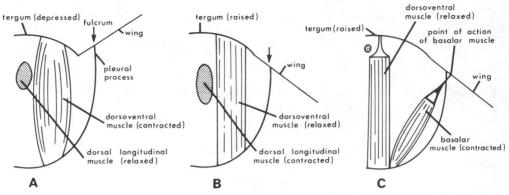

FIGURE 14.12. Diagrammatic transverse sections of thorax to show muscles used in upstroke and downstroke. (A) Use of indirect muscles to raise wing, (B) use of indirect muscles to lower wing, and (C) use of direct muscles to lower wing. [After R. F. Chapman, 1971, *The Insects: Structure and Function.* By permission of Elsevier North-Holland Inc. and the author.]

some power for a downstroke is also obtained by contraction of the basalar and subalar muscles and in Odonata and Blattodea, this power is derived entirely from contraction of these direct muscles. The points of articulation of these muscles with the wing sclerites lie outside the pleural wing process so that when the muscles contract the wing is lowered (Figure 14.12C). In all insects, however, the basalar and subalar muscles are important in wing twisting, that is, altering the angle at which the wing meets the air, thereby affecting the values of the lift and thrust generated. Contraction of the basalar muscles causes the anterior edge of the wing to be pulled down (pronation), whereas contraction of the subalar induces supination (the pulling down of the posterior edge of the wing). Contraction of the other direct muscle, the wing flexor, causes the third axillary sclerite to twist and to be pulled inward and dorsally. This pulls the vannal area of the wing (see Figure 3.27) up over the body and enables the wing to fold along predetermined lines (usually the anal veins, and vannal and jugal folds). Unfolding (extension) of the wing occurs when the basalar muscle contracts.

Early descriptions of the role of muscles in wing flapping, for example, that of Chabrier (1822) (cited from Pringle, 1957) envisioned the muscles as supplying power directly to the wings to effect the wing beat. Thus, the speed with which muscles could contract would determine the speed at which a wing was lowered or raised. In turn, this determined the value of the lift generated. Not until the 1950s was it realized that a large proportion of the energy produced during the early stages of contraction of the power-producing muscles is stored as elastic energy within the lateral walls of the pterothorax and especially the hinge between the pleural wing process and second axillary sclerite. Then, at a critical point, both during a downstroke and during an upstroke, the stored energy is suddenly released, thereby enabling the wings to move with a greatly enhanced velocity so that the maximum lift might be generated. This system of wing movement is known as the "click mechanism."

Mechanical details of the click mechanism vary among different groups of Insecta but commonly the system is bistable; that is, the wing may rest in either the up or the down position. Approximately midway between these two positions, the wing is in a position of maximum instability and, depending on which muscles are contracting, will click into the up or down position. The system is shown diagrammatically in Figure 14.13. It can be seen that the system depends essentially on a triple hinge (three interconnected points of articulation, X, Y, and Z). During an upstroke, contraction of the muscles lowering the tergum produces a laterally directed force which attempts to push the side wall of the pterothorax outward. This is resisted by the elasticity of the wall components. The size of the lateral force (and, therefore, the extent of the elastic resistance) increases until it is maximal when the three points X, Y, and Z are in a straight line. The slightest additional downward movement—bringing Z below Y—produces an unstable situation so that the elastic energy is suddenly released, forcing Y upward, bringing X and Z closer together, and causing the wing to click rapidly into the up position. Conversely, to effect a downstroke, contraction of the dorsal longitudinal muscles, causing the center of the tergum to be raised, will again create a laterally directed force whose magnitude is increased (like that of the opposing elastic resistance) as Z comes into line with X and Y. Immediately Z is raised above Y, the energy is released, forcing X and Z closer, and the wing clicks downward.

In the above description, it has been assumed that the vertical plane

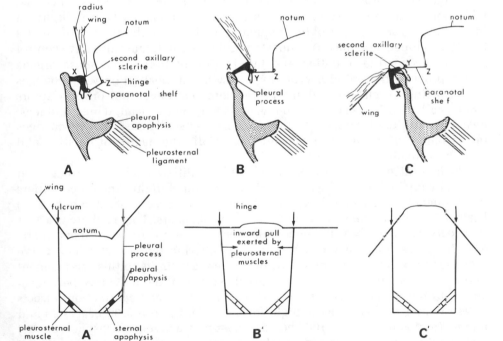

FIGURE 14.13. (A–C) Sections through wing base and (A′–C′) diagrammatic sections through thorax of a fly to illustrate the click mechanism. (A, A′) Wing in stable "up" position; (B, B′) wing in unstable position, with hinges (X, Y, Z) in a straight line; and (C, C′) wing in stable "down" position. [After R. F. Chapman, 1971, *The Insects: Structure and Function*. By permission of Elsevier North-Holland Inc. and the author.]

through which the hinge Z moves either remains constant through the stroke, or perhaps even moves slightly inward because of the elastic resistance. In fact, in many insects the shape of the tergum is such that during both an upstroke and a downstroke, muscle contraction causes a lateral extension of the tergum (that is, increases the distance between Z on each side) so that the value of the outwardly exerted lateral force is even greater.

Raising the wings during flight is aided by the pressure of the onrushing air on the underside of the wing, so that only a small amount of the elastic energy stored in the skeleton is used in an upstroke. The remainder (about 86% in the desert locust) can be used in the following downstroke (Pringle, 1974).

Maximum wing velocity (and, therefore, maximum power) is achieved shortly after a wing "clicks." If a wing were allowed to swing through its complete amplitude, a significant time would elapse during which little or no power was being produced (viz., at the beginning or end of each stroke when the wing velocity is low). Therefore, it becomes more efficient for an insect to reduce the amplitude of wing beat by means of cuticular "wing stops" or by initiating contraction of the antagonistic muscles before the end of each stroke. In this way, wing-beat frequency can be increased.

3.3.4. Control of Wing Movements

Flight is a complex process and at any point in time an insect is concerned with monitoring, and varying if necessary, many parameters. During flight, an insect must be able to control the frequency of wing beat, the amount of lift and thrust developed, and the direction (stability) of flight. It must also have control mechanisms for the initiation and termination of flight. The sense organs which provide the central nervous system with information on what changes are occurring in relation to flight are the compound eyes and proprioceptors strategically distributed over the body, especially on the head, wings, and legs. Responses to stimuli received by these organs are usually mediated via changes in the nature of the wing movements (especially the degree of twisting and wing-beat frequency).

Wing-beat frequency varies widely among different insects as has been noted already in the introduction to this discussion of flight. Among poor fliers the frequency is generally low, for example, in the desert locust about 15–20 beats/sec, and neurogenic (synchronous) control occurs. That is, there is a 1 : 1 ratio between wing-beat frequency and nervous input. In such fibers, therefore, wing-beat frequency can be varied by altering the rate at which nerve impulses arrive at the muscles. However, there are limits to this arrangement (maximum frequency about 100 beats/sec) because of the refractory period required to return the muscle to its resting state after each contraction. Insects whose wing-beat frequency is high, for example, the bee (190 beats/sec) and midge *Forcipomyia* (up to 1000 beats/sec), employ a myogenic (asynchronous) system where the frequency of muscle contraction is much greater than that of nervous input. Such a system is possible only where there are antagonistic muscles which contract regularly and alternately. These muscles have the property of contracting autonomously when the tension developed in them as a result of stretching reaches a critical value. In the case of the flight musculature, the alternating contractions are, in a sense, self-perpetuating, though their initiation and cessation are under nervous control. There is also evidence to suggest that changes in the frequency at which nerve impulses arrive at these muscles can modify their frequency of contraction. However, as contractions are tension-dependent, it follows that their frequency could also be altered by modifying the elastic resistance in the exoskeleton. In Diptera, for example, this is achieved by contraction or relaxation of the pleurosternal muscles which serves to move the pleural wing process (X) closer to or farther from the tergal hinge (Z) (Figure 14.13).

As noted earlier, during a beat a wing does not make simple up and down movements, but rather twists about the vertical axis so that its tip describes an ellipse (*Schistocerca*) or figure eight (*Apis, Musca*). The size and direction of the twisting force (torque), which effectively measure the lift and thrust developed, are monitored by campaniform sensilla at the base of each wing (halteres in Diptera—see below). Input from the sensilla initiates reflex excitation of the basalar and subalar muscles, which regulate the extent to which the wing twists. This mechanism is known as the lift control reflex. *Schistocerca* employs such a reflex to control the value of lift and, in flight, holds its body at a fairly steady angle to the horizontal. Other insects vary the angle at which the body is held in flight in order to alter the lift and thrust components.

The direction (and correlated with it, the stability) of flight necessitates the monitoring and regulation of movement in three dimensions (Figure 14.14): movement of the body in the horizontal, transverse axis is pitching; rotation about the vertical axis is yawing; and rotation around the longitudinal axis is rolling.

In *Schistocerca gregaria* pitching is controlled via the lift control reflex outlined above. In this species the fore and hind wings beat in antiphase. The hind wings are not capable, however, of twisting and, therefore, the value of lift generated by them is a function of the body angle (the angle at which the body is held relative to the direction of airflow). Any change in body angle, altering the lift generated by the hind wings, is compensated for by appropriate twisting of the fore wings so that the total lift produced remains constant. Other insects control pitching by varying the amplitude of the wing beat, the stroke plane (the angle at which the wings move relative to the body), or the stroke rhythm (by delaying or enhancing the moment at which pronation of the wings occurs during the beat).

The control of yawing involves sensory input from both the compound eyes and mechanoreceptors. As an insect turns, the visual field will rotate. Also, in some insects such as *Schistocerca*, this movement results in unequal stimulation of mechanosensory hairs at each side of the head. Yawing may be corrected by varying the angle of attack, the amplitude, or the frequency of beat, between the wings on each side of the body. All these variables will alter the thrust generated.

Rolling is largely monitored as a result of certain visual stimuli. Commonly, the head is rotated so that the dorsal ommatidia receive maximal illumination, the source of which is normally directly overhead. Early and late in the day, however, when the latter is not the case, use is also made of the marked contrast in light intensity between the earth and sky at the horizon.

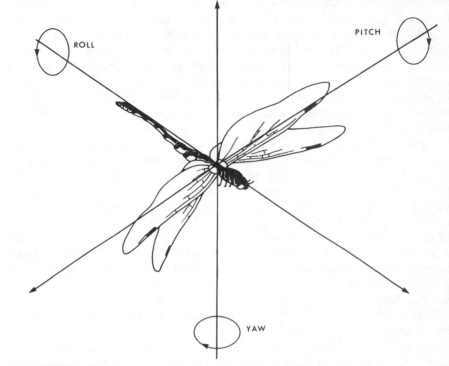

FIGURE 14.14. Diagram showing axes about which a flying insect may rotate.

Goodman's (1965) study of the importance of the horizon in orientation of the desert locust showed that the insect rotates its head so as to maintain the horizon horizontally across the visual field of the compound eyes. Her study also demonstrated that the dorsal light reaction overrides the "horizon response" at higher light intensity, and vice versa. Alignment of the thorax and abdomen with the head is achieved by means of proprioceptive hairs on each side of the neck. When the thorax is out of line with the head, the hairs on each side are differentially stimulated. This information induces appropriate wing twisting to compensate.

Wing movements are initiated in many insects by means of the "tarsal reflex," that is, loss of contact between the substrate and tarsi, as well as by a variety of nonspecific stimuli. It seems probable that sensory input to inhibit the tarsal reflex is fed in via campaniform sensilla on the femur and trochanter which are sensitive to stresses in the cuticle in the standing insect (see Chapter 12, Section 2.2). The reflex stimulates unfolding of the wings (in some species this involves the same muscles as are used in flapping the wings), stiffening of the pleural wall (contraction of the pleurosternal muscles) to facilitate the click mechanism, and raising the legs to the flying position. The tarsal reflex is not found in insects which capture prey during flight, in insects which vibrate their wings during "warm-up" prior to takeoff, in bees which employ wing movements in hive ventilation, or in many beetles which must raise the elytra and unfold the wings prior to flight.

Some insects, for example, *Drosophila*, will fly until exhausted, provided their tarsi no longer make contact with the substrate. For most species, however, additional stimuli are necessary to maintain wing movements, especially airflow, which stimulates hairs on the upper part of the head, Johnston's organs in the antennae, and proprioceptors at the base of the wings. In the honeybee, flight is maintained as a result of visual stimuli received as the insect moves forward over the ground (Heron, 1959, cited from Pringle, 1974).

Visual signals experienced by the compound eyes just prior to landing stimulate leg extension. Goodman (1960), working on the fly *Lucilia sericata*, concluded that the response was based on three variables acting singly or in combination: (1) decrease in light intensity experienced by successive ommatidia as the fly approaches the substrate; (2) number of ommatidia stimulated; and (3) rate of successive stimulation of ommatidia (based on the increase in angular velocity which occurs as the fly comes closer to the substrate).

Halteres. Deserving of special mention are the much modified hind wings—halteres of Diptera (Figure 14.15), which function as gyroscopic organs of balance. The halteres vibrate in the vertical plane only, at the same frequency as the wings with which they are in antiphase. These vertical vibrations, combined with the movement of the insect through the air, cause the development of torque at the base of the halteres, which is monitored by various groups of campaniform sensilla oriented in different planes. Any

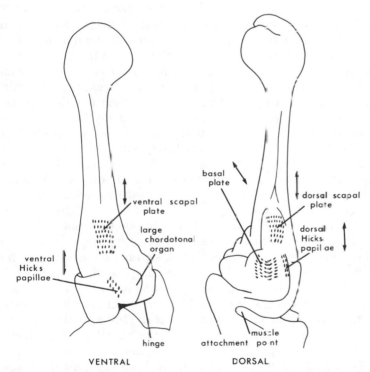

FIGURE 14.15. Haltere of *Lucilia* (Diptera). The orientation of the various groups of campaniform sensilla is indicated by double-ended arrows. [After J. W. S. Pringle, 1948. The gyroscopic mechanism of the halteres of Diptera, *Philos. Trans. R. Soc. London*, Ser. B 233:347–384. By permission of The Royal Society, London, and the author.]

change in the orientation of the insect will result in modification of the torques generated at the base of each haltere. Yawing can be detected by differential stimulation of the groups of sensilla in each haltere, whereas the detection of pitching and rolling requires the combined input of both halteres, which is assessed in the thoracic ganglion. To correct for instability in any of the planes of motion, the insect alters the extent to which each wing is twisted during beating. For a detailed discussion of the gyroscopic nature and function of halteres, see Pringle (1948, 1957).

3.4. Orientation

The locomotor behavior which follows receipt of a stimulus by an insect may be described as a KINESIS or a TAXIS. A kinesis is simply induced or enhanced activity without spatial reference to the source of the stimulus. Increased movement due to light (photokinesis) is common in many insects. For example, many butterflies fly only when the light intensity is above a certain threshold; the speed at which locusts walk is increased as the light intensity rises. A hygrokinetic response is shown by locusts whose activity is increased in moist air compared with dry. In contrast, wireworms (Elateridae) are inactive in air saturated with moisture yet become active at lower humidity. Insects that normally secrete themselves in cracks or crevices, for example, earwigs and bedbugs, exhibit stereokinesis, an inhibition of movement which occurs when a sufficient number of mechanoreceptors on the body surface are stimulated.

A taxis is orientation in response to a stimulus, the location of which can be perceived by an insect. To locate the stimulus, an insect uses either paired sense organs (compound eyes, antennae, and tympanal organs) or moves one sense organ from side to side, and compares the strength of the stimulus received by each organ or by the single organ in each position. In its simplest form, an insect's response is to move toward or away from the stimulus. For example, the bloodsucking insects *Rhodnius* and *Cimex* possess the ability to orient to a heat source (positive thermotaxis) which enables them to locate a host. Many insects exhibit chemotaxis in locating food, a mate, or a host. Locusts are attracted to a light source (positive phototaxis). A more complex form of phototaxis is where an insect orients itself and moves with its body at a constant angle to the light source. As the insect moves, therefore, it receives a constant visual stimulus. Such orientation is called menotaxis. The stimulus may be a simple point source of light, for example, the sun which forms the basis of the "light–compass reaction" of bees and other insects which navigate. Alternately, the stimulus may be a more complex visual pattern. Marching behavior in groups of juvenile locusts (hoppers) is based on menotaxis, involving both a simple and a complex visual stimulus. The simple stimulus, the sun, gives rise to a light–compass reaction so that the individual marches in a fixed direction. The more complex stimulus is the visual pattern created by the presence of adjacent hoppers. Thus, as one hopper moves, neighboring hoppers will also be stimulated to move in order to keep constant the visual pattern "seen" by each eye. This ensures that the group marches as a whole; that is, menotaxis is the basis of their gregarious behavior.

4. Summary

All insect muscles are striated, though their structural details vary according to their function. Skeletal muscles can be categorized as tubular (leg and body segment muscles, and wing muscles of Odonata and Dictyoptera), close-packed (wing muscles of Orthoptera and Lepidoptera), and fibrillar (flight muscles of most insects). All skeletal and many visceral muscles are innervated. Each muscle fiber may receive one to three functionally distinct axons, one of which is always excitatory. Excitatory axons may be subdivided into "fast" axons, whose impulses each initiate strong, rapid contractions and "slow" axons, whose impulses individually cause a weak contraction but are additive in effect, facilitating a graded response in the muscle. Some fibers also receive an inhibitory axon. Most muscles operate in pairs, with one antagonistic to the other so that as one is stimulated to contract the other is passively stretched. Where there is no antagonistic partner, a muscle may be stretched due to the elasticity of the cuticle to which it is attached.

Most insects use six legs during walking (hexapodal gait). Others use only two pairs of legs (quadrupedal gait) but may use the tip of the abdomen as a point of support. On all occasions, however, there are at least three points of contact with the substrate, forming a triangle of support. Stepping movements are coordinated endogenously, though external stimuli clearly can affect the basic rhythm. Jumping, in most insects that practice it, is a function of the hindlegs, which are elongate, muscular, and capable of great extension due to the wide angle that can be developed between femur and tibia. In order to obtain sufficient power for takeoff, the energy of muscle contraction must be stored temporarily as elastic energy. This energy, at a critical point, is suddenly and rapidly released to produce the acceleration necessary to overcome gravity. Most soft-bodied larvae which crawl over, or burrow through, the substrate depend on synchronized contraction and relaxation of muscles to effect changes in body shape, the legs and accessory locomotory appendages serving simply as points of friction between the body and substrate. In these larvae the body fluids serve as a hydrostatic skeleton.

Insects that move slowly over the surface of, or through, water typically use a hexapodal gait. More rapidly moving species, which may be streamlined, usually employ a rowing motion of the midlegs, occasionally the hindlegs. The legs are often modified to increase the surface area presented during the active stroke and their point of insertion on the body is designed so as to obtain maximum power from the stroke. Some aquatic insects swim by other means, for example, body curling, jet propulsion, or flapping the wings.

Primitively, the direct muscles (those that connect directly with the wing articulations) are used both for supplying power for flight and for controlling the nature of the wing beat. However, in most flying insects efficiency is increased by separating the supply of power (the role of large indirect muscles in the thorax) from the control of wing beat (which remains the function of the direct muscles). Poor fliers generally have a low wing-beat frequency and, in them control of muscle contraction is synchronous (neurogenic); that is there is a 1 : 1 ratio between wing-beat frequency (=frequency of wing-muscle contraction) and the number of nerve impulses arriving at the muscle. The fibrillar muscles of good fliers are asynchronous (myogenic) and always operate in antagonistic pairs. Their rhythm of contraction originates endogenously and is initiated when a muscle is stretched to a critical tension. Thus, the contractions of an antagonistic pair of fibrillar muscles are self-perpetuating, though their initiation and termination are under nervous control. The indirect muscles serve to change the shape of the pterothorax, which thus acts as an elastic box, these changes in shape causing the wings to be moved up and down. As in jumping, in order to obtain sufficient power to fly, the energy of muscle contraction is briefly stored as elastic energy, in the wall of the pterothorax, then suddenly released by means of the click mechanism.

Enhanced locomotor activity, which follows receipt of a stimulus, but whose direction is without spatial reference to that stimulus, is known as a kinesis. When the direction of movement is with reference to the source of the stimulus, for example, attraction to an odor and the light–compass reaction, the movement is described as a taxis.

5. Literature

The nature and properties of insect muscle are discussed by Smith (1972) and by various authors in the treatise edited by Usherwood (1975). Walking is dealt with by Wilson (1966) and by Hughes and Mill (1974), who also review jumping, crawling, and burrowing. Nachtigall (1974) has reviewed locomotion of aquatic insects. Pringle's (1957) monograph provides an excellent introduction to flight. More recent reviews of flight, based on his monograph, are given by Pringle (1968, 1974). Insect flight has also been the subject of a symposium, the proceedings of which were edited by Rainey (1976). Wilson (1968) has examined the nervous control of flight.

Brinkhurst, R. O., 1959, Studies on the functional morphology of *Gerris najas* De Geer (Hem. Het. Gerridae), *Proc. Zool. Soc. London* **133**:531–559.

Darnhofer-Demar, B., 1969, Zur Fortbewegung des Wasserläufers *Gerris lacustris* L. auf der Wasseroberfläche, *Zool. Anz. Suppl.* **32**:430–439.

Goodman, L. J., 1960, The landing responses of insects. 1. The landing responses of the fly, *Lucilia sericata*, and other Calliphorinae, *J. Exp. Biol.* **37**:854–878.

Goodman, L. J., 1965, The role of certain optomotor reactions in regulating stability in the rolling plane during flight in the desert locust, *Schistocerca gregaria*, *J. Exp. Biol.* **42**:385–408.

Heitler, W. J., 1974, The locust jump. Specializations of the metathoracic femoral-tibial joint, *J. Comp. Physiol.* **89**:93–104.

Hoyle, G., 1974, Neural control of skeletal muscle, in: *The Physiology of Insecta*, 2nd ed., Vol IV (M. Rockstein, ed.), Academic Press, New York.

Hughes, G. M., and Mill, P. J., 1974, Locomotion: Terrestrial, in: *The Physiology of Insecta*, 2nd ed., Vol. III (M. Rockstein, ed.), Academic Press, New York.

Nachtigall, W., 1974, Locomotion: Mechanics and hydrodynamics of swimming in aquatic insects, in: *The Physiology of Insecta*, 2nd ed., Vol. III (M. Rockstein, ed.), Academic Press, New York.

Neville, A. C., 1975, *Biology of the Arthropod Cuticle*, Springer-Verlag, New York.

Pringle, J. W. S., 1948, The gyroscopic mechanism of the halteres of Diptera, *Philos. Trans. R. Soc. London, Ser. B* **233**:347–384.

Pringle, J. W. S., 1957, *Insect Flight*, Cambridge University Press, Cambridge.

Pringle, J. W. S., 1968, Comparative physiology of the flight motor, *Adv. Insect Physiol.* **5**:163–227.

Pringle, J. W. S., 1974 Locomotion: Flight, in: The Physiology of Insecta, 2nd ed., Vol. III (M. Rockstein, ed.), Academic Press, New York.

Rainey, R. C. (ed.), 1976, Insect flight, Symp. R. Entomol. Soc. 7:287 pp.

Rothschild, M., Schlein, Y., Parker, K., and Sternberg, S., 1972, Jump of the oriental rat flea Xenopsylla cheopis (Roths.), Nature (London) 239:45–48.

Smith, D. S., 1972, Muscle, Academic Press, New York.

Usherwood, P. N. R. (ed.), 1975, Insect Muscle, Academic Press, New York.

Wilson, D. M., 1962, Bifunctional muscles in the thorax of grasshoppers, J. Exp. Biol. 39:669–677.

Wilson, D. M., 1966, Insect walking, Annu. Rev. Entomol. 11:103–122.

Wilson, D. M., 1968, The nervous control of insect flight and related behavior, Adv. Insect Physiol. 5:289–338.

15

Gaseous Exchange

1. Introduction

In all organisms gaseous exchange, the supply of oxygen to and removal of carbon dioxide from cells, depends ultimately on the rate at which these gases diffuse in the dissolved state. The diffusion rate is proportional to (1) the surface area over which diffusion is occurring, and (2) the diffusion gradient (concentration difference of the diffusing material between the two points under consideration divided by the distance between the two points). Diffusion alone, therefore, as a means of obtaining oxygen or excreting carbon dioxide can be employed only by small organisms whose surface area/volume ratio is high (i.e., all cells are relatively close to the surface of the body) and organisms whose metabolic rate is low. Organisms which are larger and/or have a high metabolic rate must increase the rate at which gases move between the environment and the body tissues by improving (1) and/or (2) above. In other words, specialized respiratory structures with large surface areas and/or transport systems which bring large quantities of the gas closer to the site of use or disposal (thereby improving the diffusion gradient) have been developed. For most terrestrial animals prevention of desiccation is another important problem, and this has had a major influence on the development of their respiratory surfaces through which considerable loss of water might occur. Typically, respiratory surfaces of terrestrial animals are formed as invaginated structures within the body so that evaporative water loss is greatly reduced.

In insects the tracheal system, a series of gas-filled tubes derived from the integument, has evolved to cope with gaseous exchange. Terminally the tubes are much branched, forming tracheoles which provide an enormous surface area over which diffusion can occur. Furthermore, tracheoles are so numerous that gaseous oxygen readily reaches most parts of the body, and, equally, carbon dioxide easily diffuses out of the tissues. Thus, in most insects, in contrast to many other animals, the circulatory system is unimportant in gas transport. Because they are in the gaseous state within the tracheal system, oxygen and carbon dioxide diffuse rapidly between the tissues and site of uptake or release, respectively, on the body surface. Oxygen, for example, diffuses 3 million times faster in air than in water (Mill, 1972). Again, because the system is gas-filled, much larger quantities of oxygen can reach the tissues in a given time. (Air has about 25 times more oxygen per unit volume than does water.)

The eminent suitability of the tracheal system for gaseous exchange is illustrated by the fact that, for most small insects and many large insects at rest, simple diffusion of gases into/out of the tracheal system entirely satisfies their requirements. In large, active insects the gradient over which diffusion occurs is increased by means of ventilation; that is, air is actively pumped through the tracheal system.

2. Organization and Structure of Tracheal System

A tracheal system is present in all Insecta and in other hexapodous arthropods with the exception of the Protura and many Collembola. It arises during embryogenesis as a series of segmental invaginations of the integument. Up to 12 (3 thoracic and 9 abdominal) pairs of spiracles may be seen in embryos, though this number is always reduced prior to hatching, and further reduction may occur in endopterygotes during metamorphosis. Various terms are used to describe the number of pairs of functional spiracles, for example, holopneustic (10 pairs, located on the mesothorax and metathorax and 8 abdominal segments), amphipneustic (2 pairs, on the mesothorax and at the tip of the abdomen), and apneustic (no functional spiracles). The latter condition is common in aquatic larvae which are said, therefore, to have a closed tracheal system.

2.1. Tracheae and Tracheoles

In apterygotes other than lepismatid Zygentoma, the tracheae which run from each spiracle do not anastomose either with those from adjacent segments or with those derived from the spiracle on the opposite side. In the Lepismatidae and Pterygota both longitudinal and transverse anastomoses occur, and, though minor variations can be seen, the resultant pattern of the tracheal system is often characteristic for a particular order or family. Generally, a pair of large-diameter, longitudinal tracheae (the lateral trunks) run along the length of an insect just internal to the spiracles. Other longitudinal trunks are associated with the heart, gut, and ventral nerve cord. Interconnecting the longitudinal tracheae are transverse commissures, usually one dorsal and another ventral, in each segment (Figure 15.1). Parts of the tracheal system, for example, that of the pterothorax, may be effectively isolated from the rest of the system by reduction of the diameter or occlusion of certain longitudinal trunks. This arrangement is associated with the use of autoventilation as a means of improving the supply of oxygen to wing muscles during flight (see Section 3.3). Also, tracheae are often dilated to form large thin-walled air sacs which have an important role in ventilation (Section 3.3) and other functions.

Numerous smaller tracheae branch off the main tracts and undergo progressive subdivision until at a diameter of about 2–5 μm they form a number of fine branches each 1 μm or less across known as tracheoles. Tracheoles are intracellular, being enclosed within a very thin layer of cytoplasm from the tracheoblast (tracheal end cell) (Figure 15.2), and ramify throughout most tissues of the body. They are especially abundant in metabolically active tissues,

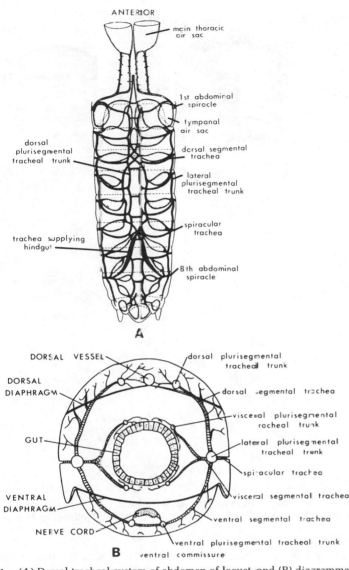

FIGURE 15.1. (A) Dorsal tracheal system of abdomen of locust, and (B) diagrammatic transverse section through abdomen of a hypothetical insect to illustrate main tracheal branches. [A, from F. O. Albrecht, 1953, *The Anatomy of the Migratory Locust*. By permission of the Athlone Press, B, from R. E. Snodgrass, *Principles of Insect Morphology*. Copyright 1935 by McGraw-Hill, Inc. Used with permission of McGraw-Hill Book Company.]

and in flight muscles, fat body, and testes, for example, tracheoles indent individual cells, so that gaseous oxygen is brought into extremely close proximity with the energy-producing mitochondria.

As derivatives of the integument, tracheae comprise cuticular components and epidermis, though they lack a basement membrane (Figure 15.2). Adjacent to the spiracle, the tracheal cuticle includes epicuticle, exocuticle, and endocuticle; in smaller tracheae and tracheoles only epicuticle is present. Pro-

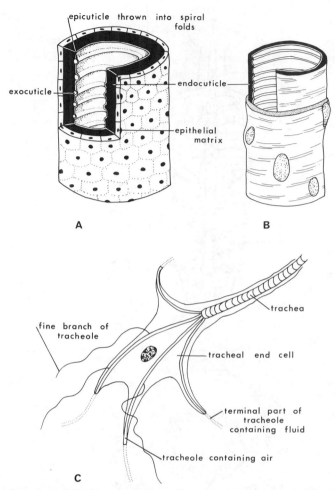

FIGURE 15.2. Structure of (A) large and (B) small tracheae; (C) origin of tracheole. [After V. B. Wigglesworth, 1965, *The Principles of Insect Physiology*, 6th ed., Methuen and Co. By permission of the author.]

viding the system with strength yet flexibility, tracheal cuticle has internal ridges which may be either separate (annuli) or form a continuous helical fold (taenidium). In large tracheae the ridges include some procuticle, but this is absent from the taenidia of tracheoles. Taenidia are absent from, or poorly developed in, air sacs. The epicuticle of tracheae comprises the same layers as that of the integument. In tracheoles, however, the wax layer is absent and, furthermore, the cuticulin layer contains fine pores. These two features may be associated with movement of liquid into and out of tracheoles in connection with gaseous exchange (see Section 3.1) (Locke, 1966).

2.2. Spiracles

Only in some apterygotes do tracheae originate at the body surface. Normally, they arise slightly below the body surface from which they are separated by a small cavity, the atrium (Figure 15.3A). In this arrangement, the term "spiracle" generally includes both the atrium and the spiracle *sensu stricto*, i.e., the tracheal pore. Except for those of a few insects which live in humid microclimates, spiracles are equipped with various valves for prevention of water loss. The valves may take the form of one or more cuticular plates which can be pulled over or across a spiracle by means of a closer muscle (Figure 15.3B–D). Opening of the valve(s) is effected either by the natural elasticity of the surrounding cuticle or by an opener muscle. Alternatively, the valve may be a cuticular lever which by muscle action constricts the trachea adjacent to the atrium (Figure 15.3E,F). In lieu of, or in addition to, the valves, there may be hairs lining the atrium or a sieve plate (a cuticular pad penetrated by many fine pores) covering the atrial pore. It is commonly assumed that an important function of these hairs and sieve plates is to prevent dust entry. However, as Miller (1974) notes, sieve plates are not better developed on inspiratory than on expiratory spiracles and several other functions can be suggested: (1) they may prevent waterlogging of the tracheal system during rain, in aquatic insects, and in species which live in moist soil, rotting vegetation, etc.; (2) they may prevent entry of parasites, especially mites, into the tracheal system; and (3) they may reduce bulk flow of gases through the system, thereby reducing evaporative water loss. This would be disadvantageous in insects which ventilate the tracheal system, and it is of interest, therefore, that those spiracles which are important in ventilation commonly lack a sieve plate or have a plate which is divided down the middle so that it may be opened during ventilation.

3. Movement of Gases within Tracheal System

Gaseous exchange between tissues and the tracheal system occurs almost exclusively across the walls of tracheoles, for it is only their walls which are sufficiently thin as to permit a satisfactory rate of diffusion. It is necessary, therefore, to ensure that a sufficient concentration of oxygen is maintained in tracheoles to supply tissue requirements and, at the same time, that carbon dioxide produced in metabolism is removed quickly, preventing its buildup to toxic levels. In small insects and inactive stages of larger insects, diffusion of gases between the spiracle and tracheoles is sufficiently rapid that these requirements are met. However, if the spiracles are kept permanently open, the amount of water lost via the tracheal system may become important. Thus, many insects utilize passive (suction) ventilation as a means of reducing this loss. The needs of large, active insects can be satisfied only by shortening the distance over which diffusion must occur. This is achieved by active ventilation movements.

3.1. Diffusion

The absence of obvious breathing movements led many nineteenth-century scientists to assume that insects obtained oxygen by simple diffusion. It was not, however, until 1920 that Krogh (cited in Miller, 1974) calculated, on the basis of (1) measurements of the average tracheal length and diameter, (2) measurements of oxygen consumption, and (3) the permeability constant for oxygen, that for *Tenebrio* and *Cossus* (goat moth) larvae at rest with the spira-

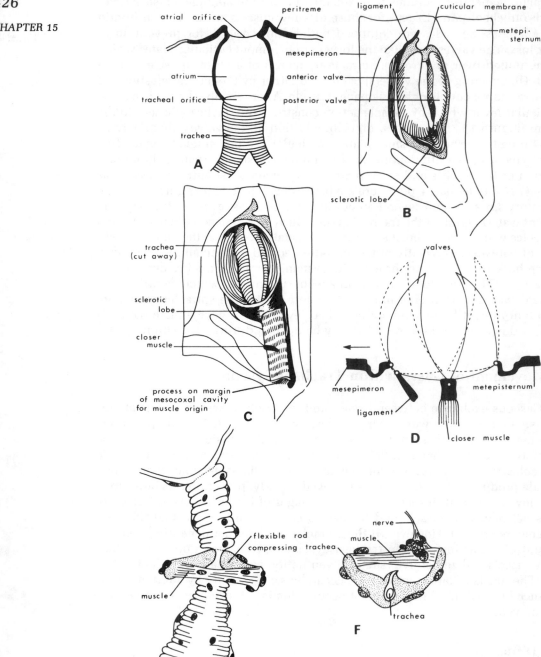

FIGURE 15.3. Spiracular structure. (A) Section through spiracle to show general arrangement; (B, C) outer and inner views of second thoracic spiracle of grasshopper; (D) diagrammatic section through spiracle to show mechanism of closure—the valve is opened by movement of the mesepimeron, closed by contraction of the muscle; (E) closing mechanism on flea trachea; and (F) section through flea trachea at level of closing mechanism. [A–C, from R. E. Snodgrass, *Principles*

cles open, the oxygen concentration difference between the spiracles and tissues is only about 2% and easily maintainable by diffusion. Even in large active insects which ventilate, diffusion is a significant process, because the ventilation movements serve only to move the air in the larger tracheae. For example, in the dragonfly *Aeshna*, oxygen reaches the flight muscles by diffusion between the primary (ventilated) air tubes and tracheoles, a distance of up to 1 mm. Even in flight, when the oxygen consumption of the muscle reaches 1.8 ml/g per min and the difference in oxygen concentration between the primary tube and tracheoles is from 5 to 13%, diffusion is quite adequate (Weis-Fogh, 1964).

Diffusion is also important in moving gases between the tracheoles and mitochondria of the tissue cells. Because diffusion of dissolved gases is relatively slow, the distance over which it can function satisfactorily (in structural terms, half the distance between adjacent tracheoles) is directly related to the metabolic activity of the tissue. In highly active flight muscles of Diptera and Hymenoptera, for example, it has been calculated that the maximum theoretical distance between tracheoles is 6–8 μm. In practice tracheoles, which indent the muscle cells, are within 2–3 μm of each other, allowing a significant "safety margin" (Weis-Fogh, 1964).

In many insects, distal parts of tracheoles are not filled with air but liquid under normal resting conditions. During activity, however, the tracheoles become completely air-filled; that is, fluid is withdrawn from them only to return when activity ceases. Wigglesworth (1953) suggested that the level of fluid in tracheoles depends on the relative strengths of the capillary force drawing fluid along the tube and the osmotic pressure of the hemolymph. During metabolic activity, the osmotic pressure increases as organic respiratory substrates are degraded to smaller metabolites, causing fluid to be withdrawn from the tracheoles (perhaps via the pores mentioned earlier) and, therefore, bringing gaseous oxygen closer to the tissue cells (Figure 15.4). As the metabolites are fully oxidized and removed, the osmotic pressure will fall, and once again the capillary force will draw fluid along the tracheoles.

Though carbon dioxide is more soluble, and has a greater permeability constant, than oxygen in water and could conceivably move by diffusion through the hemolymph to leave the body via the integument, this route does not normally eliminate a significant quantity of the gas (e.g., 2–10% of the total in some dipteran larvae with thin cuticles). The great majority of carbon dioxide leaves by gaseous diffusion via the tracheal system.

3.2. Passive (Suction) Ventilation

To reduce water loss which accompanies the diffusion of gases out of the tracheal system, many insects, especially species or stages that do not have

---

of *Insect Morphology*. Copyright 1935 by McGraw-Hill, Inc. Used with permission of McGraw-Hill Book Company. D, after P. L. Miller, 1960, Respiration in the desert locust. II. The control of the spiracles, *J. Exp. Biol.* **37**:237–263. By permission of Cambridge University Press, New York. E, F, after V. B. Wigglesworth, 1965, *The Principles of Insect Physiology*, 6th ed., Methuen and Co. By permission of the author.]

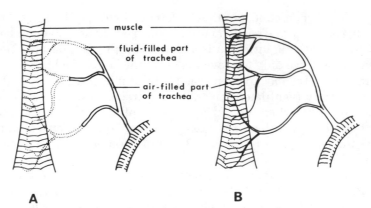

muscle

fluid-filled part
of trachea

air-filled part
of trachea

A

B

FIGURE 15.4. Changes in level of tracheolar fluid as a result of muscular activity. (A) Resting muscle and (B) active muscle. [After V. B. Wigglesworth, 1965, *The Principles of Insect Physiology*, 6th ed., Methuen and Co. By permission of the author.]

access to liquid water to replace that lost (e.g., pupae), practice suction ventilation. This makes use of the high solubility of carbon dioxide in water. In suction ventilation the spiracular valves are kept almost closed. As oxygen is used in metabolism, the carbon dioxide so produced is stored, largely as bicarbonate in the hemolymph and tissues but partially also in the gaseous state in the tracheal system. Thus, a slight vacuum is created within the tracheal system which sucks in more air. This inward movement of air effectively prevents the outward flow of water vapor. In *Hyalophora cecropia* pupae (Lepidoptera), when the concentration of carbon dioxide in the tracheae reaches about 6.5%, at which time the oxygen concentration is only about 3.5%, the valves are opened and remain open between 15 and 30 minutes. During this period there is rapid diffusion of carbon dioxide out of the tracheal system and massive release of carbon dioxide from the hemolymph. When the concentration of carbon dioxide in the tracheal system falls to 3.0%, the valves reclose (Figure 15.5). Experiments in which gases of different composition are perfused through the tracheal system or over the segmental ganglia have shown that there is dual control over the opening of the valves. In *Hyalophora* carbon dioxide in sufficient quantity directly stimulates relaxation of the valve closer muscle (the valve opens as a result of cuticular elasticity), whereas hypoxia (insufficient oxygen) acts at the level of the ganglia. In *Hyalophora* pupae the periods between bursts of carbon dioxide release may be as long as 7 hours, so that the time over which water vapor may be lost is very limited.

3.3. Active Ventilation

By alternately decreasing and increasing the volume of the tracheal system through compression and expansion of larger air tubes, a fraction of the air in the system is constantly renewed and the diffusion gradient between tracheae and tissues kept near the maximum. The volume changes are normally effected by contraction of abdominal dorsoventral and/or longitudinal muscles, which increases the hemolymph pressure, thus causing the tracheae to flatten or collapse. In some species both inspiration and expiration are brought about by

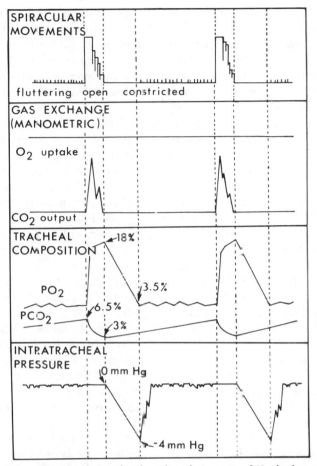

FIGURE 15.5. Discontinuous release of carbon dioxide in pupa of *Hyalophora cecropia* in relation to spiracular valve opening and closing. [After R. I. Levy and H. A. Schneiderman, 1966, Discontinuous respiration in insects. IV, *J. Insect Physiol.* **12**:465–492. By permission of Pergamon Press Ltd., Oxford.]

muscles; in others only expiration is under muscular control and inspiration occurs as a result of the natural elasticity of the body wall. During high metabolic activity, supplementary ventilation movements may occur. For example, the desert locust normally ventilates by means of dorsoventral movements of the abdomen but can supplement these by "telescoping" the abdomen and protraction/retraction of the head and prothorax (Miller, 1960).

The diffusion gradient can be further improved by increasing the volume of air in the tracheal system, which is renewed during each stroke (the tidal volume). This is achieved through the development of large, compressible air sacs. However, simple tidal flow (pumping of air in and out of all spiracles) is still somewhat inefficient because a considerable volume of air (the dead air space) remains within the system at each stroke. The size of the dead air space is greatly reduced by practicing unidirectional ventilation in which air is made to flow in one direction (usually anteroposteriorly) through the tracheal system. Unidirectional air flow is achieved by synchronizing the opening and closing of spiracular valves with ventilation movements. In the resting desert locust, for example, the first, second, and fourth spiracles are inspiratory, while the tenth (most posterior) is expiratory. When the insect becomes more active, the first four spiracles become inspiratory, the remainder expiratory. The spiracular valves do not form an airtight seal, however, so that a proportion of the inspired air continues to move tidally rather than unidirectionally (20% in the resting desert locust).

During flight, the oxygen consumption of an insect increases enormously (up to 24 times in the desert locust), almost entirely due to the metabolic activity of the flight muscles. To facilitate this activity, a massive exchange of air occurs in the pterothorax, made possible by certain structural features of the pterothoracic tracheal system and by changes in the body's normal (resting) ventilatory pattern. As noted earlier, the tracheal system of the pterothorax is effectively isolated from that of the rest of the body by reduction in the diameter or occlusion of the main longitudinal tracheae. Autoventilation of flight muscle tracheae also occurs. This is ventilation which results from movements of the nota and pleura during wing beating, and it serves to effect a considerable flow of air into and out of the thoracic tracheae. During autoventilation, normal unidirectional flow, where such occurs, becomes masked by the massive increase in tidal flow in the pterothorax. To achieve this tidal flow, in the desert locust, spiracles 2 and 3 remain permanently open. Spiracles 1 and 4–10, however, continue to open and close in synchrony with abdominal ventilation so that some unidirectional flow occurs. The rate of abdominal ventilatory movements also increases during flight to about four times the resting value, but these movements probably serve primarily to increase the rate of flow of hemolymph around the body, bringing fresh supplies of metabolites to the flight musculature.

Autoventilation is practiced by many Odonata, Orthoptera, Dictyoptera, Isoptera, Hemiptera, Lepidoptera, and Coleoptera, but is of little importance in Diptera and Hymenoptera. Its significance can be broadly correlated with body size, the type of muscles used in flight (see Chapter 14, Section 2.1), and the extent of movements of the thorax during flight. Odonata, for example, are generally large; movements of their thorax during wing beating are pronounced and their flight muscles are of the tubular type which lack indented tracheoles. Therefore, autoventilation is extremely important in this order. In contrast, in Hymenoptera and Diptera, movements of the thorax in flight are relatively slight and, therefore, the volume change that could be achieved in the tracheal system is probably not significant. However, the fibrillar flight muscles of these insects are much indented with tracheoles so that gaseous oxygen is brought close to the mitochondria. Thus, in Hymenoptera simple telescopic abdominal ventilation normally effects sufficient air exchange in the thorax. In Diptera, abdominal ventilation movements are weak or nonexistent, and here diffusion alone satisfies the insects' oxygen requirements in flight (Miller, 1974).

Ventilation movements are initiated within and controlled via the central nervous system. Some isolated abdominal segments, provided that they contain a ganglion, can carry out normal respiratory movements, though usually an appropriate stimulus such as carbon dioxide or hypoxia is necessary to initiate the movements. The coordination of these autonomous ventilation movements and, where unidirectional airflow occurs, of spiracular valve opening and closing, is achieved by a pacemaker center situated usually in the metathoracic or first abdominal ganglion. The nature of the center is not understood. However, in the desert locust, the pacemaker appears to send bursts of impulses to each ganglion, which both excite the motor neurons to the muscles used in expiration and inhibit those going to inspiratory muscles. The activity of the pacemaker center itself is modified by sensory input. For example, the brain and anterior thoracic ganglia, which are directly sensitive to carbon dioxide, can stimulate pacemaker activity so that ventilation is increased.

4. Gaseous Exchange in Aquatic Insects

Perhaps not surprisingly, in view of the rapid rate at which gases can diffuse within it, a gas-filled tracheal system has been retained by almost all aquatic forms in their evolution from terrestrial ancestors. Only rarely, for example, in the early larval stages of *Chironomus* and *Simulium* (Diptera), and *Acentropus* (Lepidoptera) is the system filled with liquid.

Oxygen may enter the tracheal system in gaseous form, that is, via functional spiracles (the "open" tracheal system) or may pass, in solution, directly across the body wall to the tracheal system, in which arrangement the spiracles are sealed (nonfunctional), and the tracheal system is said to be "closed." Aquatic insects with open tracheal systems exchange the gas within the system either by periodically visiting the water surface, by obtaining gas from gas-filled spaces in aquatic plants, or through the use of a "gas gill" (a bubble or film of air that covers the spiracles, into or out of which oxygen and carbon dioxide, respectively, can diffuse from/to the surrounding water). A significant amount of gas exchange may occur by direct diffusion across the body surface (cutaneous respiration) in larvae with an open system whose integument is thin, for example, mosquito larvae. Cutaneous respiration may entirely satisfy the requirements of insects with closed tracheal systems. However, in many species supplementary respiratory surfaces, tracheal gills, have evolved, though these often become important only under oxygen-deficient conditions.

4.1. Closed Tracheal Systems

For small aquatic insects or those with a low metabolic rate, diffusion of gases across the body wall provides an adequate means of obtaining oxygen and excreting carbon dioxide. In larger and/or more active forms, all or part of the body wall has a very thin cuticle and becomes richly tracheated to facilitate rapid entry and exit of gases from the tracheal system. Some tube-dwelling species show rhythmic movements which create water currents over the body so that water in the tube is periodically renewed.

In many aquatic larvae there are richly tracheated outgrowths of the body wall (hindgut in dragonfly larvae) collectively known as tracheal gills. Whether these function as accessory respiratory structures is controversial because even without them an insect may survive perfectly well and its oxygen consumption may not change. Included in the term "tracheal gills" are caudal lamellae (in Zygoptera), lateral abdominal gills (in Ephemeroptera, and some Zygoptera, Plecoptera, Neuroptera, and Coleoptera), rectal gills (in Anisoptera), and fingerlike structures, often found in tufts on various parts of the body (for example, in some Plecoptera and case-bearing Trichoptera).

Juvenile Zygoptera have three caudal lamellae (see Figure 6.10) which were believed by some early entomologists to function as rudders during swimming. However, swimming is a rare occurrence and, in any case, is not affected by loss of the lamellae. Other nineteenth-century authors, who observed their well-tracheated nature, suggested a respiratory function for them, though the ability of lamellaeless larvae to survive normally casts doubts on this suggestion. Experimentally, it has been demonstrated that the lamellae normally are major sites of gas exchange and in oxygen-deficient water become especially important. For example, in *Coenagrion* up to 60% of oxygen uptake may occur via the lamellae (Harnisch, 1958; cited from Mill, 1974). In certain *Enallagma* species normal larvae can survive in water with an oxygen content of only 2.4% saturation, whereas the minimum for lamellaeless larvae is 14.5% saturation. Above this value, lamellaeless larvae live apparently normally and obtain sufficient oxygen by cutaneous diffusion (Pennak and McColl, 1944).

In larval Ephemeroptera the lateral abdominal gills are paired, segmental, normally platelike structures (see Figures 6.2–6.6) whose size is inversely related to the oxygen content of the surrounding water. In other orders, the gills are filamentous (Mill, 1974). In burrowing mayfly larvae and larvae of other species which live in an oxygen-poor environment, the gills are an important site of gas exchange, and about one half of an insect's oxygen requirement may be obtained via this route. In addition, their rhythmic beating serves to move a current of water over the body to facilitate cutaneous respiration. The frequency at which the gills beat depends both on their size and on the oxygen content of the water. In species which inhabit fast-flowing water, the gills do not beat, and, further, removal of them does not affect oxygen consumption, indicating that most gaseous exchange occurs cutaneously.

The anterior part of the rectum of larval Anisoptera is enlarged to form a branchial chamber whose walls are folded and richly tracheated. Water is periodically drawn into and forced out of the rectum via the anus as a result of muscular movements of the abdomen, comparable to those which effect ventilation in terrestrial species. Contraction of dorsoventral muscles in the posterior abdominal segments decreases the abdominal volume and results in water being forced out of the rectum. A muscular diaphragm in the fifth segment may prevent hemolymph from being forced anteriorly. Water enters the rectum as the volume of the abdomen is increased, as a result of the contraction of two transverse muscles, the diaphragm and subintestinal muscles, which pull the tergal walls inward and force the sterna downward. This is aided by the natural elasticity of the body wall and by relaxation of the dorsoventral muscles (Hughes and Mill, 1966; Mill, 1977). The rate of ventilation of the branchial chamber varies with the oxygen content of the water and metabolic rate of the insect. Cutaneous respiration is probably not significant in Anisoptera, whose body wall cuticle is generally thick.

Among insects which have an open tracheal system can be traced a series of stages leading from complete dependence on atmospheric air (that is, where an insect must frequently visit the water surface to exchange the gas in its tracheal system) to the stage where an insect maintains around its body a supply of atmospheric gas into which oxygen can diffuse from the surrounding water at a sufficient rate to totally satisfy the insect's needs. Thus, the insect is completely independent of atmospheric air.

Surface-breathing insects must solve two problems. First, they must prevent waterlogging of the tracheal system when they are submerged, and second, they must be able to overcome the surface tension force at the air–water interface. Both problems are solved by having hydrofuge (water-repellent) structures around the spiracles. In many dipteran larvae special epidermal glands (perispiracular/peristigmatic glands) secrete an oily material at the entrance of the spiracle. The spiracles of some other aquatic insects are surrounded by hydrofuge hairs, which, when submerged, close over the opening but when in contact with the water surface spread out to permit exchange of air (Figure 15.6). In addition, insects may possess other modifications to the tracheal system to help cope with these problems, for example, reduction of the number of functional spiracles and restriction of the spiracles to special sites, typically at the tip of a posterior extension of the body (postabdominal respiratory siphon), as occurs in mosquito larvae. In a few species of Coleoptera and Diptera, whose larvae live in mud, the siphon, which is long and pointed, is forced into air spaces in the roots of aquatic plants.

Insects have gained variable degrees of independence from atmospheric air by holding a gas store about their body. The gas store may be subelytral or may occur as a thin film of gas over certain parts of the body, held in place by a mat of hydrofuge hairs (Figure 15.7A). A third arrangement is for the gas to be held as a layer adjacent to the body by cuticular extensions of the body wall adjacent to the spiracles, known as spiracular gills (Figure 15.7B).

The degree of independence from atmospheric air (measured as the length of time that an insect is able to remain submerged between visits to the surface) depends on a number of factors. These include (1) the metabolic rate of the insect, which itself is temperature-dependent; (2) the volume, shape, and location of the gas store, which determine the surface area of the store in contact

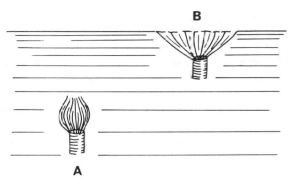

FIGURE 15.6. Hydrofuge hairs surrounding a spiracle. (A) Position when submerged and (B) position when at water surface. [After V. B. Wigglesworth, 1965, *The Principles of Insect Physiology*, 6th ed., Methuen and Co. By permission of the author.]

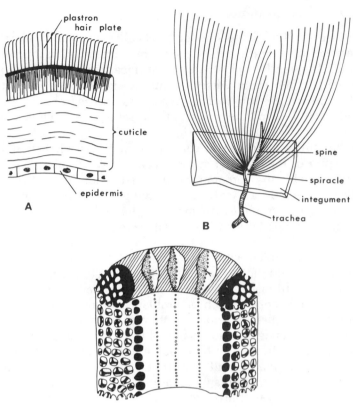

FIGURE 15.7. Examples of gas stores. (A) Hair pile on abdominal sternum of *Aphelocheirus* (Hemiptera), (B) spiracular gill of pupa of *Psephenoides gahani* (Coleoptera) comprising about 40 hollow branches, and (C) part of wall of spiracular gill branch of *P. gahani* showing cuticular struts that support the plastron. [A, after W. H. Thorpe and D. J. Crisp, 1947, Studies on plastron respiration. I, *J. Exp. Biol.* **24**:227–269. By permission of Cambridge University Press, New York. B, C, after H. E. Hinton, 1968, Spiracular gills, *Adv. Insect Physiol.* **5**:65–162. By permission of Academic Press Ltd., London.]

with the water; and (3) the oxygen content of the water. Factors (2) and (3) relate particularly to use of the gas store as a physical (gas) gill; that is, a structure which can take up oxygen from the surrounding water. When the oxygen used by an insect can be only partially replaced by diffusion into the gas store of oxygen from the water, the volume of the gas store will decrease and, eventually, the insect must return to the surface to renew the gas store. This is known as a temporary (compressible) gas gill. When an insect's oxygen requirements can be fully satisfied by diffusion of oxygen into the gill, whose volume will therefore remain constant, the gill is described as a permanent (incompressible) gas gill or plastron.

In a compressible gill the pressure of the gas will be equal to that of the surrounding water. The gas in the gill may be considered to include only oxygen and nitrogen, as the carbon dioxide produced in metabolism will readily dissolve in the water. Immediately after a visit to the surface, the composition of the gas in the gill will approximate that of air, that is, about 20%

oxygen and 80% nitrogen and will be in equilibrium with the dissolved gases in the surrounding water. As oxygen is used by an insect, a diffusion gradient will be set up so that this gas will tend to move into the gill. At the same time, the proportion of nitrogen in the gas will have increased and, therefore, nitrogen will tend to diffuse out of the gill in order to restore equilibrium. However, oxygen diffuses into the gill about three times as rapidly as nitrogen diffuses out. Thus, inward movement of oxygen will be more important than outward movement of nitrogen in restoring equilibrium. The effect of this is to prolong considerably the life of the gas store and, therefore, the duration over which an insect can remain submerged. The rate of diffusion, which is a function of the surface area of the gill exposed to the water, will obviously be greater in the case of films of gas spread over the body surface than in subelytral gas stores whose contact with the surrounding water is relatively slight. Ultimately, however, in a compressible gill the rate of oxygen use by the insect will reduce the proportion of oxygen in the gill below a critical level, and the insect will be stimulated to return to the surface.

For example, adult water beetles of the family Dytiscidae and back swimmers (Notonecta spp.: Hemiptera) are generally medium-to-large insects, which in summer, when water temperatures may approximate 20–25°C, show great activity. Under these conditions, the gas stored beneath the elytra (and also on the ventral body surface in Notonecta) must be regularly exchanged by visits to the water surface. However, in winter, when the water is cold (or may even be continuously frozen for several months as occurs, for example, on the Canadian prairies), the insects are more or less inactive. During this period, the gas store satisfies most or all of the insects' oxygen requirements.

In contrast, the volume of a plastron is constant but small. Hence, a plastron does not serve as a store of oxygen but solely as a gas gill. In certain adult Hemiptera, for example Aphelocheirus, and Coleoptera, for example Elmis, the plastron is held in place by dense mats of hydrofuge hairs. The hairs number about $200 \times 10^6/cm^2$ in Aphelocheirus, are bent at the tip and are slightly thickened at the base. As a result, they can resist becoming flattened (which would destroy the plastron) by the considerable pressure differences that may arise between the gas in the plastron and the surrounding water, as the insect moves into deeper water or uses up its oxygen supply. Spiracular gills are mostly found on the pupae of certain Diptera (for example, Tipulidae and Simuliidae) and Coleoptera (for example, Psephenicae) but occasionally occur on beetle larvae (Hinton, 1968). In almost all instances they include a plastron. In many dipteran pupae the gill is a long, hollowed-out structure (see Figure 15.7) which carries a plastron over its surface. The plastron is held in place by means of rigid cuticular struts and connects via fine tubes with the gas-filled center of the gill (and hence the spiracle).

Because the volume of the plastron remains constant, the nitrogen does not diffuse out. However, the rate and direction of diffusion of oxygen will depend on differences in the oxygen content of the plastron (always somewhat less than maximum because of use of oxygen by the insect) and the surrounding water. Therefore, to ensure that oxygen always diffuses into the plastron, the water around should be saturated with the gas. If the water is not saturated with oxygen, then the latter will either diffuse into the plastron relatively slowly or may even diffuse out, leading eventually to asphyxiation of the insect. It is not surprising, then, to discover that plastron respiration is employed by aquatic insects living in fast-moving streams, at the edges of shallow lakes, and in the intertidal zone. Plastron-bearing species known to inhabit water whose oxygen content may fluctuate daily, for example, that of marshes which drops greatly at night, may employ behavioral means to overcome the danger of asphyxiation, such as moving closer to the water surface or even climbing out of the water. Hinton (1968) also points out that the waters occupied by plastron-bearers are often subject to changes in level, leaving the insects periodically exposed to atmospheric air. However, desiccation does not present a problem to these insects because the connection between the plastron and internal tissues may be quite restricted. In addition, bulk flow of gas within the plastron is negligible, so that evaporative water loss is small.

5. Gaseous Exchange in Endoparasitic Insects

It is probably not surprising that endoparasitic insects, since they too are surrounded by fluid, show many parallels with aquatic insects in the way that they obtain oxygen. Most endoparasites satisfy a proportion of their requirements by cutaneous diffusion. In some first-instar larvae of Hymenoptera and Diptera the tracheal system may be liquid-filled, but generally it is gas-filled with closed spiracles and includes a rich network of branches immediately beneath the integument. Many endoparasitic forms, especially larval Braconidae, Chalcidae, and Ichneumonidae (Hymenoptera), and some Diptera, possess "tails" filled with hemolymph or can evaginate the wall of the hindgut through the anus. It has been suggested that these structures may facilitate gaseous exchange though the evidence on which this suggestion is based is, in most instances, not strong.

Endoparasites with greater oxygen requirements usually are in direct contact with atmospheric air either via the integument of the host or via the host's tracheal system. In larvae of many Chalcidoidea, for example, only the posterior spiracles are functional, and these open into an air cavity formed at the base of the egg pedicel which penetrates the host's integument (Figure 15.8A). Many larval Tachinidae (Diptera) become enclosed in a "respiratory funnel" produced by the host in an attempt to encapsulate the parasite (Figure 15.8B). The funnel is produced by inward growth of the host's integument or tracheal wall. Within it, the parasite attaches itself by means of mouth hooks while retaining contact with atmospheric air via the entrance of the funnel.

6. Summary

The tracheal system is a system of gas-filled tubes which develops embryonically as a series of segmental invaginations of the integument. The invaginations anastomose and branch and eventually form tracheoles across which the vast majority of gaseous exchange occurs. The external openings of

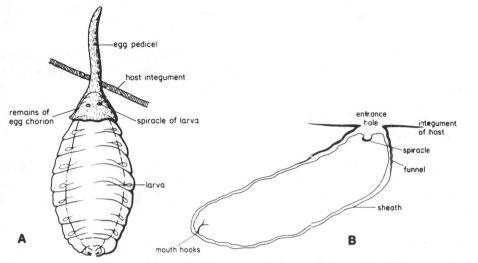

FIGURE 15.8. Respiratory systems of endoparasites. (A) Larva of *Blastothrix* (Hymenoptera) attached posteriorly to remains of egg, thereby maintaining contact with the atmosphere via the egg pedicel; and (B) larva of *Thrixion* (Diptera) surrounded by the respiratory funnel formed by ingrowth of the host's integument. [From A. D. Imms, 1937, *Recent Advances in Entomology*, 2nd ed. By permission of Churchill-Livingstone, Publishers.]

the tracheal system (spiracles) are generally equipped with valves, hairs, or sieve plates, whose primary function is probably prevention of water loss.

Because oxygen can diffuse more rapidly in the gaseous state and is in higher concentration in air than in water, the requirements of most small insects and many large resting insects can be satisfied entirely by diffusion. To reduce water loss, larger insects, especially species or stages that do not have access to liquid water, may use passive (suction) ventilation. In suction ventilation, the spiracles are kept almost closed and carbon dioxide is temporarily stored, largely as bicarbonate in the hemolymph. As oxygen is used, a slight vacuum is created in the tracheal system that sucks in more air and reduces outward diffusion of water vapor. Periodically, the spiracular valves are opened for a short time when a massive release of carbon dioxide occurs.

To increase the diffusion gradient between the tracheal system and tissues, many large insects ventilate the system by alternately increasing and decreasing its volume. Frequently the volume of tidal air moved during ventilation is increased through the development of large compressible air sacs, and by practicing unidirectional airflow. Autoventilation of flight muscles, based on movements of the pterothorax during wing beating, may occur.

Aquatic insects have either an open tracheal system with functional spiracles or a closed tracheal system in which the spiracles are sealed. In many insects with a closed tracheal system, cutaneous diffusion may entirely satisfy oxygen requirements. Accessory respiratory structures (tracheal gills) may be present though these are often important only under oxygen-deficient conditions. Some insects with open tracheal systems obtain oxygen by periodic visits to the water surface, or from air spaces in plants. Others hold a store of gas (gas gill) about their body. This may be either temporary, when an insect must visit the water surface to renew the oxygen content of the gill, or permanent (a plastron), when oxygen is renewed by diffusion into the gill from the surrounding medium.

For many endoparasitic insects, cutaneous diffusion is sufficient to satisfy requirements. In others there are special structural adaptations that ensure the parasite remains in contact with atmospheric air.

7. Literature

Gaseous exchange is reviewed by Buck (1962), Mill (1972, 1974), and Miller (1974). Hinton (1968) discusses spiracular gills. The regulation of gaseous exchange is examined by Miller (1966). Whitten (1972) provides structural details of the tracheal system.

Buck, J., 1962, Some physical aspects of insect respiration, *Annu. Rev. Entomol.* **7**:27–56.

Hinton, H. E., 1968, Spiracular gills, *Adv. Insect Physiol.* **5**:65–162.

Hughes, G. M., and Mill, P. J., 1966, Patterns of ventilation in dragonfly larvae, *J. Exp. Biol.* **44**:317–334.

Locke, M., 1966, The structure and formation of the cuticulin layer in the epicuticle of an insect, *Calpodes ethlius, J. Morphol.* **118**:461–494.

Mill, P. J., 1972, *Respiration in the Invertebrates*, Macmillan, London.

Mill, P. J., 1974, Respiration: Aquatic insects, in: *The Physiology of Insecta*, 2nd ed., Vol. VI (M. Rockstein, ed.), Academic Press, New York.

Mill, P. J., 1977, Ventilation motor mechanisms in the dragonfly and other insects, in: *Identified Neurons and Behavior of Arthropods* (G. Hoyle, ed.), Plenum Press, New York.

Miller, P. L., 1960, Respiration in the desert locust. I.–III., *J. Exp. Biol.* **37**:224–236, 237–263, 264–278.

Miller, P. L., 1966, The regulation of breathing in insects, *Adv. Insect Physiol.* **3**:279–344.

Miller, P. L, 1974, Respiration–aerial gas transport, in: *The Physiology of Insecta*, 2nd ed., Vol. VI (M. Rockstein, ed.), Academic Press, New York.

Pennak, R. W., and McColl, C. M., 1944, An experimental study of oxygen absorption in some damselfly naiads, *J. Cell. Comp. Physiol.* **23**:1–10.

Weis-Fogh, T., 1964, Diffusion in insect wing muscle, the most active tissue known, *J. Exp. Biol.* **41**:229–246.

Whitten, M. J., 1972, Comparative anatomy of the tracheal system, *Annu. Rev. Entomol.* **17**:373–402.

Wigglesworth, V. B., 1953, Surface forces in the tracheal system of insects, *Q. J. Micros. Sci.* **94**:507–522.

16

Food Uptake and Utilization

1. Introduction

Insects feed on a wide range of organic materials. The majority of species are phytophagous, and these form an important link in the transfer of energy from primary producers to second-order consumers. Others are carnivorous, omnivorous, or parasitic on other animals. In accord with the diversity of feeding habits, the means by which insects locate their food, the structure and physiology of their digestive system, and their metabolism are highly varied.

The feeding habits of insects take on special significance for Man on the one hand, because of the enormous damage that feeding insects do to his food, clothing, and health, and, on the other, because of the massive benefits that insects provide as plant pollinators during their search for food (see also Chapter 24). In addition, because many species are easily and cheaply mass-cultured in the laboratory, they have been used widely in research on digestion and absorption, as well as in the elucidation of basic biochemical pathways, the role of specific nutrients, and other aspects of animal metabolism.

2. Food Selection and Feeding

Visual, chemical, and mechanical stimuli may act as cues which enable an insect to locate and/or select material suitable for ingestion. As food is ingested, it may be tasted, and this determines whether or not feeding continues. The extent of stimulation required to initiate feeding varies with the state of an insect. Starved insects, for example, may become highly sensitive to odors or tastes associated with their normal food. In extreme cases, they may become quite indiscriminate in terms of what they ingest. Feeding is terminated either by mechanical stimuli, for example, stretching of the crop wall, or by chemical means, for example, by adaptation of chemoreceptors so that the feeding reflex is no longer initiated (Dethier, 1966).

In some plant-feeding (phytophagous) species visual stimuli such as particular patterns (especially stripes) or colors may serve to initially attract an insect to a potential food source Usually, however, the initial orientation, where this occurs, is dependent on olfactory stimuli. In many larval forms there appear to be no specific orienting stimuli because, under normal circumstances, larvae remain on the food plant selected by the mother prior to oviposition. When feeding begins, food is tasted by receptors located either on the mouthparts, especially the palps, or within the pharynx. Tasting determines whether or not intake continues and depends on the presence in the food of specific stimulating or inhibitory substances. These substances may have nutritional value to the insect or may be nutritionally unimportant ("token stimuli"). Nutritional factors are almost always stimulating in effect. Sugars, especially sucrose, are important phagostimulants for most phytophagous insects. Amino acids, in contrast, are generally by themselves weakly stimulating or nonstimulating, though may act synergistically with certain sugars or token stimuli. For example, Heron (1965) showed in the spruce budworm (*Choristoneura fumiferana*) that, whereas sucrose and L-proline in low concentration were individually only weak phagostimulants, a mixture of the two substances was highly stimulating. In addition to sugars and amino acids, other specific nutrients may stimulate feeding in a given species. Such nutrients include vitamins, phospholipids, and steroids. Token stimuli may either stimulate or inhibit feeding. Thus, derivatives of mustard oil, produced by cruciferous plants, including cabbage and its relatives, are important phagostimulants for a variety of insects which normally feed on these plants, for example, larvae of the diamondback moth (*Plutella maculipennis*), the cabbage aphid (*Brevicoryne brassicae*), and the mustard beetle (*Phaedon cochleariae*) (Dethier, 1966). Indeed, *Plutella* will feed naturally only on plants which contain mustard oil compounds. Such species, whose choice of food is limited, are said to be oligophagous. In extreme cases, an insect may be restricted to feeding on a single plant species and is described as monophagous. Species which may feed on a wide variety of plants are polyphagous, though it must be noted that even these exhibit selectivity when given a choice.

In many predaceous insects, especially those which actively pursue prey, vision is of primary importance in locating and capturing food. As noted in Chapter 12 (Section 7.1.2), some predaceous insects have binocular vision which enables them to determine when prey is within catching distance. Carnivorous species, especially larval forms, whose visual sense is less well developed, depend on chemical or tactile stimuli to find prey. For example, many beetle larvae which live on or in the ground locate prey by their scent; ladybird beetle larvae, on the other hand, must make physical contact with prey in order to determine its presence. Species parasitic on other animals usually locate a host by its scent, though tsetse flies may initially orient by visual means to a potential host. For many species which feed on the blood of birds and mammals, temperature and/or humidity gradients are important in determining the precise location at which an insect alights on a host and begins to feed.

The extent of food specificity for carnivorous insects is variable. Many insects are quite nonspecific and will attempt to capture and eat any organism which falls within a given size range (even to the extent of being cannibalistic!). Others are more selective, for example, spider wasps (Pompilidae), as their name indicates, capture only spiders for provisioning their nest. Parasitic insects, too, exhibit various degrees of host specificity. Thus, certain sarcophagid flies parasitize a range of grasshopper species; the common cattle grub (*Hypoderma lineata*) is typically found on cattle or bison, rarely on horses and man; lice are extremely host-specific, as would be expected of sedentary species.

Apart from the specific cues outlined above that facilitate location and selection of food, there are other factors that influence feeding activity. Typically insects do not feed shortly before and after a molt or when there are mature eggs in the abdomen. In addition, a diurnal rhythm of feeding activity may occur, in response to a specific light, temperature, or humidity stimulus. For example, the red locust (*Nomadacris septemfasciata*) feeds in the morning and evening, and many mosquitoes feed during the early evening (though this may change in different habitats). Pupae, most insects when in diapause, and some adult Ephemeroptera, Lepidoptera, and Diptera, do not feed.

3. The Alimentary System

The gut and its associated glands (Figure 16.1) triturate, lubricate, store, digest, and absorb food material and expel the undigested remains. Structural differences throughout the system reflect regional specialization for performance of these functions and are correlated also with feeding habits and the

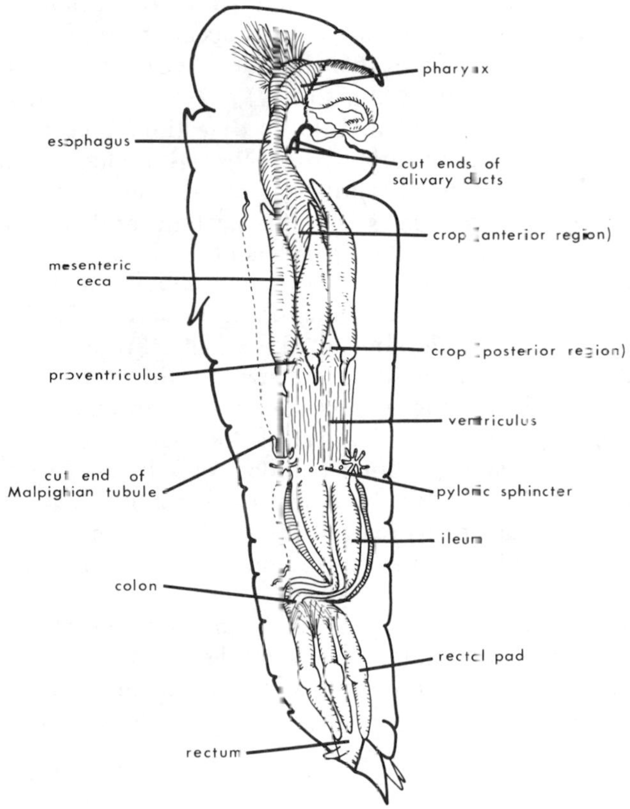

FIGURE 16.1. Alimentary canal and associated structures of a locust. [After C. Hodge, 1939, The anatomy and histology of the alimentary tract of *Locusta migratoria* L. (Orthoptera: Acrididae), *J. Morphol.* **64**:375–399. By permission of The Wistar Press.]

nature of normal food material. The structure of the system may vary at different stages of the life history because of the different feeding habits of the larva and adult of a species. The gut normally occurs as a continuous tube between the mouth and anus, and its length is broadly correlated with feeding habits, being short in carnivorous forms where digestion and absorption occur relatively rapidly, and longer (often convoluted) in phytophagous forms. In a few species which feed on fluids, such as larvae of Neuroptera and Hymenoptera Apocrita and some adult Heteroptera, there is little or no solid waste in the food, and the junction between the midgut and hindgut is occluded.

As Figure 16.1 indicates, food first enters the buccal cavity, which is enclosed by the mouthparts and is not strictly part of the gut. It is into the buccal cavity that the salivary glands release their products. The gut proper comprises three main regions: the foregut, in which the food may be stored, filtered, and partially digested; the midgut, which is the primary site for digestion and absorption of food; and the hindgut, where some absorption and feces formation occur.

3.1. Salivary Glands

Salivary glands are present in most insects, though their form and function are extremely varied. Frequently they are known by other names according to either the site at which their duct enters the buccal cavity, for example, labial glands and mandibular glands, or their function, for example, silk glands and venom glands. Typically, saliva is a watery, enzyme-containing fluid which serves to lubricate the food and initiate its digestion. Like that of man, the saliva generally contains only carbohydrate-digesting enzymes (amylase and invertase), though there are exceptions to this statement. For example, the saliva of some carnivorous species contains protein- and/or fat-digesting enzymes only; that of bloodsucking species has no enzymes.

Other substances which may occur in saliva, though having no direct role in digestion, are important in food acquisition. The saliva of aphids contains a pectinase that facilitates penetration of the stylets through the intercellular spaces of plant tissues. Hyaluronidase, an enzyme which breaks down connective tissue, is secreted by some insects which suck animal tissue fluids. Anticoagulants are present in the saliva of bloodsucking species such as mosquitoes. Toxins (venoms), which paralyze or kill the prey, occur in the saliva of some assassin bugs (Pentatomidae) and robber flies (Asilidae). It is also reported that substances which induce gall formation by stimulating cell division and elongation are present in the saliva of some gall-inhabiting species.

In some species the glands have taken on functions quite unrelated to digestion, for example, production of silk by the labial glands of caterpillars and caddis fly larvae, and pheromone production by the mandibular glands of the queen honeybee.

3.2. Foregut

The foregut, formed during embryogenesis by invagination of the integument, is lined with cuticle (the intima) which is shed at each molt. Surrounding the intima, which may be folded to enable the gut to stretch when filled, is a thin epidermis, small bundles of longitudinal muscle, a thick layer of circular muscle, and a layer of connective tissue through which run nerves and tracheae (Figure 16.2). The foregut is generally differentiated into pharynx, esophagus, crop, and proventriculus. Attached to the pharyngeal intima are dilator muscles. These are especially well developed in sucking insects and form the pharyngeal pump (see Chapter 3, Section 3.2.2). The esophagus is usually narrow but posteriorly may be dilated to form the crop where food is stored. During storage the food may undergo some digestion in insects whose saliva contains enzymes or which regurgitate digestive fluid from the midgut. In some species the intima of the crop forms spines or ridges which probably aid in breaking up solid food into smaller particles and mixing in the digestive fluid (Figure 16.2A). The hindmost region of the foregut is the proventriculus, which may serve as a valve regulating the rate at which food enters the midgut,

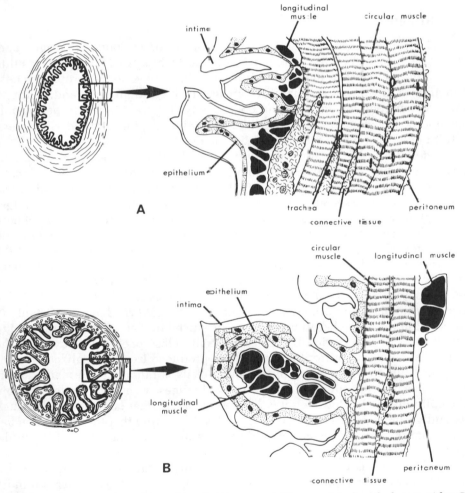

FIGURE 16.2. Transverse sections through (A) crop and (B) proventriculus of a locust. [After C. Hodge, 1939, The anatomy and histology of the alimentary tract of *Locusta migratoria* L. [Orthoptera: Acrididae], *J. Morphol.* **64**:375–399. By permission of The Wistar Press.]

as a filter separating liquid and solid components, or as a grinder to further break up solid material. Its structure is, accordingly, quite varied. In species where it acts as a valve the intima of the proventriculus may form longitudinal folds and the circular muscle layer is thickened to form a sphincter. When a filter, the proventriculus contains spines which hold back the solid material, permitting only liquids to move posteriorly. Where the proventriculus acts as a gizzard, grinding up food, the intima is formed into strong, radially arranged teeth, and a thick layer of circular muscle covers the entire structure (Figure 16.2B).

Posteriorly the foregut is invaginated slightly into the midgut to form the esophageal (=stomodeal) invagination (Figure 16.3). Its function is to ensure that food enters the midgut within the peritrophic membrane. It also appears to assist in molding the peritrophic membrane into the correct shape in some insects.

3.3. Midgut

The midgut (=ventriculus=mesenteron) is of endodermal origin and, therefore, has no cuticular lining. In most insects, however, it is lined by a thin peritrophic membrane made of chitin (and, in some species, protein) whose function is to prevent mechanical damage to the midgut cells and, perhaps, entry of microorganisms into the body cavity (Figure 16.4). The peritrophic membrane is generally absent in fluid-feeding insects, for example, Hemiptera, adult Lepidoptera, and bloodsucking Diptera.

The peritrophic membrane may be formed either by delamination of successive concentric lamellae throughout the length of the midgut (in Odonata, Ephemeroptera, Phasmida, some Orthoptera, some Coleoptera, and larval Lepidoptera), or by secretion from special cells at the anterior end of the midgut (in Diptera, and perhaps Dermaptera and Isoptera), or by a combination of both methods as seems to occur in Dictyoptera, other Orthoptera, Hymenoptera, and Neuroptera. In Diptera the esophageal invagination presses firmly against the anterior wall of the midgut so that the originally viscous secretion of the peritrophic membrane-producing cells, as it hardens, is squeezed to form the tubular membrane. The membrane is made up of a meshwork of microfibrils between which is a thin proteinaceous film. The microfibrils have a constant 60° orientation to each other in membranes produced by delamination, thought to result from their secretion by the hexagonally close-packed microvilli of the epithelial cells. In peritrophic membranes produced from specialized anterior midgut cells, the orientation of the microfibrils is random. The membrane is permeable to the products of digestion and to digestive enzymes released from the epithelial cells. However, it is not permeable to other large molecules, such as undigested proteins and polysaccharides, indicating that the membrane has a distinct polarity and is not merely an ultrafilter (see Richards and Richards, 1977).

The midgut is usually not differentiated into structurally distinct regions apart from the development, at the anterior end, of a variable number of blindly ending ceca, which serve to increase the surface area available for enzyme secretion and absorption of digested material. In many Heteroptera, however,

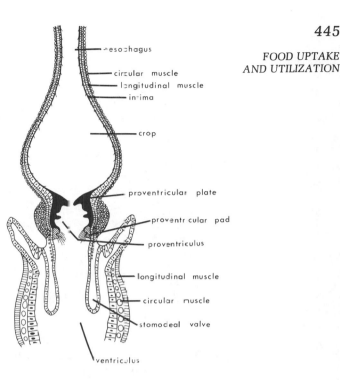

FIGURE 16.3. Longitudinal section through crop, proventriculus, and anterior midgut of a cockroach. [From R. E. Snodgrass, *Principles of Insect Morphology.* Copyright 1935 by McGraw-Hill, Inc. Used with permission of McGraw-Hill Book Company.]

the midgut is divided into three or four easily visible regions. In the chinch bug (*Blissus leucopterus*) four such regions occur (Figure 16.5). The anterior region is large and saclike, and serves as a storage region (no crop is present). The second region serves as a valve to regulate the flow of material into the third region where digestion probably occurs. Ten fingerlike ceca filled with bacteria

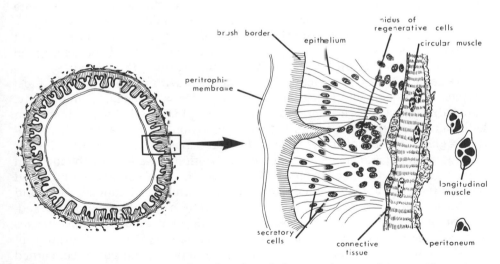

FIGURE 16.4. Transverse section through midgut of a locust. [After C. Hodge, 1939, The anatomy and histology of the alimentary tract of *Locusta migratoria* L. (Orthoptera: Acrididae), *J. Morphol.* **64:**375–399. By permission of The Wistar Press.]

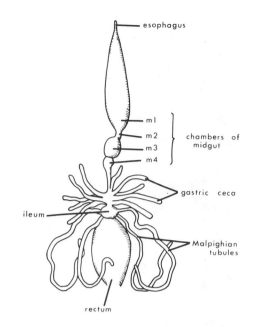

esophagus

m1

m2

m3   chambers of
midgut

m4

gastric ceca

ileum

Malpighian
tubules

rectum

FIGURE 16.5. Alimentary canal of chinch bug (*Blissus leucopterus*) showing regional differentiation of midgut. [After A. D. Imms, 1957, *A General Textbook of Entomology*, 9th ed. (revised by O. W. Richards and R. G. Davies). By permission of Chapman and Hall Ltd.]

are attached to the fourth region, which may be absorptive in function. The role of the bacteria is not known.

In many Homoptera, which feed on plant sap, the midgut is modified both morphologically and anatomically so that excess water present in the food can be removed, thus preventing dilution of the hemolymph. Though details vary among different groups of Homoptera, the anterior end of the midgut (or, in some species, the posterior part of the esophagus) is brought into close contact with the posterior region of the midgut (or anterior hindgut), and the region of contact becomes enclosed within a sac called the "filter chamber" (Figure 16.6). Such an arrangement facilitates rapid movement of water by osmosis from the lumen of the anterior midgut across the wall of the posterior midgut and possibly also the Malpighian tubules. Thus, relatively little of the original water in the food actually passes along the full length of the midgut.

Despite the lack of structural differentiation within the midgut of most insects, functional differentiation occurs, and this is reflected histologically. Specialization of certain anterior cells for peritrophic membrane production in Diptera was noted earlier. In addition, differentiation into digestive and absorptive regions occurs in many species. In tsetse flies the cells of the anterior midgut are small and are concerned with absorption of water from the ingested blood. They produce no enzymes and digestion does not begin until food reaches the middle region whose cells are large, rich in ribonucleic acid, and produce enzymes. In the posterior midgut the cells are smaller, closely packed, and probably concerned with absorption of digested food. In some species different regions of the midgut are apparently adapted to the absorption of particular food materials. In *Aedes* larvae the anterior midgut is concerned with fat absorption and storage, whereas the posterior portion absorbs carbohydrates and stores them as glycogen.

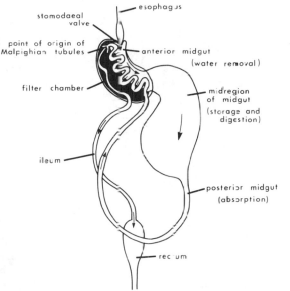

esophagus

stomodaeal
valve

point of origin of
Malpighian tubules

anterior midgut

(water removal)

filter chamber

midregion
of midgut
(storage and
digestion)

ileum

posterior midgut
(absorption)

rectum

FIGURE 16.6. Alimentary canal of cercopid (Homoptera) showing filter chamber arrangement. [From R. E. Snodgrass, *Principles of Insect Morphology.* Copyright 1935 by McGraw-Hill, Inc. Used with permission of McGraw-Hill Book Company.]

3.4. Hindgut

The hindgut is an ectodermal derivative and, as such, is lined with cuticle, though this is thinner than that of the foregut, a feature related to the absorptive function of this region. The epithelial cells which surround the cuticle are flattened except in the rectal pads (see below) where they become highly columnar and filled with mitochondria. Muscles are only weakly developed and, usually, the longitudinal strands lie outside the sheet of circular muscle.

In the hindgut the following regions usually can be distinguished: pylorus, ileum, and rectum. The pylorus may have a well-developed circular muscle layer (pyloric sphincter) and regulate the movement of material from midgut to hindgut. Also, the Malpighian tubules characteristically enter the gut in this region. The ileum (Figure 16.7A) is generally a narrow tube which serves to conduct undigested food to the rectum for final processing. In some insects, however, some absorption of ions and/or water may occur in this region. In a few species production and excretion of nitrogenous wastes occur in the ileum (see Chapter 18, Section 2.2). In many wood-eating insects, for example, species of termites and beetles. the ileum is dilated to form a "fermentation pouch" housing bacteria or protozoa which digest wood particles. The products of digestion, when liberated by the microorganisms, are absorbed across the wall of the ileum. The most posterior part of the gut, the rectum, is frequently dilated. Though for the most part thin-walled, the rectum includes six to eight thick-walled rectal pads (Figure 16.7B) whose function is to absorb ions, water, and small organic molecules (see Chapter 18, Section 4). As a result, the feces of terrestrial insects are expelled as a more or less dry pellet. Frequently, the pellets are ensheathed within the peritrophic membrane which continues into the hindgut.

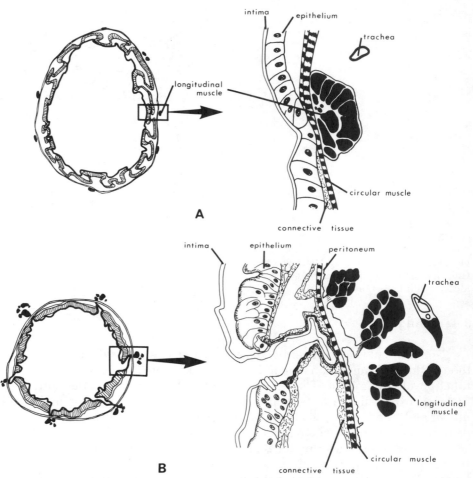

FIGURE 16.7. Transverse sections through (A) ileum and (B) rectum of a locust. [After C. Hodge, 1939, The anatomy and histology of the alimentary tract of *Locusta migratoria* L. (Orthoptera: Acrididae), *J. Morphol.* **64**:375–399. By permission of The Wistar Press.]

4. Gut Physiology

The primary functions of the alimentary canal are digestion and absorption. For these processes to occur efficiently, food is moved along the canal. In some species, enzyme secretions are moved anteriorly so that digestion can begin some time before food reaches the region of absorption.

4.1. Gut Movements

Rhythmic waves of peristalsis move the food posteriorly through the gut. The contractions are myogenic, that is, originate within the muscles themselves rather than occurring as a result of nervous stimuli, and myogenic centers have been located in the esophagus, crop, and proventriculus, in *Galleria*, for example. In insects which form a peritrophic membrane at the anterior end of the midgut, backward movement of food may be aided by "growth" of the membrane. Antiperistaltic movements also occur in some species and serve to move digestive fluid forward from the midgut into the crop.

The rate at which food moves through the gut is not uniform. It varies according to the "physiological state" of an insect; for example, it is greater when an insect has been starved previously or is active. The rate may also differ between sexes and with age. Another important variable is the nature of the food. Some insects are able to move some components of the diet rapidly through the gut while retaining others for considerable periods. Within the gut, food moves at variable rates in different regions.

The proventricular and pyloric valves are important regulators of food movement, though little is known about how their opening and closing are controlled. In *Periplaneta* opening of the proventriculus has been shown to depend on the osmotic pressure of ingested fluid (Davey and Treherne, 1963). As the concentration is increased, the proventriculus opens less often and less widely, and vice versa. Davey and Treherne suggest that osmoreceptors in the pharynx provide information on the osmotic pressure of the food and this information travels via the frontal ganglion to the ingluvial ganglion which controls the proventriculus. Removal of the frontal ganglion results in a massive buildup of food in the crop in *Schistocerca* (Highnam *et al.*, 1966).

4.2. Digestion

As noted above, digestion may be initiated by enzymes present in the saliva either mixed with the food as it enters the buccal cavity or secreted onto the food prior to ingestion. Most digestion is dependent, however, on enzymes secreted by the midgut epithelium. Digestion mostly occurs in the lumen of the midgut, though regurgitation of digestive fluid into the crop is important in some species. In wood-eating forms, digestion is largely carried out by microorganisms in the hindgut.

4.2.1. Digestive Enzymes

A wide variety and large number of digestive enzymes have been reported to be produced by insects. In many instances, however, enzymes have been characterized (and named) on the basis of their activity on "unnatural" substrates, that is, materials which do not occur in the normal diet of the insect. This is because many digestive enzymes, especially carbohydrases, are "group-specific"; that is, they will hydrolyze any substrate which includes a particular bond between two parts of the molecule. For example, α-glucosidase splits all α-glucosides, including sucrose, maltose, furanose, trehalose, and melezitose. Further, in preparing enzyme extracts for analysis, either gut contents or midgut tissue homogenates are typically used. As House (1974) notes, the former may include enzymes derived from the food per se, while the latter contains endoenzymes (intracellular enzymes) which have no digestive function. Thus, reports on digestive enzyme activity must be examined cautiously.

As would be expected, the enzymes produced reflect both qualitatively and quantitatively the normal constituents of the diet. Omnivorous species produce enzymes for digesting proteins, fats, and carbohydrates. Carnivorous species produce mainly lipases and proteases; in some species these may be highly specific in action. Blowfly larvae (*Lucilia cuprina*), for example, produce large amounts of collagenase. The nature of the enzymes produced may change at different stages of the life history as the diet of an insect changes. For example, caterpillars feeding on plant tissue secrete a spectrum of enzymes, whereas nectar-feeding adult Lepidoptera produce only invertase.

A wide range of carbohydrates can be digested by insects, even though only a few distinct enzymes may be produced. As noted earlier, α-glucosidase will effect hydrolysis of all α-glucosides. Likewise, β-glucosidase facilitates splitting of cellobiose, gentiobiose, and phenylglucosides; β-galactosidase hydrolyses β-galactosides such as lactose. In some species, however, there appear to be carbohydrate-digesting enzymes which exhibit absolute specificity. Thus, adult *Lucilia cuprina* produce an α-glucosidase, trehalase, which splits only trehalose. The normal polysaccharide-digesting enzyme produced is amylase for hydrolysis of starch, though particular species may produce enzymes for digestion of other polysaccharides. For example, wood-boring Cerambycidae and Anobiidae (Coleoptera) produce a cellulase, though in most insects production of this enzyme is restricted to microorganisms present in the hindgut; Scolytinae (Coleoptera) produce a hemicellulase; chitinase is reported to occur in the intestinal juice of *Periplaneta*; and lichenase is produced by some herbivorous Orthoptera.

As in other organisms, the protein-digesting enzymes produced by the midgut are divisible into two types, endopeptidases, which effect the initial splitting of proteins into polypeptides, and exopeptidases, which bring about degradation of polypeptides by the sequential splitting off of individual amino acids from each end of a molecule. Exopeptidases can be further categorized into carboxypeptidases, which remove amino acids from the carboxylic end of a polypeptide, and aminopeptidases, which cause hydrolysis at the amino end of a molecule. A dipeptidase also is frequently present. In some species only endopeptidases occur in the midgut lumen, indicating that the final stages of protein digestion may be intracellular. Specific enzymes are produced by some insects for the digestion of particularly resistant structural proteins. Collagenase has been mentioned already. Keratin, the primary constituent of wool, hair, and feathers, is a fibrous protein whose polypeptide components lie side by side linked by highly stable disulfide bonds between adjacent sulfur-containing amino acids, such as cystine and methionine. A keratinase has been identified in clothes moth larvae (*Tineola*) and may also occur in other keratin-digesting species, such as dermestid beetles and Mallophaga. The keratinase is active only under anaerobic (reducing) conditions and, in this context, it is interesting to note that the midgut of *Tineola* is poorly tracheated.

Dietary fats of either animal or plant origin are almost always triglycerides, that is, glycerol in combination with three fatty acid molecules. The latter may range from unsaturated to fully saturated. Lipases, which hydrolyse fats to the constituent fatty acids and glycerol, exhibit low specificity. Therefore, the presence of one such enzyme will normally satisfy an insect's requirements. In a few species, however, at least two lipases have been identified, having different pH optima and acting upon triglycerides of different sizes.

According to House (1974), three factors markedly affect digestion in insects: pH, buffering capacity, and redox potential of the gut.

The pH determines not only the activity of digestive enzymes, but also the nature and extent of microorganisms in the gut and the solubility of certain materials in the gut lumen. The latter affects the osmotic pressure of the gut contents and, in turn, the rate of absorption of molecules across the gut wall. Analyses of the pH in various regions of the gut have been made for a wide range of species, and various authors have attempted to correlate these with the feeding habits or phylogenetic position of an insect. At best, these correlations are only broadly correct, and many exceptions are known. Swingle (1931, cited from House, 1974) observed that, in most insects studied by him, the gut is slightly acidic or slightly alkaline throughout its length. Further, the pH generally increases from foregut to midgut, then decreases from midgut to hindgut. Though the latter is true for most phytophagous species, in many omnivorous and carnivorous species the pH of the hindgut is greater than that of the midgut. Many variables affect the pH of different regions of the gut. Generally the pH of the crop is the same as that of the food, though in some species it is consistently less than 7 due to the digestive activity of microorganisms or regurgitation of digestive juice from the midgut. The pH of the midgut differs among species but tends to be constant for a given species because of the presence in this region of buffering agents. In a few species there are local variations in pH within the midgut, which are perhaps related to changes in digestive function from one part to another. The hindgut typically has a pH slightly less than 7, presumably resulting from the presence of the nitrogenous waste product, uric acid (see Chapter 18, Section 3.2). The hindgut contents of some phytophagous species may be quite acidic due to the formation of organic acids from cellulose by symbiotic microorganisms.

The relatively constant pH found in the midgut results from the presence in the lumen of both inorganic and organic buffering agents. In some species, inorganic ions, especially phosphates, but including aluminum, ammonium, calcium, iron, magnesium, potassium, sodium, carbonate, chloride, and nitrate, seem to offer sufficient buffering capacity. In other species organic acids, including amino acids and proteins, tend to supplement or replace the buffering effect of the inorganic ions.

The redox potential, which measures ability to gain or lose electrons, that is, to be reduced or oxidized, respectively, is an important factor in digestion in some insects. Normally, the redox potential of the gut is positive, which is indicative of oxidizing (aerobic) conditions. However, in species able to digest keratin the redox potential of the midgut fluid is strongly negative. It has been suggested that such an anaerobic (reducing) environment is necessary to effect splitting (reduction) of the disulfide bonds (see House, 1974). Subsequently, a "normal" protease facilitates hydrolysis of the polypeptides. A reducing environment was assumed to be provided by the presence in the gut of reducing agents such as ascorbic acid, glutathione, or riboflavin. However, Gilmour (1965) has pointed out that the protease present in the gut is unusual in that it functions under anaerobic conditions. Further, the cystine released by hydrolysis could be converted intracellularly to cysteine, which itself could yield hydrogen sulfide. Both these substances, which are known to be present in the gut during digestion of keratin, are reducing agents and would facilitate breaking of disulfide bonds. Thus, the negative value for redox potential is, on this proposal, consequent upon, rather than necessary for, the digestion of keratin.

4.2.3. Control of Enzyme Synthesis and Secretion

Apparently, digestive enzymes are not stored in the midgut cells but are liberated immediately into the gut lumen. In many insects which feed more or less continuously, digestive enzyme synthesis and secretion seem not to be regulated but continue even during nonfeeding periods, including starvation. In other species production of enzymes is clearly related quantitatively to food intake and secretagogue, neural, and hormonal control mechanisms have been suggested by different authors. Unfortunately, for most species, the evidence presented in support of one mechanism or another is equivocal. In a secretagogue system enzymes are produced in response to food present in the midgut. Presumably the amount produced is directly influenced by the concentration of food in the lumen. Where neural or hormonal control of enzyme production has been proposed, the amount of food passing along the foregut, measured as the degree of stretching of the gut wall, is believed to result in the production of a proportionate amount of enzyme. In systems under neural control, information passes directly to the midgut via the stomatogastric nervous system. Where hormonal control exists, feeding stimulates the release of neurosecretion which travels via the hemolymph to the midgut cells. Insects do not appear able to regulate enzyme production qualitatively, that is, to alter the relative proportions of enzymes which they produce in accord with a change in diet. This suggests that a midgut epithelial cell produces a complete package of digestive enzymes.

4.2.4. Digestion by Microorganisms

Microorganisms (bacteria, fungi, and protozoa) may be present in the gut, but for only a few species has there been a convincing demonstration of their importance in digestion. In many insects microorganisms appear to have no role, as the insects can be reared equally well in their absence. In other species microorganisms may be more important with respect to an insect's nutrition than digestion *per se*. Where a role for microorganisms in digestion has been demonstrated, the relationship between the microorganisms and insect host is not always obligate, but may be facultative or even accidental.

Bacteria are important cellulose-digesting agents in many phytophagous insects, especially wood-eating species whose hindgut may include a fermentation pouch in which the microorganisms are housed. In other species, for example, the wood-eating cockroach *Panesthia*, bacteria in the crop are essential for cellulose digestion. In larvae of the wax moth *Galleria*, bacteria normally present in the gut undoubtedly aid in the digestion of beeswax, yet bacteriologically sterile larvae produce an intrinsic lipase capable of degrading certain wax components. Finally, many insects feed on decaying vegetation and must, therefore, ingest a large number of saprophytic bacteria which, temporarily at least, would continue their degradative activity in the gut. In this sense, therefore, though the relationship is accidental, the microorganisms are assisting in digestion.

In lower termites and some primitive wood-eating cockroaches (Cryptocercus), flagellate and ciliate protozoa occur in enormous numbers in the hindgut. The relationship between the insects and protozoa is mutualistic; that is, in return for a suitable, anaerobic environment in which to live, the protozoa phagocytose particles of wood eaten by the insects, fermenting the cellulose and releasing large amounts of glucose (in Cryptocercus) or organic acids (in termites) for use by the insects.

Fungi rarely play a direct role in the digestive process of insects, though it is reported that yeasts capable of hydrolyzing carbohydrates occur in the gut of some leafhoppers (Homoptera). However, a mutualistic relationship has evolved between many fungi and insects in which the fungi convert wood into a more usable form, while the insects serve to transport the fungi to new locations. Some ants and higher termites, for example, culture ascomycete or basidiomycete fungi in special regions of the nest called fungus gardens. Chewed wood or other vegetation is brought to the fungus garden and becomes the substrate on which the fungi grow, forming hyphae to be eaten by the insects. Certain wood-boring insects, for example, bark beetles (Scolytinae) inoculate their tunnels with fungal cells when they invade a new tree. The fungal mycelium which develops, along with partially decomposed wood, can then be used as food by the insects.

4.3. Absorption

Contrary to the views of some early entomologists, the foregut, specifically the crop region, is not the site of absorption of digestive products. The majority of absorption occurs in the midgut, especially the anterior portion, including the mesenteric ceca, and to a lesser extent the hindgut. The latter region is, however, primarily of importance as the site of water or ion resorption in connection with osmoregulation (see Chapter 18), though in insects which have symbiotic microorganisms in the hindgut it may also be an important site for absorption of small organic molecules, especially carboxylic and amino acids.

Most absorption of organic molecules across the midgut wall is passive, that is, from a higher to a lower concentration, though the rapid rate at which some molecules are absorbed suggests that their movement is facilitated by special carriers. The absorption rate is enhanced by a steep concentration gradient maintained between the midgut lumen and hemolymph. This may be achieved by absorption of water from the gut lumen so that the hemolymph becomes more dilute or by rapid conversion of the absorbed molecules to a more complex form. The absorption of organic molecules across the gut wall in Schistocerca and Periplaneta formed the subject of a series of papers by Treherne in the late 1950s (see Treherne, 1967, for details).

Using isotopically labeled monosaccharides, Treherne demonstrated that nearly all sugars are absorbed in the anterior region of the midgut, especially the ceca. Further, the monosaccharides are converted rapidly to the disaccharide trehalose in the fat body. Interestingly, in *Schistocerca* much of the fat body is in close proximity to the midgut wall. Of the monosaccharides studied, glucose was found to be absorbed most rapidly. Fructose and mannose are absorbed relatively slowly because of their accumulation in the hemolymph. The latter is related to the lower rate at which they are converted to trehalose. Apparently no mechanism for active uptake of monosaccharides occurs (or is necessary) in *Schistocerca* because its principal "blood sugar" is trehalose. In certain insects, for example the honeybee, a considerable amount of glucose is normally present in the hemolymph and, in such species, active transport systems may be necessary for sugar absorption.

In *Schistocerca*, amino acids, like sugars, are apparently absorbed passively through the wall of the mesenteric ceca and anterior midgut. The observation that their absorption is passive is of interest, since it is known that the concentration of amino acids in the hemolymph is normally very high. Treherne discovered that, prior to amino acid absorption, there is rapid movement of water from the gut lumen to the hemolymph, which establishes a favorable concentration gradient for passive absorption of amino acids. When certain amino acids were fed to starved *Aedes*, glycogen rapidly appeared in some of the cells of the midgut and ceca. However, it is not known whether the glycogen was formed directly from these amino acids.

Early histochemical studies demonstrated that droplets of lipid are present in the epithelial cells of the crop and gave support to the idea that the crop was the site of lipid absorption. However, work by Treherne, again involving labeled compounds, showed convincingly that, in *Schistocerca*, absorption of lipid occurs not across the crop wall but via the anterior midgut and ceca. In other insects the occurrence of lipophilic cells in the middle and posterior regions of the midgut suggests that these may be sites of lipid absorption. Absorption of lipids is again a passive process.

5. Metabolism

Substances absorbed through the gut wall (occasionally the integument, for example, certain insecticides) seldom remain unchanged in the hemolymph for any length of time but are quickly converted into other compounds. Metabolism comprises all the chemical reactions that occur in a living organism. It includes anabolism (reactions that result in the formation of more complex molecules and are, therefore, energy-requiring) and catabolism (reactions from which simpler molecules result and energy is released). Anabolic reactions include, for example, the formation of structural proteins or enzymes from amino acids, and the formation from simple sugars of polysaccharides which serve as an energy store. Many catabolic reactions have evolved for the specific purpose of producing the large quantities of energy required by the organism for performance of work.

The metabolism of insects generally resembles that of mammals, details of which can be found in standard biochemical texts. The present account, therefore, will be largely comparative in nature.

5.1. Sites of Metabolism

Chemical reactions are carried out by all living cells, though they are usually limited in number and, of course, are related to the specific function of the cell in which they occur. For example, in midgut epithelial cells, metabolism is directed largely toward synthesis of specific proteins, the enzymes used in digestion. Metabolism in muscle cells is specifically concerned with production of large amounts of energy, in the form of ATP, for the contraction process. In epidermal cells reactions leading to the production of chitin and certain proteins, the components of cuticle, are predominant. Certain tissues, however, are not so specialized and in them a multitude of biochemical reactions, involving the three major raw materials (sugars, amino acids, and lipids), are carried out. In vertebrates the liver performs these multiple functions. The analogous tissue in insects is the fat body (Kilby, 1965).

5.1.1. Fat Body

The fat body is derived during embryogenesis from the mesodermal walls of the coelomic cavities. In other words, it is initially a segmentally arranged tissue though this becomes obscured as the hemocoel develops. Nevertheless, and contrary to what a casual examination may suggest, the fat body does have a definite arrangement in the hemocoel characteristic of the species. Typically, there is a layer of fat body beneath the integument and, often, a sheath surrounding the gut. In addition, sheets or cords of cells occur in specific locations.

The fat body is composed mainly of cells called trophocytes, though in some species urate cells and/or mycetocytes also can be seen scattered throughout the tissue. In embryos, early postembryonic stages, and starved insects the individual trophocytes are easily distinguishable, their nucleus is rounded, and their cytoplasm contains few inclusions. As such, they closely resemble hemocytes, with which they probably have a close phylogenetic relationship. In later larval stages and adults the trophocytes enlarge and become vacuolated. The vacuoles contain reserves of fat, protein, and glycogen. The trophocyte nuclei are proportionately large and frequently become elongate and much branched. During metamorphosis in endopterygotes, the reserves are liberated into the hemolymph. In some Diptera and Hymenoptera the majority of trophocytes also disintegrate at this time, and the fat body appears to be completely reformed in the adult from the few cells which remain.

Urate cells are found in the fat body of insects whose Malpighian tubules are nonfunctional (in terms of uric acid production), for example, larval Apocrita (Hymenoptera) and *Blatta*, and in Collembola which lack Malpighian tubules. Accumulation of uric acid in the urate cells is, therefore, a means of storage excretion. It has been suggested that the uric acid may also act as a reserve of nitrogen, perhaps for purine synthesis. However, to date, the enzymes necessary for this process have not been demonstrated.

Mycetocytes are cells within the fat body (and other tissues) which contain microorganisms. In cockroaches the cells appear to be modified trophocytes which contain bacteria. It is thought that the bacteria provide an insect with certain nutrients, though proof of this has been obtained in very few cases.

5.2. Carbohydrate Metabolism

As in other animals, simple sugars provide a readily available substrate which can be oxidized for production of energy. However, in contrast to vertebrates where glucose in the blood is the sugar of importance as an energy source, in insect hemolymph glucose and other monosaccharides usually are present only in minimal amounts. An exception to this statement is the worker honeybee whose hemolymph glucose may reach a concentration of almost 3 g/100 ml and is used as the energy source during flight. In most insects, a disaccharide, trehalose, is the immediate energy source. Trehalose consists of two glucose molecules joined through a $\alpha1,1$-linkage. Its level in the hemolymph is constant and in a state of dynamic equilibrium with glycogen stored in the fat body (Friedman, 1978). In this respect, therefore, the situation is similar to that in vertebrates whose blood glucose level is in equilibrium with liver glycogen. The similarity goes further. Just as the conversion of liver glycogen to blood glucose is promoted by the hormone glucagon, which stimulates glycogen phosphorylase activity, in insect fat body the formation of trehalose from glycogen is promoted by a hyperglycemic hormone released from the corpora cardiaca. This hormone activates the phosphorylase which removes a glucose unit from the glycogen. Because of its highly polar, polyhydroxyl nature, trehalose does not easily penetrate the muscle cell membrane. Therefore before it can be oxidized by muscle it must first be converted into glucose by a hydrolyzing enzyme, trehalase, present in the muscle cell membrane. Hemolymph trehalose is also the source of the glucose which is converted by epidermal cells into acetylglucosamine during production of the nitrogenous polysaccharide chitin (see Chapter 11).

Glycogen is an important reserve substance in almost all insects and is found in high concentration in the fat body, muscle, especially flight muscle, and sometimes the midgut epithelium. In the mature bee larva, for example, glycogen makes up about one-third of the dry weight. It is produced principally from glucose and other monosaccharides absorbed from the gut following digestion. In some insects glycogen may also be synthesized from amino acids. As noted above, fat body glycogen (and perhaps also that stored in the midgut epithelium) is used to maintain a constant level of trehalose in the hemolymph. The glycerol produced as an antifreeze in the hemolymph of some insects which must withstand extremely low winter temperatures is probably also derived from fat body glycogen. Glycogen in muscle is used directly as an energy source, being degraded as in mammalian tissue via the glycolytic pathway, Krebs cycle, and respiratory chain, with resultant production of ATP.

Glycogen also is a significant component of the yolk in the eggs of some insects. Its use as an energy source in this situation is, however, secondary to its importance as a provider of glucose units for chitin synthesis in the developing embryo.

In insects the major functions of the pentose cycle are (1) production of reducing equivalents (as NADP) that are used, for example, in lipid synthesis, and (2) production of 5-carbon sugars (pentoses) for nucleic acid synthesis. In addition, through its ability to interconvert sugars containing from three to seven carbon atoms, the pentose cycle can change "unusual" sugars produced during digestion into 6-carbon derivatives and hence into glycogen.

For most insects, fats stored in the fat body are the primary energy reserve and, like those of other animals, are mostly triglycerides. They may be formed directly by combination of the fatty acids and glycerol produced during digestion or from amino acids and simple sugars. Typically, fat is stored throughout the juvenile period, especially in endopterygotes where at metamorphosis it may make up between one-third and one-half of the dry weight of an insect. Large amounts of fat also accumulate in the egg during vitellogenesis. Fats are used as an energy source during "longer-term" energy-requiring events, for example, embryogenesis, metamorphosis, starvation, and sustained flight. On a weight-for-weight basis, fats contain twice as much energy as carbohydrate; they are therefore more economical to store.

Insects that fly for long periods, for example, during migration, typically use carbohydrate initially (for the first few minutes of flight) but then convert to the use of fat. When flight begins in the migratory locust, for example, hemolymph trehalose is the substrate oxidized, and its concentration steadily decreases for about 10 minutes. During this period, no conversion of stored carbohydrate to trehalose appears to occur, presumably because hyperglycemic hormone is not released from the corpora cardiaca. Instead, flight triggers the release of adipokinetic hormone from these glands which promotes mobilization of the stored lipid in the fat body. Specifically, the hormone triggers the conversion of stored triglycerides into diglycerides which are released into the hemolymph for use by the flight muscles (see Goldsworthy, 1976). In addition, the hormone probably induces a switch in metabolism of the flight muscles, stimulating them to use lipid rather than carbohydrate substrate. To this end, the muscles contain high concentrations of lipases and other enzymes required for hydrolysis of lipid and degradation of free fatty acids to 2-carbon fragments which can be oxidized by the Krebs cycle to carbon dioxide and water.

In addition to the fats just described which serve solely as energy reserves or sources of carbon, many other lipids having structural or metabolic functions occur in insects. Waxes are mixtures of long-chain alcohols or acids, their esters, and paraffins. The number of carbon atoms that form the chain ranges between 12 and 36, and both unsaturated and saturated compounds have been identified. It seems that the various components of wax are synthesized from fatty acid precursors. The paraffins and, possibly, some acids are produced by oenocytes, whereas alcohol and ester synthesis occurs in the fat body. Compound lipids are fatty acids combined with a variety of organic or inorganic residues, for example, carbohydrates, nitrogenous bases, amino acids, phosphate, and sulfate. The metabolism of these lipids in insects is for the most part poorly known. Certain of them, for example, choline, a phospholipid, cannot be synthesized, however, by insects and must be included in the diet. Likewise, sterols are essential components of the diet in almost all insects.

5.4. Amino Acid and Protein Metabolism

In growing insects a large proportion of the amino acids that result from digestion is used directly in the formation of new tissue proteins, both structural and metabolic. Within the fat body especially, but also in other tissues, a variety of transaminations also occur; that is, the amino group from an amino acid can be transferred to a keto acid to form a new amino acid. Such transaminations are especially important when an insect's diet contains insufficient amounts of particular amino acids. As in vertebrates, not all amino acids can be synthesized in insect tissues. Those which cannot must be included in the diet or provided by symbiotic microorganisms. During starvation or when present in excess, amino acids may undergo oxidative deamination (that is, be oxidized and simultaneously lose their α-amino group) within the fat body, resulting in the formation of the corresponding keto acids. The latter can then be further oxidized via the Krebs cycle and respiratory chain to provide energy or may be converted into carbohydrate or fat reserves. The ammonia produced during deamination is normally converted into uric acid.

The fat body is an important site of hemolymph protein synthesis, especially in the late juvenile stages of endopterygotes and in adult female insects. In silk moth (*Bombyx*) caterpillars, for example, the hemolymph protein concentration increases sixfold from the fourth to the final instar, in preparation for spinning the cocoon and metamorphosis. When the adult emerges, the concentration has fallen to about one third the value at pupation. In the female of many insect species, the importance of the fat body in producing one or more "female-specific" proteins is now well documented. These proteins, whose synthesis is regulated by juvenile hormone or ecdysone, are accumulated in large amounts by the developing oocytes (see Chapter 19 for further discussion and references). In the male of some species the fat body produces proteins which are accumulated by the accessory reproductive glands, probably for use in spermatophore production.

5.5. Metabolism of Insecticides

Insecticides now may be regarded as a "normal" environmental hazard for insects, survival over which has been achieved through natural selection of resistant strains. Though the purpose of this section is to outline the biochemical pathways by which insecticides are rendered harmless, it should be realized that resistance can also be developed, solely or partially, as a result of "physical" rather than "metabolic" changes in a species. This is because, ultimately, the degree of resistance is dependent on the rate at which an insect can degrade the toxic material so that lethal quantities do not accumulate at the site of action. Among the physical alterations which may lead to increased resistance are (1) a decline in the permeability of the integument (achieved by increasing cuticle thickness or the extent of tanning, or by modifying the composition of the cuticle); (2) a change in the pH of the gut, resulting in a decrease in solubility and, therefore, rate of absorption of an insecticide; (3) an increase in the amount of fat stored (most insecticides are fat-soluble and, therefore, accumulate in fatty tissues in which they are ineffective); and (4) a decrease in permeability of the membranes surrounding the target tissue (usually the nervous system).

Insects have various methods for detoxifying potentially harmful substances, many of which parallel those found in vertebrates. Hydrolysis, hydroxylation, sulfation, methylation, acetylation, and conjugation with cysteine, glycine, glucose, glucuronic acid, or phosphate are examples of the methods employed. Hydroxylation and conjugation are also important in making the normally fat-soluble insecticides water-soluble so that they can be excreted. Each of these processes is enzymatically controlled, and it is not surprising to find, therefore, that metabolic resistance to insecticides most often results from qualitative or quantitative changes in the enzymes concerned so that the rate of detoxication is increased. In some resistant strains, an increase in the amount of enzyme responsible for detoxication has been observed; in others, it appears that the enzyme has changed so that it is now more specific toward its "new" substrate, the insecticide. In a few species, resistance seems to have developed as a result of an increase in the quantity, or decrease in the sensitivity, of the enzyme normally affected by the insecticide, specifically cholinesterase in the nervous system.

Mechanisms for detoxication vary among the different categories of insecticides. It is appropriate, therefore to examine separately the metabolism of compounds in these categories. Three categories of insecticides will be considered: chlorinated hydrocarbons, organophosphates, and carbamates.

Among the chlorinated hydrocarbon insecticides are DDT, lindane (γ-BHC), chlordane, heptachlor, aldrin, and isodrin.* Though their precise mode of action is not yet understood, chlorinated hydrocarbons undoubtedly act on the nervous system to prevent normal impulse transmission. A current hypothesis proposes that the insecticides become distributed within the lipid layer of axonal membranes in such a manner that sodium ions can readily diffuse across the membrane. As a result, the membrane becomes permanently depolarized, because as rapidly as active transport attempts to restore the normal ionic balance, sodium ions will diffuse back in the opposite direction.

DDT was the first synthetic insecticide to be developed and "appropriately" was the first to which insects developed resistance! Resistance to DDT is most often due to the presence of an enzyme which dechlorinates the compound, forming the less toxic dichloroethylene derivative, DDE. Some species, however, convert DDT into other less harmful materials such as DDA (the acetic acid derivative), dicofol (kelthane; the trichloroethanol derivative), and DDD (the dichloroethane derivative). Many comparisons of the activity of the DDT-degrading enzyme in resistant and susceptible strains of a species have shown, however, that metabolic resistance alone is often insufficient to account for the full extent of resistance; specifically, the known maximum rate of degradation of DDT measured in vitro is not high enough to account for the high tolerance shown by the resistant strain. In such cases, further work has usually shown the importance also of physical resistance mechanisms of the type outlined earlier. Like DDT, lindane and chlordane are rapidly degraded in resistant strains to less harmful forms.

The cyclodiene compounds, heptachlor, aldrin, and isodrin, are of interest from several viewpoints. In themselves they are not toxic but are oxidized within an insect's tissues to the highly toxic epoxy derivatives, heptachlor-epoxide, dieldrin, and endrin, respectively, a process known as "autointoxica-

*These are approved common names. For the chemical names, see Perry and Agosin (1974).

tion." Because of this conversion, insects treated with these compounds show no symptoms for 1 or 2 hours after treatment, in contrast to insects treated with other insecticides which react within a matter of minutes. The resistance exhibited by certain strains is not because they no longer convert an insecticide to its toxic form as might be anticipated. Further, the toxic derivatives appear to have great stability, remaining unchanged even in resistant insects for several days. Thus, resistance is not due to an increased ability to detoxicate these compounds. Indeed, the reason for resistance to cyclodiene compounds is not yet established, though decreased sensitivity of nervous tissue, perhaps due to changes in structure of the myelin sheath seems most probable.

In contrast to the situation for chlorinated hydrocarbons, the mode of action of organophosphates (for example, parathion, malathion, diazinon, and dimethoate) is well known. These phosphate esters bind covalently with and inhibit the action of cholinesterase, the enzyme which normally degrades acetylcholine at the synapse, though there are reports that their toxicity is partially due also to inhibition of other tissue esterases. As with chlorinated hydrocarbons, resistance to organophosphates may be developed as a result of physical change, but generally is metabolic. Like cyclodienes, many organophosphates are "activated" (rendered more toxic) as a result of oxidation; for example, parathion is converted to paraoxon. Thus, resistance may be due to a decrease in the rate of activation and/or an increase in the rate of conversion of the compound to a nontoxic form. (Apparently, resistance does not develop as a result of decreased sensitivity of the cholinesterase to an insecticide.) Many reports have shown that resistant strains are more able to carry out conjugation, especially with glutathione, or hydrolysis of the insecticide than are susceptible insects. This ability is due to the presence of either greater quantities of esterifying enzymes or enzymes, which, through mutation and natural selection, have become more specific for an insecticide.

Carbamates, for example furadan, sevin, pyrolan, and isolan, are substituted esters of carbamic acid, which, like organophosphates, attack cholinesterase. Resistance to these insecticides also is very similar to that for organophosphates. Some resistance can be achieved by physical changes, but most is due to increased rates of degradation, especially through oxidation and hydrolysis.

Shortly after the discovery that most resistance is metabolic, that is, due to increased quantities or specificity of particular enzymes which cause more rapid breakdown of an insecticide, it was realized that the phenomenon of synergism might be explored to advantage in the use of insecticides. Synergism describes the situation in which the combined effect of two substances is much greater than the sum of their separate effects. In practical terms, in the present context, it means that appropriate substances (synergists), when mixed with an insecticide, would increase the latter's effectiveness by combining with (and inhibiting) the enzymes which normally degrade the insecticide. The synergists used may be quite unrelated chemically to the insecticide but, most often, are analogues. The principle of synergism has been applied with limited success in the case of pyrethrin insecticides and DDT. For example, in the early 1950s DMC, the ethanol derivative of DDT, was found to be an effective synergist for DDT in DDT-resistant houseflies. However, perhaps not surprisingly, by 1955 the flies had developed resistance to the combination!

461

FOOD UPTAKE
AND UTILIZATION

6. Summary

Visual, tactile. or chemical cues stimulate food location and/or selection in most insects. The stimuli may be general, for example, color, pattern, and size, or highly specific, such as the particular odor or taste of a chemical. The chemicals that promote feeding (phagostimulants) may have no nutritional value for an insect.

The gut includes three primary subdivisions, foregut, midgut, and hindgut, and these are typically differentiated into regions of differing function. The foregut is concerned with storage and trituration of food, the midgut with digestion and absorption of small organic molecules, and the hindgut with absorption of water and ions, though some absorption of small organic molecules may occur across the hindgut wall, especially in insects with symbiotic microorganisms in their hindgut.

The digestive enzymes produced match qualitatively and quantitatively the normal composition of the diet. The enzymes may have low specificity, enabling an insect to digest a variety of molecules of a given type, or may be highly specific, for example, when a species feeds solely on a particular food. Gut fluid is buffered within a narrow pH range to facilitate digestion and absorption. Enzymes are released as soon as they are synthesized. Synthesis is regulated so that an appropriate amount of enzyme is produced for the food consumed. Microorganisms in the gut may be important in digestion, especially in wood-eating species, where they degrade cellulose.

Absorption of digestion products occurs mostly in the anterior midgut and mesenteric ceca and is a passive process. The rate at which sugars are absorbed is linked to the rate at which they are converted to trehalose and, hence, glycogen. Amino acid absorption is preceded by absorption of water across the midgut wall to produce a favorable gradient for diffusion.

The fat body is the primary site of intermediary metabolism as well as a site for storage of metabolic reserves. In most insects, trehalose in the hemolymph is the sugar of importance as an energy reserve. Its concentration in the hemolymph is constant and is in dynamic equilibrium with glycogen stored in the fat body. Lipids in the fat body form the major energy reserve molecules and are used in long-term energy requiring processes such as flight, metamorphosis, starvation, and embryogenesis. The fat body is important in protein metabolism, including amino acid transamination and synthesis of some specific proteins.

The development of resistance to insecticides is normally due to increased ability of an insect to degrade the insecticides to less harmful and excretable products, but may be due also to increased physical resistance, that is, to structural changes that prevent insecticides from reaching their site of action. Metabolic resistance normally develops through the production of more specific or greater quantities of insecticide-degrading enzymes.

7. Literature

A large body of literature exists on gut physiology and metabolism. The following selection is representative of some of the more recent reviews of the subject. Dethier (1966) deals with feeding behavior. The regulation of food intake is examined by Gelperin (1971), Bernays and Chapman (1974), and Barton Browne (1975). Digestion is covered by Dadd (1970) and House (1974), and absorption by Treherne (1967). Waldbauer (1968) discusses food consumption and utilization. General metabolism is the subject of Gilmour's (1965) text, while the hormonal control of metabolism is reviewed by Steele (1976) and Keeley (1978). Metabolism of insecticides is reviewed by Perry and Agosin (1974) and in the volume edited by Wilkinson (1976).

Barton Browne, L., 1975, Regulatory mechanisms in insect feeding, *Adv. Insect Physiol.* **11**:1–116.

Bernays, E. A., and Chapman, R. F., 1974, The regulation of food intake by acridids, in: *Experimental Analysis of Insect Behaviour* (L. Barton Browne, ed.), Springer Verlag, New York.

Dadd, R. H., 1970, Digestion in insects, *Chem. Zool.* **5**:117–145.

Davey, K. G., and Treherne, J. E., 1963, Studies on crop function in the cockroach. I and II, *J. Exp. Biol.* **40**:763–773, 775–780.

Dethier, V. G., 1966, Feeding behaviour, *Symp. R. Entomol. Soc.* **3**:46–58.

Friedman, S., 1978, Trehalose regulation, one aspect of metabolic homeostasis, *Annu. Rev. Entomol.* **23**:389–407.

Gelperin, A., 1971, Regulation of feeding, *Annu. Rev. Entomol.* **16**:365–378.

Gilmour, D., 1965, *The Metabolism of Insects*, Oliver and Body, Edinburgh.

Goldsworthy, G. J., 1976, Hormones and flight in the locust, *Perspect. Exp. Biol.* **1**(Zoology):167–177.

Heron, R. J., 1965, The role of chemotactic stimuli in the feeding behavior of spruce budworm larvae on white spruce, *Can. J. Zool.* **43**:247–269.

Highnam, K. C., Hill, L., and Mordue, W., 1966, The endocrine system and oocyte growth in *Schistocerca* in relation to starvation and frontal ganglionectomy, *J. Insect Physiol.* **12**:977–994.

House, H. L., 1974, Digestion, in: *The Physiology of Insecta*, 2nd ed., Vol. V (M. Rockstein, ed.), Academic Press, New York.

Keeley, L. L., 1978, Endocrine regulation of fat body development and function, *Annu. Rev. Entomol.* **23**:329–352.

Kilby, B. A., 1965, Intermediary metabolism and the insect fat body, *Symp. Biochem. Soc.* **25**:39–48.

Perry, A. S., and Agosin, M, 1974, The physiology of insecticide resistance by insects, in: *The Physiology of Insecta*, 2nd ed., Vol. VI (M. Rockstein, ed.), Academic Press, New York.

Richards, A. G., and Richards, P. A., 1977, The peritrophic membranes of insects, *Annu. Rev. Entomol.* **22**:219–240.

Steele, J. E., 1976, Hormonal control of metabolism in insects, *Adv. Insect Physiol.* **12**:239–323.

Treherne, J. E., 1967, Gut absorption, *Annu. Rev. Entomol.* **12**:43–58.

Waldbauer, G. P., 1968, The consumption and utilization of food by insects, *Adv. Insect Physiol.* **5**:229–288.

Wilkinson, C. F. (ed.), 1976, *Insecticide Biochemistry and Physiology*, Plenum Press, New York.

17

The Circulatory System

1. Introduction

The circulatory system of insects, like that of all arthropods, is of the "open" type; that is, the fluid which circulates is not restricted to a network of conducting vessels as, for example, in vertebrates, but flows freely among the body organs. An open system results from the development, in evolution, of a hemocoel rather than a true coelom. A consequence of the open system is that insects have only one extracellular fluid, hemolymph, in contrast to vertebrates which have two such fluids, blood and lymph. The occurrence of an open system does not mean that hemolymph simply bathes the organs it surrounds because usually thin granular membranes separate the tissues from the hemolymph itself. Insects generally possess pumping structures and various diaphragms to ensure that hemolymph flows through the body along a definite route. As the only extracellular fluid, it is perhaps not surprising that hemolymph, in general, serves the functions of both blood and lymph of vertebrates. Thus, plasma is important in providing the correct milieu for body cells and is the transport system for nutrients, hormones, and metabolic wastes, while hemocytes provide the major defence mechanism against foreign organisms which enter the body and are important in wound repair and in the metabolism of specific compounds.

2. Structure

The primary pump for moving hemolymph around the body is a middorsal vessel which runs more or less the entire length of the body (Figure 17.1). The posterior portion of the vessel has ostia (valves) and is known as the heart, whereas the cephalothoracic portion, which is often a simple tube, is termed the aorta. In some insects the heart is the only part which contracts, but in many others the entire vessel is contractile. The vessel is held in position by connective tissue strands attached to the dorsal integument, tracheae, gut, and other organs and by a series of paired, usually fan-shaped alary muscles. Normally, the vessel is a straight tube, though in some species the aorta may loop laterally or vertically. Anteriorly the aorta runs ventrally to pass between the corpora

463

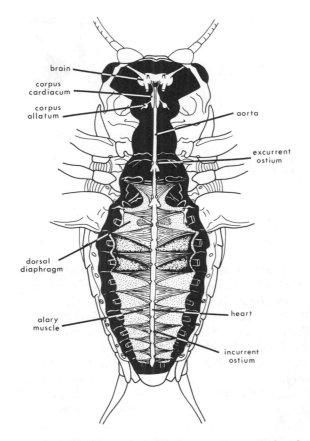

brain

corpus
cardiacum

corpus
allatum

aorta

excurrent
ostium

dorsal
diaphragm

heart

alary
muscle

incurrent
ostium

FIGURE 17.1. Ventral dissection of the field cricket, *Acheta assimilis*, to show dorsal vessel and associated structures. [After W. L. Nutting, 1951, A comparative and anatomical study of the heart and accessory structures of the orthopteroid insects, *J. Morphol.* **89:**501–597. By permission of The Wistar Press.]

cardiaca and under the brain. Posteriorly the vessel is usually closed off, though in some Diptera one or two ostia may occur.

In most insects the dorsal vessel is well tracheated. The heart may not be innervated or may receive paired lateral motor nerves from the brain and/or segmental ventral ganglia. Ostia may be simple, slitlike valves or deep, funnel-shaped structures in the wall of the heart, or internal flaps (Figure 17.2). Their position and number are equally varied. They may be lateral, dorsal, or ventral and may be as numerous as 12 pairs (in cockroaches) or as few as one pair (in some dragonflies). Ostia are usually incurrent, that is, they open to allow hemolymph to enter the heart but close to prevent backflow. In some orthopteroid insects, however, some ostia are excurrent. Histologically, the dorsal vessel in its simplest form comprises a single layer of circular muscle fibers, though more often longitudinal and oblique muscle layers also occur.

Assisting in directing the flow of hemolymph are various diaphragms (septa) (Figure 17.3), which include both connective tissue and muscular elements. The spaces they enclose are known as sinuses. The pericardial septum (dorsal diaphragm) lies immediately below the dorsal vessel and spreads between the alary muscles. Laterally, it is attached at intervals to the terga so that the pericardial sinus is in effect continuous with the perivisceral sinus. Normally, the septum forms a continuous sheet, but in some species it is fenestrated. Ventrally, a perineural septum (ventral diaphragm) may occur, which

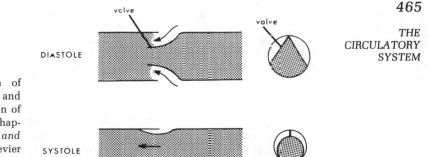

DIASTOLE

valve

valve

SYSTOLE

cuts off the perineural sinus from the perivisceral sinus. It is capable of performing posteriorly directed undulations and may be fenestrated. It may receive motor nerves from segmental ganglia, which regulate the rate at which it undulates, though the undulations originate myogenically. Contrary to what is stated in early reports, lateral extensions of the ventral diaphragm do not pass into each leg (J. C. Jones, personal communication).

To further facilitate hemolymph flow, especially through appendages, accessory pulsatile organs (accessory hearts) may occur. These have been identified in the head, antennae, thorax, legs, and wings. In many species they are saclike structures which have a posterior incurrent ostium and an anteriorly extended vessel. In antennal pulsatile organs the vessel may run the length of the appendage but is perforated at intervals to permit exit of hemolymph. The wall of the sac may be muscular, so that constriction of the sac is the active phase, and dilation is due to elasticity of the wall, or the sac may have attached to it a discrete dilator muscle, and constriction is due to the sac's elasticity. Normally, accessory hearts are quite separate from the dorsal vessel, though in

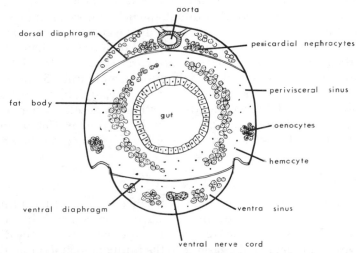

aorta

dorsal diaphragm

pericardial nephrocytes

perivisceral sinus

fat body

gut

oenocytes

hemocyte

ventral diaphragm

ventral sinus

ventral nerve cord

FIGURE 17.3. Diagrammatic transverse section through abdomen to show arrangement of septa. [From R. E. Snodgrass, *Principles of Insect Morphology*. Copyright 1935 by McGraw-Hill, Inc. Used with permission of McGraw-Hill Book Company.]

some Odonata they are connected via short vessels with the aorta into which they pump hemolymph. Most accessory pulsatile organs are not innervated.

At specific locations in the circulatory system are sessile cells, usually conspicuously pigmented, called athrocytes. They may occur singly, in small groups, or may form distinct lobes. In most species athrocytes are situated on the surface of the heart (occasionally also along the aorta), and these are referred to as pericardial cells. They may also be found as scattered cells in the fat body (in *Lepisma*), in clusters at the bases of legs (in *Gryllus* and *Periplaneta*), or as a garland of cells around the esophagus (in some larval Diptera). When mature they may contain several nuclei, as well as mitochondria, Golgi apparatus, and pigment granules or crystals of various colors. The cells are able to accumulate colloidal particles, for example, certain dyes, hemoglobin, and chlorophyll, which led to an early suggestion that they segregated and stored waste products (hence their alternate name of nephrocytes). The current view is that the cells accumulate and degrade large molecules such as proteins, peptides, and pigments, and the products are then used or excreted.

3. Physiology

3.1. Circulation

Contractions of the dorsal vessel and accessory pulsatile organs, along with movements of other internal organs and, in some species, abdominal ventilatory movements, serve to move hemolymph around the body. In *Periplaneta* larvae, for example, circulation time is 3 to 6 minutes; in *Tenebrio* the time for complete mixing of injected radioisotope is reported as 8 to 10 minutes. Generally hemolymph is pumped rapidly through the dorsal vessel but moves slowly and discontinuously through sinuses and appendages.

The normal direction of hemolymph flow is indicated in Figure 17.4. Hemolymph is pumped anteriorly through the dorsal vessel from which it exits via either excurrent ostia of the heart or mainly the anterior opening of the aorta in the head. The resultant pressure in the head region forces hemolymph posteriorly through the perivisceral and perineural sinuses. Undulations of the ventral diaphragm aid the backward flow of hemolymph. Relaxation of the heart muscle results in an increase in heart volume, and, by negative pressure, hemolymph is sucked in via incurrent ostia. As noted earlier, circulation through appendages is aided by accessory pulsatile organs. Hemolymph enters the wings via the anterior veins and returns to the thorax via the anal veins.

3.2. Heartbeat

Contraction of the heart (systole) is followed, as in other animals, by a phase of relaxation (diastole) during which muscle cell membranes become repolarized. A third phase, diastasis, may follow diastole, when the diameter of the dorsal vessel suddenly enlarges due to the influx of hemolymph. Diastole in many insects seems to be passive, that is, due to natural elasticity of the heart

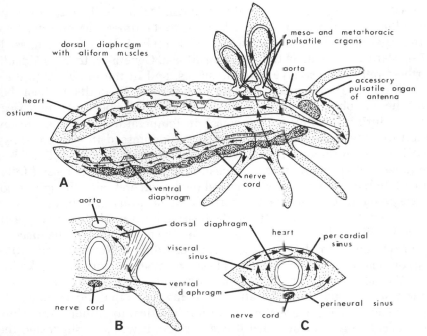

FIGURE 17.4. Diagrams showing direction of hemolymph flow. (A) Longitudinal section, (B) transverse section through thorax, and (C) transverse section through abdomen. Arrows indicate direction of flow. [After V. B. Wigglesworth, 1965, *The Principles of Insect Physiology*, 6th ed., Methuen and Co. By permission of the author.]

muscle. Though alary muscles may be quite well developed in these species, they apparently have no role in the relaxation process. They have been shown to be electrically inexcitable, and cutting them has no effect on the rate and strength of the heartbeat. In a few species structural integrity of the heart and alary muscles is vital, and cutting the alary muscles terminates the heartbeat.

Contraction of the dorsal vessel normally begins at the posterior end and passes forward as the peristaltic wave. However, contraction can be induced artificially at any point along the length of the vessel and individual semiisolated segments (portions of the heart with tergum still attached) continue to beat rhythmically. These observations suggest that the heartbeat is normally coordinated by a pacemaker located posteriorly. Reversal of heartbeat is also known to occur and is characteristically seen in pupae and adults of Lepidoptera and Diptera.

As noted earlier, insect hearts may or may not be innervated. In aneural (noninnervated) hearts, obviously, the initiation and control of the heartbeat must be a property of the heart muscle itself. When initiation of heartbeat resides in the muscle, the heart is said to be myogenic. Even innervated hearts are myogenic, though control of the frequency and amplitude of heartbeat is a property of the cardiac neurons. In other words, insects do not have neurogenic hearts and thus differ in this respect from Arachnida and Crustacea.

The rate at which the heart beats varies widely both among species and even within an individual under different conditions. In the pupa of *Anagasta kuhniella,* for example, the heart beats 6 to 11 times per minute. In larval *Blattella germanica* rates of 180 to 310 beats/min have been recorded (see Jones, 1974). Many factors affect the rate of heartbeat. Generally, there is a decline in heartbeat rate in successive juvenile stages, and in the pupal stage the heart beats slowly or even ceases to beat for long periods. In adults the heart beats at about the rate observed in the final larval stage. Heartbeat rate increases with activity, during feeding, with increase in temperature or in the presence of carbon dioxide in low concentration, but is depressed in starved or asphyxiated insects. Hormones, too, may affect heartbeat rate. Roussel (1969–1971) (see Jones, 1974) reports that in *Locusta* heartbeat rate is increased following implantation of corpora allata or several pars intercerebralis, though he considers the latter to operate via the corpora allata. Roussel believes the corpus allatum secretion to act directly on the heart, but his work does not rule out the strong likelihood that the hormone modifies the metabolism of the insect. Ablation or implantation of corpora cardiaca, according to Roussel, had little or no effect on heartbeat rate in *Locusta.* These observations contrast markedly with those of other authors who have reported the presence of cardioaccelerators in extracts of these glands. Originally, it was suggested that the cardioaccelerator might be an orthodiphenol similar but not identical to epinephrine. However, more recent work indicates that it is a peptide. It was also suggested that the factor actually stimulates pericardial cells to release the *true* heart stimulant which was believed to be an amine. However, involvement of the pericardial cells has been disputed, and it now seems likely that the corpus cardiacum material acts directly on the heart.

4. Hemolymph

Hemolymph, like the blood of vertebrates, includes a cellular fraction, the hemocytes, and a liquid component, the plasma, whose functions are broadly comparable with those found in vertebrates. Several of the features of hemolymph, however, contrast markedly with what is seen in vertebrates. First, associated with the evolution of a tracheal system, hemolymph has no gas transporting function, except perhaps in some chironomid larvae. In addition, the composition of hemolymph (especially in the more advanced endopterygotes) is both very different from that of blood and is much more variable on a day-to-day basis. Among the trends seen in the evolution of the higher endopterygotes are substitution of organic molecules for the predominant inorganic ions (sodium and chloride), an increase in the proportion of divalent to monovalent cations, and an increase in the importance (quantitatively) of organic phosphate. Further, as noted in the previous chapter, monosaccharide sugars are generally of little importance in the hemolymph and are replaced by the disaccharide trehalose. In many insects the hemolymph osmotic pressure is held reasonably constant over a range of environmental conditions. In other species, the osmotic pressure changes in parallel with the environmental conditions, yet the body cells are able to tolerate these changes (see Chapter 18).

4.1. Plasma

4.1.1. Composition

Plasma contains a large variety of components both organic and inorganic whose relative proportions may differ greatly both among species and within an individual under different physiological conditions. Despite this variability, some general statements may be made.

In primitive orders, the predominant cation is sodium, with potassium, calcium, and magnesium present in low proportions. The major anion is chloride, though plasma also contains small amounts of phosphate and bicarbonate. These inorganic constituents are the major contributors to the hemolymph osmotic pressure (Figure 17.5A,B).

In higher orders certain trends can be observed. The relative importance of sodium decreases at the expense of potassium and, especially, magnesium. Chloride also decreases in importance and is replaced by organic anions, especially amino and carboxylic acids. Finally, the relative contribution that inorganic ions make to the hemolymph osmotic pressure declines, and organic constituents become the major osmotic effectors (Figure 17.5D,E).

Superimposed on these phylogenetic relationships may be dietetic and ontogenetic considerations, especially with respect to the cationic components of hemolymph. Thus, zoophagous species generally have a larger proportion of sodium in the hemolymph, in contrast to phytophagous species where magnesium (derived from chlorophyll) and potassium are the major cations [Figure 17.5C]. The ionic composition may also change with stage of development, in endopterygotes at least, though whether this is due to a change of diet which, of course, may also occur from the juvenile to the adult stage, does not seem to have been considered. For example, in the exopterygotes *Aeshna cyanea*

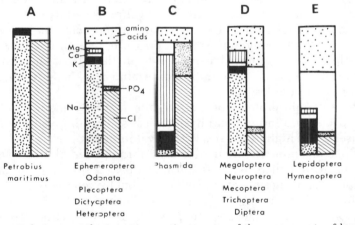

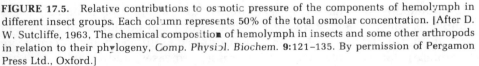

FIGURE 17.5. Relative contributions to osmotic pressure of the components of hemolymph in different insect groups. Each column represents 50% of the total osmolar concentration. [After D. W. Sutcliffe, 1963, The chemical composition of hemolymph in insects and some other arthropods in relation to their phylogeny, *Comp. Physiol. Biochem.* **9:**121–135. By permission of Pergamon Press Ltd., Oxford.]

(Odonata), *Periplaneta americana* (Dictyoptera), and *Locusta migratoria* (Orthoptera) and the endopterygote *Dytiscus* (Coleoptera), the composition of the hemolymph is similar in larvae and adults, but so, too, is the diet. In contrast, in a few endopterygote species, specifically Lepidoptera and Hymenoptera, for which data are available, larval hemolymph is of the high magnesium type, whereas adult hemolymph has a much greater sodium content. Again, however, the fact that the diet of the adult (if, indeed, it feeds at all) is typically different from that of the larva apparently has not received consideration.

As Florkin and Jeuniaux (1974) point out, insects whose hemolymph contains such large quantities of magnesium and potassium must have become adapted so that physiological processes, especially neuromuscular function, can be carried out normally because these ions are detrimental. Hoyle (1954, cited in Florkin and Jeuniaux, 1974) suggested that the high magnesium–potassium type of hemolymph characteristic of phytophagous endopterygote larvae might reduce, through an effect on the nervous system, locomotor activity of larvae, so that they would tend to remain close to their food. The adult, in contrast, is usually much more active and possesses the more primitive high sodium type of hemolymph.

Organic acids are important hemolymph constituents, especially in juvenile endopterygotes. Carboxylic acids (citric, α-ketoglutaric, malic, fumaric, succinic, and oxaloacetic), which in the hemolymph are anionic, are present in large amounts and may neutralize almost 50% of the inorganic cations. They are apparently synthesized by insects (or by symbiotic bacteria), as their levels in the hemolymph are independent of diet. Whether these acids, which are components of the Krebs cycle, also have a metabolic function in the hemolymph, is not known. Insects, especially endopterygotes, also characteristically have high concentrations of amino acids in their hemolymph. The proportions of amino acids vary among species and within an individual according to diet and developmental and physiological state, though glutamic acid, glycine, histidine, lysine, proline, and valine generally each constitute at least 10% of the total amino acid pool. The amino acids make a significant contribution to the hemolymph osmotic pressure, though whether they function as cations or as anions depends both on the pH of the fluid (usually between 6.0 and 7.0) and on the individual amino acid. In addition, some have important metabolic roles.

The hemolymph protein concentration is generally about 1 to 5% but varies with species and individual physiological states. For example, it is low in starved insects and high in females with developing oocytes. In endopterygotes the protein concentration often increases through larval life, especially in the final instar, but then declines during pupation. Many proteins occur in hemolymph, including a number of enzymes, but for relatively few of these has the function been determined. Some appear to be involved in prevention of infection. Others, the female-specific proteins or vitellogenins, are selectively accumulated by oocytes during yolk formation. Yet other proteins may serve simply as concentrated stores of nitrogen which can be degraded for use in growth and metabolism. Among the enzymes identified in hemolymph are hydrolases (for example, amylase, esterases, proteases, and trehalase), dehydrogenases and oxidases important in carbohydrate metabolism and tyrosinase.

The principle carbohydrate in the hemolymph of most insects is the disaccharide trehalose, which serves as a source of readily available energy (see Chapter 16, Section 5.2). Monosaccharides and polysaccharides normally occur in only small amounts. Glycerol and sorbitol are sometimes present in high concentration in the hemolymph of overwintering stages where they serve as antifreezes.

Free lipids seldom occur in high concentrations in hemolymph, except in some species after feeding, during flight, or at metamorphosis when they are being transported to sites of use or storage.

Other constituents of hemolymph include phosphate esters; urate salts and traces of other nitrogenous waste products; amino sugars such as acetylglucosamine, which are produced during digestion of the cuticle; pigments (often conjugated with protein); and hormones, which may be transported to their sites of action in combination with protein

4.1.2. Functions

Apart from the specific functions of particular components, which were outlined in the above consideration of its composition, the plasma has some important general functions. It serves as the medium in which nutrients, hormones, and waste materials can be transported to sites of use, action, and disposal, respectively. It is an important site for the storage, usually temporary, of metabolites. Plasma is the source of cell water, and during periods of desiccation its volume may decline at the expense of water entering the tissues. By virtue of certain of its components (proteins, amino acids, carboxylic acids, bicarbonate, and phosphate) it is a strong buffer and resists changes in pH which might occur as a result of metabolism. As a liquid, it is also used by insects for transmitting pressure changes from one part of the body to another. It is used hydrostatically, for example, to maintain the turgor necessary for movement in soft-bodied insects, split the old exocuticle during ecdysis, expand appendages after ecdysis, evert structures such as the penis, and extend the labium of larval Odonata.

4.2. Hemocytes

4.2.1. Number and Form

In many species all or nearly all the hemocytes are in circulation; in some species very few hemocytes circulate, the great majority remaining loosely attached to tissue surfaces. In adult mosquitoes there are no circulating hemocytes (Jones, 1977). Hemocyte counts are normally highly variable within the same species, as well as differing among species. Thus, in *Periplaneta americana*, counts ranging from 45,000 to 120,000 hemocytes/μl may be measured. As the insect has about 170 μl of hemolymph, it will have between 7 and 20 million circulating cells.

According to Crossley (1975) there is evidence that the number of circulating hemocytes may depend on the hormone titer of the hemolymph. Both ecdysone and juvenile hormone are said to stimulate an increase in the number of circulating hemocytes, especially plasmatocytes. In some instances, the increase seems to be due to increased mobility of preexisting cells rather than an increased rate of cell multiplication, though how hormones bring about this effect is not known.

Though several types of hemocyte have been recognized, which differ in size, stainability, function, and cytology (including fine structure) (Figure 17.6), their classification has proven difficult. This difficulty stems partly from the natural structural variability and multifunctional nature of some hemocytes both within and among species, and partly from the differences in methodology and criteria used to distinguish different hemocytes (see Arnold, 1974; Crossley, 1975; Jones, 1977). Notwithstanding these difficulties, it is apparent that three types of hemocyte are common to almost all insects, though one or more additional types may also occur in a given species. The relationships of these various types are, in most instances, unclear. In this account, the scheme of Arnold (1974) is followed.

The three types common to most insects are prohemocytes, plasmatocytes, and granular hemocytes, Prohemocytes (stem cells) are small (10 μm or less in diameter), spherical, or ellipsoidal cells whose nucleus fills almost the entire cytoplasm. They are frequently seen undergoing mitosis and are assumed to be the primary source of new hemocytes and the type from which other forms differentiate. Plasmatocytes (phagocytes) are cells of variable shape and size, with a centrally placed, spherical nucleus surrounded by well-vacuolated cytoplasm. In the cytoplasm is a well-developed Golgi complex and endoplasmic reticulum, as well as many lysosomes. The cells are capable of amoeboid movement and are phagocytic. Granular hemocytes are usually round or disc-

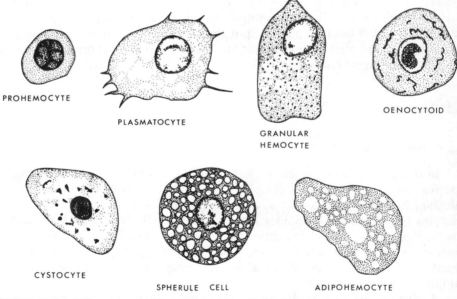

PROHEMOCYTE

PLASMATOCYTE

GRANULAR
HEMOCYTE

OENOCYTOID

CYSTOCYTE

SPHERULE CELL

ADIPOHEMOCYTE

FIGURE 17.6. Different types of hemocytes. [After R. F. Chapman, 1971, *The Insects: Structure and Function*. By permission of Elsevier North-Holland, Inc., and the author.]

shaped, with a relatively small nucleus surrounded by cytoplasm filled with
prominent granules. In some species they are amoeboid and phagocytic which,

473

THE
CIRCULATORY
SYSTEM

together with the occurrence of intermediate forms suggests that they may be
derived from plasmatocytes. More often, they are nonmotile and appear to be
involved in intermediary metabolism.

Other types of hemocytes include adipohemocytes, oenocytoids, spherule
cells, and cystocytes. As their name indicates, the adipohemocytes are cells
whose cytoplasm contains droplets of lipid. In addition to lipid droplets, the
cytoplasm may have nonlipid vacuoles and granules. The cells, which occasionally may be phagocytic, are considered by some authors to be a form of
granular hemocyte. Oenocytoids are spherical or ovoid cells with one, occasionally two, relatively small, eccentric nuclei. They are almost never phagocytic, as the absence of lysosomes, small Golgi complexes, and poorly developed
endoplasmic reticulum attest. Spherule cells are readily identifiable cells
whose central nucleus is often obscured by the mass of dense spherical inclusions occupying most of the cytoplasm. They are especially common in higher
Diptera and Lepidoptera, though their function is not known. Cystocytes
(coagulocytes) are spherical cells in whose small central nucleus the chromatin
is so arranged as to give the nucleus a "cartwheellike" appearance. The cytoplasm contains granules which, when liberated from these fragile cells, cause
the surrounding plasma to precipitate. Thus, the cells, which are again a
specialized kind of granular hemocyte, play a major role in hemolymph coagulation.

4.2.2. Functions

The functions of hemocytes may be discussed under four headings: endocytosis, encapsulation, coagulation, and metabolic and homeostatic functions.

Endocytosis. This is a process whereby a cell plasma membrane folds
around a substance which is thus ingested by the cell without rupture of the
membrane. It includes pinocytosis ("cell drinking") in which the material
ingested is in solution and phagocytosis where the material engulfed is par-
ticulate. Once it reaches the interior of the cell, the membrane-bound vesicle
containing the substance fuses with enzyme-containing lysosomes and degradation of the substance ensues. In insects the primary phagocytic cells are plasmatocytes, though, as noted above, other types of hemocytes may also engulf
material. Like other endocytotic cells, hemocytes are apparently selective with
regard to what they ingest though the basis of this selectivity is not known.
(See, however, encapsulation below.)

Endocytosis, especially phagocytosis, is important in both metamorphosis
and defense against disease, as well as in routine cleaning up of dead or damaged cells. In some insects (including Lepidoptera and higher Diptera), during
metamorphosis, phagocytic hemocytes invade many larval tissues and bring
about their rapid histolysis. However, invasion does not occur (at least in muscle which has been well studied) until autolysis has begun in tissue cells.
Autolysis is believed to be under hormonal control and presumably renders
tissue "foreign" to the hemocytes. In other insects (including Coleoptera and

mosquitoes) the hemocytes never invade larval tissues during metamorphosis, and the tissues disappear strictly by autolysis. A variety of microorganisms (viruses, bacteria, fungi, rickettsias, and protozoa) are known to be phago-cytozed by hemocytes, and in most instances insects have excellent resistance to infection as a result. However, for some viruses, phagocytosis is not followed by digestion and therefore does not prevent infection. Rather, the virus uses this process as a means of entering a cell prior to replication. Furthermore, habitual protozoan and fungal parasites often escape phagocytosis, presumably because they are not recognized as foreign by the hemocytes.

Encapsulation. Objects too large to be phagocytozed may become sur-rounded by layers of hemocytes, especially plasmatocytes, in a process called encapsulation (see Nappi, 1975). This reaction on the part of hemocytes thus forms an insect's most important defense mechanism against many metazoan endoparasites such as nematodes and insects. Typically, a capsule is 50 or more cells thick and includes, especially in the layers adjacent to the encapsulated object, intercellular secretions, in particular, mucopolysaccharide. Melanin may be present also in the capsule, though whether the hemocytes are involved in its production is not known. The primary effects of encapsulation on a foreign organism would appear to be starvation and asphyxiation. Though an organism does not always die as a result of its entombment, its development may be severely inhibited.

As in phagocytosis, recognition of an object as "foreign" is a key event in the encapsulation process, yet how this takes place remains largely a mystery. Salt (1970) has pointed out that it is impossible for all encapsulated objects to have something in common which is recognized by hemocytes. Rather, they are alike only in that they all lack a particular feature, specifically, Salt suggests, the connective tissue sheath which covers an insect's tissues and organs. Though Salt has provided evidence in support of his proposal, Crossley (1975) notes that (1) hemocytes themselves have no such sheath, yet do not attack each other; (2) at specific times individual tissues lack a sheath, yet are not attacked; and (3) conversely, occasionally tissues are phagocytozed even though their sheath appears intact. Clearly, the true recognition signal remains in doubt.

Though we do not know how hemocytes recognize foreign matter, many parasites obviously do because on entering a host they do not elicit an encapsu-lation reaction! This is especially true of many parasitic larval Hymenoptera. In such species it seems likely that a parasite's outer covering contains material identical to that surrounding the host's tissues. However, an integral part of the "recognition" process is adhesion to the foreign surface. Thus, another possi-bility is that a parasite secretes a substance which prevents hemocytes from attaching themselves to it. Also, prior to encapsulation, hemocytes are stimu-lated to move from their normal resting sites to the site of ensheathment and to divide. Therefore, the possibility that secretions from the parasite prevent mobilization and/or mitosis of hemocytes should not be ignored.

Coagulation. Studies on hemolymph coagulation in insects lag behind those on vertebrate blood clotting for which a mass of physiological and biochemical information is available. According to Grégoire (1974), this is due largely to the extremely rapid, even instantaneous nature of hemolymph coagulation, which hampers studies on the relative roles of hemocytes and plasma and the biochemical events which occur.

Although some early reports indicated that coagulation could occur in cell-free preparations, it is now accepted that all forms of coagulation involve the participation of a specific form of hemocyte, the coagulocyte. Among orthopteroid insects most Hemiptera, some Coleoptera and some Hymenoptera, coagulation is initiated by a sometimes "explosive" discharge of small pieces of cytoplasm from the coagulocytes, though the cells remain intact. As a result, plasma surrounding the cells forms granular precipitates which gradually increase in size and density. In some Scarabaeidae (Coleoptera) and larval Lepidoptera, the coagulocytes extrude long, threadlike pseudopodia which interweave with and stick to each other to form a cytoplasmic network in which other hemocytes become trapped. Concurrently, the plasma itself gels, forming transparent elastic sheets between the pseudopodia. In some Homoptera, other Coleoptera, for example, Tenebrionidae, and other Hymenoptera, clotting appears to combine the features of the types described above; that is, the coagulocytes send out pseudopodia, yet the plasma forms a granular precipitate around the cells as well as transparent sheets between pseudopodia. It is not known whether the pseudopodia-producing coagulocytes and those which induce granular precipitation of the plasma are identical.

In some insects (some Heteroptera, Curculionidae (Coleoptera), and Neuroptera) hemocytes morphologically identical to coagulocytes are present, but these do not release material to induce clotting. They do, however, accumulate at wounds, and Crossley (1975) suggests that they may release bacteriostatic substances.

Few biochemical details of the clotting process are known. For example, the nature of the coagulation-initiating substance released from the coagulocytes and its precise function are purely speculative. Perhaps the cells release a pharmacological trigger. Alternately, they may liberate a precursor substance directly involved in coagulation (Crossley, 1975). Specific hemolymph proteins are assumed to be involved in clot formation, though there appears to be almost no direct evidence in support of this assumption.

Metabolic and Homeostatic Functions. Hemocytes have been implicated in a variety of metabolic and homeostatic functions, though for most of these convincing evidence is not available. Largely on the basis of electron microscopy and histochemical studies, in which the cells have been shown to contain, for example, glycogen, mucopolysaccharide, lipid, and protein, it has been suggested that hemocytes are important in storage of nutrients and their distribution to growing tissues connective tissue formation, chitin synthesis, and hemolymph sugar level maintenance. Recent work cited in Crossley (1975) has shown that certain amino acids, including glutamate, can be actively accumulated by hemocytes, suggesting that the cells might be important in hemolymph amino acid homeostasis. The ability to accumulate and store glutamate (thereby effectively removing it from the hemolymph) may be of special significance in view of this substance's probable role as a transmitter substance at the neuromuscular junction (see Chapter 13). Certain hemocytes (spherule cells in Diptera, oenocytoids in other insects) contain enzymes for metabolism of tyrosine, derivatives of which are important in tanning and/or darkening of cuticle (Chapter 11, Section 3.3) and have a bacteriostatic effect.

Various authors also have suggested that the hemocytes may have a significant role in detoxicating poisons, though again convincing evidence is lacking.

5. Resistance to Disease

Resistance to disease may be considered to include two components, prevention of entry of the disease-causing organisms and rendering harmless organisms which do manage to reach the body cavity. The insect cuticle, covering the entire body surface and tracheal system presents a major obstacle to the entry of such organisms,* leaving virtually only the midgut and surface wounds as potential sites of invasion. A considerable number of potentially dangerous microorganisms must be ingested during feeding, yet normally these have no detrimental effect on the insect. What little work has been done suggests that special substances occur in the gut which deal with these organisms. Bacteria (especially gram-positive forms) appear to be digested due to the presence of enzymes called "lysozymes." Substances with antiviral activity are also known to occur in the gut, though their nature and mode of action has not been studied.

Hemolymph plays a major role in an insect's resistance to disease. It is important, through its role in wound healing, in preventing entry of pathogenic organisms, and in the destruction of organisms which manage to enter the body cavity.

5.1. Wound Healing

Wounding leads to a rapid increase in the number of circulating hemocytes and their aggregation in the vicinity of the wound. The nature of the factors that cause mobilization and aggregation are unknown, though it has been proposed that the damaged tissue cells release chemicals (wound hormones) which "attract" hemocytes. At the site of damage, hemocytes may become involved in various ways. The coagulocytes may initiate clotting; the plasmatocytes and perhaps other types of hemocytes may phagocytoze or encapsulate dead tissue cells and foreign organisms which have entered the wound; and hemocytes may arrange themselves in sheets to form a scaffold on which damaged tissue can regenerate.

5.2. Immunity

Immunity in the present discussion may be defined as the ability of an insect to resist the pathogenic effects of microorganisms which have gained entry into the body cavity. As in vertebrates, two forms of immunity may be distinguished, innate (natural) and acquired (induced).

---

*Some fungi enter the host via the integument (see Chapter 23, Section 4.2.4).

Innate immunity refers to resistance due to factors already present in an organism, that is, prior to any stimulation resulting from appearance of a pathogen. Innate immunity in Insecta appears to be principally a cellular phenomenon, comprising phagocytosis and encapsulation on the part of hemocytes. However, the plasma may contain powerful agglutinating and lytic substances (Jones, 1977).

Acquired immunity is resistance which is developed only when a pathogen enters a host; that is, the pathogen (or toxins produced by it) stimulates the development of this resistance. Unfortunately, despite more than 50 years of study of acquired immunity in insects, we can say with certainty little more than that it occurs! The existence of the phenomenon was first demonstrated by Paillot and Metalnikov, working independently, in 1920 (cited from Whitcomb et al., 1974). These authors found that insects previously "vaccinated" with killed bacteria were immune to live pathogens injected subsequently. The immunity which develops is different, however, to the antigen–antibody system of vertebrates in several respects. First, the immunity is short-lived, lasting only a matter of days, indicating that the cells which produce the immunogen (the substance which reacts against the pathogen) have no "memory," as do the antibody-producing lymphoid cells in vertebrates Second the response is nonspecific, which is in marked contrast to the highly specific interaction between antibody and antigen in vertebrates. Third, in many species, the immunogens are not proteins, since they are dialyzable, heat-stable, and unaffected by trypsin. Some circumstantial evidence suggests that phenolic compounds derived from tyrosine might be the immunogens. These compounds could be produced either within hemocytes, which are known to contain enzymes for the metabolism of tyrosine, or within the plasma after release of enzymes by hemocytes. Probably, plasmatocytes are important in immunogen production as their number increases significantly shortly after vaccination. Nonspecific proteins in the form of lysozymes may also be important, as there are reports which indicate that vaccination induces a marked increase in concentration of these enzymes. Again, the response is nonspecific, as increments can also be obtained by injecting sterile water or various dyes.

6. Summary

The insect circulatory system includes a contractile middorsal vessel, divisible into a posterior portion, the heart, equipped with valves (ostia) through which hemolymph enters the vessel, and an anterior aorta which delivers hemolymph to the head region. Muscular diaphragms and, in many species, accessory pulsatile organs assist in directing the hemolymph around the body. The heartbeat is myogenic in all insects, though its rate may be regulated neuronally in some species. The heartbeat is seen as a forwardly moving wave of peristalsis that begins near the posterior end of the heart.

Hemolymph is composed of cellular components, hemocytes, and a noncellular fraction, plasma. In primitive insects, inorganic ions, especially sodium and chloride, are the major osmotic effectors in hemolymph, but, through evolution, the tendency is for sodium and chloride to be replaced by magnesium, potassium, and organic anions, and for organic components to become the dominant contributors to osmotic pressure. However, diet may considerably influence the composition of plasma. Plasma, especially of larval endopterygotes, contains high and relatively constant concentrations of amino and carboxylic acids. Other organic constituents of plasma include enzymes and other proteins, trehalose and other carbohydrates, hormones and lipids, but, with the exception of trehalose, their concentration is variable. Apart from the specific functions of its components, plasma is important as a transport system, a source of cell water, and a hydrostatic system for transmitting pressure changes from one part of the body to another.

In many species virtually all the hemocytes are in circulation; in some other species most hemocytes do not normally circulate through the body but remain loosely attached to tissue surfaces to be mobilized under specific conditions such as wounding, entry of foreign organisms, and ecdysis. Several types of hemocytes occur of which three, prohemocytes, plasmatocytes, and granular hemocytes, are common to most insects. Prohemocytes are the stem cells from which other types differentiate; plasmatocytes are phagocytic and encapsulating agents; granular hemocytes are probably important in intermediary metabolism. In addition, in many species, cystocytes are present which participate in hemolymph coagulation.

Hemolymph plays an important role in insect immunity to disease. In insects innate immunity appears to be principally a cellular phenomenon, though in some species agglutinating and lytic substances occur in the plasma. Acquired immunity is short-lived and nonspecific. The immunogens may be either nonproteinaceous, perhaps phenolic compounds, or nonspecific enzymes known as lysozymes.

7. Literature

General descriptions of the circulatory system are given by Jones (1964, 1977). The forms and functions of hemocytes are discussed by Arnold (1974), Crossley (1975), and Jones (1975). Florkin and Jeuniaux (1974) deal with the composition of the hemolymph, and Grégoire (1970, 1974) with coagulation. Salt (1970) and Whitcomb et al. (1974) discuss the mechanisms by which insects resist disease.

Arnold, J. W., 1974, The hemocytes of insects, in: The Physiology of Insecta, 2nd ed., Vol. V (M. Rockstein, ed.), Academic Press, New York.

Crossley, A. C., 1975, The cytophysiology of insect blood, Adv. Insect Physiol. 11:117–221.

Florkin, M., and Jeuniaux, C., 1974, Hemolymph: Composition, in: The Physiology of Insecta, 2nd ed., Vol. V (M. Rockstein, ed.), Academic Press, New York.

Grégoire, C., 1970, Hemolymph coagulation in arthropods, Symp. Zool. Soc. London 27:45–74.

Grégoire, C., 1974, Hemolymph coagulation, in: The Physiology of Insecta, 2nd ed., Vol. V (M. Rockstein, ed.), Academic Press, New York.

Jones, J. C., 1964, The circulatory system of insects, in: The Physiology of Insecta, 1st ed., Vol. III (M. Rockstein, ed.), Academic Press, New York.

Jones, J. C., 1974, Factors affecting heart rates in insects, in: The Physiology of Insecta, 2nd ed., Vol. V (M. Rockstein, ed.), Academic Press, New York.

Jones, J. C., 1975, Forms and functions of insect hemocytes, in: *Invertebrate Immunity* (K. Maramorosch and R. E. Shope, eds.), Academic Press, New York.

Jones, J. C., 1977, *The Circulatory System of Insects*, Thomas, Springfield, Illinois.

Nappi, A. J., 1975, Parasite encapsulation in insects, in: *Invertebrate Immunity* (K. Maramorosch and R. E. Shope, eds.), Academic Press, New York.

Salt, G., 1970, *The Cellular Defence Reactions of Insects*, Cambridge University Press, Cambridge.

Whitcomb, R. F., Shapiro, M., and Granados, R. R., 1974, Insect defense mechanisms against microorganisms and parasitoids, in: *The Physiology of Insecta*, 2nd ed., Vol. V (M. Rockstein, ed.), Academic Press, New York.

18

Nitrogenous Excretion and Salt and Water Balance

1. Introduction

Enzymatically controlled reactions occur at the optimum rate within a narrow range of physical conditions. Especially important are the pH and ionic content of the cell fluid, since these factors most easily affect the active site on an enzyme. Since the conditions existing within cells and tissues are necessarily dependent on the nature of the fluid that bathes them—in insects, the hemolymph—it is the regulation of this fluid that is important. By regulation we mean the removal of unwanted materials and the retention of those that are useful to maintain as nearly as possible the best cellular environment. Regulation is a function of the excretory system and is of great importance in insects because they occupy such varied habitats and, therefore, have different regulatory requirements. Terrestrial insects lose water by evaporation through the integument and respiratory surfaces and in the process of nitrogenous waste removal. Brackish-water and saltwater forms also lose water as a result of osmosis across the integument; in addition, they gain salts from the external medium. Insects inhabiting fresh water gain water from and lose salts to the environment. The problem of osmoregulation is complicated by an insect's need to remove nitrogenous waste products of metabolism, which in some instances are very toxic. This removal utilizes both salts and water, one or both of which must be recovered later from the urine.

2. Excretory Systems

2.1. Malpighian Tubules—Rectum

The Malpighian tubules and rectum, functioning as a unit, form the major excretory system in most insects. Details of the rectum are given in Chapter 16, and only the structure of the tubules is described here.

The blindly ending tubules, which usually lie freely in the hemocoel, open into the alimentary canal at the junction of the midgut and hindgut (Figure

18.1A). They usually enter the gut individually but may fuse first to form a common sac or ureter that leads into the gut. Their number varies from two to several hundred and does not appear to be closely related to either the phylogenetic position or the excretory problems of an insect. Malpighian tubules are absent in Collembola, some Diplura, and aphids (Homoptera); in other Diplura, Protura, and Strepsiptera there are papillae at the junction of the midgut and hindgut. With the tubules are associated tracheoles and, usually, muscles (Figure 18.1E). The latter take the form of a continuous sheath, helical strips, or circular bands and are situated outside the basement membrane. They enable the tubules to writhe, which ensures that different parts of the hemolymph are exposed to the tubules and assists in the flow of fluid along the tubules.

A tubule is made up of a single layer of epithelial cells, situated on the

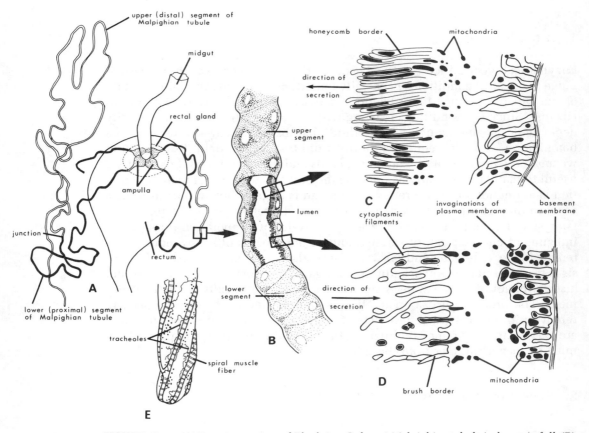

FIGURE 18.1. (A) Excretory system of *Rhodnius*. Only one Malpighian tubule is drawn in full. (B) Junction of proximal and distal segments of a Malpighian tubule of *Rhodnius*. Part of the tubule has been cut away to show the cellular differentiation. (C, D) Sections of the wall of the distal and proximal segments, respectively, of a tubule. (E) Tip of Malpighian tubule of *Apis* to show tracheoles and spiral muscles. [A, B, E, after V. B. Wigglesworth, 1965, *The Principles of Insect Physiology*, 6th ed., Methuen and Co. By permission of the author. C, D, from V. B. Wigglesworth and M. M. Saltpeter, 1962, Histology of the Malpighian tubules in *Rhodnius prolixus* Stal. (Hemiptera), *J. Insect Physiol.* **8**:299–307. By permission of Pergamon Press Ltd., Oxford.]

inner side of a basement membrane (Figure 18.1B–D). In many species where the tubules have only a secretory function (see Section 3.2) the histology of the tubules is constant throughout their length and basically resembles that of the distal part of the tubule of Rhodnius (Figure 18.1C). The inner (apical) surface of the cells takes the form of a brush border (microvilli). The outer (basal) surface is also extensively folded. Both these features are typical of cells involved in the transport of materials and serve to increase enormously the surface area across which transport can occur. Numerous mitochondria occur, especially adjacent to or within the folded areas, to supply the energy requirements for active transport of certain ions across the tubule wall. Adjacent cells are closely apposed near their apical and basal margins, though not necessarily elsewhere.

In some insects (for example, Rhodnius) two distinct zones can be seen in the Malpighian tubule (Figure 18.1C,D). In the distal (secretory) zone the cells possess large numbers of closely packed microvilli, but very few infoldings of the basal surface. Mitochondria are located near or within the microvilli. In the proximal (absorptive) part of the tubule the cells possess fewer microvilli, yet show more extensive invagination of the basal surface. The mitochondria are correspondingly more evenly distributed. In the flies Dacus and Drosophila, where pairs of Malpighian tubules unite to form a ureter prior to joining the gut, the ultrastructure of the ureter resembles that of the proximal part of the Rhodnius tubule, suggesting that the ureter may be a site of resorption of materials from the urine.

Yet other species have even more complex Malpighian tubules in which up to four distinct regions may be distinguished on histological or ultrastructural grounds. On the basis of the structural features of their cells, these regions have been designated as secretory or absorptive, though it must be emphasized that physiological evidence for these proposed functions is lacking. For a survey of insects whose tubules show regional differentiation and a discussion of tubule function in such species, see Jarial and Scudder (1970).

A cryptonephridial arrangement of Malpighian tubules is found in larvae and adults of many Coleoptera, some larval Hymenoptera and Neuroptera, and nearly all larval Lepidoptera (Figure 18.2). Here the distal portion of the Malpighian tubules is closely apposed to the surface of the rectum and enclosed within a perinephric membrane. The system is particularly well developed in insects living in very dry habitats, and its function is to improve water resorption from the material in the rectum (see Section 4.1).

2.2. Other Excretory Structures

In a few insects the labial glands may function as excretory organs. In apterygotes which lack Malpighian tubules the glands can accumulate and eliminate dyes such as ammonia carmine and indigo carmine from the hemolymph, but there is no evidence that they can deal similarly with nitrogenous or other wastes. The labial glands of saturniid moths excrete copious amounts of fluid just prior to emergence from the cocoon, and it may well be that the primary function of the glands is to reduce hemolymph volume and hence body weight, which, in such large flying insects, needs to be kept as low

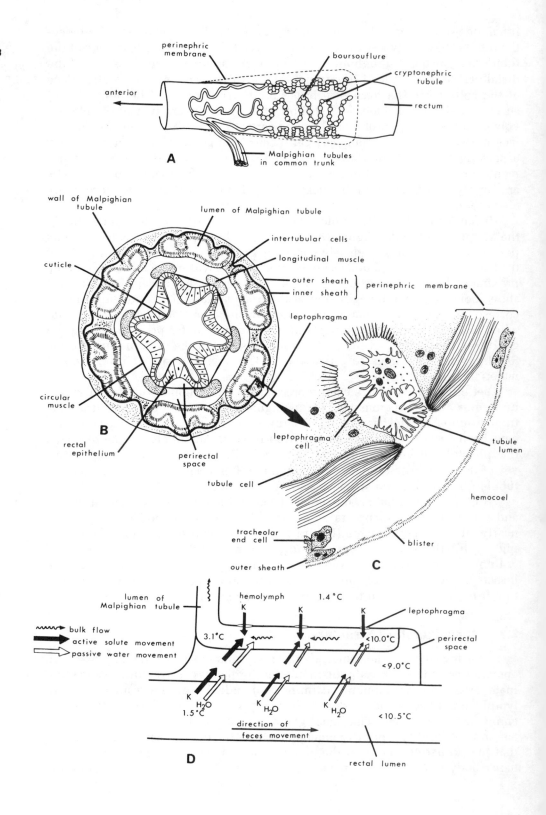

A

B

C

D

as possible. The midgut of silkmoth larvae actively removes potassium from the hemolymph, thus protecting the tissues from the very high concentration of potassium ions present in the leaves eaten by these insects.

In a few insects it appears that the Malpighian tubules, though present, play no part in nitrogenous excretion. In *Periplaneta americana*, for example, uric acid is not found in the tubules but does occur in small amounts in the hindgut, which may excrete it directly from the hemolymph. In *P. americana* much uric acid is stored in urate cells in the fat body, and the major form of excreted nitrogen in this species is ammonia. How this reaches the hindgut lumen in *P. americana* is unclear. However, in the flesh fly *Sarcophaga bullata*, ammonia, the primary excretory product, is actively secreted as ammonium ions into the lumen across the hindgut wall.

In males of some species of cockroaches, for example, *Blattella germanica*, a considerable amount of uric acid (as much as 5% of the live weight of the insect) is found in the utriculi majores (part of the accessory reproductive gland complex). The uric acid becomes part of the wall of the spermatophore and is, in a sense, "excreted" during copulation.

3. Nitrogenous Excretion

3.1. The Nature of Nitrogenous Wastes

In nitrogenous wastes structural complexity, toxicity, and solubility go hand in hand. The simplest form of waste (ammonia) is highly toxic and very water-soluble. It contains a high proportion of hydrogen which can be used in production of water. It is generally found as an excretory product, therefore, only in those insects which have available large amounts of water, for example, freshwater forms and the larvae of meat-eating flies. Generally, however, in insects, as in other terrestrial organisms, water must be conserved, and more complex nitrogenous wastes must be produced, which are both less toxic and less soluble. In the egg and pupal stage the problem is accentuated because water lost cannot be replaced, and nitrogenous wastes must remain in the body in the absence of a functional excretory system. Most insects, then, excrete their waste nitrogen as uric acid. This is only slightly water-soluble, relatively nontoxic, and contains a smaller proportion of hydrogen compared with ammonia. It should be realized, however, that uric acid is not the only form of nitrogenous waste. Usually, traces of other materials (especially the related compounds

---

FIGURE 18.2. Cryptonephridial arrangement of Malpighian tubules in *Tenebrio* larva. (A) General appearance. Note that only three of the six tubules are drawn fully and that in reality the tubules are much more convoluted and have more boursouflures than are shown. (B) Cross section through posterior region of cryptonephridial system, (C) details of a leptophragma, and (D) diagram illustrating proposed mode of operation of system. Solid arrows indicate movements of potassium, hollow arrows indicate movements of water. Numbers indicate osmotic concentration (measured as freezing-point depression) of fluids in different compartments. [After A. V. Grimstone, A. M. Mullinger, and J. A. Ramsay, 1968, Further studies on the rectal complex of the mealworm *Tenebrio molitor* L. (Coleoptera, Tenebrionidae), *Philos. Trans. R. Soc. London, Ser. B* **253**:343–382. By permission of the Royal Society, London, and Professor J. A. Ramsey.]

allantoin and allantoic acid) can be detected, and in many species one of these has become the predominant excretory product (Bursell, 1967). Urea is only rarely a major constituent of insect urine. Usually, it represents less than 10% of the nitrogen excreted. Traces of amino acids can be found in the excreta of many insects, but their presence should be regarded as accidental loss rather than deliberate excretion by an insect (Bursell, 1967). Only rarely has the excretion of particular amino acids been authenticated; for example, the clothes moth *Tineola* and the carpet beetle *Attagenus* excrete large amounts of the sulfur-containing amino acid cystine. Although in tsetse flies uric acid is the primary excretory product, two amino acids, arginine and histidine, are important components of the urine. These make up about 10% of the protein amino acids in human blood; because their nitrogen content is high, it is probably uneconomical to degrade them, and they are therefore excreted unchanged (Bursell, 1967). The amino acids excreted in vast quantity by plant-sucking Hemiptera must, of course, be considered as fecal and not metabolic waste products. Because of the large amount of water taken in by aphids, it has been suggested that they might produce ammonia as their nitrogenous waste. Indeed, uric acid, allantoin, and allantoic acid cannot be detected in the excreta. However, ammonia makes up only 0.5% of the total nitrogen excreted, which has led to the suggestion that it is used (and detoxified) by the symbiotic microorganisms residing in the mycetome. Table 18.1 contains selected examples to show the variety of nitrogenous wastes produced by insects.

As can be seen in Figure 18.3, uric acid and the other nitrogenous waste

TABLE 18.1. Nitrogenous Excretory Products of Various Insects[a,b]

	Uric acid	Allantoin	Allantoic acid	Urea	Ammonia	Amino acids
Odonata						
Aeshna cyanea (larva)	0.08	—	0.00	—	1.00	—
Dictyoptera/Phasmida						
Periplaneta americana	1.00	0.00	0.00	—	—	—
Blatta orientalis	0.64	0.64	1.00	—	—	—
Dixippus morosus	0.69	1.00	0.44	—	—	—
Hemiptera						
Dysdercus fasciatus	0.00	1.00	0.00	0.26	—	0.24
Rhodnius prolixus	1.00	—	—	0.33	—	trace
Coleoptera						
Melolontha vulgaris	1.00	0.00	0.00	—	—	—
Attagenus piceus	0.72	—	—	1.00	0.57	0.50
Diptera						
Lucilia sericata	1.00	0.30	—	—	0.30	—
Lucilia sericata (pupa)	1.00	0.00	—	—	0.15	—
Lucilia sericata (larva)	0.05	0.02	—	—	1.00	—
Lepidoptera						
Pieris brassicae	1.00	0.04	0.01	—	—	—
Pieris brassicae (pupa)	1.00	0.03	0.05	—	—	—
Pieris brassicae (larva)	0.28	0.16	1.00	—	—	—

[a] From Bursell (1967), after various authors.
[b] The quantity of nitrogen excreted in the different products is expressed as a proportion of the nitrogen in the predominant end product.

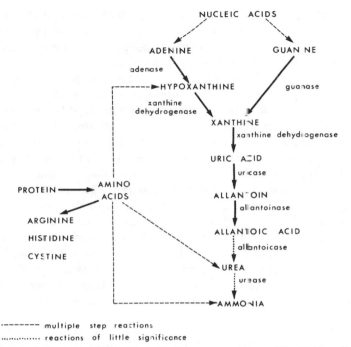

FIGURE 18.3. Metabolic interrelationships of nitrogenous wastes. [After E. Bursell, 1967, The excretion of nitrogen in insects, *Adv. Insect Physiol.* 4:33–67. By permission of Academic Press Ltd., London, and the author.]

products are derived from two sources, nucleic acids and proteins. Degradation of nucleic acids is of minor importance; most nitrogenous waste comes from protein breakdown followed by synthesis of hypoxanthine from amino acids. The biochemical reactions which lead to synthesis of this purine appear to be similar to those found in other uric acid-excreting organisms [Bursell, 1967; Barrett and Friend, 1970].

In addition to the enzymes for uric acid synthesis there are also uricolytic enzymes which catalyze the degradation of this molecule in many insects (Figure 18.3). Uricase has a wide distribution within the insect class. Active preparations of allantoinase have been obtained from many species, but the distribution of this enzyme appears to be rather restricted compared with uricase. Although there are reports that indicate the occurrence of allantoicase and urease in tissue extracts from a few insects, their presence should not be regarded as having been established unequivocally. In other words, when urea and ammonia are produced in significant amounts, they are probably derived in a manner other than by the degradation of uric acid. The existence of an ornithine cycle for urea production, such as is found in vertebrates, has not been proved conclusively, even though the constituent molecules of the cycle (arginine, ornithine, and citrulline) and the enzyme arginase have been identified in several species. Similarly, the way in which ammonia is produced (especially in those insects in which it is a major excretory molecule) is poorly understood. It is generally assumed to result from deamination of amino acids, but the precise way in which this occurs is unclear.

It has been suggested that the most primitive state was that in which the complete series of uricolytic enzymes was present, and ammonia was the excretory material. As insects became more independent of water, selection pressures led to loss of the terminal enzymes and production of more appropriate excretory molecules. This simple view should be regarded with caution. Thus, in some caterpillars diet can affect the nature of the nitrogenous waste. In certain insects substantial quantities of a particular nitrogenous waste molecule are produced, yet the appropriate enzyme in the uricolytic pathway has not been demonstrated, and vice versa; that is, the effects of other metabolic pathways may override the uricolytic system. In many insects (especially endopterygotes) the predominant nitrogenous excretory product changes during development. For example, in *Pieris brassicae* (Lepidoptera) the major excretory product in the pupa and adult is uric acid; in the larva this compound constitutes only about 20% of the nitrogenous waste, allantoic acid being the predominant end product. Indeed, in some Lepidoptera, the ratio of uric acid to allantoin may fluctuate widely from day to day (Razet, 1961, cited from Bursell, 1967). Of great interest will be determination of factors which stimulate inhibition or activation (degradation or synthesis?) of uricolytic enzymes so that the most suitable form of nitrogenous waste is produced under a given set of conditions.

3.2. Physiology of Nitrogenous Excretion

Uric acid is produced in the fat body and/or Malpighian tubules (occasionally the midgut) and released into the hemolymph. It is secreted into the lumen of the tubules as the sodium or potassium salt, along with other ions, water, and various low molecular weight organic molecules. In *Dixippus* secretion occurs along the entire length of the tubule. No resorption of materials takes place across the tubule wall, and urate leaves the tubule in solution. In the rectum resorption of water and sodium and potassium ions occurs, and the pH of the fluid decreases from 6.8–7.5 to 3.5–4.5. The combined effect of water resorption and pH change is to cause massive precipitation of uric acid. Useful organic molecules such as amino acids and sugars are also resorbed through the rectal wall. The Malpighian tubule–rectal wall excretory system thus shows certain functional analogies with the vertebrate nephron. The excretion of uric acid in *Dixippus* is summarized in Figure 18.4A.

In *Rhodnius*, whose tubules show structural differentiation along their length, the process of excretion is basically the same as in *Dixippus*. However, in *Rhodnius* only the distal portion of the tubule is secretory and resorption of water and cations begins in the proximal part. Slight change in pH occurs (from 7.2 to 6.6) as the fluid passes along the tubule and this is sufficient to initiate uric acid precipitation. Further water and salt resorption occurs in the rectum (pH 6.0), causing precipitation of the remaining waste (Figure 18.4B).

Although allantoin is the major nitrogenous waste in many insects, its mode of excretion appears to have been studied in only one species, *Dysdercus fasciatus* (Hemiptera) (Berridge, 1965). This insect is required, because of its diet, to excrete large quantities of unwanted ions (magnesium, potassium, and phosphate). This, combined with the insect's inability to actively resorb water

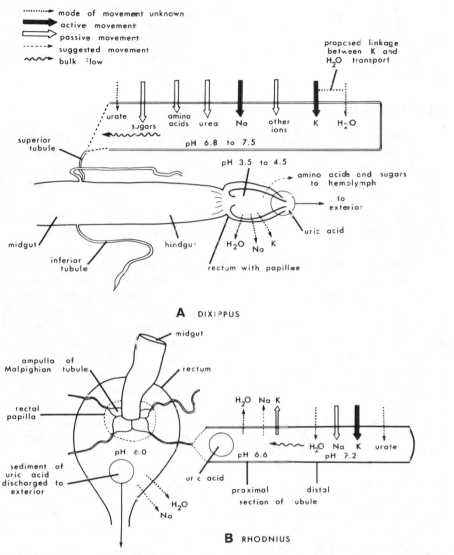

FIGURE 18.4. Movements of water, ions, and organic molecules in the excretory systems of (A) *Dixippus* and (B) *Rhodnius*. [After R. H. Stobbart and J. Shaw. 1974. Salt and water balance; excretion, in: *The Physiology of Insecta*, 2nd ed., Vol. V (M. Rockstein, ed.). By permission of Academic Press Inc. and the authors.]

from the rectum, results in the production of a large volume of urine. Because no resorption or acidification occurs which could cause precipitation of uric acid, this molecule is no longer used as an excretory product. Thus, allantoin, which is 10 times more soluble than uric acid (yet of equally low toxicity), is preferred. However, the insect does not possess a mechanism for actively transporting this molecule from the hemolymph to tubule lumen; that is, allantoin only moves passively across the wall of the tubule. It is therefore maintained in high concentration in the hemolymph to achieve a sufficient rate of diffusion into the tubule. Whether a similar mechanism occurs in other allantoin-excreting insects remains to be seen. It may be significant that many other allantoin producers are herbivorous and have the problem of removing large quantities of unwanted ions.

The physiological mechanisms for excretion of other nitrogenous wastes are not known.

3.3. Storage Excretion

Storage excretion is the retention of waste material within the body, rather than its removal through the excretory system. Thus, in Collembola and other apterygotes which lack Malpighian tubules, in those larval Hymenoptera whose tubules are not fully developed, in many endoparasites, and in *Periplaneta* where the tubules appear not to function in nitrogenous waste removal, there are specialized urate cells in the fat body in which uric acid concretions develop. These cells replace, functionally speaking, the Malpighian tubule system.

At other times storage of urate occurs even when the tubules are working normally and may be regarded as a supplementary excretory mechanism for occasions when the tubules cannot cope with all the waste that is being produced. In the larval stages of many species uric acid crystallizes out in ordinary fat body cells and epidermis, even though the Malpighian tubules are functional. It appears that this is caused by the metabolic activity of the cells themselves (that is, they are not accumulating uric acid from the hemolymph), and crystallization occurs by virtue of the particular conditions (pH, ionic content, etc.) existing in the cells. During the later stages of pupation the crystals disappear, the uric acid apparently having been transferred to the meconium via the excretory system. It is worth noting that in many species the Malpighian tubules are entirely reconstituted during the pupal stage. Thus, storage of uric acid in fat body and epidermal cells is of great importance at this time (Wigglesworth, 1965). Often the uric acid is stored in specific regions of the body. In *Dysdercus*, for example, the Malpighian tubules cease to function toward the end of each instar. At this time uric acid is deposited in the epidermal cells of the abdomen forming distinct, white transverse bands. The uric acid is stored permanently in these cells and thus serves an important pigmentary function (Berridge, 1965).

Storage excretion of material other than uric acid takes place. Calcium salts (especially carbonate and oxalate) are found in the fat body of many plant-eating insect larvae. During metamorphosis they are released and dissolved, to be excreted via the Malpighian tubules in the adult. Dyes present in food are often accumulated in fat body cells where they appear to become associated with particular proteins. These proteins are then transferred to the egg during vitellogenesis and the dyes subsequently "excreted" during oviposition.

Nephrocytes (Chapter 17) accumulate a variety of substances, especially pigments, and it is thought by many workers that storage excretion is one of their major functions. According to Wigglesworth (1965), this view is mistaken; if they are involved in excretion it is probably in the intermediary metabolism of waste materials.

4. Salt and Water Balance

491

NITROGENOUS
EXCRETION AND
SALT AND WATER
BALANCE

Salt and water balance involves more than simply the control of hemolymph osmotic pressure; the relative proportions of the ions that contribute to this pressure must be maintained within narrow limits. The osmotic pressure of the hemolymph is generally within the same limits as that of the blood of other organisms, but it can be increased considerably under specific conditions (by the addition, for example, of glycerol which serves as an antifreeze during hibernation). Regulation of the salt and water content is obviously related to the nature of the external environment. Different osmotic problems are faced by insects in different habitats. Nevertheless, they have been solved using the same basic mechanism, namely, the production of a "primary excretory fluid" in the Malpighian tubules followed by differential resorption from or secretion into this fluid when it reaches the rectum. For clarity we shall consider separately the problems of insects living on land, in brackish or saline water, or in fresh water; considerable similarity in the solution of these problems will, however, be seen.

4.1. Terrestrial Insects

Although only a limited amount of data is available, it appears that terrestrial insects can regulate their hemolymph osmotic pressure over a wide range of conditions. For example, in *Tenebrio* the hemolymph osmotic pressure varies only from 223 to 365 mM/liter (measured as the equivalent of a sodium chloride solution) over a range of relative humidity from 0 to 100% (Marcuzzi, 1956, cited in Stobbart and Shaw, 1974). In starving *Schistocerca* there is only a 30% difference in hemolymph osmotic pressure between animals kept in air at 100% relative humidity and given only tap water and those kept in air at 70% relative humidity and given saline (osmotic pressure equivalent to 500 mM/liter sodium chloride) to drink (Phillips, 1964a).

In terrestrial insects water is lost (1) by evaporation across the integument, although this is considerably reduced by the presence of the wax layer in the epicuticle (Chapter 11); (2) during respiration through the spiracles [many insects possess devices both physiological and structural for reducing the loss (Chapter 15)]; and (3) during excretion. The major source of water for most terrestrial insects is obviously food and drink. Some insects eat excessively solely for the water content of the food. Where sufficient water cannot be obtained by drinking or in food, the insect must obtain it by other means. One source is the water produced during metabolism. Absorption of water vapor from the atmosphere is a method employed by a few insects (for example, *Thermobia* and *Tenebrio*) which are normally found in very dry conditions. Interestingly, the site of absorption is the rectum, which, as is noted below, is the site of uptake of liquid water in other terrestrial and brackish-water insects.

Small amounts of ions are lost from the body via the excretory system, and these are readily made up by absorption across the midgut wall. Indeed, in terrestrial insects the usual problem is removal of unwanted ions present in the diet. The food often contains ions in concentrations which are widely different from those of the hemolymph. It is probable that these ions enter the hemo-lymph passively in the same proportions as they occur in the diet, and excesses are subsequently expelled via the excretory system. In other words, the midgut does not act as a selectively permeable barrier to the entry of ions (Stobbart and Shaw, 1974).

The role of the Malpighian tubules and rectum has been investigated by examination of the ionic composition of the fluids within them and, more recently, by the use of radioisotopes to measure the direction and rate of movement of individual ions. The studies of Ramsay in the 1950s (see reviews for references) revealed that the fluid in the tubules is isosmotic with the hemolymph (Table 18.2) but has a very different ionic composition. Particu-larly obvious is the difference in potassium ion concentration which is several times higher in the tubule fluid than in the hemolymph. The sodium ion con-centration is usually lower in the fluid than in the hemolymph, as is the case with most other ions (except phosphate). The tubule fluid, which is produced continuously, contains a number of low molecular weight organic molecules, for example, amino acids and sugars; thus, it is broadly comparable with the glomerular filtrate of the vertebrate kidney, though it is not produced by hy-drostatic pressure. The high potassium concentration in the tubule fluid and the demonstration that the rate at which tubule fluid is formed depends on the hemolymph potassium concentration led Ramsay to suggest that the active

TABLE 18.2. The Osmotic Pressure and Concentration (mM/liter) of Some Ions in the Hemolymph (H), Malpighian Tubule Fluid (MT), and Rectal Fluid (R) in Insects from Different Habitats[a]

Habitat	Species (stage and conditions)	Fluid	Osmotic pressure ($\equiv$ NaCl solution)	Ions		
				Na^+	K^+	Cl^-
Terrestrial	*Schistocerca gregaria* (adult, water-fed)	H	214	108	11	115
		MT	226	20	139	93
		R	433	1	22	5
	Dixippus morosus (adult, feeding)	H	171	11	18	87
		MT	171	5	145	65
		R	390	18	327	—
	Rhodnius prolixus (adult, 19–29 hr after meal)	H	206	174	7	155
		MT	228	114	104	180
		R	358	161	191	—
Salt water	*Aedes detritus* (larvae, in seawater)	H	157	—	—	—
		MT	—	—	—	—
		R	537	—	—	—
	Aedes detritus (larvae, in distilled water)	H	97	—	—	—
		MT	—	—	—	—
		R	56	—	—	—
Fresh water	*Aedes aegypti* (larvae, in distilled water)	H	138	87	3	—
		MT	130	24	88	—
		R	12	4	25	—

[a] Data mainly from Stobbart and Shaw (1974).

transport of potassium ions is fundamental to the production and flow of the fluid. The other ions and organic molecules probably enter the tubule fluid passively.

It is clear that, since the tubule fluid and hemolymph are isosmotic, the tubules are not directly concerned with regulation of hemolymph osmotic pressure. Experimental proof for this was obtained by Ramsay, who showed, using isolated tubules, that the isosmotic condition is retained over a wide range of external concentrations. Indirectly, however, the tubules are important in regulation, since the rate at which ions and water are excreted from the body is the difference between their rate of secretion into the tubule lumen and their rate of resorption by the rectum.

As noted in Section 3.2, in some insects the tubules show regional differentiation, secretion taking place in the distal part and resorption beginning in the proximal part of the tubule. In most species, however, resorption occurs mainly in the rectum, though the anterior part of the hindgut may also modify the fluid. In the rectum major changes occur in the osmotic pressure and composition of the urine (Table 18.2). Generally the urine becomes greatly hypertonic to the hemolymph, but when much water is available a hypotonic fluid may be excreted.

Wigglesworth (1931) first noted that the rectum is the main site of water resorption in many terrestrial insects. Water is resorbed against a concentration gradient; that is, it is an active process and energy is expended. Phillips (1964a) showed that in *Schistocerca* the rate of water movement across the rectal wall is independent of the rate of salt accumulation. The rate at which water is resorbed depends on the osmotic gradient across the wall, and, as the gradient increases during resorption, the point is reached at which the rate of active accumulation is balanced by the rate of passive diffusion back into the rectal lumen; that is, the concentration of the rectal fluid reaches a maximum value. However, this value varies according to the water requirements of the insect. Insects that have been kept in a dry environment and given strong saline to drink have a rectal fluid whose osmotic pressure is about twice that of insects with access to tap water. The physiological basis of this increased ability to concentrate the urine is not known.

The precise mechanism of water uptake is still unclear. Though models have been proposed in which water *per se* is actively transported across the rectal wall, there is now direct evidence that water movements occur as a result of active movements of inorganic ions, especially sodium, potassium, and chloride (secondary transport of water) (see Phillips, 1977). Fine-structural studies of the rectal wall of *Calliphora* (Berridge and Gupta, 1967), *Periplaneta* (Oschman and Wall, 1969) and other insects, and the elegant work of Wall and Oschman (1970) and Wall et al. (1970) who used micropuncture to obtain fluid samples from the subepithelial sinus and intercellular spaces in the rectal epithelium led to the following scheme for water absorption from the rectum (Figure 18.5). Ions are actively secreted into the intercellular space between the highly convoluted plasma membranes of adjacent epithelial cells so that local pockets of high salt content are formed. Thus, an osmotic gradient is developed down which water flows from the rectal lumen to the intercellular spaces via the cytoplasm of the epithelial cells. The entry of water into the spaces pro-

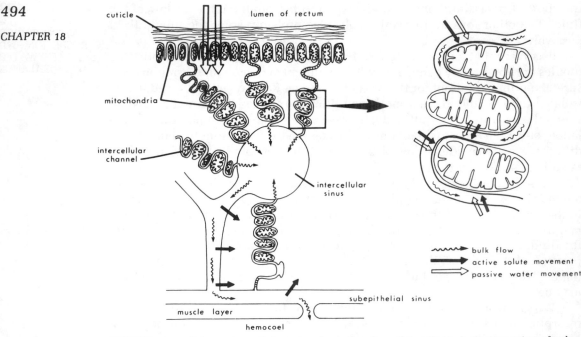

FIGURE 18.5. Scheme to explain water absorption from the rectum. Active secretion of solute into the intercellular channels induces passive movement of water into the channels from the epithelial cells and into the cells from the rectal lumen. The intercellular fluid thus formed flows toward the hemocoel, and, as it moves through the sinuses, solute is actively resorbed by the cells for recycling. For further details, see text. [After S. H. P. Maddrell, 1971, The mechanisms of insect excretory systems, *Adv. Insect Physiol.* **8:**199–331. By permission of Academic Press Ltd., London, and the author.]

duces a hydrostatic pressure which forces the ions and water toward the hemolymph. As the fluid moves through the larger (inner) intercellular spaces and subepithelial sinus, active resorption of ions occurs across the epithelial cell membrane. However, relatively little water moves into the cells because the spaces have a low surface area/volume ratio (i.e., the plasma membrane of the cells is not convoluted in these regions as it is in the distal intercellular spaces).

Ramsay's early work on *Rhodnius* and *Dixippus* provided a strong indication that the rectum is also capable of resorbing salts, and this has been confirmed by Phillips (1964b) in *Schistocerca*. This author showed that sodium, potassium, and chloride ions can be accumulated against a concentration gradient and independently of the movement of water. Furthermore, the rate of accumulation of these ions depends on their concentrations in the rectal fluid and the hemolymph. In this way the requirements of the insect can be satisfied. In water-fed locusts ions are resorbed from the rectum as quickly as they arrive in the fluid from the tubules, and low rectal concentrations are found (Table 18.2). At the other extreme, in saline-fed animals, the rates of resorption are low and a greatly hyperosmotic fluid is produced.

It appears from the limited amount of experimental work carried out that the cryptonephridial arrangement of Malpighian tubules increases the power of the rectal wall to resorb water against high concentration gradients. The system is particularly well developed in insects that inhabit dry environments. According to Ramsay (1964), the perinephric membrane is impermeable to water, and, under dry conditions, the osmotic pressure of the perinephric cavity is raised mainly due to the presence of some unknown electrolyte. Thus, the concentration gradient across the rectal wall is reduced, facilitating water uptake. Ramsay suggested that potassium and chloride ions are actively transported into the lumen of the perirectal tubules (which contain many mitochondria), with an accompanying movement of water. Resorption of ions and water occurs across the wall of the parts of the tubule bathed in hemolymph. An apparent lack of mitochondria in the perinephric membrane and leptophragma cells led Ramsay to conclude that ions move passively across the membrane in order to balance those removed by the tubule. However, a fine-structural and experimental study (Grimstone et al., 1968) has shown this conclusion to be wrong. These authors found that the leptophragma cells have a normal complement of mitochondria. Active transport of potassium ions occurs across the cells, which are, however, impermeable to water. Chloride ions follow passively. The scheme is summarized in Figure 18.2D.

4.2. Brackish-Water and Saltwater Insects

The habitat occupied by brackish-water and saltwater insects can vary widely in ionic content and osmotic pressure. During periods of warm, dry weather the salinity may increase severalfold. Conversely, after heavy rains or the melting of snow in spring, the salinity may approach that of fresh water. It is not surprising, therefore, to find experimentally that such insects can regulate their hemolymph osmotic pressure over a wide range of external salt concentrations (Figure 18.6). Larvae of *Aedes detritus* and *Ephydra riparia*, inhabitants of salt marshes, can survive in media containing the equivalent of 0 to about 7–8% sodium chloride. Over this range of concentrations the hemolymph osmotic pressure changes by only 40–60%. Larvae of *Ephydra cinerea* are found in the Great Salt Lake of Utah where the salinity may exceed the equivalent of 20% sodium chloride.

At low salinities the hemolymph is hyperosmotic to the medium and a very dilute urine is excreted. Salts are actively resorbed through the rectal wall. At high salinities the hemolymph osmotic pressure is less than that of the environment. Water will therefore tend to move out of the insect as a result of osmosis. The water loss is counterbalanced by ingestion of the medium during feeding, but this in itself creates a problem because of the high salt content of the ingested fluid. Until recently, it was assumed that (1) the midgut wall was highly impermeable to salts, most of which therefore did not enter the hemolymph, and (2) as in terrestrial insects, most of the water in the tubule fluid was resorbed in the rectum, forming a greatly hyperosmotic urine (Stobbart and Shaw, 1974). However, the work of Phillips and co-workers (see Phillips et al., 1978, for references) has shown that in saline-water mosquitoes, at least, these assumptions are incorrect. The midgut wall is markedly permeable both to ions and to water; indeed, almost all ions and water ingested are absorbed from the midgut into the hemolymph, a feature which Phillips et al. (1978) suggest may serve to concentrate food in the gut prior to digestion. The

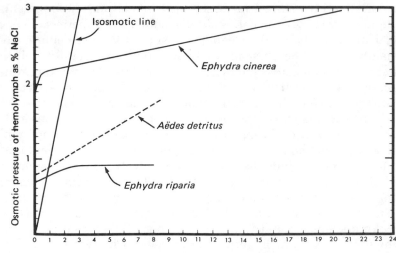

FIGURE 18.6. The relationship between osmotic pressure of the hemolymph and that of the external medium in some saltwater larvae. [After J. Shaw and R. H. Stobbart, 1963, Osmotic and ionic regulation in insects, *Adv. Insect Physiol* 1:315–399. By permission of Academic Press Ltd., London, and the authors.]

excess water taken into the hemolymph is lost either by osmosis through the body wall or in the formation of Malpighian tubule fluid. As in insects from other habitats, the tubule fluid is isosmotic with the hemolymph, and the main function of the tubules in saline-water species appears to be excretion of sulfate ions which may be present in the medium in high concentration. Other ions in excess in the hemolymph are actively secreted into the lumen of the rectum by cells of the posterior rectal segment (which, interestingly, is not present in strictly freshwater mosquitoes) to form a greatly hyperosmotic urine.

4.3. Freshwater Insects

The regulatory problems facing freshwater insects are the opposite of those in insects from saline conditions. Water enters the body osmotically [despite the relatively impermeable cuticle (Chapter 11)] and must be removed, and salts will be lost from the body and must be replaced if the hyperosmotic condition of the hemolymph is to be maintained. Freshwater insects can regulate their hemolymph osmotic pressure successfully to the point at which the external environment becomes isosmotic with the hemolymph (Figure 18.7). This is achieved by the production of urine which is hypoosmotic to the hemolymph. Beyond this point regulation breaks down because freshwater insects are not able to produce a hyperosmotic urine; that is, they cannot resorb water against a concentration gradient (Stobbart and Shaw, 1974).

As in terrestrial insects, the Malpighian tubules produce a fluid which is isosmotic with the hemolymph but of different ionic composition. Particularly obvious is the great difference in the potassium ion concentration between the two solutions (Table 18.2). When the fluid enters the rectum resorption of ions occurs. The osmotic pressure of the fluid that finally leaves the body is much

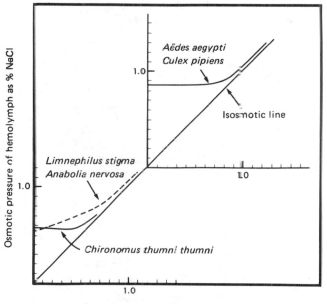

FIGURE 18.7. The relationship between osmotic pressure of the hemolymph and that of the external medium in some freshwater insects. [After J. Shaw and R. H. Stobbart, 1963, Osmotic and ionic regulation in insects, *Adv. Insect Physiol* 1:315–399. By permission of Academic Press Ltd., London, and the authors.]

lower than that of the hemolymph but not as low as would be expected from knowledge of the extent of ionic resorption in the rectum. This is because large quantities of ammonium ions appear in the rectal fluid. These ions cannot be detected in the Malpighian tubules, and it is presumed that they are secreted directly across the rectal wall, as occurs in larvae of *Sarcophaga bullata* (see Section 2.2).

In freshwater insects food is the usual source of ions which are absorbed through the midgut wall. However, in some forms ions are accumulated through other parts of the body, for example, the gills of caddis fly larvae, the rectal respiratory chamber of dragonfly and mayfly larvae, the anal gills of syrphid larvae, and the anal papillae of mosquito and midge larvae (Wigglesworth, 1965). The role of the anal papillae in ionic regulation has been particularly well studied. In mosquito larvae a pair of papillae is located on each side of the anus (Figure 18.8A). They communicate with the hemocoel and are well supplied with tracheae. Their walls are one cell thick and covered with a thin cuticle (Figure 18.8B). Koch (1938) and Wigglesworth (1938) were the first to demonstrate that mosquito larvae can accumulate chloride ions against a large concentration gradient and that the mechanism for doing so is located in the papillae. Later workers showed that sodium, potassium, and phosphate ions are also actively transported into the papillae. The ability to accumulate ions varies with the habitat in which an insect is normally found. Thus, *Culex pipiens*, which is found in contaminated water, is less efficient at collecting ions than *Aedes aegypti*, which typically lives in fresh rainwater pools. Indeed, normal larvae of the latter species can maintain a constant hemolymph sodium

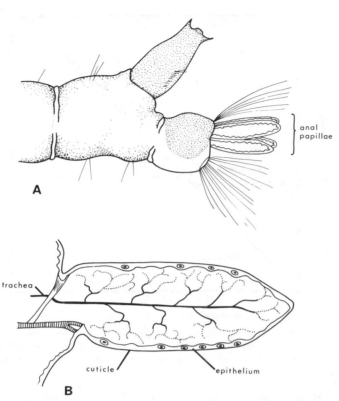

FIGURE 18.8. (A) Posterior end of *Aedes aegypti* to show anal papillae, and (B) structural details of a single anal papilla. [A, after V. B. Wigglesworth, 1965, *The Principles of Insect Physiology*, 6th ed., Methuen and Co. By permission of the author. B, after V. B. Wigglesworth, 1933, The effect of salts on the anal glands of the mosquito larva, *J. Exp. Biol.* **10**:1–15. By permission of Cambridge University Press, New York.]

concentration when the sodium concentration of the external medium is only 6 μM/liter. The hemolymph sodium concentration under these conditions is only about 5% below the normal level (Shaw and Stobbart, 1963).

4.4. Hormonal Control

As we have seen above, insects possess, in the form of the Malpighian tubules and rectum, an excellent mechanism for the regulation of the salt and water content of the hemolymph. However, such a mechanism is useful only if it can be told when to start, stop, accelerate, or decelerate to suit the needs of an insect. This coordination is effected by hormones.

The occurrence of diuretic hormones produced by neurosecretory cells of the brain and other ganglia is firmly established. Antidiuretic factors are also thought to occur in some insects, but the evidence for them is rather circumstantial at present. A diuretic hormone, which stimulates excretion of water, is released following feeding in many terrestrial insects. In *Rhodnius* stretching of the abdominal wall brings about hormonal release (Maddrell,

1964). In *Schistocerca, Dysdercus,* and other insects that feed more or less continuously, it is probably the stretching of the foregut that causes release of hormone (Mordue 1969; Berridge, 1966). The hormone appears to act primarily on the Malpighian tubules, stimulating them to secrete potassium ions at a greater rate, thereby creating an enhanced flow of water across the tubule wall (Pilcher, 1970). However, Mordue (1969) reports that in *Schistocerca* the hormone has a dual action, causing accelerated secretion through the tubules and a slowing down of water resorption through the rectal wall.

Because, presumably, there is a direct relationship between the amount of food consumed, the amount of hormone released and the quantity of water removed across the tubules, it is difficult to see how such a simple arrangement will work other than in insects such as *Rhodnius* whose food has a constant water content. The physical stimulus of "stretching" alone would not provide a precise enough mechanism for water regulation in insects whose food differs in water content. Some other control system must therefore operate. Perhaps an insect can monitor the water content of its food or, alternatively, the resorptive power of the rectal wall may be controlled directly by the hemolymph itself. Phillips (1964b) showed that the rate at which sodium and potassium ions are resorbed is dependent on the ionic concentration of the hemolymph.

Almost no work has been done on the hormonal control of salt and water balance in aquatic insects. However, as a result of his experiments on *Aedes aegypti,* Stobbart (1971) has suggested that accumulation of sodium ions across the wall of the anal papillae is under endocrine control. It appears that when the sodium concentration of the hemolymph changes an abdominal monitoring center passes information to a center in the thoracic ganglia, which causes a change in the rate of hormone production.

5. Summary

The removal of nitrogenous wastes and maintenance of a suitable hemolymph salt and water content are two closely linked processes. In most insects the predominant nitrogenous waste is uric acid, which is removed from the hemolymph via the Malpighian tubules as the soluble sodium or potassium salt. Precipitation of uric acid occurs usually in the rectum as a result of resorption of ions and water from, and acidification of, the urine. Allantoin and allantoic acid are excreted in quantity by some insects and may be the major nitrogenous waste. Urea is of little significance as a waste product, and ammonia is generally produced only in aquatic species.

Insects are usually able to regulate the salt and water content of the hemolymph within narrow limits. In all insects, a primary excretory fluid, isosmotic with hemolymph but differing in ionic composition, is produced in the Malpighian tubules. When this fluid reaches the posterior rectum, it is modified according to an insect's needs. In terrestrial and aquatic insects selective resorption of ions and/or water occurs. In saline-water species ions are actively secreted into the fluid across the rectal wall. Salt and water balance is regulated hormonally.

6. Literature

For additional information, readers should consult the reviews of Maddrell (1971) and Stobbart and Shaw (1974) [general]; Craig (1960) and Bursell (1967) [nitrogenous waste excretion]; and Barton Browne (1964), Beament (1964), Shaw and Stobbart (1963, 1972), and Phillips (1977) [salt and water balance].

Barrett, F. M., and Friend, W. G., 1970, Uric acid synthesis in *Rhodnius prolixus*, *J. Insect Physiol.* **16:**121–129.

Barton Browne, L., 1964, Water regulation in insects, *Annu. Rev. Entomol.* **9:**63–82.

Beament, J. W. L., 1964, The active transport and passive movement of water in insects, *Adv. Insect Physiol.* **2:**67–130.

Berridge, M. J., 1965, The physiology of excretion in the cotton stainer, *Dysdercus fasciatus* Signoret. III. Nitrogen excretion and excretory metabolism, *J. Exp. Biol.* **43:**511–521.

Berridge, M. J., 1966, The physiology of excretion in the cotton stainer, *Dysdercus fasciatus* Signoret. IV. Hormonal control of excretion, *J. Exp. Biol.* **44:**553–566.

Berridge, M. J., and Gupta, B. L., 1967, Fine-structural changes in relation to ion and water transport in the rectal papillae of the blowfly, *Calliphora*, *J. Cell Sci.* **2:**89–112.

Bursell, E., 1967, The excretion of nitrogen in insects, *Adv. Insect Physiol.* **4:**33–67.

Craig, R., 1960, The physiology of excretion in the insect, *Annu. Rev. Entomol.* **5:**53–68.

Grimstone, A. V., Mullinger, A. M., and Ramsay, J. A., 1968, Further studies on the rectal complex of the mealworm *Tenebrio molitor* L. (Coleoptera, Tenebrionidae), *Philos. Trans. R. Soc. London, Ser. B* **253:**343–382.

Jarial, M. S., and Scudder, G. G. E., 1970, The morphology and ultrastructure of the Malpighian tubules and hindgut of *Cenocorixa bifida* (Hung.) (Hemiptera, Corixidae), *Z. Morphol. Tiere* **68:**269–299.

Koch, H. J., 1938, The absorption of chloride ions by the anal papillae of Diptera larvae, *J. Exp. Biol.* **15:**152–160.

Maddrell, S. H. P., 1964, Excretion in the blood-sucking bug, *Rhodnius prolixus* Stål. III. The control of the release of the diuretic hormone, *J. Exp. Biol.* **41:**459–472.

Maddrell, S. H. P., 1971, The mechanisms of insect excretory systems, *Adv. Insect Physiol.* **8:**199–331.

Mordue, W., 1969, Hormonal control of Malpighian tube and rectal function in the desert locust, *Schistocerca gregaria*, *J. Insect Physiol.* **15:**273–285.

Oschman, J. L., and Wall, B. J., 1969, The structure of the rectal pads of *Periplaneta americana* L. with regard to fluid transport, *J. Morphol.* **127:**475–510.

Phillips, J. E., 1964a, Rectal absorption in the desert locust, *Schistocerca gregaria* Forskål. I. Water, *J. Exp. Biol.* **41:**14–38.

Phillips, J. E., 1964b, Rectal absorption in the desert locust, *Schistocerca gregaria* Forskål. II. Sodium, potassium and chloride, *J. Exp. Biol.* **41:**39–67.

Phillips, J. E., 1977, Excretion in insects: Function of gut and rectum in concentrating and diluting the urine, *Fed. Proc.* **36:**2480–2486.

Phillips, J. E., Bradley, T. J., and Maddrell, S. H. P., 1978, Mechanisms of ionic and osmotic regulation in saline-water mosquito larvae, in: *Comparative Physiology—Water, Ions and Fluid Mechanics* (K. Schmidt-Nielson, L. Bolis, and S. H. P. Maddrell, eds.), Cambridge University Press, Cambridge.

Pilcher, D. E. M., 1970, The influence of the diuretic hormone on the process of urine secretion by the Malpighian tubules of *Carausius morosus*, *J. Exp. Biol.* **53:**465–484.

Ramsay, J. A., 1964, The rectal complex of the mealworm *Tenebrio molitor* L. (Coleoptera, Tenebrionidae), *Philos. Trans. R. Soc. London, Ser. B* **248:**279–314.

Shaw, J., and Stobbart, R. H., 1963, Osmotic and ionic regulation in insects, *Adv. Insect Physiol.* **1:**315–399.

Shaw, J., and Stobbart, R. H., 1972, The water balance and osmoregulatory physiology of the desert locust (*Schistocerca gregaria*) and other desert and xeric arthropods, *Symp. Zool. Soc. London* **31:**15–38.

Stobbart, R. H., 1971, The control of sodium uptake by the larva of the mosquito *Aedes aegypti* (L.), *J. Exp. Biol.* **54:**29–66.

Stobbart, R. H., and Shaw, J., 1974, Salt and water balance: Excretion, in: *The Physiology of Insecta*, 2nd ed., Vol. V (M. Rockstein, ed.), Academic Press, New York.

Wall, B. J., and Oschman, J. L., 1970, Water and solute uptake by the rectal pads of *Periplaneta americana, Am. J. Physiol.* **218:**1208–1215.

Wall, B. J., Oschman, J. L., and Schmidt-Nielson, B., 1970, Fluid transport: Concentration of the intercellular compartment, *Science* **167**:1497–1498.

Wigglesworth, V. B., 1931, The physiology of excretion in a blood-sucking insect, *Rhodnius prolixus* (Hemiptera, Reduviidae). I. The composition of the urine, *J. Exp. Biol.* **8:**411–427.

Wigglesworth, V. B., 1938, The regulation of osmotic pressure and chloride concentration in the haemolymph of mosquito larvae, *J. Exp. Biol.* **15:**235–247.

Wigglesworth, V. B., 1965, *The Principles of Insect Physiology*, 6th ed., Methuen, London.

Reproduction and Development

19

Reproduction

1. Introduction

As was discussed in Chapter 2 (Section 4.1), an important factor in the success of the Insecta is their high reproductive capacity, the ability of a single female to give rise to many offspring, a relatively large proportion of which may reach sexual maturity under favorable conditions. Since reproduction is almost always sexual in insects, there arise within insect populations large numbers of genetic combinations, as well as mutations, which can be tested out in the prevailing environmental conditions. As these conditions change with time, insects are able to adapt readily, through natural selection, to a new situation. Over the short term their high reproductive capacity enables insects to exploit temporarily favorable conditions, for example, availability of suitable food plants. The latter requires that both the timing of mating, egg production and hatching, and the location of a suitable egg-laying site must be carefully "assessed" by an insect.

Like other terrestrial animals insects have had to solve two major problems in connection with their reproductive biology, namely, the bringing together of sperm and egg in the absence of surrounding water and the provision of a suitable watery environment in which an embryo can develop. The solution to these problems has been the evolution of internal fertilization and an egg surrounded by a waterproof cover (chorion), respectively. The latter has itself created two secondary problems. First, because of the generally impermeable nature of the chorion, structural modifications have had to evolve to ensure that adequate gaseous exchange can occur during embryonic development. Second, the chorion is formed while an egg is still within the ovarian follicle, that is, prior to fertilization, which has necessitated the development of special pores (micropyles) to permit entry of sperm.

2. Structure and Function of the Reproductive System

The external structure of male and female reproductive systems has been dealt with in Chapter 3 (Section 5.2.1), so that the structure of internal reproductive organs only will be described here.

2.1. Female

Functions of the female reproductive system include production of eggs, including yolk and chorion formation, reception and storage of sperm, sometimes for a considerable period, and coordination of events that lead to fertilization and oviposition.

Though details vary, the female system (Figure 19.1) essentially includes a pair of ovaries from each of which runs a lateral oviduct. The lateral oviducts fuse in the midline, and the common oviduct typically enters a saclike structure, the vagina. In some species the vaginal wall evaginates to form a pouchlike structure, the bursa copulatrix, in which spermatophores and/or seminal fluid is deposited during copulation. Also connected with the vagina are the spermatheca in which sperm are stored and various accessory glands. In some species part of the spermatheca takes the form of a diverticulum, the spermathecal gland.

The ovaries are usually dorsolateral to the gut, and each comprises a number of tubular ovarioles ensheathed by a network of connective tissue in which numerous tracheoles and muscles are embedded. The number of ovarioles per ovary, though approximately constant within a species, varies widely among species. For example, in some viviparous aphids and in dung beetles (Coprinae) there is one ovariole per ovary in contrast to the more than

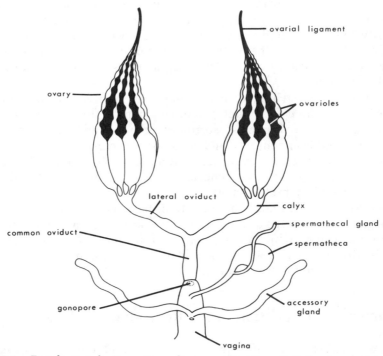

FIGURE 19.1.   Female reproductive system, diagrammatic. [From R. E. Snodgrass, *Principles of Insect Morphology*. Copyright 1935 by McGraw-Hill Inc. Used with permission of McGraw-Hill Book Company.]

2000 ovarioles per ovary in some higher termite queens. The wall of each ovariole includes an outer epithelial sheath and an inner acellular, elastic layer, the tunica propria. Each ovariole (Figure 19.2) consists of a terminal filament, germarium, vitellarium, and pedicel (ovariole stalk). The terminal filaments may fuse to form a sheet of tissue attached to the dorsal body wall or dorsal diaphragm by which an ovary is suspended within the abdominal cavity. Within the germarium, oogonia, derived from primary germ cells, give rise to oocytes and, in some types of ovarioles, also to nutritive cells (see below). As oocytes mature and enter the vitellarium they tend in most insects to become arranged in a linear sequence along the ovariole. Each oocyte also becomes enclosed in a one-cell thick layer of follicular epithelium derived from meso-dermal prefollicular tissue located at the junction of the germarium and vitellarium. As its name indicates, the vitellarium is the region in which an

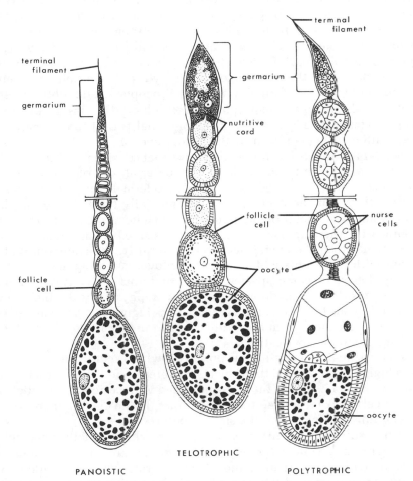

FIGURE 19.2. Types of ovarioles. The upper portion of each figure is enlarged to a greater extent than the lower in order to make details of germarial structure clear. [After A. P. Mahowald, 1972. Oogenesis, in: *Developmental Systems: Insects*, Vol. I (S. J. Counce and C. H. Waddington, eds). By permission of Academic Press Ltd., London, and the author.]

oocyte accumulates yolk, a process known as vitellogenesis (see Section 3.1.1). Normally vitellogenesis occurs only in the terminal oocyte, that is, the oocyte closest to the lateral oviduct, and during the process the oocyte's volume may increase enormously, for example, by as much as 10^5 times in *Drosophila*. Each ovariole is connected to a lateral oviduct by a thin-walled tube, the pedicel, whose lumen is initially occluded by epithelial tissue. This plug of tissue is lost during ovulation (movement of a mature oocyte into a lateral oviduct) and replaced by the remains of the follicular epithelium which originally covered the oocyte. Ovarioles may join a lateral oviduct linearly, as in some apterygotes and Ephemeroptera or, more often, open confluently into the distal expanded portion of the oviduct, the calyx.

Three types of ovarioles can be distinguished (Figure 19.2). The most primitive type, found in Thysanura, Paleoptera, most orthopteroid insects, and Siphonaptera, is the panoistic ovariole in which specialized nutritive cells (trophocytes) are absent. Trophocytes occur in the two remaining types, the polytrophic and telotrophic ovarioles, which are sometimes grouped together as meroistic ovarioles. In polytrophic ovarioles, a number of trophocytes (nurse cells) are enclosed in each follicle along with an oocyte. The trophocytes and oocyte originate from the same oogonium. Polytrophic ovarioles are found in most endopterygotes, and in Dermaptera, Psocoptera, and Phthiraptera. In Hemiptera and Coleoptera telotrophic (acrotrophic) ovarioles occur in which the trophocytes form a syncytium in the proximal part of the germarium and connect with each oocyte by means of a trophic cord.

The lateral oviducts are thin-walled tubes that consist of an inner epithelial layer set on a basement membrane and an outer sheath of muscle. In many species they include both mesodermal and ectodermal components. In almost all insects they join the common oviduct medially beneath the gut, but in Ephemeroptera the lateral oviducts remain separate and open to the exterior independently. The common oviduct, which is lined with cuticle, is usually more muscular than the lateral oviducts. Posteriorly, the common oviduct is confluent with the vagina which, as noted above, may evaginate to form the bursa copulatrix. In some species the bursa forms a diverticulum off the oviduct. In nearly all Lepidoptera the bursa is physically distinct from the oviduct and opens to the outside via the vulva (Figure 19.3). A narrow sperm duct connects the bursa with the oviduct and forms the route along which the sperm migrate to the spermatheca.

Usually a single spermatheca is present in which sperm are stored, though in some higher Diptera up to three such structures occur. The spermatheca and the duct with which it joins the bursa are lined with cuticle. The cuticle overlays a one-cell thick layer of epithelium whose cells are glandular and assumed to secrete nutrients for use by the stored sperm.

Various accessory glands may be present, which usually open into the bursa, though in Acrididae (Orthoptera) the glands are anterior extensions of the lateral oviducts. Normally, the glands secrete materials which form a protective coating around the eggs or stick the eggs to the substrate during oviposition. However, in Hymenoptera, the glands may produce the poison used in the sting, secrete trail-marking pheromones or lubricate the ovipositor valves.

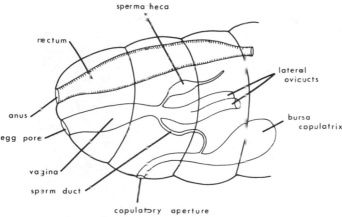

FIGURE 19.3. Reproductive system of female Lepidoptera–Ditrysia. [After A. D. Imms, 1957, *A General Textbook of Entomology,* 9th ed. (revised by O. W. Richards and R. G. Davies). By permission of Chapman and Hall Ltd.]

2.2. Male

Functions of the male reproductive system include production, storage, and, finally, delivery to the female of sperm. In some species, the system produces substances transferred during copulation that regulate female receptivity and fecundity. An additional, perhaps incidental, function may be to supply the female with nutrients which can be incorporated into developing oocytes, thereby increasing the rate and number of eggs produced.

The male system includes paired testes (in Lepidoptera these fuse to form a single median organ), paired vasa deferentia and seminal vesicles, a median ejaculatory duct, and various accessory glands (Figure 19.4). The testes, which lie either above or below the gut, comprise a variable number of tubular follicles bound together by a connective tissue sheath. The follicles may open into the vas deferens either confluently or in a linear sequence. The wall of each follicle is a layer of epithelium set on a basement membrane. Within the follicles several zones of maturation can be readily distinguished (Figure 19.5). The distal zone is the germarium in which spermatogonia are produced from germ cells. In Orthoptera, Dictyoptera, Hemiptera, and Lepidoptera a prominent apical cell is also present whose presumed function is to supply nutrients to the spermatogonia. As each spermatogonium moves proximally into the zone of growth, it becomes enclosed within a layer of somatic cells, forming a "cyst." Within the cyst, the cell divides mitotically to form a variable number (usually 64–256) of spermatocytes. In the zone of maturation, the spermatocytes undergo two maturation divisions, so that from each spermatocyte four haploid spermatids are formed. In the proximal part of the follicle, the zone of transformation, spermatids differentiate into flagellated spermatozoa. At this time the cyst wall normally has ruptured, though often the sperm within a bundle (spermatodesm) remain held together by a gelatinous cap which covers their anterior end. This cap may be lost as the sperm enter the vas deferens or persist until the sperm have been transferred to the female. The sperm are moved to the seminal vesicles by peristaltic contractions of the vas deferens. The seminal

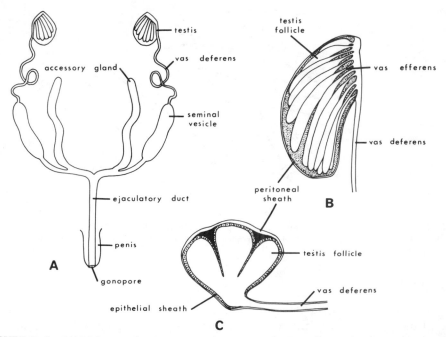

FIGURE 19.4. (A) Male reproductive system, diagrammatic; (B) structure of testis; and (C) section of testis and vas deferens. [From R. E. Snodgrass, *Principles of Insect Morphology*. Copyright 1935 by McGraw-Hill Inc. Used with permission of McGraw-Hill Book Company.]

vesicles, where sperm are stored, are dilations of the vasa deferentia. Their walls are well tracheated and frequently glandular, which may indicate a possible nutritive function.

The vasa deferentia enter the anterior tip of the ejaculatory duct, an ectodermally derived tube lined with cuticle whose walls normally are heavily muscularized. Posteriorly, the ejaculatory duct may run through an evagination of the body wall, which thus forms an intromittent organ. In insects which form a complex spermatophore, subdivision of the ejaculatory duct into specialized regions may occur. In Ephemeroptera no ejaculatory duct is present, and each vas deferens opens directly to the exterior.

The accessory glands may be either mesodermal (mesadenia) or ectodermal (ectadenia) in origin and are connected with either the lower part of the vasa deferentia or the upper end of the ejaculatory duct. In some species considerable morphological and functional differentiation of the glands occurs. Essentially, however, their secretions may contribute to the seminal fluid and/or form the spermatophore. In some species the glands produce substances which, when transferred to the female during insemination, cause increased egg production and/or decreased receptivity (willingness to mate subsequently).

3. Sexual Maturation

Most male insects eclose (emerge as adults) with mature sperm in their seminal vesicles. Indeed, in a few insects, for example, some Lepidoptera,

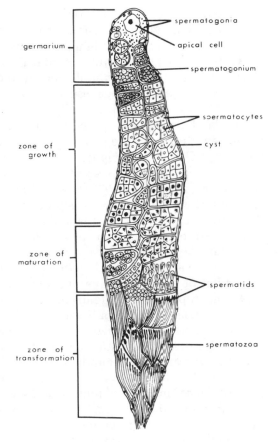

FIGURE 19.5. Zones of maturation in testis follicle. [After V. B. Wigglesworth, 1965, *The Principles of Insect Physiology*, 6th ed., Methuen and Co. By permission of the author.]

Plecoptera, and Ephemeroptera, both egg and sperm production occur in the final larval or pupal instar to enable mating and egg laying to take place within a few hours of eclosion. Generally, however, after eclosion, a period of sexual maturity is required in each sex during which important structural, physiological, and behavioral changes occur. This period may extend from only a few days up to several months in species which have a reproductive diapause (see Section 3.1.3).

3.1. Female

Among the processes that occur as a female insect becomes sexually mature are vitellogenesis, development of characteristic body coloration, maturation of pheromone-producing glands, growth of the reproductive tract, including accessory glands, and an increase in receptivity. These processes are controlled by the endocrine system whose activity, in turn, is influenced by various environmental stimuli.

3.1.1. Vitellogenesis

As noted above, vitellogenesis occurs, by and large, only in the terminal oocyte within an ovariole, yet in many species the process is highly synchronized among ovarioles and between ovaries; that is, the eggs are produced in batches. Why vitellogenesis does not occur to any great extent in the more distal oocytes is unclear, though various suggestions have been made. One suggestion is that the terminal oocyte, as the first to mature, that is, to become capable of vitellogenesis, simply outruns the competition. In other words, once the oocyte begins vitellogenesis and increases in size, its increasing surface area enables it to capture virtually all the available nutrients. This, however, cannot be the complete answer because in many female insects vitellogenesis in the penultimate oocyte appears to be inhibited even after the terminal oocyte has completed its yolk deposition and become chorionated, provided that the mature egg is not laid. Two explanations have been proposed. Adams and co-workers (see Adams, 1970) have proposed that, in *Musca domestica* at least, an ovary containing mature eggs produces an oostatic hormone which inhibits the activity of the corpora allata, thereby preventing further yolk production and deposition. In contrast, in *Rhodnius prolixus* an antigonadotropic hormone may be produced by the pedicel region of an ovariole when the latter contains a mature egg. The function of this hormone, it is proposed, is to block the action of the corpus allatum hormone on the follicle cells (see Section 3.1.3), again preventing vitellogenesis (Huebner and Davey, 1973).

As yolk appears, it can be seen to be made up almost entirely of roundish granules or vacuoles known as yolk spheres. Within the yolk spheres, protein, lipid, or carbohydrate, can be detected. The membrane-bound protein yolk spheres are most abundant, followed by lipid droplets which are not membrane-bound. Relatively few glycogen-containing yolk spheres are usually present. Small amounts of nucleic acids are normally detectable, but these are not within the yolk spheres. The source of some of these materials is different in the various types of ovarioles.

In all ovarioles, however, almost all yolk protein is extraovarian in origin, being derived especially from the hemolymph. The source of these proteins, as was noted in Chapter 16, is the fat body which, during vitellogenesis, synthesizes* and releases large quantities of a few specific proteins (vitellogenins or female-specific proteins) that are selectively accumulated by the terminal oocytes. Shortly before vitellogenesis, a space appears between the follicle cells and the terminal oocytes, and intercellular spaces develop in the follicular epithelium, so that the oocytes become bathed in hemolymph. The tunica propria appears to be freely permeable to all solutes within the hemolymph. Electron microscopic and other studies have shown that extensive pinocytosis occurs in the oocyte plasma membrane, resulting in the accumulation of yolk protein in membrane-bound vacuoles. Though most yolk protein is produced in the fat body, small contributions may be made by follicular epithelial cells, nurse cells where present, or the terminal oocyte. Studies have shown, for example, that in some species active RNA synthesis occurs in the follicular epithelium during vitellogenesis, especially the early stages, and that isotopically labeled amino acids are first taken up by follicle cells, to appear later in

---

*In insects which have fully developed eggs at eclosion, the proteins are synthesized (and stored) by the fat body during larval development, to be released during the pupal stage when vitellogenesis occurs.

protein spheres within the oocytes. In telotrophic and polytrophic ovarioles some protein may be transferred during early vitellogenesis to the oocyte from the nurse cells. However, the latter appear to be more important as suppliers of nucleic acids to the developing oocyte and several elegant autoradiographic and electron microscopic studies have shown the movement of labeled RNA or ribosomes down the trophic cord in telotrophic ovarioles or across adjacent nurse cells into the oocyte in polytrophic ovarioles. It is presumed that this RNA is then associated with protein synthesis within the oocyte. In addition to the RNA derived from nurse cells, RNA may also be produced by the oocyte nucleus for use in protein synthesis. In several species, active incorporation of labeled RNA precursors into the oocyte nucleus has been observed to occur early in vitellogenesis, concomitant with the accumulation of protein (non-membrane-bound) adjacent to the nuclear envelope

Studies on the accumulation of the lipid and carbohydrate components of yolk are few. Though lipid may make up a considerable proportion of the yolk, its source remains doubtful in most species. An apparent association between the Golgi apparatus and the accumulation of lipid led early authors to suggest that the lipid was synthesized by the oocyte per se, though this has not been confirmed. Another suggestion which requires further study is that the follicle cells contribute lipid to the oocyte. In the polytrophic ovariole of *Drosophila*, the nurse cells supply lipids to the oocyte, though this apparently is not the case in *Culex*. In telotrophic ovarioles, lipid may be derived both from the nurse cells (early in vitellogenesis) and from the follicle cells.

Glycogen usually can be detected only in small amounts and, in meroistic ovarioles, after degeneration of the nurse cells. In *Apis* and *Musca*, labeled glucose injected into the hemolymph is rapidly accumulated by oocytes in late vitellogenesis and apparently converted to glycogen (Engelmann, 1970).

3.1.2. Vitelline Membrane and Chorion Formation

When vitellogenesis is completed, the vitelline membrane and, later, the chorion (eggshell) are formed. Early observations suggested that the vitelline membrane was produced by the oocyte itself, perhaps as a modification of the existing plasma membrane, and this may be the case in some species, for example, grasshoppers. However, more recent studies on Diptera have demonstrated that the follicle cells secrete droplets of material which coalesce to form the vitelline membrane.

The chorion is usually secreted entirely by the follicle cells and can be seen to comprise two main layers, an endochorion adjacent to the vitelline membrane and an exochorion (Figure 19.6). In some insects, for example, Acrididae, the shell takes on a third layer, the extrachorion, as an oocyte moves through the common oviduct. Interestingly, though the follicle cells are mesodermal derivatives, the chorion is cuticlelike in nature and contains layers of protein and lipoprotein, some of which are tanned by polyphenolic substances released by the cells.

After chorion formation the oocyte secretes a wax layer between the vitelline membrane and chorion which renders it waterproof.

The chorion is not produced as a uniform layer over the oocyte. For exam-

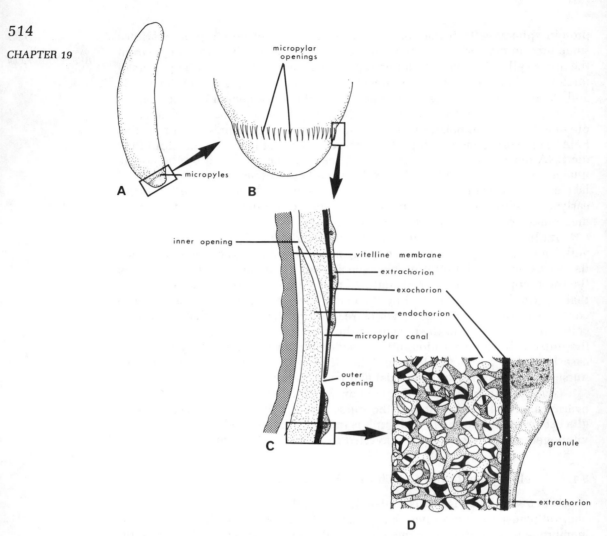

FIGURE 19.6. Egg of *Locusta*. (A) General view, (B) enlargement of posterior end, (C) section through chorion along micropylar axis, and (D) details of chorion structure. [A, after R. F. Chapman, 1971, *The Insects: Structure and Function*. By permission of Elsevier North-Holland, Inc., and the author. B, C, after M. L. Roonwal, 1954, The egg-wall of the African migratory locust, *Locusta m. migratorioides* R. & F. (Orthoptera: Acrididae), *Proc. Natl. Inst. Sci. India* **20**:361–370. By permission of the Indian National Science Academy. D, after J. C. Hartley, 1961, The shell of acridid eggs, *Q. J. Microsc. Sci.* **102**:249–255. By permission of Cambridge University Press, New York.]

ple, in some species a ring of follicle cells near the anterior end of the oocyte secrete no exochorion, so that a line of weakness is created at this point which facilitates hatching. Also, certain follicle cells appear to have larger than normal microvilli which, when withdrawn after chorion formation, leave channels (micropyles) to permit entry of sperm (Figure 19.6C,D) The aeropyles (air canals) appear to be formed in a similar way. They connect with a network of minute air spaces within the endochorion, which facilitates gaseous exchange between the oocyte and atmosphere without concomitant loss of water during embryonic development (Figure 19.6D).

The internal structure of a mature (i.e., chorionated) egg is shown diagrammatically in Figure 19.7.

3.1.3. Factors Affecting Sexual Maturity in the Female

Environmental. A number of environmental factors have been observed to influence the rate at which a female becomes sexually mature, for example, quantity and quality of food consumed, population density and structure, mating, photoperiod, temperature, and humidity. For most factors there is evidence that they exert their influence by modifying the activity of the endocrine system, though availability of food and temperature also have obvious direct effects on the rate of egg development.

Many reports indicate that the qualitative nature of food may have a marked effect on the number of eggs matured. However, it is often not clear whether the observed differences in egg production are due to differences in palatability (more palatable foods might be eaten in larger quantities) or to differences in the nutritive value of the food. For most insects, dietary proteins are essential for maturation of eggs. Many Diptera, for example, may survive for several weeks on a diet that contains carbohydrate but no protein, yet will mature no eggs. Equally, different proteins may have different nutritional values related, presumably, to their amino acid composition and possibly, to their digestibility. Carbohydrates, too, are important, especially in the provision of energy for synthesis of yolk components. As noted above, lipids may be a significant component of the yolk and, therefore, an essential part of the diet, though some may be formed from carbohydrates. Water, vitamins, minerals, and, for some species, specific growth substances, are also necessary constituents of the diet if maximum egg production is to be achieved.

Quantitative influences of feeding are much more easily documented. Anautogenous mosquitoes and many other bloodsucking insects, for example, do not mature eggs until they have fed. Furthermore, the number of eggs pro-

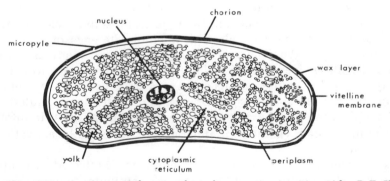

FIGURE 19.7. Diagrammatic sagittal section through an egg at oviposition. [After R. F. Chapman, 1971, *The Insects: Structure and Function.* By permission of Elsevier North-Holland, Inc. and the author.]

duced is, within limits, proportional to the quantity of blood ingested. However, it must be noted that in endopterygotes many of the materials to be used in egg production are laid down during larval development, and therefore the nutrition of the larva must also be considered when attempting comparisons.

In addition to its direct function of providing raw materials for egg maturation, feeding also has an important indirect effect, namely, to stimulate endocrine activity. In continuous feeders (those which take a series of small meals) stretching of the foregut as food is ingested results in information being sent via the stomatogastric nervous system to the brain/corpora cardiaca complex where it stimulates synthesis and release of neurosecretion. In occasional feeders such as *Rhodnius* and anautogenous mosquitoes, which require a single, large meal in order to mature a batch of eggs, it is said that stretching of the abdominal wall is the stimulus, sent to the brain via the ventral nerve cord, that triggers endocrine activity. In both arrangements the degree of endocrine activity and, in turn, the number of eggs developed, will be proportional (other things being equal) to the amount of food ingested.

In some insects, especially gregarious species, population density and structure may influence sexual maturation. In some species of *Drosophila*, *Locusta migratoria*, and *Nomadacris septemfasciata* (the red locust), for example, the greater the population density, the slower the rate of egg maturation. Though there are some obvious potential reasons for this, such as interference with feeding, other effects of this stress, perhaps manifest through a decrease in endocrine activity, may also be important. However, in the desert locust, *Schistocerca gregaria*, crowded females mature eggs faster than isolated individuals due, it appears, to increased endocrine activity in the former group. In addition, in *S. gregaria* and *S. paranensis* (the Central American locust) there is evidence that egg development in females is promoted in the presence of older males, which are believed to secrete a maturation-accelerating pheromone. Conversely, in colonies of social insects, secretion of a maturation-inhibiting pheromone by the queen prevents development of the reproductive system of other females, the workers. Again, the effects of these pheromones are probably mediated through the endocrine system, though the evidence for this proposal is mostly circumstantial.

Mating is, for many species, a most important factor in sexual maturation. For example, in some bloodsucking Hemiptera and some cockroaches almost no eggs develop in virgin females. In other insects eggs of virgin females mature more slowly than those of mated females, and many are eventually resorbed if mating does not take place. The stimulus given to a female is normally physical in nature, and its effect appears to be enhancement of endocrine activity. In some cockroaches, for example, mechanical stimulation of the genitalia or the presence of a spermatophore in the bursa results in information being sent to the brain via the ventral nerve cord, followed by activation of the corpora allata.

Like all metabolic processes, egg development is affected by temperature and occurs at the maximum rate at a specific, optimum temperature, whose value presumably reflects the normal temperature conditions experienced by a species during reproduction. On each side of this optimum egg maturation is decreased, in the normal temperature-dependent manner of enzymatically controlled reactions. Sometimes superimposed on this basic effect, however, are more subtle effects of temperature, of both a direct and an indirect nature. For example, in *Locusta migratoria* regular temperature fluctuations (provided these are not too extreme) appear to stress the insect, causing release of neurosecretion and enhanced rates of development. In some other species mating occurs only within a certain temperature range, yet, as noted earlier, may have an important influence on egg development. It follows that, in this situation, temperature can have an important indirect effect on maturation.

Few direct observations have been made on the effects of humidity on egg maturation, though it is known that humidity may determine whether or not oviposition occurs. In many species eggs are laid only when the relative humidity is high (80–90%), and oviposition is increasingly retarded as the environment becomes drier. Engelmann (1970) suggests a possible explanation for this may be that as increasing amounts of water are lost from the body by evaporation, insufficient remains for use in egg development whose rate is therefore decreased.

Photoperiod, the earth's naturally recurring alternation of light and darkness, is probably the best studied environmental factor that influences egg maturation. The effect of photoperiod is long-term (seasonal) and serves to correlate egg development with the availability of food, suitable egg-laying conditions, and/or suitable conditions for the eventual development of the larvae. Implicit in this statement is the idea that an insect, by having its reproductive activity seasonal in nature, is able to overcome adverse environmental conditions. Commonly, an insect survives these adverse conditions by entering a specific physiological condition known as diapause, whose onset and termination are induced by changes in daylength (sometimes acting in conjunction with temperature). (For a general discussion of diapause, see Chapter 22, Section 3.2.3.) Essentially diapause is a phase of arrested development, and, in the context of adult (reproductive) diapause, this means that the eggs do not mature. In different species diapause may be induced by increasing daylength (number of hours of light in a 24-hour period), which enables an insect to overcome hot and/or dry summer conditions (aestivation), or by decreasing daylength prior to the onset of cold winter conditions (hibernation). For example, in the Egyptian grasshopper *Anacridium aegyptium* reproductive diapause is induced by the decreasing daylengths experienced in fall and is maintained for about 4 months during which no eggs are produced. Termination of diapause, brought about by increasing daylengths in spring, is correlated with renewed availability of oviposition sites and food for the juvenile stages. Likewise, hibernation in newly eclosed adult Colorado potato beetles (*Leptinotarsa decemlineata*) is induced by the short daylengths of fall (and also by a lack of food at this time). At the onset of diapause, the beetles become negatively phototactic and bury themselves under several inches of soil. In this situation, it is perhaps not surprising that diapause is terminated not by changing daylength but by the increasing temperature of the soil experienced in the spring.

Aestivation is seen in *Schistocerca gregaria* and *Hypera postica*, the alfalfa weevil. Sexual maturation in *S. gregaria* is retarded by long and promoted by short daylengths This observation can be correlated with the availability of food and oviposition sites in its natural habitat, arid areas of Africa and Asia, where in the summer the weather is hot and dry, but rain falls intermittently during winter. *Hypera postica* adults emerge in late spring and undergo reproductive diapause before laying eggs in late summer and fall. Low winter temperatures prevent the eggs from hatching until the following spring when new alfalfa foliage on which the larvae feed has begun to appear.

The effects of photoperiod on egg maturation are mediated via the endocrine system, though for many species the evidence for this statement is largely circumstantial, for example, differences in the histological appearance of the endocrine glands in diapausing and nondiapausing insects. In diapausing adult insects the corpora allata are small, and the neurosecretory system is typically full of stainable material, which are taken to indicate inactivity of these glands. When diapause is terminated and egg development begins, the corpora allata increase in volume and the amount of stainable material in the neurosecretory system decreases. In the beetle *Galeruca tanaceti* autoradiographic studies have shown that in postdiapause beetles the rate of incorporation of labeled cystine into neurosecretory cells (taken to be a measure of their synthetic activity) is high compared to that of aestivating insects. In other species diapause may be terminated by treating insects with juvenile hormone or its mimics.

Endocrine. The endocrine control of sexual maturation in female insects, especially the control of oocyte development, has been one of the most popular areas of research in insect physiology over the past two decades. Despite this popularity and the wealth of literature that has resulted, several major aspects of hormonal control remain unclear. It is apparent, however, that among the Insecta the relative importance of the various endocrine centers in reproduction may differ, as might be anticipated in a group of such diverse habits. The following account is therefore generalized, though the major points of contention and differences among insects will also be outlined (see also Figure 19.8).

Essentially, the two endocrine components involved are the corpora allata and cerebral neurosecretory system, though recent work indicates that in some insects ecdysone, oostatic hormone, or antigonadotropic hormone also may be important.

The importance of the corpora allata in egg development first became apparent in 1936 when Wigglesworth and Weed-Pfeiffer (cited in de Wilde and de Loof, 1974b) demonstrated independently that in *Rhodnius* and *Melanoplus*, respectively, allatectomy (removal of the corpora allata) prevented vitellogenesis. Since this date, authors, by removing the allata, followed by replacement therapy (implantation of "active" glands from other insects, or treatment with juvenile hormone or its mimics) have confirmed the importance of these glands as the source of a gonadotropic hormone in most but not all insects. In the flesh fly, *Sarcophaga bullata*, for example, vitellogenesis occurs only when the median neurosecretory cells are present and is, apparently, quite independent of the corpora allata (Figure 19.8D). In mosquitoes, corpus allatum hormone controls only the previtellogenic stages of egg development and is not required for vitellogenesis (Figure 19.8C).

Originally, it was believed that corpus allatum hormone probably triggered the synthesis of yolk precursors in the follicle cells, which then passed these materials on to developing oocytes. However, recent studies have shown that

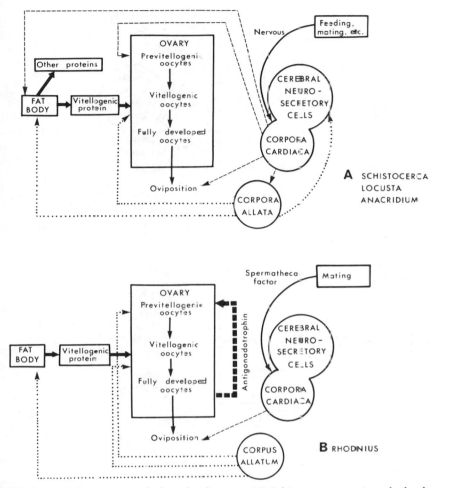

FIGURE 19.8. Endocrine control of egg development. (A) *Schistocerca gregaria* and other locusts, (B) *Rhodnius prolixus*, (C) *Aedes aegypti* and other mosquitoes, and (D) *Sarcophaga bullata*. [After K. C. Highnam and L. Hill, 1977, *The Comparative Endocrinology of the Invertebrates*, 2nd ed. By permission of Edward Arnold Publishers Ltd.]

most yolk is of extraovarian origin, and the hormone regulates not the synthetic activity but rather the permeability of the follicular epithelium. In the presence of gonadotropic hormone, the epithelium differentiates from a flattened cuboidal type lacking intercellular spaces to one comprising tall, columnar cells, between which are prominent intercellular channels. Proteins have been shown to move along these channels, to be accumulated pinocytotically by the oocyte during vitellogenesis. The precise way in which the hormone stimulates the follicle cells to change their shape is unknown, though an effect on the arrangement of microtubules within the cells has been demonstrated. (Evidence from other noninsectan systems strongly implicates the involvement of these organelles in the regulation of cell shape.)

As was noted earlier, the source of most yolk components is extraovarian, specifically the hemolymph, which serves as a reservoir for materials synthe-

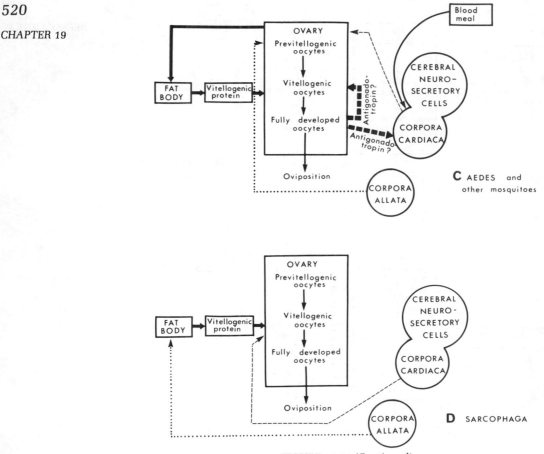

FIGURE 19.8. *(Continued).*

sized in the fat body. In the early to mid-1960s, a considerable body of information was collected by Highnam, Hill, and colleagues (see references in Highnam and Hill, 1977) which indicated that a hormone from the median neurosecretory cells of the brain regulated the protein synthetic activity of the fat body in some insects. In a series of papers, these authors reported that in *S. gregaria* (1) cycles of synthesis and release of neurosecretion were correlated with the development of successive egg batches, (2) cautery of the median neurosecretory cells prevented vitellogenesis due to the inability of the fat body to synthesize certain yolk proteins (an effect which could be reversed by implantation of corpora cardiaca), and (3) stimuli known to enhance the rate of vitellogenesis (e.g., presence of mature males, feeding, or artificial electrical stimulation) exerted their effect by promoting the release of neurosecretion from the corpora cardiaca.

Thus, for a time, an attractive proposition appeared to be that neurosecretion regulated protein production in the fat body, while corpus allatum hormone controlled protein uptake in the ovary. However, in the light of later work, this proposal has required significant modification. In a large number of insect species (including *S. gregaria*) it has been demonstrated that synthesis of vitellogenins in the fat body, as well as their accumulation in the ovary, is under the control of corpus allatum hormone. In these species the function of neurosecretion in egg maturation appears to be mainly allatotropic, that is, to activate the corpora allata, though it may exert an overall control of protein synthesis in the fat body (see Highnam and Hill, 1977) (Figure 19.8A). In some insects, for example, *Rhodnius*, neurosecretion is not involved in vitellogenesis, though it has other important functions in the reproductive process (see Section 6.2) (Figure 19.8B).

A different arrangement occurs in mosquitoes in which neurosecretion is released for only the first few hours after the blood meal, and, as noted earlier, the corpora allata are not important in vitellogenesis. In these insects neurosecretion appears to act directly on the ovary, stimulating it to release ecdysone, which regulates vitellogenin synthesis in the fat body (Hagedorn et al., 1975) (Figure 19.8C). Of particular interest is the involvement of ecdysone, a hormone previously associated with molting and known to be produced by glands which disappear in all but a very few adult insects. Ecdysone has been detected in significant quantities in adults of other insect species and, in some at least, can be implicated in egg development (see Hagedorn et al., 1975). Whether the ovary actually synthesizes ecdysone or accumulates it during the juvenile stage is not known. Nor is the role of ecdysone in egg maturation known in species other than mosquitoes. Clearly, however, future work in this area may result in considerable revision of the present concepts of the endocrine control of oocyte development.

An aspect of egg maturation which requires much more study is the interrelationship of the endocrine glands, ovary, and fat body; that is, how is endocrine activity increased or decreased during each cycle of egg development so that the system operates most efficiently. For most insects, the neurosecretory system acts as the center for translating external stimuli into endocrine language. As a result of these stimuli, production and release of allatotropic hormone will be altered, causing, in turn, a change in corpus allatum activity. However, corpus allatum activity can be modified in other ways, as is obviously the situation in insects whose neurosecretory system is not involved in egg development. For example, in the allatic nerves (see Figure 13.4) there are nonneurosecretory axons which conceivably could carry neural information from the brain to activate or inhibit the corpora allata. Another possibility is that the corpora allata respond directly to the nutritional milieu in which they are bathed. In an actively feeding insect whose hemolymph contains large quantities of nutrients, the gland could be expected to be active to facilitate production and accumulation of yolk. As oocytes mature, feeding activity declines, so that there are fewer nutrients in the hemolymph, which might lead to a reduction in corpus allatum activity. An important question is, therefore, "What causes feeding activity to decline as vitellogenesis nears completion?" To date, there is no clear answer to this question, though some authors have suggested that distension of the oviducal or abdominal wall by mature eggs might inhibit feeding behavior via the central nervous system.

A further possibility which may account for the cyclic nature of egg production involves a common endocrine principle, namely, feedback inhibition.

There is evidence that a high level of circulating corpus allatum hormone may inhibit synthesis and release of neurosecretion. In turn, this will result in a decline in corpus allatum activity, a drop in the level of gonadotropic hormone, and, eventually, a renewal of neurosecretory activity.

A relatively unexplored aspect of these interrelationships is whether the ovary may function as an endocrine structure whose secretion may competitively inhibit or bind with, for example, corpus allatum hormone, or may inhibit production of hormone by a gland. As noted earlier in this chapter and also in Chapter 13 (Section 3.4), the ovaries of *Musca* and *Rhodnius* are reported to produce an oostatic and an antigonadotropic hormone, respectively, and it will be interesting to see whether comparable situations are widespread among the Insecta.

3.2. Male

Compared to that of females, sexual maturation in male insects has received only slight attention and, consequently, is relatively poorly understood.

In species in which males after eclosion live for only a brief time, spermatogenesis is completed during the late larval and/or pupal stages. Presumably, also, in such short-lived species, when sperm is transferred to the female in a spermatophore, the accessory glands (which produce the spermatophore components) must become active in the juvenile stages. In many male insects, especially those that mate frequently, spermatogenesis continues at a low rate after eclosion. During sexual maturation, other events may also occur, such as synthesis and accumulation of accessory gland secretions for use in spermatophore or seminal fluid formation. Pheromone-producing glands may become functional, and the male may develop characteristic behavior patterns and coloration. All these processes appear to be influenced by the endocrine system.

Involvement of the endocrine system in sperm production has only recently been clarified. Several early workers observed that precocious adult males, obtained by removing the corpora allata from larvae, had testes that contained sperm capable of fertilizing eggs and concluded that juvenile hormone inhibited differentiation of sperm. Conversely, other authors had noted that in diapausing larvae or pupae differentiation within the testis ceased and suggested that ecdysone promoted this process, an effect which has been confirmed in several species following the commercial availability of hormone. If, however, the hormones regulate differentiation, then an adult insect, which lacks ecdysone and has active corpora allata, presumably presents an unfavorable environment for spermatogenesis. Yet, as noted above, this process occurs after eclosion in some species. Dumser and Davey (1975 and earlier), working on *Rhodnius*, have resolved this paradox by proposing that the hormones affect only the rate of spermatogonial mitosis and not differentiation *per se*. The authors propose that in the testes there is a basal (endogenous) rate of mitosis which occurs even in the absence of hormones. Ecdysone increases the rate of division, whereas juvenile hormone depresses it, though never below the basal level. Differentiation of the germ cells follows division, but its rate is constant (not directly affected by hormones) and species-specific. Thus, in early juvenile stages, the rate of division will be low, due to the presence in the hemolymph of juvenile hormone as well as ecdysone. However, in the final larval and/or pupal stages when the corpora allata are inactive, ecdysone will accelerate spermatogonial division (reflected in the enormous growth of the testes seen at this time), which results in production of large numbers of mature sperm. In adults, spermatogenesis continues at the basal rate, despite the presence of juvenile hormone in the hemolymph.

The male accessory reproductive glands are fully differentiated at eclosion. During sexual maturation they become active and increase greatly in size as a result of the synthesis and accumulation of materials used in spermatophore and/or seminal fluid formation (see Section 4.3.1). In some species, for example, the cockroach *Leucophaea maderae*, it is reported that the activity of the accessory glands is not under endocrine control, and allatectomized adult males are able to produce spermatophores throughout their life. In contrast, in many other species there is a clear correlation between the onset of corpus allatum activity and development of secretory activity in the accessory glands. Furthermore, accessory glands of allatectomized males of these species remain small and show only weak secretory activity, effects that can be reversed by treatment with juvenile hormone compounds. Though Engelmann (1970) suggested that the effect of juvenile hormone on the secretory activity of the accessory glands may be indirect, stemming from a general endocrine control of protein metabolism, other authors for example, Odhiambo (1966) and Gillott and Friedel (1976), have argued that juvenile hormone has a specific role in controlling accessory gland activity. Indeed, the studies of Friedel and Gillott (1976) on the production of accessory gland secretion in the male migratory grasshopper, *Melanoplus sanguinipes*, have revealed some interesting parallels with the process of vitellogenesis. Male *M. sanguinipes* are promiscuous insects. They may copulate several times on a single day and, on each occasion, transfer several spermatophores. Such promiscuity requires either extremely active accessory glands or the participation of other tissue(s) in the production of the necessary materials. The authors showed that the fat body produces specific proteins which are accumulated by the accessory glands. Removal of the accessory glands led to accumulation of protein in the fat body and hemolymph. Furthermore, both synthesis of these proteins in the fat body and their accumulation by the accessory glands were prevented by allatectomy, an effect which could be reversed by treating operated insects with a juvenile hormone mimic. Whether extraglandular synthesis of accessory gland materials, under the control of juvenile hormone, is of widespread occurrence among Insecta remains to be seen (Figure 19.9).

The development and control of sexual behavior in male insects have been studied in detail in relatively few species, mostly Acrididae. As Engelmann (1970) notes, the best-known species in this regard is the desert locust S. gregaria, which begins to show elements of sexual behavior between 6 and 12 days after eclosion. Paralleling the onset of this behavior is a change in body color, from the greyish-brown-pink of newly emerged insects to the uniform yellow of mature males, and the beginning of secretion from epidermal glands of a maturation-accelerating pheromone. The work of Loher, Odhiambo, and Pener (see references in Pener, 1974) has established that the changes are con-

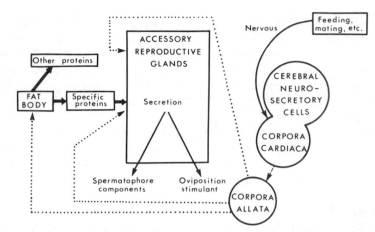

FIGURE 19.9 Endocrine relationships in male insects. Note that some of these relationships may not exist in all species.

trolled by the corpora allata in this species. Allatectomy shortly after eclosion, for example, leads to retention of immature coloration and a lack of desire to copulate, both of which effects may be reversed by implantation of corpora allata from mature individuals. Similar procedures have demonstrated corpus allatum-controlled sexual behavior in the males of some other species. However, in male migratory locusts, L. migratoria ssp. migratorioides, the corpora allata control only the development of the mature body color; mating behavior is controlled primarily by "C-type" median neurosecretory cells in the brain. The control exerted by the cells is direct, that is, not mediated through the corpora allata.

4. Mating Behavior

Mating behavior, which can be considered as the events which encompass and relate to the act of mating, may be subdivided into four components, some or all of which may be observed in a given species. The components are (1) location and recognition of a mate, (2) courtship, (3) copulation, and (4) postcopulatory behavior. The primary function of mating behavior is to ensure the transfer of sperm from male to female, though as is discussed below, there are additional functions of this behavior which optimize the reproductive economy of the species.

4.1. Mate Location and Recognition

A prerequisite to internal fertilization for almost all terrestrial animals, insects included, is the coming together of male and female so that sperm can be transferred directly into a female's reproductive tract. Interestingly, however, in some apterygotes indirect sperm transfer occurs. In some Collembola, for example, males deposit packages of sperm onto the substrate, though at the time no females may be in the vicinity. Females apparently find the sperm packages by chance and take them up into the reproductive tract. In Thysanura, also, droplets of sperm are placed on the substrate, though only when a female is present.

Among pterygote insects, a variety of stimuli alone or in combination may signal the location of a potential mate. These may be visual, olfactory, auditory, or tactile. Visual cues are used mostly, though not solely, by many diurnal species. Movement, color, form, and size may attract one individual to another, though final determination of the suitability of the proposed partner is usually made by tactile or chemical stimuli. Many male Diptera, for example, are attracted randomly to dark objects of a given size range but do not make copulatory movements unless they receive the correct tactile stimulus on contact. Male butterflies may fly toward objects whose movements follow a particular sequence and which are of a particular color pattern. Again, however, final contact and attempts to mate depend on the odor of the object. The best-known nocturnal insects which use a visual cue to locate (or attract) a mate are fireflies (Lampyridae: Coleoptera), both sexes or only females of which may produce light by means of bioluminescent organs located on the posterior abdominal segments. In some species a female "glows" continuously to attract males which land in her general vicinity, then locate her exact position by olfaction. In others, a male in flight flashes regularly (!), and a female responds (begins to flash) when he comes within a certain range. The time interval between the partners' flashes is critical and determines whether a male is attracted to a responding female. Again, final contact is made by means of touch and/or smell.

Pheromones (see also Chapter 13, Section 4.1) are probably the most common signal used by insects in mate location. In contrast to visual stimuli, volatile chemical attractants are both highly specific and capable of exerting their effect over a considerable distance (in some Lepidoptera several kilometers). Pheromones are usually produced by females and are employed by both diurnal and nocturnal species.

Auditory stimuli, like pheromones, can exert their effect over some distance and may be species-specific. Because transmission of sound is not seriously impeded by vegetation, auditory stimuli are particularly useful cues for insects that live among grass, etc., for example, many Orthoptera. Among the Gryllidae, only males produce sounds which serve to orient and attract potential mates from distances up to 30 meters in some species. In some Acrididae, in contrast, both males and females produce sounds which attract the opposite sex.

"Accidental" sex attraction occurs when all members of a species are attracted by the topography or scent of a particular site or when individuals of both sexes produce aggregation sounds, for example, cicadas.

4.2. Courtship

Recognition and attraction is immediately followed, in some insects, by copulation. In other species, however, copulation may be preceded by more or less elaborate forms of courtship for which several functions have been suggested (Manning, 1966). Specialized courtship behavior may prevent interspecific mating, especially between closely related species, though Manning (1966) argues that there is little evidence to support this proposal. In species where a female is normally aggressive (even predaceous to the point of consuming a potential "lover"), courtship may serve to appease her so that she becomes willing to copulate. One of the best-known examples of such appeasement is found in the dance flies (Empididae: Diptera), so-called because in many species a male has an elaborate courtship dance in which he presents a female with a silken ball. In some species the ball contains prey which is actually eaten by the female during copulation. In others, it contains nutritionally useless material, though by the time a female discovers this it is too late! Finally, in some species the procedure becomes entirely ritualized, and a male simply presents his partner with an empty ball.

As an antithesis to female appeasement, courtship may also be a necessary prerequisite for bringing a male into a suitable state for insemination. In other words, the process serves to synchronize the behavior of the pair, thereby increasing the chance of successful sperm transfer. In some species, courtship has a very obvious function, namely, bringing the partners together in the correct physical relationship for insemination. For example, in his courtship display, the male cockroach *Byrsotria fumigata* raises his wings to expose on the metanotum a gland whose secretion is attractive to the female. At the same time, the male turns away from the female, who, in order to reach the secretion and feed on it, must mount the male from the rear. In this way the female comes to take up the appropriate position for copulation.

4.3. Copulation

Receptivity (willingness of a female to copulate) depends not only on a male's efforts to seduce her but also on the female's physiological state. Age and the presence of semen in the spermatheca are the two most important factors governing receptivity, though external influences may also be important. For many species an obvious correlation exists between receptivity and state of egg development in virgin females, and it appears that the level of circulating juvenile hormone governs receptivity. Many females mate only once or a few times, and after mating become unreceptive to males for a variable length of time. The switching off of receptivity has been related to the presence of semen in the spermatheca. In some species it appears that stretching of the spermathecal wall may lead to nervous inhibition of receptivity. In others, it is clear that inhibition is pheromonal, the seminal fluid containing a "receptivity-inhibiting" substance which either directly, or by causing the spermatheca to liberate a hormone into the hemolymph, acts on the brain to render the female unreceptive (see Gillott and Friedel, 1977).

In many species, mating occurs only at a certain time of the day. For example, certain fruit flies (*Dacus* spp.) mate only when the light intensity is decreasing. Other species have built-in circadian rhythms of mating.

Some Lepidoptera mate only in the vicinity of the larval food plant, the odor of the plant stimulating release of sex attractant by the female.

During copulation, some of the behavioral elements introduced during courtship may be continued, presumably to keep the female pacified until insemination is completed. Sometimes pacification continues after insemination and is thought to prevent the female from ejecting the spermatophore until the sperm have migrated from it.

4.3.1. Insemination

As noted earlier, indirect sperm transfer occurs in some apterygotes, but in almost all Pterygota, sperm is transferred during copulation directly to the female reproductive tract. Primitively, sperm are enclosed in a special structure, a spermatophore, which may be formed some time before copulation (in Gryllidae and Tettigoniidae) or, more often, as copulation proceeds. Spermatophores are not produced by males of many species of endopterygotes, especially higher Diptera, or by some male Hemiptera.

Spermatophore Production. Production of a spermatophore has been studied in relatively few species, and it is difficult, therefore, to generalize. Essentially, however, the structure is formed by secretions of the accessory glands and sometimes also the ejaculatory duct. The secretions from different gland components may mix or remain separate, so that the wall of the spermatophore is formed of a series of layers which surround a central mass of sperm.

Gerber (1970) proposed that four general methods of spermatophore formation occur which form a distinct evolutionary series. In the most primitive method (first male-determined method), found in many orthopteroid species, the spermatophore is complex and formed either at the anterior end of the ejaculatory duct or within the male copulatory organ. After transfer to a female, the spermatophore is usually held between her external genital plates and only its anterior, tubelike portion enters the vagina or bursa. In the second male-determined method, the spermatophore has a less complex structure and is formed within a special spermatophore sac of the copulatory organ. The thin-walled sac is everted into the bursa, which therefore essentially determines the final shape and size of the spermatophore. After spermatophore formation, the sac is withdrawn Rhodnius, some Coleoptera, and some Diptera form spermatophores in this way. The two remaining methods are female-determined. In the first, seen in Trichoptera, Lepidoptera, some Diptera, some Coleoptera and a few Hymenoptera, male accessory gland secretions empty directly into the vagina or bursa in a definite sequence, either before or after transfer of sperm which they encapsulate. The spermatophore takes up the shape of the genital duct. The spermatophore produced by the second female-determined method is the least complex. Male accessory gland secretions are produced concurrently with or immediately after sperm transfer and often do not encapsulate the sperm; rather, they harden to form a mating plug which prevents backflow and loss of semen (and also, further mating). This method is seen in mosquitoes, the honeybee, and some Lepidoptera. Gerber (1970) speculates that the next step in the evolutionary sequence would be complete loss of the spermatophore and, concurrently, the development of a more elongate penis for depositing sperm close to the spermatheca.

The number of spermatophores formed, the length of time the spermatophore remains with a female, and its fate are variable. In some species a male produces a single spermatophore during each copulation, and this may remain in the female's genital tract or between the genital plates for several hours to ensure complete evacuation of semen. The empty spermatophore may then be discarded and/or eaten by the female, or partially to completely digested within her genital tract. In cockroaches the digestive fluid appears to be secreted by the female's accessory glands.

Precisely how sperm move from the spermatophore to the spermatheca is for most species unclear. In some species the anterior end of the spermatophore is open to facilitate the escape of sperm. Where the spermatophore completely encloses the sperm, its wall may be either ruptured by spines protruding from the wall of the bursa or digested by secretions of the bursa wall or accessory glands. Transfer of sperm into the spermatheca is achieved normally as a result of rhythmic contractions of the reproductive tract. In *Rhodnius* the contractions are promoted by a substance contained within the seminal fluid and produced in the male's accessory glands. In some species sperm may migrate actively into the spermatheca, possibly in response to a chemotactic stimulus released by the storage organ.

Insemination without a Spermatophore. In species where spermatophores are not used in insemination, the penis may be rigid and erection achieved by means of muscles. In such instances the organ penetrates only a short distance into the female's genital tract. Alternatively and more commonly, the penis is thin-walled and erected by hydrostatic pressure, either of the hemolymph or of fluid contained within a special reservoir off the ejaculatory duct. It extends far into the genital tract of the female and terminates adjacent to the spermatheca.

Hemocoelic Insemination. This most unusual form of insemination, practiced by Cimicoidea (Hemiptera) and Strepsiptera, refers to the injection of sperm into the body cavity of a female from which they migrate to specialized storage sites, conceptacula seminales (not homologous with the spermathecae of other Insecta), adjacent to the oviducts. Comparative studies of the phenomenon among various Cimicoidea have led authors to propose a possible evolutionary sequence (Hinton, 1964; Carayon, 1966). Primitively, the penis is placed in the vagina but penetrates its wall, thereby injecting semen into the body cavity. At a more advanced stage, the penis penetrates the integument, though not at any predefined site, and sperm are still injected into the hemolymph. Next, the site of penetration becomes fixed, and beneath it a special structure, the spermalege, develops to receive sperm. However, the sperm must still migrate via the hemolymph to the conceptacula seminales. At the most advanced level, insemination into a spermalege occurs, and sperm move to the conceptacula along a solid core of cells. What is the functional significance of this arrangement? In all forms of hemocoelic insemination a proportion of the sperm are phagocytosed either by hemocytes or by cells of the spermalege. As a result, it has been suggested that in Cimicoidea hemocoelic insemination is a method of providing nutrients to the female, enabling her to survive for longer periods in the absence of suitable food. It should be remembered that many Cimicoidea are semiparasitic or parasitic, and the chances of locating a host are slight. Interestingly, some species are apparently homosexual; that is, males inseminate other males, enabling the recipients to resist starvation for longer periods, while reducing the donors' own viability, surely a truly noble and altruistic act!

4.4. Postcopulatory Behavior

In many species characteristic behavior follows copulation. This may be a continuation of the events which occur during copulation, for example, antennation or palpation of the female by the male, or feeding by the female on special secretions (nuptial gifts) offered by the male, so that the female remains quiet, enabling sperm to be evacuated from the spermatophore which, as noted above, may then be eaten. A female often becomes unreceptive after copulation, that is, fails to respond to a male's courtship display, and may vigorously reject his advances, especially where this involves physical contact. In Odonata, egg laying immediately follows copulation, often while the male is still grasping the female.

5. Entry of Sperm into the Eggs and Fertilization

In almost all insects sperm enter the eggs as the latter pass through the common oviduct during oviposition. However, in C. micoidea and some scale insects sperm enter the eggs in the ovary, and in Strepsiptera sperm entry occurs as the eggs float within the hemocoel. Two problems associated with the entry of sperm are (1) release of sperm from the spermatheca in synchrony with movement of an egg through the oviduct, and (2) location by sperm of the micropyles. The solution to the first problem is not known with certainty, though in some species the presence of muscle in the wall of the spermatheca and its duct indicates that sperm are squeezed toward the oviduct rather than moving of their own accord. In Periplaneta, it has been suggested that a nervous pathway might exist between sensory hairs found within the oviduct and the spermathecal muscle which could synchronize these events. Other insects have spermathecae with rigid walls, and in these a chemotactic stimulus for induction of sperm movement seems more likely.

Various mechanisms ensure that sperm can locate and enter the micropyles. In many species, an egg is precisely oriented as it moves along the oviduct so that the micropylar region directly faces the opening of the spermathecal duct. Furthermore, movement of an egg along the oviduct may stop briefly at this point. Where an egg's orientation is less precise, there may be a large number of micropyles. For example, in Periplaneta, up to 100 of these funnel-shaped structures occur in a cluster at the cephalic end of the egg. In Rhodnius the micropyles lie within a groove that encircles the anterior end of the egg. As an egg moves along the oviduct, the groove comes to lie opposite the openings of the paired spermathecae (Davey, 1965). Actual entry of sperm into the micropyles, whose diameter at the inner end may be only a fraction of a micrometer, may be due to release of a chemical attractant by the oocyte, though there is little evidence to support this suggestion.

Polyspermy (the entry of two or more sperm into an egg) is common in insects, though only one sperm normally undergoes subsequent transformation into a pronucleus, and the remainder degenerate.

Fertilization *sensu stricto*, that is, the fusion of male and female pronuclei, does not occur until after oviposition. Indeed, completion of the meiotic divisions of the oocyte nucleus, which give rise to three polar body nuclei and the female pronucleus, is inhibited until after an egg is laid. Whether or not entry of sperm is responsible for removal of this inhibition is uncertain. Pronuclear fusion may occur at a relatively fixed site to which both pronuclei migrate or may occur randomly within an oocyte, depending on the rate and direction of movement of the pronuclei. The polar body nuclei normally migrate to the periphery of the oocyte and eventually degenerate.

6. Oviposition

Oviposition is an extremely important phase in an insect's life history for, unless it is carried out at the correct time and in a suitable location, the chances of the eggs developing and of the larvae reaching adulthood are slim. Except in normally parthenogenetic species, unfertilized eggs are generally inviable, and it is important, therefore, that oviposition occurs only after mating. The eggs when laid must be protected from desiccation and predation. Further, because larvae are relatively immobile, it is necessary to lay eggs close to or even on/in the larva's food. This is especially true when the food is highly specialized and/or in limited supply as, for example, in many parasitic species.

Conversely, it is desirable that the female not expend energy searching for and testing potential oviposition sites until she is ready to lay. Accordingly, oviposition behavior characteristically does not begin until the eggs are more or less mature and may be induced by the hormonal balance in the female at this time.

6.1. Site Selection

Only a few examples are known of insects which apparently show no site selection behavior. Phasmids simply release their eggs, which fall among the dead vegetation beneath the host plant, though even this may be adaptive, as the eggs presumably will be hidden and protected from heat, desiccation, and predation. Many species attach their eggs, either singly or in batches, to an appropriate surface (often the food source) using secretions of the accessory glands. Such species typically lack an ovipositor. Other insects lay their eggs in crevices, plant or animal tissues, and, for this purpose, an ovipositor may be required which may be formed by modification of either the terminal abdominal segments *per se* or the appendages of these segments (see Chapter 3, Section 5).

The location and final selection of an egg-laying site may be more or less specific. The initial location of a site depends on rather general stimuli, often visual, which tend to attract the female. For example, *Pieris* (Lepidoptera), which is attracted by blue or yellow objects when sexually immature, shows a preference for green when ready to lay. However, final choice of an oviposition site depends on more specific environmental cues. Mosquitoes, for example, are attracted to water by light reflected from its surface. However, whether oviposition occurs subsequently is determined by water quality, measured by sensilla on the tarsi. Salinity, pH, and amount of organic matter present are factors which influence oviposition in these insects. Many phytophagous species are stimulated to oviposit by odors characteristic of the host plant. Pieris, for example, is stimulated by the odor of oil of mustard, which is released by cabbage and its relatives. For grasshoppers and locusts, the texture, moisture content, and salt content of the soil determine whether a female lays eggs. In these insects, sense organs receptive to these stimuli appear to be located at the tip of the abdomen.

6.2. Mechanics and Control of Oviposition

Expulsion of eggs via the genital pore results from rhythmic peristaltic contractions of the walls of the oviducts. Concurrently, muscles may move the ovipositor, where present, so that a suitable egg-laying cavity is formed. In some species, oviposition appears to be controlled neurally; in others, pheromones and hormones appear to regulate the process. In Locusta migratoria ssp. man lensis, oviposition is reported to be controlled via the terminal abdominal ganglion. Removal of the ganglion prevents egg laying, though cutting the ventral nerve cord anterior to this center does not. In some other species, including Rhodnius prolixus and the acridids, Schistocerca gregaria and Melanoplus sanguinipes, there is evidence for the production, by median neurosecretory cells, of an oviposition-stimulating (myotropic) hormone. The action of this hormone is to enhance the frequency and amplitude of contractions of the oviducal muscles.

Release of an oviposition-stimulating hormone or initiation of motor impulses in species where egg laying is controlled neurally depends on sensory input received by the central nervous system, and examples were given above of chemical and physical information on the oviposition site which was acquired in this way. However, another important consideration for the female "contemplating" oviposition, is whether or not she has been inseminated, since, as noted above, unfertilized eggs are normally inviable. In some insects, such as the cockroaches Pycnoscelus and Leucophaea, filling the spermatheca with semen stimulates neural pathways which terminate in the muscles used in oviposition. In others, it is possible that filling the spermatheca may cause this structure to release a hormone which, directly or indirectly, triggers oviposition. In a number of Diptera and Orthoptera, and other insects, there is evidence that semen contains a pheromone (fecundity-enhancing substance) which stimulates release of an oviposition hormone by the brain (see Gillott and Friedel, 1977).

6.3. Oothecae

Many orthopteroid insects, for example, locusts and grasshoppers, mantids, and cockroaches, surround their eggs, which are laid in batches, with a protective coat, the ootheca, produced from secretions of the accessory glands and thought to prevent desiccation and, perhaps, parasitism. The egg pod of Acrididae consists of eggs surrounded by a hard, frothy mass (Figure 19.10A). The egg "cocoon" of mantids is also a hard but vacuolated sheath (Figure 19.10B,C), and, in these examples, there is clearly no hindrance to gaseous exchange. In contrast, the ootheca of cockroaches is a hardened shell (Figure 19.10D) formed by tanning of the proteins secreted from the accessory glands and for the most part is virtually impermeable to gases. Accordingly, the ootheca contains an air-filled cavity along one side which connects with the exterior via small pores (Figure 19.10E). Oothecae are restricted to the lower insect orders and Davey (1965) speculates that their absence in higher insects might represent an economy measure which permits members of these groups to produce more eggs.

7. Summary

Almost all species of insects practice internal fertilization and lay eggs that contain much yolk. The female reproductive system basically includes paired

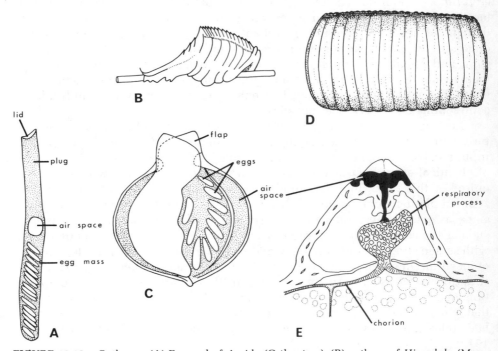

FIGURE 19.10.   Oothecae. (A) Egg pod of *Acrida* (Orthoptera), (B) ootheca of *Hierodula* (Mantodea), (C) transverse section through ootheca of *Hierodula*, (D) ootheca of *Blattella* (Blattodea), and (E) transverse section through crista. [A, after R. F. Chapman and I. A. D. Robertson, 1958, The egg pods of some tropical African grasshoppers, *J. Entomol. Soc. South. Afr.* **21**:85–112. By permission of the Entomological Society of Southern Africa. B, C, after A. D. Imms, 1957, *A General Textbook of Entomology*, 9th ed. (revised by O. W. Richards and R. G. Davies). By permission of Chapman and Hall Ltd. E, after V. B. Wigglesworth, 1965, *The Principles of Insect Physiology*, 6th ed., Methuen and Co. By permission of the author.]

ovaries and lateral oviducts, a common oviduct, bursa copulatrix, spermatheca, and accessory glands. Each ovary includes ovarioles, which may be panoistic (lacking nurse cells) or meroistic, where nurse cells either are enclosed in each follicle with an oocyte (polytrophic type) or form a syncytium in the germarium and connect with an oocyte via a trophic cord (telotrophic type). The male system includes paired testes, vasa deferentia and seminal vesicles, a median ejaculatory duct, and accessory glands. Each testis is composed of follicles in which zones of maturation of germ cells occur.

In most insects a period of sexual maturation is required after eclosion. In females maturation may include vitellogenesis (formation of yolk), development of characteristic body coloration, maturation of pheromone-producing glands, growth of the reproductive tract, and an increase in receptivity. Sexual maturation is affected by quality and quantity of food eaten, population density and structure, mating, temperature, humidity, and photoperiod. These factors exert their influence by modifying endocrine activity. The two primary endocrine components in most insects are the neurosecretory system and corpora allata, though in a few species hormones also may be released from the ovaries. The neurosecretory system liberates an allatotropic hormone, the corpora allata release juvenile hormone, which has both a gonadotropic effect (stimulates differentiation of the follicular epithelium so that it becomes permeable to proteins) and a metabolic effect [promotes synthesis of specific yolk proteins (vitellogenins) in the fat body].

In male insects sexual maturation may include synthesis and accumulation of accessory gland secretions, and development of pheromone-producing glands, mature body coloration, and courtship behavior. As in females, maturation is controlled hormonally, especially by juvenile hormone.

Mating behavior serves to ensure that sperm is transferred from male to female under the most suitable conditions and, perhaps, to prevent interspecific mating. It includes mate location and/or recognition, courtship, copulation, and postcopulatory behavior. Mate location and recognition are achieved through visual, chemical, auditory, and tactile stimuli. Courtship may synchronize the behavior of male and female and appease a normally aggressive female. Copulation may depend on the receptivity of a female and may occur only under specific conditions, for example, at a set time of the day, near the food plant, or immediately after feeding.

Primitively, sperm are transferred in a spermatophore produced from secretions of the accessory glands and formed within the male genital tract. At a more advanced stage, the spermatophore is formed in the female genital tract and may simply serve as a plug to prevent loss of semen. Some species do not form spermatophores but use a penis for depositing sperm in the female genital tract. Sperm normally reach the spermatheca as a result of peristaltic movements of the genital tract, though in some species active migration may occur in response to a chemical attractant.

In almost all insects sperm enter eggs via micropyles as the eggs move down the common oviduct during oviposition. Release of sperm from the spermatheca is closely synchronized with movements of the eggs. Eggs may be precisely oriented in the oviduct or the micropyles may be arranged in a cluster or a groove to ensure location of the micropyles by sperm. Polyspermy is common, though usually only one sperm undergoes transformation into a pronucleus. Fusion of male and female pronuclei does not occur until after eggs are laid.

Insects may show great selectivity in their choice of oviposition sites. Eggs may be attached to surfaces by secretions of the female accessory glands or buried using an ovipositor. Eggs may be covered with an ootheca, again formed from accessory gland secretions, which may prevent desiccation and/or parasitism.

8. Literature

Insect reproduction is the subject of texts by Davey (1965) and Engelmann (1970), of a symposium edited by Highnam (1964), and of a review by de Wilde and de Loof (1974a). The endocrine control of reproduction, mainly of female insects, is discussed by Engelmann (1968), de Wilde and de Loof (1974b), and Highnam and Hill (1977). Reviews of other aspects of reproduction are given by Phillips (1970) and Baccetti (1972) [sperm]; Barth and Lester (1973) [endocrine control of sexual behavior]; King (1970), Mahowald (1972), and Telfer (1975) [ovarian development]; and Leopold (1976) [male accessory reproductive glands].

Adams, T. S., 1970, Ovarian regulation of the corpus allatum in the housefly, *Musca domestica, J. Insect Physiol.* **16**:349–360.

Baccetti, B., 1972, Insect sperm cells, *Adv. Insect Physiol.* **9**:315–397.

Barth, R. H., and Lester, L. J., 1973, Neuro-hormonal control of sexual behavior in insects, *Annu. Rev. Entomol.* **18**:455–472.

Carayon, J., 1966, Traumatic insemination and the paragenital system, *The Thomas Say Foundation, Publ.* **7**:81–166.

Davey, K. G., 1965, *Reproduction in the Insects*, Oliver and Boyd, Edinburgh.

Dumser, J. B., and Davey, K. G., 1975, The *Rhodnius* testis: Hormonal effects on germ cell division, *Can. J. Zool.* **53**:1682–1689.

Engelmann, F., 1968, Endocrine control of reproduction in insects, *Annu. Rev. Entomol.* **13**:1–26.

Engelmann, F., 1970, *The Physiology of Insect Reproduction*, Pergamon Press, Oxford.

Friedel, T., and Gillott, C., 1976, Extraglandular synthesis of accessory reproductive gland components in male *Melanoplus sanguinipes, J. Insect Physiol* **22**:1309–1314.

Gerber, G. H., 1970, Evolution of the methods of spermatophore formation in pterygotan insects, *Can. Entomol.* **102**:358–362.

Gillott, C., and Friedel, T., 1976, Development of accessory reproductive glands and its control by the corpus allatum in adult male *Melanoplus sanguinipes, J. Insect Physiol.* **22**:365–372.

Gillott, C., and Friedel, T., 1977, Fecundity-enhancing and receptivity-inhibiting substances produced by male insects: A review, *Adv. Invert. Reprod.* **1**:199–218.

Hagedorn, H. H., O'Connor, J. D., Fuchs, M. S., Sage, B., Schlaeger, D. A., and Bohm, M. K. 1975, The ovary as a source of α-ecdysone in an adult mosquito, *Proc. Natl. Acad. Sci. USA,* **72**:3255–3259.

Highnam, K. C. (ed.), 1964, Insect reproduction, *Symp. R. Entomol. Soc.* **2**:120 pp.

Highnam, K. C., and Hill, L., 1977, *The Comparative Endocrinology of the Invertebrates*, 2nd ed., Edward Arnold, London.

Hinton, H. E., 1964, Sperm transfer in insects and the evolution of haemocoelic insemination, *Symp. R. Entomol. Soc.* **2**:95–107.

Huebner, E., and Davey, K. G., 1973, An antigonadotropin from the ovaries of the insect *Rhodnius prolixus* Stål, *Can. J. Zool.* **51**:113–120.

King, R. C., 1970, *Ovarian Development in Drosophila melanogaster*, Academic Press, New York.

Leopold, R. A., 1976, The role of male accessory glands in insect reproduction, *Annu. Rev. Entomol.* **21**:199–221.

Mahowald, A. P., 1972, Oogenesis, in: *Developmental Systems: Insects*, Vol. I (S. J. Counce and C. H. Waddington, eds.), Academic Press, New York.

Manning, A., 1966, Sexual behavior, *Symp. R. Entomol. Soc.* **3**:59–68.

Odhiambo, T. R., 1966, Growth and the hormonal control of sexual maturation in the male desert locust, *Schistocerca gregaria* (Forskål), *Trans. R. Entomol. Soc.* **118**:393–412.

Pener, M. P., 1974, Neurosecretory and corpus allatum controlled effects on male sexual behaviour in acridids, in: *Experimental Analysis of Insect Behavior* (L. Barton Browne, ed.), Springer-Verlag, New York.

Phillips, D. M., 1970, Insect sperm: Their structure and morphogenesis, *J. Cell Biol.* **44**:243–277.

Telfer, W. H., 1975, Development and physiology of the oocyte–nurse cell syncytium, *Adv. Insect Physiol.* **11**:223–319.

de Wilde, J., and de Loof, A., 1974a, Reproduction, in: *The Physiology of Insecta*, 2nd ed., Vol. I (M. Rockstein, ed.), Academic Press, New York.

de Wilde, J., and de Loof, A., 1974b, Reproduction—Endocrine control, in: *The Physiology of Insecta*, 2nd ed., Vol. I (M. Rockstein, ed.), Academic Press, New York.

20

Embryonic Development

1. Introduction

Embryonic development begins with the first mitotic division of the zygote nucleus and terminates at hatching. Not surprisingly, in view of their diversity of form, function, and life history, insects exhibit a variety of embryonic developmental patterns though, as more knowledge is obtained, certain evolutionary trends are becoming apparent. Eggs of most species contain a considerable amount of yolk. In exopterygote eggs there is such a preponderance of yolk that the egg cytoplasm is readily obvious only when it forms a small island surrounding the nucleus. In eggs of endopterygotes, the yolk/cytoplasm ratio is much lower than that of exopterygotes and, as a result, the cytoplasm can be seen as a conspicuous network connecting the central island with a layer of periplasm lying beneath the vitelline membrane. This trend toward reduction in the relative amount of yolk in the egg, carried to an extreme in certain parasitic Hymenoptera and viviparous Diptera (Cecidomyiidae), whose eggs are yolkless and receive nutrients from their surroundings, has some important consequences. Broadly speaking, the eggs of endopterygotes are smaller (size measured in relation to the body size of the laying insect) and develop more rapidly than those of exopterygotes. The increased quantity of cytoplasm leads to the more rapid formation of more and larger cells at the yolk surface which, in turn, facilitates the formation of a larger embryonic area from which development can take place. Compared with that of exopterygotes, development of endopterygotes is streamlined and simplified. There has been, as Anderson (1972b, p. 229) puts it "reduction or elimination of ancestral irrelevancies," which when taken to an extreme, seen in the apocritan Hymenoptera and cyclorrhaph Diptera, results in the formation of a structurally simple larva which hatches within a short time of egg laying. However, superimposed on this process of short-circuiting may be developmental specializations associated with an increasing dissimilarity of juvenile and adult habits.

2. Cleavage and Blastoderm Formation

As it moves toward the center of an egg after fusion, the zygote nucleus begins to divide mitotically. The first division occurs at a predetermined site, the cleavage center (Figure 20.1), located in the future head region, which cannot be recognized morphologically but which appears to become activated either when sperm enter an egg or when an egg is laid. Early divisions are synchronous, and as nuclei are formed and migrate through the yolk toward the periplasm, each becomes surrounded by an island of cytoplasm (Figure 20.2A). Each nucleus and its surrounding cytoplasm are known as a cleavage energid. In eggs of endopterygotes and possibly exopterygotes, but not those of apterygotes, the energids remain interconnected by means of fine cytoplasmic bridges.

The rate at which nuclei migrate to the yolk surface and the method of colonization are variable. In eggs of some species nuclei appear in the periplasm as early as the 64-energid state (after 6 divisions); in others, nuclei are not seen in the periplasm until the 1024-energid stage. In eggs of most endopterygotes and in those of paleopteran and hemipteroid exopterygotes, the periplasm is invaded uniformly by the energids. However, in eggs of orthopteroid insects the periplasm at the posterior pole of the egg receives energids first, after which there is progressive colonization of the more anterior regions.

In eggs of most insects not all cleavage energids migrate to the periphery but continue to divide within the yolk to form primary vitellophages, phagocytic cells whose function is to digest the yolk (Figure 20.2B). In eggs of Lepidoptera, Diptera, and some orthopteroid insects, however, all the energids migrate to the periplasm and only later do some of their progeny move back into the yolk as secondary vitellophages (Figure 20.2F). Secondary vitellophages are also produced in eggs of other insects to supplement the number of primary vitellophages. So-called tertiary vitellophages are produced in eggs of some cyclorrhaph Diptera and apocritan Hymenoptera from the anterior and posterior midgut rudiments.

After their arrival at the periplasm, the energids continue to divide, often synchronously, until the nuclei become closely packed (the syncytial blastoderm stage), after which cell membranes form (the uniform blastoderm stage) (Figure 20.2C–F).

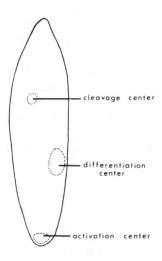

FIGURE 20.1. Positions of cleavage center, activation center, and differentiation center in eggs of *Platycnemis* (Odonata). [After D. Bodenstein, 1953, Embryonic development, in: *Insect Physiology* (K. D. Roeder, ed.). Copyright © 1953, John Wiley and Sons, Inc. Reprinted by permission of John Wiley and Sons, Inc.]

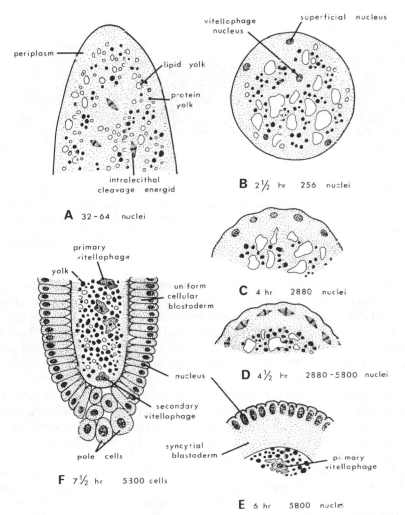

FIGURE 20.2. Stages in cleavage and blastoderm formation in egg of *Dacus tryoni* (Diptera). (A) Frontal section through anterior end during sixth division, (B) transverse section after eighth division, (C) transverse section after twelfth division, (D) transverse section during thirteenth division, (E) transverse section at syncytial blastoderm stage, and (F) frontal section through posterior end after formation of uniform cellular blastoderm. [After D. T. Anderson, 1972b, The development of holometabolous insects, in: *Developmental Systems: Insects*, Vol. I (S. J. Counce and C. H. Waddington, eds.). By permission of Academic Press Ltd., London, and the author.]

3. Formation and Growth of Germ Band

The next stage is blastoderm differentiation, giving rise to the embryonic primordium (an area of closely packed columnar cells from which the future embryo forms) and the extraembryonic ectoderm from which the extraembryonic membranes later differentiate (Figure 20.3). Differentiation is controlled by two centers, as the classic experiments of the German embryologist Seidel first demonstrated (see Bodenstein, 1953, and Counce, 1973). As energids move toward the posterior end of the egg they "interact" with a so-called

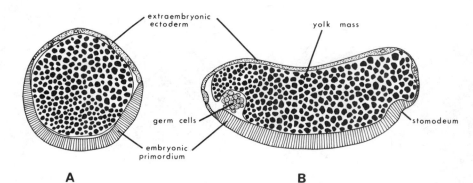

FIGURE 20.3. Diagrammatic transverse (A) and sagittal (B) sections of egg of *Pontania* (Hymenoptera) to show differentiation of blastoderm into embryonic primordium and extraembryonic ectoderm. Note also the germ (pole) cells at the posterior end. [After D. T. Anderson, 1972b, The development of holometabolous insects, in: *Developmental Systems: Insects*, Vol. I (S. J. Counce and C. H. Waddington, eds.). By permission of Academic Press Ltd., London, and the author.]

"activation center" (Figure 20.1) and differentiation subsequently occurs. Seidel's experiments showed that neither an energid nor the activation center alone could stimulate differentiation. It is presumed that the center is caused to release an as yet unidentified chemical which diffuses anteriorly. This diffusion is seen morphologically as a clearing and slight contraction of the yolk. As the chemical reaches the future prothoracic region of the embryo (the "differentiation center") (Figure 20.1), the blastoderm in this region gives a sharp twitch and becomes slightly invaginated. Blastoderm cells aggregate within this invagination and differentiate into the embryonic primordium. (Later in embryogenesis, other processes, for example, mesoderm formation and segmentation, begin at the differentiation center and spread anteriorly and posteriorly from it.)

As a result of the differing amounts of yolk that exopterygote and endopterygote eggs contain, important differences occur in the formation of the embryonic primordium. In exopterygote eggs where there is initially little cytoplasm, the embryonic primordium is normally relatively small, and its formation depends on the aggregation and, to some extent, proliferation of cells. In these eggs it usually occupies a posterior midventral position (Figure 20.4A–D). In contrast, in endopterygote eggs with their greater quantity of cytoplasm, the primordium forms as a broad monolayer of columnar cells that occupies much of the ventral surface of the yolk (Figure 20.5A,B). In other words, the primordium in endopterygote eggs does not require to undergo much increase in size, as is necessary in eggs of exopterygotes, so that tissue differentiation can occur directly and embryonic growth more rapidly. At its extreme, seen in eggs of some Diptera and Hymenoptera, the primordium occupies both ventral and lateral areas of the egg, with the extraembryonic ectoderm covering only the dorsal surface (Figure 20.5C).

The shape of the primordium is variable, though in most insects the anterior region is expanded laterally as a pair of head lobes (=protocephalon), behind which is a region of variable length, the protocorm (postantennal region) (Figure 20.4). In eggs of Paleoptera, hemipteroid insects, and some or-

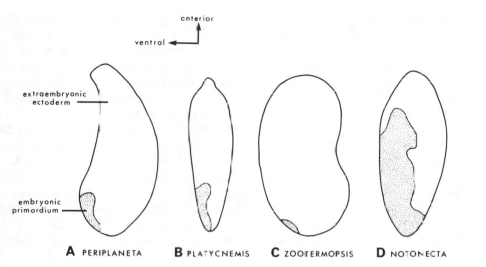

FIGURE 20.4. Form and position of embryonic primordium in exopterygotes. [After D. T. Anderson, 1972a, The development of hemimetabolous insects, in: *Developmental Systems: Insects*, Vol. I (S. J. Counce and C. H. Waddington, eds.). By permission of Academic Press Ltd., London, and the author.]

thopteroid species, the protocorm is semilong and at its formation includes the mouthpart-bearing segments, the thoracic segments, and a posterior growth region from which the abdominal segments arise. In eggs of other orthopteroid insects the postantennal region consists initially only of the growth zone. Though the protocorm in most endopterygote embryos is long, it also includes

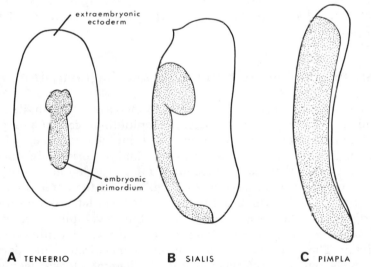

FIGURE 20.5. Form and position of embryonic primordium in endopterygotes. [After D. T. Anderson, 1972a,b, The development of hemimetabolous insects, and The development of holometabolous insects, in: *Developmental Systems: Insects*, Vol. I (S. J. Counce and C. H. Waddington, eds.). By permission of Academic Press Ltd., London, and the author.]

a posterior growth zone from which rudimentary abdominal segments proliferate. As the embryonic primordium elongates and begins to differentiate, it becomes known as the germ band. During elongation and differentiation, the abdomen grows around the posterior end and forward over the dorsal surface of the egg (Figure 20.6). In eggs of some higher endopterygotes (Hymenoptera–Apocrita and Diptera–Cyclorrhapha), there is no posterior growth zone and the abdominal segments arise directly from the primordium.

It is during the differentiation and elongation of the germ band that the primordial germ cells first become noticeable in most endopterygote eggs, though in those of some Coleoptera they are distinguishable even as the syncytial blastoderm is forming. They are largish, rounded cells in a distinct group at the posterior pole of the yolk, and accordingly are referred to as pole cells (see Figure 20.3). In eggs of Dermaptera, Psocoptera, Thysanoptera, and Homoptera also, the germ cells differentiate early at the posterior end of the primordium. In those of most exopterygotes, however, they are not apparent until gastrulation or somite formation has occurred.

As the germ band elongates and becomes broader, segmentation and limb-bud formation are apparent externally and are accompanied internally by mesoderm and somite formation. Growth of the germ band may occur either on the surface of the yolk (superficial growth) as seen in eggs of Dictyoptera, Dermaptera, Isoptera, some other orthopteroid insects and all endopterygotes (Figure 20.6), or by immersion into the yolk (immersed growth) as occurs in eggs of Paleoptera, most Orthoptera, and hemipteroid insects (Figure 20.7). Immersion of the germ band (anatrepsis) forms the first of a series of embryonic movements, collectively known as blastokinesis. The reverse movement (katatrepsis), which brings the embryo back to the surface of the yolk, occurs later (see Section 6). Anatrepsis has developed secondarily (i.e., superficial growth is the more primitive method) and convergently among those exopterygotes in which it occurs. Its functional significance is, however, not yet clear (Anderson, 1972a).

4. Gastrulation, Somite Formation, and Segmentation

As the embryonic primordium begins to increase in length, its midventral cells sink inward to form a transient, longitudinal gastral groove (Figure 20.8A). The invaginated cells soon separate from the outer layer which closes to obliterate the groove. It is from the anterior and posterior points of closure of the gastral groove that the stomodeum and proctodeum, respectively, develop. The outer layer can now be distinguished as the embryonic ectoderm. The invaginated cells, which proliferate and spread laterally, form the mesoderm (Figure 20.8B,C) except adjacent to the developing stomodeum and proctodeum where they become the anterior and posterior midgut rudiments, respectively. The mesodermal cells become concentrated into paired longitudinal tracts which soon separate into segmental blocks, leaving only a thin longitudinal strip, the median mesoderm, from which hemocytes later differentiate. From these segmental blocks, paired hollow somites usually arise (Figure 20.8E). Somite formation is initiated and occurs more or less simultaneously in the gnathal and thoracic segments, spreading anteriorly and pos-

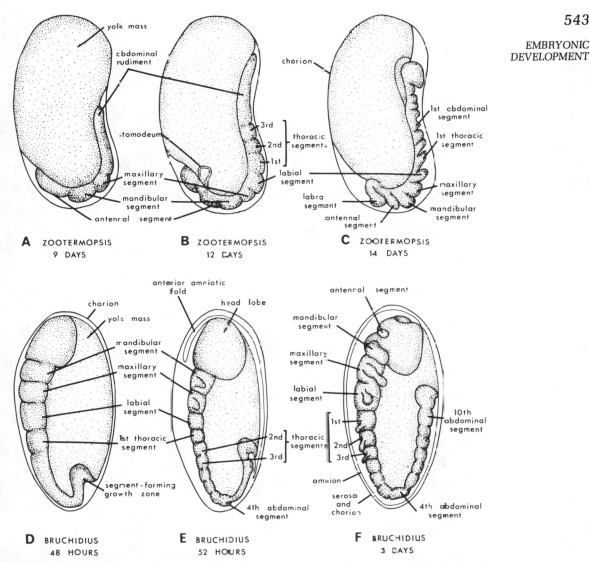

FIGURE 20.6. Stages in elongation and segmentation of germ band in *Zootermopsis* (Isoptera) (A–C) and *Bruchidius* (Coleoptera) (D–F). [After D. T. Anderson, 1972a,b, The development of hemimetabolous insects, and The development of holometabolous insects, in *Developmental Systems: Insects*, Vol. I (S. J. Counce and C. H. Waddington, eds.). By permission of Academic Press Ltd., London, and the author.]

teriorly after gastrulation takes place. Formation of the coelom (the cavity within a somite) may occur in one of two ways, by internal splitting of a somite or by median folding of the lateral part of each somite. In embryos of a given species, one or both methods may be seen in different segments. For example, internal splitting of the somites occurs in all segments of embryos of Phasmida, most hemipteroid insects, and most endopterygotes, and in the abdominal segments of *Locusta* embryos. Median folding is the method used in all segments in embryos of Odonata, Dictyoptera, and Mallophaga, and in the gnathal

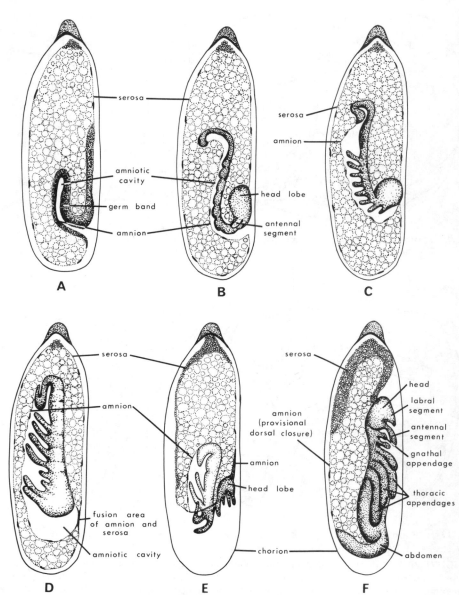

FIGURE 20.7. Early embryonic development in *Calopteryx* to show anatrepsis and katatrepsis. [A–E, after O. A. Johannsen and F. H. Butt, 1941, *Embryology of Insects and Myriapods*. By permission of McGraw-Hill Book Co. Inc., New York. F, after R. F. Chapman, 1971, *The Insects: Structure and Function*. By permission of Elsevier North-Holland, Inc., and the author.]

and thoracic segments of those of *Locusta*, and some Coleoptera, Lepidoptera, and Megaloptera. In exopterygote embryos, all somites usually develop a central cavity, though this may be only temporary. Among endopterygotes, members of more primitive orders retain a full complement of somites in their embryos and the latter usually develop a coelom. In embryos of some species, however, cavities may not form, and somite formation may be suppressed in the head segments. In embryos of Diptera and Hymenoptera, no distinct head

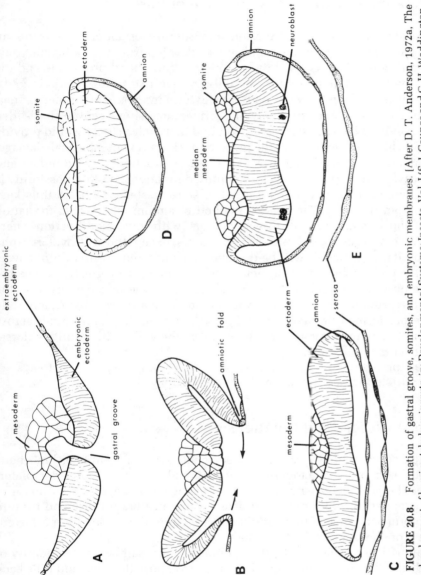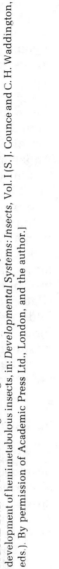

FIGURE 20.8. Formation of gastral groove, somites, and embryonic membranes. [After D. T. Anderson, 1972a. The development of hemimetabolous insects. in: *Developmental Systems: Insects*, Vol. I (S. J. Counce and C. H. Waddington, eds.). By permission of Academic Press Ltd., London, and the author.]

somites appear, and in those of some Cyclorrhapha and Apocrita, somite formation is entirely suppressed, so that mesodermal derivatives are produced directly from a single midventral mass.

5. Formation of Extraembryonic Membranes

Simultaneously with gastrulation and somite formation, two extraembryonic membranes, the amnion and serosa, develop from the extraembryonic
ectoderm (Figure 20.8). Cells at the edge of the germ band proliferate and the
tissue formed on each side folds ventrally to give rise to the amniotic folds.
These meet and fuse in the ventral midline to form inner and outer membranes,
the amnion and serosa, respectively, the former enclosing a central fluid-filled
amniotic cavity. Many authors have suggested that such a cavity would provide
space in which an embryo could grow and also prevent physical damage.
Anderson (1972a) considers, however, that these functions are redundant and
that the cavity must have an as yet unidentified function. Another possibility is
that the amnion and its cavity are used to store wastes which are thus kept
separate from the yolk. The general method of amnion and serosa formation
outlined above is found in all insect embryos (with some modification where
immersion of the germ band into the yolk occurs) except those of Cyclorrhapha
and Apocrita, in which, it will be recalled, the embryonic primordium covers
most of the yolk surface. In these, embryonic membranes are greatly reduced or
lost. In embryos of Apocrita the extraembryonic ectoderm separates from the
edge of the primordium and grows ventrally to form a serosa; that is, amniotic
folds are not formed. In embryos of Cyclorrhapha neither amnion nor serosa
forms and the extraembryonic ectoderm covers the yolk until definitive dorsal
closure occurs (see below).

After the embryonic membranes form, the serosa in most insect eggs secretes a cuticle that is often as thick as the chorion.

6. Dorsal Closure and Katatrepsis

When germ band elongation and segmentation are complete, limb buds
develop, the embryonic ectoderm becomes broader, that is, grows dorsolaterally over the yolk mass, and internally organogenesis begins. This phase of
growth is ended abruptly as the extraembryonic membranes fuse and rupture
and the germ band reverts to its original (preanatreptic) position (in most exopterygotes) or shortens (endopterygotes).

In embryos of most insects, the amnion and serosa fuse in the vicinity of
the head, and the combined tissue then splits to expose the head and rolls back
dorsally over the yolk (Figure 20.9A). As a result, the serosa is reduced to a
small mass of cells, the secondary dorsal organ, and the amnion becomes
stretched over the yolk, forming the provisional dorsal closure (Figure 20.9B).
In some endopterygote embryos, variations of this process can be seen. In those
of Nematocera (Diptera) and Symphyta (Hymenoptera), for example, it is the
amnion which ruptures and is reduced, leaving the serosa intact. As noted

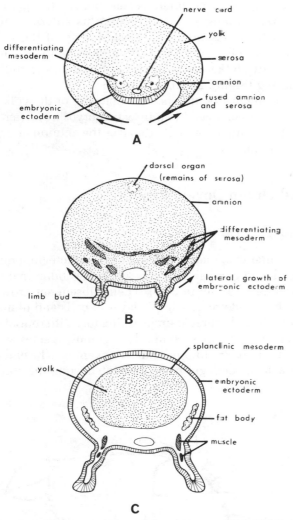

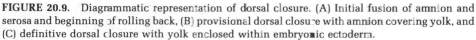

FIGURE 20.9. Diagrammatic representation of dorsal closure. (A) Initial fusion of amnion and serosa and beginning of rolling back, (B) provisional dorsal closure with amnion covering yolk, and (C) definitive dorsal closure with yolk enclosed within embryonic ectoderm.

above, in eggs of Apocrita only a serosa is formed, and this persists until definitive dorsal closure occurs, and in those of Cyclorrhapha no extraembryonic membranes develop, and the yolk remains covered by the extraembryonic ectoderm until definitive dorsal closure.

Except in dictyopteran embryos where the germ band remains superficial and ventral during elongation, extensive movement of the germ band now occurs in exopterygote eggs which serves (1) to bring an immersed germ band back to the surface of the yolk, and (2) to restore the germ band to its preanatreptic orientation, that is, on the ventral surface of the yolk with the head end facing the anterior pole of the egg. This movement, the reverse of anatrepsis, is known as katatrepsis (Figure 20.7).

At the beginning of provisional dorsal closure the germ band of most endopterygotes is quite long so that, although its anterior end is ventral, its posterior component passes round the posterior tip of the yolk and forward along the dorsal side (Figure 20.6F). During closure, the germ band shortens and broadens rapidly so that its posterior end now comes to lie near the posterior end of the egg (Figure 20.10A).

Definitive dorsal closure, that is, the enclosing of the yolk within the embryo, then occurs. It is achieved in all insect embryos by a lateral growth of the embryonic ectoderm which gradually replaces the amnion or, rarely, the serosa (Figures 20.9C and 20.10B).

7. Tissue and Organ Development

7.1. Appendages

Paired segmental evaginations of the embryonic ectoderm appear on the thoracic, antennal, and gnathal segments while the abdominal part of the germ band is still forming (see Figure 20.6). Their subsequent growth results from proliferation of the ectoderm as a single layer of cells and of mesodermal cells within. The cephalic and thoracic limbs ultimately differentiate into their specific form, except in eggs of secondarily apodous species where they soon shorten or become reduced to epidermal thickenings. In embryos of Cyclorrhapha, the thoracic appendages never develop beyond the epidermal thickening stage.

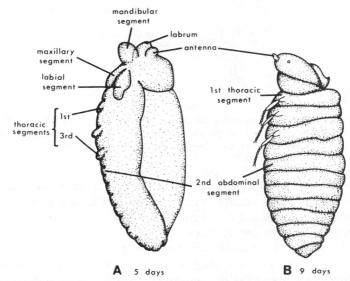

FIGURE 20.10. (A) Five-day embryo of *Bruchidius* (Coleoptera) after shortening of germ band; compare this figure with Figure 20.6F. (B) Embryo of *Bruchidius* at hatching stage (9 days). [After D. T. Anderson, 1972b, The development of holometabolous insects, in: *Developmental Systems: Insects,* Vol. I (S. J. Counce and C. H. Waddington, eds.). By permission of Academic Press Ltd., London, and the author.]

In Paleoptera and orthopteroid insects, 11 pairs of abdominal appendages evaginate before provisional dorsal closure. In most hemipteroid embryos, no sign of abdominal limbs is evident, though in those of Hemiptera and Thysanoptera appendages develop on the first and last abdominal segments. Ten pairs of abdominal evaginations develop in most endopterygote embryos. The fate of the abdominal appendages varies, and some or all of them may disappear before embryonic development is completed. The first (most anterior) pair disappears after blastokinesis in embryos of Paleoptera and some orthopteroid insects, but remains as glandular pleuropodia in those of Dictyoptera, Phasmida, Orthoptera, Hemiptera, and some Coleoptera and Lepidoptera. The function of the pleuropodia is uncertain, though some authors have suggested that in orthopteran embryos they secrete an enzyme which brings about dissolution of the serosal cuticle. The pleuropodia are resorbed or discarded before hatching. The appendages of the second through seventh abdominal segments are resorbed, except in some endopterygotes where they persist as larval prolegs. Pairs 8 to 10 may differentiate into the external genitalia or disappear, while the last pair either persists as cerci or disappears.

7.2. Integument and Ectodermal Invaginations

Soon after definitive dorsal closure, the outer embryonic ectoderm differentiates into epidermis, which in embryos of most insects then secretes the first instar larval cuticle. In some insects, however, one or more embryonic cuticles are produced which may be shed before or at hatching.

Concurrently with the formation of abdominal appendages a number of ectodermal invaginations develop from which differentiate endoskeletal components, various glands, the tracheal system, and certain parts of the reproductive tract (for the latter, see Section 7.6). From ventrolateral invaginations at the junctions of the antennal/mandibular segments and the mandibular/maxillary segments are derived the anterior and posterior arms of the tentorium. Paired mandibular apodemes differentiate from invaginations near the bases of the mandibles. The apodemes of the trunk region arise from intersegmental invaginations in the thorax and abdomen.

Salivary glands develop from a pair of invaginations near the bases of the labial appendages. When the appendages fuse the invaginations merge to form a common salivary duct that opens midventrally on the hypopharynx.

The corpora allata develop from a pair of ventrolateral invaginations at the junction of the mandibular/maxillary segments. Initially, they exist as hollow vesicles, though these fill in as they move dorsally to their final position adjacent to the stomodeum. The molt glands also originate as paired ventral ectodermal invaginations, usually on the prothoracic segment. Other invaginations on the head may give rise to specialized exocrine glands on the mandibles or maxillae.

Elements of the tracheal system can be seen first as paired lateral invaginations on each segment from the second thoracic to the eighth (ninth in a few Thysanura) abdominal. However, not all of these invaginations develop completely into tracheae. Those which do, bifurcate and anastomose with branches from adjacent segments and from their opposite partner of the same segment. The cells differentiate as tracheal epithelium and then secrete a cuticular lining. After cuticle secretion but before hatching gas is secreted into the tracheal system. Some of the invaginated ectodermal cells differentiate into oenocytes. These may remain closely associated with the tracheal system, form definite clusters in specific body regions, or become embedded as single cells in the fat body.

7.3. Central Nervous System

Soon after somite formation has commenced, specialized ectodermal cells on each side of the midventral line, the neuroblasts (Figure 20.8E), begin to proliferate, resulting in the formation of paired longitudinal neural ridges separated by a neural groove. As proliferation occurs, the cells move slightly inward so that they become separated from the ectoderm. They then begin to divide vertically, unequally and repeatedly, the small daughter cells eventually developing into ganglion cells (Figure 20.11). With the onset of segmentation, neuroblasts in the intersegmental regions become less active, so that paired segmental swellings, the future ganglia, now become apparent. As the ganglion cells take on the appearance of neurons, their cell bodies become arranged peripherally around the central axons (neuropile). Subsequent growth of the axons leads to formation of longitudinal connectives and transverse commissures. The neurilemma is also formed from ganglion cells. As embryogenesis continues, fusion of ganglia occurs in the head region to form the brain and subesophageal ganglion and at the posterior end of the abdomen where ganglia from segments 8 to 11 form a composite structure. In embryos of species belonging to different orders of Insecta, varying degrees of fusion of other ganglia may subsequently occur.

7.4. Gut and Derivatives

As noted earlier, the stomodeum and proctodeum arise at the anterior and posterior ends of the gastral groove, respectively. Both develop as hollow invaginated tubes, the stomodeum slightly earlier than the proctodeum, and as they differentiate into the subdivisions of the foregut and hindgut, respectively, various associated structures arise.

On the roof of the stomodeum, evaginations or thickenings give rise to the frontal ganglion, hypocerebral ganglion, intrinsic cells of the corpora cardiaca, and ingluvial ganglia. At its distal end, the proctodeum develops pouches which are the rudiments of the Malpighian tubules. (In many insects additional tubules develop during larval life.)

The anterior and posterior midgut rudiments which appeared at gastrulation begin to proliferate at about the time of provisional dorsal closure to form the midgut. Each rudiment proliferates a pair of strands which grow caudad or cephalad, respectively, between the nerve cord and yolk. After strands from each rudiment meet in the middle of the embryo, they grow laterally and dorsally so as to eventually enclose the yolk mass. The cells then differentiate as midgut epithelium. Prior to this, in embryos of some species, the vitellophages form a temporary "yolk sac" around the yolk mass.

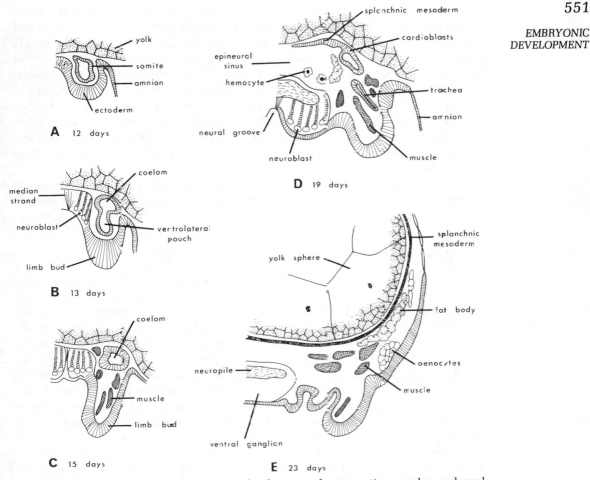

FIGURE 20.11. Transverse sections to show development of nervous tissue and mesodermal derivatives in prothoracic segment of *Tachycines* (Orthoptera). In (A–D), which are prakatatreptic stages, the serosa is omitted. [After D. T. Anderson, 1972a, The development of hemimetabolous insects, in: *Developmental Systems: Insects*, Vol. I (S. J. Counce and C. H. Waddington eds.). By permission of Academic Press Ltd., London, and the author.]

7.5. Circulatory System, Muscle, and Fat Body

The heart, aorta, musculature, fat body, lining of the hemocoel, and some components of the reproductive system are derived from the somites and median mesoderm formed after gastrulation. As noted earlier, the median mesoderm gives rise to hemocytes. In most embryos, each somite becomes hollow and forms three interconnected chambers, the anterior, posterior, and ventrolateral pouches. The latter grows into the adjacent ectodermal limb bud and breaks up to form intrinsic limb muscles (Figure 20.11). The splanchnic walls (i.e., those facing the yolk) of the two remaining pouches spread round the gut, forming the gut musculature, some fat body and part of the reproductive system (see Section 7.6). The somatic walls (those facing the ectoderm) of the anterior and posterior pouches give rise to extrinsic limb muscles, dorsal and ventral longitudinal muscles, and more fat body. Thus in insects as in other arthropods, the breaking up of the somite walls into discrete tissues means that there is no true coelom. Rather, the latter merges with the epineural sinus (the space between the dorsal surface of the embryo and the yolk) and is correctly called a mixocoel (hemocoel). From mesodermal cells at the dorsal junction of the somatic and splanchnic walls of the labial to the tenth abdominal somites, a sheet of cardioblasts develops. As the mesoderm grows around the gut, the sheets on each side become apposed to form the heart. Other somatic mesoderm cells adjacent to the cardioblasts differentiate as alary muscles, pericardial septum, and pericardial cells. The aorta develops from the median walls of the antennal somites, which become apposed and grow posteriorly to meet the heart.

In embryos of insects where the somites remain solid or are not formed as discrete segmental structures, the mesoderm still gives rise to the same components.

7.6. Reproductive System

The reproductive system includes both mesodermal and ectodermal components. In female exopterygotes, the vagina and spermatheca develop after hatching as midventral ectodermal invaginations of the seventh or eighth abdominal segment. In males, the ejaculatory duct and ectadenes (ectodermal accessory glands) are formed from a similar midventral invagination of the ectoderm of the ninth or tenth abdominal segment.

The paired genital ducts and mesadenes (mesodermal accessory glands) arise in exopterygotes from mesoderm of the splanchnic walls of certain abdominal somites which first thickens then hollows out to form coelomoducts. Some of these soon disappear, but those of the seventh and eighth somites (in females) or ninth and tenth somites (in males) enlarge to form the ducts and/or accessory gland components. In endopterygotes, the genital ducts are formed during postembryonic development. In *Drosophila* and other cyclorrhaph Diptera the reproductive system (excluding the gonads) develops from a single or pair of imaginal discs during metamorphosis (see also Chapter 21, Section 4.2).

Development of the gonads varies, though two related trends can be seen, namely, earlier segregation of the primordial germ cells and restriction of these cells to fewer abdominal segments. In the most primitive arrangement, seen in some thysanuran and orthopteran embryos, the germ cells do not become distinguishable until they appear in the splanchnic walls of several abdominal somites. In *Locusta* embryos, for example, they are found initially in the somites of abdominal segments 2 through 10, though they remain only in segments 3 through 6. Eventually they fuse longitudinally to form a compact gonad on each side. Such a segmental arrangement is presumably primitive as it is seen also in adult Annelida, Onychophora, Myriapoda, and noninsectan apterygotes (Anderson, 1972a).

In embryos of other orthopteroid insects (Dictyoptera, Phasmida, and Embioptera) and some hemipteroid insects (Heteroptera) the germ cells become apparent early in gastrulation. Nevertheless, they still become associated with the splanchnic mesoderm of several anterior abdominal segments.

In embryos of Dermaptera, Psocoptera, Thysanoptera, Homoptera, and endopterygotes, the germ cells differentiate as the blastoderm forms (see Figure 20.3B). After somite formation, they migrate to the third and fourth abdominal segments in exopterygote embryos or fifth and sixth abdominal segments in endopterygote embryos where they divide into left and right halves and become surrounded by splanchnic mesoderm.

553

EMBRYONIC
DEVELOPMENT

8. Special Forms of Embryonic Development

The great majority of insect species are bisexual and females lay eggs that contain a considerable amount of yolk. However, in some species, males may be rare and females may produce viable offspring from unfertilized eggs (parthenogenesis). In another form of asexual reproduction, polyembryony, which is characteristic of some parasitic Hymenoptera and Strepsiptera, several embryos develop from one fertilized egg. In other insects, fertilized eggs may be retained within the female reproductive tract for variable periods of time so that a young insect may hatch from the egg almost as soon as or even before the latter is laid (viviparity). In a few species paedogenesis may occur where mature larvae are able to produce, parthenogenetically and usually viviparously, a further generation of young.

8.1. Parthenogenesis

The ability of unfertilized eggs to develop is common to many insect species, and in some is the normal mode of reproduction under certain conditions. In all insects except Lepidoptera and Trichoptera, the female is the homogametic sex (that is having two X sex chromosomes) and the male, heterogametic (XY or XO). Unfertilized eggs, therefore, will contain only X chromosomes. However, whether they contain one or two such chromosomes and, therefore, the sex of parthenogenetic offspring, depends on the behavior of the chromosomes during meiosis in the oocyte nucleus (Sucmalainen, 1962; White, 1973).

In some species no meiotic division occurs during oogenesis. Therefore, offspring have the same genetic makeup as the mother; that is, they are diploid and female (ameiotic parthenogenesis). In meiotic parthenogenesis, the typical reduction division is followed by nuclear fusion so that a diploid chromosome complement is retained. Again, therefore, the offspring are female. Haploid parthenogenesis, where the oocyte nucleus undergoes meiosis which is not followed by nuclear fusion, is of relatively rare occurrence, though typical of Hymenoptera, Thysanoptera, and some Homoptera and Coleoptera, and results in the production of males. In Hymenoptera, haploid parthenogenesis is facultative; that is, a female determines whether or not an egg will be fertilized. In the honeybee, for example, a queen normally lays fertilized eggs which develop into workers (diploid females). However, under certain conditions, for example, when the hive is crowded, and the workers construct larger than normal drone cells on the honeycomb, she will lay unfertilized eggs from which haploid males develop, as a preliminary to swarming.

Parthenogenesis, producing in most species female offspring, may confer two advantages. In a species whose population density may be (temporarily) low, the ability of an isolated female to reproduce parthenogenetically may ensure survival of her genotype until the population density increases and males are again likely to be encountered. More often, however, parthenogenesis is employed as a mechanism that provides a rapid mode of reproduction, to enable a species to take full advantage of temporarily ideal conditions. Thus, a parthenogenetic female, who does not require to locate or be located by a male, can devote her time and energy to egg production. Further, all her offspring are female, so that her maximum reproductive potential can be realized. The disadvantage of parthenogenesis is that the genotype of successive generations remains more or less constant so that adaptation of a species to changing environmental conditions is very slow. To counteract this, many species alternate one or more parthenogenetic generations with a normal sexual generation. Aphids, for example, reproduce for most of the year by ameiotic parthenogenesis (see Figure 8.11). However, toward fall (and affected by changing environmental conditions) there occurs, during maturation of some oocytes, a separation of the two X chromosomes, one of which migrates to the polar body and is destroyed. From such eggs (with an XO constitution) males will develop. As spermatogenesis occurs in these individuals, spermatocytes containing either one X chromosome or no X chromosome are produced. However, the latter do not mature, so that only sperm with an X chromosome result. Therefore, the overwintering eggs produced as a result of mating will have an XX sex chromosome complement and give rise the following spring only to females.

8.2. Polyembryony

Polyembryony, the development of several embryos from one egg, is known to be a normal occurrence in about 30 species of parasitic Hymenoptera (mostly Encyrtidae, Platygasteridae, and Braconidae) and one species of Strepsiptera (Ivanova-Kasas, 1972). In these insects it is always associated with either parasitism or viviparity and is presumed to have evolved in conjunction with the abundance of food offered by these two modes of life. Characteristically, the eggs of polyembryonic species are minute and devoid of yolk. Because they depend on an external (host or maternal) source of nutrients the chorion, which is initially thin and permeable, soon disappears. Further, in Hymenoptera, the serosa becomes modified for the uptake of nutrients and is known as a "trophamnion."

Both the number of embryos formed and the point in development at which they become discernible vary. In *Platygaster hiemalis*, a parasite of Hessian fly larvae, for example, at the 4-cell stage, the cells may separate into two groups so that twin embryos are formed. In contrast, in the chalcid *Litomastix truncatellus*, which parasitizes larvae of the moth genus *Plusia*, formation of embryos does not begin until the 220–225-blastomere stage. At this stage, certain of the blastomeres become spindle-shaped and fuse to form a syncytial sheath which divides the remaining blastomeres into groups, the primary embryonic masses. In due course, secondary, tertiary, etc., embryonic masses form so that the final number of potential embryos may exceed 1000. The early development of *Litomastix* is summarized in Figure 20.12. Eventu-

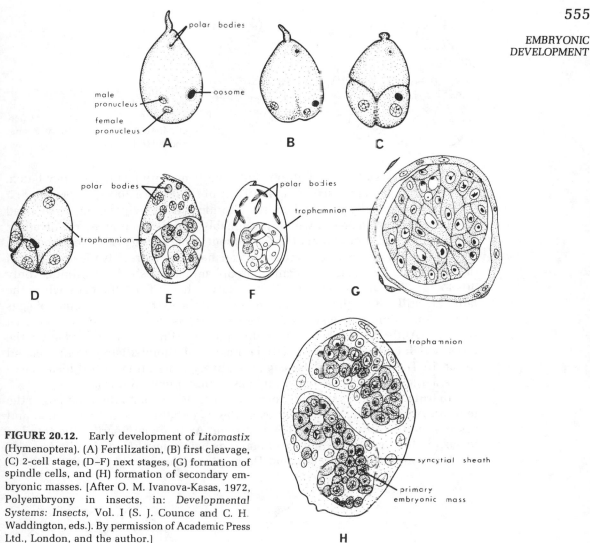

FIGURE 20.12. Early development of *Litomastix* (Hymenoptera). (A) Fertilization, (B) first cleavage, (C) 2-cell stage, (D–F) next stages, (G) formation of spindle cells, and (H) formation of secondary embryonic masses. [After O. M. Ivanova-Kasas, 1972, Polyembryony in insects, in: *Developmental Systems: Insects*, Vol. I (S. J. Counce and C. H. Waddington, eds.). By permission of Academic Press Ltd., London, and the author.]

ally the polygerm (the total embryonic mass within the trophamnion) disintegrates, and each embryo develops into a larva. The larvae feed within a host until all the soft parts are consumed and then pupate. At this point the host is nothing more than a cuticular bag full of parasites (Figure 20.13).

8.3. Viviparity

Viviparity, the retention of developing offspring within the maternal genital tract, is found in a range of complexity within the Insecta. It is seen in different forms in species from several orders, but among Diptera the entire range of variation may occur.

In its simplest form (ovoviviparity) the eggs retain their full complement of yolk for nourishment of the embryo. The eggs may be retained within the

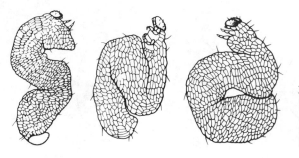

FIGURE 20.13. Caterpillars parasitized by *Litomastix*. [From R. R. Askew, 1971, *Parasitic Insects.* By permission of Heinemann Educational Books Ltd.]

mother for a variable period of time but usually are laid just before they hatch. Such an arrangement is seen in many Tachinidae (Figure 20.14) where the first instar larvae actually escape from the chorion during oviposition. Associated with retention of the eggs, which presumably affords them greater protection, is a trend toward production of fewer of them. As fewer eggs are produced each can acquire more yolk so that larvae can hatch at more advanced stages of development. For example, many Sarcophagidae (flesh flies) produce only 40–80 eggs but are larviparous; that is, larvae hatch from the eggs while the latter are still within the reproductive tract. At the extreme, the number of eggs that mature simultaneously is reduced to one, as, for example, in *Hylemya strigosa* (Anthomyiidae) and *Termitoxenia* sp. (Phoridae). In *Hylemya* the larva which emerges from a newly laid egg molts immediately to the second instar; in *Termitoxenia*, whose egg is relatively larger, it is a third instar larva that emerges from an egg and it pupates within a few minutes.

In truly viviparous species, developing offspring obtain their food from the mother. Accordingly, the structures of the maternal reproductive system and egg are modified to facilitate this exchange. As in ovoviviparity, the trend is toward reduction of the number of embryos being developed simultaneously.

Some aphids, Psocoptera, and Dermaptera (*Hemimerus*) exhibit pseudo-

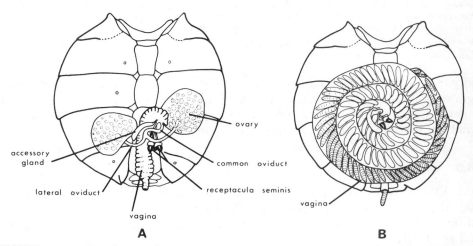

FIGURE 20.14. Female reproductive system of the tachinid *Panzeria* (Diptera). (A) Newly emerged fly and (B) mature female, with greatly enlarged vagina forming a brood chamber. An egg containing a fully formed embryo is being laid. [After V. B. Wigglesworth, 1965, *The Principles of Insect Physiology*, 6th ed., Methuen and Co. By permission of the author.]

placental viviparity (Hagan, 1951). Eggs of these insects contain little or no yolk and lack a chorion. They develop within the ovariole, where the follicle cells supposedly supply at least some nourishment to the embryo. (In species with meroistic ovarioles, the nurse cells are also important.) In *Hemimerus*, for example, follicle cells adjacent to the anterior and posterior ends of an oocyte proliferate and become connected with the embryonic membranes forming pseudoplacentae. Later, the follicle cells degenerate because, it is assumed, they are supplying nutrients to the developing embryo (Figure 20.15).

In *Glossina* spp. and pupiparous Diptera adenotrophic viviparity occurs. In this arrangement, an egg is normal, that is, contains yolk and possesses a chorion, yet is retained within the expanded bursa, the so-called uterus. Embryonic development, is, therefore, correctly described as ovoviviparous. However, after hatching, the larva remains within the uterus and feeds on secretions (uterine milk) of the enormous accessory glands which ramify through the abdomen (Figure 20.16). One larva at a time develops and pupation occurs shortly after birth.

In hemocoelic viviparity, practiced by Strepsiptera and some paedogenetic Cecidomyiidae (Diptera), oocytes are released from the ovarioles into the maternal hemocoel. In Strepsiptera fertilization occurs within the maternal body cavity, and, during embryonic development, nutrients are absorbed directly from the hemolymph. After hatching, the larvae escape from the female's body via the genital pores. In some cecidomyiids, oocytes develop parthenogenetically. Initially, development occurs within the ovarioles, but the larvae on hatching escape into the hemocoel. The larvae remain within the mother and feed on her tissues until just prior to pupation when they exit via the body wall.

8.4. Paedogenesis

Though not actually a form of embryonic development, paedogenesis, that is, precocious sexual maturation of juvenile stages, is conveniently mentioned

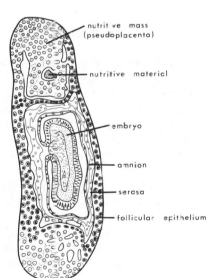

FIGURE 20.15. Longitudinal section through ovarian follicle of *Hemimerus* (Dermaptera) to show pseudoplacentae. [After V. B. Wigglesworth, 1965, *The Principles of Insect Physiology*, 6th ed., Methuen and Co. By permission of the author.]

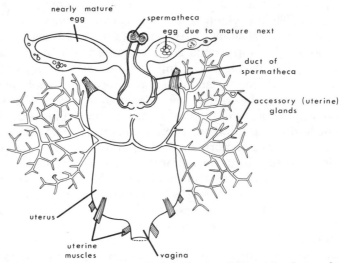

FIGURE 20.16. Female reproductive system of *Glossina* (Diptera) to show enlarged accessory glands. Note also that only one egg at a time is maturing. [After V. B. Wigglesworth, 1965, *The Principles of Insect Physiology*, 6th ed., Methuen and Co. By permission of the author.]

here. Paedogenesis is usually associated with both parthenogenesis and viviparity, and, probably, is best exemplified in certain Cecidomyiidae, though it is known to occur also in some Chironomidae (Diptera), Coleoptera, and Hemiptera. In some cecidomyiids, the oocytes develop viviparously in the hemocoel of the last larval or pupal instar. In some chironomids, embryonic development begins in female pupae. In the viviparous hemipteran *Hesperoctenes* female larvae which have been inseminated hemocoelically develop embryos within their ovarioles. In an extreme situation seen in some aphids, development of young may begin in the mother while she herself is still in her own mother's reproductive system!

9. Factors Affecting Embryonic Development

Temperature is probably the single most important environmental variable affecting embryonic development. For eggs of most species, there are upper and lower temperature limits, outside which development is greatly retarded or completely inhibited. Within these limits, however, an inverse but linear relationship exists between temperature and time required to complete development; that is, the total heat requirement (temperature above minimum required × duration of exposure to this temperature) is constant for a given species. This heat requirement is typically measured in degree-days. Outside these developmental limits, yet within the limits of viability, an egg may survive but does not develop. Under these conditions, it is said to be quiescent and in this state may survive for a considerable length of time. In a quiescent state, an egg is always ready to take advantage of favorable conditions, even if only temporary, to continue its development. However, quiescence is a relatively sensitive developmental state; that is, outside certain temperature limits, an egg will be killed. For many species, therefore, which exist in habitats exposed to climatic extremes, especially of temperature but also of precipitation, a more resistant state of developmental arrest, diapause, has evolved to permit their survival. Diapause is discussed at greater length in Chapter 22 (Section 3.2.3), though in the present context it is worth noting that diapause may occur at different stages of embryonic development and with variable strength in different species. In all instances, however, it is characterized by a cessation of morphogenesis and a considerable lowering of the metabolic rate. Also, the water content of an egg is often low at this time. In *Bombyx*, diapause, which begins in overwintering eggs almost as soon as they are laid, is extremely strong; that is, even when eggs are experimentally maintained at 15–20°C from the time of laying they will not develop. Development begins only after they have been exposed to a temperature of about 0°C for several months. In eggs of the damselfly, *Lestes congener*, diapause is also strong but does not commence until after anatrepsis. Diapause in eggs of the grasshopper *Melanoplus differentialis* also occurs after anatrepsis but is weak. Should the temperature to which eggs are exposed be maintained at summer levels (around 25°C), some of the eggs will develop directly, though more slowly than those which undergo chilling. In eggs of some insects, for example, certain mosquitoes (*Aedes* spp.) and the damselfly *Lestes disjunctus*, embryonic development is almost completed before diapause is initiated.

Water is another important requirement and in eggs of many species must be acquired from the external environment before embryonic development can begin. When it is available to an egg in insufficient quantity, the embryo becomes quiescent or remains in diapause (though this was not induced by the lack of moisture). Some species can obtain sufficient water from moisture in the air. For example, eggs of the beetle *Sitona*, when kept at 20°C and 100% relative humidity, hatch in 10.5 days; at the same temperature, but only 62% relative humidity, development takes twice as long. In other species contact of the egg with liquid water is necessary for continued development. Such is the case in the damselfly eggs mentioned above which pass the winter in snow-covered, dried-out *Scirpus* stems and do not continue their development until the stems become waterlogged following the spring thaw.

10. Hatching

To escape from the egg, a larva must break through the various membranes which surround it. These include the chorion, vitelline membrane, and, in eggs of some species, serosal cuticle. Further, in many exopterygotes and some endopterygotes a newly hatched larva is surrounded by embryonic cuticle which also must be shed before the insect is truly free.

The general mechanism of hatching is as follows. An insect first swallows amniotic fluid,* followed usually by air, which diffuses into the amniotic cavity. The abdomen is then contracted to force hemolymph into the head and tho-

---

*This fluid is, of course, no longer in the amniotic cavity whose membranes were destroyed during dorsal closure.

rax, which enlarge and cause the egg membranes to rupture. To facilitate rupture the chorion may have predetermined lines of weakness which run longitudinally or transversely, the latter separating an anterior egg cap from the more posterior portion of the egg. In many species, egg bursters, in the form of hard cuticular spines or plates, or thin eversible bladders, may develop on the head, thorax, or abdomen. In Acrididae and those Hemiptera in which pleuropodia develop, it is believed that these glands secrete enzymes which dissolve the serosal cuticle as an aid to hatching. Larvae of Lepidoptera simply eat their way out of the egg.

Where an embryonic cuticle is present, this may be shed concurrently with the other enclosing membranes, or may ensheath a larva until it has completely escaped from the egg, as in Odonata, Orthoptera, and some Hemiptera. In these insects the embryonic cuticle underwent apolysis some time prior to hatching, and the first instar larval cuticle is already formed beneath. Thus, the insect hatches as a pharate first instar larva. In Orthoptera and endophytic Odonata, the embryonic cuticle presumably protects a larva until it reaches the surface of the substrate in which the egg was laid. In other species, however, its function is unclear. It is shed a few minutes after a larva has reached the surface, a process called the intermediate molt.

11. Summary

Most insects are oviparous and therefore lay eggs that contain much yolk. However, there is an evolutionary trend toward reduction of the yolk : cytoplasm ratio to permit more rapid embryonic development.

Cleavage begins at a predetermined site, the cleavage center and early divisions are synchronous. Most energids migrate through the yolk to the periplasm and form the blastoderm; some remain in the yolk as vitellophages which supply nutrients to the embryo. Energids that move posteriorly interact with the activation center to stimulate differentiation of part of the blastoderm into the embryonic primordium. Differentiation begins at a predetermined site, the differentiation center, located in the region of the future prothorax. The embryonic primordium of exopterygotes is usually small and grows by aggregation and proliferation of cells, whereas that of most endopterygotes is large to permit rapid tissue differentiation and embryonic growth. Elongation and differentiation of the primordium (now known as the germ band) occur, and externally segmentation and appendage formation are obvious; internally somites form and mesoderm differentiates. Simultaneously, in embryos of most species the amnion and serosa develop from proliferating extraembryonic cells at the margins of the germ band. Anatrepsis, movement of the germ band into the yolk core, occurs in eggs of most exopterygotes at this time.

At the end of germ band formation the amnion and serosa fuse, then break in the head region, and the combination rolls back dorsally over the yolk which is left covered by only the amnion (provisional dorsal closure). Katatrepsis now takes place in eggs with immersed germ bands, so that the embryo is returned to the yolk surface with its head facing the anterior pole of the egg. Embryonic ectoderm now extends around the yolk to replace the amnion (definitive dorsal closure).

Paired segmental appendages develop from evaginations of the embryonic ectoderm but may become reduced or disappear. Shortly after definitive dorsal closure, the embryonic ectoderm differentiates into epidermis and secretes a cuticle. Invaginations of the ectoderm give rise to the endoskeleton, tracheal system, salivary glands, corpora allata, molt glands, exocrine glands, and, in females, the vagina and spermatheca, in males, the ejaculatory duct and ectadenes. The foregut and hindgut develop from ectodermal invaginations at the anterior and posterior ends, respectively, of the gastral groove. The midgut is formed from anterior and posterior midgut rudiments that grow toward each other and on meeting extend dorsolaterally to enclose the yolk. The central nervous system arises from neuroblasts in the midventral line. The stomatogastric nervous system develops from evaginations in the roof of the stomodeum.

The heart, aorta, septa, muscle, fat body, paired genital ducts, and mesadenes are mesodermal derivatives. Gonads arise from primordial germ cells that become enclosed in mesoderm.

Parthenogenesis, the development of unfertilized eggs, may be ameiotic (no meiosis in oocyte nucleus) or meiotic (meiosis is followed by nuclear fusion), both of which result in diploid (female) offspring, or haploid (meiosis is not followed by nuclear fusion) from which males arise.

Polyembryony, the formation of many embryos in a single, small, yolkless egg, is restricted to a few parasitic or viviparous Hymenoptera and Strepsiptera.

Viviparity occurs in several forms. Ovoviviparity is retention of yolky eggs in the genital tract. In true viviparity developing offspring receive their nourishment directly from the mother. In pseudoplacental viviparity the follicle cells and embryonic membranes become closely apposed, and nourishment appears to be derived largely from the degeneration of follicle cells and from trophocytes. In adenotrophic viviparity, eggs are yolky, but larvae are retained in the uterus and feed on secretions of the accessory glands. Hemocoelic viviparity is where embryos receive nutrients directly from the hemolymph.

Paedogenesis is precocious sexual maturation of juvenile stages and is normally associated with parthenogenesis and viviparity.

Within species-specific limits the rate of embryonic development is inversely related to temperature. Outside these limits, an embryo may survive but not develop; that is, it is quiescent. Survival of an embryo at extreme temperatures may be achieved through diapause. Eggs of many species must take up water from the environment before embryonic development can begin.

At hatching, hemolymph is forced into the head and thorax as a result of abdominal muscle contraction. As the anterior end of the embryo increases in volume, the chorion is split. Hatching may be facilitated by lines of weakness in the chorion, by egg bursters or eversible bladders on the head or thorax, or by secretion of pleuropodial enzymes that dissolve the serosal cuticle.

12. Literature

General reviews of insect embryology are given by Johannsen and Butt (1941), Anderson (1972a,b, 1973), and Counce (1973). Works dealing with specific aspects of insect development include those by Hagan (1951) [embryology of viviparous insects], Anderson (1966) [embryology of Diptera], Jura (1972) [apterygote development], Ivanova-Kasas (1972) [polyembryony], White (1973) [parthenogenesis and sex determination], and Matsuda (1976) [embryogenesis of abdomen, gonads, and germ cells]. Several chapters of the symposium proceedings edited by Lawrence (1976) deal with experimental embryogenesis. For readers who may wish to trace the embryology of a species as an aid to understanding the chronology of embryogenesis, the series of papers by Rempel, Church, and co-workers (see Rempel *et al.*, 1977, for earlier references) on the beetle, *Lytta viridana*, will be most useful.

Anderson, D. T., 1966, The comparative embryology of the Diptera, *Annu. Rev. Entomol.* **11**:23–46.

Anderson, D. T., 1972a, The development of hemimetabolous insects, in: *Developmental Systems: Insects*, Vol. 1 (S. J. Counce and C. H. Waddington, eds.), Academic Press, New York.

Anderson, D. T., 1972b, The development of holometabolous insects, in: *Developmental Systems: Insects*, Vol. 1 (S. J. Counce and C. H. Waddington, eds.), Academic Press, New York.

Anderson, D. T., 1973, *Embryology and Phylogeny in Annelids and Arthropods*, Pergamon Press, Oxford.

Bodenstein, D., 1953, Embryonic development, in: *Insect Physiology* (K. D. Roeder, ed.), Wiley, New York.

Counce, S. J., 1973, The causal analysis of insect embryogenesis, in: *Developmental Systems: Insects*, Vol. 2 (S. J. Counce and C. H. Waddington, eds.), Academic Press, New York.

Hagan, H. R., 1951, *Embryology of the Viviparous Insects*, Ronald Press, New York.

Ivanova-Kasas, O. M., 1972, Polyembryony in insects, in: *Developmental Systems: Insects*, Vol. 1 (S. J. Counce and C. H. Waddington, eds.), Academic Press, New York.

Johannsen, O. A., and Butt, F. H., 1941, *Embryology of Insects and Myriapods*, McGraw-Hill, New York.

Jura, C., 1972, Development of apterygote insects, in: *Developmental Systems: Insects*, Vol. 1 (S. J. Counce and C. H. Waddington, eds.), Academic Press, New York.

Lawrence, P. A. (ed.), 1976, Insect development, *Symp. R. Entomol. Soc.* **8**:240 pp.

Matsuda, R., 1976, *Morphology and Evolution of the Insect Abdomen*, Pergamon Press, Oxford.

Rempel, J. G., Heming, B. S., and Church, N. S., 1977, The embryology of *Lytta viridana* Le Conte (Coleoptera; Meloidae). IX. The central nervous system, stomatogastric nervous system, and endocrine system, *Quaest. Entomol.* **13**:5–23.

Suomalainen, E., 1962, Significance of parthenogenesis in the evolution of insects, *Annu. Rev. Entomol.* **7**:349–366.

White, M. J. D., 1973, *Animal Cytology and Evolution*, 3rd ed., Cambridge University Press, Cambridge.

21

Postembryonic Development

1. Introduction

During their postembryonic growth period insects pass through a series of stages (instars) until they become adult, the time interval (stadium) occupied by each instar being terminated by a molt. Apterygotes continue to grow and molt as adults, periods of growth alternating with periods of reproductive activity. In these insects structural differences between juvenile and adult instars are slight, and their method of development is thus described as ametabolous. Among the Pterygota, which with rare exceptions do not molt in the adult stage, two forms of development can be distinguished. In almost all exopterygotes the later juvenile instars broadly resemble the adult, except for their lack of wings and incompletely formed genitalia. Such insects, in which there is some degree of change in the molt from juvenile to adult, are said to undergo partial (incomplete) metamorphosis, and their development is described as hemimetabolous. Endopterygotes and a few exopterygotes have larvae whose form and habits, by and large, are very different from those of the adults. As a result they undergo striking changes (complete metamorphosis), spread over two molts, in the formation of the adult (holometabolous development). The final juvenile instar has become specialized to facilitate these changes and is known as the pupa [see also Chapter 2, Section 3.3).

In insect evolution increasing functional separation has occurred between the larval phase, which is concerned with growth and accumulation of reserves, and the adult stage, whose functions are reproduction and dispersal. Associated with this trend is a tendency for the greater part of an insect's life to be spent in the juvenile phase, which is in contrast to the situation in many other animals. Thus, in apterygotes, the adult stage may be considerably longer than the juvenile stage. Furthermore, feeding (in the adult) serves to provide raw materials both for reproduction and for growth. In exopterygotes and primitive endopterygotes adults may live for a reasonable period, but this is not usually as long as the larval phase. Feeding in the adult stage is primarily associated with reproductive requirements, though in some insects it provides nutrients for an initial, short "somatic growth phase' in which the flight muscles, gut, and cuticle become fully developed. Many endopterygotes live for a relatively short time as adults and may feed little or not at all because sufficient reserves have been acquired during larval life to satisfy the needs of reproduction.

2. Growth

2.1. Physical Aspects

Growth in insects and other arthropods differs from that of, say, mammals in various respects. In insects growth is almost entirely restricted to the larval instars, though in some species there is a short period of somatic growth in newly eclosed adults when additional cuticle may be deposited, and growth of flight muscles and the alimentary canal may occur. As a consequence the length of the juvenile stage is considerably longer than that of the adult. An extreme example of this is seen in some mayfly species whose aquatic juvenile stage may require 2 or 3 years for completion, yet give rise to an adult which lives for only a few hours or days. Growth in many animals is discontinuous or cyclic; that is, periods of active growth are separated by periods when little or no growth occurs. Nowhere is discontinuous growth better exemplified than in arthropods, which must periodically molt their generally inextensible cuticle in order to significantly increase their size (volume). It should be appreciated, however, that, though increases in volume are discontinuous, increases in weight are not (Figure 21.1). As an insect feeds during each stadium, reserves are laid down in the fat body, whose weight and volume increase. This increase in volume may be compensated for by a decrease in the volume occupied by the tracheal system or by extension of the abdomen.

For many insects grown under standard conditions the amount of growth which occurs is predictable from one instar to the next; that is, it obeys certain "growth laws." Dyar's law, based on measurements of the change in width of the head capsule which occurs at each molt, states that growth follows a geometric progression; that is, the proportionate increase in size for a given structure is constant from one instar to the next. Mathematically expressed, the law states x/y = constant (value usually 1.2–1.4), where x = size in a given instar, and y = size in previous instar (Figure 21.1). Thus, when the size of a structure is plotted logarithmically against instar number, a straight line is obtained, whose gradient is constant for a given species (Figure 21.2). In those insects where it applies Dyar's law can be used to determine how many instars there are in the life history. However, so many factors affect growth rates and the frequency of ecdysis that the law is frequently inapplicable. In any event, the law requires that the interval between molts remains constant, but this is rarely the case.

An extension of Dyar's law, Przibram's rule, states that the progression ratio (x/y) for all parameters should be the same, and have a value of 1.26 (i.e., $\sqrt{3/2}$). Przibram proposed this rule following his observations on the mantid *Sphrodomantis* whose linear measurements for a variety of structures closely approximate this value and whose weight doubles at each instar. Przibram's proposal was that during each instar the mass of cells doubles (each cell divides once and grows to its original size), thereby doubling the volume of the

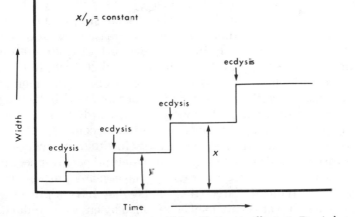

FIGURE 21.1. Change in head width with time to illustrate Dyar's law.

insect. Each linear dimension would therefore increase by the value of $\sqrt{3}/2$. Przibram showed as support for his rule, that the density of epidermal cells beneath the cuticle remained constant from one instar to the next. However, it is now known, for *Rhodnius* at least, that it is the increase in area of cuticle which determines the increase in number of epidermal cells. Furthermore, in many insects growth is not simply a matter of cell division; a good deal of histolysis and rebuilding of tissues occurs in each molt cycle. In some Diptera cell division does not occur in the larval instars; rather, the cells simply in-

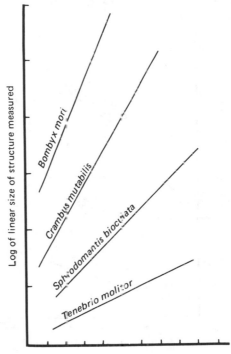

FIGURE 21.2. Linear size of structure plotted logarithmically against instar number in various species. [After V. B. Wigglesworth, 1965. *The Principles of Insect Physiology*, 6th ed., Methuen and Co. By permission of the author.]

crease their size, though the chromosomes divide, giving rise to a polyploid condition. It is now appreciated that most insects simply do not conform to Przibram's rule; their linear dimensions may increase by a ratio very different from 1.26, and their weight may increase several fold from one instar to the next. The greatest deficiency in Przibram's rule is that it requires growth in an insect to be isogonic (harmonic); that is, all parts of the organism grow at the same rate. In fact, growth in insects, as in other animals, is heterogonic (disharmonic, allometric), each part having its own growth rate (Figure 21.3). Thus, during growth changes in body proportions occur.

The laws outlined above also do not apply in situations where the number of instars is variable. This variability may be a natural occurrence, especially in primitive insects such as mayflies which have many instars. In addition, females which are typically larger than males may have a greater number of instars than males. Or variability may be induced by environmental conditions. For example, rearing insects at abnormally high temperature often increases the number of instars, as does semistarvation. In contrast, in some caterpillars crowding leads to a decrease in the number of molts.

2.2. Biochemical Changes during Growth

Like the physical changes noted above, biochemical changes which occur during postembryonic development may also be described as heterogonic. That is, the relative proportions of the various biochemical components change as growth takes place. These changes are especially noticeable in endopterygotes during the final larval and pupal stages. At hatching, the fat content of a larva is

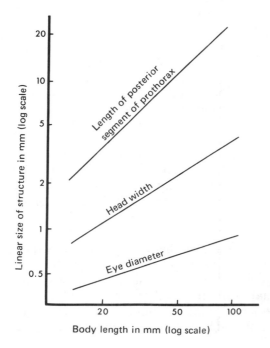

FIGURE 21.3. Heterogonic growth in *Carausius* (Phasmida). [After V. B. Wigglesworth, 1965, *The Principles of Insect Physiology*, 6th ed., Methuen and Co. By permission of the author.]

typically low (less than 1% in the caterpillar *Malacosoma*, for example) and remains at about this level until the final larval stadium when fat is synthesized and stored in large quantity, reaching about 30% of the dry body weight. Though fat is the typical reserve substance in most insects, members of some species store glycogen. Again, this usually occurs in small amounts in newly hatched insects, but its proportion increases steadily through larval development, and at pupation glycogen may be a significant component of the dry weight (one-third in the honeybee). Like fat, glycogen is stored in the fat body.

In contrast, the proportions of water, protein, and nucleic acids generally decline during larval development. However, this is often not the situation in larvae which require large amounts of protein for specific purposes, for example, spinning a cocoon. In *Bombyx mori*, for example, the hemolymph protein concentration increases sixfold in late larval development, and about 50% of the total protein content of a mature larva is used in cocoon formation. The great increase in concentration of hemolymph protein often can be accounted for almost entirely by synthesis, in the fat body, of a few specific proteins. In the fly *Calliphora stygia*, for example, the protein "calliphorin" makes up 75% (about 7 mg) of the hemolymph protein by the time a mature larva stops feeding. The calliphorin is used in the pupa as a major source of nitrogen (in the form of amino acids) for formation of adult tissues and as a source of the energy required in biosynthesis. Thus, at eclosion, the hemolymph calliphorin content has fallen to 0.03 mg, and, one week after emergence, the protein has entirely disappeared.

During metamorphosis some of the above trends may be reversed. The proportions of fat and/or glycogen decline as these molecules are utilized in energy production. In *Calliphora* the fat content decreases from 7 to 3% of the dry weight through the pupal period. In the honeybee, which mainly uses glycogen as an energy source, the glycogen content drops to less than 10% of its initial value as metamorphosis proceeds. For most insects there is little change in the net protein content during pupation, though major qualitative changes occur as adult tissues develop. In members of a few species a significant decline in total protein content occurs during metamorphosis as protein is used as an energy source. The moth *Celerio*, for example, obtains only 20% of its energy requirements in metamorphosis from fat, the remaining 80% coming largely from protein.

Superimposed on the overall biochemical changes from hatching to adulthood are changes which occur in each stadium, related to the cyclic nature of growth and molting. Factors to be considered include the phasic pattern of feeding activity throughout the stadium, synthesis of new and degradation of old cuticle, and net production of new tissues (though some histolysis also occurs in each instar).

Measurement of oxygen consumption shows that it follows a U-shaped curve through each stadium with maximum values being obtained at the time of molting. The maxima are correlated with the great increase in metabolic activity at this time, associated especially with the synthesis of new cuticle and formation of new tissues. In *Locusta* larvae there are significant decreases in the carbohydrate and lipid contents of the fat body and hemolymph at ecdysis, probably correlated with the use of these substrates to supply energy (Hill and Goldsworthy, 1968). Conversely, as feeding restarts after a molt, these materials are again accumulated.

Changes in the amount of protein in the fat body and hemolymph of *Locusta* are also cyclical, with maximum values occurring in the second half of each stadium (Hill and Goldsworthy, 1968). The early increase in protein content is due to renewed feeding activity after the molt. Feeding activity reaches a peak in the middle of the stadium, providing materials for growth of muscles (and presumably other tissues, though these were not studied by Hill and Goldsworthy) and for synthesis of cuticle. Excess material is stored in the fat body and hemolymph. In the second half of the stadium feeding activity declines, and this is followed ·by a decrease in the level of protein in the hemolymph and fat body. Hill and Goldsworthy (1968) suggest that the latter probably reflects the utilization of protein in the synthesis of new cuticle. It should be remembered, however, that digested protein from the old cuticle could account for most (about 80% in *Locusta*) of the protein content of the new cuticle.

3. Forms of Development

Through insect evolution there has been a trend toward increasing functional and structural divergence between juvenile and adult stages. Juvenile insects have become more concerned with feeding and growth, whereas adults form the reproductive and dispersal phase. This specialization of different stages in the life history became possible with the introduction into the life history of a pupal instar, though the latter's original function was probably related specifically to evagination of the wings and development of the wing musculature (see Chapter 2, Section 3.3).

In modern insects three basic forms of postembryonic development can be recognized, described as ametabolous, hemimetabolous, and holometabolous, according to the extent of metamorphosis from juvenile to adult (Figure 21.4).

3.1. Ametabolous Development

In Thysanura (and other primitive hexapods), which as adults remain wingless, the degree of change from juvenile to adult form is slight and is manifest primarily in increased body size and development of functional genitalia. Juvenile and adult apterygotes inhabit the same ecological niche, and the insects continue to grow and molt after reaching sexual maturity. The number of molts through which an insect passes is very high and variable. For example, in the firebrat, *Thermobia domestica*, between 45 and 60 molts have been recorded.

3.2. Hemimetabolous Development

Exopterygotes usually molt a fixed number of times, but, with the exception of Ephemeroptera, which pass through a winged subimago stage, never as

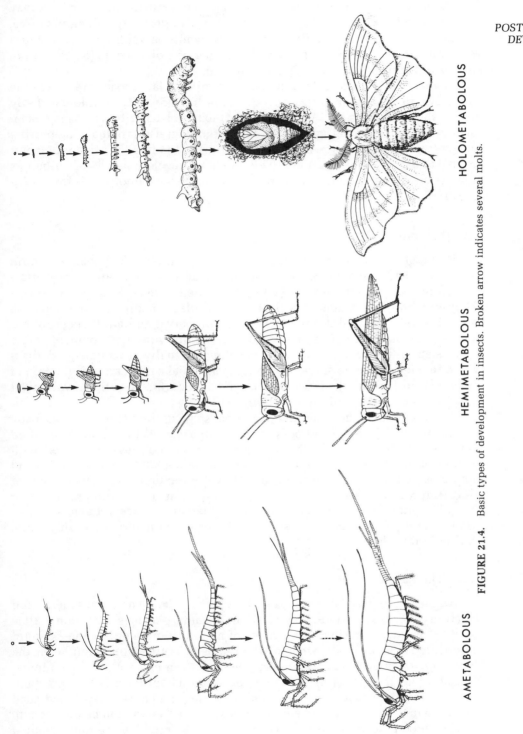

HOLOMETABOLOUS

HEMIMETABOLOUS

AMETABOLOUS

FIGURE 21.4. Basic types of development in insects. Broken arrow indicates several molts.

adults. In species where the female is much larger than the male, she may undergo an additional larval molt. The number of molts is typically four or five, though in some Odonata and Ephemeroptera whose larval life may last 2 or 3 years a much greater and more variable number of molts occurs (e.g., 10–15 in species of Odonata, 40–45 in Ephemeroptera).

In almost all exopterygotes the later juvenile instars broadly resemble the adult, except that their wings and genitalia are not fully developed. Early instars show no trace of wings, but, later, external wing buds arise as sclerotized, nonarticulated evaginations of the tergopleural area of the wing-bearing segments. Wings develop within the buds during the final larval stadium and are expanded after the last molt. This mode of development is described as hemimetabolous and includes a partial (incomplete) metamorphosis from larva to adult.

3.3. Holometabolous Development

Holometabolous development, in which there is a marked change of form from larva to adult (complete metamorphosis), occurs in endopterygotes and a few exopterygotes, for example, whiteflies (Aleurodidae: Hemiptera), thrips (Thysanoptera), and male scale insects (Coccidae: Hemiptera). Perhaps the most obvious structural difference between the larval and adult stages of endopterygotes is the absence of any external sign of wing development in the larval stages. The wing rudiments develop internally from imaginal discs which in most larvae lie at the base of the peripodial cavity, an invagination of the epidermis beneath the larval cuticle, and are evaginated at the larval–pupal molt (see Section 4.2 and Figure 21.10).

As noted above, the evolution of a pupal stage in the life history has made holometabolous development possible. The pupa is probably a highly modified final juvenile instar which, through evolution, became less concerned with feeding and building up reserves (this function being left to earlier instars) and more specialized for the breakdown of larval structures and construction of adult features. In other words, the pupa has become a nonfeeding stage; it is generally immobile as a result of histolysis of larval muscles; it broadly resembles the adult and thereby serves as a mold for the formation of adult tissues, especially muscles.

3.3.1. The Larval Stage

Among endopterygotes the extent to which the larval and adult habits and structure differ [and therefore the extent of metamorphosis (see Section 4.2)] is variable. Broadly speaking, in members of more primitive orders the extent of these differences is small, whereas the opposite is true, for example, in the Hymenoptera and Diptera. Endopterygote larvae can be arranged in a number of basic types (Figure 21.5). The most primitive larval form is the oligopod. Larvae of this type have three pairs of thoracic legs and a well-developed head with chewing mouthparts and simple eyes. Oligopod larvae can be further subdivided into (1) scarabaeiform larvae (Figure 21.5A), which are round-bodied and have short legs and a weakly sclerotized thorax and abdomen, features

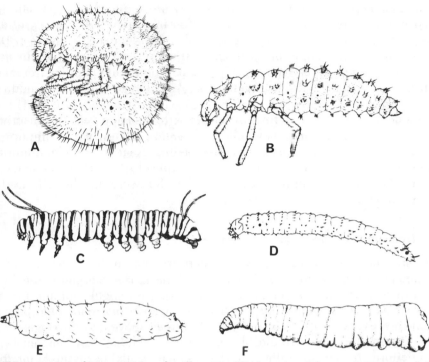

FIGURE 21.5. Larval types. (A) Scarabaeiform (*Popillia japonica*, Coleoptera), (B) campodeiform (*Hippodamia convergens*, Coleoptera), (C) eruciform (*Danaus plexippus*, Lepidoptera), (D) eucephalous (*Bibio* sp., Diptera), (E) hemicephalous (*Tanyptera frontalis*, Diptera), and (F) acephalous (*Musca domestica*, Diptera). [A–E, from A. Peterson, 1951, *Larvae of Insects*. By permission of Mrs. Helen Peterson. F, from V. B. Wigglesworth, 1959, Metamorphosis, polymorphism, differentiation. Copyright © February 1959 by Scientific American, Inc. All rights reserved.]

associated with the habit of burrowing into the substrate, and (2) campodeiform larvae (Figure 21.5B), which are active, predaceous surface-dwellers with a dorsoventrally flattened body, long legs, strongly sclerotized thorax and abdomen, and prognathous mouthparts. Scarabaeiform larvae are typical of the Scarabaeidae and other beetle families; campodeiform larvae occur in Neuroptera, Coleoptera–Adephaga, and Trichoptera.

Polypod (eruciform) larvae (Figure 21.5C) have, in addition to thoracic legs, a variable number of abdominal prolegs. The larvae are generally phytophagous and relatively inactive, remaining close to or on their food source. The thorax and abdomen are weakly sclerotized in comparison with the head, which has well-developed chewing mouthparts. Eruciform larvae are typical of Lepidoptera, Mecoptera, and some Hymenoptera [sawflies (Tenthredinidae)].

Apodous larvae, which lack all trunk appendages, occur in various forms in many endopterygote orders. The variability of form concerns the extent to which a distinct head capsule is developed. In eucephalous larvae (Figure 21.5D), characteristic of some Coleoptera (Buprestidae and Cerambycidae), aculeate Hymenoptera and more primitive Diptera (suborder Nematocera), the head is well sclerotized and bears normal appendages. The head and its appendages of hemicephalous larvae (Figure 21.5E) are reduced and partially retracted into the thorax. This condition is seen in crane fly larvae (Tipulidae: Nematocera) and in the larvae of Diptera, suborder Brachycera. Larvae of Diptera, suborder Cyclorrhapha, are acephalous (Figure 21.5F); no sign of the head and its appendages can be seen apart from a pair of minute papillae (remnants of the antennae) and a pair of sclerotized hooks believed to be much modified maxillae.

Frequently a larva in the final instar ceases to feed and becomes inactive a few days before the larval–pupal molt. Such a stage is known as a prepupa. In some species, the entire instar is a nonfeeding stage in which important changes related to pupation occur. For example, in the prepupal instar of sawflies, the salivary glands become modified for secreting the silk used in cocoon formation.

3.3.2. Heteromorphosis

In most endopterygotes the larval instars are more or less alike. However, in some species of Neuroptera, Coleoptera, Diptera, Hymenoptera, and in all Strepsiptera, a larva undergoes characteristic changes in habit and morphology as it grows, a phenomenon known as heteromorphosis (hypermetamorphosis). In such species several of the larval types described above may develop successively (Figure 21.6). For example, blister beetles (Meloidae) hatch as free-living campodeiform larvae (planidia, triungulins) which actively search for food (grasshopper eggs and immature stages, or food reserves of bees or ants). At this stage the larvae can survive for periods of several weeks without food. Larvae which locate food soon molt to the second stage, a caterpillarlike (eruciform) larva. The insect then passes through two or more additional larval instars which may remain essentially eruciform or may become scarabaeiform. Some species overwinter in a modified larval form known as the pseudopupa or

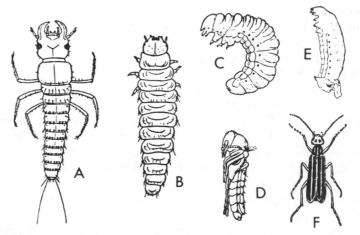

FIGURE 21.6. Heteromorphosis in *Epicauta* (Coleoptera). (A) Triungulin, (B) caraboid second instar, (C) final form of second instar, (D) coarctate larva, (E) pupa, and (F) adult. [From A. D. Imms, 1957, *A General Textbook of Entomology*, 9th ed. (revised by O. W. Richards and R. G. Davies). By permission of Chapman and Hall Ltd.]

coarctate larva, so-called because the larva remains within the cuticle of the previous instar. The pseudopupal stage is followed the next spring by a further larval feeding stage, which then molts into a pupa

3.3.3. The Pupal Stage

The pupa is a nonfeeding, generally quiescent instar which serves as a mold in which adult features can be formed. For many species it is also the stage in which an insect survives adverse conditions by means of diapause (see Chapter 22, Section 3.2.3). The terms "pupa" and "pupal stage" are commonly used to describe the entire preimaginal instar. This is, strictly speaking, incorrect because for a variable period prior to eclosion (emergence of the adult) the insect is in fact a "pharate adult," that is, an adult enclosed within the pupal cuticle. The insect thus becomes an adult immediately after apolysis of the pupal cuticle and formation of the adult epicuticle. The distinction between the true pupal stage and the pharate adult condition becomes important in consideration of so-called "pupal movements," including locomotion and mandibular chewing movements (used in escaping from the protective cocoon or cell in which metamorphosis took place). In most instances these movements result from the activity of muscles attached to the adult apodemes which fit snugly around the remains of the pupal apodemes (Figure 21.7).

Pupae may be arranged in the following categories: according to whether or not the mandibles are functional and whether or not the remaining appendages are sealed closely against the body (Figure 21.3). Decticous pupae, found in more primitive endopterygotes [Neuroptera, Mecoptera, Trichoptera, and Lepidoptera (suborders Zeugloptera and Dacnonypha)], have well-developed, articulated mandibles (moved by means of the pharate adult's muscles) with which an insect can cut its way out of the cocoon or cell. Decticous pupae are always exarate; that is, the appendages are not sealed against the body so that

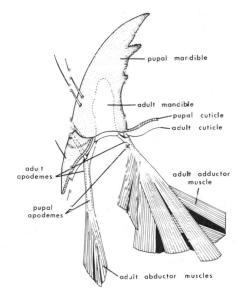

FIGURE 21.7. Section through mandible of a decticous pupa to show adult apodemes around remains of pupal apodemes. [After H. E. Hinton, 1946, A new classification of insect pupae, Proc. Zool. Soc. London 116:282–328. By permission of the Zoological Society of London.]

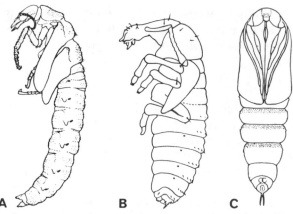

FIGURE 21.8. Pupal types. (A) Decticous (*Chrysopa* sp., Neuroptera), (B) exarate adecticous (*Brachyrhinus sulcatus*, Coleoptera), and (C) obtect adecticous (*Heliothis armigera*, Lepidoptera). [From A. Peterson, 1951, *Larvae of Insects*. By permission of Mrs. Helen Peterson.]

they may be used in locomotion. Some neuropteran pupae, for example, can crawl and some pupae of Trichoptera swim to the water surface prior to eclosion. Adecticous pupae, whose mandibles are nonfunctional and often reduced, may be either exarate or obtect. In the latter condition the appendages are firmly sealed against the body and are usually well sclerotized. Adecticous exarate pupae are characteristic of Siphonaptera, brachycerous and cyclorrhaphous Diptera, most Coleoptera and Hymenoptera, and Strepsiptera. In nematocerous Diptera, Lepidoptera (suborders Monotrysia and Ditrysia), and in a few Coleoptera and Hymenoptera, pupae are of the adecticous obtect type.

Clearly, an immobile pupa is vulnerable to attack by predators or parasites and to severe changes in climatic conditions, particularly as the pupal stadium may last for a considerable time. To obtain protection against such adversities the pupa of many species is enclosed within a cocoon or subterranean cell constructed by the previous larval instar. The cocoon may comprise various kinds of extraneous material, for example, soil particles, small stones, leaves or other vegetation, or may be made solely of silk. In cyclorrhaphous Diptera a special protective capsule, the puparium, has evolved. It is formed from the cuticle of the last larval instar, which becomes thickened and tanned. At the larval–pupal molt, the cuticle is not shed but remains as a rigid coat around the pupa. In some endopterygotes the pupa is not surrounded by a protective cocoon but obtains protection by being well sclerotized and/or by taking on the color of its surroundings. Many parasitic species remain within, and are thus protected by, the host's body in the pupal stage.

4. Histological Changes during Metamorphosis

Though we have distinguished, in the preceding discussion, between hemimetabolous development (where partial metamorphosis occurs in the molt from larva to adult) and holometabolous development (in which metamorphosis is striking and requires two molts, larval–pupal and pupal–imaginal, for completion) it is important to realize that the distinction is primarily useful in discussions of insect evolution; in a physiological sense the difference between partial and complete metamorphosis is a matter of degree rather than kind. Indeed, as is described in Section 6.1, the endocrine basis of growth, including molting and change of form, is common to all insects.

4.1. Exopterygote Metamorphosis

Among most exopterygotes the larval and adult forms of a species occupy the same habitat, eat the same kinds of food (though specific preferences may change with age) and are subject to the same environmental conditions. Accordingly, most organ systems of a juvenile exopterygote are smaller and/or less well-developed versions of those found in an adult and simply grow progressively during larval life to accommodate changing needs. Even larval Odonata and Ephemeroptera which are aquatic and possess transient adaptive features, for example, gills or caudal lamellae, broadly resemble the adult stage. The system which undergoes the most obvious change at the final molt is the flight mechanism. In the last larval instar wings develop within the wing buds as much folded sheets of integument, and, concurrently, the articulating sclerites differentiate. Direct flight muscle rudiments are present in larval instars and are attached to the integument at points corresponding to the future locations of the sclerites. Some of these (bifunctional muscles) may be important in leg movements during larval life (see Chapter 14, Section 3.3.1). Like the direct flight muscles, the indirect flight muscles grow progressively through larval life but remain unstriated and nonfunctional until the adult stage.

4.2. Endopterygote Metamorphosis

In more primitive endopterygotes such as Neuroptera and Coleoptera, as in exopterygotes, a good deal of progressive development of organ systems occurs during larval life so that metamorphosis, relatively speaking, is slight and concerns, again, mainly the flight mechanism. At the opposite extreme, seen in many Diptera and Hymenoptera, most larval tissues are histolyzed, with adult tissues being formed anew, often from specific groups of undifferentiated cells, the imaginal discs or buds. The imaginal discs occur as thickened regions of epidermis whose cells remain embryonic; that is, in the larval instars their differentiation is suppressed, probably by the hormonal milieu which exists in an insect at this time. At metamorphosis striking changes occur in the concentration of certain hormones, as a result of which the cells can multiply and differentiate into adult tissues. Experiments in which cells have been selectively destroyed by X irradiation have shown that formation of imaginal discs occurs very early in embryogenesis and at specific sites. Furthermore, each imaginal disc differentiates in a predetermined manner. During larval development, the discs grow exponentially in relation to general body growth and, typically, come to lie within an invagination, the peripodial cavity, beneath the cuticle (Figure 21.10). With the evolution of imaginal discs the way was open for the development of a larva whose form is highly different from that of the adult, and capable of existing in a different habitat from that of the adult, thus avoiding competition for food and space.

To discuss the histological changes which occur in endopterygote metamorphosis it is convenient to consider separately the various organ systems.

The cuticle of most adult endopterygotes is produced by epidermal cells carried over from the larval stage. However, in Hymenoptera–Apocrita and Diptera–Cyclorrhapha the larval epidermis is more or less completely histolyzed and replaced by cells derived from imaginal discs. Appendage formation is also variable. In lower endopterygotes formation of adult mouthparts, antennae, and legs begins early in the final larval stadium from larval epidermis. Certain predetermined areas of the epidermis thicken, then proliferate and differentiate so that, at pupation, the basic form of the adult appendages is evident. During the pupal stadium the final form of the adult appendages is expressed (Figure 21.9). In contrast, where the larval appendages are very different from those of the adult, or are absent, the adult structures develop from imaginal discs which undergo marked proliferation and differentiation in the

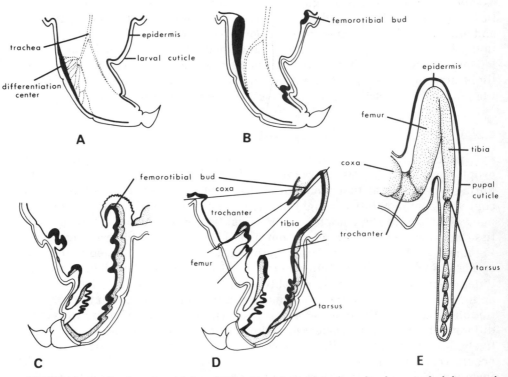

FIGURE 21.9. Sections through leg of *Pieris* (Lepidoptera) to show develoment of adult appendage. (A) Leg of last instar larva 3 hours after ecdysis, (B) same as (A) but 1 day after ecdysis, (C) same as (A) but 3 days after ecdysis, (D) leg at beginning of prepupal stage showing presumptive areas of adult leg, and (E) leg of pupa. [After Chang-Whan Kim, 1959, The differentiation centre inducing the development from larval to adult leg in *Pieris brassicae* (Lepidoptera), *J. Embryol. Exp. Morphol.* **7**:572–582. By permission of Cambridge University Press, New York.]

last larval instar and are evaginated from the peripodial cavity at the larval–pupal molt. Wings are formed in all endopterygotes from imaginal discs. In most species their early development is similar to the development of paired segmental appendages outlined above; that is, the wing rudiments form in a peripodial cavity and become everted at the larval–pupal molt (Figure 21.10). The forming wing bud in the peripodial cavity is initially a hollow, fingerlike structure, but this becomes flattened so that the central cavity is more or less obliterated, leaving only small lacunae (Figure 21.11A,B). A nerve and trachea which have been associated with the imaginal disc now grow along each lacuna occasionally branching in a predetermined pattern. At the larval–pupal molt, hemolymph pressure forces the sides of a wing bud apart so that there is sufficient space within it for development of an adult wing (Figure 21.11C,D). During the pupal stadium extensive proliferation of the epidermal cells within the wing bud occurs, as a consequence of which the epidermis becomes folded and closely apposed over most of the wing surface. The epidermal layers remain separate adjacent to the nerve and trachea, forming the definitive wing veins (Figure 21.11E–H).

The gut of endopterygotes typically changes its form markedly during

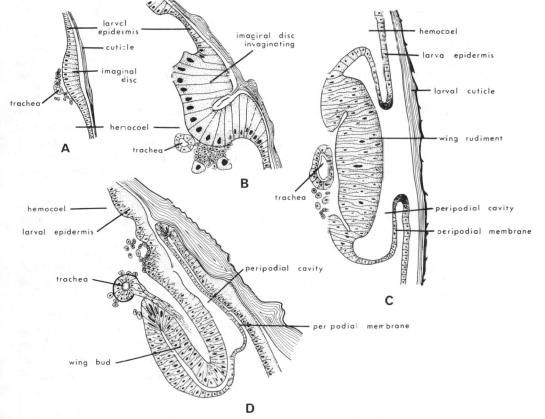

FIGURE 21.10. Sections through developing wing bud of first four larval instars of *Pieris* (Lepidoptera). [After J. H. Comstock, 1918, *The Wings of Insects*, Comstock, New York.]

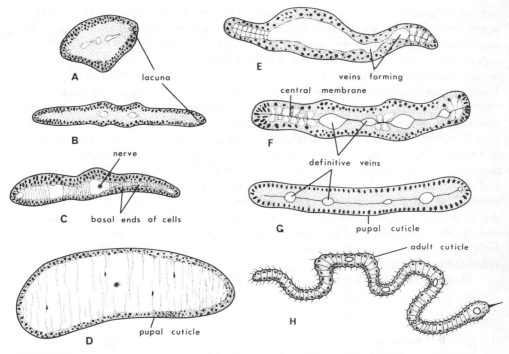

FIGURE 21.11. Transverse sections of developing wing of *Drosophila*. (A,B) Successive stages in pharate pupa, (C–G) stages in pupa, and (H) pharate adult. [After C. H. Waddington, 1941, The genetic control of wing development in *Drosophila*, *J. Genet.* **41**:75–139. By permission of Cambridge University Press, New York.]

metamorphosis. In Coleoptera the foregut and hindgut undergo relatively slight modification, this being achieved by the activity of larval cells. In higher endopterygotes these regions are partially or entirely renewed from imaginal discs located at the junctions of the foregut and midgut and midgut and hindgut, and adjacent to the mouth and anus. The larval midgut of all endopterygotes is fully replaced as a result of the activity of either regenerative cells from the larval midgut, or undifferentiated cells at the junction of the midgut and hindgut, or both. In either arrangement, the histolyzed larval cells eventually are surrounded by adult tissue.

Larval Malpighian tubules may be retained in some adult Diptera, but in other endopterygotes they are partially or completely replaced at metamorphosis from special cells located either along the length of each tubule or at the anterior end of the hindgut.

Generally, the larval tracheal system is carried over to the adult with little modification, except as required by the development of new tissues such as flight muscles and reproductive organs. However, in Diptera, considerable replacement of the larval system occurs from scattered cells in the larval tracheae.

The central nervous system of most endopterygotes becomes more concentrated at metamorphosis due to shortening of the interganglionic connectives and the forward migration of ganglia. Concurrently, some neurons enlarge and additional ones differentiate. Breakdown and rebuilding of the perineurium and neural lamella are necessitated by these changes.

During metamorphosis the muscular system undergoes considerable modification. The muscles of an adult insect arise in several ways: (1) from larval muscles which remain unchanged, (2) from partially histolyzed and reconstructed larval muscles, (3) from previously inactive imaginal nuclei within the larval muscles, (4) from myoblasts which previously adhered loosely to the surface of the larval muscles, or (5) from rudimentary, nonfunctional fibers present in the larva. The point at which histolysis of larval muscles begins is variable and somewhat dependent on their function. For many muscles histolysis begins in the final larval instar and continues in the pupa. Other muscles, however, have particular functions at the larval–pupal molt and beyond and do not histolyze until later.

The extent of histolysis of the fat body is quite variable and depends on the overall degree of metamorphosis in an insect. In more primitive endopterygotes where metamorphosis is relatively slight, much of the larval fat body is carried over unchanged into the adult stage. However, in cyclorrhaph Diptera, for example, the larval tissue is completely broken down, and the adult fat body is formed from mesenchyme cells associated with imaginal discs.

The heart and aorta are normally not histolyzed and continue to contract in the pupa.

5. Adult Emergence

For exopterygotes adult emergence (eclosion) consists solely of escape from the cuticle of the previous instar. Many endopterygotes must, in addition, force their way out of the cocoon or cell in which pupation occurred and, in some species, to the surface of the substrate in which they have been buried. Some aquatic species which pupate under water have special devices to enable the adult to reach the water surface.

Eclosion is accomplished in a manner similar to larval–larval molts. A pharate adult swallows air to increase its body volume and, by contraction of abdominal muscles, forces hemolymph anteriorly. As the hemolymph pressure increases the pupal cuticle splits along an ecdysial line on the thorax and/or head. In obtect pupae the pupal mouth is sealed over, but the adult swallows air which enters the pupal case via the tracheal system. Some adult spiracles remain in contact with those of the pupa, whereas others become separated so that a channel is open along which air can move into the pupal case.

Among more primitive endopterygotes an insect escapes from its cocoon or cell as a pharate adult using the mandibles of the decticous pupa to force an opening in the wall. Pharate adults of some species also have backwardly facing spines on the pupal cuticle, which enable them to wriggle out of the cell and through the substrate. Many primitive Lepidoptera and Diptera, whose pupae are adecticous, also escape as pharate adults, frequently making use of special spines (cocoon cutters) on the pupal cuticle. In higher Lepidoptera, adults may eclose while still in the cocoon. In such species the cocoon may possess an "escape hatch" or part of it may be softened by special salivary secretions. In cyclorrhaph Diptera an eversible membranous sac on the head, the ptilinum, can be expanded by hemolymph pressure. This enables an adult to push off the tip of the puparium and tunnel to the surface of the substrate in which it has been buried. Adult Coleoptera, Hymenoptera, and Siphonaptera eclose while in the cocoon or cell, then use their mandibles or special cocoon cutters to cut their way out. In some species this is the sole function of the mandibles, which, like cocoon cutters, are shed after eclosion.

6. Control of Development

Despite the apparently wide differences in the pattern of development seen in Insecta, the physiological system which regulates growth, molting, and metamorphosis is common to all members of the class, namely, the endocrine system. Variations in the hormonal milieu (the relative levels of different hormones) in an insect's body determine the nature and extent of tissue differentiation which is expressed at the next molt. In other words, it is the hormonal balance which determines, in a holometabolous insect, for example, whether the next molt is larval–larval, larval–pupal, or pupal–adult. Hormones also coordinate the sequence of events in the growth and molting cycle and in some species ensure that an adult ecloses under suitable environmental conditions. The hormones act by regulating genetic activity. In a particular hormonal milieu, the genes which are active may be responsible for expression of larval characters; under other hormonal conditions genes for pupal or imaginal features are activated.

Many environmental factors can modify developmental patterns. Some of these factors, for example, temperature, may act directly to affect development; most factors, however, exert their effect indirectly via the endocrine system.

6.1. Endocrine Regulation of Development

Postembryonic development is controlled by three endocrine centers, the brain–corpora cardiaca complex, corpora allata, and molt glands (see Chapter 13, Section 3, for a description of their structure). The median neurosecretory cells in the brain produce a thoracotropic hormone, which activates the molt glands, and an allatotropic hormone, which stimulates the corpora allata. Experimental evidence indicates that in most species the tropic hormones reach the target gland via the hemolymph, having been released from their site of storage in the corpora cardiaca. In some insects, however, some neurosecretory axons do not terminate in the corpora cardiaca but continue to the corpora allata where they are presumed to release their product. (The molt glands also receive nerves, but these very largely comprise nonneurosecretory neurons thought to inhibit rather than promote glandular activity.) The molt glands secrete molting hormone (ecdysone) (or perhaps its precursor which is converted to ecdysone en route to target tissues), which initiates the molting cycle. The corpora allata produce juvenile hormone, which can exert an influence on development only in the presence of molting hormone, that is, after a molting cycle has begun. It is the level of circulating juvenile hormone during a critical period of the stadium which determines the nature of the succeeding molt.

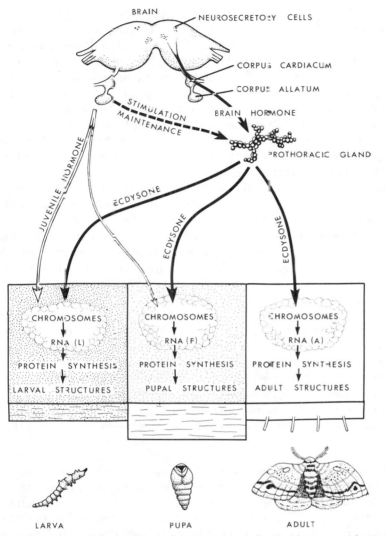

FIGURE 21.12. Schematic representation of endocrine control of development. [After L. I. Gilbert, 1964, Physiology of growth and development: Endocrine aspects, in: *The Physiology of Insecta*, 1st ed., Vol. I (M. Rockstein, ed.). By permission of Academic Press and the author.]

When the concentration of juvenile hormone is high,* the next molt will be larval–larval. At intermediate concentrations a larval–pupal molt will ensue, and when there is little or no circulating juvenile hormone due to inactivity of the corpora allata, an adult insect will emerge at the next molt (Figure 21.12). Apart from permitting the expression of adult characters the low concentration of juvenile hormone has another major effect: it leads to degeneration of the molt glands, which disappear within a few days of eclosion in most insects,

---

*Terms such as "high" and "low" are relative, and absolute concentration values will vary among species.

exceptions being apterygotes, which, as noted earlier, continue to grow and molt as adults and solitary locusts whose corpora allata apparently do not become completely inactive at the final molt. Though degeneration of the molt glands is of critical importance in the life history of an insect, when viewed in the perspective of metamorphosis the phenomenon becomes simply an example of programmed cell (tissue) death. In other words, the molt glands, like many other structures in juvenile insects, especially endopterygotes, are larval tissues whose structural well-being is dependent on juvenile hormone. In the absence of the hormone, at metamorphosis, they histolyze. Conversely, the development of adult tissues, for example, imaginal discs, is suppressed by juvenile hormone during larval life.

The precise modes of action of the hormones responsible for development are uncertain at the moment. One of the earliest observable effects of thoracotropic hormone is renewed RNA synthesis in the molt glands. However, this effect could be achieved in a variety of ways and indicates nothing about the site of action of the hormone. Activation of the corpora allata in juvenile insects has hardly been studied. In adult insects, however, activation of the corpora allata is accompanied by increased RNA synthesis in the glands. The major obstacle to studies on the mode and site of action of the tropic hormones continues to be the lack of pure preparations. For the molting hormone and juvenile hormone such preparations are available, and consequently much more is known about their effects close to the site of action. Unfortunately, as more becomes known, the more complex appears the situation, and it is not possible in a text of this nature to enter into a lengthy discussion of the subject. Gilbert and King (1973), however, have written an excellent review for those who wish to delve further into the matter.

Several major levels exist at which the hormones might act: (1) at the cell membrane (perhaps inducing cyclic AMP formation), (2) at the level of translation (decoding of messenger RNA for specific protein synthesis in the cytoplasm), (3) at the nuclear membrane (perhaps through an effect on membrane permeability so that the ionic content of the nucleoplasm is altered, leading to gene activation), and (4) at the level of transcription (gene activation) so that specific messenger RNAs are produced. For each of these levels there are several steps, one or more of which could conceivably be a specific site of action. In the case of ecdysone, with which most work has been done, effects at all these levels have been reported, and, indeed, there is no reason to suspect that the hormone cannot act at several locations. Data on juvenile hormone are fewer but, on the grounds that the hormone modifies the effects of ecdysone, it might be expected to have (a) similar site(s) of action.

6.2. Factors Initiating and Terminating Molt Cycles

Compared with the enormous volume of literature on the endocrine interactions that regulate growth and molting, relatively little is known about the external factors that initiate or terminate molting cycles.

A number of environmental variables have been shown to affect growth and molting, and some of these clearly exert their effect via the neurosecretory system, that is, they stimulate or depress the synthesis/release of tropic hormones from the brain. Probably the best studied of these variables are feeding and photoperiod.

In *Rhodnius* and *Cimex*, for example, engorgement with the host's blood, causing distension of the abdominal wall, initiates a molt cycle. Information from stretch receptors in the wall passes along the ventral nerve cord to the brain whose neurosecretory cells are activated. In insects which feed more or less continuously through the stadium, also, the intake of food is probably an important stimulus for the release of neurosecretion. *Locusta migratoria*, for example, has stretch receptors in the wall of the pharynx, which are stimulated as food passes through the foregut. Information from the receptors reaches the brain–corpora cardiaca complex via the stomatogastric nervous system (Clarke and Langley, 1963). Thus, in a continuously feeding insect such as the locust, neurosecretion will be released throughout the stadium, and presumably therefore will have functions in addition to activation of the molt glands and corpora allata. Clarke and Langley (1963) have suggested that the normal function of neurosecretion in a continuous feeder is to promote tissue growth (protein synthesis). Just prior to ecdysis, however, an insect stops feeding and instead begins to swallow air in preparation for splitting the remains of the old cuticle. Thus, synthesis and release of neurosecretion continue, and its concentration in the hemolymph will rise because at this point it is no longer being used in protein synthesis, there being no raw materials available as the insect is not feeding. This increase in the hemolymph concentration of neurosecretion, which becomes maximal at ecdysis, serves to activate the molt glands and initiate another growth and molting cycle. As an insect begins to feed after ecdysis neurosecretion will again be used in protein synthesis, and its concentration will fall. In support of their proposal Clarke and Langley observed histologically that the median neurosecretory cells showed no signs of cyclic activity (i.e., they released hormone continuously), and, furthermore, coincident with ecdysis, the molt glands exhibited considerable mitotic activity.

Photoperiod is another important environmental factor in the regulation of growth and molting, particularly in relation to diapause, a more or less prolonged condition of arrested development, which enables insects to survive periods of adverse conditions (see Chapter 22, Section 3.2). Members of most species studied enter diapause when the daily amount of light to which they are exposed falls below a certain value (usually 14 to 16 hours). In diapause, an insect is physiologically "turned-off"; generally, it does not feed or move actively, and its metabolic rate is abnormally low. These effects result from inactivity of the endocrine system. In a manner which is not yet clear, short daylengths lead to reduced neurosecretory activity which, in turn, results in inactivity of molt glands and corpora allata. Conversely, diapause is terminated as the daylength increases beyond a certain point in spring, due to renewed endocrine activity. In members of some species, however, the neurosecretory system must be exposed to low temperatures for a critical length of time during diapause before it can respond to increasing daylength (see Chippendale, 1977).

In some insects, the "feel" of the surroundings is important for continued normal development. For example, larvae of the wheatstem sawfly, *Cephus cinctus*, will not pupate if removed from the cavity at the base of the stem.

Larvae of the squash fly, *Zeugoducus depressus*, live in the cavity of squash where the carbon dioxide concentration is initially about 4–6%. Pupation is delayed by this concentration of gas and will not occur until the level falls to about 1%, some 6 months later. Presumably, this delay serves to synchronize the emergence of adult flies with the opening of the squash flowers (in which eggs are laid) the following season.

Very little is known about the inhibition of endocrine activity and other events which bring a molting cycle to a close with the shedding of the old cuticle. Probably negative feedback pathways exist so that when the concentration of circulating hormone reaches a critical level, the activity of the gland producing it is depressed. The pathway may be direct, that is, the hormone itself may depress glandular activity. Alternatively, in analogy with the situation in vertebrates, circulating ecdysone and juvenile hormone may inhibit the activity of the thoracotropic hormone- and allatotropic hormone-producing cells, respectively. A third possibility is that hormonal levels are monitored by chemoreceptors which send the information via sensory neurons to the brain. Reduction in activity of the molt glands and/or corpora allata might then be brought about via the nerves to the glands.

What causes an insect to stop feeding and begin swallowing air prior to ecdysis has not been determined. It is known, however, that hormones can greatly modify behavior, and it seems possible that the particular hormonal milieu just before molting is responsible for this changeover. Truman and his co-workers have established that the factor which initiates eclosion behavior in the Chinese oak silk moth, *Antheraea pernyi*, is a specific neurosecretion, the eclosion hormone (see Truman, 1973). Like many other insects, adult *A. pernyi* emerge during a characteristic short period of the day, specifically, the early evening. Truman (1973) showed that the hormone is produced throughout most of the pharate adult stage and stored in the brain and corpora cardiaca. Coincident with the onset of characteristic eclosion movements, hormone is liberated from the corpora cardiaca into the hemolymph. Truman (1973) showed that the hormone is not produced in the larval instars of *Antheraea* and could not be detected in larvae of two hemimetabolous species, *Leucophaea maderae* and *Pyrrhocoris apterus*. He concludes that the hormone is used only in the molt from pupa to adult and probably, therefore, is restricted to holometabolous insects. Whether other insects (including hemimetabolous forms such as Odonata) which emerge over a restricted period of the day use a system comparable to that of *Antheraea* is not known.

7. Polymorphism

Polymorphism, the existence of several distinct forms of the same life stage of an organism,* though not a common phenomenon in insects, occurs in representatives of several widely different orders. The phases of locusts (Or-

*Wigglesworth (1961) defines polymorphism more broadly as simply the occurrence of different forms of an individual. Thus, in his view, the larva, pupa, and adult are forms of a polymorphic organism, and all insects are accordingly polymorphic. Normally the term "polyeidism" is used to describe the condition in which an individual exists in different forms during its development.

thoptera) and some caterpillars (Lepidoptera), castes of social insects (Isoptera and Hymenoptera), alary polymorphism in crickets (Orthoptera), aphids and other Hemiptera, and color polymorphism of mimetic butterflies (Lepidoptera) are examples of insect polymorphism. Though these examples refer only to difference of form, it should be appreciated that the physiology, behavior, and ecology of these forms are also different.

Polymorphism, like differentiation, has a genetic basis. In some examples of polymorphism such as the color forms of adult butterflies, the genetic system is relatively little influenced by environmental conditions [Wigglesworth (1961) describes the constituent genes as "strong"], and the relative proportions of the various forms of the species can be determined from normal genetic considerations. However, many examples of polymorphism are known whose genetic basis is "weak," that is, greatly influenced by prevailing environmental conditions. Furthermore, as with differentiation, these conditions often exert their influence via the endocrine system. It is, in other words, changes in the hormonal milieu which lead to development of polymorphism. Under conditions where the juvenile hormone titer is greater than normal, adults will retain certain juvenile features, for example, lack of wings; when juvenile hormone secretion is depressed, the final larval instar may develop adultoid characters and become prematurely sexually mature.

Aphid polymorphism is a complex phenomenon for, in addition to extensive structural polymorphism (some species include as many as eight distinct forms) and the physiological polymorphism which accompanies it, there is also "temporal" or "successive" polymorphism in which these structural and physiological features gradually change from generation to generation (Lees, 1966). Aphids reproduce parthenogenetically (and in some species paedogenetically) for a large part of the year, giving rise to large numbers of wingless individuals which can exploit the rich food supplies available in spring and summer. However, to avoid starvation due to overcrowding, to move to nutritionally more valuable food sources, and to reproduce sexually, it is necessary for winged individuals (alates) to develop. The development of alates is influenced by many environmental factors, for example, photoperiod, temperature, population density, and water content of food plants, all of which ultimately appear to bring about changes in endocrine activity. In some species these factors act directly on an individual to modify its form, whereas in others the effect is not seen until the following generation. For example, when first or, to a lesser extent, second instar larvae of Myzus persicae are kept under crowded conditions, they develop into winged forms. In contrast, in Megoura viciae, crowding young larvae does not induce wing development either in the larvae or in their progeny. In this species, sensitivity to crowding develops in the fourth (final) instar and is retained through adulthood. Thus, when older larvae or wingless adults are crowded, their progeny are winged. Experiments have indicated that the stimulus is not visual or chemical but due to repeated physical contact between individuals. (Interestingly, it is known that ants which tend aphids for their honeydew have a "tranquilizing" effect on the aphids. This effect apparently leads to reduction in the amount of physical contact between the aphids, which thus remain apterous, to the ants' obvious advantage.) It is presumed the crowding stimulus operates via the brain, which somehow reduces corpus allatum activity. The nature of the link between the brain and corpora allata is not known but is likely to be hormonal because in species such as *M. viciae* the crowding stimulus is received by the maternal brain, but its effect is made apparent in the progeny. In other words, it is the corpora allata of the developing embryos within the mother whose activity is modified, yet obviously there is no nervous connection between these glands and the mother's brain (Lees, 1966).

In colonies of social insects, individuals fall into a number of functionally and, usually, structurally distinct castes. In lower termites, for example, there is a pair of primary reproductives (king and queen) which found the colony, supplementary (replacement) reproductives which develop as the colony reaches a certain size and eventually take over the reproductive function, soldiers, nymphs (juveniles with wing buds from which primary reproductives develop), and larvae (juveniles which lack wing buds). Each caste contains members of both sexes. The number of individuals belonging to each caste is normally maintained as a fixed proportion of the total number of insects in the colony by means of inhibitory pheromones secreted by already differentiated individuals (see Chapter 13, Sections 4.1 and 4.2). When the concentration of a pheromone falls below a certain level, due, for example, to growth of the colony or death of the pheromone-producing individuals, inhibition no longer occurs, and differentiation of new individuals restores the correct proportion. The relationship of the castes and the course of development in a lower termite, *Zootermopsis angusticollis*, is indicated in Figure 21.13. It can be seen that young larvae pass through several progressive molts (that is, grow and differentiate) until they become pseudergates (false workers) comparable with the true workers of higher termites. Pseudergates may undergo additional progressive molts to form specific castes, or may molt without differentiation (stationary molts) (Yin and Gillott, 1975a).

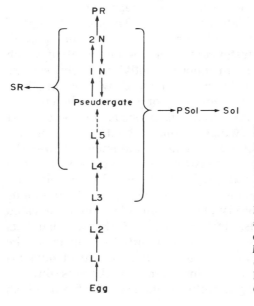

FIGURE 21.13. Course of development in *Zootermopsis angusticollis* (Isoptera). Broken arrow indicates the potentiality for several molts. Abbreviations: L1–L5, first to fifth instar larvae; 1N, 2N, first and second stage nymphs; PR, primary reproductive; SR, supplementary reproductive; PSol, presoldier; Sol, soldier. [After C.-M. Yin and C. Gillott, 1975a, Endocrine activity during caste differentiation in *Zootermopsis angusticollis* Hagen (Isoptera): A morphometric and autoradiographic study, *Can. J. Zool.* **53**:1690–1700. By permission of the National Research Council of Canada.]

The inhibitory pheromones influence caste differentiation by modifying the activity of the endocrine system, especially the corpora allata. Most of our knowledge of the endocrine control of caste formation has resulted from the extensive studies of Lüscher and his associates, which began in the 1950s. As a result of this early work, Lüscher (1960, 1965) developed an elaborate hypothesis for hormonal control, included in which was the suggestion that the corpora allata were capable of producing two, possibly three, distinct hormones. In parallel with the situation in nonpolymorphic insects, Lüscher proposed that in the differentiation of adult termites (larvae or nymphs → supplementary reproductives; nymphs → primary reproductives) the corpora allata stopped secreting juvenile hormone, permitting the expression of adult features. In larval → larval or larval → nymphal molts, the corpora allata continued to secrete juvenile hormone. Soldier development was promoted, according to Lüscher (1965), by activation of the corpora allata, which secreted a soldier-inducing hormone distinct from both juvenile hormone and gonadotropic hormone. The latter hormone, Lüscher considered, was produced in adult insects in relation to oocyte development.

Though many aspects of Lüscher's original proposal are still accepted, the hypothesis has required substantial revision in the light of more recent studies carried out by Lüscher's group and other authors, especially experiments involving application of juvenile hormone analogues. A major modification has come about with the realization that the corpora allata do not produce more than one hormone. In other words, all developmental possibilities result from variations in the concentration of circulating juvenile hormone at a critical time during a stadium. Yin and Gillott (1975b) have presented a modified version of Lüscher's hypothesis in which they propose that there are two critical juvenile hormone levels (Figure 21.14). When the concentration of hormone falls below the lower level, an adult emerges at the next molt. At intermediate levels, that is, between the upper and lower thresholds, a molt may be progressive (larva → nymph), regressive (nymph → larva), or stationary (larva → larva). Soldiers differentiate when the juvenile hormone concentration exceeds and remains above the upper threshold.

In contrast to the situation in aphids and termites, whose different forms are clearly distinct (discontinuous polymorphism), in some locusts a spectrum of slightly different forms (continuous polymorphism) can be obtained by varying the population density over a number of generations. At opposite ends of the spectrum are the solitary form (phase), produced when the population density remains continuously low for a long period of time, and the gregarious phase, typical of high-density populations. It is not feasible here to describe in detail the many differences which occur between the two phases. Suffice it to say there are significant differences in their structure, color, physiology (growth rates, occurrence of diapause, reproductive capacity), and behavior (feeding activity, migratory habits), which enable populations to develop and reproduce at the optimum rate according to the habitat in which they occur. For example, female solitary locusts, for which food is abundantly available, are larger, may lay up to five times more eggs, and have slightly shorter wings than gregarious females. Because of the greater population density, gregarious females must have the capacity to migrate to new food supplies (see also Chap-

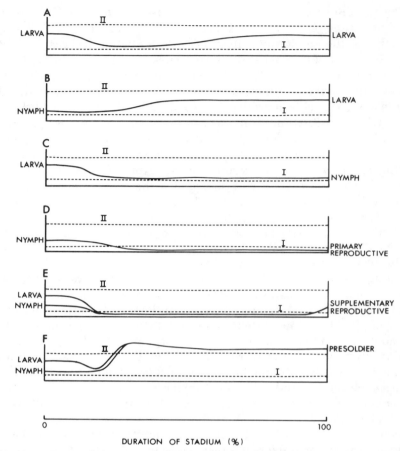

FIGURE 21.14. Proposed changes in hemolymph juvenile hormone titer during caste differentiation in *Zootermopsis angusticollis* (Isoptera). Broken lines indicate upper and lower threshold titers. [From C.-M. Yin and C. Gillott, 1975b, Endocrine control of caste differentiation in *Zootermopsis angusticollis* Hagen (Isoptera), *Can. J. Zool.* **53**:1701–1708. By permission of the National Research Council of Canada.]

ter 22, Section 5.1) for which smaller body size and larger wings will be advantageous. In addition, they accumulate more fat (used as an energy source during migration) than solitary females, an aspect which may be correlated with production of fewer eggs [see Kennedy (1961b) and Uvarov (1966) for further information]. There is convincing experimental evidence that the influence of population density and other environmental factors on phase determination is exerted via the endocrine system and especially the corpora allata. For example, adults with many of the features of solitary locusts can be produced by implanting corpora allata or applying juvenile hormone to gregarious larvae. Solitary adults are structurally more juvenile than gregarious adults, and, as noted above, solitary females lay more eggs than their gregarious counterparts, both of which features are probably related to the enhanced activity of the corpora allata. Interestingly, because of this above-normal activity of the corpora allata, the molt glands of solitary individuals do not degenerate at eclosion, though they never trigger a further molt.

8. Summary

Among the evolutionary trends that may be seen in insect postembryonic development are (1) increasing separation of the processes of growth and accumulation of reserves (functions of the juvenile stage) from reproduction and dispersal, which are functions of the adult stage; (2) the spending of a greater proportion of the organism's life in the juvenile stage; and (3) an increasing degree of difference between larval and adult habits and form. The latter has been accompanied by modification of the final larval instar into a pupa in which the considerable changes from larval to adult form can occur.

Insects may be arranged in three basic groups in terms of the pattern of postembryonic development which they display. Apterygotes are ametabolous; that is, the changes from juvenile to adult form are very slight. Adults continue to molt, and the number of instars is both large and variable. Almost all exopterygotes are hemimetabolous. Juveniles broadly resemble adults and undergo only a partial metamorphosis. The number of instars is generally four or five and constant for a species. The major event in exopterygote metamorphosis is the full development of wings and genitalia. Internal organs grow progressively through larval life. Endopterygotes and a very few exopterygotes are holometabolous. Juveniles and adults are normally strikingly different and major changes (complete metamorphosis) occur in the pupa. In primitive endopterygotes most organs grow progressively during larval life, and metamorphosis consists mainly of the development of the flight mechanism. In most endopterygotes considerable differentiation of adult tissues occurs during metamorphosis, often from imaginal discs, groups of cells that remain embryonic through larval life, probably because of the hormonal milieu in juvenile instars.

Several types of endopterygote larvae may be distinguished, oligopod (including scarabaeiform and campodeiform larvae), which are most primitive, polypod (eruciform), and apodous (including eucephalous, hemicephalous, and acephalous larvae).

Pupae may be decticous (having functional mandibles) and exarate (appendages not sealed against the body), or adecticous and exarate or obtect (appendages sealed against the body). For protection, a pupa may be enclosed within a cocoon or cell, or may be heavily sclerotized and/or camouflaged, or in cyclorrhaph Diptera remain inside the cuticle of the last larval instar (the puparium).

Adult emergence (eclosion) is achieved by swallowing air to increase the body volume and thus split the pupal cuticle. When an insect pupates in a cocoon, etc., it may chew or tear its way out, either as a pharate adult or after eclosion, using mandibles, spines, or cocoon cutters. The cocoon of some species is equipped with an escape hatch.

Development is regulated by hormones. A molt cycle is initiated with the release of thoracotropic hormone from the brain, which stimulates release of ecdysone (or its precursor) from the molt glands. The nature of a molt is determined by the concentration of juvenile hormone at a critical period in the stadium. When the concentration is high, a larval–larval molt follows; at lower concentrations the molt is larval–pupal; when very little or no juvenile hormone is present, an adult is produced.

Polymorphism, the existence of several distinct forms of the same stage of a species, is in many insects the result of variation in the hormonal milieu, especially the concentration of juvenile hormone. Examples of hormonally controlled polymorphism are caste differentiation in social insects, polymorphism in aphids, and continuous polymorphism of locusts.

9. Literature

Postembryonic development continues to be one of the most popular subjects of entomological research, and an abundance of literature is available covering a range of aspects. The following list includes some of the more recent publications. Wigglesworth (1972) provides basic information on insect growth, while the same author (Wigglesworth, 1964, 1970), Novák (1975), Doane (1973), Gilbert and King (1973), and Highnam and Hill (1977) discuss the endocrine regulation of postembryonic development. Chippendale (1977) has reviewed the hormonal control of larval diapause. Biochemical aspects of development are dealt with by Agrell and Lundquist (1973). Polymorphism is the subject of the volumes edited by Kennedy (1961a) and Lüscher (1976). Caste determination is discussed by Weaver (1966), while Lees (1966) reviews the control of polymorphism in aphids. Imaginal discs are discussed in a series of papers edited by Ursprung and Nöthiger (1972).

Agrell, I. P. S., and Lundquist, A. M., 1973, Physiological and biochemical changes during insect development, in: The Physiology of Insecta, 2nd ed., Vol. I (M. Rockstein, ed.), Academic Press, New York.

Chippendale, G. M., 1977, Hormonal regulation of larval diapause, Annu. Rev. Entomol. 22:121–138.

Clarke, K. U., and Langley, P. A., 1963, Studies on the initiation of growth and moulting in Locusta migratoria migratorioides R. and F. IV. The relationship between the stomatogastric nervous system and neurosecretion, J. Insect Physiol. 9:423–430.

Doane, W. W., 1973, Role of hormones in insect development, in: Developmental Systems: Insects, Vol. 2 (S. J. Counce and C. H. Waddington, eds.), Academic Press, New York.

Gilbert, L. I., and King, D. S., 1973, Physiology of growth and development: Endocrine aspects, in: The Physiology of Insecta, 2nd ed., Vol. I (M. Rockstein, ed.), Academic Press, New York.

Highnam, K. C., and Hill, L., 1977, The Comparative Endocrinology of the Invertebrates, 2nd ed., Edward Arnold, London.

Hill, L., and Goldsworthy, G. J., 1968, Growth, feeding activity, and the utilisation of reserves in larvae of Locusta, J. Insect Physiol. 14:1085–1098.

Kennedy, J. S. (ed.), 1961a, Insect polymorphism, Symp. R. Entomol. Soc. 1:115 pp.

Kennedy, J. S., 1961b, Continuous polymorphism in locusts, Symp. R. Entomol. Soc. 1:80–90.

Lees, A. D., 1966, The control of polymorphism in aphids, Adv. Insect Physiol. 3:207–277.

Lüscher, M., 1960, Hormonal control of caste differentiation in termites, Ann. N.Y. Acad. Sci. 89:549–563.

Lüscher, M., 1965, Functions of the corpora allata in the development of termites, Proc. 16th Int. Congr. Zool. 4:244–250.

Lüscher, M. (ed.), 1976, Phase and Caste Determination in Insects, Pergamon Press, New York.

Novák, V. J. A., 1975, Insect Hormones, 2nd Engl. ed., Chapman and Hall, London.

Truman, J. W., 1973, Physiology of insect ecdysis. II. The assay and occurrence of the eclosion hormone in the Chinese oak silkmoth, Antheraea pernyi, Biol. Bull. 144:200–211.

Ursprung, H., and Nöthiger, R. (eds.), 1972, The Biology of Imaginal Disks, Springer-Verlag, New York.

Uvarov, B. P., 1966, Grasshoppers and Locusts. A Handbook of General Acridology, Vol. 1, Cambridge University Press, Cambridge.

Weaver, N., 1966, Physiology of caste determination, *Annu. Rev Entomol.* 11:79–102.

Wigglesworth, V. B., 1961, Insect polymorphism—A tentative synthesis, *Symp. R. Entomol. Soc.* 1:103–113.

Wigglesworth, V. B., 1964, The hormonal regulation of growth and reproduction in insects, *Adv. Insect Physiol.* 2:247–336.

Wigglesworth, V. B., 1970, *Insect Hormones*, Oliver and Boyd, Edinburgh.

Wigglesworth, V. B., 1972, *The Principles of Insect Physiology*, 7th ed., Chapman and Hall, London.

Yin, C.-M., and Gillott, C., 1975a, Endocrine activity during caste differentiation in *Zootermopsis angusticollis* Hagen (Isoptera): A morphometric and autoradiographic study, *Can. J. Zool.* 53:1690–1700.

Yin, C.-M., and Gillott, C., 1975b, Endocrine control of caste differentiation in *Zootermopsis angusticollis* Hagen (Isoptera), *Can. J. Zool.* 53:1701–1708.

IV

Ecology

22

The Abiotic Environment

1. Introduction

The development and reproduction of insects are greatly influenced by a variety of abiotic factors. These factors may exert their effects on insects either directly or indirectly (through their effects on other organisms) and in the short or long term (light, for example, may exert an immediate effect on the orientation of an insect as it searches for food, and may induce changes in an insect's physiology in anticipation of adverse conditions some months in the future). Another abiotic factor to which insects are now routinely subjected (deliberately or otherwise) are pesticides. Apart from the obvious effect of lethal doses of such chemicals, pesticides may have more subtle, indirect effects on the distribution and abundance of species, for example, alteration of predator–prey ratios and, in sublethal doses, changes in fecundity or rates of development.

   Under natural conditions organisms are subject to a combination of environmental factors, both biotic and abiotic, and it is this combination that ultimately determines the distribution and abundance of a species. Frequently the effect of one factor modifies the normal response of an organism to another factor. For example, light, by inducing diapause, may make an insect unresponsive to (unaffected by) temperature fluctuations. As a result an insect is not harmed by abnormally low temperatures, but nor is it "fooled" into reactivity by temporary periods of warmer weather which may occur in the middle of winter.

2. Temperature

2.1. Effect on Developmental Rate

   The body temperature of insects, as poikilothermic animals, normally follows closely the temperature of the surroundings. Within limits, therefore, metabolic rate is proportional to ambient temperature. Consequently, the rate of development is inversely proportional to temperature (Figure 22.1). Outside these temperature limits the rate of development no longer bears an inversely linear relationship to temperature, presumably because of the deleterious ef-

595

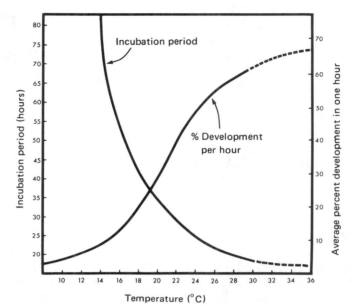

FIGURE 22.1. Relationship between temperature and rate of development in eggs of *Drosophila melanogaster* (Diptera). [After H. G. Andrewartha, 1961, *Introduction to the Study of Animal Populations*, University of Chicago Press. By permission of the author.]

fects of extreme temperatures on the enzymes which regulate metabolism, and eventually temperatures are reached (the so-called upper and lower lethal limits) where death occurs.

Within the range of linearity the product of temperature multiplied by time required for development will be constant. This constant, known as the thermal constant or heat budget, is commonly measured in units of degree-days. This relationship will hold even when the temperature fluctuates, provided that the fluctuations do not exceed the range of linearity.

The temperature limits outside which development ceases and the rate of development at a given temperature vary among species, points which, though seemingly obvious, were apparently overlooked in some early attempts at biological control of insect pests. A predator which, on the basis of laboratory tests and short-term field trials, had good control potential, was found to exert little or no control of the pest under natural conditions. Further study showed this to be due to the differing effects of temperature on development, hatching, and activity between the pest and its predator.

A broad correlation exists between the temperature limits for development and the habitat occupied by members of a species. For example, many Arctic insects which overwinter in the egg stage complete their entire development (embryonic + postembryonic) in the temperature range 0–4°C, whereas in the Australian grasshopper, *Austroicetes cruciata*, development ceases below 16°C. This means that the distribution of a species will be limited by the range of temperature experienced in different geographic regions, as well as by other factors. However, the distribution of a species may be significantly greater than that anticipated on the basis of temperature data for the following reasons; (1) temperature adaptation may occur; that is, genetically different strains may evolve, each capable of surviving within a different temperature range; (2) the temperature limits of development may differ among developmental stages [this also serves as an important developmental synchronizer in some species (see Section 2.3)]; and (3) the insect may have mechanisms for surviving extreme temperatures (see Section 2.4).

Because of the ameliorating effects of the water surrounding them aquatic insects are not normally exposed to the temperature extremes experienced by terrestrial species. Further, because ice is a good insulator, development may continue through the winter in some aquatic species in temperature climates, though air temperatures render development of terrestrial species impossible. Indeed, through evolution there has been a trend in some insects (for example, species of Ephemeroptera and Plecoptera) to restrict their period of growth to the winter, passing the summer as eggs in diapause. Such species, whose developmental threshold is usually only slightly above 0°C, appear to gain at least two advantages from this arrangement. First, through the winter there is an abundance of food in the form of rotting vegetation, yet relatively little competition for it. Second, they are relatively safe from predators (fish) which are sluggish and feed only occasionally at these temperatures (Hynes, 1970b). Such a life cycle may also allow some species to inhabit temporary or still bodies of water that dry up or become anaerobic during summer.

2.2. Effect on Activity and Dispersal

Through its effect on metabolic rate temperature clearly will affect the activity of insects. Many of the generalizations made above with regard to the influence of temperature on development have their parallel in relation to activity. Thus, there is a range of temperature within which activity is normal, though this range may vary among different strains of the same species. The temperature range for activity is correlated with a species' habitat; for example, in the Arctic, chironomid larvae are normally active in water at 0°C, and adults can fly at temperatures as low as 3.5°C (Downes, 1964).

By affecting an insect's ability to fly temperature may have a marked effect on a species' dispersal and, therefore, distribution. Further, because flight is of such importance in food and/or mate location and ultimately, reproduction, temperature is of great consequence in determining the abundance of species. Insects utilize various means of raising their body temperature to that at which flight is possible even when the ambient temperature is low. For example, they may be darkly colored so as to absorb solar radiation or they may bask on dark surfaces, again using the sun's heat. Some moths and bumblebees beat their wings while at rest in order to increase their body temperature. A dense coat of hairs or scales covers the body of some insects, which, by its insulating effect, will retard loss of heat generated or absorbed.

In extremely cold climates, such as that of the Arctic, these physiological, behavioral, or structural features may no longer be sufficient to enable flight to occur, especially in a larger-bodied egg-carrying female. Thus, different temperature-adaptation strategies are employed, some of which are exemplified especially well by Arctic blackflies (Simuliidae: Diptera). Typical southern species are, as adults, active insects which mate in flight and may, in the case of females, fly considerable distances in search of a blood meal necessary for egg maturation. In contrast, females of Arctic species seldom fly. Their mouthparts are reduced and eggs mature from nutrients acquired during larval life. Mating occurs on the ground as a result of chance encounters close to the site of adult emergence. In two species parthenogenesis has evolved, thereby overcoming the difficulty of being located by a mate (Downes, 1964).

Temperature change, through its effect on the solubility of oxygen in water, may markedly modify the activity and, ultimately, the distribution and survival of aquatic insects. Members of many aquatic species are restricted to habitats whose oxygen content remains relatively high throughout the year. Such habitats include rivers and streams which are normally well oxygenated because of their turbulent flow and lower summer temperature, and high altitude or latitude ponds and lakes, which generally remain cool through the summer. Alternatively, as noted in the previous section, the life cycle of some species is such that the warmer (oxygen-deficient) conditions are passed through in a resistant, diapausing, egg stage.

2.3. Temperature-Synchronized Development and Emergence

Many species of insects exhibit highly synchronized larval development (all larvae are more or less at the same developmental stage) and/or synchronized eclosion, especially those which live in habitats where the climate is suitable for growth and reproduction for a limited period each year. Synchronized eclosion increases the chances of finding a mate. It may also increase the probability of finding suitable food or oviposition sites, or of escaping potential predators. Synchronized larval development also may be related to the availability of food, and in some situations it may be necessary in order to avoid interspecific competition for the same resource. For certain carnivorous species, such as Odonata, synchronized development may help reduce the incidence of cannibalism among larvae.

Perhaps not surprisingly in view of its effects on rate of development and activity, temperature is an important synchronizing factor in the life of insects. Its importance may be illustrated by reference to the life history of *Coenagrion angulatum*, which, along with several other species of damselflies (Odonata: Zygoptera), is found in or around shallow ponds on the Canadian prairies (Sawchyn and Gillott, 1975). For these insects the season for growth and reproduction lasts from about mid-May to mid-October. For the remaining 7 months of the year *Coenagrion angulatum* exists as more or less mature larvae, which, between about November and April, are encased in ice as the ponds freeze to the bottom. (The larvae themselves do not freeze, as the ice temperature seldom falls more than a few degrees Celsius below zero as a result of snow cover.) In *C. angulatum* both larval development and eclosion are synchronized by temperature. Synchronized development is achieved by means of different temperature thresholds for development in different instars; that is, younger larvae can continue to grow in the fall after the growth of older larvae has been arrested by decreasing water temperatures and by a photoperiodically induced diapause. Thus, samples collected in mid-September include larvae of the last seven instars, whereas those from early October are composed almost entirely of larvae of the last three instars. Conversely, after the ice melts the following April, younger larvae can continue their development earlier than their more mature relatives, so that by mid-May more than 90% of the larvae are in the final instar. After their release from the ice larvae migrate into shallow water at the pond margin whose temperature parallels that of the air. Thus, larvae are able to "monitor" the air temperature and determine when it is suitable for emergence. Emergence occurs when the air temperature is 20–21°C (and the water temperature is about 12°C). It begins normally during the last week of May and reaches a peak within 10 days. Emergence of *C. angulatum* follows that of various chironomids and chaoborids (Diptera), which form the main food of the adult damselflies during the period of sexual maturation. The development and emergence of other damselfly species which inhabit the same pond are also highly synchronized but occur at different times of the growing season. This enables the species to occupy the same pond and make use of the same resources, yet avoid interspecific competition. This point will be discussed further in Chapter 23 (Section 3.2.1).

Temperature may also have an important influence on diapause and other photoperiodically induced phenomena, and this aspect is dealt with below (Section 3.2).

2.4. Survival at Extreme Temperatures

In many tropical areas climatic conditions are suitable for year-round development and reproduction in insects. In other areas of the world, the year is divisible into distinct seasons, in some of which growth and/or reproduction is not possible. One reason for this arrest of growth and reproduction may be the extreme temperatures which occur at this time and are potentially lethal to an insect. In many instances shortage of food would also occur under these conditions.

To avoid the detrimental effects of periods of moderately low (down to freezing) or high temperature, insects may employ a number of behavioral and physiological mechanisms. First, the life history of many species is so arranged that the period of adverse temperature is passed as the immobile, nonfeeding egg or pupa. Second, prior to the advent of adverse conditions [and it should be realized that an insect "anticipates" the onset of these conditions (see Section 3.2)] an insect may actively seek out a habitat in which the full effect of the detrimental temperature is not felt. For example, it may burrow or oviposit in soil, litter, or plant tissue, which acts as an insulator. Third, it may enter diapause where its physiological systems are largely inactive and resistant to extremes of temperature.

2.4.1. Cold-Hardiness

For insects in environments which experience temperatures below 0°C, an additional problem presents itself, namely, how to avoid being damaged by freezing of the body cells. The formation of ice crystals within cells causes irreversible damage and frequently death of an organism (1) by physical disruption of the protoplasm, and (2) by dehydration, reduction of the liquid water content which is essential for normal enzyme activity. Insects which survive freezing temperatures fall into two categories: (1) freezing-susceptible species which avoid freezing by lowering the freezing point of the body fluids and by supercooling, and (2) freezing-tolerant (=freezing-resistant=frost-resistant) species whose extracellular body fluids can freeze without damage to the insect. In both systems, one or more cryoprotectants (substances which protect against freezing) are normally involved. Cryoprotectants identified to date include the polyhydric alcohols, glycerol and sorbitol, and the disaccharide trehalose, each of which contains a number of hydroxyl groups, the significance of which is discussed below. However, in order to appreciate the mode of action of these cryoprotectants, it is necessary first to understand the process of freezing. When water is cooled the speed at which individual molecules move decreases, and the molecules aggregate. As cooling continues there is an increased probability that a number of aggregated molecules will become so oriented with respect to each other as to form a minute rigid latticework, that is, a crystal. Immediately this minute crystal (nucleator) is formed the rest of the water freezes rapidly as additional molecules bind to the solid frame now available to them. Freezing of a liquid does not always depend on the formation of a nucleator, but can be induced by foreign nucleating agents such as dust particles or, in the present context, particles of food in the gut or a rough surface such as that of the cuticle.

Hence, the problem for a freezing-susceptible species is to reduce as much as possible the chance that its hemolymph, and subsequently its intracellular fluid, will freeze. For many species, this includes emptying the gut of food; for others, it may be an additional reason for overwintering as a nonfeeding pupa. By selecting dry locations in which to overwinter and/or by building structures that prevent contact with moisture an insect reduces the possibility of nucleation on its body surface that eventually may cause internal freezing. The presence of an hydrophobic wax layer also reduces this possibility. The most important mechanism, however, for avoiding freezing is the production of cryoprotectants (antifreezes). These molecules not only increase the concentration of solutes in the body fluid (an effect which is achieved in some species by active excretion of water) so that the freezing point is depressed, but by their chemical nature they facilitate considerable supercooling; that is, the body fluids remain liquid at temperatures much below their normal freezing point. The known cryoprotectants, because of their hydroxyl groups, are capable of extensive hydrogen bonding with the water within the body. The binding of the water has two important effects with respect to supercooling. First, it greatly reduces the chance of the water molecules to aggregate and form a nucleating crystal, and second, even if an ice nucleus is formed, the rate at which freezing spreads through the body is greatly retarded because of the increased viscosity of the fluid.

A remarkable degree of supercooling can be achieved through the use of cryoprotectants. In the overwintering larva of the parasitic wasp *Bracon cephi*, for example, glycerol makes up 25% of the fresh body weight and lowers the supercooling point of the hemolymph to $-47°C$ (Salt, 1958; cited in Asahina, 1969). Perhaps a disadvantage to the use of supercooling as a means of overwintering is that the probability of freezing occurring increases both with duration of exposure and with the degree of supercooling so that, for example, an insect might freeze in 1 minute at $-19°C$ but survive for 1 month at $-10°C$. Thus, to ensure survival an insect must have the ability to remain supercooled at extreme temperatures for significant periods of time, even though the average temperatures to which it is exposed may be 10–15°C higher. In other words, it may have to produce much more antifreeze in anticipation of those extremes than would be judged necessary on the basis of the average temperature.

The alternative method, employed by freezing-tolerant species, is to permit (be able to withstand) a limited amount of freezing within the body. Freezing must be restricted to the extracellular fluid, as intracellular freezing damages cells. Ice formation in the extracellular fluid, which is accompanied by release of heat (latent heat of fusion), will therefore reduce the rate at which the body's tissues cool as the ambient temperature falls. Thus, it will be to an insect's advantage to have a large volume of hemolymph (and there is evidence that this is characteristic of pupae) and to be able to tolerate freezing of a large proportion of the water within it. The problems for an insect are twofold. First, it must be able to prevent freezing from extending to the cell surfaces (and hence into the cells), and second, it must prevent damage to cells as a result of dehydration (as water in the extracellular fluid freezes, the osmotic pressure of the remaining liquid will increase, so that water will be drawn out of the cells by osmosis). Both problems are overcome by the use of polyhydroxyl cryoprotectant molecules with their ability to bind extensively with water. Extracellular cryoprotectant will retard the rate at which freezing spreads, while intracellular cryoprotectants will hold water within cells, to counteract the outwardly pulling osmotic force. It has also been suggested that the cryoprotectants may bind with plasma membranes to reduce their permeability to water.

Of interest is the evolutionary selection of glycerol as a cryoprotectant because in high concentration this molecule is toxic at above freezing temperatures. Thus, insects which use this molecule should possess biochemical mechanisms for synthesizing it in increasing amounts as the temperature falls progressively below 0°C and, equally, for degrading it when temperature increases. Such has been shown to be the case in *Pterostichus brevicornis*, an Arctic carabid beetle which overwinters as a freezing-tolerant adult. In *P. brevicornis* glycerol synthesis begins when an insect is exposed to a fall temperature of 0°C, and by the following December–January the concentration of this molecule may reach or exceed 30 g%, sufficient to enable an insect to withstand the -40 to $-50°C$ temperatures to which it may be exposed at this time. Conversely, as temperatures increase toward 0°C with the advent of spring, the glycerol concentration falls and the cryoprotectant disappears from the hemolymph by about the end of April, coincident with the return of above freezing average temperatures (Baust and Morrissey, 1977). A comparable situation is observed in *Eurosta solidagensis*, a gall-forming fly which overwinters as a freezing-tolerant third instar larva. The larva has a three-phase cryoprotectant system which comprises glycerol, sorbitol, and trehalose. Production of the molecules begins somewhat above 0°C but is probably triggered by declining temperatures At temperatures below 0°C, production of glycerol and sorbitol is greatly enhanced. With the return of warm weather in spring, the concentration of the three molecules rapidly declines (Baust and Morrissey, 1977).

3. Light

Light exerts a major influence on the ability of almost all insects to survive and multiply. A well-developed visual system enables insects to respond immediately and directly to light stimuli of various kinds in their search for food, a mate, a "home," or an oviposition site, and in avoidance of danger. This aspect has been discussed already in Chapter 12 (Section 7). But light influences the biology of many insects in another manner which stems from the earth's rotation about its axis, resulting in a regularly recurring 24-hour cycle of light and darkness, the photoperiod.* Because the earth's axis is not perpendicular to the plane of the earth's orbit around the sun, and because the orbit varies throughout the year, the relative amounts of light and darkness in the photoperiod change seasonally and from point to point over the earth's surface.

Photoperiod influences organisms in two ways: it may either induce short-term (diurnal) behavioral responses which occur at specified times in the 24-hour cycle, or bring about long-term (seasonal) physiological responses which keep organisms in tune with changing environmental conditions. In both situations, however, a key feature is that the organisms which respond have the ability to measure time. In short-term responses the time interval between the onset of light or darkness and commencement of the activity is important; for seasonal responses, the absolute daylength (number of hours of light in a 24-hour period) is usually critical, though in some species it is the day-to-day increase or decrease in the light period which is measured. In other words, organisms which exhibit photoperiodic responses are said to possess a "biological clock," the nature of which is unknown, though its effects in animals are frequently manifest through changes in endocrine activity.

3.1. Daily Influences of Photoperiod

Various advantages may accrue to members of a species through the performance of particular activities at set times of the photoperiod. It may be advantageous for some insects to become active at dawn, dusk, or through the night when ambient temperatures are below the upper lethal limit, chances of predation are reduced, and the rate of water loss through the cuticle is lessened by the generally greater relative humidity which occurs at these times. For other insects, in which visual stimuli are important, activity during specific daylight hours may be advantageous; for example, food may be available for only a limited part of the day, or conversely, other, detrimental factors may restrict feeding to a specific period. For many species it is clearly beneficial for its members to show synchronous activity as this will increase the chance of

---

*As Beck (1968) notes, some authors use this term to describe the light portion of a light–dark cycle (i.e., synonymously with daylength).

contact between sexes. "Activity" in this sense is not restricted to locomotion, however. For example, in many species of moths, it is by and large only the males who exhibit daily rhythms of locomotor activity. The females are sedentary, but, in their virgin condition, have daily rhythms of "calling" (secretion of male-attracting pheromones) which enable males to locate them.

3.1.1. Circadian Rhythms

In a few species daily rhythms of activity are triggered by environmental cues and are therefore of exogenous origin. For example, the activity of the stick insect *Carausius morosus* is directly provoked by daily changes in light intensity. However, in most species these rhythms are not simply a response to the onset of daylight or darkness; that is, dawn or dusk do not act as a trigger that switches the activity on or off. Rather, the rhythms are endogenous (originate within the organism itself) but are subject to modification (regulation) by photoperiod and other environmental factors. That the rhythm originates internally may be demonstrated by placing the organism in constant light or darkness. The organism continues to begin its activity at approximately the same time of the 24-hour cycle, as it did when subject to alternating periods of light and darkness. Because the rhythm has an approximately 24-hour cycle, it is described as a circadian rhythm. When the rhythm is not influenced by the environment, that is, when environmental conditions are kept constant, the rhythm is said to be free-running. When environmental conditions vary regularly in each 24-hour cycle, and the beginning of the activity occurs at precisely the same time in the cycle, the rhythm is said to be entrained. For example, if a cockroach begins its locomotor activity 2 hours after darkness, this activity is described as being photoperiodically entrained. The role of photoperiod is therefore to adjust (phase set) the endogenous rhythm so that the activity occurs each day at the same time in relation to the onset of daylight or darkness. Though photoperiod is probably the most important regulator of circadian rhythms in insects, other environmental factors such as temperature, humidity, and light intensity, as well as physiological variables such as age, reproductive state, and degree of desiccation or starvation may modify behavior patterns. Photoperiodically entrained daily rhythms are known to occur in relation to locomotor activity, feeding, mating behavior (including swarming), oviposition, and eclosion, examples of which are given below.

Many examples are known of insects which actively run, swim, or fly during a characteristic period of the 24-hour cycle, this activity usually occurring in relation to some other rhythm such as feeding or mate location. In *Periplaneta* and other cockroaches activity begins shortly before the anticipated onset of darkness, reaches a peak some 2 to 3 hours after dark, and declines to a low level for the remaining period of darkness and during most of the light period (Figure 22.2A). *Drosophila robusta* (Figure 22 2B) flies actively during the last 3 hours of the light phase but is virtually inactive for the rest of the 24-hour period. Male ants of the species *Camponotus clarithorax* are most active during the first few hours of the light period but show little activity at other times (Figure 22.2C). The above examples show a well-defined single peak (unimodal rhythm) of activity. Other species, however, exhibit bimodal or

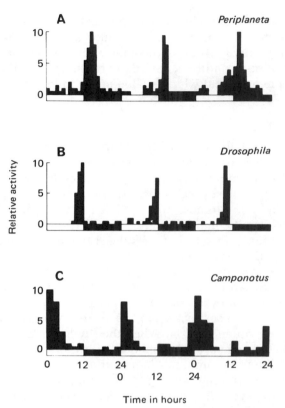

FIGURE 22.2. Locomotor activity rhythms in insects, illustrating photoperiodic entrainment. [From S. D. Beck, 1968, *Insect Photoperiodism.* By permission of Academic Press Inc., New York, and the author.]

trimodal rhythms. For example, females of the silver-spotted tiger moth, *Halisidota argentata*, show two peaks of flight activity during darkness, the first shortly after darkness begins, the second about midway through the dark period (Figure 22.3A). Males of this species, in contrast, have a trimodal rhythm of flight activity (Figure 22.3B) (Beck, 1968).

Rhythmic feeding activity is apparent in larvae of some Lepidoptera, for example, *H. argentata*, which feed almost exclusively during darkness. Female mosquitoes, too, show peaks of feeding activity either at dawn or dusk, or during both these periods, though there is some argument with regard to whether feeding activity is endogenous or simply a direct response to a particular light intensity.

Several good examples may be cited to illustrate the importance of photoperiod in entraining daily endogenous rhythms of mating behavior. Many virgin female Lepidoptera begin to secrete male-attracting pheromones shortly after the onset of darkness and are maximally receptive to males about midway through the dark period. Equally, males show maximum excitability to these pheromones in the early part of the dark period. The males of certain ant species undertake mating flights at characteristic times within the light period, typically near dawn or dusk. Mosquitoes and other Nematocera form all-male

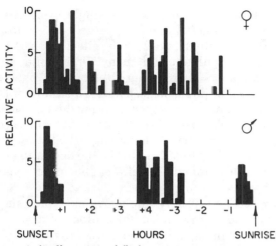

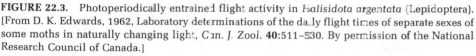

FIGURE 22.3. Photoperiodically entrained flight activity in *Halisidota argentata* (Lepidoptera). [From D. K. Edwards, 1962, Laboratory determinations of the daily flight times of separate sexes of some moths in naturally changing light, *Can. J. Zool.* **40**:511–530. By permission of the National Research Council of Canada.]

swarms into which females fly for insemination. Formation of these swarms, which occurs both at dawn and dusk, is an endogenous rhythm, entrained by photoperiod, though temperature and light intensity are also involved (Beck, 1968).

For some insect species, egg-laying has been shown to be a photoperiodically entrained endogenous rhythm. In the mosquitoes *Aedes aegypti* and *Taeniorhynchus fuscopennatus*, for example, oviposition is concentrated in the period immediately after sunset and before dawn, respectively. In other mosquitoes, however, no oviposition rhythm exists and egg-laying appears to be dependent on light intensity.

Many examples are known of insects which molt to adults during a characteristic period of the day. Many tropical Odonata exhibit mass eclosion during the early evening and are able to fly by the following morning. Corbet (1963) suggests that this may minimize the effects of predators such as birds and other dragonflies which hunt by sight. In temperate climates where nighttime temperatures are generally too low for emergence, there may be a switch to emergence during certain daylight hours. In more rigorous climates temperature appears to override photoperiod* as a factor regulating emergence, which occurs opportunistically at any time of the day provided the ambient temperature is suitable (see Section 2.3). Species of Ephemeroptera and Diptera also have daily emergence patterns, which may be associated with immediate mating and oviposition. Though many insect species are known which have a daily emergence rhythm, for only a few of these, mainly Diptera, is experimental evidence available that proves the endogenous nature of the rhythm. In contrast to the previously described rhythmic processes, emergence occurs but once in the life of an insect and results in the appearance of a very different develop-

---

*In the Arctic summer there are, of course, 24 hours of light per day, and photoperiod cannot serve as an entraining factor for diurnal rhythms.

mental stage, the adult. Nevertheless, this single event, like daily repeated processes, is an endogenous rhythm, entrained by environmental stimuli, especially photoperiod, which exert their effect in earlier developmental stages. For example, populations of many *Drosophila* species emerge at maximum rates 1 to 2 hours after dawn on the basis of photoperiodic entrainment either in the larval or pupal stage. Thus, if a culture of *Drosophila* larvae of variable ages is maintained in darkness from the egg stage except for one brief period of light (a flash lasting as little as 1/2000 of a second is sufficient) the adults will emerge at regular 24-hour intervals, based on the onset of the light period being equivalent to dawn; that is, the beginning of the light period serves as the reference point for entraining the insects' emergence rhythm (Beck, 1968).

Though described only briefly, the above examples should demonstrate the importance of circadian rhythms in the ecology of insects. Beck (1968) and Saunders (1974) discuss in detail insect circadian rhythms in relation to photoperiod.

3.2. Seasonal Influences of Photoperiod

Photoperiod affects a variety of long-term physiological processes in insects and, in doing so, allows a species to (1) exploit suitable environmental conditions, and (2) survive periods when climatic conditions are adverse. Some of the ways in which species are enabled to exploit a suitable environment include being in an appropriate developmental stage as soon as the suitable conditions appear, and growing or reproducing at the maximum rate while conditions last. Obviously, to survive adverse conditions, members of a species must already be in an appropriate physiological state when the conditions develop. In other words, organisms must be able to anticipate the arrival of inclement climatic conditions. Thus, among the processes known to be affected by photoperiod are the nature (qualitative expression) and rate of development, reproductive ability and capacity, synchronized adult emergence, induction of diapause, and possibly cold-hardiness. Several of these processes are closely related and are therefore affected simultaneously. Other environmental factors, especially temperature, may modify the effects of photoperiod.

3.2.1. Nature and Rate of Development

In some species larval growth rates are affected by photoperiod. For some species growth is accelerated under long-day conditions (when there are 16 or more hours of light in each 24-hour cycle) and inhibited in photoperiods that contain 12 or fewer hours of light; for other species, the converse is true. Often the effect of photoperiod on growth rate is correlated with the nature of diapause induction; that is, species which grow more slowly under short-day conditions tend also to enter diapause as a result of short days. However, it should be noted that the growth rate of many species which enter a photoperiodically controlled diapause is not affected by photoperiod.

Exposure to different photoperiods such as occur in different seasons may result in the development of distinct forms of a species, that is, polymorphism.

The physiological (endocrine) basis of polymorphism has been outlined in Chapter 21 (Section 7), and the present discussion will be restricted to a consideration of its induction by photoperiod. Sometimes the forms which develop are so strikingly different that they were described originally as separate species. Beck (1968) cites the instance of the European butterfly *Araschnia levana*, described originally as two species, *A. levana* and *A. prorsa*, but which is now known to be a seasonally dimorphic species. Caterpillars reared under long-day conditions metamorphose into the nondiapausing, black-winged (prorsa) form; when they have developed at short daylengths the caterpillars emerge as red-winged (levana) adults which overwinter in diapause. This example shows a typical feature of most dimorphic Lepidoptera, namely, that one form is characteristically found in summer and is nondiapausing, whereas the alternate form is the diapausing, overwintering stage.

Photoperiodically influenced polymorphism is also exemplified by the seasonal occurrence of normal-winged, brachypterous, and/or apterous forms of species of Orthoptera and Hemiptera. But perhaps the best-known example of the effects of photoperiod on development is that of temporal polymorphism in aphids. The life cycle of aphid species is complex and variable but shows beautifully how an insect takes full advantage of suitable conditions for growth and reproduction. A key feature of the life cycle is the occurrence within it of wingless, neotenic females which reproduce viviparously and parthenogenetically. In many species the offspring are entirely female. This combination of features enables aphids to reproduce rapidly and build up massive populations in the spring and summer when weather conditions are good and food is abundant. As a result of the crowding which results from this reproductive activity, winged migratory forms develop, and a part of the population moves on to alternate host plants. From these migratory forms several more generations of female aphids (alienicolae) are produced (again through viviparity and parthenogenesis) which may be winged or apterous. Eventually the alienicolae give rise to winged sexuparae (all female) that migrate back to the original host plant, and whose progeny may be either winged males or wingless females (oviparae). These reproduce sexually and lay eggs which pass the winter in diapause on the host plant. The following spring each egg gives rise to a female individual, the "stem mother" or fundatrix, normally wingless, that reproduces asexually, and from which several generations of neotenic females (fundatrigeniae) arise (see Figure 8.11). There are many variants of this generalized life cycle, most often through its simplification; that is, one or more of the life stages is omitted as, for example, in species which do not alterate hosts when migrants and whose offspring do not appear as distinct forms. Indeed, in some species sexual forms have never been described and reproduction appears to be strictly parthenogenetic.

As noted in the previous chapter, the development of seasonally occurring aphid forms is influenced by a variety of environmental factors, including photoperiod. Crowding is the major factor that influences production of summer migrants, whereas the shorter days of late summer and early fall induce development of sexuparae and oviparae. For some species there is a critical daylength for induction of oviparous forms. In *Megoura viciae*, for example, which does not alternate host plants (i.e., it has no migrant form, and the oviparae are produced directly from fundatrigeniae), the critical daylength is 14 hours 55 minutes at 15°C. At greater daylengths continuous production of viviparous, parthenogenetic females occurs; when the daylength is below this critical value oviparae are produced.

In some species production of males also is induced by short-day photoperiods, though temperature and age exert a strong influence. For example, in the pea aphid *Acyrthosiphon pisum* male offspring are not produced by young females or by females reared under long-day conditions. Old females reared at short daylengths and temperatures from 13 to 20°C produced a large proportion of males. Outside this temperature range the proportion of males declined.

3.2.2. Reproductive Ability and Capacity

The effects of photoperiod on reproductive processes are almost all indirect, that is, result from other photoperiodically induced phenomena, especially adult diapause (see below). By its effect on the nature of development, as in aphids, photoperiod may indirectly modify the fecundity of a species. Beck (1968) notes one example of an apparently direct effect of photoperiod on fecundity. In *Plutella maculipennis*, the diamondback moth, egg production in individuals reared under long-day photoperiods averaged 74 eggs/moth, whereas egg production under short-day conditions was only half this value.

3.2.3. Diapause

Beck (1968, p. 135) describes diapause as a "genetically determined state of suppressed development, the manifestation of which may be induced by environmental factors." It is a physiological state in which insects can survive cyclic, usually long, periods of adverse conditions, unsuited to growth and reproduction, including high summer or low winter temperatures, drought, and absence of food. An insect enters diapause usually some time in advance of the adverse conditions and terminates diapause after the conditions have ended. In other words, not only does the insect leave itself a margin of safety, but it anticipates the arrival of the adverse conditions. Furthermore, the factor which leads to the induction of diapause (most often photoperiod) is not in itself an adverse condition. Thus, diapause differs markedly from quiescence, which is a temporary, nonadaptive form of dormancy, usually induced directly by the arrival of adverse conditions.

Occurrence and Nature. Diapause may occur at any stage of the life history, egg, larva, pupa, or adult, though this stage is usually species-specific. Only rarely does diapause occur at more than one stage in the life history of a species. Such may be the case in species that require two or more years in which to complete their development. Anticipation of the arrival of adverse conditions means that the environmental stimuli which induce diapause must exert their influence at an earlier stage in development. Thus, egg diapause is the result of stimuli which affect the parental generation. These stimuli act on the female parent either in the adult stage or, more often, during her embryonic or larval development. In *Bombyx mori*, for example, the daylength experienced by developing female embryos determines whether or not these insects will lay eggs that enter diapause. Specifically, exposure of embryos to long daylengths results in females which lay diapausing eggs, and vice versa. For *B. mori* good evidence exists for the production of a diapause hormone by females exposed during embryogenesis to long-day conditions. This hormone, synthesized in the subesophageal ganglion, has as its target organ the ovary which is caused to produce diapause eggs. Whether this scheme is applicable to the induction of egg diapause in other species is not known.

Diapause may occur at any larval stage, though the instar in which it is present is characteristic for a species. In many species it occurs in the final instar and, upon termination, is immediately followed by pupation. In this situation it is referred to as prepupal diapause. In the induction of larval diapause environmental stimuli normally exert their influence at an earlier larval stage, though species are known in which the environmental stimulus is given in the egg stage or during the previous generation. For example, in the pink bollworm *Pectinophora gossypiella* the photoperiod experienced by the eggs, as well as that during larval life, is important in determining whether or not prepupal diapause occurs. In *Nasonia vitripenns*, a parasitic hymenopteran, induction of larval diapause is dependent on the age of the female parent at oviposition, as well as on the photoperiod and temperature to which she is exposed early in adult life.

The pupa is the stage in which a large number of species enter diapause. The environmental signal that induces diapause is generally given during larval development, though for some species the influence is exerted in the parental generation. In the Chinese oak silkworm, *Anthercea pernyi* for example, the last two larval instars are sensitive to photoperiod whereas in the horn fly, *Haematobia irritans*, pupal diapause results when the female parent has been exposed to short daylengths.

Diapause may also occur in adult insects when it is known as reproductive diapause. Either young adults or larval instars are the stages sensitive to environmental stimuli. Newly emerged adult Colorado potato beetles (*Leptinotarsa decemlineata*), when subjected to short daylengths, will enter diapause. In the boll weevil, *Anthonomus grandis*, short-day conditions experienced by larvae will induce diapause in the adult stage.

Mansingh (1971) has subdivided diapause and the events surrounding it into a number of phases (Figure 22.4). This arrangement is convenient for a description of the sequence in which various processes occur, though it must be realized that these phases normally are not clearly separated in time but merge gradually with one another. In the preparatory phase environmental factors induce changes in metabolic activity in anticipation of diapause, resulting generally in accumulation of reserves, especially fats, but including carbohydrates and in some species cryoprotectants. During this phase the metabolic rate remains normal. Entry into the first phase (induction phase) of diapause is signaled by a great decline in metabolic rate and, in postembryonic stages, in the activity of the endocrine system. For example, in diapausing *Hyalophora cecropia*, a saturniid moth, the rate of oxygen consumption (a measure of metabolic rate) is only about 2% of the prediapause value; in larval European corn borers (*Ostrinia nubilalis*), which have a "weak" diapause, the rate of oxygen consumption falls to about one quarter of the prediapause level. In the

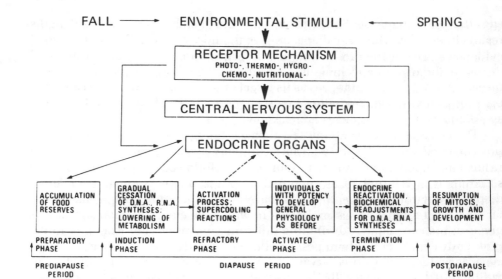

FALL ⟶ ENVIRONMENTAL STIMULI ⟵ SPRING

FIGURE 22.4. Phases before, during, and after diapause in overwintering insects. Probable (solid arrows) and possible (broken arrows) relationships between the environment, endocrine system, and the various phases are indicated. [From A. Mansingh, 1971, Physiological classification of dormancies in insects, *Can. Entomol.* **103**:983–1009. By permission of the Entomological Society of Canada.]

induction phase continued production of certain reserves, especially cryo-protectants, probably occurs. What causes the decline in metabolic rate associated with the beginning of diapause is uncertain. It may be due to the continued effects of environmental stimuli, or it may result from the changed metabolism of an insect. For most insects that overwinter in diapause a period of exposure to low temperature is necessary before development can continue, that is, before diapause can be terminated. This is the refractory phase (phase of diapause development) and is perhaps the least understood aspect of the diapause condition. Some authors have suggested that low temperature is necessary for breakdown of diapause-inducing substances (perhaps hormones) or growth-inhibiting substances produced in earlier phases. Others have proposed that this phase is necessary for reactivation of specific systems, for example, the endocrine system, important in postdiapause development. The refractory phase is followed by the activated phase, a period in which insects are capable of terminating diapause but do not do so because of prevailing environmental conditions (especially low temperature). Certain authors consider that once insects reach this stage, when their dormancy is (often) simply temperature-dependent, they must be considered as being quiescent, that is, no longer in diapause. Mansingh (1971), however, points out that, although insects in this phase are capable of continued development, several aspects of their physiology are similar to those of the refractory phase, for example, greatly depressed respiratory rate and presence of cryoprotectants. He believes, therefore, that activated insects should be considered to be still in diapause. In Mansingh's scheme the final phase of diapause is the termination phase which ensues as environmental conditions become favorable for development. In this phase the metabolic rate returns to normal, the endocrine system once more becomes active, body tissues again become capable of nucleic acid and protein synthesis, and any cryoprotectants present gradually disappear. As a result of these changes postdiapause development can begin.

In view of the varying degrees of severity of climatic conditions that insects in different geographic regions may encounter, it is perhaps not surprising to find that the intensity (duration and stability) of diapause varies. This variability, which is both interspecific and intraspecific, is manifest as a broad spectrum of dormancy that ranges from a state virtually indistinguishable from quiescence to one of great stability in which an insect can resist extremely unfavorable conditions. In each situation the strength of diapause is precisely adjusted through natural selection to provide an insect with adequate protection against the adverse conditions, yet to continue growth and reproduction as soon as an amenable climate returns. Broadly speaking, insects from less extreme climates exhibit "weak" diapause [called oligopause by Mansingh (1971)], in which development may not be completely suppressed; the insects may continue to grow slowly (and even molt) and feed when conditions permit during the period of generally adverse climate. In weak diapause the induction phase is relatively short, since the biochemical adjustments which an insect makes in order to cope with the adverse conditions are relatively simple. As a corollary of this, insects that overwinter in weak diapause are not, for example, very cold-tolerant. The refractory phase is short so that the activated phase is entered relatively soon after diapause has begun, and diapause is quickly terminated when environmental conditions return to normal. Conversely, in "strong" diapause, which is the rule in insects from severe climates, there is a lengthy induction phase, after which development is fully suppressed. The refractory phase usually lasts for several weeks or months, and the activation phase usually does not begin until diapause is more than half over. The termination phase is relatively slow, normally spanning 2 or 3 weeks after the return of suitable climatic conditions. Frequently insects which overwinter in strong diapause are very cold-hardy.

Diapause was formerly subdivided into facultative and obligate diapause. Facultative diapause described the environmentally controlled diapause of bivoltine and multivoltine species (having two or more generations per year) in which the members of certain generations had no diapause in their life history. Obligate diapause referred to the diapause found in univoltine species (those with one generation per year) in which every member of the species undergoes diapause. It was incorrectly assumed that in univoltine species diapause was not induced by environmental factors. Careful experimental work on a number of univoltine species has now revealed that in these species diapause is environmentally controlled. Further study may well demonstrate that this is always the case and render invalid the distinction between obligate and facultative diapause.

Induction and Termination. Various factors may influence the course of diapause. Photoperiod is especially important in the induction of diapause, though ambient temperatures and diet during the preparatory and induction phases may influence the incidence (proportion of individuals entering diapause) and intensity of dormancy. As noted earlier, the refractory phase commonly requires that an insect be chilled for a certain length of time. For some species, however, exposure to long daylengths may supplement or re-place the temperature treatment. Diapause is normally terminated spontane-ously; that is, as soon as temperatures return to reasonable levels after the activated phase has been entered, development continues. In some species, however, photoperiod or availability of moisture are important determinants in the onset of postdiapause development.

For the great majority of insects that exhibit a photoperiodically induced diapause it is the absolute daylength which is critical rather than daily changes in daylength. Most insects studied to date show a long-day response to photo-period (Figure 22.5A). That is, when reared under long-day conditions, they show continuous development, whereas at short daylengths diapause is in-duced. Between these extremes is a critical daylength at which the incidence of diapause changes abruptly. Examples of insects that show a long-day response are the Colorado potato beetle, *Leptinotarsa decemlineata*, and the pink bollworm, *Pectinophora gossypiella*. In a number of species, including the silkworm, *Bombyx mori*, diapause is induced when the daylength is long, while at short daylengths development is continuous. Such insects are said to show a short-day response (Figure 22.5B). The European corn borer, *Ostrinia nubilalis*, and the imported cabbage worm, *Pieris brassicae*, have a short-day–long-day response to photoperiod; that is, the incidence of diapause is low at short and long daylengths, but high at intermediate daylengths (14–16 hours of light per day) (Figure 22.5C). The ecological significance of such a response is unclear, since under natural conditions, insects would already be overwinter-ing in diapause when the daylength was short. A few northern species of Lepidoptera behave in the opposite manner, namely, show a long-day–short-

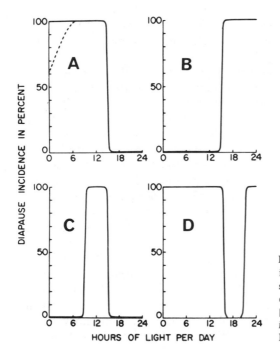

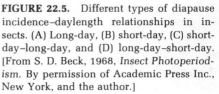

FIGURE 22.5. Different types of diapause incidence–daylength relationships in in-sects. (A) Long-day, (B) short-day, (C) short-day–long-day, and (D) long-day–short-day. [From S. D. Beck, 1968, *Insect Photoperiod-ism*. By permission of Academic Press Inc., New York, and the author.]

day response to photoperiod (Figure 22.5D). All photoperiods except those with 16 to 20 hours of light per day, induce diapause. Again, however, the ecological value of such a response is uncertain.

The precise value of the critical daylength for a species varies with latitude (Figure 22.6). For example, the sorrel dagger moth, *Acronycta rumicis*, studied in Russia by Danilevskii [1961], is a long-day insect which, near Leningrad (latitude about 60°N), has a critical daylength of about 19 hours. In more southerly populations the critical daylength is gradually reduced and is, for example, only 15 hours on the Black Sea coast (43°N).

In the dragonfly, *Anax imperator*, and perhaps a few other insects, diapause is induced by daily changes in daylength rather than by absolute number of hours of light per day. *Anax* larvae that enter the final instar by the beginning of June are able to metamorphose the same year. Those which reach the final instar after this date enter diapause and do not emerge until the following spring. It seems that larvae are able to determine the extent by which the daylength increases. When the daily increment is 2 minutes or more per day larvae can develop directly, whereas at smaller increments or decreases in daylength diapause is induced (Corbet, 1963).

Temperature may profoundly modify or overrule the normal effect of photoperiod on diapause induction. For example, the critical photoperiod depends on the particular (constant) temperature at which insects are maintained: in *A. rumicis* a 5°C difference in temperature results in a 1-hour difference in the critical daylength. At extreme values the effects of temperature may overcome those of photoperiod with reference to induction of diapause. In long-day insects exposure to constant high temperature may completely avert diapause induction regardless of photoperiod. Conversely, in short-day insects high temperature induces diapause, even under long-day conditions.

In nature temperatures normally fluctuate daily about a mean value. This daily fluctuation (thermoperiod) also may modify the influence of photoperiod according to whether or not it is in phase with the light–dark cycle. For example, in *A. rumicis* the incidence of diapause was increased by low nighttime temperatures and vice versa, though at very long daylengths (18 hours or more) temperature had little effect.

In most species studied diet influences the induction of diapause only slightly or not at all. In *P. gossypiella*, for example, the incidence of diapause

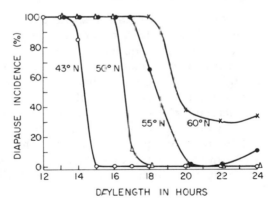

FIGURE 22.6. Effect of photoperiod on diapause incidence in *Acronycta rumicis* (Lepidoptera) populations from different northern latitudes. [From S. D. Beck, 1968, *Insect Photoperiodism*. By permission of Academic Press Inc., New York, and the author.]

induction may be increased by feeding the larvae on cotton seeds whose water content is low and/or oil content high, provided that the daylength is not much greater than the critical value.

For most insects diapause termination is not under photoperiodic control but occurs, under natural conditions, when suitable temperatures for development return in spring. In a few species, however, exposure to appropriate photoperiods will terminate diapause. For example, in *H. cecropia* and *P. gossypiella* long-day conditions terminate diapause. Conversely, in some *Limnephilus* species (Trichoptera), which in the adult stage have a summer diapause, diapause is ended by short daylengths.

In some species, especially those which overwinter in the egg stage or in a partially dehydrated condition, contact with liquid water is necessary for the initiation of postdiapause development. In diapausing larvae of *O. nubilalis*, for example, whose water content falls by midwinter to about 50% of the prediapause level, uptake of water (by drinking) is essential before the insect can continue its development (Beck, 1968). *Lestes congener*, a damselfly found on the Canadian prairies, oviposits in late summer in dried-out, dead stems of *Scirpus* (bulrush). The eggs begin to develop immediately but only to the end of anatrepsis and then enter diapause. Postdiapause development in the spring will not begin until the eggs are wetted, regardless of temperature. Wetting is achieved under natural conditions as the level of the water rises during snow melt and also as a result of wind action, which causes ice movements and subsequent breaking and submersion of the plant stems (Sawchyn and Gillott, 1974a).

4. Water

Water, an essential constituent of living organisms, is obviously an important determinant of their distribution and abundance. Active organisms must retain within their body a certain proportion of water in order for metabolism to occur normally. Deviation from this proportion for any length of time results in injury or death.

4.1. Terrestrial Insects

For terrestrial organisms, the problem generally is to reduce water loss from the body, which occurs as a result of surface evaporation and during excretion of metabolic wastes. Surface evaporation is especially important in small organisms, including insects whose surface area is relatively large in relation to body volume. That insects have been able to solve this problem is one of the main reasons for their success as a terrestrial group. Not only do insects generally possess a highly impermeable cuticle (see Chapter 11) and various devices for reducing water loss from the respiratory system (Chapter 15), but they also have an efficient method of excretion, that is, one which uses a minimum of water (Chapter 18). Such water loss as does occur is normally made up by drinking or from water in the food, though members of a few species from very dry habitats are able to take up water from moist air should the opportunity arise, or use water produced in metabolism.

The importance of water is not restricted to postembryonic stages; during embryogenesis, also, the correct proportion of water must be present within the egg. Again, the primary problem is to prevent water loss (unlike postembryonic stages of most species, eggs cannot move in search of water or into habitats where loss is reduced!). To facilitate this a female may oviposit in a moist medium and/or surround the eggs with protective material (Chapter 19). In addition, the eggshell (chorion) is highly impermeable to water. As a result the egg is very resistant to desiccation and is frequently the stage in which periods of drought are overcome.

Dormant stages, especially those in diapause, are sometimes able to withstand considerable water loss without damage, though the deficiency must be remedied before development can continue. As noted above, renewal of the normal water content may serve as the signal which initiates postdiapause development. The example of O. nubilalis larvae has already been cited. Likewise, yellow woollybear caterpillars (Diacrisia virginica) enter diapause as mature larvae weighing about 600 mg. During diapause their weight falls to about 200 mg, mainly as a result of the loss of water. This loss must be made up in the spring before the larvae can pupate.

In view of the importance of water, it is not surprising to find that many terrestrial insects behave in a characteristic manner with respect to moisture in the surrounding air or substrate. The response may have immediate survival value for the individual concerned or may confer a long-term advantage on the individual and, ultimately on the species. The ability to recognize and respond to slight differences in relative humidity enables an insect to move into a region of preferred humidity. Not only does this have immediate survival value, but because other individuals of the species will tend to respond similarly it may also increase the chances for perpetuation of the species. Some insects seek out sites with a preferred humidity, in which to enter diapause. Though this behavior is of no immediate value to the insect, it increases the chances of survival of the dormant stage. Similarly, female grasshoppers about to oviposit dig "test holes" with their ovipositor to determine the moisture content (and probably other features) of the soil. Eggs are normally laid in moist soil, and a female may retain the eggs in the oviducts for some time if she does not immediately find a suitable site. Again, this behavior has no immediate value to the female but certainly increases the eggs' chances of survival.

Thus far, the discussion has emphasized the harmful effects of too little water and the mechanisms by which terrestrial insects avoid this problem. On occasions, however, too much moisture may be equally detrimental to insects' survival. The effects of excessive moisture may be direct (viz., causing drowning) but more often are indirect. For example, insects that normally develop cold-hardiness partially as a result of dehydration may be less cold-hardy and therefore less capable of surviving low temperatures of winter if this has been preceded by a wet fall. Wet conditions may also affect a species' food supply. However, the most important way in which excessive moisture affects insect populations is by stimulating the development and spread of pathogenic microorganisms (bacteria, protozoa, fungi, and viruses). For example, in the summer of 1963 in Saskatchewan (Canada) the weather was abnormally humid, with above average rainfall in some areas of the province. These conditions appeared ideal for the fungus *Entomophthora grylli*, which underwent a widespread epizootic, causing high mortality in populations of several species of grasshoppers, especially *Camnula pellucida*, the clear-winged grasshopper, and to a lesser extent *Melanoplus bivittatus* (two-striped grasshopper) and *M. packardii* (Packard's grasshopper). Such was the effect of the fungus on *C. pellucida* that by the fall of 1963 its proportion in the grasshopper species complex had fallen to 7% compared with 64% the previous year (Pickford and Riegert, 1964).

Finally, the beneficial effects of snow on the survival of insects must be noted. Snow is an excellent insulator and in extremely cold climates serves to reduce considerably the rate of heat loss from the substrate. Thus, the substrate remains considerably warmer than the air above the snow. For example, with an air temperature of $-30°C$ and a snow depth of 10 cm, the temperature of soil about 3 cm below its surface is about $-9°C$. In the absence of snow the soil temperature at this depth is only a degree or two higher than that of the air. This means that species with only limited cold-hardiness may be able to survive the winter in cold climates provided that there is ample snow cover. In other words, because of snow a species may be able to extend its geographical range into areas with low winter temperatures. In Saskatchewan, for example, the damselflies *Lestes disjunctus* and *L. unguiculatus* overwinter as eggs (in diapause) laid in emergent stems of *Scirpus*. The eggs can tolerate exposure to temperatures as low as $-20°C$ and remain viable. Below this temperature mortality increases significantly (Sawchyn and Gillott, 1974b). At Saskatoon, where this study was carried out, the *mean* temperature for January is, however, about $-22°C$, though the temperature frequently falls well below this value (the record low being about $-48°C$!). Field collection of eggs throughout the winter showed that, whereas the viability of eggs from beneath the snow remained near 100%, no eggs collected from exposed stems survived. Thus, the insulating effect of snow is essential to the survival of these species in this region of Canada. In addition, the snow cover may also prevent desiccation.

4.2. Aquatic Insects

The most important features of the surrounding medium that affect the distribution and abundance of aquatic insects appear to be its temperature, oxygen content, ionic content, and rate of flow. The influence of temperature on development and activity (through its effect on oxygen content) has already been outlined in Sections 2.1 and 2.2.

The ability of insects to regulate both the total ionic concentration and the level of individual ions in the hemolymph is a major determinant of their distribution. Typical freshwater insects are restricted to waters of low ionic content because, although they are capable of excreting excess water that enters their body osmotically, they have no mechanism for removing excess ions which enter the body when the insect is in a saline medium; that is, they cannot produce a hyperosmotic urine (see Chapter 18, Section 4.3). Further, members of some species may be unable to colonize some freshwater habitats because these contain certain ions such as Mg^{2+} and Ca^{2+} in too high a concentration.

In contrast, members of many species that normally inhabit saline environments appear to be able to regulate their hemolymph osmotic pressure and ionic content over a wide range of external salt concentrations. In other words, they can produce hyperosmotic urine when it is necessary, in a saline medium, to excrete excess ions, or hypoosmotic urine, when in freshwater when excess water must be removed from the body (see Chapter 18, Section 4.2). As they are normally found only in saline habitats, it must be assumed that their distribution is governed by other environmental factors.

The insect fauna of an aquatic habitat may vary with the speed at which the water is moving. Insects in still or slowly moving water are not prevented from moving, for example, in search of food or to the surface for gaseous exchange. In contrast, rheophilic insects (those that live in swiftly moving streams or rivers) have of necessity evolved structural, behavioral, and physiological adaptations to survive in these habitats. Among the structural adaptations which may be found in rheophilic insects are flattening or streamlining of the body, and the development of friction discs or hydraulic suckers (Hynes, 1970a,b). Flattening may take on differing significance among species, though ultimately its function is to enable insects to avoid being washed downstream by the current. In members of some species, which live on exposed surfaces, flattening enables them to remain within the so-called "boundary layer," a thin layer of almost static water covering the substrate. For members of most species flattening is associated with their cryptic habit, permitting them to live under stones, in cracks, crevices, etc. Streamlining, too, is a modification mainly used by insects to avoid currents by burrowing into the substrate, though members of a few streamlined species, for example, most species of *Baetis* and *Centroptilum* (mayflies), do live on exposed surfaces and are able to swim against quite strong currents (Hynes, 1970a,b).

The major physiological adaptation of rheophilic species is related to gaseous exchange. Because of the danger of being washed downstream, insects in moving water cannot come to the surface to obtain oxygen; they rely on oxygen dissolved in the medium. Through evolution, members of rheophilic species have become adapted to a medium with a high oxygen content and conduct most or all gaseous exchange directly across the body wall. Further, they depend on the water current to renew the oxygen supply at their body surface. As a result, in many species, gills, if present, are reduced, and the ability to ventilate, by flapping the gills or undulating the abdomen, has been lost.

Their relative inability to move because of the current has been paralleled, in many rheophilic insects, by the evolution of devices which enable them to obtain food passively; that is, they depend on the current bringing food (especially microorganisms and detritus) to them. These devices include the nets built by many trichopteran larvae, fringes of hairs on the forelegs and/or mandibles of some larval Plecoptera, the fans on the premandibles of blackfly larvae, and the sticky strings of saliva produced by the chironomid *Rheotanytarsus* (Hynes, 1970a,b).

An important factor in the distribution of aquatic insects, and one which is related to the extent of water movement, is the substratum. Members of many species of stream insects are characteristically associated with particular types of substratum. For some insects the significance of this association is easily understood. For example, water pennies [larvae of Psephenidae (Coleoptera)], found in fast-moving waters, require largish rocks to which they can become attached. Similarly, larval Blepharoceridae (Diptera) need smooth rocks, not covered with silt or algal growth, to which to attach their suckers. And some Leuctridae (Plecoptera) require gravel of the correct texture in which to burrow.

5. Wind

Because of their weight and relatively large surface area/volume ratio, insects may be profoundly affected by wind. By altering the rate of evaporation of water from the body surface wind may be important in the water relations of the insect. Flight activity (whether or not flight occurs, the direction of movement and the distance traveled) is also directly related to the strength and direction of the wind. Wind action may also exert indirect effects on insects, for example, by causing erosion of soil or snow so that the insects (or their eggs) are exposed to predators, extremes of temperature, or desiccation.

Through its effect on the flight activity of winged insects and because insects by virtue of their weight are easily transported on wind currents, wind is an important factor in dispersal, the movement away from a crowded habitat so that scattering of a population results. Though a good deal of insect dispersal is of no benefit, for some species the dispersal is adaptive, that is, confers a long-term advantage on the species by transferring some adult members to new breeding sites. Because of its advantageous nature, physiological, structural, and behavioral features which facilitate adaptive dispersal (=migration) will become fixed in a population through natural selection.

5.1. Migration

Johnson (1969, p. 8) describes migration as "essentially a transference of adults of a new generation from one breeding habitat to others." In many species migration begins shortly after the molt to the adult and mass migrations are frequently preceded by highly synchronized adult emergence. In a sense, therefore, migration forms part of a species' development just as do mating and oviposition.

The form of migration varies widely among species. Some of the variables are the proportion of the population which migrates, whether migration occurs in every generation or only in certain generations, the distance traveled, and the nature of the migratory movements (wind-dependent or wind-independent, feeding en route or proceeding directly). For example, the swarming flights of social insects, such as ants and termites, involve only a fraction of a colony's population, may be completed in a matter of minutes, and may take the migrating individuals only a few yards from the original colony. In contrast, the migrations of locusts are undertaken by all members of a population and may cover several thousand miles. The migrations extend over a number of weeks and are interspersed with short periods of feeding activity. Johnson (1969) suggests that all forms of migration may be arranged in three major categories, though there is gradation both within and between each of them. In the first category are included species which, as adults, migrate from the emergence site to a new breeding site where they oviposit, then die. Johnson includes in this group species such as the housefly, all of whose members, at emergence, leave the old habitat and disperse randomly. After a period of maturation, they seek out new breeding sites. The category also contains species which seasonally produce populations of weakly flying, migratory individuals, for example, aphids, termites, and ants. Migrations are largely wind-dependent and may occur in every direction from the emergence site. The migration time and distance are usually short, and once individuals reach a suitable breeding site, they remain there for the rest of their lives. Indeed, on reaching such a site, some species characteristically shed their wings.

Also placed in the first category, but having migrations which are of a much grander scale, are the migratory locusts. Desert locusts, *Schistocerca gregaria*, for example, may travel thousands of miles as they move from one breeding area to another as each becomes unsuitable because of drought (Figure 22.7). Like those of aphids, etc., the migrations of desert locusts are wind-dependent. Because they inhabit fairly dry regions, breeding in the desert locust is synchronized with the arrival of a rainy season. As rain comes to different parts of the inhabited area at different times of the year, adults migrate in order to continue breeding activity. It is the winds on which locusts migrate that also bring rain to the new areas. During spring breeding occurs in north and northwest Africa (in conjunction with rain in the Mediterranean region), in the Middle East across to Pakistan, and to the south in East Africa. In the latter regions, local seasonal rains occur at this time. As northern Africa and the Middle East become dry in early summer, locusts migrate southward on prevailing winds to an area which runs across Africa, from east to west, lying just south of the Sahara desert, then northward across southern Arabia and into Pakistan. This area is closely associated with the Inter-Tropical Convergence Zone (ITCZ), where hot, northbound air from the equatorial region meets cooler air flowing south. The mixing of these air masses within the ITCZ results in the production of rain and a reduction of wind speed so that the locusts again become earthbound. Locusts from East Africa, south of the ITCZ, move north and east on prevailing winds to be deposited in southern Arabia and India. These summer migrations may take locust swarms several hundred or even thousands of miles in a relatively short time. In contrast, the fall and winter movements which bring locusts to their spring breeding sites generally consist of a number of shorter migrations made over a longer period of time, mainly because the air temperatures at this time of the year are only intermittently suitable for migration. The northward migration results largely from cyclonic weather disturbances that move eastward across Africa every few days. These disturbances bring with them warm southerly winds on which locusts may be carried. As the winds push northward they mix with cooler air, which results in rainfall and temporary cessation of migration. Successive waves of warm air gradually bring the locusts to their spring habitat. The above summary of the annual movements of locusts is of necessity extremely simplified. Neverthe-

seasonal movements (———▶) breeding areas (▨▨)

limit of invasion area (————)

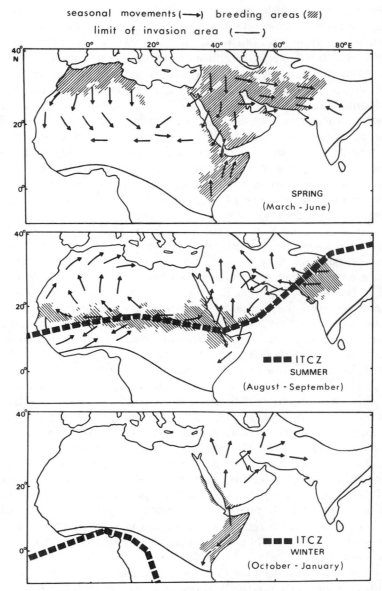

FIGURE 22.7. Major movements of swarms of *Schistocerca gregaria* from spring, summer, and winter breeding areas, in relation to the position of the Inter-Tropical Convergence Zone (ITCZ). [From Z. Waloff, 1966, *Antilocust Memoir* **8**. Crown copyright 1966. Reproduced with permission of the Controller of Her Britannic Majesty's Stationery Office.]

less, it shows how migration, based on wind currents, has evolved as an integral part of locusts' life history to facilitate year-round breeding activity through the exploitation of temporarily suitable habitats.

Finally, category 1 includes some species (mainly Lepidoptera) whose migrations are independent of wind currents. That is, the insects do not rely on wind to make the migrations, though their movements are undoubtedly influenced by wind speed and direction. An important difference, therefore, between these "active" migrations and the "passive" movements of the form described, say, for locusts, is that, in the former, insects must have a "sense of direction." In other words, they must be able to orient themselves with reference to environmental cues, especially the sun. Because the sun's position changes through the day the insects must also have a sense of time so that they may make corrections to account for this change (the light–compass reaction) (see Chapter 12, Section 7.1.4). A much-studied example of a species that migrates under its own power is the great southern white butterfly *Ascia monuste* (Pieridae), whose migrations up and down the Florida coastline are well known. In Florida the species breeds year round but not in all localities simultaneously. Periodically populations comprising immature females and males of all ages make migratory flights over distances up to 100 miles or more to new areas where *Batis maritima* (maritime salt wort), the primary host plant, is abundant. In contrast to the situation in locust migration, it is not the arrival of adverse conditions that stimulates migration in *A monuste*. The migrations occur from areas where food and oviposition sites are still abundant. Though the sun has been suggested as a reference point by which the butterflies orient themselves during flight, local cues are also important. For example, the insects may closely follow the shoreline, roads, railway tracks, or telephone lines.

In his second category Johnson (1969) includes species whose migration is in two parts, an emigration to feeding sites where sexual maturation occurs, followed by a return flight to the original (or a similar) site of emergence where the insects oviposit. Many Odonata, for example, do not remain in the area of the pond from which they emerge but migrate to nearby woods or hedgerows to prey on other insects until mature. They then return to water, mate, and lay eggs. Some species regularly return to the feeding habitat between each oviposition period. Some species of mosquitoes also have a two-part migration, first to find a host on which to feed and later to locate an egg-laying site. In some cases the initial part of the migration is considerably longer than the second. Like that of Odonata, the migratory flight of some mosquitoes is wind-independent. For most migratory species, however, wind determines the direction and distance traveled.

The third category includes species which again have a two-part migration. The initial migratory flight takes the species to suitable hibernating or aestivating sites where they enter diapause, after which they return to the region in which they emerged and reproduce. Within this type of migration three subcategories can be recognized. In the first the sites of diapause are within the general breeding area of the species. Species which adopt this arrangement include the Colorado potato beetle, *Leptinotarsa decemlineata*, and the corn thrips, *Limothrips cerealium*.

Belonging to the second subcategory are species that migrate to a climatically different region prior to diapause. Especially common is migration between warmer lowland areas where the insects have emerged and mountainous regions, either to avoid summer heat or to overwinter. Such migrations are seen in various noctuid moths, for example, the army cutworm, *Euxoa auxiliaris*, in Montana, which moves southwest to the Rocky Mountains, and in some coccinellid beetles, such as the convergent lady beetle *Hippodamia convergens*, in northern California. Adult beetles first appear in early May and soon most migrate, using prevailing winds, to mountain canyons in the Sierra Nevada range where they aggregate under stones, litter, etc., and enter diapause. Diapause lasts for about 9 months, and the following February and March adults, again windborne, return to the valleys where a new generation is produced. Breeding activity is thus closely correlated with mass emergence of spring-breeding aphids. (Interestingly, as a result of Man's agricultural activities, aphids are now available on a year-round basis, and some populations of *H. convergens* no longer migrate but through the summer produce several generations of progeny, the last one of which overwinters in diapause at the breeding site.)

Migrations included in the third subcategory differ from those of the second subcategory only in terms of the distance covered, especially during prediapause movements. The classic example of a species in this group is the monarch butterfly, *Danaus plexippus*, some populations of which spend early summer in southern Canada, then migrate southward during August to October to overwintering sites in southern United States. Mark and recapture experiments have demonstrated that the butterflies may travel upward of 2000 miles at average speeds of about 20 miles per day. Since on certain days the butterflies may not migrate because of poor weather conditions, the distances traveled on suitable days may be considerably in excess of the average value. Females which migrate south are sexually immature and remain in reproductive diapause until about the end of February when oogenesis begins and butterflies begin to move slowly northward. However, there is some dispute as to whether overwintering insects make the return journey to their original habitat. Some authors believe this to be the case, whereas others believe that the northward migration occurs in a series of shorter steps, each made by successive generations. The latter arrangement would place *D. plexippus* in the same category as the migratory locusts. Like those of locusts the migrations (southward and northward) depend very largely on wind currents.

As the above examples demonstrate, migration can take on many forms among different species, yet its common purpose is to improve a species' ability to survive and multiply. For most species, wind supplies the power for migration, and their physiology and behavior have so evolved as to make best use of this.

6. Summary

The distribution and abundance of insects are markedly affected by temperature, photoperiod, water, and wind.

As poikilotherms, insects have a metabolic rate that within species- and stage-specific limits is proportional to temperature. Their rate of development within these limits is inversely proportional to temperature. Outside these limits insects will survive, but their development is retarded or prevented. The temperature extremes for survival are known as the upper and lower lethal limits. Survival at extreme ambient temperatures may be accomplished by (1) behavioral means, such as burrowing or ovipositing in a substrate; and/or (2) entering a physiologically dormant condition (diapause). At below-freezing temperatures insects may also become freezing-tolerant, that is, capable of withstanding freezing of their extracellular fluids, or, when they are freezing-susceptible, become supercooled. In both arrangements, polyhydroxyl cryo-protectants such as glycerol, sorbitol, and trehalose are important. In species from habitats whose climate is suitable for development and/or reproduction for a limited period each year, temperature may be an important synchronizer of development and/or eclosion.

Photoperiod, the naturally occurring 24-hour cycle of light and darkness, exerts both short-term and long-term effects on behavior and physiology which keep insects in tune with changing environmental conditions. In a few species daily activities are triggered by changing light intensity, that is, are of exogenous origin. However, in most species, diurnal rhythms of activity, for example, locomotor activity, feeding, mating behavior, oviposition, and eclosion, originate endogenously and are phase set by photoperiod.

By responding to seasonal changes in photoperiod insects can exploit suitable environmental conditions for development and reproduction and survive periods when climatic conditions are adverse. Among the long-term processes affected by photoperiod are the nature and rate of development, reproductive ability and capacity, synchronized eclosion, diapause, and possibly cold-hardiness.

Diapause is a genetically determined state of suppressed development. It may occur at any stage of the life history, though this is usually species-specific. Photoperiod exerts its influence at a stage earlier than the one in which diapause occurs. In this way insects are able to anticipate the onset of adverse conditions. Induction of diapause is, in almost all species, a response to the absolute daylength (number of hours of light in a 24-hour cycle) rather than daily differences in the daylength. For a species, there is a critical day-length at which the incidence of diapause (proportion of individuals that enter diapause) changes markedly. Long-day insects develop continuously at all daylengths above the critical daylength (usually about 16 hours of light per day) but enter diapause at shorter daylengths. In short-day insects, development is continuous at daylengths below the critical value (usually about 12 hours). In short-day–long-day insects, development is continuous at short and long daylengths, but at intermediate daylengths (about 14–16 hours of light per day) the incidence of diapause is high. Long-day–short-day insects develop continuously within a narrow range of daylengths (16–20 hours light per day) and enter diapause at all other daylengths. The value of the critical daylength for a species may change with temperature and latitude.

Water is an important determinant of the distribution and abundance of insects. A problem for most terrestrial species is to reduce water loss from the integument and tracheal system and in excretion. In postembryonic stages this is achieved by means of a relatively impermeable cuticle, valves and/or hairs that reduce water vapor movement out of the tracheae, production of highly concentrated urine, as well as by active selection of more humid microclimates. Eggs are covered with a cuticlelike chorion and may be laid in an ootheca and/or substrate. For some species in cold climates snow cover may be important as an insulator and in preventing desiccation.

In addition to temperature and light, important abiotic factors affecting the distribution and abundance of aquatic insects are oxygen content, ionic content, and rate of movement of the surrounding water.

Because of their size, insects may be greatly affected by wind. Wind affects the rate of water loss from the body and is an important agent of dispersal in many terrestrial species. Dispersal by wind may be adaptive, that is, advantageous to a species which can migrate to new breeding sites. The form of migration varies among species, in terms of the proportion of the population that migrates, whether migration occurs in all or selected generations, the distance traveled, and whether the migration is continuous or intermittent.

7. Literature

Information on the effects of abiotic factors on the distribution and abundance of insects is provided by Andrewartha and Birch (1954), Andrewartha (1961), and Varley et al. (1973). Beck (1968) discusses insect photoperiodism in detail. Tauber and Tauber (1976) and Jungreis (1978) have reviewed insect diapause. Insect migration and dispersal is the subject of Johnson's (1969) text. Salt (1961), Asahina (1966, 1969), Baust and Morrissey (1977), and Danks (1978) review insect cold-hardiness. Factors affecting the distribution of aquatic insects are discussed by Macan (1962, 1974), Hynes (1970a,b), and by various authors in the text edited by Merritt and Cummins (1978). The latter contains an extensive bibliography. Cloudsley-Thompson (1975) reviews the adaptations of insects in arid environments.

Andrewartha, H. G., 1961, Introduction to the Study of Animal Populations, University of Chicago Press, Chicago, Ill.

Andrewartha, H. G., and Birch, L. C., 1954, The Distribution and Abundance of Animals, University of Chicago Press, Chicago, Ill.

Asahina, E., 1966, Freezing and frost resistance in insects, in: Cryobiology (H. T. Meryman, ed.), Academic Press, New York.

Asahina, E., 1969, Frost resistance in insects, Adv. Insect Physiol. 6:1–49.

Baust, J. G., and Morrissey, R. E., 1977, Strategies of low temperature adaptation, Proc. XV Int. Congr. Entomol., pp. 173–184.

Beck, S. D., 1968, Insect Photoperiodism, Academic Press, New York.

Cloudsley-Thompson, J. L., 1975, Adaptations of Arthropoda to arid environments, Annu. Rev. Entomol. 20:261–283.

Corbet, P. S., 1963, A Biology of Dragonflies, Quadrangle Books, Chicago, Ill.

Danilevskii, A. S., 1961, Photoperiodism and Seasonal Development of Insects (Engl. transl., 1965), Oliver and Boyd, Edinburgh.

Danks, H. V., 1978, Modes of seasonal adaptation in the insects. I. Winter survival, Can. Entomol. 110:1167–1205.

Downes, J. A., 1964, Arctic insects and their environment, Can. Entomol. 96:280–307.

Hynes, H. B. N., 1970a, The Ecology of Running Waters, University of Toronto Press, Toronto.

Hynes, H. B. N., 1970b, The ecology of stream insects, Annu. Rev. Entomol. 15:25–42.

Johnson, C. G., 1969, Migration and Dispersal of Insects by Flight, Methuen, London.

Jungreis, A. M., 1978, Insect dormancy, in: Dormancy and Developmental Arrest (M. E. Clutter, ed.), Academic Press, New York.

Macan, T. T., 1962, Ecology of aquatic insects, Annu. Rev. Entomol. 7:261–288.

Macan, T. T., 1974, Freshwater Ecology, 2nd ed., Wiley, New York.

Mansingh, A., 1971, Physiological classification of dormancies in insects, Can. Entomol. 103:983–1009.

Merritt, R. W., and Cummins, K. W. (eds.), 1978, An Introduction to the Aquatic Insects of North America, Kendall Hunt, Dubuque, Iowa.

Pickford, R., and Riegert, P. W., 1964, The fungous disease caused by Entomophthora grylli Fres., and its effects on grasshopper populations in Saskatchewan in 1963, Can. Entomol. 96:1158–1166.

Salt, R. W., 1961, Principles of insect cold-hardiness, Annu. Rev. Entomol. 6:55–74.

Saunders, D. S., 1974. Circadian rhythms and photoperiodism in insects, in: The Physiology of Insecta, 2nd ed., Vol. II (M. Rockstein, ed.), Academic Press, New York.

Sawchyn, W. W., and Gillott, C., 1974a., The life history of Lestes congener (Odonata: Zygoptera) on the Canadian Prairies, Can. Entomol. 106:367–376.

Sawchyn, W. W., and Gillott, C., 1974b, The life histories of three species of Lestes (Odonata: Zygoptera) in Saskatchewan, Can. Entomol. 106:1283–1293

Sawchyn, W. W., and Gillott, C., 1975, The biology of two related species of coenagrionid dragonflies (Odonata: Zygoptera) in Western Canada, Can. Entomol. 107:119–128.

Tauber, M. J., and Tauber, C. A., 1976, Insect seasonality: Diapause maintenance, termination, and postdiapause development, Annu. Rev. Entomol. 21:81–107.

Varley, G. C., Gradwell, G. R., and Hassell, M. P., 1973, Insect Population Ecology: An Analytical Approach, Blackwell, Oxford.

23

The Biotic Environment

1. Introduction

This chapter will deal with the biotic environment of insects, which is composed of all other organisms that affect insects' ability to survive and multiply. In other words, the interactions of insects with other organisms (of the same and other species) will be discussed. As most insects eat living organisms, food is probably the most obvious and important biotic factor. However, other interactions are known which, though not as easily recognized as feeding, are nonetheless important regulators of insect distribution and abundance.

2. Food

2.1. Quantitative Aspects

Though the amount of food available might be considered as an important regulator of insect abundance, it has been found in natural communities that populations do not normally use more than a small fraction of the total available food. This is primarily because other components of the environment, especially weather but including, for example, predators, parasites, or pathogens, usually have a significant adverse effect on growth and reproduction. Other features of insects may, however, be important in this regard. Many species, especially plant feeders, are polyphagous. Thus, when the preferred food plant is in limited quantity, alternate choices can be used. Among endopterygotes, larvae and adults of a species may eat quite different kinds of food, and in some species such as mosquitoes the food of the adult female differs from that of the adult male.

Two situations may occur in which the quantity of food limits insect distribution and abundance. In the first, there is no absolute shortage of food, but only a proportion of the total is available to a species. Thus, there is said to be a "relative shortage" of food. Various reasons may account for the food not being available. (1) The food may be concentrated within a small area so that it is available to relatively few insects. As an interesting example of this Andrewartha (1961) cites the Shinyanga Game-Extermination Experiment in East

627

Africa in which, over the course of about 5 years, the natural hosts of tsetse flies were virtually exterminated over an area of about 800 square miles. At the end of this period one small elephant herd and various small ungulates remained in the game reserve. However, almost no tsetse flies could be found, despite the fact that, collectively, the mammals which remained could supply enough blood to feed the entire original population of flies. The distribution of the food was now so sparse that the chance of flies obtaining a meal was practically nil. (2) The food may be randomly distributed but difficult to locate. Thus, only a fraction of the individuals searching ever find food. Such is probably the situation in many parasitic or hyperparasitic species whose host is buried within the tissues of plants or other animals. (3) A proportion of the food may occur in areas which for other reasons are not normally visited by the consumer so that, in effect, it is not available.

In the second situation, food may become a limiting factor in population growth when a species' numbers are not kept in check by other influences, especially natural enemies. This may happen, for example, when a species is accidently transferred (often as a result of Man's activity) from its original environment to a new geographic area where its natural enemies are absent. Under these conditions, the population may grow unchecked, and its final size is limited only by the amount of food available. Occasionally, even in a species' natural habitat, food may limit population growth, for example, when weather conditions are favorable for development of a species but not for development of those organisms which prey on or parasitize it.

2.2. Qualitative Aspects

The nature of the food available may have striking effects on the survival, rate of growth, and reproduction potential of a species, and much work has been done on insects in this regard. For example, of the insect fauna associated with stored products, the sawtoothed grain beetle, *Oryzaephilus surinamensis*, can survive only on foods with a high carbohydrate content such as flour, bran, and dried fruit, whereas species of spider beetles, *Ptinus* spp., and flour beetles, *Tribolium* spp., have no such carbohydrate requirement and are consequently cosmopolitan, occurring in animal meals and dried yeast, in addition to plant products. For some phytophagous insects, a combination of plants of different kinds appears necessary for survival and/or normal rates of juvenile development. In the migratory grasshopper, *Melanoplus sanguinipes*, for example, a smaller percentage of insects survive from hatching to adulthood, and the development of those that do survive is slower when the grasshoppers are fed on wheat (*Triticum aestivum*) alone compared with wheat plus flixweed (*Descurainia sophia*) or dandelion (*Taraxacum officinale*) (Pickford, 1962).

Both the rate of egg production and number of eggs produced may be markedly affected by the nature of the food available. Many common flies, for example, species of *Musca, Calliphora,* and *Lucilia*, may survive as adults for some time on a diet of carbohydrate. However, for females to mature eggs a source of protein is essential. Pickford (1962) showed that *M. sanguinipes* females fed a diet which included wheat and wild mustard (*Brassica kaber*) or wheat and flixweed produced far more eggs (579 and 467 eggs per female, respectively) than females fed on wheat (243 eggs/♀), wild mustard (431 eggs/♀), or flixweed (249 eggs/♀), alone. These differences in egg production resulted largely from variations in the duration of adult life, though differences in rate of egg production were also evident. For example, percent survival of females fed wheat plus mustard after 1, 2, and 3 months was 93, 60, and 13% respectively. These females produced, on average, 8.4 eggs/female-day. The corresponding figures for females fed on wheat alone were (1) survival: 87, 27, and 0% over 1, 2, and 3 months, respectively; and (2) rate of egg production: 4.6 eggs/female-day. The metabolic basis for these differences was not determined.

2.3. Insect–Plant Interactions

Deserving of special mention in a discussion of insects and their food are the relationships which have evolved between insects and higher plants. As might be anticipated in view of the length of time over which they have evolved, some of these relationships are extremely intimate and refined, though essentially the relationships have a common theme. Insects gain energy (food) at the expense of plants, whereas plants attempt to defend themselves (conserve their energy) or at least to obtain something in return for the energy which insects take from them. Though the theme remains constant through time, the relationships themselves are always changing as a result of natural selection. Insects strive to improve their energy-gathering efficiency (most often by concentrating on energy in a particular form and from a restricted source and by specialization of the method used to collect the energy) while plants concurrently improve their defenses. There is, as Price (1975a) puts it, "constant warfare" between the two opponents, and this forms the basis of their coevolution.

The most common method used by plants as defense against insects (and other herbivorous animals) is production of toxic metabolites. It is possible that originally some of these metabolites were simply short-lived intermediates in normal biochemical pathways within plants and/or provided a means of storing chemical energy for later use by the plant. In other words, the original function(s) of these compounds may have been unrelated to the occurrence of herbivores. An example of such a compound might be nicotine produced by the tobacco plant (*Nicotiana* spp.). Radioisotope studies have shown that, although about 12% of the energy trapped in photosynthesis is used for nicotine production, the nicotine has a relatively short half-life, 40% of it being converted to other metabolites (possibly sugars, amino acids, and organic acids) within 10 hours.

Another possibility is that the chemicals arose as by-products of a plant's primary metabolism, and the plant, being unable to excrete the molecules, simply retained them within its tissues.

Regardless of the origin and earlier function(s) of these chemicals, the evolution of animals that fed upon the plants which produced them would create strong selection pressure for the production of greater quantities of a chemical or more toxic derivatives of it. This pressure might be especially great on longer-lived plants. Selection would also favor production of greater quantities of a chemical or more toxic derivatives in the reproductive parts of plants, as these parts represent concentrated stores of energy and are therefore especially attractive to herbivores. Price (1975a) cites several examples in support of this proposal, including *Hypericum perforatum* (Klamath weed), which, like other members of the genus, produces the toxicant hypericin. The concentration of the toxicant is 30 μg/g wet weight in the lower stem, 70 μg/g in the upper stem and 500 μg/g in the flower.

Thus animals that become adapted to feeding on plants that produce toxins will be at a considerable advantage over animals which do not. Among herbivores, insects show the greatest ability to cope with the toxins. In part, this arises from the enormous period of time over which coevolution of insects and plants has occurred, but it is also due to insects' high reproductive rate and short generation time which facilitate rapid adaptation to changes that occur in the host plant. Through evolution, many insect species have not only developed increasing tolerance to a host plant's toxins but are now attracted by them. In other words, insects locate food plants by the scent or taste of their toxic substance and frequently are restricted to feeding on such plants. For example, certain species of flea beetles, *Phyllotreta* spp., and cabbageworms, *Pieris* spp., feed exclusively on plants such as Cruciferae that produce mustard oils. Colorado potato beetles, *Leptinotarsa decemlineata*, and various hornworms, *Manduca* spp., feed only on Solanaceae, the family that includes potato (*Solanum tuberosum*) (produces solanine), tobacco (*Nicotiana* spp.) (nicotine), and deadly nightshade (*Atropa belladonna*) (atropine) (Price, 1975a).

The method most often used to overcome the potentially harmful effects of these chemicals is to convert them into nontoxic or less toxic products. Especially important in such conversions is a group of enzymes known as mixed function oxidases which, as their name indicates, catalyze a variety of oxidation reactions. The enzymes are located in the microsome fraction* of cells and occur in particularly high concentrations in fat body and midgut. (Interestingly, it is these same enzymes that are largely responsible for the resistance of insects to man-made insecticides.)

Some insects are able to feed on potentially dangerous plants as a result of either temporal or spatial avoidance of the toxic materials. For example, the life history of the winter moth, *Operophtera brumata*, is such that the caterpillars hatch in the early spring and feed on young leaves of oak (*Quercus* spp.) which lack the highly toxic tannins. Though, later in the season, weather conditions are suitable and food is still apparently plentiful, a second generation of winter moths does not develop because by this time large quantities of tannins are present in the leaves. Spatial avoidance is possible for many Hemiptera whose delicate suctorial mouthparts can bypass localized concentrations of toxin in the host plant. Some aphids feed on senescent foliage where the concentration of toxin is less than that of younger, metabolically active tissue (Price, 1975a).

Price (1975a) proposes that at least four advantages may accrue to an insect

---

*The microsome fraction is obtained by differential high-speed centrifugation of homogenized cells and consists of fragmented membranes of endoplasmic reticulum, ribonucleoproteins, and vesicles.

able to feed on potentially toxic plants. First, competition with other herbivores for food will be much reduced. Second, the food plant can be located easily. Related to this, as members of a species will tend to aggregate on or near the food plant, the chances of finding a mate will be increased. Third, if an insect can store within its tissues the toxin ingested as it feeds, it may gain protection from would-be predators. This appears to be the situation with most of the insect fauna associated with plants of the family Asclepiadaceae (milkweeds), many of which produce cardiac glycosides, substances that at sublethal levels, induce vomiting in vertebrates. Most insect species that feed on milkweed are aposematically (brightly and distinctly) colored, a feature commonly indicative of a distasteful organism and one which makes them stand out against the background of their host plant. On sampling such insects, a would-be vertebrate predator discovers their unpalatability and quickly learns to avoid insects having a particular color pattern. Interestingly, a few insect predators have evolved tolerance to the plant-produced toxins stored by their insect prey and are, themselves, unpalatable to predators further up the food chain! The fourth advantage to be gained by tolerance to these plant products is protection against pathogenic microorganisms. For example, the cardiac glycosides present in the hemolymph of the large milkweed bug, *Oncopeltus fasciatus*, have a strong antibacterial effect.

The channeling of energy into production of toxic or at least repellent substances is the most often used method by which plants may obtain protection, though others are known. A few plants expend this "energy of protection" on formation of structures which prevent or deter feeding, or even harm would-be feeders. For example, passion flower plants (*Passiflora adenopoda*) have minute hooked hairs that grip the integument of caterpillars which attempt to feed on them. The hairs both impede movement and tear the integument as the caterpillars struggle to free themselves so that the insects die from starvation and/or desiccation (Gilbert, 1971). Leguminous plants have evolved a variety of physical (as well as chemical) mechanisms to protect their seeds from Bruchidae (pea and bean weevils). These include production of gum as a larva penetrates the seed pod so that the insect is drowned or its movements hindered, production of a flakey pod surface which is shed, carrying the weevil's eggs with it, as the pod breaks open to expose its seeds, and production of pods which open "explosively" so that seeds are immediately dispersed and, therefore, not available to females which oviposit directly on seeds (Center and Johnson, 1974).

In a curious evolutionary twist, some plants use insects to gain protection from herbivores, in return for which they provide the insects with food and shelter. A well-known example of such a mutualistic relationship is that which has evolved between the bull's-horn acacias (*Acacia* spp.) and ants, *Pseudomyrmex* spp. The aggressive ants guard the plant against herbivores, while the plant produces nectar (in petioles) and protein (in special "Beltian bodies" formed at the tips of new leaves) on which the ants feed (Price, 1975a).

A mutualistic relationship of a very different kind is that in which the plants supply food to insects usually in the form of nectar and pollen, and, in return, insects provide the transport system necessary for effective cross-pollination. The success (importance) of insects as pollinators compared with pollinators from other groups such as birds and bats is presumably a result of their much longer evolutionary association with plants. Most of the modern insect orders were well established by the time the earliest flowering plants appeared about 225 million years ago. Thus, insects were able to gain a considerable head start as pollinators over birds and bats, the earliest fossil records for which date back about 150 and 60 million years, respectively (Price, 1975a).

To achieve effective cross-pollination, two important factors must be taken into consideration in an evolutionary sense. First, plants must produce precisely the right amount of nectar to make an insect's visit energetically worthwhile, yet stimulate visits to other plants, and second, plants of the same species must be easily recognized by an insect. If too much energy is made available by each plant, then insects need visit fewer plants and the extent of cross-pollination is reduced. If a plant produces too little food (to ensure that an insect will visit many plants) there is a risk that the insect will seek more accessible sources of food. Natural selection determines the precise amount of energy which each plant must offer to an insect, and this amount depends on a number of factors. The amount of energy gained by an insect during each visit to a flower is related to both quantity and quality of available food. Thus, until recently, it was considered that many adult insects obtained their carbohydrate requirements from nectar and their protein requirements from other sources such as pollen, vegetative parts of the plant (as a result of larval feeding), or other animals. Baker and Baker (1973) have shown, however, that the nectar of many plants contains significant amounts of amino acids, so that insects can concentrate their efforts on nectar collection. This not only increases the extent of cross-pollination by inducing more visits to flowers, but it may also lead to economy in pollen production, since pollen becomes less important as food for the insects. The amount of nectar produced is a function of the number of flowers per plant. Thus, it is important for plants which have a number of flowers blooming synchronously, that each flower produces only a small amount of nectar and pollen, so that an insect must visit other plants to satisfy its requirements.

More nectar is produced by plant species whose members typically grow some distance apart, so that it is still energetically worthwhile for an insect species to concentrate on these plants. Related to this is the fact that the insects which forage over greater distances are larger species such as bees, moths, and butterflies whose energy requirements are high. When nectar is produced in large amounts, it is typically accessible only to larger insects which are strong enough to gain entry into the nectar-producing area or have sufficiently elongate mouthparts. This ensures that nectar is not wasted on smaller insects who lack the ability to carry pollen to other members of the plant species.

Temperature also affects the amount of nectar produced, as it is related to the energy expended by insects in flight and to the time of day and/or season. For example, in temperate regions and/or at high altitudes, flowers which bloom early in the day or at night, or early or late in the season, when temperatures may not be much above freezing, must provide a large enough reward as to make foraging profitable at these temperatures. An alternative to production of large amounts of nectar by individual flowers, is for plants which bloom at lower temperatures to grow in high density and flower synchronously (Heinrich and Raven, 1972).

Beyond a certain distance between plants, however, the amount of nectar which an insect requires to collect at each plant (in order to remain "interested" in that species) exceeds the maximum amount that the plant is able to produce. Thus, the plant must adopt a different strategy. Among orchids, for example, about one half of the species produce no nectar, but rely on other methods to attract insects, especially deception by mimicry. The flowers may resemble (1) other nectar-producing flowers, (2) female insects so that males are attracted and attempt pseudocopulation, (3) hosts of insect parasitoids, or (4) insects which are subsequently attacked by other territorial insects. These somewhat risky methods of attracting insects are offset by the evolution of highly specific pollen receptors (so that only pollen from the correct species is acquired) and a high degree of seed set for each pollination (Price, 1975a).

It is important for both plants and insects that insects visit members of the same plant species. The chances of this occurring are greatly increased (1) when the plant species has a restricted period of bloom, in terms of both season and/or time of day; (2) where members of a species grow in aggregations, though this is counterbalanced by a restriction of gene flow if pollinators work within a particular plant population; and (3) when the flowers are easily recognized by an insect which learns to associate a given plant species with food. Recognition is achieved as a result of flower morphology (and related to this is accessibility of the nectar and pollen), color, and scent. The advantage to an insect species when its members can recognize particular flowers is that, through natural selection, the species will become more efficient at gathering and utilizing the food produced by those flowers.

The degree of influence that these variables exert is manifest as a spectrum of intimacy between plants and their insect pollinators. At one end of the spectrum, the plant–insect relationship is nonspecific; that is, a variety of insect species serve as pollinators for a variety of plants. Neither insects nor flowers are especially modified structurally or physiologically. At the opposite extreme, the relationship is such that a plant species is pollinated by a single insect species. Flower morphology is precisely complemented by structural features of the pollinator; the plant's blooming period is synchronized with the life history and diurnal activity of the insect; and, where present, nectar is produced in exactly the right quantity and quality to satisfy the insect's requirements.

3. Interactions between Insects and Other Animals

Interactions between insects and other animals (including other members of the same species) take many forms, though most are food-related. Insects may be predators (which require more than one prey individual in order to complete development), parasitoids or parasites (which need only one host to complete development). Parasitoids differ from parasites in that they ultimately kill their "host" which is typically another arthropod. Alternatively, insects may serve as prey or host for other animals. In a third form of interaction, insects may compete either with other members of the species or with other animals for the same resource, for example, food, breeding or egg-laying sites, overwintering sites, or resting sites. Between opposite sexes of the same species, the interaction may be for a very obvious purpose, propagation of the species.

3.1. Intraspecific Interactions

The nature and number of interactions among members of the same species will depend on the density of the population. These interactions may be either beneficial or harmful. It follows that there will be an optimal range of density for a given population, within which the net effect of these interactions will be most beneficial. Outside this range of density, that is, when there is underpopulation or overpopulation (crowding), the net result of these interactions will be less than optimal for perpetuation of the species. Interestingly, animals have evolved various regulatory mechanisms which serve either to maintain this optimal density or to alter the existing density so as to bring it within the optimal range. In the discussion which follows we shall see how these mechanisms operate in some insects.

3.1.1. Underpopulation

Probably the most obvious detrimental effect of underpopulation is the increased difficulty of locating a mate for breeding purposes. For most species, mate location requires an active search on the part of the members of one sex which, under conditions of underpopulation, might present a special problem for weakly flying insects. To alleviate this, many species have evolved highly refined mechanisms (for example, production of pheromones, sounds, or light) which facilitate aggregation or location of individuals of the opposite sex. The relatively slight chance of finding a mate is offset to some extent by the fact that one mating may suffice for fertilization of all eggs a female may produce, that is, sperm may remain viable for a considerable period, and a female may produce a large number of eggs.

Andrewartha and Birch (1954) suggested that on some occasions the effect of nonspecific predators might be much greater when prey density was lower than normal because the chance of an individual prey organism being eaten is increased. In support of this suggestion, they cite observations on the Australian plague locust, *Austroicetes cruciata*, whose population density was high in the period 1935–1940. Drought conditions in the winter of 1940 resulted in a great shortage of grasshopper food and a decline in population density. Only a few small areas of land remained moist enough to support growth of grass and surviving grasshoppers congregated in these areas. But so did the birds which normally fed on the insects and they reduced the population density to an extremely low level.

Lower-than-normal densities may also have a serious effect in species which modify their environment, for example, social insects. The temperature and humidity within the nest are normally quite different from those outside, being regulated by the activity of members of the colony. If a proportion of the population is removed or destroyed, those individuals which remain may no longer be able to keep the temperature and humidity at the desired level and the colony may die. Another interesting example noted by Andrewartha and Birch (1954) is that of the lesser grain borer, *Rhizopertha dominica* (Coleoptera), which is a serious pest of stored grain in the United States. In damaged (cracked) grain beetles can survive and reproduce even at low population density. However, in sound grain only cultures whose density is quite high will survive because the insects themselves, through their chewing activity, can cause sufficient damage to the grain that it becomes suitable for reproduction. The nature of this suitability is unknown.

Below a certain level of population the so-called "threshold density," the chances of survival for a population are slim because of the unlikelihood of a meeting between insects of opposite sex and in reproductive condition. This fact is important in two areas of applied entomology, namely, quarantine service and biological control. Quarantine regulations are designed so that for a given pest the number which enter a country over a period of time is sufficiently low that the chances of the pest establishing itself are very slim. In biological control using insects as the controlling agents experience has taught that it is wiser to release the insects in a restricted area, especially if they are limited in number, rather than distributing them sparsely, in order to improve the chances of establishing a breeding population (Andrewartha and Birch, 1954).

Populations of many insect species may be considered as "self-regulating," that is, should the density of the population fall below normal (though not below threshold) it will, in the course of time, return to its original level. There may be various reasons for this. As a species' density falls, its predators may experience greater difficulty in finding food so that they migrate elsewhere or produce fewer young. As a result of the decline in predator density a larger proportion of the prey species may survive to reproduce. If this continues for several generations the original population density may be reestablished. Another possibility is that with a decrease in density, there will be a greater choice of egg-laying and, perhaps, resting sites. Selection of the best of these sites will again increase the chances of survival of an insect or its progeny and lead to a population increase. Some species have rather more specific mechanisms for overcoming the disadvantages of underpopulation. For example, females of some species practice facultative parthenogenesis in the absence of males, which serves not only to maintain continuity of the population, but, as the progeny are generally all female, any offspring that do find a mate can make a substantial contribution to the next generation. In the desert locust adult females in the solitary phase (see Chapter 21, Section 7) live longer, so that the chance of encountering a male is increased and, further, produce up to four times as many eggs compared to gregarious females.

3.1.2. Overpopulation

As population density rises beyond the normal level members of a species will increasingly compete with each other for such resources as oviposition sites, overwintering sites, resting places, and, occasionally, food. Such competition may itself have a regulatory effect, as a proportion of the population will have to be satisfied with less than optimal conditions. Thus, if oviposition sites are marginally suitable, few or no progeny may result. In less than adequate overwintering sites, insects may die if the weather is severe. If insects cannot find proper resting places, their chances of discovery by predators or parasitoids are increased, as are their chances of dying due to unfavorable weather conditions. As noted earlier (Section 2.1) food is seldom limiting, though under unusual circumstances it may become so.

In addition to the general regulating mechanisms just mentioned, some insects regulate population density in more specific ways. Migration, discussed in Chapter 22, Section 5.1, is a means by which a species may reduce its population density. In such species, it is crowding that induces the necessary physiological and behavioral changes which put an insect into migratory condition. Crowding may also lead to reduction in fecundity. For example, in *Schistocerca gregaria* gregarious females lay fewer eggs than solitary females because (1) their ovaries contain fewer ovarioles, (2) a smaller proportion of the ovarioles produce oocytes in each ovarian cycle [as a result of (1) and (2), each egg pod contains fewer eggs], and (3) they have fewer ovarian cycles (Kennedy, 1961). Likewise, in the migratory grasshopper, *Melanoplus sanguinipes*, mating frequency, which is a function of population density, is inversely related to number of eggs produced and longevity (Pickford and Gillott, 1972).

A few species employ a very obvious means of reducing crowding, namely, cannibalism of either the same or a different life stage. Among larval Zygoptera, for example, cannibalism of earlier instars is common under crowded conditions. For species whose larvae inhabit small and/or temporary ponds with limited food resources, cannibalism may be important in ensuring that at least a proportion of the population reaches the adult stage. Another consequence is that stragglers are eliminated, which results in greater synchrony of adult emergence (see also Chapter 22, Section 2.3). In the confused flour beetle, *Tribolium confusum*, and some other beetle species, all of whose life stages are spent in grain or its products, egg cannibalism occurs. Adults eat any eggs they find, and, therefore, the higher the adult population density, the greater the number of eggs consumed.

In some species of insects that inhabit a homogeneous environment, population density is regulated by making the environment less suitable for growth. For example, *T. confusum* larvae and adults "condition" the flour in which they live, and as a result a smaller proportion of the larvae survive to maturity, and the duration of the larval stage is increased. The nature of this conditioning is not known.

In some species, regulation of population density is achieved by having individuals that dominate others so that the reproductive capacity of the latter is either reduced or totally suppressed. This is seen most clearly in social Hymenoptera where one individual, the queen, dominates the other members of the colony, which are mostly female. In more primitive species, dominance is achieved initially by physical aggression, though in time the subordinates recognize the queen by scent and consequently avoid her. In highly social forms, such as the honeybee, dominance is asserted entirely through the release of pheromones.

In many species, dominance has taken on another form, namely, territoriality, the defense of a particular area. The size of the area defenced (territory) may vary, but not below a minimum value, so that a maximum population density is attained. Territoriality is exhibited by insects that belong to a number of orders, both primitive and advanced, and is typically associated with some aspect of reproduction. Most often males establish territories and defend them against other males, usually by chasing and fighting but occasionally by nonaggressive means such as chirping in some Orthoptera. Females enter males' territories for mating and oviposition. Among Odonata, where mating immediately preceeds egg-laying, males also protect females as they oviposit.

Territoriality with respect to food availability may be seen in some species, especially parasitoids and social insects. For example, female ichneumons, braconids, chalcids, and scelionids (Hymenoptera) may mark the host either chemically or physically as they oviposit so that other females of the species do not lay in the same host. Such behavior ensures that the offspring will have adequate food for complete development. (The marks may also be the means by which hyperparasites locate a host!) Social insects defend both their nest and foraging sites against members of other colonies.

3.2. Interspecific Interactions

3.2.1. Competition and Coexistence

An important form of interaction is when an insect competes with other organisms for the same resources. Grasshoppers, sheep, and rabbits all eat grass, and if this is in short supply the presence of the mammalian herbivores will have a very obvious effect on the distribution and abundance of grasshoppers living in the same area. However, as noted earlier, food is seldom a limiting factor as far as the abundance of animals is concerned, and the other requirements of these three species are so different that the species can coexist perfectly well. The collection of requirements that must be satisfied in order for a species to survive and reproduce under natural conditions is described as a niche. Thus, a niche includes both physical and biotic requirements, and its complexity varies with the environment in which a species finds itself. For example, as was noted in Chapter 22, Section 3.2.3, the critical daylength for induction of diapause in a species may vary with latitude. Equally, with reference to biotic requirements, the complexity of a niche will differ according to the number and nature of other species utilizing the same resources. The more closely two species are related, the more nearly identical will be their requirements, that is, their niche, and the greater will be the degree of competition between them where the two species coexist. Normally, in this situation the less well-adapted species becomes extinct or restricted to areas where it can again compete favorably with the other species as a result of different environmental conditions, a phenomenon known as competitive exclusion or displacement. In the absence of competition, a species' niche will be broader (less complex); that is, a species' requirements will be less stringent and form the so-called "fundamental" niche. Conversely, the niche occupied by a species that coexists with others is known as the "realized" niche.

A well-documented example of competitive exclusion in insects involves three species of chalcid, belonging to the genus *Aphytis*, which are parasitoids of the California red scale, *Aonidiella aurantii*, found on citrus fruits. In the early 1900s, the golden chalcid *Aphytis chrysomphali* was accidentally introduced into southern California, probably along with red scale on nursery stock imported from the Mediterranean region, though it is a native of China. During the next 50 years, *Aphytis chrysomphali* spread along with its host throughout the citrus-growing area and exerted a reasonable degree of control over red scale, particularly in the milder coastal areas. However, in 1948, a second species, also Chinese, *Aphytis lingnanensis*, was introduced in the hope of obtaining even better control of the pest. During the 1950s, *A. lingnanensis* gradually displaced *A. chrysomphali* so that, by 1961, the latter was virtually extinct, being restricted to a few small areas along the coast. However, *A. lingnanensis* was ineffective as a control agent of red scale in the inland citrus-growing areas around San Fernando, San Bernadino, and Riverside, where annual climatic changes are greater. It was found for example, that periods of cool weather (18°C or less for 1 or 2 weeks) or several nights of hard frost caused high mortality of all stages. Further, even light overnight frosts (−1°C for 8 hours) killed sperm in the spermathecae of females, who did not mate again, and rendered males sterile. Exposure of females to a temperature of 15°C for 24 hours led to an increase in proportion of male progeny. These factors caused a reduction in "effective progeny production" (number of female offspring produced per female) from 21.4 to 4.5 and a resultant inability of the species to control red scale populations. Consequently, a third species, *A. melinus*, was introduced from India and Pakistan in 1956 and 1957. This species rapidly displaced *A. lingnanensis* from these inland areas, and by 1961 virtually the entire population of chalcids in these areas was made up of *A. melinus* (DeBach, 1974). Changes in the distribution of the three *Aphytis* species between 1948 and 1965 are summarized in Figure 23.1.

Competitive displacement does not always occur, however, because closely related organisms have evolved mechanisms that enable them to occupy almost but not quite the same niche. These mechanisms include habitat selection (spatial selection), microhabitat selection, temporal (diurnal and seasonal) segregation, and dietary differences. Two or more of these mechanisms may operate simultaneously to prevent competition between species.

As an example of spatial segregation, we may cite the distribution of *A. lingnanensis* and *A. melinus* in southern California which DeBach (1974) considers to have stabilized, with *A. lingnanensis* occupying the milder (less climatically extreme) coastal districts and *A. melinus* the interior. Spatial separation is also seen in damselflies (Zygoptera), which inhabit prairie ponds. For example, two species of Coenagrionidae, *Coenagrion resolutum* and *Enallagma boreale*, hatch, develop, and emerge as adults almost synchronously. However, the species can coexist because larval *E. boreale* are restricted to

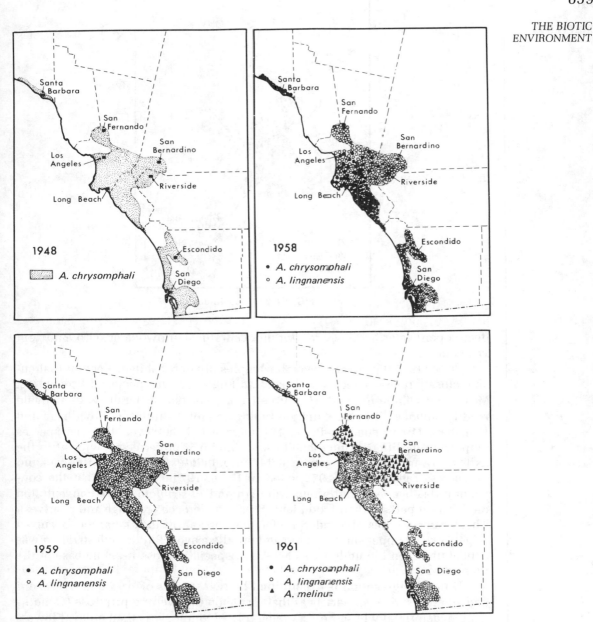

FIGURE 23.1. Changes in the distribution of *Aphytis chrysomphali*, *A. lingnanensis* and *A. melinus* in Southern California between 1948 and 1965. [After P. Debach and R. A. Sundby, 1963, Competitive displacement between ecological homologues, *Hilgardia* **34**:105–166. By permission of Agricultural Sciences Publications, University of California; and P. DeBach, D. Rosen, and C. E. Kennett. 1971, Biological control of coccids by introduced natural enemies, in: *Biological Control* (C. B. Huffaker, ed.). By permission of Plenum Publishing Corporation and the authors.]

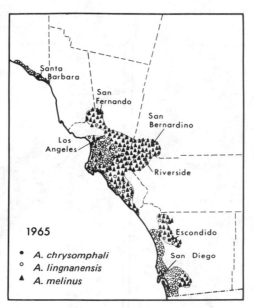

FIGURE 23.1 (Continued).

deep, open water, while *C. resolutum* occurs in shallow water with emergent vegetation.

Price (1975a) provides several examples of microhabitat selection that enable closely related species to coexist. In England, two species of Psocoptera, *Mesopsocus immunis* and *M. unipunctatus*, coexist on larch twigs with no readily obvious differences in their biology. Careful studies by Broadhead and Wapshere (1966, cited in Price, 1975a) revealed, however, that the species oviposited in different microhabitats. *M. immunis* preferred to oviposit in the axils of dwarf side shoots, whereas *M. unipunctatus* selected girdle scars and leaf scars. Tahvanainen (1972, cited in Price, 1975a) observed that the congeneric flea beetles, *Phyllotreta crucifera* and *P. striolata*, were concentrated on different parts of their food plant, *Brassica oleracea* (cabbage and relatives). The former species showed a preference for sunny locations and occurred largely on the upper surface of top and middle leaves. *Phyllotreta striolata* was concentrated on the underside of leaves, especially those near the base of the plant.

Diurnal segregation is exemplified by two species of *Andrena*, *A. rozeni* and *A. chylismiae*, solitary bees that forage on the evening primrose (*Oenothera clavaeformis*) whose flowers remain in bloom for less than a day. Flowers open in late afternoon and are visited by *A. rozeni* between about 4:00 P.M. and 7:00 P.M. *Andrena chylismiae* is an early morning forager and visits flowers between 5:00 A.M. and 8:00 A.M., that is, just before they wilt (Linsley *et al.*, 1963b, cited in Price, 1975a).

Excellent seasonal segregation is shown by the damselflies studied by Sawchyn and Gillott (1974a,b; 1975), who were able to arrange the damselflies into three types according to their seasonal biology. Type A species, which included *Coenagrion resolutum*, *Enallagma boreale*, and other Coenagrionidae, overwinter in diapause as well-developed larvae and emerge highly synchronously between the last week of May and mid-June. Sexual maturation takes about 1 week and the oviposition period extends to the end of July. Females lay eggs in the submerged parts of floating plants. Embryogenesis is direct and requires less than 3 weeks; half-grown larvae may be collected before the end of July and mature larvae by mid-September. Included in Type B are three species of *Lestes*, *L. unguiculatus*, *L. disjunctus*, and *L. dryas*, which overwinter in diapause as well-developed embryos. Eggs hatch synchronously during early May, but the very young larvae are not preyed on by the larger larvae of Type A species either because they are too small, that is, outside the range of prey size, or because the Type A larvae have ceased to feed in preparation for the final molt. Type B larvae develop rapidly and synchronized adult emergence begins in early July and is completed within 2 weeks. Adult maturation requires 16–18 days, and females oviposit in green emergent stems of *Scirpus* (bulrush), which may relate to the requirement of water for embryogenesis. Adults are not normally seen after the end of August, though in mild years they may survive into October. In Type C is included one species, *Lestes congener*, which is characterized by the lateness of its seasonal chronology. *Lestes congener* overwinters in diapause at an early (preblastokinetic) stage of embryogenesis. Embryonic development continues in the spring after the eggs are wetted and hatching occurs at the end of May. However, the young larvae are too small to serve as prey for the Type B species. Larval development is rapid in *L. congener* so that synchronized emergence begins in late July and continues for about 3 weeks. Thus, larvae of Type A species generally are not eaten by the much larger larvae of *L. congener*. Sexual maturation in *L. congener* takes about 3 weeks. Oviposition begins in mid-August and copulating adults may be seen until early October. Female *L. congener* oviposit only in dry stems of *Scirpus*, a feature associated with the lack of prediapause embryonic development observed in this species.

Thus, the occurrence of seasonal segregation between types and of microhabitat segregation (e.g., deep versus shallow water for larvae, and oviposition in floating vegetation, or emergent green or dry stems in adults) both between and within types, enables a number of species of Zygoptera to coexist and make use of the rich food supply (in the form of *Daphnia*, *Diaptomus*, and dipteran larvae) which is found in prairie ponds.

Dietary differences also enable closely related species to coexist. For example, larvae of the caddis flies *Pycnopsyche gentilis* and *P. luculenta* are able to coexist in woodland streams in Quebec because the former prefers fallen leaves, whereas *P. luculenta* feeds on submerged twigs or, if these are not available, on detritus or well-rotted leaves (MacKay and Kalff, 1973).

3.2.2. Predator–Prey Relationships

It will be abundantly clear that the distribution and abundance of a species will be greatly affected by those organisms that use it as food and that the reverse is also true, namely, that the distribution and abundance of prey will determine the distribution and abundance of predators.

Most Insecta feed on plant material in one form or another, that is, are primary consumers, and therefore play a major role in the flow of energy stored in plants to higher trophic levels. However, another large group, probably numbering about 10% of known species, feed on other animals, especially insects. Some of these are typical predators or parasites, but the majority are parasitoids that belong especially to the Tachinidae (Diptera), Strepsiptera, and so-called "parasitic" Hymenoptera (for details of the latter, see Chapter 10, Section 7). A parasitoid may be defined as "an insect that requires and eats only one animal in its life span, but may be ultimately responsible for killing many" (Price, 1975a, p. 55). Typically, a female parasitoid deposits a single egg or larva on each host, which is then gradually eaten as the offspring develop. Adult parasitoids are free-living and either do not feed or subsist on nectar and/or pollen. Thus, a parasitoid differs from a typical predator which feeds on many organisms during its life and a parasite which may feed on one to several host individuals but does not kill them. However, as Price (1975a) points out, the distinction between predator and parasitoid is not always clear. For example, a bird that captures insects as food for its offspring is comparable with a parasitoid that lays its egg on a freshly killed or paralyzed host. Further, a predator and parasitoid face the same problem, namely, location of prey (host), and may solve the problem in an identical manner. Of course, from the prey's point of view, it matters not whether the aggressor is predator or parasitoid; for either, it must take appropriate steps to avoid being eaten! In the final analysis, the population dynamics of predator–prey and parasitoid–host relationships will be identical, and it is therefore appropriate to discuss these relationships under the same heading. In the remainder of this section, therefore, the terms "predator" and "prey" should be taken to include "parasitoid" and "host," respectively, except where specifically stated otherwise.

Let us consider first the strategies employed by prey species in order to reduce the chances of their members being eaten. Probably the most obvious strategy is for insects to avoid detection. This they may do in various ways—by burrowing into a substrate which frequently also serves as food, by hiding, for example, on the underside of leaves, by becoming active for a restricted period of the day, or through camouflage where their color pattern merges with the background on which they normally rest, or they precisely resemble a twig or leaf of their food plant. Other prey species have opted for other protective mechanisms which depend on initial recognition of the prey by the predator for their effectiveness. Such mechanisms include being distasteful, a feature usually accompanied by aposematic (warning) coloration so that a predator soon learns to recognize that species are distasteful. Related to this is Müllerian mimicry, in which distantly related, distasteful species resemble each other, so that if a predator recognizes their pattern of coloration all species are protected. Another form of mimicry is Batesian, in which an edible species (the mimic) comes to resemble a distasteful species (the model). The success of this method of avoiding predation relies on the probability of the predator selecting the distasteful model rather than the edible mimic; that is, the population density of the model must greatly outweigh that of the mimic. Another chemical method of defense is to secrete obnoxious liquid or vapor whose odor repels predators. Other species release poisons, which on contact with skin or when injected by means of spines, hairs, or sting, injure or kill the attacker.

Some insects, especially species of butterflies, practice intimidation displays aimed at frightening would-be (vertebrate) predators. The butterflies normally rest with their wings closed vertically above the body. On being disturbed, the butterflies rapidly open their wings to reveal a striking color pattern, often including large "eyespots," intended to evoke prompt retreat of the enemy.

Predators make use of a variety of stimuli in order to locate prey. Some may attempt to capture and eat anything that moves within a certain size range and employ only simple visual or mechanical cues for detection of prey. Most species are, however, relatively prey-specific (feed on only a few or a single species of prey), and prey location is therefore a much more elaborate process. For many of these more specialized predators, the first step is location of the prey's habitat, and this is frequently achieved as a result of attraction to odors released from the food of the prey. For example, females of the ichneumon fly, *Itoplectis conquisitor*, a parasitoid, are attracted by the odor of pine oil, especially that of Scots pine (*Pinus sylvestris*), on which one of its preferred hosts, caterpillars of the European pine shoot moth, *Rhyacionia buoliana*, are found. For some predators, attraction is greater after the food of the prey has been damaged by the prey. The hymenopteran *Nasonia vitripennis*, for example, is attracted to meat, especially when this has been contaminated by the parasitoid's hosts, various muscid flies. Similarly, the ichneumon *Nemertis canescens* is more attracted to oatmeal contaminated by its host, larvae of the Mediterranean flour moth, *Ephestia kuhniella*, than to clean oatmeal (Vinson, 1975).

Having been attracted to the habitat of its prey, a predator must now specifically locate the prey. For many species this involves a systematic search, though this behavior is initiated only after receipt of an appropriate signal which indicates the likelihood of prey in the immediate area. Usually, such signals are again chemical in nature and include, for example, odors from the prey's feces or from the damaged tissues of the prey's food plant. Final location of prey is commonly achieved by means of its odor (though such odors are effective only over a very short range) or, more often, taste. Some parasitoids locate hosts by the vibrations or sounds the latter make as they burrow through the substrate.

In some species, location of prey is followed immediately by feeding or, in parasitoids, oviposition or larviposition. In others, additional stimuli must be received before prey is deemed acceptable. These appear to be especially critical in parasitoids for which selection of hosts of the correct age (judged by size, color, shape, texture, or taste) may be important. For example, some parasitoids accept only hosts above a certain size; cylindrical host shape improved acceptance in the ichneumon *Pimpla instigator*; hairiness of the host (by preference, caterpillars of the gypsy moth, *Lymantria dispar*) is an important determinant of acceptibility in the braconid *Apanteles melanoscelus*; female *Itoplectis conquisitor* probe host larvae with their ovipositor but will lay eggs only if the host's hemolymph has a suitable taste. Movement may be an important stimulant or deterrent of acceptibility. Some parasitoids oviposit only if a larval host moves, whereas movement of an embryo within an egg (indicative of the egg's age) may inhibit oviposition by egg parasitoids (Vinson, 1975, 1976). A special feature of many parasitoids is their ability to discriminate between non-parasitized and parasitized hosts on the basis of physical markings, odors or tastes left by the original parasitoid as it oviposited or larviposited. Such marks render a host unacceptable to the parasitoids that locate it subsequently and ensure that the parasitoid larva, on hatching, has sufficient food for its complete development. Other parasitoids leave trail-marking pheromones as they search for hosts, which inhibit researching of an area and thereby facilitate dispersal of the species.

When given a choice a predator may consistently select a particular species for prey. However, in its natural habitat its survival does not normally depend on availability of that species, and in its absence other species are acceptable as prey. Special mention is made of this point, as it has sometimes been overlooked in attempts at biological control of pests. Some attempts have failed because the introduced predator reduced the pest to low density and then died out because alternate prey was not available in the new habitat. As a result, the pest was able to rebound to an economically important level. In other words, secondary prey species form an important reservoir for the predator at times when the density of the primary prey is low.

4. Insect Diseases

In insects, as well as in almost all other organisms, the great majority of individuals (80–99.99% of those born) never survive to reproduce. Fifty percent or more of this mortality may be due to predation, while the remainder results from unsuitable weather conditions, perhaps starvation, and especially disease. Diseases may be subdivided into noninfectious and infectious categories. The former includes those that result from physical or chemical injury, nutritional diseases (due to deficiencies of specific nutrients), genetic diseases (inherited abnormalities), and physiological, metabolic, or developmental disturbances. Infectious diseases, which can be spread rapidly within a population of organisms, are caused by microorganisms, including viruses, rickettsias, spirochaetes, bacteria, fungi, protozoa, and nematodes. Though infections of many of these pathogens may be directly fatal, other pathogens simply "weaken" an insect, rendering it more susceptible to predation, parasitism, or other pathogens, to chemical and other means of control, or altering its growth rate and reproductive capacity. Most of the time in natural populations, the effects of pathogens are not readily obvious. This is described as the enzootic stage. On occasion, however, conditions are such that the pathogens can reproduce and spread rapidly to decimate the host population. This is known as the epizootic phase, and the outbreak is described as an epizootic, comparable to an epidemic within a population of humans. Study of the factors that lead to epizootics, epizootiology, is of interest not only from a purely ecological point of view but also in view of the potential use of microorganisms in the biological control of insect pests.

This section will outline the important factors in outbreaks of infectious diseases and survey the major groups of insect pathogens.

4.1. Epizootics

Essentially, there are four primary components in the development of an epizootic: the pathogen population, the host population, an efficient means of pathogen transmission, and the environment, all of which are closely interrelated.

Key features of a pathogen are its virulence (disease-producing power), infectivity (capacity to spread among hosts), and ability to survive. Clearly, pathogens (or specific strains of a pathogen) which have both high virulence and high infectivity are the ones that most often cause epizootics, though the susceptibility of the host is also important. Some pathogens may be highly virulent but of low infectivity and, as a result, have a low potential for causing epizootics. *Bacillus thuringiensis*, for example, though pathogenic for many Lepidoptera, seldom causes an epizootic under natural conditions because of its poor powers of dispersal. Indeed, the inability to disperse, together with limited capacity to survive outside a host, are probably the main reasons why epizootics are, relatively speaking, of rare occurrence. Dispersal may be effected either by abiotic or biotic agents in the environment, including wind, rain, running water, snow, host organisms (both healthy and infected), and their predators (both vertebrate and invertebrate) or parasites. Host organisms may disperse the pathogen as a result of defecation, regurgitation, oviposition (that is, the pathogen occurs either on or within the eggs), disintegration of the body after death, or cannibalism. Predators commonly distribute pathogens via their feces, though some insect parasitoids transfer the microorganisms via their ovipositor when they either sting the host or lay an egg on or in it. Pathogens may survive in either the host or the environment, sometimes for considerable periods. Those which survive in the environment typically have a highly resistant resting stage, such as spores (bacteria, fungi, and protozoa), inclusion bodies (viruses), or cysts (nematodes).

Members of the host population pick up pathogens as a result of physical contact with contaminated surfaces, or eating contaminated food (including cannibalism), or receive pathogens directly from the mother via transovarian transmission. Contact with a pathogen, however, does not necessarily result in ill effects for the host, as the latter has various means of defending itself. Further, even when some members of a population are susceptible to a pathogen, an epizootic does not always follow because of the difficulty of dispersal of the pathogen referred to above and because other members of the population may have varying degrees of resistance.

Density, distribution, and mobility of hosts are important factors in the development of an epizootic. Generally speaking, epizootics are more likely to occur at high densities, even distribution, and high mobility of hosts, as the chances of dispersal of the pathogen are greater under these conditions. On occasion an epizootic may develop at low host density, as a result of widely dispersed but long-lived pathogens that remain from a previous high-density outbreak. Even at high host-population density, an epizootic may not develop if the host population has a discontinuous distribution and/or poor mobility.

The importance of the environment, physical and biotic, in the dispersal and survival of pathogens has already been noted. Environmental factors are important in other ways in relation to epizootics. For example, factors which induce stress in an insect, especially extremes of temperature, high humidity, and inadequate food, may lower its resistance to a pathogen.

In conclusion it should now be apparent that whether or not an epizootic occurs depends on a variety of conditions relating to pathogen, host, and environment. Only when knowledge of all these factors is available for a given host–pathogen interaction can an accurate forecast of a potential disease outbreak be made. Such knowledge is of critical importance in determining the success or otherwise of biological control using pathogenic microorganisms.

4.2. Types of Pathogens

It is natural that the best-known pathogens are those which cause epizootics in economically important insects and show potential for use in microbial control of pest species. In this short account, the major features of some of these pathogens will be outlined, as a basis for the discussion on microbial control presented in Chapter 24.

4.2.1. Bacteria

Many species of bacteria have been shown experimentally to be highly pathogenic should they gain access to the hemocoel. However, under natural conditions, the majority of these never cause epizootics (even though their infectivity is high) because of the barrier presented by the integument and lining of the gut. Further, the gut may be unsuitable for survival and multiplication of the bacteria because of its pH or redox potential, or because of the presence within it of antibacterial substances or antagonistic microorganisms. Infection of an individual insect occurs when the integument or gut is physically damaged. The commonest route of invasion appears to be via the midgut. Bacteria are sometimes able to slip between the peritrophic membrane and the midgut epithelium at either the anterior or the posterior end of the midgut, or through ruptures in the membrane caused by the passage of food.

The subsequent mode of action of the bacteria, that is, how they cause a pathological condition, is variable. Invasion and destruction of the midgut epithelium is typically the first step, and for some bacteria this may be their only activity, their host dying of starvation. More often, bacteria not only destroy the midgut epithelium but then grow rapidly in the hemolymph and other tissues, causing a massive septicemia. Others liberate toxins that kill the host's cells. Other bacteria become pathogenic only when they are able to enter the hemocoel via lesions in the midgut epithelium caused by other microorganisms such as protozoa, viruses, and nematodes.

Bacteria known to be naturally pathogenic in insects can be arranged in two groups, the non-spore-formers and the spore-formers.* Included among the non-spore-forming bacteria is *Serratia marcescens*, varieties of which attack a range of insect species and whose presence is recognized by the reddish color of the dead host. Outbreaks of *S. marcescens* are common in high-density laboratory cultures of insects. According to Bailey (1968), at least five species of bacteria are involved in European foulbrood disease of larval honeybees, though a non-spore-former, *Streptococcus pluton*, is the causative agent. The remaining species are secondary pathogens or saprophytes.

The spore-formers are the most important group from the point of view of

---

*This is not a taxonomic arrangement but one of convenience used by insect pathologists.

epizootics and for their potential importance in biological control, largely because they remain viable for a considerable time outside their host. Further, some species are highly pathogenic should the sporangia be ingested because their sporangia include, in addition to a spore, a crystalline structure, the parasporal body, that contains various toxins and enzymes such as hyaluronidase and phospholipase for digestion of the cement holding together gut epithelial cells. Thus, the bacteria may be pathogenic due to the parasporal toxins, even though, initially, the spore may not germinate because of prevailing conditions of the gut. Bacteria that produce a parasporal body are known as crystalliferous bacteria.

Two of the better known, spore-forming, noncrystalliferous bacteria are *Bacillus cereus*, which has been isolated from a range of host species, and *B. larvae* which is the cause of American foulbrood in bee larvae. Only young larvae (up to 55 hours old) are susceptible, suggesting that only in early larval life are gut conditions suitable for spore germination. Despite the availability of antibiotics for treatment of American foulbrood, it continues to be a common disease. Though this is partially due to the long-lived nature of *B. larvae* spores (especially in honey), its appearance is frequently the result of poor management on the part of beekeepers who expose contaminated yet unoccupied supers.* Bees from adjacent hives visit the supers to steal honey and in doing so spread the disease.

Of the crystalliferous species, *B. thuringiensis* *B. popilliae*, and *B. lentimorbus* have been the subject of considerable research, pure and applied, during the past 30 years. *Bacillus thuringiensis* includes an enormous number of highly pathogenic varieties which attack Lepidoptera (more than 100 susceptible species) and representatives of several other orders, including some economically important species such as mosquitoes and houseflies. Honeybees are not, however, susceptible. Extensive field testing has shown that *B. thuringiensis* has desirable attributes for successful biological control of insect pests and commercial preparations are now available. This aspect will be taken up again in Chapter 24.

Bacillus popilliae and *B. lentimorbus* cause "milky disease" in larvae of some scarabaeid beetles, including the Japanese beetle, *Popillia japonica*, an important pest in the eastern United States, that feeds on roots of grasses, vegetables, and nursery stock. Mass-produced spore preparations of *B. popilliae* are now marketed for control of this pest.

4.2.2. Rickettsias

Only a few species of rickettsias are pathogenic in insects, though some infect pests and may have potential as biological control agents, for example, *Rickettsiella popilliae*, which attacks Japanese beetles. However, almost no work has been done to test this possibility perhaps, as St. Julian et al. (in Bulla, 1973) suggest, because some rickettsias of insects are also pathogenic in mammals.

---

*Supers are the boxes, without top or bottom, that contain the frames of honeycomb. Placed one on top of the other, they constitute the hive.

4.2.3. Viruses

According to David (1975), more than 700 species of insects are known to be susceptible to viral diseases. Of these, about 80% are Lepidoptera, 10% Diptera, and 5% Hymenoptera. Many of these species are economically important, and much research has been and is being conducted to determine the value of viruses in biological control.

Viruses may be arranged in two categories, according to whether the nucleic acid they contain is in the form of DNA or RNA. Viruses in both groups use raw materials in host cells for replicating their nucleic acids, and, by producing viral mRNA, they use the host's ribosomes for synthesis of viral proteins, thereby causing disruption of the host cell's metabolism. The DNA viruses include nuclear polyhedrosis viruses (NPVs), granulosis viruses (GVs), and iridescent viruses (IVs), while among the RNA-containing viruses the best known are the cytoplasmic polyhedrosis viruses (CPVs). The characteristics of these various forms are given in Steinhaus (1963, Vol. I), by David (1975), and in standard texts on virology or microbiology.

Like bacteria, viruses mainly infect insects when ingested on contaminated food or during cannibalism, though some may be transmitted on the surface of or within eggs. Initially, they invade midgut epithelial cells, and, for some, this is the only tissue affected. More often, they move into the hemocoel and subsequently attack fat body, hemocytes, and, to a lesser extent, other tissues.

Some viruses are relatively short-lived outside the insect host and their survival from year to year requires that at least some members of the host population survive. Others, such as NPVs and CPVs are able to survive outside the host for a considerable time under suitable conditions. Cabbage looper NPV, for example, may persist for 9 years in soil. Viruses are relatively stable within the temperature and humidity range that they normally experience. However, they are sensitive to sunlight, especially its shorter (ultraviolet) wavelengths, and are inactivated by a few hours of continuous exposure. Thus, those which are disseminated on the host's food plant may have limited viability. In commercial preparations viability may be increased by additives which screen out ultraviolet radiation. Viruses are also sensitive to pH and need to be maintained in conditions close to neutrality. Again, in commercial preparations such sensitivity may be partially overcome by the inclusion of buffers.

The best-studied insect viruses are the NPVs and GVs. The former have been isolated from more than 240 Lepidoptera, and from representatives of other orders, whereas the GVs apparently are restricted to Lepidoptera. Bergold (in Steinhaus, 1963, Vol. I) presents a partial list of species susceptible to NPVs, included in which are many pests, among them *Heliothis zea* (corn earworm), *H. viriscens* (tobacco budworm), *Trichoplusia ni* (cabbage looper), *Prodenia* and *Pseudaletia* spp. (armyworms), and *Neodiprion* spp. (sawflies), against which commercial NPV preparations have been developed. Examples of pests with highly pathogenic GVs, which are now being developed commercially and/or field tested, are *Pieris* spp. (cabbageworms), *Carpocapsa pomonella* (codling moth), and *Choristoneura murinana* (fir budworm).

Iridescent viruses have been mainly isolated from Diptera, including mosquitoes and blackflies. Though highly infective when injected experimentally into the hemocoel, IVs do not appear to have much future as biological control agents in Diptera, at least, because they are rapidly destroyed when ingested. Further, the peritrophic membrane of dipteran larvae does not have gaps like that of larval Lepidoptera (David, 1975).

Almost all of the approximately 150 species of insect from which CPVs have been isolated are Lepidoptera, and a number of these are pests. Relatively little work appears to have been carried out on the potential of CPVs as control agents, perhaps because many of the pests in which they occur are also susceptible to the better-known NPVs and because they are relatively slow-acting.

4.2.4. Fungi

Studies on entomogenous fungi, including their potential as control agents, tend to be overshadowed by the enormous volume of work being carried out on bacteria and viruses. However, as Roberts (in Bulla, 1973) points out, the first attempt at microbial control used the green-muscardine fungus, *Metarrhizium anisopliae*, and the first large-scale program, which covered almost the whole of Kansas, used a fungus, *Beauveria bassiana*.

Unlike bacteria, viruses, and protozoa, fungi normally enter insects via the integument rather than the gut, whose conditions (especially pH) are unsuitable for fungal spore germination. Apart from suitable temperature and pH (in soil-dwelling fungi), high humidity or even liquid water is essential for spore germination, an observation which accounts for the frequent occurrence of fungal epizootics during periods of rainy or humid weather. After germination, access through the cuticle is probably achieved through the secretion of chitinase and proteinase enzymes by elongating hyphae. Initially, the hyphae which penetrate colonize the epidermis but then spread to other specific body tissues. Some tissues are not attacked until after the host's death when the fungus becomes saprophytic. Death may result from the mere physical presence of a mass of hyphae which disrupt tissues or inhibit hemolymph circulation. Alternatively, fungal secretions may cause histolysis or be toxic to the host.

At the end of an epizootic a fungus may survive in various ways. It may persist at low incidence in remaining members of the host population or it may infect other, less susceptible, species. It may enter a facultatively saprophytic phase, for example, in soil. Or, most commonly, it may produce spores that can survive outside the host for some time. The longevity of spores varies among species and in relation to weather conditions. Low humidity and temperatures a few degrees above freezing point appear to result in greatest longevity. Spores are also susceptible to sunlight, though whether this is due to ultraviolet radiation is uncertain.

Several hundred species of fungi are known to attack insects. Some of these are very common and can cause epizootics. It is natural, therefore, that these are the best studied, with emphasis on their potential as microbial control agents. Species of *Metarrhizium* (especially *M. anisopliae* which has over 200 host species, mostly soil inhabitants) and *Beauveria* (*B. bassiana* and *B. tenella* in particular) appear to be the best prospects for control agents. Spores of *M. anisopliae* and *B. bassiana* are readily producible on a commercial scale, and field testing is underway. *Coelomomyces* species are aquatic fungi which are obligate parasites almost exclusively of larval mosquitoes, and which, under natural conditions, are probably important regulators of mosquito population density. Roberts (in Bulla, 1973) notes, however, that laboratory culture of these fungi is proving difficult, so that studies on them are hindered. The genus *Entomophthora* includes more than 100 species of fungi, many of which cause epizootics among grasshoppers, aphids, caterpillars, mosquitoes, and houseflies. Though *Entomophthora* species are usually highly specific with reference to their insect host, many are facultative parasites and are amenable to culture on a mass scale.

4.2.5. Protozoa

Protozoa pathogenic to insects are typically acquired by hosts as they ingest food, though a few species may be passed on to subsequent generations in the host's eggs. In some species the midgut epithelium is the first tissue attacked, and only later are other tissues, especially fat body, invaded. In other species the protozoa migrate through the midgut wall to specific tissues in the hemocoel. Disease is caused by the general debilitating effect of the protozoa as they reproduce, by toxins released from the protozoa, or as a result of secondary invasions of viruses or bacteria. Protozoa are disseminated as spores or cysts, some of which may survive for long periods outside the host.

Most disease-causing protozoa in insects are Sporozoa, including Coccidia, Gregarinia, and especially Microsporidia for which more than 200 host species are known. The protozoa probably play an important role in regulation of insect populations, though in only a few instances has this been authenticated. Species of the microsporidian *Nosema* are agents for a number of well-known insect diseases, two examples of which are pébrine disease of silkworms caused by *N. bombycis* and nosema disease of honeybee (pathogen *N. apis*). Other species of *Nosema* known to be highly pathogenic include *N. lymantriae* [primary host, the gypsy moth (*Lymantria dispar*)], *N. polyvora* [imported cabbageworm (*Pieris brassicae*)], and *N. locustae* (various species of grasshoppers).

Though pathogenic protozoa would seem to be good candidates for biological control agents, research into this possibility is proceeding slowly, primarily because, with few exceptions, the pathogens cannot be cultured outside their host. As a result obtaining sufficient spore material for testing is somewhat slow.

4.2.6. Nematodes

Though obviously not microorganisms *sensu stricto*, nematodes are generally included in this term in discussions of insect pathogens. Nematodes enter into a variety of interactions with insects, but those that are pathogenic largely belong to two families, Mermithidae and Neoaplectanidae. Nematode pathogens exert various effects on insects but in most instances do not directly cause death but, rather, protracted larval development, abnormal morphology (including wing shortening), and reduced fecundity. Many of the effects noted can be attributed to debilitation of the host as the nematode feeds on its tissues, especially fat body. Other effects result from more ' specific'' activities on the part of the pathogen. For example, some morphological abnormalities probably result from endocrine imbalances, induced perhaps by toxins released by the parasite. Sterility results in some insects because the parasite selectively feeds on the gonads. *Neoaplectana carpocapsae* carries a pathogenic bacterium, *Achromobacter nematophilus*, which is released when juvenile nematodes enter a host. As the bacterium develops within the host (whose death it rapidly causes as a result of septicemia), it is fed upon by the nematode which ultimately produces up to 10^5 progeny.

Infection of new hosts may occur passively when nematode eggs are ingested on contaminated food and hatch within the gut, or actively where juvenile nematodes search out a suitable host whose body they enter via the integument.

Pathogenic nematodes have been found in representatives of at least 15 insect orders, including mosquitoes, blackflies, grasshoppers, weevils, ants, and various caterpillars (Pramer and Al-Rabiai, in Bulla, 1973), and probably they are important natural regulators of insect populations. However, because they are obligate parasites, their large-scale culture has proved difficult, and consequently few studies have been made on biological control using nematodes. Some field tests have been made using N. *carpocapsae* against codling moth, corn earworm, Colorado potato beetle, tobacco budworm, and other pests, but the results were variable.

5. Summary

The biotic environment of insects is composed of all other organisms that affect insects' survival and multiplication.

Food is not normally an important regulator of insect abundance because other environmental factors have a significant adverse effect on insect growth and reproduction. In addition, many insects are polyphagous, and larvae and adults may eat different foods. In two situations the quantity of food may be limiting: (1) when only a proportion of the total food is available, and (2) when insect population density is not kept in check by other factors. The nutritional quality of food also markedly affects survival, rate of growth, and fecundity of insects.

Through evolution complex interactions between plants and insects have developed based on the theme that insects feed on (gain energy from) plants, while plants attempt to defend themselves (conserve this energy) or obtain service (most often cross-pollination, rarely protection) from insects in exchange. Some plants protect themselves by producing toxins. However, some insects have become able to cope with these toxins and may even accumulate them for protection against predators and, possibly, microorganisms. Other insects have become adapted to feeding on parts of plants that lack toxins (spatial avoidance) or when plants have a low toxin content (temporal avoidance). To obtain effective cross-pollination: (1) plants must produce the correct amount of nectar to maintain insects' "interest," yet stimulate visits to other plants of the same species; and (2) insects must be able to recognize members of the same plant species. The quantity of nectar produced in each flower depends on factors such as number of flowers per plant, synchronous or asynchronous blooming of flowers on a plant, population density of plants (average distance between plants), air temperature when flowers are blooming, and the size and foraging capacity of the pollinators.

Interactions between insects and other animals may be intraspecific or interspecific. Intraspecific interactions include those due to underpopulation and those that result from overpopulation. Populations are generally self-regulating, that is, maintain their density within a suitable range. As a prey species' density falls, predators may experience greater difficulty in locating food and may migrate or produce fewer progeny. Thus, more prey may survive to reproduce, leading to restoration of the original population density. In some species females produce more eggs or reproduce parthenogenetically in under-populated conditions. Overpopulation results in competition for resources such as oviposition, overwintering, and resting sites, and, occasionally, food, and renders a greater proportion of the population susceptible to predation, the effects of weather, and disease. When overcrowded, part of a population may migrate, cannibalism may increase, females may lay fewer eggs, and the rate of larval development may be reduced. Some species regulate breeding population density by territoriality.

When two species that coexist (live in the same habitat) require a common resource, they are said to compete for that resource. The more closely related are the species, the more nearly identical will be their total requirements (niche) and the greater will be the competition between the species, leading ultimately to the competitive exclusion (displacement) of one species from that habitat. To avoid displacement closely related species evolve mechanisms which make their niches sufficiently different that both can occupy the same habitat. The mechanisms include spatial segregation, microhabitat selection, temporal (diurnal or seasonal) segregation, and dietary differences.

Insects which are preyed on by other organisms may reduce their chances of being eaten by avoiding detection by burrowing into a substrate, hiding in vegetation, becoming active during a restricted period, or through camouflage. Some species are aposematically colored and distasteful. Distasteful species may resemble each other (Müllerian mimicry), so that when a predator learns their pattern of coloration, all species are protected. An edible species (the mimic) may resemble an inedible (distasteful) species (the model) to avoid detection (Batesian mimicry), though this method requires that the population density of the model is much greater than that of the mimic.

Most insect predators, parasitoids, and parasites are relatively or highly prey- (host-) specific. They may find suitable prey in a sequence of steps: (1) location of the prey's habitat, often by the odor of the prey's food (especially if this has been damaged by the prey); (2) search for and location of prey, stimulated by specific odors, for example, that of the prey's feces; and (3) acceptance of prey, which may be dependent on its size, color, shape, texture, or taste.

Diseases caused by pathogenic microorganisms, particularly bacteria, viruses, fungi, protozoa, and nematodes, are important regulators of insect populations. Often the incidence of disease in a population is low and the disease is said to be in the enzootic stage. When conditions are such that a disease can spread rapidly through a population, the disease is described as in the epizootic stage. The occurrence of an epizootic depends on a pathogen's virulence, infectivity, and viability, on the host's density, distribution, and mobility, and on abiotic factors such as temperature, humidity, light, and wind. The normal route of entry of pathogens is via the midgut, though fungi commonly enter via the integument. How microorganisms cause a pathological condition is variable. Bacteria may damage the midgut epithelium, causing starvation, or may invade other tissues causing septicemia and/or liberating toxins. Viruses disrupt the metabolism of the host's cells. Fungi may physically disrupt tissues, or may secrete histolyzing or toxic substances. Protozoa have a generally debilitating effect and may release toxins. Nematodes also have a debilitating effect, causing protracted or morphologically abnormal larval development, or reduced fecundity. Some feed on selected organs, for example, the gonads, and thereby exert specific effects on the host.

6. Literature

Much useful information on the nature and influence of the interactions of insects with other organisms is included in the texts by Andrewartha and Birch (1954), Andrewartha (1961), and Price (1975a). Wallace and Mansell (1975) have edited a series of papers that deal with the biochemical interactions between plants and insects. Several articles on insect–plant coevolution are in the text edited by Gilbert and Raven (1975). The biology of parasitoids is dealt with by Vinson (1976) and in several of the papers edited by Price (1975b). Rettenmeyer (1970) discusses insect mimicry. The relationships between insects and pathogens are examined in the two-volume treatise edited by Steinhaus (1963). In addition, information on pathogens in relation to the biological control of insects is given in the texts edited by Burges and Hussey (1971) and Bulla (1973).

Andrewartha, H. G., 1961, Introduction to the Study of Animal Populations, University of Chicago Press, Chicago, Ill.
Andrewartha, H. G., and Birch, L. C., 1954, The Distribution and Abundance of Animals, University of Chicago Press, Chicago, Ill.
Bailey, L., 1968, Honey bee pathology, Annu. Rev. Entomol. 13:191–212.
Baker, H. G., and Baker, I., 1973, Amino-acids in nectar and their evolutionary significance, Nature (London) 241:543–545.
Bulla, L. A., Jr. (ed.), 1973, Regulation of insect populations by microorganisms, Ann. N.Y. Acad. Sci. 217:243 pp.
Burges, H. D., and Hussey, N. W. (eds.), 1971, Microbial Control of Insects and Mites, Academic Press, New York.
Center, T. D., and Johnson, C. D., 1974, Coevolution of some seed beetles (Coleoptera: Bruchidae) and their hosts, Ecology 55:1096–1103.
David, W. A. L., 1975, The status of viruses pathogenic for insects and mites, Annu. Rev. Entomol. 20:97–117.
DeBach, P., 1974, Biological Control by Natural Enemies, Cambridge University Press, London.
Gilbert, L. E., 1971, Butterfly–plant coevolution: Has Passiflora adenopoda won the selectional race with heliconiine butterflies? Science 172:585–586.

Gilbert, L. E., and Raven, P. H., 1975, *Coevolution of Animals and Plants*, University of Texas Press, Austin, Texas.

Heinrich, B., and Raven, P. H., 1972, Energetics and pollination ecology, *Science* **176**:597–602.

Kennedy, J. S., 1961, Continuous polymorphism in locusts, *Symp. R. Entomol. Soc.* **1**:80–90.

MacKay, R. J., and Kalff, J., 1973, Ecology of two related species of caddis fly larvae in the organic substrates of a woodland stream, *Ecology* **54**:499–511.

Pickford, R., 1962, Development, survival and reproduction of *Melanoplus bilituratus* (Wlk.) (Orthoptera: Acrididae) reared on various food plants, *Can. Entomol.* **94**:859–869.

Pickford, R., and Gillott, C., 1972, Coupling behaviour of the migratory grasshopper, *Melanoplus sanguinipes* (Orthoptera: Acrididae), *Can. Entomol.* **104**:873–879.

Price, P. W., 1975a, *Insect Ecology*, Wiley, New York.

Price, P. W. (ed.), 1975b, *Evolutionary Strategies of Parasitic Insects and Mites*, Plenum Press, New York.

Rettenmeyer, C. W., 1970, Insect mimicry, *Annu. Rev. Entomol.* **15**:43–74.

Sawchyn, W. W., and Gillott, C., 1974a, The life history of *Lestes congener* (Odonata: Zygoptera) on the Canadian Prairies, *Can. Entomol.* **106**:367–376.

Sawchyn, W. W., and Gillott, C., 1974b, The life histories of three species of *Lestes* (Odonata: Zygoptera) in Saskatchewan, *Can. Entomol.* **106**:1283–1293.

Sawchyn, W. W., and Gillott, C., 1975, The biology of two related species of coenagrionid dragonflies (Odonata: Zygoptera) in Western Canada, *Can. Entomol.* **107**:119–128.

Steinhaus, E. A. (ed.), 1963, *Insect Pathology—An Advanced Treatise*, Vols. 1 and 2, Academic Press, New York.

Vinson, S. B., 1975, Biochemical coevolution between parasitoids and their hosts, in: *Evolutionary Strategies of Parasitic Insects and Mites* (P. W. Price, ed.), Plenum Press, New York.

Vinson, S. B., 1976, Host selection by insect parasitoids, *Annu. Rev. Entomol.* **21**:109–133.

Wallace, J. W., and Mansell, R. L. (eds.), 1975, Biochemical interaction between plants and insects, *Recent Adv. Phytochem.* **10**:425 pp.

24

Insects and Man

1. Introduction

In this final chapter we shall focus on those insects that Man describes, in his economically minded way, as beneficial or harmful, though it should be appreciated from the outset that these constitute only a very small fraction of the total number of species. Further, it must also be realized that the ecological principles that govern the interactions between insects and Man are no different from those between insects and any other living species, even though Man with his modern technology can modify considerably the nature of these interactions.

Of an estimated 3 million species of insects, probably not more than about 15,000 (0.5%) interact, directly or indirectly, with Man. Perhaps some 3000 species constitute pests that, either alone or in conjunction with microorganisms, cause significant damage or death to Man, agricultural products, and manufactured goods (Williams, 1967). In the United States alone, for example, it is estimated that harmful insects annually do about 5 billion dollars' worth of damage (Borror et al., 1976). On the other hand, according to Borror et al., the value of benefits derived from insects each year amounts to more than 7 billion dollars. That insects do more good than harm probably would come as a surprise to lay persons whose familiarity with insects is normally limited to mosquitoes, houseflies, cockroaches, various garden pests, etc., and to farmers who must protect their livestock and crops against a variety of pests. If asked to prepare a list of useful insects, many people most likely would not get farther than the honeybee and, perhaps, the silkmoth, and would entirely overlook the enormous number of species that act as pollinating agents or prey on harmful insects which might otherwise reach pest proportions.

Man has long recognized the importance of insects in his well-being. Insects and/or their products have been eaten by Man for thousands of years. Production of silk from silkworm pupae has been carried out for almost 5000 years. Locust swarms which originally may have been an important seasonal food for Man, took on new significance as he became a farmer rather than a hunter. However, except for the honeybee and silkworm, whose management is relatively simple and labor-intensive, until recently Man neither desired nor was able, due to his lack of basic knowledge as well as of technology, to attempt large-scale modification of the environment of insects, either to increase the number of beneficial insects or to decrease the number of those designated as pests.

Several features of Man's recent evolution have made such attempts imperative. These include a massive increase in human population, a trend toward urbanization, increased geographic movement of people and agricultural products, and, associated with the need to feed more people, a trend toward monoculture as an agricultural practice.

The relatively crowded conditions of urban areas enable insects parasitic on humans both to locate a host (frequently a prerequisite to reproduction) and to transfer between host individuals. Thus, urbanization facilitates the spread of insect-borne human diseases such as typhus, plague, and malaria whose spectacular effects on human population are well documented. For example, in the sixth century A.D. plague was responsible for the death of about 50% of the population in the Roman Empire, and "Black Death" killed a similar proportion of England's population in the mid-1300s (Southwood, 1977).

An increasing need to produce more and cheaper food led, through agricultural mechanization, to the practice of monoculture, the growing of a crop over the same large area of land for many years consecutively. However, two faults of monoculture are (1) the ecosystem is simplified, and (2) as the crop plant is frequently graminaceous (a member of the grass family, including wheat, barley, oats, rice, and corn), the ecosystem is artificially maintained at an early stage of ecological succession. By simplifying the ecosystem, Man encourages the buildup of populations of the insects that compete with him for the food being grown. Further, as the competing insects are primary consumers, that is, near the start of the food chain, they typically have a high reproductive rate and short generation time. In other words, populations of such species have the potential to increase at a rapid rate.

A massive increase in Man's geographic movements and a concomitant increase in trade, led to the transplantation of a number of species, both plant and animal, into areas previously unoccupied by them. Some of these were able to establish themselves and, in the absence of normal regulators of population (especially predators), increased rapidly in number and became important pests. Sometimes, as Man colonized new areas, some of the cultivated plants he introduced proved to be an excellent food for species of insects endemic to these areas. For example, the Colorado beetle, *Leptinotarsa decemlineata*, was originally restricted to the southern Rocky Mountains and fed on wild Solanaceae. With the introduction of the potato by settlers, the beetle had an alternate, more easily accessible source of food, as a result of which both the abundance and distribution of the beetle increased and the species became an important pest. Likewise, the apple maggot, *Rhagoletis pomonella*, apparently fed on hawthorn until apples were introduced into the eastern United States (Horn, 1976).

2. Beneficial Insects

Insects may benefit Man in various ways, both directly and indirectly. The most obvious of the beneficial species are those whose products are commercially valuable. Considerably more important, however, are the insects that pollinate crop plants. Other beneficial insects are those which are used as food, in biological control of pest insects and plants, and in research. For some of these useful species, Man modifies their environment in order to increase their distribution and abundance so that he may gain the benefits.

2.1. Insects Whose Products are Commercially Valuable

The best-known insects that fall in this category are the honeybee (*Apis mellifera*), silkworm (*Bombyx mori*), and lac insect (*Laccifer lacca*).

The honeybee originally occupied the African continent, most of Europe (except the northern part), and western Asia, and within this area the usefulness of its products, honey and beeswax, has been known for many thousands of years. Though the discovery of sugar in cane (in India, about 500 B.C.) and in beet (in Europe, about 1800 A.D.) (Southwood, 1977) led to a decline in the importance of honey, it is nevertheless still a very valuable product.

Bee management was probably first practiced by the ancient Egyptians. Honeybees were brought to North America by colonists in the early 1600s, and today honey and beeswax production is a multimillion dollar industry. Good bee management aims to maintain a honeybee colony under optimum conditions for maximum honey and wax production. Management details vary according to the climate and customs of different geographical areas but may include (1) moving hives to locations where nectar-producing plants are plentiful; (2) artificial feeding of newly established, spring colonies with sugar syrup in order to build up colony size in time for the summer nectar flow; (3) checking that the queen is laying well and, if not, replacing her; (4) checking and treating colonies for diseases such as foulbrood and nosema; and (5) increasing the size of a hive as the colony develops, in order to prevent swarming.

Silk production has been commercially important for about 4700 years. The industry originated in East Asia and spread into Europe (France, Italy, and Spain) after eggs were smuggled from China to Italy in the sixth century A.D. The development of cheap synthetic fibers has had a severely deleterious effect on the silk industry which is labor-intensive, making production costs high. Nevertheless, the annual world production is about 30–35 million kilograms (Borror *et al.*, 1976).

The lac insect is a scale insect endemic to India and southeast Asia that secretes about itself a coating of lac, which may be more than 1 cm thick. The twigs on which the insects rest are collected and either used to spread the insects to new areas or ground up and heated in order to separate the lac. The lac is a component of shellac, though its importance has declined considerably with the development of synthetic materials.

2.2. Insects as Pollinators

As was noted in Chapter 23, Section 2.3, an intimate, mutualistic relationship has evolved between many species of insects and plants, in which plants produce nectar and pollen for use by insects, while the latter provide a transport system to ensure effective cross-pollination. Though some crop plants are wind-pollinated, for example, graminaceous species, a large number, including fruits, vegetables, and field crops such as clovers, rape, and sunflower, require the service of insects. In addition, ornamental flowers are almost all insect-pollinated.

The best-known, though by no means the only, important insect pollinator is the honeybee, and it is standard practice in parts of the world for fruit and vegetable growers either to set up their own beehives in their orchards and fields or to contract this job out to beekeepers in the vicinity. Indeed, it is apparent that under such conditions the value of the bee as a pollinator may be 15 to 20 times its value as a honey producer. Certainly, the presence of bees in orchards, alfalfa fields, etc., may result in a three- or fourfold increase in the yield of fruit or seed. Borror *et al.* (1976) estimate the annual value of insect-pollinated crops to be about 6 billion dollars.

2.3. Insects as Agents of Biological Control

It is only relatively recently that Man has gained an appreciation of the importance of insects in the regulation of populations of potentially harmful insects and plants. In many instances, this appreciation was gained only when, as a result of Man's own activities, the natural regulators were absent, a situation which was rapidly exploited by these organisms whose status was soon elevated to that of pest. In the first three examples given below (taken from DeBach, 1974), none of the organisms is a pest in its country of origin due to the occurrence there of various insect regulators. The discovery of these regulators, followed by their successful culture and release in the area where the pest occurs, constitutes biological control.

Probably the best-known examples of an introduced plant pest are the prickly pear cacti (*Opuntia* spp.) taken into Australia as ornamental plants by early settlers. Once established, the plants spread rapidly so that by 1925 some 60 million acres of land were infested, mostly in Queensland and New South Wales. Surveys in both North and South America, where *Opuntia* spp. are endemic, revealed about 150 species of cactus-eating insects, of which about 50 were judged to have biological control potential and were subsequently sent to Australia for culture and trials. Larvae of one species, *Cactoblastis cactorum*, a moth, brought in from Argentina in January 1925, proved to have the required qualities and within 10 years had virtually destroyed the cacti. Perhaps the most remarkable feature of this success story is that only 2750 *Cactoblastis* larvae were brought to Australia, of which only 1070 became adults. From these, however, more than 100,000 eggs were produced, and in February–March of 1926 more than 2.2 million eggs were released in the field! Additional releases, and redistribution of almost 400 million field-produced eggs until the end of 1929, ensured the project's success.

The classical example of an insect pest brought under biological control is the cottony-cushion scale, *Icerya purchasi*, which was introduced into California, probably from Australia, in the 1860s. Within 20 years, the scale had virtually destroyed the recently established, citrus-fruit industry in Southern California. As a result of correspondence between American and Australian entomologists and of a visit to Australia by an American entomologist, Albert Koebele, two insect species were introduced into the United States as biological control agents for the scale. The first, in 1887, was *Cryptochaetum iceryae*, a parasitic fly, about which little is heard, though DeBach (1974) considers that it had excellent potential for control of the scale had it alone been imported. However, the abilities of this species appear to have been largely ignored with the discovery by Koebele of the vedalia beetle, *Vedalia cardinalis*, feeding on the scale. In total, only 514 vedalia were brought into the United States, between November 1888 and March 1889, to be cultured on caged trees infested with scale. By the end of July 1889, the vedalia had reproduced to such an extent that one orchardist, on whose trees about 150 of the imported beetles had been placed for culture, reported having distributed 63,000 of their descendents since June 1! By 1890, the scale was virtually wiped out. Similar successes in controlling scale by means of vedalia or *Cryptochaetum* have been reported from more than 50 countries (DeBach, 1974).

As a third example of an introduced pest being brought under control by biological agents, we may cite the winter moth, *Operophtera brumata*, which, though endemic to Europe and parts of Asia, was accidentally introduced into Nova Scotia in the 1930s. Its initial colonization was slow, and it did not reach economically significant proportions until the early 1950s, and by 1962 it had spread to Prince Edward Island and New Brunswick. The larvae of the winter moth feed on the foliage of hardwoods such as oak and apple. Though more than 60 parasites of the winter moth are known in western Europe, only six of these were considered to be potential control agents and introduced into eastern Canada between 1955 and 1960. Only two of these, *Cyzenis albicans*, a tachinid, and *Agrypon flaveolatum*, an ichneumonid, became established, but between them they brought the moth under control by 1963. Embree (in Huffaker, 1971) notes that the two parasites are both compatible and supplementary to each other. When the density of moth larvae is high, *C. albicans*, which is attracted to feeding damage caused by the larvae, is a more efficient parasite than *A. flaveolatum*. However, once in the vicinity of damage, it does not specifically seek out winter moth larvae. Thus, at lower density, it wastes eggs on nonsusceptible defoliators such as caterpillars of the fall cankerworm, *Alsophila pometaria*. Hence, at low host densities, *A. flaveolatum* is more effective because it oviposits specifically on winter moth larvae.

The three examples described above indicate one method whereby the importance of biological control can be demonstrated, namely, by introduction of potential pests into areas where natural regulators are absent. Another way of demonstrating the same phenomenon is to destroy the natural regulators in the original habitat, which enables potential pests to undergo a population explosion. This has been achieved frequently through the use of nonselective insecticides such as DDT. For example, the use of DDT against the codling moth, *Carpocapsa pomonella*, in the walnut orchards of California, led to outbreaks of native frosted scale, *Lecanium pruinosum*, which was unaffected by DDT, whereas its main predator, an encyrtid, *Metaphycus californicus*, suffered high mortality (Hagen et al., in Huffaker, 1971). Another *Lecanium* scale, *L. coryli*, introduced from Europe in the 1600s, is a potentially serious pest of apple orchards in Nova Scotia but is normally regulated by various natural parasitoids (especially the chalcidoids *Blastothrix sericea* and *Coccophagus* sp.) and predators (especially mirid bugs). Experimentally it was clearly demonstrated in the 1960s that application of DDT destroyed a large proportion of the *Blastothrix* and mirid population, and this was followed in the next 2 years by medium to heavy scale infestations. Recovery of the parasite and predators was rapid, however, and by the third year after spraying the scale population density had been reduced to its original value (MacPhee and MacLellan, in Huffaker, 1971).

2.4. Insects as Food

In addition to the many predaceous insects which feed on other insects, many vertebrates are insectivorous to a variable degree. Some of these vertebrates, especially some freshwater fish and game birds, in turn, may be eaten by humans. In some areas of the world, insects "in season" routinely form part of the human diet. Locusts, termites, caterpillars, and grubs of various kinds are eaten in Africa; edible caterpillars are sold fresh or in cans in Mexico; and Australian aborigines are reported to eat roasted bugong moths (*Agrotis infusa*). Most North Americans and Europeans have not yet been educated to the delights (and nutritional value) of insects, and despite the efforts of authors such as Taylor and Carter (1976) to increase the popularity of insects as food, at present cans of edible insects are restricted to the gourmet counter of food stores.

2.5. Soil-Dwelling and Scavenging Insects

By their very habit the majority of soil-dwelling insects are ignored by Man. Only those which adversely affect his well-being, for example, termites, wireworms, and cutworms, normally "merit" his attention. When placed in perspective, however, it seems probable that the damage done by such pests is greatly outweighed by the benefits which soil-dwelling insects as a group confer. The benefits include aeration, drainage, and turnover of soil as a result of burrowing activity. Many species carry animal and plant material underground for nesting, feeding, and/or reproduction, which Metcalf *et al.* (1962) compare to ploughing in a cover crop.

Many insects, including a large number of soil-dwelling species, are scavengers; that is, they feed on decaying animal or plant tissues, including dung, and thus accelerate the return of elements to food chains. In addition, through their activity they may prevent use of the decaying material by other, pest insects, for example, flies. Perhaps of special interest are the dung beetles (Scarabaeidae), most species of which bury pieces of fresh dung for use as egg-laying sites. Generally, the beetles are sufficiently abundant that a pat of fresh dung may completely disappear within a few hours, thus reducing the number of dung-breeding flies that can locate it. Furthermore, the chances of fly eggs or larvae surviving within the dung are very low because the dung is ground into a fine paste as the beetles or their larvae feed. Likewise, the survival of the eggs of tapeworms, roundworms, etc., present in the dung producer, is severely reduced by this activity.

In Australia, which has a large population of cattle (derived largely from animals taken in by early settlers), cow dung presents a serious problem because, until recently, the beetle species that normally dispose of this kind of dung were absent. (Australia does have dung beetles but these are specialized for dealing with the dung of kangaroos and other marsupials.) In addition to providing food for a massive number of fly larvae when freshly deposited, the dung, because of the dry climate, soon dries and then may remain unchanged for a considerable time. Rank herbage grows around each pad of dung, but this is not normally eaten by cattle. Thus, at any time, a considerable proportion of all grazing land (estimated at 20% or 6 million acres total) is not usable. In 1963, it was decided to initiate a program of biological control of dung using various South African species of dung beetles. After extensive research, in 1967 the first beetles were released in northern Australia. Several species have become established and are spreading rapidly down the eastern side of Australia. In general, the beetles are doing an excellent job of "cleaning up," except for the cooler months when they are inactive. Additional species, which are active at lower temperatures are being tested so as to obtain year round dung removal (Waterhouse, 1974).

2.6. Other Benefits of Insects

Their relatively simple food and other requirements, short generation time, and high fecundity enable many insects to be reared cheaply and easily under laboratory conditions and, consequently, make them valuable in teaching and research. The fruit fly, *Drosophila melonagaster*, with its array of mutants, is familiar to all who take an elementary genetics class, though it must also be appreciated that the insect continues to have an important role in advanced genetic research. Studies on other insects have provided us with much of our basic knowledge of animal and cell physiology, particularly in the areas of nutrition, metabolism, endocrinology, and neuromuscular physiology. Grain beetles are good subjects for the study of population ecology.

Many insects give us pleasure through their aesthetic value. Because of their beauty, certain groups, especially butterflies, moths, and beetles, are sometimes collected as a hobby. Some are embedded in clear materials from which jewelry, paperweights, bookends, place mats, etc., are made. Others are simply used as models on which paintings and jewelry are based.

3. Pest Insects

Since Man evolved, insects have fed on him, competed with him for food and other resources, and acted as vectors of microorganisms that cause diseases in Man or the organisms of value to him. However, as was noted in the Introduction, the impact of such insects increased considerably as the human population grew and became more urbanized. Urbanization presented easy opportunities for the dissemination of Man's insect parasites and the diseases they carry. Large-scale and long-term cultivation of the same crop over an area facilitated rapid population increases in certain plant-feeding species and the spread of plant diseases. Modern transportation, too, encourages the spread of pest insects and insect-borne diseases. Further, as described in Section 2.3, some of Man's attempts at pest eradication have backfired, resulting in even greater economic damage.

3.1. Insects Affecting Man Directly

A large number of insect species may be external, or temporary internal, parasites of Man. Some of these are specific to Man, for example, the body louse (*Pediculus humanus*) and pubic louse (*Phthirus pubis*), but most have a variable number of alternate hosts which compounds the problem of their eradication. With rare exceptions, for example, some myiasis-causing flies, insect parasites are not fatal to humans. In large numbers, insect parasites may generally weaken their host, making him more susceptible to the attacks of disease-causing organisms. Or the parasites, as a result of feeding, may cause irritation or sores which may then become infected.

But by far the greatest importance of insect parasites of Man is their role as vectors of pathogenic microorganisms (including various "worms") some well-known examples of which are given in Table 24.1. The pathogen is picked up when a parasitic insect feeds and may or may not go through specific stages of its life cycle in the insect. Bacteria and viruses are directly transmitted to new hosts, an insect serving as a mechanical vector, whereas for protozoa and worms (tapeworms and nematodes), an insect serves as an intermediate host in which an essential part of the parasites' life cycle occurs. In the latter arrangement the insect is known as a biological vector.

A pathogen may reside (and multiply) in alternate vertebrate hosts which are immune to or only mildly infected by it. For example, the bacterium *Pasteurella pestis*, which causes bubonic plague (Black Death), is endemic in wild rodent populations. However, in domestic rats and humans, to which it is transmitted by certain fleas, it is highly pathogenic. Similarly, in South America, yellow fever virus, transmitted by mosquitoes, is found in monkeys though these are immune to it. Such alternate hosts are thus an important reservoir of disease.

Transmission of human disease-causing microorganisms is not, however, entirely the domain of parasitic insects. Many insects, especially flies, may act as mechanical vectors, contaminating human food as they rest or defecate on it, with pathogens picked up during contact with feces or other organic waste. Examples of such insects and the disease transmitted by them are listed in Table 24.1.

A third category of insects that directly affect Man includes those which may bite or sting when accidental contact is made with them, for example, bees, wasps, ants, some caterpillars (with poisonous hairs on their dorsal surface), and blister beetles. Normally, the effect of the bite or sting is temporary and nothing more than skin irritation, swelling, or blister formation. Bee stings, however, may cause anaphylaxis or death in some sensitive individuals.

3.2. Pests of Domesticated Animals

A range of insect parasites may cause economically important levels of damage to domestic animals. The majority of these parasites are external and include bloodsucking flies (for example, mosquitoes, horseflies, deerflies, blackflies, and stable flies), biting and sucking lice and fleas. Other parasites are internal for part of their life history, for example, bot-, warble, and screwworm flies, which as larvae live in the gut (horse bot), under the skin (warble and screwworm of cattle), or in head sinuses (sheep bot). Other examples are given in the chapters that deal with the orders of insects. In addition to insects, other arthropods are also important livestock pests, for example, various mites and ticks. For further details of the pests which attack particular domesticated animals, readers may consult Metcalf et al. (1962) and Pfadt (1971).

Generally the effect of such parasites is to cause a reduction in the health of the infected animal. In turn, this results in a loss of quality and/or quantity of meat, wool, hide, milk, etc., produced. When severely infected by parasites, an animal may eventually die. In addition to their own direct effect on the host, some parasites are vectors of livestock diseases, examples of which are included in Table 24.1.

3.3. Pests of Cultivated Plants

Damage to crops and other cultivated plants by insects is enormous and is valued at about 3 billion dollars annually in the United States (Borror et al., 1976). Damage is caused either directly by insects as they feed (by chewing or sucking) or oviposit, or by viral, bacterial, or fungal diseases, for which insects serve as vectors. Especially important as "direct damagers" of plants are Orthoptera, Lepidoptera, Coleoptera, and Hemiptera, and the reader should consult the chapters that deal with these orders for specific examples of such pests. Several hundred diseases of plants are known to be transmitted by insects (for examples see Table 24.2) including about 300 which are caused by viruses (Eastop, 1977). Especially important in disease transmission are Hemiptera, particularly leafhoppers and aphids. Three aspects of the behavior of these insects facilitate their role as disease vectors: (1) they make brief but frequent probes with their mouthparts into host plants; (2) as the population density reaches a critical level, winged migratory individuals are produced; and (3) in many species, winged females deposit a few progeny on each of many plants, from which new colonies develop. On the basis of their method of transmission and viability (persistence in the vector), viruses may be arranged in three categories. The nonpersistent (stylet-borne) viruses are those believed to be transmitted as contaminants of the mouthparts. Such viruses remain infective in a vector for only a very short time, usually an hour or less. Semipersistent viruses are carried in the anterior regions of the gut of a vector, where they may multiply to a certain extent. Vectors do not normally remain infective after a molt, presumably because the viruses are lost when the foregut intima is shed. Persistent (circulative or circulative-propagative) viruses are those which, when acquired by a vector, pass through the midgut wall to the salivary glands from where they can infect new hosts. Such viruses may multiply within tissues of a vector which retains the ability to transmit the virus for a considerable time, in some instances for the rest of its life. Persistent viruses, in contrast to those in the first two categories, may be quite specific with respect to the vectors capable of transmitting them (Matthews, 1970).

TABLE 24.1. Examples of Insects Which Serve as Vectors for Diseases of Man and Domestic Animals[a]

Insect vector	Pathogen	Disease	Host	Distribution
ANOPLURA				
Pediculus humanus (body louse)	*Rickettsia prowazekii* (rickettsian)	Epidemic typhus (Brills disease)	Man, rodents	Worldwide
	Pasteurella tularensis (bacterium)	Tularemia	Man, rodents	N. America, Europe, the Orient
HEMIPTERA				
Rhodnius spp. (assassin bugs)	*Trypanosoma cruzi* (protozoan)	Chagas' disease	Man, rodents	S. America, Central America, Mexico, Texas
DIPTERA				
Phlebotomus spp. (sand flies)	*Leishmania donovani* (protozoan)	Kala-azar (Dumdum fever)	Man	Mediterranean region, Asia, S. America
	L. tropica	Oriental sore	Man	Africa, Asia, S. America
	L. braziliensis	Espundia	Man	S. America, Central America, N. Africa, southern Asia
	(virus)	Pappataci fever (Sand fly fever)	Man	Mediterranean region, India, Sri Lanka
Anopheles spp. (mosquitoes)	*Plasmodium vivax* (protozoan)	Malaria	Man	Worldwide in tropical, subtropical, and temperate regions
	P. malariae	Malaria	Man	
	P. falciparum	Malaria	Man	
Aedes spp. (mosquitoes)	(virus)	Yellow fever	Man, monkeys, rodents	American and African tropics and subtropics
	(virus)	Dengue	Man	Worldwide in tropics and subtropics
	(virus)	Encephalitis	Man, horses	N. America, S. America, Europe, Asia
Culex spp. (mosquitoes)	*Wucheria bancrofti* (nematode)	Filariasis (Elephantiasis)	Man	Worldwide in tropics and subtropics
	(virus)	Dengue	Man	Worldwide in tropics and subtropics

Vector	Pathogen	Disease	Host	Distribution
	(virus)	Encephalitis	Man, horses	N. America, S. America, Europe, Asia
Tabanus spp. (horse flies)	*Wucheria bancrofti* (nematode)	Filariasis (Elephantiasis)	Mar	Worldwide in tropics and subtropics
	Bacillus anthracis (bacterium)	Anthrax	Man, other animals	Worldwide
Chrysops spp. (deer flies)	*Pasteurella tularensis* (bacterium)	Tularemia	Man, rodents	N. America, Europe, the Orient
	Loa loa (nematode)	Loiasis (Calabar swelling)	Man	Africa
Glossina spp. (tsetse flies)	*Trypanosoma rhodesiense* (protozoan)	Sleeping sickness	Man, other animals	Equatorial Africa
	T. gambiense	Sleeping sickness	Man, other animals	Equatorial Africa
	T. brucei	Nagana	Cattle, wild ungulates	Equatorial Africa
SIPHONAPTERA				
Xenopsylla cheopsis (oriental rat flea)	*Pasteurella pestis* (bacterium)	Bubonic plague (Black Death)	Man, rodents	Worldwide
Xenopsylla spp.	*Rickettsia typhi* (rickettsian)	Endemic (murine) typhus	Man, rodents	Worldwide
Xenopsylla cheopsis	*Hymenolepis nana* (cestode)	Tapeworm	Man	Europe and N. America
	H. diminuta (cestode)	Tapeworm	Man	Worldwide
Nosopsyllus fasciatus (northern rat flea)	*Pasteurella pestis* (bacterium)	Bubonic plague (Black Death)	Man, rodents	Worldwide
	Rickettsia typhi (rickettsian)	Endemic (murine) typhus	Man, rodents	Worldwide
	Hymenolepis diminuta (cestode)	Tapeworm	Man	Worldwide
Ctenocephalides canis (dog flea)	*Dipylidium caninum* (cestode)	Tapeworm	Man, dogs, cats	Worldwide
	Hymenolepis nana (cestode)	Tapeworm	Man	Europe, N. America
Pulex irritans (human flea)	*Hymenolepis nana* (cestode)	Tapeworm	Man	Europe, N. America

[a] Data from Askew (1971) and Borror et al. (1976).

TABLE 24.2. Examples of Plant Diseases Transmitted by Insects[a]

	Disease	Important hosts	Vectors	Distribution
Viruses	Alfalfa mosaic	Alfalfa, tobacco, potato, beans, peas, celery, zinnia, petunia	Aphids (at least 16 spp.) incl. *Acyrthosiphon primulae, A. solani, Aphis craccivora, A. fabae, A. gossypii, Macrosiphum euphorbiae, M. pisi, Myzus ornatus, M. persicae, M. violae*	Worldwide
	Barley yellow dwarf	Barley, oat, wheat, rye, wild and tame grasses	Numerous aphids incl. *Macrosiphum granarium, M. miscanthi, Myzus circumflexus, Rhopalosiphum padi, R. maidis*	North America, Australia, Denmark, Holland, U.K.
	Bean common mosaic	Beans	Aphids (at least 11 spp.) esp. *Aphis rumicis, Macrosiphum pisi, M. gei*	Worldwide
	Beet yellows	Sugarbeet, spinach	Aphids esp. *Aphis fabae, Myzus persicae*	Wherever sugarbeet is grown
	Cauliflower mosaic	Cauliflower, cabbage, Chinese cabbage	Aphids esp. *Brevicoryne brassicae, Rhopalosiphum pseudobrassicae, Myzus persicae*	Europe, U.S.A., New Zealand
	Dahlia mosaic	Dahlia, zinnia, calendula	Aphids esp. *Myzus persicae, Aphis fabae, A. gossypii, Macrosiphum gei, Myzus convolvuli*	Wherever dahlias are grown
	Lettuce mosaic	Lettuce, sweet pea, garden pea, endive, aster, zinnia	Aphids esp. *Myzus persicae, Aphis gossypii, Macrosiphum euphorbiae*	Europe, U.S.A. (esp. California), New Zealand

Pea mosaic	Garden pea, sweet pea, broadbean, lupin, clovers	Aphids: *Acyrthosiphon pisi, Myzus persicae, Aphis fabae, A. rumicis*	Europe, U.S.A., New Zealand, Australia, Japan
Potato virus Y	Potato, tobacco, tomato, petunia, dahlia	Aphids esp. *Myzus persicae, M. certus, M. ornatus, Macrosiphum euphorbiae*	U.K., France, U.S.A.
Soybean mosaic	Soybean	Aphids incl. *Myzus persicae, Macrosiphum pisi*	Wherever soybean is grown
Sugarcane mosaic	Sugarcane, corn, sorghum, other tame and wild grasses	Numerous aphids incl. *Rhopalosiphum maidis, Aphis gossypii, Schizaphis graminum, Myzus persicae*	Wherever sugarcane is grown
Tomato spotted wilt	Tomato, tobacco, dahlia, pineapple	Thrips: *Thrips tabaci, Frankliniella schultzeri, F. fusca, F. occidentalis*	Africa, Asia, Australia, Europe, North and South America
Turnip yellow mosaic	Turnip, cauliflower, Chinese cabbage, kohlrabi, cabbage, broccoli	Flea beetles (*Phyllotreta* spp.); Mustard beetle (*Phaedon cochleariae*); Grasshoppers (*Leptophyes punctatissima, Chorthippus bicolor*); Earwig (*Forficula auricularia*)	U.K., Germany, Portugal, North America
Aster yellows	Aster, celery, carrot, squash, cucumber, wheat, barley	Numerous leafhoppers incl. *Gyponana hasta, Scaphytopius acutus, S. irroratus, Macrosteles fascifrons, Paraphlepsius apertinus, Texananus* (several species)	Worldwide
Mycoplasmas			

(Continued)

TABLE 24.2. (Continued)

	Disease	Important hosts	Vectors	Distribution
	Clover phyllody	Most clovers	Leafhoppers incl. *Aphrodes albifrons, Macrosteles cristata, M. fascifrons, M. viridigriseus, Euscelis lineolatus, E. plebeja*	U.S.A., U.K.
	Corn stunt	Corn	Leafhoppers esp. *Dalbulus elimatus, D. maidis, Graminella nigrifrons*	North and Central America
Bacteria	Stewarts bacterial wilt	Corn	Corn flea beetle (*Chaetocneme pulicara*); toothed flea beetle (*C. denticulata*)	U.S.A.
	Cucurbit wilt	Cucumber, muskmelon	Cucumber beetles (*Diabrotica vittata, D. duodecimpunctata*)	U.S.A., Europe, South Africa. Japan
	Potato blackleg	Potato	Seedcorn maggot (*Hylemya cilicrura*), *H. trichodactyla*	North America
	Fire blight	90 spp. of orchard trees and ornamentals, esp. apple, pear, quince	Wide range of insect vectors, esp. bees, wasps, flies, ants, aphids	North America, Europe
Fungi	Dutch elm disease	Elm	Elm bark beetles, esp. *Scolytus multistriatus, S. scolytus, Hylurgopinus rufipes*	Asia, Europe, North America
	Ergot	Cereals and other grasses	About 40 spp. of insects esp. flies, beetles, aphids	Worldwide

[a] Data from various sources.

Almost any stored material, whether of plant or animal origin, may be subject to attack by insects, especially species of Coleoptera (larvae and adults) and Lepidoptera (larvae only). Among the products that are frequently damaged are grains and their derivatives, beans, peas, nuts, fruit, meat, dairy products, leather, and woolen goods. In addition, wood and its products may be spoiled by termites or ants. Again the reader is referred to the appropriate chapters describing these groups for specific examples.

Borror *et al.* (1976) cite a figure of about 1 billion dollars for the estimated value of losses caused by pests of stored products. The nature of the damage caused by these pests varies. Pests of household goods, including clothes, furniture, and various grain-based foods, cause damage as a result of spoilage, for example, by tunneling through or defecation on/in their food. Grain and other seed pests eat economically valuable quantities of food.

4. Pest Control

As will be apparent from what was said above, pests are organisms which damage, to an economically significant extent, Man or his possessions, or which in some other way are a source of "annoyance" to Man. Implicit in the above description are value judgments which may vary according to who is making them, as well as where and when they are being made. Nevertheless, in a given set of circumstances, there will be an economic injury (annoyance) threshold, measured in terms of a species' population density, above which it is desirable (profitable) to take control measures that will reduce the species' density. As the margin between economic injury threshold and actual population density widens, the desirability (profitability) of control increases. Pest control is, then, essentially a sociological problem—a matter of economics, politics, and psychology.

A range of methods is available for the control of insect pests. Each of these methods has its advantages and disadvantages and these must be balanced against each other in determining which (combination of) method(s) is most appropriate in a given instance. Some of these methods are spectacular but short-term and will be appropriate, for example, where massive outbreaks of pests are relatively sudden, yet unpredictable and temporary. Others are more slowly acting but relatively permanent in effect, and may be used for pests that are more or less permanent but whose populations are relatively stable.

Conway (1976) and Southwood (1977) have proposed that pests can be arranged in a spectrum according to their "ecological strategies" and that the principle (best) method of control is based on their position in the spectrum (Table 24.3). At either end of the spectrum are the so-called "r pests" and "K pests," respectively, with in between, the intermediate pests.

The r pests are characterized by their potentially high rates of population increase (due to the high fecundity and short generation time), well-developed powers of dispersal (migration) and ability to locate new food sources, and rather general food preferences. These features enable r pests to colonize tem-

TABLE 24.3. Principal Control Methods in Relation to Ecological Strategies of Pests[a]

	r Pests	Intermediate pests	K Pests
Control method:	Pesticides ⟶		
		Biological control ⟶	
			Cultural control ⟵
			Genetic control ⟵
Examples (with important features):	*Schistocerca gregaria* (desert locust) Fecundity/♀ = 400 eggs Generation time = 1–2 months Migratory, defoliates many crops	(Most deciduous forest pests, fruit pests, and some vegetable pests)	*Oryctes rhinoceros* (rhinoceros beetle) Fecundity/♀ = 50 eggs Generation time = 3–4 months Feeds on apical growing points of coconuts
	Aphis fabae (black bean aphid) Fecundity/♀ = 100 eggs Generation time = 1–2 weeks Feeds on wide range of plants		*Glossina* spp. (tsetse fly) Fecundity/♀ = 10 eggs Generation time = 2–3 months Feeds on narrow range of hosts
	Musca domestica (housefly) Fecundity/♀ = 500 eggs Generation time = 2–3 weeks Feeds on organic waste		*Carpocapsa pomonella* (codling moth) Fecundity/♀ = 40 eggs Generation time = 2–6 months Larvae feed on apple and some other fruits
	Agrotis ipsilon (black cutworm) Fecundity/♀ = 1500 eggs Generation time = 1–1½ months Feeds on seedlings of most crops		*Melophagus ovinus* (sheep ked) Fecundity/♀ = 15 eggs Generation time = 1–2 months External parasite of sheep

[a] Data from Conway (1976) and Southwood (1977).

porarily suitable habitats, in which there is typically little interspecific competition for the resources available. Because r pests may occur in such large but unpredictable numbers and rapidly change their location, predators (of which there may be many) have relatively little effect on their population. Further, although like other organisms r pests are subject to disease, the latter is slow to take effect, by which time significant damage may have been done. Finally, because of their high reproductive potential, r pests are able to tolerate mass mortality and rapidly regenerate their original density. Hence, biological control, which is a relatively slow but long-term method, is of little use against r pests. For such pests specific insecticides, which can be stored for application at short notice, continue to be the most important tool in their control. Included in the r-pest group are the "classic" pests: locusts, aphids, mosquitoes, and houseflies (Table 24.3).

K pests, on the other hand, have lower fecundity and longer generation time, poor ability to disperse, relatively specialized food preferences, and are found in habitats that remain stable over long periods of time. Under natural conditions, insects with the features of K strategists seldom become pests. If, however, probably as a result of Man's activity, their niche is expanded (e.g., their food plant becomes an important crop), or if they can occupy a new niche (e.g., feeding on domestic cattle rather than wild ungulates) they may become a pest. Once established, such pests are often difficult to eradicate over the short term, for example, through the use of insecticides. Insecticides are frequently not feasible tools because the K pests attack the fruit rather than the foliage of crop plants, or because the cost is prohibitive in view of the low density of the pest population. (In some instances, however, where even at low population density a pest may cause considerable damage, for example, codling moth on apple, insecticidal control may be profitable.) Nor is biological control an appropriate method because K pests have few natural enemies, a feature probably related to their low density under natural conditions. For K pests, the best methods of control are those which disturb a pest's habitat, for example, the breeding of resistant strains of plant(s) or animal(s) attacked by it, and cultural practices. Examples of K pests are given in Table 24.3.

The majority of pests are classified as intermediate pests in the Conway scheme because they exhibit a mixture of the features of r and K pests. For some of these, with a relatively high reproductive potential, insecticidal control may be necessary under certain conditions, and conversely, for pests approaching the K end of the spectrum, cultural control sometimes may be adequate. However, the most important feature of intermediate pests is the relatively large number of natural enemies that they have. These enemies, under normal circumstances, are important regulators of the pest population. In addition, intermediate pests are frequently foliage- or root-damaging pests, for example, spruce budworm and some scale insects, and, therefore, the economic injury threshold is reasonably high; that is, a fair amount of damage can be tolerated without economic loss. Hence, for these pests, biological control would appear to be the single most appropriate method of control, which can be supplemented as necessary with insecticidal and other methods. The latter is, in other words, an integrated control program.

With these general considerations in mind, it is now appropriate to consider in more detail the methods available for pest control.

4.1. Legal Control

Also known as regulatory control, legal control is based primarily on the old adage "Prevention is better [in this instance, cheaper] than cure." Legal control is the enactment of legislation to prevent or control damage by insects. It includes, therefore, establishment of quarantine stations at major ports of entry into an area. Usually the stations are located at international borders, though in some instances domestic quarantines are necessary, for example, when certain parts of a country are widely separated from the rest (Hawaii and continental United States). At quarantine stations people and goods are inspected to prevent the accidental introduction of potential insect pests and plant and animal diseases. Prior to the introduction of quarantine legislation in the United States early this century (Plant Quarantine Act of 1912) a number of insect species had been accidentally introduced and become established as plant pests, for example, the cottony-cushion scale discussed in Section 2.3. Quarantine has severely reduced the number of pest introductions, though several species still managed to enter the United States, for example, the Mediterranean fruit fly (*Ceratitis capitata*) which became temporarily established in Florida and Texas on several occasions in the 1950s and 1960s, and the cereal leaf beetle (*Oulema melanopus*), found in Michigan in 1962 (Horn, 1976). The Mediterranean fruit fly was successfully exterminated using insecticides following each of its invasions, though at a cost of several million dollars. However, the cereal leaf beetle is now established in several midwestern states (primarily Michigan, Ohio, and Indiana) and efforts must now be directed toward restricting its dispersal.

Thus, another aspect of legal control is the setting up of surveillance systems for monitoring the insect population in a given area so that, should an outbreak occur, it can be dealt with before it has a chance to spread. Such surveillance is an important duty of state/provincial entomologists, in cooperation with local agriculture representatives and crop and livestock producers.

As an adjunct to quarantine, many countries (or areas within countries) have legislation that requires international or interstate shipments of animals or plants, or their products, to be certified as disease- or insect-free by qualified personnel prior to shipment.

Another aspect of legal control, and one which is becoming increasingly important, is the licensing of insecticides and the establishment of (1) regulations regarding their use, and (2) monitoring systems to assess their total impact on the environment. In the United States the Environmental Protection Agency is responsible for assessing the effectiveness of pesticides, as well as their possible hazardous effects on humans, wildlife, and other organisms, including bees, other pollinating species, and beneficial parasitoids. As noted earlier (Section 2.3), indiscriminate use of insecticides can result in greatly increased rather than decreased pest damage.

4.2. Chemical Control

The use of chemicals either to kill or to repel insect pests is the oldest method of pest control. Fronk (in Pfadt, 1971) notes that sulfur was used by the

Greeks against pests almost 3000 years ago and the Romans used asphalt fumes to rid their vineyards of insect pests. The Chinese used arsenic compounds against garden pests before 900 A.D , though arsenic was not used in the Western world until the second half of the seventeenth century.

Until about 1940, insecticides belonged to two major categories, the "inorganics" and the "botanicals." Among the inorganic insecticides are arsenic and its derivatives (arsenicals), including Paris green (copper acetoarsenite), which was the first insecticide to be used on a large scale in the United States—against Colorado potato beetle in 1865. Other inorganics include fluoride salts (developed at about the end of the nineteenth century, following realization that toxic residues were left by arsenicals), sulfur, borax, phosphorus, mercury salts, and tartar. These inorganic insecticides were typically sprayed on the pest's food plant or mixed with a suitable bait. In other words, all are "stomach poisons" that require ingestion and absorption to be effective. Thus, they were unsatisfactory pesticides for sucking insects for which "contact poisons," absorbed through the integument or tracheal system are necessary.

The "botanicals" are organic contact poisons produced by certain plants in which they serve as protectants against insects (see Chapter 23, Section 2.3). The group includes (1) nicotine alkaloids, derived from certain species of *Nicotiana*, including *N. tabaca* (tobacco) (family Solanaceae); (2) rotenoids extracted from the roots of derris (*Derris* spp.) and cubé (*Lonchocarpus* spp.); and (3) pyrethroids, produced by plants in the genus *Pyrethrum* (*Chrysanthemum*) (family Compositae). The pyrethroids, when first available commercially, were an important group of insecticides for use in the home, as livestock sprays, and against stored-product, vegetable, or fruit pests, primarily because of their low toxicity to mammals. The great disadvantage or pyrethroids has been their photolabile nature (instability in light) and the need, therefore, to reapply them frequently made them expensive to use. Thus, they were largely replaced by cheaper, synthetic insecticides in the 1940s, an important consequence of which is that very few insects (6 of 364 species with insecticide resistance of some type) are presently resistant to them. This feature, in conjunction with the recent synthesis of photostable pyrethroids, may lead to renewed importance for this group of compounds (Elliott et al., 1978).

Though two synthetic organic insecticides had been commercially available prior to 1939 (dinitrophenols in Germany, first used in 1892; organic thiocyanates in the United States from 1932 on), it is generally acknowledged that this is the date at which the synthetic organic insecticide industry took off. After several years of research for a better mothproofing compound, Müller, who worked for the Geigy company in Switzerland, discovered the value of DDT as an insecticide. In the next few years, production of DDT began at the company's plants in the United Kingdom and United States, although because of the second world war, knowledge of DDT was kept a closely guarded secret. In early 1944, DDT was first used on a large scale, in a delousing program in Naples where typhus had recently broken out. Some 1.3 million civilians were treated with DDT, and within 3 weeks the epidemic was controlled (Fronk, in Pfadt, 1971). Later that year the identity of the "miracle cure" was revealed, and the world soon became convinced that with DDT (and other recently developed insecticides) pest insects would become a thing of the past. In 1948 Müller was awarded a Nobel prize, though, interestingly, the first example of insect resistance to DDT had been reported two years earlier!

Through the 1940s and into the 1950s, much research was carried out in Western Europe and the United States for other insecticides as effective as DDT. Three major groups of compounds emerged: (1) chlorinated hydrocarbons, such as DDT, lindane, chlordane, aldrin, dieldrin, endrin, and heptachlor; (2) organophosphates, like TEPP, diazinon, dichlorvos, parathion, and malathion; and (3) carbamates, for example, sevin, isolan, and furadan. For details of structure, physical properties, formulation, lethal doses, usage, etc., consult Pfadt (1971) and Metcalf *et al.* (1962).

The search for suitable synthetic insecticides continues today, though at a somewhat reduced rate because the profitability of such ventures for industrial concerns has greatly diminished, for a variety of interrelated reasons. The primary reasons are (1) the time and cost of producing and testing an insecticide, estimated at an average of 7 years and more than 10 million dollars (Brown, 1977), with no guarantee of its final approval for use; (2) a general unwillingness of government agencies to grant approval for use of new insecticides following public outcry against the environmental hazards of such compounds; if approval is granted, it may restrict the use (and hence saleability and profitability) of an insecticide; and (3) the relatively short "life" of an insecticide due to its being no longer effective and/or becoming an environmental hazard. As a result, some of the largest companies in the chemical industry have either considerably reduced or abandoned research into the development of new insecticides (Brown, 1977).

Three major problems have arisen as a result of the massive use of insecticides over the past 70 years. First, many insects and mites have developed resistance to one or more of the chemicals (Tables 24.4 and 24.5). (See Chapter

TABLE 24.4. Number of Species of Insects and Mites in Which Resistance to One or More Chemicals Has Been Documented[a]

Year	Species
1908	1
1928	5
1938	7
1948	14
1954	25
1957	76
1960	137
1963	157
1965	185
1967	224
1975	364[b]

[a] Data from Georghiou and Taylor (1977).
[b] Includes 59 species reported on basis of field tests or observations; of these 3 are of medical/veterinary importance and 56 of agricultural importance.

TABLE 24.5. Number of Species of Arthropoda with Reported Cases of Resistance to Pesticides as of 1975[a]

	Pesticide group					Nature of pest		
	DDT	Cyclo-dienes	Organo-phosphates	Carba-mates	Other	Medical/veterinary	Agricultural	Total
Acarina	21	10	32	6	13	10	33	43
Anoplura	5	3	2			5		5
Coleoptera	26	48	18	7	8		56	56
Dermaptera	1						1	1
Diptera	91	100	40	6	4	110	23	133
Ephemeroptera	2						2	2
Hemipt./Het.	4	12	3			4	10	14
Hemipt./Hom.	10	11	28	4	4		41	41
Hymenoptera	1	1					2	2
Lepidoptera	31	32	22	12	4		52	52
Mallophaga		2				2		2
Orthoptera	3	1	1	1		3		3
Siphonaptera	5	3	1			5		5
Thysanoptera	3	2			2		5	5
Total	203	225	147	36	35	139	225	364[b]

[a] Data from Georghiou and Taylor (1977).
[b] Includes 59 species reported on basis of field tests or observations; of these 3 are of medical/veterinary importance and 56 of agricultural importance. Note that many species are resistant to pesticides in two or more groups.

16, Section 5.5 for a discussion of the mechanism of resistance.) *Quadraspidiotus perniciosus* (San Jose scale) takes credit for being the first recorded species to develop resistance to an insecticide (lime sulfur) in 1908, less than 30 years after use of the insecticide began. The housefly was the first species resistant to a synthetic insecticide (DDT in 1946), and by the end of 1975 the total had reached 364, of which 139 species are of medical or veterinary importance and 225 are agricultural pests (Georghiou and Taylor, 1977). Depending on the method of resistance developed by a species against an insecticide, the species may also have resistance to other chemicals of the same group (class resistance) or even to chemicals of other groups (cross resistance). The development of resistance necessitates the use of ever-increasing doses of insecticide to achieve the same benefits, with a concomitant enhancement of the second and third problems described below, as well as increased cost to the user. Alternatively, the user may turn to newly developed insecticides though, again, the cost will be high.

The second problem associated with insecticide use is one already mentioned in Section 2.3, namely, the nonspecificity of action of these chemicals, with the result that beneficial as well as pest species are destroyed. (Indeed, some were developed precisely because of their "general purpose" nature!) As pest species typically can recover from insecticide application more rapidly than their natural enemies (because of their greater reproductive potential), they rebound with even greater force, necessitating additional insecticidal treatment and increasing costs to the user.

The third problem, the potential health hazard of many of the synthetic insecticides to humans and wildlife, is of especial interest because it is based on a feature of the synthetics that was once considered highly beneficial, namely, their highly persistent (indestructible) nature. To the user this means that one or a few insecticide treatments will suffice for the entire season. For example, DDT is highly stable and only slowly degraded in the presence of sunlight and oxygen. Thus, a single spraying in a house or barn may remain effective up to a year, and even outdoor applications (on foliage) may be stable through an entire growing season. Unfortunately, this stability is retained following ingestion or absorption of DDT by living organisms. Thus, DDT tends to be stored in fatty tissue, due to its lipid solubility, and is concentrated as it is transferred from organism to organism in food webs. Recognition of the phenomenon of bioconcentration (biological magnification) via food webs and observation of harmful effects of insecticides in the terminal members of food chains (especially predatory birds but not, so far, humans) led to enormous public outcry against insecticides. As a result, governments have been forced to examine carefully the balance between the benefits gained and the risks entailed in the use of insecticides, and where necessary enact legislation to protect human interest. One result of this was the banning in 1972 of DDT use in the United States, in all except a very few situations where benefits clearly outweighed risks (Whittemore, 1977).

Despite this rather gloomy picture, it must be strongly emphasized that synthetic insecticides have saved and will continue to save millions of human lives and billions of dollars' worth of food and organic manufactured goods. They are still by far the principal method for control of insect pests, though there is a slowly growing realization among users that insecticides are probably more valuable (i.e., cheaper and having a longer period of service) when used selectively, in conjunction with other methods such as biological control, rather than in a "blanket" manner as in the past (see Doutt and Smith, in Huffaker, 1971). This is associated with an appreciation that actual extermination of a pest is almost never achievable, or is even necessary in most situations; that is, a certain amount of pest damage can be tolerated without suffering economic loss.

The chemicals discussed above operate on the principle of control through rapid death of (most) members of a pest population. Recently, however, great interest has been shown in other chemicals which, though widely different in nature, are collectively known as insectistatics because they suppress insects' growth and reproduction (Levinson, 1975). They include (1) substances that inhibit chitin synthesis or tanning, rendering an insect more susceptible to microbial (especially fungal) infection or preventing normal activity because the muscles do not have a firm structural base; (2) antagonists or analogues of essential metabolites (e.g., essential amino acids and vitamins) whose effect is to prolong larval life and/or retard egg production, frequently resulting in death; (3) insect growth regulators (IGRs), substances with juvenile-hormone or ecdysonelike activity; and (4) sex attractants.

Some 600 compounds are known to mimic juvenile hormone to various degrees and in different species. Work on the possible use of IGRs in pest control is still largely at the stage of laboratory experiment or small-scale field trial, though early results indicate considerable promise for IGRs against pests of some orders (Staal, 1975). Among the effects noted are (1) interference with embryogenesis, followed by death, at IGR doses about 1000th the value of conventional ovicides; (2) abnormal development of the integument in post-embryonic stages, leading to inability to molt properly and impaired sensory function (hence inability to locate food, mates, oviposition sites, etc.); (3) improper metamorphosis of internal organs or external genitalia, causing sterility and/or inability to mate; (4) interference with diapause, so that an insect becomes seasonally maladjusted; and (5) abnormal polymorphism in aphids. Despite the seemingly bright picture for IGRs at the present time, they might, if used indiscriminately, soon become subject to the same criticisms that have been leveled at conventional insecticides.

Another area of chemical control that holds considerable promise is the use of sex attractants (see also Chapter 13, Section 4.1). Considerable progress has been made in the elucidation of their structure and artificial synthesis. Though the substances exert their effect at extremely low concentrations, the high cost of manufacture and technical problems associated with their application presently prohibit large-scale use. However, it may be anticipated that pheromones may find the following uses in pest control: (1) estimation of population density (followed by treatment, if necessary, with a suitable control method); (2) attraction of pests into traps in which they are then killed; and (3) permeation of an environment with sex attractant, so that individuals are unable to locate mates.

4.3. Biological Control

Biological control, in the sense used here, may be described as the regulation of pest populations by natural enemies (parasites, predators, and pathogens). It includes both naturally occurring control and control achieved as a result of Man's augmentation of the natural enemy component (through importation, rearing, and release of selected enemies).

As in the history of chemical control, several early successes with biological control led some scientists to believe that this method might be the one to solve the world's pest problems. However, with an increase in knowledge of ecology, specifically predator–prey and host–parasite relationships, and in the length of the list of "failures" in biological control projects, this view has been markedly tempered. It is now realized that for some pest species, biological control will not work, and that for some others, biological control has been or will be a highly effective method. Between these two extremes, and forming the majority of pests, are those for which biological control will be an effective tool when used in conjunction with other methods of pest control, that is, as a component of integrated pest control. For many years, biological control was a "poor relation" to chemical control, even though several outstanding successes of biological control were recorded before the advent of synthetic insecticides. However, with the increasing appreciation of the problems caused by synthetic insecticides, biological control began to receive a greater share of the attention of applied entomologists, industrial concerns, and government agencies.

The importance of naturally occurring biological control must be emphasized. DeBach (1974) suggests (p. 59) that "Upon it rests our entire ability to successfully grow crops because without it, the potential pests would overwhelm us." He estimates (p. 60) that "99 percent or more of potential pests [are] under natural biological control." However, by its very effect, namely, the prevention of species from becoming sufficiently populous to be designated as pests, it is easily overlooked. Only by very careful study of ecosystems in equilibrium or by disturbance of ecosystems can its value be appreciated. Two common (man-made) methods of ecosystem disruption are (1) transfer of a species from its original habitat where it is not a pest to a new habitat where, in the absence of natural enemies, it flourishes and becomes a pest, and (2) indiscriminate use of broad-spectrum insecticides which decimate both pest and natural enemy populations. As noted above, the pest normally recovers more rapidly than its predators or parasitoids and becomes even more destructive than before. Examples of both forms of disruption were given in Section 2.3. DeBach (1974) lists 20 cases of naturally occurring control of homopteran pests (mostly scale insects), and further examples involving pest insects of other orders are given in the papers by Hagen *et al.*, Rabb, and MacPhee and MacLellan, in Huffaker (1971).

Man's first recorded attempts at biological control were made by the ancient Chinese who used predatory ants in their citrus orchards against caterpillars and boring beetles. This practice is still used in parts of Burma and perhaps China (DeBach, 1974). In the eighteenth century, mynah birds imported from India were successfully used in the control of red locusts *(Nomadacris septemfasciata)* in Mauritius. However, full appreciation of the potential that insect parasitoids and predators might have as control agents developed only in the middle of the nineteenth century, following careful studies of the biology of such insects in the early 1800s. During the latter half of the nineteenth century, many well-known entomologists, both in Europe and North America, studied biological control and extolled its virtues, though these were largely unheeded until cottony-cushion scale was spectacularly controlled by the vedalia beetle in 1888–1889 (see Section 2.3). More or less concurrently, based on studies of insect diseases (especially those of the silkworm), several authorities, most notably Pasteur and an American entomologist Le Conte, suggested that microorganisms might be used in the control of pest species (DeBach, 1974). However, for reasons outlined below, successful biological control through the use of microorganisms has been achieved relatively infrequently.

Biological control, when successful, has several advantages over control through the use of insecticides. First, it is persistent; that is, once a control agent is established, it will exert a continuing influence on the population density of the pest. In part related to this, biological control is cheap because one application of the control agent is sufficient. Furthermore, the control agent is "ready-made"—it does not have to go through an extended and costly phase of research and development, though determination of the most suitable control agent(s) may take some time. As an example of the cheapness of biological control, DeBach (1974) notes that the cottony-cushion scale project in 1888–1890 cost less than 5000 dollars, yet it has saved the California citrus-fruit industry millions of dollars each year since. Because of escalating costs of insecticides, biological control may be the only (or principal) method of pest control in underdeveloped countries. A third advantage is that biological control does not stimulate "genetic counterattack" by the pests. Fourth, it does not result in the growth to pest status of species that are economically unimportant. And finally, it is a method that does not endanger humans or wildlife through pollution of the environment (Huffaker, 1971).

If biological control offers so many advantages over insecticides, why has it, in most instances, come only a distant second to these compounds in terms of usage? What are its disadvantages? Perhaps the main one is psychological because users like to see immediate results (profits?) for their efforts. In biological control it may take some time (even years) for a control agent to subdue a pest, by which time a user's patience (and profit margin) have worn thin, especially under the considerable and continuous advertising pressure of insecticide producers. Further, the new equilibrium density that a pest attains when controlled biologically is almost certainly higher than the density immediately after insecticide treatment, which again may be unsatisfactory so far as a user is concerned, especially if the crop being grown is of high unit value, for example, fresh fruit. Consumers, too, are involved here, since they have become accustomed to "blemish-free" produce and may not buy even slightly damaged material.

Biological control has not always been successful, and proponents of insecticidal control (which include a significant number of applied entomologists as well as industrial concerns) have been quick to point this out. As a result many governments and universities have been loath to invest money and manpower in research on biological control. According to DeBach (1974), had such research been possible, a number of major pests could have been controlled (biologically) long before they were. In the few countries or states (most notably, California) where significant research effort has been put into biological control, the success rate for this method is high. In other words, in many instances, failure to control a pest by biological measures has stemmed from incomplete or poorly performed study of the pest and its predators or parasites, not from the unsuitability of biological control per se.

The greatest success has been achieved through the use of insects, especially parasitoids, as control agents and some examples are given in Table 24.6. According to DeBach (1974), up to 1970 biological control using insects as control agents had been attempted against 223 pest species throughout the world. Forty-two of these species were completely controlled; that is, their population densities were reduced, then permanently maintained, below the economic threshold, making insecticide treatment only rarely or never necessary. Substantial control was achieved over 48 species. Substantial control includes instances where the pest or crop is not of major importance, where control is only partial in some areas of infestation, and where occasional insecticide treatment is required. Another 30 species fell into the "partial control" category, where pest density was reduced but not consistently below the economic threshold, making regular, though less frequent, insecticidal treatment necessary. In short, some degree of success was achieved against more than one half of the species against which the method was used. Indeed, the total number of successful projects up to 1970 was 253 on a world basis, since some of the pests were controlled in several different countries.

Not surprisingly, in view of the "unprofitability" of successful (long-term)

TABLE 24.6. Examples of Successful Biological Control Projects
Using Insects as Control Agents[a]

Pest	Primary control agent[b] and source	Location and date of project
Icerya purchasi (cottony-cushion scale)	*Vedalia cardinalis* (vedalia beetle) (Australia)	California (1888–1889)
Perkinsiella saccharicida (sugarcane leafhopper)	*Paranagrus optabilis* (mymarid) (Australia) *Cytorhinus mundulus* (mirid) (Australia)	Hawaii (1904–1920)
Levuana irridescens (coconut moth)	*Ptychomyia remota* (tachinid) (Malaysia)	Fiji (1925)
Aleurocanthus woglumi (citrus blackfly)	*Eretmocerus serius* (aphelinid) (Malaysia)	Cuba (1930), other Caribbean islands and Central American countries (1931 on)
	Amitus hesperidum (platygasterid) *Prospaltella opulenta* (aphelinid) *P. clypealis* (aphelinid) (India and Pakistan)	Mexico (1949)
Planococcus kenyae (coffee mealybug)	*Anagyrus* nr. *kivuensis* (encyrtid) (Uganda)	Kenya (1938)
Pseudococcus citriculus (citriculus mealybug)	*Clausenia purpurea* (encyrtid) (Japan)	Israel (1939–1940)
Dacus dorsalis (oriental fruit fly)	*Opius oophilus* (braconid) (Phillipines and Malaysia)	Hawaii (1947–1951)
Lepidosaphes beckii (purple scale)	*Aphytis lepidosaphes* (eulophid) (China)	California (1948 on), Texas, Mexico, Greece, Brazil, Peru (1952–1968)
Antonina graminis (rhodesgrass scale)	*Neodusmetia sangwani* (encyrtid) (India)	Texas, Florida, Brazil (1949 on)
Operophtera brumata (winter moth)	*Cyzenis albicans* (tachinid) (Europe) *Agrypon flaveolatum* (ichneumon) (Europe)	Nova Scotia (1955–1960)
Chrysomphalus aonidum (Florida redscale)	*Aphytis holoxanthus* (eulophid) (Hong Kong)	Israel, Lebanon, Florida, Mexico, South Africa, Brazil, Peru (1956 on)
Chrysomphalus dictyospermi (dictyospermum scale)	*Aphytis melinus* (eulophid) (California)	Greece (1962)

TABLE 24.6. (Continued)

681

INSECTS
AND MAN

Pest	Primary control agent[b] and source	Location and date of project
Nezara viridula (green vegetable bug)	Trissolcus basalis (scelionid) (Egypt and Pakistan)	Australia (1933–1962), New Zealand (1949–1952), Fiji and other Pacific islands (1941–1953)
	T. basalis and Trichopoda pennipes var. pilipes (tachinid) (Antigua and Monserrat)	Hawaii (1962–1965)
Opuntia spp. (prickly pear cactus)	Cactoblastis cactorum (moth) (Argentina)	Australia (1920–1925)
Hypericum perforatum (Klamath weed)	Chrysolina quadrigemina (chrysomelid beetle) (Australia)	California (1944–1946) and other western states

[a] Data from DeBach (1974).
[b] In many projects, a variety of control agents were introduced and became established, but assist in control of the
the pest to only a minor degree.

biological control using insect agents, industry has shown little interest in the method. Research and development in this area have been the domains of government and university scientists.

Biological control using microorganisms (microbial control), specifically viruses, bacteria, protozoa, fungi, and nematodes has not met with an equivalent degree of success (see Burges and Hussey, 1971; Bulla, 1973; Angus, 1977). This stems largely from the nature of the control agents, many of which are obligate pathogens, relatively labile (short-lived) outside the host, have poor powers of dispersal, and are active only in certain environmental conditions. Added to these problems, are technical difficulties that have hampered research in this area, especially the inability to mass-produce microorganisms on a year-round basis for laboratory study as well as field trials. This is slowly being overcome through the use of artificial insect diets, enabling cultures of the host insect to be maintained year-round, and tissue culture. However, these methods are not, by and large, satisfactory for large-scale commercial production because of their labor-intensive nature and high costs. Another major technical problem that requires solution is the development of suitable protectants against sunlight to which most microorganisms are especially sensitive. The poor viability of most microorganisms outside their host means that, to be effective, they must be reintroduced at least on an annual basis, if not more frequently. In this regard they are, of course, comparable with chemical insecticides, though they do not pose the same environmental hazards. The continuing need for reapplication of microbial agents has not gone unnoticed by industrial concerns, and a number of commercial preparations are now available or under development (Table 24.7).

Four examples are known of microorganisms that exert a permanent regu-

TABLE 24.7. Examples of Insect Pathogens Developed by Industry and Various Agencies[a]

Group	Pathogen	Product name	Source
Bacteria	*Bacillus lentimorbus*	Japidemic	Ditman Corp., U.S.A.
	B. popilliae	Doom	Fairfax Biological Labs., U.S.A.
	B. sphaericus	—	International Minerals Chemical Corp. (IMC), U.S.A.
	B. thuringiensis	Agritrol®	Merck and Co., U.S.A.
		Bakthane® L69	Rohm and Haas Co., U.S.A.
		Bactospeine	Pechiney Progil Lab. Roger Bellon, France
		Bathurin	Chemapol, Biokrma, Czechoslovakia
		Biospor 2802	Farbwerke Hoechst, Germany
		Biotrol® BTB	Nutrilite Products, Inc., U.S.A.
		Dendrobacillin	Moskovs. zavod. bakt. prepavatov. atov., U.S.S.R.
		Entobakterin 3	All-Union Institute Plant Protection, U.S.S.R.
		HD-1 (Experimental)	Abbott Laboratories, U.S.A.
		Parasporin®	Grain Processing Corp., U.S.A.
		Sporeine	Laboratoire L.I.B.E.C., France
		Thuricide®	IMC, U.S.A.
Fungi	*Beauveria bassiana*	—	IMC, U.S.A.
		Biotrol FBB	Nutrilite Products, Inc., U.S.A.
	Metarrhizium anisopliae	—	IMC, U.S.A.
Polyhedrosis viruses	*Heliothis*	Biotrol VHZ	Nutrilite Products, Inc., U.S.A.
		Viron/H®	IMC, U.S.A.
	Neodiprion	Polyvirocide	Indiana Farm Bureau Co-op Assoc., U.S.A.
	Prodenia	Biotrol VPO	Nutrilite Products, Inc., U.S.A.
	Spodoptera	—	IMC, U.S.A.
		Biotrol VSE	Nutrilite Products, Inc., U.S.A.
	Trichoplusia	—	Biological Control Supplies, U.S.A.
		—	IMC, U.S.A.
		Biotrol VTN	Nutrilite Products, Inc., U.S.A.
Nematodes	*Neoaplectana carpocapsae* (DD-136)	Biotrol NCS	Nutrilite Products, Inc., U.S.A.

[a] Data from Falcon (in Huffaker, 1971).

latory effect on pest insects. These are *Bacillus popilliae* and *B. lentimorbus*, which cause milky disease in the Japanese beetle, *Popillia japonica*, an important pest of lawn and other grasses in the United States; a nuclear polyhedrosis virus, accidentally introduced into Canada in the early 1940s, which has kept populations of the European spruce sawfly (*Diprion hercyniae*) below the economic injury threshold for more than 30 years since the initial epizootic; and a nuclear polyhedrosis virus which exerts good control over the European pine sawfly (*Neodiprion sertifer*) in Canada (Falcon, in Huffaker, 1971).

Bacillus thuringiensis, which, as noted in the previous chapter, is pathogenic to a large number of Lepidoptera, has been available commercially in the United States since 1958. As of 1971 it was the only microbial agent registered for use against crop, as well as forest and shade-tree, pests. The great advantage of B. thuringiensis is that it can be cultured outside its hosts by fermentation and is therefore cheap to produce. The bacterium is short-lived, however, and several applications may be necessary each season.

As Table 24.7 indicates, a number of viruses (especially nuclear polyhedrosis viruses) have been developed for use against insect, especially lepidopteran, pests, though only a few of these have been registered for commercial use. Two major problems beset the use of viruses as microbial control agents, their low viability in sunlight and their culture. Although in soil viruses may survive from season to season, when exposed to ultraviolet light, their half-life is usually less than 1 day. Even when additives which screen out ultraviolet light are added to the preparation, the half-life is only extended to a few days. As of 1975 all viruses used in field trials or available commercially were produced in living insects, the mass-rearing of which may be time-consuming and expensive. This difficulty may eventually be overcome through the use of tissue culture, though at the moment this method is also relatively costly.

As was noted in Chapter 23, Section 4.2.4, it was a fungus (Metarrhizium anisopliae) that was the control agent in the first attempt at microbial control. Generally speaking, however, the potential of fungi as pest control agents has not been widely studied. Roberts (in Bulla, 1973) suggests three reasons why workers may have been unwilling to commit themselves to such a study. First, fungi appear to be unpredictable; that is, in some tests high infectivity of hosts was achieved, whereas in others the opposite occurred. Roberts considers this inconsistency to stem from inadequate information on the fungal requirements for infectivity. Second, because fungal infections depend heavily on weather conditions, which cannot be controlled, the use of fungi as control agents was thought to be impractical. However, it is the microclimate (rather than weather conditions) which is important, and this is amenable to modification under certain circumstances. Third, because some fungi that are pathogenic in insects are also known to affect vertebrates, the tendency has been to ignore the group as a whole on the grounds of safety.

Nonetheless, fungi have some qualities that indicate potential as control agents, for example, many (even some which were once thought to be obligate parasites) can be cultured on artificial media, they may be highly virulent, and among different species a range of host specificity can be found, so that a species appropriate for a given control problem may be selected (Roberts, in Bulla, 1973, see also Ferron, 1978).

Very little work has been done on the possibility of using pathogenic protozoa in the biological control of insect pests, though some, especially species of Microsporidia are probably important natural control agents (Kellen, 1974). As with other microorganisms, a serious deterrent has been the inability to culture the protozoa outside the host. However, using spores produced in mass-reared hosts, a number of Nosema spp. have been field tested with good results. For example, an extensive study is being undertaken in the United States by Henry and colleagues (see Henry and Oma, 1974) on the potential use of N. locustae, N. acridophaga, and N. cuneatum as control agents against grasshoppers. Using N. *locustae* spores mixed with bran bait, Henry and co-workers obtained significant reductions in the grasshopper populations of experimental plots. However, the timing of spore application was found to be critical. Application too early in the season, that is, when grasshoppers were in the early larval instars, caused rapid and high mortality, with the result that too few spores were produced for infection of other grasshoppers. Conversely, application to older (near adult) grasshoppers, while resulting in a high incidence of infection (most individuals were infected), caused little mortality because the grasshoppers were able to tolerate the protozoa. (It is probable, however, that the fecundity of these individuals would be reduced.) The best time of application appeared to be when the dominant, early summer species were in the third instar. Such applications caused the greatest initial mortality and highest infection in later populations. A comparable program to that initiated by Henry's group has recently begun in Western Canada.

Like protozoa, pathogenic nematodes have been little studied with regard to their potential as control agents, largely because of the difficulty of culturing them on artificial media (Dutky, 1974; Pramer and Al-Rabiai, in Bulla, 1973). As of 1974 only two species of nematodes pathogenic in insects [*Neoaplectana glaseri* and *N. carpocapsae* (=*dutkyi*)] could be cultured outside their host. *Neoaplectana carpocapsae* (also known as DD-136) appears to be a good candidate for a biological control agent. Its advantages, apart from ease of culture, include a high reproductive capacity, a potentially long-lived infective stage, and resistance of the infective stage to many insecticides (making it possible to use the nematode in conjunction with conventional chemical methods of pest control). Another possible advantage is its wide host range, which includes many pests, though this must be weighed against its detrimental effects on some beneficial insects. Its disadvantages include the need for a moist surface in order to migrate in search of a host and its sensitivity to high temperatures and desiccation. Hence, the timing of field applications is very important (Dutky, 1974).

4.4. Genetic Control

Methods for genetic control fall into two distinct categories: (1) those by which pests are rendered less capable of reproduction, and (2) those in which resistance is increased in the organism attacked by the pest.

A variety of genetic methods are potentially applicable for regulation of a pest's reproductive capability [see reviews by Proverbs (1969), and Whitten and Foster (1975)]. To date, however, only one of these, the sterile insect release method (SIRM), has been used on a full scale; the rest are either under examination in laboratories or in field trials or still at the theoretical stage. Knipling (1955) first proposed the idea of releasing sterile insects into wild populations of the pest so as to reduce total fecundity. Successive releases of sterile insects over a number of generations would lead, cumulatively, to eradication of the pest. SIRM has been used with striking success against the screwworm fly (*Cochliomyia hominivorax*) in Curacao and the southeastern United States where the fly has been eliminated, and in the southwestern United States where good control has been achieved, though not eradication because of immigration of wild flies from Mexico (DeBach, 1974). Flies are mass-cultured and sterilized by irradiation. SIRM has also been used successfully against small, isolated populations of other pests, for example, melon fly (*Dacus cucurbitae*) and Oriental fruit fly (*D. dorsalis*) in the Mariana Islands. Field tests have indicated that the method holds promise for the control of codling moth (*Carpocapsa pomonella*) in the United States and Canada, Mediterranean fruit fly (*Ceratitis capitata*) in Spain, Italy, Hawaii, and Nicaragua, and Queensland fruit fly (*D. tryoni*) in Australia. The advantages of SIRM are its specificity, the permanency of its effect (though it may take several years to achieve this), and the fact that it does not pollute the environment. An important disadvantage is its limited applicability. It is not a very suitable method for species that mate frequently as the chances of successful (fertile) matings are increased. In this regard, it is worth noting that the pests controlled by this method are mostly Diptera or Lepidoptera, in many species of which females mate only once or a very few times. Nor is it feasible to use SIRM on pests that appear sporadically or in high density; the latter would make it very difficult to achieve the necessary high ratio of sterile: wild males in the field. In addition, there are several technical problems related to the mass production of individuals all of the same sex. Presently, both males and females reared in culture are irradiated and released. Thus, irradiated females will compete with wild females for the males' attention, reducing the efficacy of the method. Two solutions have been proposed for this dilemma: (1) Incorporation of lethal sex-linked mutants into laboratory cultures. Such mutants might be, for example, temperature-sensitive, so that exposure of the population to a certain temperature would kill one sex, or sensitive to a particular insecticide. Thus, treatment of juvenile stages with the insecticide would leave only the resistant sex alive. (2) Chemosterilization of wild populations. This probably would be cheaper than mass-rearing followed by irradiation, though this would have to be weighed against the potential nonspecificity and pollution of the environment by the sterilant.

The use of irradiation or chemosterilants to achieve sterility (known as induced sterility) may not always be feasible due to health hazards, costs, etc., and a number of workers are now examining various methods of causing inherited sterility. This is sterility which results either from matings between genetically or cytoplasmically incompatible partners, with production of inviable zygotes, or because gametogenesis is abnormal in one or both sexes. For further details, Whitten and Foster (1975) should be consulted.

The development of plant varieties capable of resisting attack by pests is a well-known method of pest control. Probably the classic example of plant resistance is the resistance of wheat varieties to attack by the Hessian fly (*Mayetiola destructor*). Resistance of "Underhill" wheat was recorded in New York in 1782 (Gallun et al., 1975), and several resistant varieties are now used regularly in the United States, with enormous annual savings.

Basically, increased resistance can be achieved by introducing either physical or biochemical changes in plants. For example, wheat resistant to attack by wheat stem sawfly (*Cephus cinctus*) has solid rather than hollow stems; wheat grown in Michigan and resistant to cereal leaf beetle (*Oulema melanopus*) has "hairier" leaves than other varieties; resistance to Hessian fly is due to the presence of antibiotics in the plant tissues which cause the death of the larvae. Other examples of plants having varieties resistant to pests are corn to the European corn borer (*Ostrinia nubilalis*), and corn earworm (*Heliothis zea*); alfalfa to the pea aphid (*Acyrthosiphon pisum*); cabbage to the potato leafhopper (*Empoasca fabae*), and beans to the Mexican bean beetle (*Epilachna varivestis*). All told, plant varieties resistant to more than 100 pest species are known (DeBach, 1974).

The advantages of using resistant varieties include greater crop yield, relatively low cost of development, and absence of side effects (environmental damage, etc.). In addition, pests that feed on these plants may be more susceptible to disease, adverse weather conditions, and insecticides which can therefore be used in lesser amounts. Disadvantages include the length of time required to develop a resistant variety, ordinarily 10 to 15 years with crop plants and even longer with trees. Related to this, an extensive program of screening must be carried out to ensure that the varieties are totally satisfactory. Thus, resistance may vary according to the form of the insect, for example, between apterous and winged aphids. Or increased resistance to one pest may result in decreased resistance to another. Another problem may be psychological, that is, to persuade a grower that a new variety is superior to the one previously used.

4.5. Cultural Control

Cultural control, the use of various agricultural practices to make a habitat less suitable for reproduction and/or survival of pests, is a long-established method of pest control. Cultural control aims, therefore, to reduce rather than eradicate pest populations and is typically used in conjunction with other control methods. However, in some instances, cultural practices alone may effect almost complete control of a pest, as occurs with the tobacco hornworm (*Manduca sexta*) in North Carolina and the pink bollworm (*Pectinophora gossypiella*) on cotton in central Texas.

The agricultural practices used either may have a direct effect on the pest or may act indirectly by stimulating population buildup of a pest's predators or parasites, or by making plants more tolerant of pest attack. An essential prerequisite for effective cultural control is detailed knowledge of a pest's life history so that its most susceptible stages can be determined. Important agricultural practices include (1) crop rotation to prevent buildup of pest populations; (2) planting or harvesting out of phase with a pest's injurious stage(s), which is especially important against species that have a limited period of infestation or for plants with a short period of susceptibility; (3) use of trap crops on which a pest will concentrate, making its subsequent destruction easy; (4) soil preparation, so as to bury or expose a pest, or increase the crop's strength so that it can more easily tolerate a pest; and (5) clean culture, the removal, destruction, or ploughing under of crop remains, in or under which pests may hibernate.

4.6. Integrated Control

Integrated control, synonymous with pest management, is "a pest population management system that utilizes all suitable techniques either to reduce pest populations and maintain them at levels below those causing economic injury, or to so manipulate the populations that they are prevented from causing such injury" (van den Bosch et al., in Huffaker, 1971, p. 378). The techniques employed must be compatible and the system must be flexible to accommodate changes in an ecosystem. It is an approach to pest control that arose, of necessity, with the realization that, for almost all pests, existing methods either individually were not capable of exerting permanent control, gave rise to harmful side effects, or created, by their very use, new pest problems.

Integrated control is not a new concept. It was envisioned by C. V. Riley, former Chief Entomologist in the United States, in the 1890s and was practiced in the 1920s in California citrus groves (DeBach, 1974). However, it was not in most instances given serious consideration until the use of insecticides alone had increased to the point at which an entire crop-growing operation became unprofitable. (The importance of insecticides as environmental hazards did not enter into consideration until relatively recently.) As has been noted on previous occasions, a major fault of pesticide use has been its indiscriminate nature. Insecticides were not used only as necessary (a decision based on pest population density) but on a regular (calendar) basis as a form of "preventive medicine" or "insurance." This approach to the use of pesticides, referred to by Doutt and Smith (in Huffaker, 1971) as "the pesticide syndrome," led rapidly to the development of resistance in the pests against which it was aimed, to the use of larger doses or of different insecticides (to which also the pests soon became resistant!), to a rise to pest status of previously innocuous species, and ultimately to greatly elevated costs and decreased profits.

The primary step toward integrated control is collation of as much information as possible about the agroecosystem being studied; the more information that is available, the closer an integrated control program will come to providing maximum returns from the investment of time, effort, and money. In many early integrated control projects, relatively little information was available on which to base integrated control programs, and these were developed largely by intuition, despite which they were remarkably successful. Nowadays the considerably greater body of information generally available is stored in computers for use in constructing and testing mathematical models of the agroecosystem. Once developed, such models can be used to examine the effects of varying one or more of the factors that influence the population density of a pest and, therefore, to determine the optimum method of reducing and maintaining this density below the economic injury threshold.

Some of the important questions to be answered in an integrated control program might be as follows. (1) What is the economic injury threshold? It has been observed frequently that insecticide users were maintaining pest population density at levels much below the economic injury threshold, increasing their operational costs in both the short and the long term. (2) What are the best parasitoids and predators to use and how might their abundance be increased? (3) What are the best insecticides to use, when should they be applied, and what are suitable doses? To answer this question requires consideration of the specificity of action, potential side effects, and stability of the insecticides. (4)

Are suitable cultural procedures available? (5) Are resistant strains of the crop available? It will be apparent from these questions that pest management is interdisciplinary in nature and may require collaboration between experts in widely different areas, for example, economists, statisticians, meteorologists, plant breeders, entomologists, chemists, and toxicologists.

Many examples of pest management might be cited, that differ in complexity, geographic location, and crop pest being controlled. In its simplest (and rarest) form, pest management involves only one method of control and, strictly speaking, cannot be described as "integrated control." Thus, biological control is entirely satisfactory for control of sugarcane pests in Hawaii and coconut pests in Fiji (DeBach, 1974). More often, two or more control methods are employed. For example, at the height of the pesticide syndrome, apple growers in Nova Scotia used a battery of fungicides and insecticides (especially broad-spectrum synthetics) against apple pests. Far from achieving the desired effects, this led to even greater problems. By about 1950 earlier pests such as codling moth (*Carpocapsa pomonella*), eye-spotted bud moth (*Spilonota ocellana*), and oystershell scale (*Lepidosaphes ulmi*) had become even more serious, and other species, for example, European red mite (*Panonychus ulmi*) and fruit-tree leafroller (*Archips argyrospilus*) had become pests due to destruction of their natural enemies. The need to reduce the escalating costs of this program stimulated intensive study of apple orchard ecology, including the effects of pesticides on the natural enemy complex. The outcome of this work was the development of an integrated control program which utilized fewer but more specific insecticides (as well as different fungicides, some of which had killed some natural enemies) and allowed for recovery of many of the natural enemies. By 1955 the mite and scale insect had practically disappeared, and the density of the eye-spotted bud moth had fallen significantly.

As a final example, the integrated control program for cotton pests in the San Joaquin Valley, California will be outlined (van den Bosch *et al.*, in Huffaker, 1971). The usual picture emerges. After extensive use of organochlorine insecticides, followed by organophosphates and carbamates, there was a population resurgence and an increase in resistance of the two major pests, the corn earworm (*Heliothis zea*) and a mirid bug (*Lygus hesperus*), as well as outbreaks of secondary pests such as cabbage looper (*Trichoplusia ni*), beet armyworm (*Spodoptera exigua*), salt-marsh caterpillar (*Estigmene acraea*), and spider mites (Tetranychidae).

Early in the development of the integrated program, it was realized that no reliably established economic injury thresholds existed for either of the major pests and that, for *Lygus*, time of insecticide application was critical. For *Lygus*, it was determined that the generally accepted injury threshold of 10 bugs/50 net sweeps was invalid; rather, the major period of importance was June 1 to July 20 and only if the density exceeded 10 bugs/50 net sweeps in this period should insecticides be applied. After July 20, *Lygus* densities of 20 or more bugs/50 net sweeps would not affect the quantity or quality of cotton produced, provided that the plants had flowered normally in June and early July.

It was noted that corn earworms seldom became pests except in fields treated for *Lygus* outbreaks (using the old value for economic injury threshold) where the corn earworms' natural enemies (the bugs *Geocoris pallens* and *Nabis americoferus* and the lacewing, *Chrysopa carnea*) were destroyed by insecticide. With the new threshold for economic injury due to *Lygus*, tied in with the plant's seasonal development, the use of insecticides was reduced, especially after late July, and the earworms' natural enemies survived. Further reduction in insecticide usage came about with the raising of the economic injury threshold for corn earworms from 4 earworms/plant to 15 *treatable* earworms (first and second instar larvae)/plant. This arose, in part, from realization that larvae often feed on surplus buds, flowers, and small bolls that are not picked.

Cultural control is also used to reduce the effects of *Lygus* on the cotton crop. *Lygus* prefers alfalfa (*Medicago sativa*) to cotton and will remain on the former plant when given a choice. Traditionally, however, alfalfa fields are solid cut (all the field is cut at once) provoking mass migration of *Lygus* into adjacent cotton fields. Alfalfa growers are now being persuaded to practice strip cutting to prevent such migration, though the method is not widely adopted. Probably a better technique is for cotton growers to plant strips of alfalfa through their cotton fields, on which *Lygus* will concentrate.

Another cultural method that may become commercially feasible is the use of food sprays in attempts to prolong the seasonal effectiveness of natural enemies whose populations frequently decline after midseason. For example, in experimental plots adult *Chrysopa* were attracted to the food sprays and produced more eggs, resulting in significantly less injury to plants by corn earworms (see Hagen and Hale, 1974).

Finally, microbial control may become an important method for use against the corn earworm and other lepidopteran pests of cotton. Some growers use homemade sprays of cabbage looper nuclear polyhedrosis virus, prepared from infected larvae which they collect. *Heliothis* nuclear polyhedrosis virus is now registered for use on cotton. *Bacillus thuringiensis* is also registered for commercial use but is expensive and not sufficiently virulent against corn earworm to merit widespread application.

Integrated control results in substantial savings, both financial and in terms of pollution, through the greatly decreased use of insecticides. For example, in the San Joaquin Valley, the traditional control program for cotton pests cost about twice as much per acre as the integrated program, and even greater savings may be incurred as the integrated program is improved. Adkisson (1971, cited in DeBach, 1974) considers that insecticide use against cotton pests could be reduced by 50% throughout the United States without loss of production. As more than one half of the insecticide produced in the United States (totaling about 0.5 billion kilograms in 1971) is used on cotton, enormous savings are possible in the growing of this crop alone. Indeed, a 50–80% reduction in the use of insecticides, after the initiation of an integrated control program, has been observed in many other instances. Despite these remarkable savings, a great majority of crop growers and government officials remain convinced that the present method of pest control, the unilateral use of insecticides, is best. Persuading these individuals that to continue with this approach will lead not only to financial ruin but to long-term, perhaps irreversible, deterioration of environmental quality is perhaps the greatest challenge that scientists have ever faced.

5. Summary

Only a small proportion (perhaps 0.5%) of insect species interact directly or indirectly with Man. Most of these are beneficial to Man's existence by virtue of their pollinating activity, the materials they produce, or their role as biological control agents. Relatively few insects (about 0.1% of all species) constitute pests which alone or in conjunction with microorganisms cause damage or death to Man, crops, livestock, and manufactured goods.

Pest insects may be classified according to their ecological strategies as r pests, K pests, or intermediate pests. The r pests have potentially high rates of population increase, well-developed ability to disperse and find new food sources, and general food preferences. Though they have a large number of natural enemies, biological control would not be effective against r pests because of their reproductive potential. The use of insecticides is the best method for the control of r pests. At the opposite end of the spectrum are K pests, which have relatively low fecundity, poor powers of dispersal, specific food preferences, and few natural enemies. They occur in relatively stable habitats and are best controlled by methods which render their habitat less stable, for example, cultural practices and by the breeding of resistant strains of the organisms they attack. Intermediate pests form a continuum between r and K pests and are normally held in check (below the economic injury threshold) by natural enemies. Biological control is the primary method for control of these pests, supplemented as appropriate by chemical, genetic, and/or cultural methods.

Legal control aims to prevent or control pest damage through legislation. It includes the setting up of quarantine stations, systems for monitoring pest populations, and mechanisms for the certification of disease-free plants and animals.

Chemical control traditionally has been the use of naturally occurring or synthetic chemicals to kill pests. It has been the major method of pest control for about 70 years but has created three serious problems: (1) a great increase in the resistance of pests to the chemicals, (2) the death of many beneficial insects due to the chemicals' nonspecific activity, and (3) pollution of the environment. More recently, increased interest has been shown in insectistatics and sex attractants, which interfere with normal growth and reproduction.

Biological control is the regulation of pest populations by natural enemies. For insect pests, parasitoids, predaceous insects, and microorganisms are the major control agents. Biological control offers several advantages over control by insecticides: (1) the control agent is ready-made (does not have to be developed), cheap to produce and apply, and persistent; (2) the method does not stimulate genetic counterattack by pests; and (3) it is specific, and does not endanger humans or wildlife through pollution of the environment. Its main disadvantages are its slowness of effect and the fact that the final (equilibrium) pest population density is normally higher than that achieved after insecticide application.

Methods of genetic control include (1) those which render pests less capable of reproduction, for example, the sterile insect release method, use of chemosterilants, and sterility resulting from mating incompatibility or abnormal gametogenesis, and (2) those in which resistance is increased in the organisms attacked by pests.

Cultural control includes the long-established agricultural practices which make habitats less suitable for pests. The methods used may either directly affect a pest, stimulate an increase in density of a pest's natural enemies, or make the organisms on which a pest feeds more tolerant of attack.

Integrated control (pest management) is a combination of methods for reducing and maintaining pest populations below the economic injury threshold. To be most effective it requires the input of as much information as possible about the agroecosystem under consideration and the collaboration of experts from a variety of disciplines. If conducted properly, integrated control programs lead to considerable financial saving and a great improvement in environmental quality.

6. Literature

The literature dealing with economic entomology is voluminous, and the following list is highly restrictive. Descriptions of economically important insects (mainly North American species) may be found in standard textbooks of applied entomology, for example, Metcalf et al. (1962), Pfadt (1971), and Little (1972). Problems associated with the use of insecticides are discussed in the Proceedings of the XV International Congress of Entomology (1977), in volumes of the Annual Review of Entomology, and in the volume edited by Metcalf and McKelvey (1976). Biological control is the subject of the volumes edited by Burges and Hussey (1971), Huffaker (1971), and Huffaker and Messenger (1976), of DeBach's (1974) text, and of reviews by several authors in the Proceedings of the XV Congress. Metcalf and Luckmann (1975), Apple and Smith (1976) and various contributors to the Proceedings of the XV Congress have discussed integrated control (pest management). The genetic control of insects is reviewed by Whitten and Foster (1975). Selected aspects of genetic control are discussed in the Proceedings of the XV Congress.

Angus, T. A., 1977, Microbial control of arthropod pests, Proc. XV Int. Congr. Entomol., pp. 473–477.

Apple, J. L., and Smith, R. F. (eds.), 1976, Integrated Pest Management, Plenum Press, New York.

Askew, R. R., 1971, Parasitic Insects, American Elsevier, New York.

Borror, D. J., Delong, D. M., and Triplehorn, C. A., 1976, An Introduction to the Study of Insects, 4th ed., Holt, Rinehart and Winston, New York.

Brown, A. W. A., 1977, Epilogue: Resistance as a factor in pesticide management, Proc. XV Int. Congr. Entomol., pp. 816–824.

Bulla, L. A., Jr. (ed.), 1973, Regulation of insect populations by microorganisms, Ann. N.Y. Acad. Sci. 217:243 pp.

Burges, H. D., and Hussey, N. W. (eds.), 1971, Microbial Control of Insects and Mites, Academic Press, New York.

Conway, G., 1976, Man versus pests, in: *Theoretical Ecology: Principles and Applications* (R. M. May, ed.), Blackwell, Oxford.

DeBach, P., 1974, *Biological Control by Natural Enemies*, Cambridge University Press, London.

Dutky, S. R., 1974, Nematode parasites, in: *Proceedings of the Summer Institute on Biological Control of Plant Insects and Diseases* (F. G. Maxwell and F. A. Harris, eds.), University Press of Mississippi, Jackson, Miss.

Eastop, V. F., 1977, Worldwide importance of aphids as virus vectors, in: *Aphids as Virus Vectors* (K. F. Harris and K. Maramorosch, eds.), Academic Press, New York.

Elliott, M., Janes, N. F., and Potter, C., 1978, The future of pyrethroids in insect control, *Annu. Rev. Entomol.* **23:**443–469.

Ferron, P., 1978, Biological control of insect pests by entomogenous fungi, *Annu. Rev. Entomol.* **23:**409–442.

Gallun, R. L., Starks, K. J., and Guthrie, W. D., 1975, Plant resistance to insects attacking cereals, *Annu. Rev. Entomol.* **20:**337–357.

Georghiou, G. P., and Taylor, C. E., 1977, Pesticide resistance as an evolutionary phenomenon, *Proc. XV Int. Congr. Entomol.*, pp. 759–785.

Hagen, K. S., and Hale, R. 1974, Increasing natural enemies through use of supplementary feeding and non-target prey, in: *Proceedings of the Summer Institute on Biological Control of Plant Insects and Diseases* (F. G. Maxwell and F. A. Harris, eds.), University Press of Mississippi, Jackson, Miss.

Henry, J. E., and Oma, E. A., 1974, Effects of infection by *Nosema locustae* Canning, *Nosema acridophagus* Henry, and *Nosema cuneatum* Henry (Microsporida: Nosematidae) in *Melanoplus bivittatus* (Say) (Orthoptera: Acrididae), *Acrida* **3:**223–231.

Horn, D. J., 1976, *Biology of Insects*, Saunders, Philadelphia, Pa.

Huffaker, C. B. (ed), 1971, *Biological Control*, Plenum Press, New York.

Huffaker, C. B., and Messenger, P. S. (eds.), 1976, *Theory and Practice of Biological Control*, Academic Press, New York.

Kellen, W. R., 1974, Protozoan pathogens, in: *Proceedings of the Summer Institute on Biological Control of Plant Insects and Diseases* (F. G. Maxwell and F. A. Harris, eds.), University Press of Mississippi, Jackson, Miss.

Knipling, E. F., 1955, Possibilities of insect control or eradication through the use of sexually sterile males, *J. Econ. Entomol.* **48:**459–462.

Levinson, H. Z., 1975, Possibilities of using insectistatics and pheromones in pest control, *Naturwissenschaften* **62:**272–282.

Little, V. A., 1972, *General and Applied Entomology*, 3rd ed., Harper and Row, New York.

Matthews, R. E. F., 1970, *Plant Virology*, Academic Press, New York.

Metcalf, C. L., Flint, W. P., and Metcalf, R. L., 1962, *Destructive and Useful Insects: Their Habits and Control*, 4th ed., McGraw-Hill, New York.

Metcalf, R. L., and Luckmann, W., 1975, *Introduction to Insect Pest Management*, Wiley, New York.

Metcalf, R. L., and McKelvey, J. J., Jr. (eds.), 1976, *The Future for Insecticides: Needs and Prospects*, Wiley, New York.

Pfadt, R. E. (ed.), 1971, *Fundamentals of Applied Entomology*, 2nd ed., Macmillan, New York.

Proverbs, M. D., 1969, Induced sterilization and control of insects, *Annu. Rev. Entomol.* **14:**81–102.

Southwood, T. R. E., 1977, Entomology and mankind, *Proc. XV Int. Congr. Entomol.*, pp. 36–51.

Staal, G. B., 1975, Insect growth regulators with juvenile hormone activity, *Annu. Rev. Entomol.* **20:**417–460.

Taylor, R. L., and Carter, B. J., 1976, *Entertaining with Insects*, Woodbridge Press, Santa Barbara, Calif.

Waterhouse, D. F., 1974, The biological control of dung, *Sci. Am.* **230**(April):100–109.

Whittemore, F. W., 1977, The evolution of pesticides and the philosophy of regulation, *Proc. XV Int. Congr. Entomol.*, pp. 714–718.

Whitten, M. J., and Foster, G. G., 1975, Genetical methods of pest control, *Annu. Rev. Entomol.* **20:**461–476.

Williams, C. M., 1967, Third-generation pesticides, *Sci. Am.* **217**(July):13–17.

Author Index

Subject Index

Page numbers in **boldfaced** type indicate illustrations.

The Biology of Scorpions · Iblis

549.46
B615

SR